Nanobiotechnology for Sustainable Food Management

Among the most novel and ever-growing approaches to improving the food industry is nanobiotechnology. In this book, the prospective role of nanobiotechnology in food which includes quality control and safety through nanosensors and biosensors, targeted delivery of nutrients, controlled release of nutrients, proteins, antioxidants, and flavors through encapsulation and enzymatic reactions for food fortification of fat-soluble compounds is discussed. Along the chapters of this book, nanobiotechnological techniques are addressed in detail with specific emphasis on food science applications.

Features:

- Discusses nanobiotechnology in food for quality control and safety.
- Covers food processing and packaging for food safety.
- Explores the positive role of nanomaterials toward the sustainability of food.
- Provides efficient, real, and sustainable solutions to pertinent global problems.
- Includes case studies and research directions of nanobiotechnology.

This book is aimed at researchers and graduate students in nanotechnology and food engineering.

AF386640

Nanobiotechnology for Sustainable Food Management

Edited by
Eduardo Alberto López-Maldonado, Amber R. Solangi,
and Fabio Granados-Chinchilla

CRC Press
Taylor & Francis Group
Boca Raton London New York

CRC Press is an imprint of the
Taylor & Francis Group, an **informa** business

Designed cover image: Sergio Granados

First edition published 2025
by CRC Press
2385 NW Executive Center Drive, Suite 320, Boca Raton FL 33431

and by CRC Press
4 Park Square, Milton Park, Abingdon, Oxon, OX14 4RN

CRC Press is an imprint of Taylor & Francis Group, LLC

ISBN: 9781032622842 (hbk)
ISBN: 9781032845968 (pbk)
ISBN: 9781003514039 (ebk)

DOI: 10.1201/9781003514039

Typeset in Times
by codeMantra

Contents

Preface

Among the most novel and ever-growing approaches to improving the food industry is nanobiotechnology. Nanotechnology encompasses science, engineering, and applications of submicron materials that involve harnessing unique physical, chemical, and biological properties of nanoscale substances in fundamentally new and practical ways (Sargent, 2014). Biotechnology uses several techniques and information from biology to manipulate molecular, genetic, and cellular processes to create novel products and services (Gupta et al., 2016). Nanobiotechnology is considered a unique merger of both knowledge areas. Hence, technology can be merged with a molecular biological approach. This particular area of expertise also uses the main techniques in which nanotechnology is founded, i.e., "top-down" (mechanical processes such as fine grinding and milling), "bottom-up" (larger structures constructed from atoms/molecules), and biologically mediated nanoparticle production (Halake and Haro, 2022).

Nanobiotechnology applications have a singular place within the medical sector (e.g., diagnostic, therapy, and drug delivery) (Anjum et al., 2021). However, its role within food technology remains to be exploited despite being regarded as one of the industry sectors where it can be more relevant. Specifically, food additives and packaging seem to be the areas where nanobiotechnology has found more space (Figure P1). For example, nanofibers from polymeric, ceramic, and carbonaceous materials have been used to produce "intelligent" packaging for the food industry (Ehsani et al., 2022; Forghani et al., 2021; Kumar et al., 2019). Another example of engineered materials in the food industry is delivering functional bioactive ingredients (e.g., vitamins, probiotics, bioactive peptides, and antioxidants) through nanoencapsulation (Hamad et al., 2018). Additionally, there are many positive effects of nanomaterials in agriculture in different forms that include nanofertilizers (Nongbet et al., 2022; Toksha et al., 2021), nanopesticides (Chaud et al., 2021), sustainable agriculture and post-harvest management (Babu et al., 2022), and waste management (Nehra et al., 2021) among others.

Nanobiotechnological techniques that have found applications in food science include nanoparticles [organic (fat crystals, micelles, oil droplets, and vesicles, Pan and Zhong, 2016) and inorganic (metals and metal oxides, Ropers et al., 2017)], nanoclays (Bumbudsanpharoke and Ko, 2019), nanocapsules (González-Reza et al., 2020), nanocoating (Poonia and Mishra, 2022), electrospun nanofibers (Han et al., 2022), quantum dots (Zhang et al., 2022), physicochemical and whole-cell biosensors (McLamore et al., 2021), and bionanocomposites (Youssef and El-Sayed, 2018) to name just a few. Along the chapters of this book, several of these techniques will be addressed in detail with specific emphasis on food science applications.

Nanobiotechnology has also been applied in the detection of pollutants [e.g., pesticides, antimicrobials, additives, and even microplastics] (Yilmaz et al., 2022), detection and remediation of food contaminants [e.g., heavy metals, dioxins] (Salek Maghsoudi et al., 2021; Nasr-Eldahan et al., 2021), food spoilage and adulteration detection (Mohammadi and Jafari, 2020), bacterial pathogens and toxins (Stephen Inbaraj and Chen, 2016). On the other hand, metal nanoparticles have been introduced within the packaging to provide antimicrobial properties (Hoseinnejad et al., 2017).

Furthermore, nanobiotechnology has found a surplus of applications for water remediation, decontamination, and purification of increased contamination levels (Ajith et al., 2021). In this scenario, water is linked to human food consumption as it is a nutrient for animals and humans alike and, environmentally speaking, is used to irrigate crops.

Although several of the above target analytes can be determined by employing other analytical techniques (mainly instrumental approaches), this technology tends to be expensive, requires special training and personnel, and the upkeep is usually time-consuming. Then, biotechnological processes can serve as an alternative to test these types of analytes and even determine properties

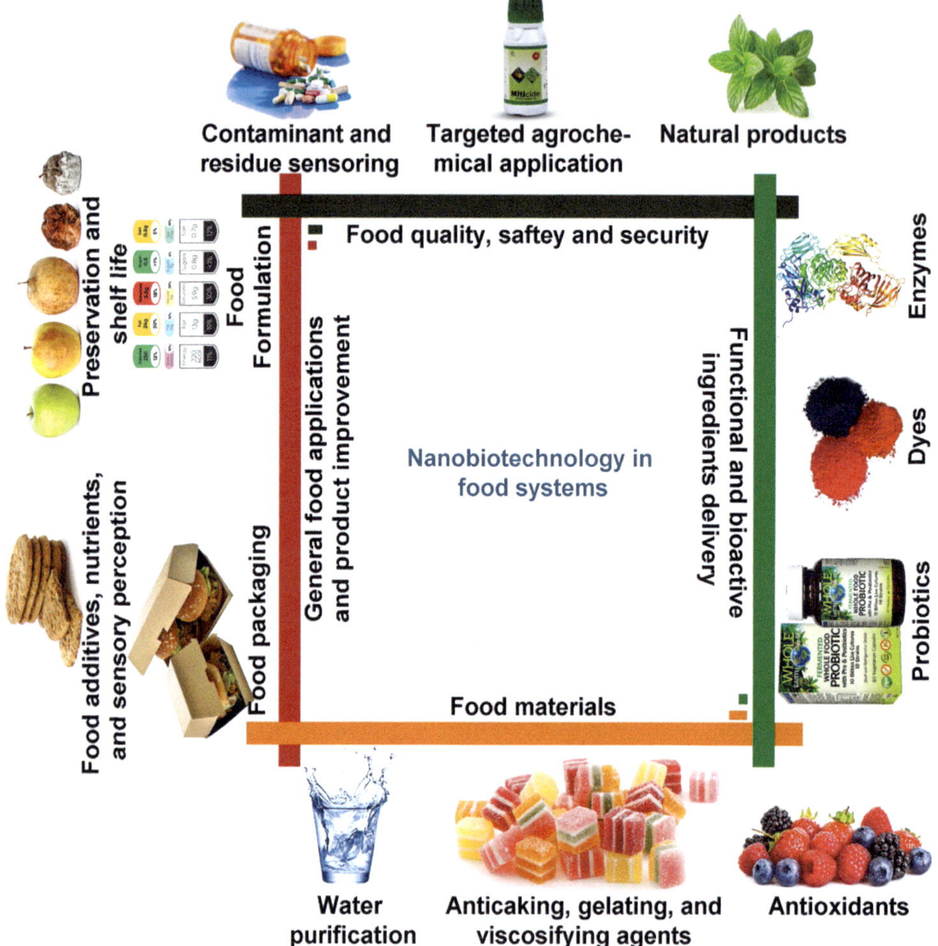

FIGURE P1 Prospective role of nanobiotechnology in food, which includes quality control and safety through nanosensors and biosensors, targeted delivery of nutrients, controlled release of nutrients, proteins, antioxidants, and flavors through encapsulation and enzymatic reactions for food fortification of fat-soluble compounds.

that clear-cut analytical techniques might not permit. Just in lieu of examples, analytical determination of aflatoxin B_1 and nitrite in foods might be analytically cumbersome and work-demanding, but the biotechnological alternatives are fast and straightforward (Xia et al., 2022; Li et al., 2022). The techniques mentioned above have a myriad of applications notwithstanding, they can be beneficial within the food industry as, on occasions, food technicians, professionals, or other stakeholders might have to make swift decisions regarding the safety of a food parcel or batch.

Finally, nanoparticles and structures used within the food industry have a unique classification (i.e., zero one, two, and three-dimensional) according to their complexity (Bhatia, 2016). They can be classified based on composition (Joudeh and Linke, 2022). Overall, organic nanoparticles are claimed to be less toxic than inorganic ones, considering their metabolism (Khan et al., 2019). Furthermore, despite their plentiful applications and advantages in the food sector, nanobiotechnological structures are not without showing some toxicity and allergenicity, which can hinder their widespread application (Khan et al., 2019; Lamas et al., 2020; Issa et al., 2022). However, in some instances, nanoparticles have been demonstrated to have less noxious effects than their traditional counterparts (Bhattacharjee et al., 2019).

As is evident, many novel developments have been put together using nanobiotechnology within the food industry. Nanobiotechnological applications have already covered the most relevant aspects of food, from food processing and packaging to food safety and quality control. In this quick summary, we intended to give the reader an overall look at some approaches and applications of nanobiotechnology in the food sector. Considering the above, continued research must be pursued to include beneficial nanobiotechnological insights into food technology. Nanobiotechnology is evident that today is more relevant than ever, as demonstrated by the winners of the 2023 Chemistry Nobel working with quantum dots. Such new developments in the area will undoubtedly open up even more untapped possibilities of nanotechnology in various fields, including food research.

Our book was finally divided into four sections and 17 chapters. As the reader will see, some analytes and methods are conserved and iterated to give the whole book continuity as a common thread. We hope this contribution summarizes the most critical advances in nanotechnology in different aspects and topics of food analysis. Our book proposal includes an analysis of "typical" food components, an analysis of chemical and microbiological contaminants, and innovative approaches that improve agricultural practices. It includes strides in food processing and packaging due to nanobiotechnology, until reaching impacts and regulatory and safety aspects of the use of nanobiotechnology.

Finally, we honestly anticipate that this book will be valuable for practitioners and scientists in food research. We would like to heartwarmingly thank all authors for contributing to this book and for the editorial support of CRC.

BIBLIOGRAPHY

Ajith, M. P., Aswathi, M., Priyadarshini, E., Rajamani, P. (2021). Recent innovations of nanotechnology in water treatment: A comprehensive review. *Bioresource Technology*, 342, 126000.

Anjum, S., Ishaque, S., Fatima, H., Farooq, W., Hano, C., Abbasi, B. H., Anjum, I. (2021). Emerging applications of nanotechnology in healthcare systems: Grand challenges and perspectives. *Pharmaceuticals*, 14(8), 707.

Babu, P. J., Saranya, S., Longchan, B., Rajasekhar, A. (2022). Nanobiotechnology-mediated sustainable agriculture and post-harvest management. *Current Research in Biotechnology*, 4, 326–336.

Bhatia, S. (2016). Nanoparticles types, classification, characterization, fabrication methods and drug delivery applications. In: *Natural Polymer Drug Delivery Systems*, https://doi.org/10.1007/978-3-319-41129-3_2

Bhattacharjee, A., Basu, A., Bhattacharya, S. (2019). Selenium nanoparticles are less toxic than inorganic and organic selenium to mice *in vivo*. *The Nucleus*, 62, 259–268.

Bumbudsanpharoke, N., Ko, S. (2019). Nanoclays in food and beverage packaging. *Journal of Nanomaterials*, 2019, Article ID 8927167.

Chaud, M., Souto, E. B., Zielinska, A., Severino, P., Batain, F., Oliveira-Junior, J., Alves, T. (2021). Nanopesticides in agriculture: Benefits and challenge in agricultural productivity, toxicological risks to a human health and environment. *Toxics*, 9(6), 131.

Ehsani, N., Rostamabadi, H., Dadashi, S., Ghanbarzadeh, B., Kharazmi, M. S., Jafari, S. M. (2022). Electrospun nanofibers fabricated by natural biopolymers for intelligent food packaging. *Critical Reviews in Food Science and Nutrition*, https://doi.org/10.1080/10408398.2022.2147900

Forghani, S., Almasi, H., Moradi, M. (2021). Electrospun nanofibers as food freshness and time-temperature indicators: A new approach in food intelligent packaging. *Innovative Food Science & Emerging Technologies*, 73, 102804.

González-Reza, R. M., Hernández-Sánchez, H., Zambrano-Zaragoza, M. L., Gutiérrez-López, G. F., Del-Real, A., Quintanar-Guerrero, D., Velasco-Bejarano, B. (2020). Influence of stabilizing and encapsulating polymers on antioxidant capacity, stability, and kinetic release of thyme essential oil nanocapsules. *Foods*, 9(12), 1884.

Gupta, V., Sengupta, M., Prakash, J., Charan Tripathy, B. (2016). An Introduction to Biotechnology. *Basic and Applied Aspects of Biotechnology*, 1–21. https://doi.org/10.1007/978-981-10-0875-7_1

Halake, N. H., Haro, J. M. (2022). Role of nanobiotechnology towards agri-food system. *Journal of Nanotechnology*, 2022, Article ID 6108610.

Hamad, A. F., Han, J-H., Kim, B-C., Rather, I. A. (2018). The intertwine of nanotechnology with the food industry. *Saudi Journal of Biological Sciences*, 25(1), 27–30.

Han, W.-H., Li, X., Yu, G-F., Wang, B-C., Huang, L-P., Wang, J., Long, Y-Z. (2022). Recent advances in the food application of electrospun nanofibers. *Journal of Industrial and Engineering Chemistry*, 110, 15–26.

Hoseinnejad, M., Jafari, S. M., Katouzian, I. (2017). Inorganic and metal nanoparticles and their antimicrobial activity in food packaging applications. *Critical Reviews in Microbiology*, https://doi.org/10.1080/1040 841X.2017.1332001

Issa, M., Rivière, G., Houdeau, E., Adel-Patient, K. (2022). Perinatal exposure to foodborne inorganic nanoparticles: A role in the susceptibility to food allergy? *Frontiers in Allergy*, https://doi.org/10.3389/falgy.2022.1067281.

Joudeh, N., Linke, D. (2022). Nanoparticle classification, physicochemical properties, characterization, and applications: A comprehensive review for biologists. *Journal of Nanobiotechnology*, 20, Article 262.

Khan, I., Saeed, K., Khan, I. (2019). Nanoparticles: Properties, applications and toxicities. *Arabian Journal of Chemistry*, 12(7), 908–931.

Kumar, T. S. M., Kumar, K. S., Rajini, N., Siengchin, S., Ayrilmis, N., Rajulu, A. V. (2019). A comprehensive review of electrospun. *Composites Part B: Engineering*, 175, 107074.

Lamas, B., Breyner, N. M., Houdeau, E. (2020). Impacts of foodborne inorganic nanoparticles on the gut microbiota-immune axis: Potential consequences for host health. *Particle and Fibre Toxicology*, 17, Article 19.

Li, Y., Zhou, H., Zhang, J., Cui, B., Fang, Y. (2022). Determination of nitrite in food based on its sensitizing effect on cathodic electrochemiluminescence of conductive PTH-DPP films. *Food Chemistry*, 397, 133760.

McLamore, E. S., Alocilja, E., Gomes, C., Gunasekaran, S., Jenkins, D., Datta, S. P. A., Li, Y., Mao, J., Nugen, S. R., Reyes-De-Corcuera, J. I., Takhistov, P., Tsyusko, O., Cochran, J. P., Tzeng, J., Yoon, J-Y., Yu, C., Zhou, A. (2021). FEAST of biosensors: Food, environmental and agricultural sensing technologies (FEAST) in North America. *Biosensors and Bioelectronics*, 178, 113011.

Mohammadi, Z., Jafari, S. M. (2020). Detection of food spoilage and adulteration by novel nanomaterial-based sensors. *Advances in Colloid and Interface Science*, 286, 102297.

Nasr-Eldahan, S., Nabil-Adam, A., Shreadah, M. A., Maher, A. M., El-Sayed-Ali, T. (2021). A review article on nanotechnology in aquaculture sustainability as a novel tool in fish disease control. *Aquaculture International*, 29(4), 1459–1480.

Nehra, M., Dilbaghi, N., Marrazza, G., Kaushik, A., Sonne, C., Kim, K-H., Kumar, S. (2021). Emerging nanobiotechnology in agriculture for the management of pesticide residues. *Journal of Hazardous Materials*, 401, 123369.

Nongbet, A., Mishra, A. K., Mohanta, Y. K., Mahanta, S., Ray, M. K., Khan, M., Baek, K-H., Chakrabartty, I. (2022). Nanofertilizers: A smart and sustainable attribute to modern agriculture. *Plants*, 11(19), 2587.

Pan, K., Zhong, Q. (2016). Organic nanoparticles in foods: Fabrication, characterization, and utilization. *Annual Reviews in Food Science and Technology*, 7, 245–266.

Poonia, A., Mishra, A. (2022). Edible nanocoatings: Potential food applications, challenges and safety regulations. *Nutrition & Food Science*, 52(3), 497–514.

Ropers, M-H., Axelos, M. A., Van de Voorde, M. H. (2017). Engineered Inorganic Nanoparticles in Food, https://doi.org/10.1002/9783527697724.ch5.

Salek Maghsoudi, A., Hassani, S., Mirnia, K., Abdollahi, M. (2021). Recent advances in nanotechnology-based biosensors development for detection of arsenic, lead, mercury, and cadmium. *International Journal of Nanomedicine*, 16, 803–832.

Sargent, J. F. (2014). *The National Nanotechnology Initiative: Overview, Reauthorization, and Appropriations Issues*. Congressional Research Service. Retrieved from https://sgp.fas.org/crs/misc/RL34401.pdf. Accessed January 9th, 2023.

Stephen Inbaraj, B., Chen, B. H. (2016). Nanomaterial-based sensors for detection of foodborne bacterial pathogens and toxins as well as pork adulteration in meat products. *Journal of Food and Drug Analysis*, 24(1), 15–28.

Toksha, B., Sonawale, V. A. M., Vanarase, A., Bornare, D., Tonde, S., Hazra, C., Kundu, D., Satdive, A., Chatterjee, A. (2021). Nanofertilizers: A review on synthesis and impact of their use on crop yield and environment. *Environmental Technology & Innovation*, 24, 101986.

Xia, M., Yang, X., Jiao, T., Oyama, M., Chen, Q., Chen, X. (2022). Self-enhanced electrochemiluminescence of luminol induced by palladium-graphene oxide for ultrasensitive detection of aflatoxin B1 in food samples. *Food Chemistry*, 381, 132276.

Yilmaz, D., Günaydın, B. N., Yüce, M. (2022). Nanotechnology in food and water security: On-site detection of agricultural pollutants through surface-enhanced Raman spectroscopy. *Emergent Materials*, 5(1), 105–132.

Youssef, A. M., El-Sayed, S. M. (2018). Bionanocomposites materials for food packaging applications: Concepts and future outlook. *Carbohydrate Polymers*, 193, 19–27.

Zhang, W., Zhong, H., Zhao, P., Shen, A., Li, H., Liu, X. (2022). Carbon quantum dot fluorescent probes for food safety detection: Progress, opportunities and challenges. *Food Control*, 133 (Part A), 108591.

Editors

Eduardo López-Maldonado, a distinguished academic, holds a BSc. in Chemical Engineering (2009) and a PhD. in Chemical Sciences (2012) from the National Technological Institute of Mexico-Tijuana, Center for Graduates and Research in Chemistry. His doctoral thesis, a groundbreaking piece of applied scientific research in collaboration with a semiconductor company in Tijuana's industrial zone, earned him the prestigious Mexican Society of Electrochemistry's Best Doctoral Thesis award in 2013. The same year, Dr. López-Maldonado joined the research staff at the Center for Research and Technological Development in Electrochemistry (CIDETEQ) S.C., Tijuana Branch, where he began his research career. His work there involved projects centered on the extraction and properties of natural biopolymers as coagulating, flocculating, and chelating agents. Dr. López-Maldonado's impressive track record includes extensive experience in basic and applied science research projects with industries at the regional and national levels. He is recognized as a member of the Researchers National System and has served as a facilitator of the Environmental Leadership Program for Competitiveness (PLAC), organized by SEMARNAT and PROFEPA. Dr. López-Maldonado has presented at national and international conferences focusing on Polymer Chemistry, Sustainable process development, Environment, and Nanotechnology. Currently, Dr. Eduardo López-Maldonado is a full-time professor-researcher in the Faculty of Chemical Sciences and Engineering at the Autonomous University of Baja California, where he leads the Biopolyelectrolyte Engineering Research line.

Amber R. Solangi is a full Professor at the National Centre of Excellence in Analytical Chemistry (NCEAC), University of Sindh, Jamshoro, Pakistan. She earned her Ph.D. from the NCEAC, University of Sindh, Jamshoro, in 2007 and a Post-Doc at the School of Chemistry, Monash University, Melbourne, Australia (2009–2010). Her field of expertise encompasses the innovative development of methodologies for identifying and removing environmental pollutants from aqueous systems. She also excels in synthesizing and utilizing nanomaterials/nanocomposites to fabricate cutting-edge sensors. Dr. Solangi has effectively supervised the research thesis of 10 Ph.D. and 30 M.Phil students. Her significant contributions include publishing 106 research papers in esteemed and reputable scientific journals with high impact factors. She has also authored four book chapters in books published with reputed publishers. She has delivered invited/oral talks at 16 international conferences abroad and 28 national conferences in Pakistan. Dr. Solangi has received seven international trainings. She has also organized six workshops/seminars at NCEAC. She maintains active membership in several distinguished national and international scientific societies, including a Fellowship of the Chemical Society of Pakistan, the Soil Science Society of Pakistan, the American Chemical Society, the American Society for Microbiology, USA, and the International Society for Development & Sustainability, Japan. Dr. Solangi's impactful achievements also encompass the successful completion of four projects funded by the Higher Education Commission of Pakistan, Islamabad, and Sindh Higher Education Commission, Karachi, Pakistan.

Fabio Granados-Chinchilla holds a B.Sc. Chemistry (in 2008) and M.Sc. Microbiology with emphasis in Bacteriology (in 2016) from the Universidad de Costa Rica. He worked extensively in feed and food research, with over 13 years of experience in food analysis, including safety, quality, natural products, liquid and gas chromatography, and mass spectrometry. He also worked in the implementation of pesticide residue analysis and the optimization and creation of methods for the determination of antimicrobials in environmental and food matrices. He has vast experience in determining mineral profiles (including heavy metal analysis) in complex matrices, including feed, foods, and food ingredients. His research also includes antimicrobial and mycotoxin analysis and method development for food and feed matrices, including cereals, milk, and compound feed. His main research interests include bacterial antibiotic resistance, chromatography, analytical chemistry, food and feed quality, food and feed safety, and contaminant and residue analysis. He is a regular member (id 57060626) of the American Society for Microbiology, ASM (since 2013) and a student member (id 30402995) of the American Chemical Society, ACS (since 2012). Currently, he fills a position as an instructor of organic chemistry laboratory undergrad courses for the Faculty of Basic Sciences and the School of Chemistry. He also occupies a research position in the Faculty of Microbiology and the Centre for Research in Tropical Diseases at the Universidad de Costa Rica. He has been a reviewer for over 30 internationally recognized and indexed prestigious scientific journals and contributed to over 50 research items (including articles and book chapters).

Contributors

Hani Nasser Abdelhamid
Advanced Multifunctional Materials
  Laboratory
Chemistry Department, Faculty of Science
Assiut University
Asyut, Egypt
and
Nanotechnology Research Centre (NTRC)
The British University in Egypt (BUE)
Cairo, Egypt

Matthew Ndubuisi Abonyi
Department of Chemical Engineering
Nnamdi Azikiwe University
Awka, Nigeria

Adenike A. Akinsemolu
Institute of Advanced Studies
University of Birmingham
Birmingham, United Kingdom

Sidra Amin
Department of Chemistry
Shaheed Benazir Bhutto University
Shaheed Benazirabad, Pakistan

Chukwunonso Onyeka Aniagor
Department of Chemical Engineering
Nnamdi Azikiwe University
Awka, Nigeria

Iqra Zubair Awan
Department of Polymer Engineering and
  Technology
University of the Punjab
Lahore, Pakistan

Tariq Aziz
Sub-Campus UAF at Depalpur Okara
University of Agriculture Faisalabad
Faisalabad, Pakistan

Madeeha Batool
Centre for Analytical Chemistry
School of Chemistry, University of the Punjab
Lahore, Pakistan

Jamil A. Buledi
National Centre of Excellence in Analytical
  Chemistry
University of Sindh
Jamshoro, Pakistan

Shazia Chohan
National Centre of Excellence in Analytical
  Chemistry
University of Sindh
Jamshoro, Sindh-Pakistan

Gabriela Flores-Rangel
Institute of Analytical and Bioanalytical
  Chemistry
Ulm University
Ulm, Germany

Nuran Gokdere
Faculty of Pharmacy, Department of Analytical
  Chemistry
Ankara University
Ankara, Türkiye
and
Graduate School of Health Sciences
Ankara University
Ankara, Türkiye

Fabio Granados-Chinchilla
Faculty of Microbiology, Research Center for
  Tropical Diseases (CIET)
Universidad de Costa Rica
San José, Costa Rica
and
Faculty of Basic Sciences, School of Chemistry
Universidad de Costa Rica
San José, Costa Rica

Zia-ul Hassan
Department of Soil Science
Sindh Agriculture University (SAU)
Sindh, Pakistan

Ali Hyder
National Centre of Excellence in Analytical
Chemistry
University of Sindh
Jamshoro, Pakistan

Martin Emeka Ibenta
Department of Polymer Engineering
Nnamdi Azikiwe University
Awka, Nigeria

Hafiz Muhammad Junaid
Institute of Chemistry
University of the Punjab
Punjab, Lahore, Pakistan

Bindia Junejo
National Centre of Excellence in Analytical
Chemistry
University of Sindh
Jamshoro, Pakistan

Hüseyin Kara
Department of Chemistry
Selcuk University
Konya, Türkiye

Nadir H. Khand
National Centre of Excellence in Analytical
Chemistry
University of Sindh
Jamshoro, Pakistan

Zahid Hussain Laghari
National Centre of Excellence in Analytical
Chemistry
University of Sindh
Jamshoro, Pakistan

Lorena Díaz de León-Martínez
Institute of Analytical and Bioanalytical
Chemistry
Ulm University
Ulm, Germany
and
LABINNOVA INC.
Life Sciences Technologies
Spring, Texas

Eduardo Alberto López-Maldonado
Faculty of Chemical Sciences and Engineering
Autonomous University of Baja California
Baja California, Mexico

Sarfaraz Ahmed Mahesar
National Centre of Excellence in Analytical
Chemistry
University of Sindh
Jamshoro, Pakistan

Hassan Karimi Maleh
School of Resources and Environment
University of Electronic Science and
Technology of China
Chengdu, P.R. China
and
School of Engineering
Lebanese American University
Byblos, Lebanon

Arfana Mallah
M.A. Kazi Institute of Chemistry
University of Sindh
Sindh, Pakistan

Najma Memon
National Centre of Excellence in Analytical
Chemistry
University of Sindh
Sindh, Pakistan

Boris Mizaikoff
Institute of Analytical and Bioanalytical
Chemistry
Ulm University
Ulm, Germany
and
Hahn-Schickard
Ulm, Germany

Muhammad Nawaz
National Centre of Excellence in Analytical
Chemistry
University of Sindh
Jamshoro, Pakistan

Emmanuel C. Nwadike
Department of Mechanical Engineering
Nnamdi Azikiwe University
Awka, Nigeria

Christopher Chiedozie Obi
Department of Polymer and Textile
Engineering
Nnamdi Azikiwe University
Awka, Nigeria

Helen Onyeaka
School of Chemical Engineering
University of Birmingham
Birmingham, United Kingdom

Henry Chukwuka Oyeoka
Department of Polymer Engineering
Nnamdi Azikiwe University
Awka, Nigeria

Ismail Murat Palabiyik
Faculty of Pharmacy, Department of Analytical
  Chemistry
Ankara University
Ankara, Türkiye

Ummugulsum Polat
Faculty of Pharmacy, Department of Analytical
  Chemistry
Ankara University
Ankara, Türkiye
and
Graduate School of Health Sciences
Ankara University
Ankara, Türkiye
and
Faculty of Pharmacy, Department of Analytical
  Chemistry
Cumhuriyet University
Sivas, Türkiye

Mauricio Redondo-Solano
Research Center for Tropical Diseases (CIET)
Faculty of Microbiology
Universidad de Costa Rica
San José, Costa Rica
and
Food Microbiology Research and Training
  Laboratory (LIMA)
Faculty of Microbiology
Universidad de Costa Rica
San José, Costa Rica

José Alejandro Roque-Jiménez
Animal Production Department
Autonomous Metropolitan University—
  Xochimilco CDMX
Mexico City, Mexico

and
Agricultural Sciences Institute
Autonomous University of Baja California
Baja California, Mexico

Huma Shaikh
National Centre of Excellence in Analytical
  Chemistry
University of Sindh
Sindh, Pakistan

Syed Tufail Hussain Sherazi
National Centre of Excellence in Analytical
  Chemistry
University of Sindh
Jamshoro, Pakistan

Hadia Shoaib
National Centre of Excellence in Analytical
  Chemistry
University of Sindh
Jamshoro, Pakistan

Ahmed Raza Sidhu
Dr. M. A. Kazi Institute of Chemistry
University of Sindh
Jamshoro, Pakistan

Amber R. Solangi
National Centre of Excellence in Analytical
  Chemistry
University of Sindh
Jamshoro, Pakistan

Sanam Iram Soomro
Department of Chemistry
Selcuk University
Konya, Türkiye

Mehrunnisa Tanzil
National Centre of Excellence in Analytical
  Chemistry
University of Sindh
Sindh, Pakistan

1 Nanobiotechnology in Food Science

*Jamil A. Buledi, Arfana Mallah,
Amber R. Solangi, and Sidra Amin*

1.1 INTRODUCTION

In the last few decades, nanotechnology has emerged as an enticing field with the potential to revolutionize the food industry. Functioning on a nanoscale, this technology involves the manipulation of atoms, molecules, and macromolecules ranging in size from approximately 1 to 100 nm. Its primary aim is to design and utilize materials with unique attributes, such as novel surface properties or internal structures within this nanoscale range. This allows for manipulating and observing matter at a minute scale (Buledi et al., 2022; Rai, Yadav, & Gade, 2009). Notably, the resulting nanomaterials exhibit distinct characteristics that distinguish them from their larger counterparts. These disparities arise from factors like the elevated surface area, introducing novel physiochemical attributes like color, solubility, and strength.

Furthermore, nanotechnology has spurred a modern industrial revolution, capturing the interest of both developed and developing nations as they invest more resources into this cutting-edge domain. Consequently, nanotechnology presents a vast array of prospects for innovating structures, materials, and systems with unprecedented features across diverse sectors, including agriculture, food production, and healthcare. This multidisciplinary field's potential is vast and continues to garner attention and investment on a global scale (Qureshi et al., 2012).

The growing concerns among consumers regarding the quality and health aspects of food products have spurred researchers to explore methods for improving food quality without compromising nutritional content. This has led to an increased interest in the utilization of nanoparticle-based materials within the food industry (Fakhouri et al., 2014). These materials are appealing due to their incorporation of essential elements and demonstrated non-toxic characteristics. Moreover, they exhibit stability under high temperatures and pressures (Sari et al., 2015).

Nanotechnology presents a comprehensive range of solutions for various stages of the food industry, encompassing production, processing, and packaging. The integration of nanomaterials introduces significant enhancements not only to food quality and safety but also to the potential health benefits provided by food consumption. Numerous entities, including research organizations and industries, actively pursue innovative techniques, methodologies, and products that leverage nanotechnology's applications within food science (Jamil Ahmed Buledi, Amin, Haider, Bhanger, & Solangi, 2021; Dasgupta et al., 2015; Hyder et al., 2022).

Nanotechnology has found significant applications within the food industry, primarily categorized into two key domains: food nanostructured ingredients and food nanosensing. Food nanostructured ingredients encompass a broad spectrum, from enhancing food processing techniques to innovations in food packaging solutions (Duncan, 2011). Within food processing, these nanostructures serve multifaceted roles, acting as versatile food additives, proficient carriers facilitating intelligent nutrient delivery, effective anti-caking agents, and potent antimicrobial agents. Furthermore, they can also serve as enhancers, bolstering packaging materials' mechanical robustness and longevity. On the other hand, food nanosensing holds promise for elevating the evaluation of food quality and safety standards, ushering in more precise and comprehensive assessments (Ezhilarasi, Karthik, Chhanwal, & Anandharamakrishnan, 2013).

DOI: 10.1201/9781003514039-1

1.2 SYNTHESIS AND CHARACTERIZATION OF NANOPARTICLES (NPs)

Different methods for synthesizing NPs have been used, but generally, the NPs are synthesized either by a top-to-bottom approach or by a bottom-up approach. The top-to-bottom approach employs a destructive method for NP synthesis, involving the breakdown of larger molecules into smaller units, which are then converted into NPs. Various techniques include grinding/milling, chemical vapor deposition (CVD), physical vapor deposition (PVD), and other decomposition methods. For instance, milling was used to produce coconut shell (CS) NPs, where raw CS powders were finely milled with a planetary mill and ceramic balls for varying durations. Analysis showed that longer milling times resulted in smaller crystallite sizes and a reduction in the NPs' brownish color. Conversely, the reverse approach involves building up NPs using simpler substances, as seen in techniques like sol-gel synthesis and biochemical synthesis, exemplified by the creation of TiO_2 anatase NPs embedded with graphene domains by Mogilevsky et al. through alizarin binding, confirmed by X-ray powder diffraction (XRD, a rapid analytical technique primarily used for phase identification of crystalline materials their unit cell dimensions) (Mogilevsky et al., 2014). Different NP synthesis routes are depicted in Figure 1.1.

Moreover, the morphology of NPs has long been a focal point of research due to its significant impact on NP properties. Various methods include scanning electron microscopy (SEM) and transmission electron microscopy (TEM). Both imaging techniques that detect reflected or transmitted electrons (electrons that are passing through the sample) and polarized optical microscopy (POM, used to highlight the characteristics of various anisotropic samples) are employed to study NP morphology. SEM uses electron scanning to provide nanoscale insights into NP features and dispersion patterns within matrices, as in Saeed and Khan's study on single-walled carbon nanotubes (SWNTs) in polymer matrices (Saeed & Khan, 2014). Conversely, TEM relies on electron transmittance and has been vital in revealing the diverse morphologies of gold NPs, as demonstrated by Khlebtsov and Dykman (2010). This technique also allows the visualization of complex multilayer structures. The TEM images of Au-NPs are shown in Figure 1.2.

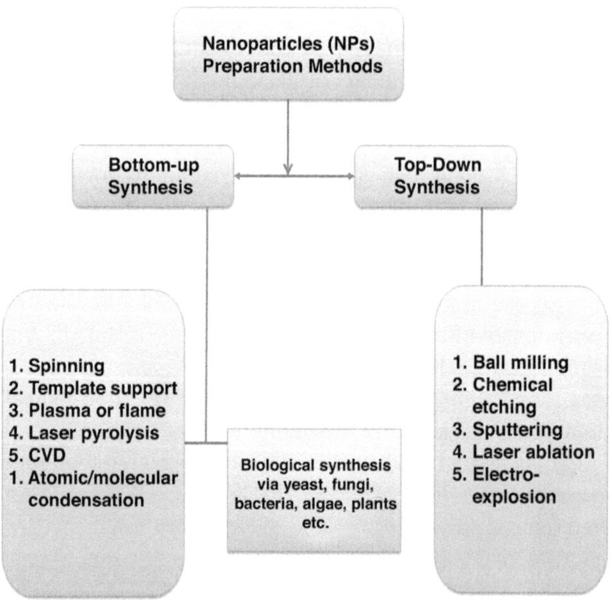

FIGURE 1.1 Bottom-up, top-down, and biological synthesis of NPs. Copyright © 2023, with permission from Elsevier (Khan, Saeed, & Khan, 2019).

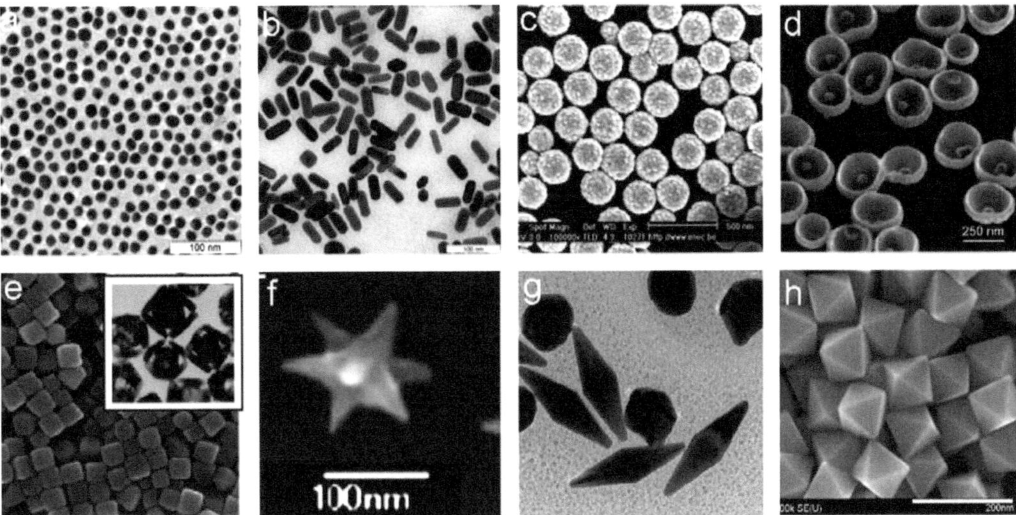

FIGURE 1.2   High-resolution TEM images of different forms of gold NPs. Copyright © 2023, with permission from Elsevier (Khlebtsov & Dykman, 2010).

In addition, studying the structural properties is essential for comprehending material composition and bonding. Various analytical methods, such as XRD, energy dispersive, and photoelectron X-ray spectroscopy (EDX and XPS, respectively, measuring X-rays or photoelectrons emitted from a sample), Infrared and Raman spectroscopy (both involve interaction of radiation with molecular vibrations but differ in the manner in which photon energy is transferred to the molecule by changing its vibrational state), Brunauer-Emmett-Teller (BET, utilized for determining specific surface areas and pore size distributions of solid materials), and Zeta potential (electrical potential at the slipping plane) analysis, are used to investigate NPs structures. XRD is vital for assessing crystallinity and phase, with the Debye-Scherer formula estimating particle size (Baksh et al., 2020; Taqvi et al., 2022). However, challenges arise with extremely small NPs and amorphous characteristics. In SEM or TEM analysis, an electron beam generates characteristic X-rays from NPs, aiding in element identification when combined with other techniques (Astefanei, Núñez, & Galceran, 2015).

Moreover, the optical properties play a pivotal role in photocatalytic applications, where researchers specializing in photochemistry have pointed their expertise to uncover the details of their photochemical processes. These investigations rely on the well-established Beer-Lambert law and fundamental principles of light, offering insights into NPs' light absorption, reflectivity, luminescence, and phosphorescence features. Metallic and semiconductor nanoparticles, known for their vibrant colors, are particularly suitable for photo-related applications. Consequently, determining the absorption and reflectivity of these materials is of great interest, enhancing our understanding of their fundamental mechanisms in diverse applications. To explore these optical characteristics, researchers employ essential tools such as Ultraviolet-Visible (UV–vis) spectroscopy, photoluminescence (PL) analysis, and null ellipsometry (Liu et al., 2015).

1.3   PROPERTIES OF NPs

The properties of NPs exhibit a strong interdependence between their optical and electronic characteristics. For example, noble metal NPs, such as gold and silver, display size-dependent optical properties that differ from the bulk metal. Specifically, they exhibit a distinct UV-visible extinction band attributed to the localized surface plasmon resonance (LSPR). This resonance occurs when incident photons match the collective excitation of conduction electrons within the NPs, resulting in wavelength-selective absorption. LSPR excitation leads to remarkable optical phenomena, including resonance Rayleigh light scattering, with an efficiency comparable to multiple fluorophores, and the creation of enhanced local electromagnetic fields near the NP surface. This enhancement enables advanced spectroscopic techniques. Notably, the peak wavelength of the LSPR spectrum is influenced by factors such as NP size, shape, interparticle spacing, dielectric properties, and local environment, including substrate, solvents, and adsorbates. An interesting outcome of these properties is the distinctive visual appearance of materials containing these NPs. For instance, gold colloidal NPs contribute to the formation of rust-colored stains on glass doors and windows, whereas silver NPs result in a characteristic yellow tint. This is attributed to freely transportable surface electrons (d electrons in silver and gold), which play a role in the coloration and interactions with light within these nanomaterials (Eustis & El-Sayed, 2006).

Moreover, magnetic NPs (NPs) have attracted significant attention across diverse scientific fields, spanning heterogeneous and homogeneous catalysis, biomedicine, magnetic fluids, data storage, magnetic resonance imaging (MRI), and environmental cleanup like water purification. Existing research highlights that NPs exhibit optimal performance when their size remains below a critical threshold, typically around 10–20 nm (Reiss & Hütten, 2005). At this diminutive scale, the magnetic characteristics of NPs take precedence, rendering these particles precious and suitable for a wide array of applications (Faivre & Bennet, 2016). The magnetic properties of NPs stem from their uneven distribution of electrons. Moreover, these properties are intricately tied to the synthetic methodology employed (Wu, He, & Jiang, 2008).

1.4 TYPES OF NANOMATERIALS USED IN FOOD INDUSTRIES

1.4.1 Organic NPs

Organic NPs have surfaced as a captivating realm of exploration within food science. These ultra-tiny particles, typically measuring between 1 and 100 nm, carry substantial potential for a transformative influence on various facets of the food industry. By capitalizing on their distinct physical and chemical attributes, organic NPs can elevate the dissolvability, robustness, and accessibility of bioactive compounds such as vitamins, antioxidants, and agents that combat microbial growth in food items. Moreover, they introduce fresh prospects for precisely managed delivery systems, enabling the gradual and targeted discharge of nutrients and aromas (Pan & Zhong, 2016). This, in turn, enhances edibles' overall sensory encounter and nutritional utility. As continuous research unveils the multifaceted utilities of organic NPs in food science, it becomes evident that their assimilation could culminate in innovative compositions, augmented food safety measures, and, ultimately, a more sustainable and health-oriented strategy toward creating and consuming sustenance (Moraru et al., 2003).

1.4.2 Polymeric NPs

Polymeric NPs utilize biocompatible and biodegradable polymers from natural and synthetic origins. The biodegradable polymers encompass synthetic variants such as poly(lactic acid), polyglycolic acid, poly(lactic-co-glycolic acid), poly(ε-caprolactone), polymethyl methacrylate, and poly(amino acid). Additionally, biodegradable polymers can be derived from natural sources, including but not limited to agarose, sodium alginate, chitosan, collagen, and fibrin (Yu et al., 2018). These polymers offer a platform for achieving controlled release of core materials, thus contributing to the growing popularity of polymeric NPs for applications like anticancer therapy and vaccine delivery. The unique chemical characteristics of polymeric NPs and their versatility make them conducive for integration with biomaterials (such as genetic material and growth factors) and enable targeted delivery to promote tissue regeneration (Saravanan et al., 2017).

1.4.3 CARBON-BASED NANOCARRIERS

Carbon nanotubes exhibit a unique carbon-based tubular morphology, resembling a rolled-up graphene sheet or capped buckyballs (Duncan & Izzo, 2005). These structures manifest in two primary configurations: single-walled nanotubes (SWNTs) and multiwalled nanotubes (MWNTs). SWNTs consist of a single graphene cylinder, whereas MWNTs encompass multiple concentric cylindrical graphene shells around a central hollow core (Xiao et al., 2016). The functionalization of nanotubes classifies them into distinct types, including target-specific, ligand-bound, solvent-dispersed, and surfactant-grafted varieties. Alongside tubular forms, fullerenes, a different carbon-based nanocarrier, display cage-like geometries comprising hexagonal and pentagonal carbon facets (Ezzati Nazhad Dolatabadi, Omidi, & Losic, 2011).

1.4.4 QUANTUM DOTS

Quantum dots represent nanoscale crystals composed of fluorescent inorganic semiconductor atoms, typically ranging in size from 2 to 10 nm. These nanocrystals utilize cadmium selenide as the semiconducting material, featuring a central core surrounded by an aqueous zinc sulfide shell. This shell acts as an insulating layer for the core, enhancing the quantum dots' optical characteristics. Notably, these engineered quantum dots can emit light across a broad spectrum, spanning from ultraviolet to infrared wavelengths (Qu et al., 2017). The emitted wavelengths exhibit considerable intensity, allowing for their detection at the subcellular level with precision. Moreover, quantum dots offer stability and inertness as carriers, facilitating the attachment of biomolecules to their outer aqueous shell. This property renders them an effective platform for delivering various biomolecules (Mo et al., 2017).

1.4.5 INORGANIC NPS

In food science, there has been a notable emergence of inorganic NPs, presenting a wealth of exciting possibilities and advantages. These minuscule materials, often comprised of metals, metal oxides, or other inorganic compounds, showcase distinct physicochemical traits that can be leveraged to elevate food quality, safety, and utility. A particularly intriguing application lies within food packaging (Buledi & Solangi, 2023; Le Guével, 2017). Integrating NPs into packaging makes it possible to establish formidable barriers against oxygen and moisture, thereby prolonging the shelf life of products and safeguarding their freshness. Furthermore, the potential of inorganic NPs extends to precisely delivering bioactive elements like vitamins or nutraceuticals. This precision facilitates enhanced nutrient absorption, culminating in overarching health benefits. It is of paramount importance, however, that as this field evolves, due consideration is accorded to safety apprehensions and regulatory frameworks. This ensures the conscientious and principled utilization of inorganic NPs within food items. The persistent advancement of research in this sphere promises a transformative trajectory for the food industry, offering innovative resolutions to the quandaries of sustenance preservation, nutritional augmentation, and advanced delivery systems (Hoseinnejad, Jafari, & Katouzian, 2018; Shah et al., 2023).

1.4.6 METAL OXIDE NPS

Metal oxide NPs have emerged as a promising and innovative food processing and safety strategy. These NPs, typically ranging from 1 to 100 nm, exhibit distinct physicochemical characteristics that render them suitable for various applications in the food industry. Metal oxide NPs are making significant advancements in food processing (Dos Santos, Ingle, & Rai, 2020; Nawaz et al., 2023). Metal oxide NPs are harnessed within food processing for their unique antimicrobial and antioxidant attributes (Nikolic, Vasiljevic, Auger, & Vidic, 2021). For instance, zinc oxide NPs have been adopted as agents to extend the shelf life of packaged foods by hindering the growth of pathogenic bacteria and spoilage microorganisms. Similarly, titanium dioxide NPs are employed as additives to augment specific products' brightness and color stability. These NPs elevate the visual allure of food items and contribute to upholding their overall quality. Furthermore, metal oxide NPs are pivotal in advancing food safety (Garcia, Shin, & Kim, 2018). The persistent concern about food contamination by harmful pathogens underscores the need for practical solutions to counteract foodborne illnesses. Metal oxide NPs, including silver and copper NPs, exhibit potent antimicrobial properties capable of combating a broad spectrum of pathogens encompassing bacteria, viruses, and fungi. Incorporating these NPs into food packaging materials or items can establish a protective barrier against microbial contamination, diminishing the peril of foodborne diseases (Zhou, Pu, & Sun, 2021).

1.4.7 Metallic NPs

Metallic NPs, such as silver, gold, zinc oxide, and titanium dioxide, exhibit remarkable potential in reshaping the landscape of food processing and safety. Through their intrinsic antimicrobial, catalytic, and photocatalytic attributes, these NPs present avenues for elevating the caliber of food products, prolonging their shelf life, and abating the presence of contaminants. Nevertheless, the conscientious assimilation of these NPs into food necessitates a comprehensive exploration of their efficiency and safety, accentuating the significance of continuous regulatory supervision and scientific inquiry (Jafarzadeh & Jafari, 2021). While the prospects presented by these metal NPs are enticing, their integration into food items and packaging mandates meticulous scrutiny of plausible hazards. The dynamic interplay between NPs and biological entities, encompassing humans, remains an ongoing realm of investigation. A stringent appraisal of their toxicity must be undertaken to ensure the judicious application of metal NPs within the food industry, aligning closely with established regulatory frameworks (Dos Santos et al., 2020). Safeguarding this amalgamation demands a rigorous evaluation of potential consequences, with a focus on potential risks. The complex interactions between NPs and living organisms serve as an ever-evolving domain of study. In this context, the assured and secure use of metal NPs within the food sector obliges a thorough examination of their safety profile, coupled with unwavering adherence to stipulated regulatory standards (Ebrahimi, Peighambardoust, Peighambardoust, & Karkaj, 2019).

1.4.8 Carbon Nanotubes

Carbon nanotubes (CNTs) exhibit an elongated, tubular morphology with diameters ranging from 1 to 2 nm (Ibrahim, 2013). The nature of their conductivity, whether metallic or semiconducting, is contingent upon their diameter (Aqel, Abou El-Nour, Ammar, & Al-Warthan, 2012). Structurally, CNTs closely resemble the act of rolling a graphite sheet into a tubular shape (refer to Figure 1.3). These rolled sheets can manifest as single-walled (SWNTs), double-walled (DWNTs), or multi-walled carbon nanotubes (MWNTs) based on the number of walls they possess. Synthesis methods encompass the deposition of carbon precursors, particularly individual carbon atoms vaporized from graphite using laser or electric arc onto metal particles. More recently, the chemical vapor deposition (CVD) technique has also been employed for their synthesis (Buledi et al., 2023; Elliott, Shibuta, Amara, Bichara, & Neyts, 2013). Leveraging their distinctive physical, chemical, and mechanical properties, these materials find utility in their pristine state and as constituents of nanocomposites for diverse commercial applications. These include deployment as fillers, a role highlighted by Saeed and Khan (2016), efficacious adsorbents for environmental remediation purposes, and as a supportive medium for various inorganic and organic catalysts.

1.4.9 Nanoclays

Nanoclays, derived from natural clay minerals at the nanoscale, have gained substantial attention in food processing, offering the potential to augment the quality and safety of food products. These

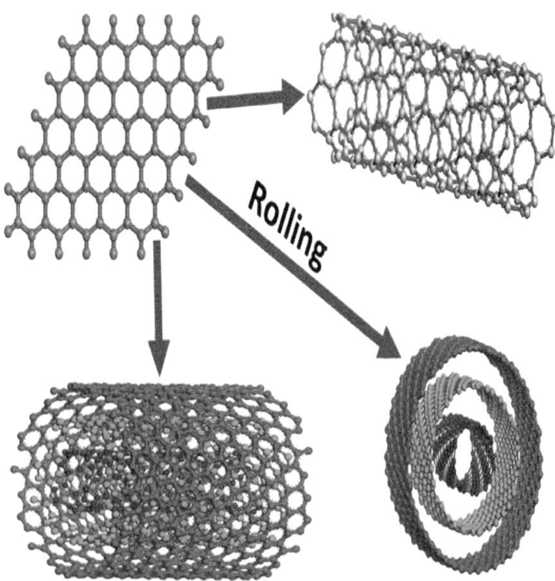

FIGURE 1.3 Rolling of graphite layer in single and multiwalled carbon nanotubes. Copyright © 2023, with permission from Elsevier (Aqel et al., 2012).

minuscule clay-derived particles possess distinctive attributes, including a considerable surface area, robust mechanical traits, and effective barrier characteristics. These attributes have positioned them as valuable augmentations in diverse food applications (Majeed et al., 2013). Nanoclays have found versatile utility in food processing as additives in packaging materials. Through their integration into packaging films or coatings, manufacturers can fabricate barriers that adeptly impede the infiltration of oxygen, moisture, and other extraneous pollutants. This, in turn, engenders an elongated shelf life, diminished spoilage, and elevated preservation standards for food items.

Furthermore, nanoclays contribute to the heightened structural integrity of packaging materials, thereby bolstering safeguarding measures during transit and storage (Bumbudsanpharoke & Ko, 2019). Moreover, nanoclays offer promising potential in the realm of food safety — their application aids in mitigating the peril of microbial contamination. By constituting a tangible barricade, nanoclays can hinder the proliferation of bacteria and pathogens on the surface of comestibles. Additionally, their proficiency in assimilating specific organic compounds facilitates the removal of undesirable impurities from food matrices. This facet is of particular significance in curtailing detrimental substances such as mycotoxins, heavy metals, and pesticides, consequently amplifying the holistic safety of consumable goods (Deshmukh, Hakim, Akhila, Ramakanth, & Gaikwad, 2023).

The primary sector in 2014, contributing USD 343.0 million, was the global nanoclay market for food packaging, with expectations of substantial growth until 2022 (Bumbudsanpharoke & Ko, 2019). Compared to alternative nanofillers such as nanosilica (Voon, Bhat, Easa, Liong, & Karim, 2012), calcium carbonate, and crystalline cellulose (Resano-Goizueta, Ashokan, Trezza, & Padua, 2018), nanoclay demonstrates equivalent or superior performance. Research by Voon and coworkers (Voon et al., 2012) showed that halloysite nanoclay outperforms nanosilica when combined with bovine gelatin polymer regarding mechanical properties while improving barrier characteristics and water solubility. Including either nanoclay or calcium carbonate enhances the mechanical strength of polypropylene film, yet the minimum required content for each compound differs significantly — 2 g/100 g for nanoclay and 8 g/100 g for calcium carbonate. The nanofiller's aspect ratio and particle shape influence the mechanical enhancement. Notably, the plate-like structure of nanoclay imparts superior mechanical attributes to bio-based polymers compared to the spherical, cubical, or acicular morphology of nanocellulose. Furthermore, nanoclays offer a cost-effective solution for functional packaging of fast-moving consumer goods, such as food and beverage items. The incorporation of a modest quantity of nanoclay (<10 g/100 g) can substantially augment a base polymer's thermal, mechanical, barrier, and degradation properties (Makaremi et al., 2017).

1.5 NANOCOATING

Nanocoatings have emerged as a groundbreaking technological advancement within food processing. These ultra-thin protective layers, typically measuring between 1 and 100 nm, are meticulously applied to food surfaces and packaging materials. The primary objective behind integrating nanocoatings into food processing procedures is to establish a formidable barrier that effectively counters pathogen growth, curbs oxidation, and combats moisture infiltration. This concerted approach plays a pivotal role in mitigating contamination and spoilage. Especially noteworthy merit of nanocoatings is their exceptional ability to hinder the proliferation of detrimental microorganisms (Jafarzadeh, Mohammadi Nafchi, Salehabadi, Oladzad-abbasabadi, & Jafari, 2021). A compelling example lies in the application of silver NPs, which possess remarkable antimicrobial attributes capable of thwarting the expansion of bacteria and viruses. This robust safeguarding measure offers invaluable protection to food items, significantly minimizing the potential for contamination. Furthermore, deploying nanocoatings is an effective deterrent against moisture migration, disrupting the optimal conditions for mold and fungi to flourish, particularly in damp environments. By erecting this safeguarding shield, nanocoatings also reduce excessive preservative usage, aligning harmoniously with the escalating consumer demand for minimally processed and more health-conscious food choices (Costa, Conte, Buonocore, Lavorgna, & Del Nobile, 2012).

In ensuring food safety, nanocoatings are pivotal in upholding packaged edibles' integrity. These specialized coatings effectively function as an impervious barrier, adeptly fending off oxygen and other gases. In doing so, they effectively stem the tide of food quality deterioration from oxidation and rancidity. Moreover, the judicious employment of nanocoatings extends the shelf life of perishable goods, resulting in a marked reduction in food wastage. This conspicuous benefit is instrumental in promoting sustainable consumption patterns and positively influences efforts to curtail unnecessary food disposal. By preserving the freshness and safety of consumable products, nanocoatings tangibly contribute to establishing a more secure and reliable food supply chain (Othman, Abd Salam, Zainal, Kadir Basha, & Talib, 2014).

1.5.1 Bio-Nanocomposite Coatings of Fruits

Bananas, mangoes, longans, and papayas stand out as among the most well-liked tropical fruits cultivated in tropical regions across the globe. These fruits boast significant nutritional value and offer numerous health benefits (Li et al., 2019). However, their brief shelf life, rapid deterioration attributed to biochemical transformations, physiological aging, and microbial contamination during storage and transportation pose a challenge. This often leads to a decline in the overall quality of these fruits, resulting in substantial annual losses of fresh produce. Researchers have proposed various strategies to preserve the quality of these fruits. These approaches encompass modified atmospheric packaging, applying protective wax coatings, and the utilization of innovative bio-nanocomposites (Shi et al., 2013). Edible coatings and bio-nanocomposite films have gained significant recognition for maintaining postharvest quality. The application of edible coatings is environmentally friendly, cost-effective, and safe, making them a promising method for preserving tropical fruits. Li and coworkers (Li et al., 2017) explored the effects of pure SPI coating and SPI/cinnamaldehyde/ZnO NP coating on bananas. The results highlighted the remarkable reduction in weight loss and ripening rate, with the nanocomposite film maintaining sensory attributes, firmness, sugar content, and acidity during storage. The SPI/CIN/ZnO NP coating exhibited stronger antifungal properties. In another investigation, Xing and coworkers (Xing et al., 2020) studied the

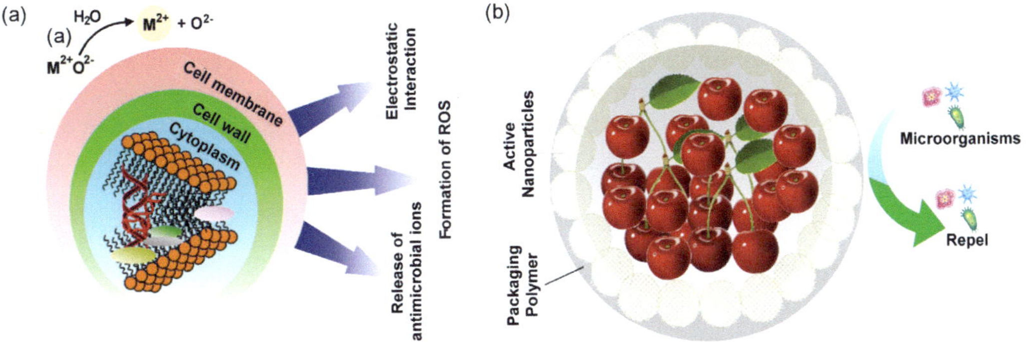

FIGURE 1.4 (a) Mechanism of antimicrobial activity of nanocomposite films, and (b) Repel mechanism of nanocomposites against microbes on fresh cherry. Copyright © 2023, with permission from Elsevier (Barikloo & Ahmadi, 2018).

influence of chitosan monolayer treatment and chitosan/nano-TiO$_2$ coating on mangoes. The nanocomposite coating demonstrated a 14.49% lower decay index and delayed respiratory activity by 5 days while enhancing firmness and altering soluble solid content. This approach effectively maintained nutrient composition and quality at a storage temperature of 13°C.

Chi and coworkers (Xing et al., 2020) introduced a poly(lactic acid) or PLA nanocomposite film incorporating nano-Ag and nano-TiO$_2$ for preserving mangoes. Throughout storage, these films minimized firmness loss and prevented undesirable changes in vitamin C, color, microbial properties, and acidity, surpassing the performance of PLA films. Mangoes coated with PLA nanocomposite films received a significantly higher acceptability index by the end of the storage period. Shi and coworkers (Shi et al., 2013) investigated nano-silica and chitosan coatings for longan fruit preservation. The semi-permeable nano-silica/chitosan films substantially extended shelf life, reduced browning, and minimized weight loss. Moreover, the peroxidase activity was notably lower in nanocomposite-coated longan. Their study highlighted the potential of nano-silica/chitosan coatings to enhance the storage quality of longan fruits (Figure 1.4).

1.5.2 COATING OF FRUITS/VEGETABLES THROUGH NPs

Although nanotechnology has made strides in food packaging, its widespread adoption has been hindered primarily due to concerns about health and environmental implications, particularly during storage. Additionally, detecting the presence of nanomaterials in food products, especially if they are released or interact with the food contents, remains a complex process due to the intricate and varied interaction and reaction profiles involved (Jafarzadeh et al., 2017; Hasan et al., 2019).

The term "nanoparticle" in food packaging encompasses a diverse range of materials, including but not limited to carbon materials, metals, mixed metal oxides, and nanolayers (Garavand et al., 2022; Pirozzi, Del Grosso, Ferrari, & Donsì, 2020). These materials find utility in nanocomposites, which involve the integration of particles with dimensions less than 100 nm as well as in coatings consisting of thin films with at least one dimension below 100 nm. This incorporation is favored due to the materials' advantageous physical, chemical, and biological properties, which can contribute to the final attributes of the products (Espitia, Otoni, & Soares, 2016). Nanocomposites fall into two categories: (i) inorganic-organic materials featuring non-covalent interactions and (ii) inorganic-organic materials exhibiting covalent interactions. Nanocomposites are fabricated through crucial chemical approaches such as sol-gel and self-assembly methods. The interplay between the nature of the polymeric materials (organic phase) and the NPs (inorganic phase) governs the ultimate properties of the resulting nanocomposites (Swaroop & Shukla, 2019).

1.6 ELECTROSPUN NANOFIBERS

Electrospun nanofibers have shown great promise in various applications, including food processing and safety. They offer several potential benefits, such as improved food packaging, controlled release of additives, enhanced shelf-life, and even new ways to deliver nutrients. However, as with any emerging technology, several critical points exist before widespread adoption in the food industry (Xue, Xie, Liu, & Xia, 2017) (Figure 1.5).

1.6.1 CHITOSAN NANOFIBERS

Chitosan, an alkaline polysaccharide with cationic properties, can be derived from the deacetylation process of chitin. This natural fiber source ranks second in size. With its excellent ability to form films, biodegradability, and antibacterial properties, chitosan is a valuable polymer in food industry applications (Shahidi, Arachchi, & Jeon, 1999). Despite its merits, chitosan poses challenges as it is not soluble in water, alkali aqueous solutions, or most mineral acidic solvents. However, certain organic solvents, such as acetic and formic acids, can dissolve it. The positively charged characteristics of chitosan and its cationic charge contribute to increased surface tension in solutions, resulting in limited spinnability. To address this, the spinnability of chitosan is often enhanced through blending with other polymers. Nevertheless, there exist instances where pure chitosan has been successfully electrospun into nanofibers using solvents like hexafluoroisopropanol, dichloromethane (DCM), and trifluoroacetate (TFA), as indicated in some reports. The electrospinnability of chitosan is significantly influenced by factors such as molecular weight, degree of deacetylation, and choice of solvent (Wang Yihan, 2014).

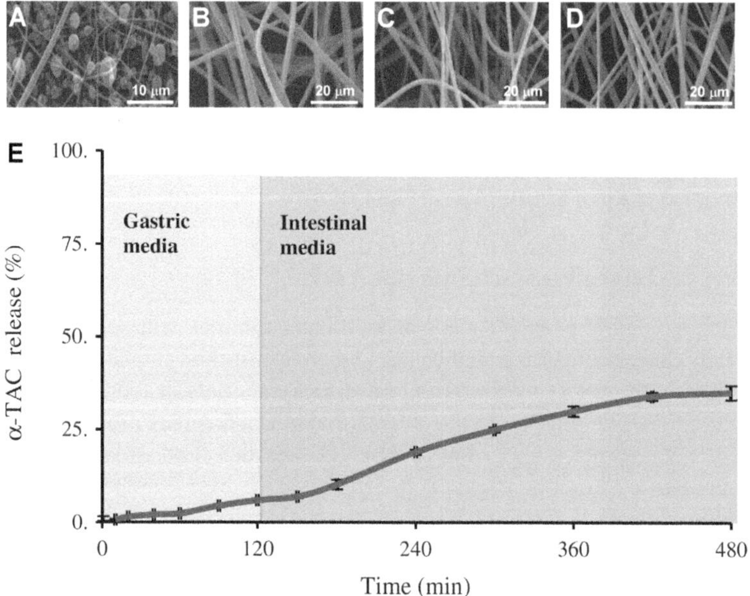

FIGURE 1.5 Microscopic images obtained through scanning electron microscopy depict various fibrous structures. (a) 0.375 g mL^{-1} dextran under the voltage of 14 kV, flow rate of 1 mL h^{-1}, and the distance of 15 cm at 25°C with a magnification of 2,000×. Vitamin E-loaded dextran nanofibers. (b) 1 g mL^{-1} dextran under 15 kV and (c) 1.25 and (d) 1 g mL^{-1} dextran under 13 kV with magnification 1,000×. (e) Release profile of vitamin E from dextran nanofibers in gastrointestinal media. Copyright © 2023, with permission from Elsevier (Fathi et al., 2017).

1.6.2 DEXTRAN

Dextran, a naturally occurring bacterial polysaccharide derived from Leuconostoc mesenteroides, exhibits a unique structure comprising α-1,6-linked D-glucopyranose leading chains with α-1,2, α-1,3, or α-1,4 linked side chains. This biopolymer is soluble in water and organic solvents, boasting biodegradability and biocompatibility. Notably, dextran forms high-viscosity solutions when dissolved in water, making it a valuable resource in the food industry for functions such as thickening and stabilizing (Aguilera-Miguel et al., 2018).

However, despite its manifold applications, the utilization of dextran nanofibers in the food sector remains relatively unexplored. The electro-spinnability of dextran (with a molecular weight range of 6,476 kDa is investigated, focusing on the influence of solvent and polymer concentrations. Within the 500–1,000 mg mL^{-1} concentration range, electrospinning at 750 mg mL^{-1} of dextran yielded the most uniform fibers, with the smallest diameters of approximately 100 nm. Intriguingly, an escalation in the electrospun solution's concentration to 1,000 mg mL^{-1} led to a dramatic increase in fiber diameters, reaching 3 mm. Furthermore, introducing 5 g/100 g bovine serum albumin (BSA) to the 750 mg mL^{-1} dextran solution showed a compelling reduction of nanofiber size. This phenomenon can be attributed to the augmented net charges, which exert enhanced forces on the jets, producing smaller fibers. A noteworthy aspect is that the inherent biological activity of BSA remained intact post-electrospinning with dextran. This finding underscores the promising potential of nanofibers as carriers for protein-based functional ingredients, marking a novel avenue in applications (Jiang, Fang, Hsiao, Chu, & Chen, 2004).

It is worth emphasizing that this condensed account sheds light on dextran's versatile attributes and its hitherto underexplored potential in food industry applications. The intricate interplay between dextran's electrospinnability and the fascinating impact of incorporating BSA holds promise for developing innovative and efficient food-related technologies (Jeong et al., 2014).

1.6.3 SILK NANOFIBERS

Silk can be categorized into mulberry and nonmulberry silks. The widely used silkworm, a prominent textile industry material, is derived from *Bombyx mori* silk and comprises two chains — heavy and light, along with a glycoprotein called P25. The ratio of heavy chain: light chain: P25 is 6:6:1. However, nonmulberry silks lack the light chain and P25 components. These structural differences lead to notable variations in bioactivity, mechanical properties, and degradation behavior between the two types of silk. Silk is characterized by biocompatibility, extensibility, biodegradability, high strength, and minimal inflammatory responses (Yang et al., 2013). Silk fibroin (SF) is extracted from Silkworms' cocoons. A solution containing 4 g/100 g SF is cast onto polystyrene Petri dishes and dried for different durations, yielding silk films primarily composed of nanospheres or nanofilaments. These films remain stable even after being dissolved in 98 mL/100 mL formic acid. Studies have shown that SF nanofilaments exhibit superior spinnability compared to SF nanospheres due to their lower viscosity. Moreover, altering the SF nanostructure and concentration allows easy control over fiber size, as evidenced by the increased average diameter of fibers with higher SF nanofilament concentrations (Lang, Neugirg, Kluge, Fery, & Scheibel, 2017).

1.6.4 GELATIN NANOFIBER

Gelatin, akin to collagen, is a related protein that undergoes partial hydrolysis. Its amino acid composition closely resembles that of collagen. In various pH environments, gelatin molecules can adopt positive ions, anions, or zwitterions. The triple-helix structure of collagen, resembling a rod, is partially disrupted through hydrolysis. Gelatin can be classified into two types based on preparation methods. Type A is derived from pig skin and acid-hydrolyzed, offering flexibility and elasticity. Type B, predominantly obtained from bones and animal skins, undergoes alkaline hydrolysis, enhancing hardness. Gelatin is popularly utilized in gel desserts due to its unique quality of "mouth-melting." It has also found utility in photographic emulsions for colloid preservation (Gomes, Rodrigues, Martins, Henriques, & Silva, 2013).

In food applications, gelatin is primarily employed for thickening and stabilizing. Similarly, electrospun gelatin nanofibers find application, possessing potent effects despite their minimal quantities. The concentration of type B gelatin powder from bovine skin and the applied voltage in electrospun gelatin nanofibers were examined. Findings suggest that electrospun gelatin nanofibers achieve a nanoscale status at 20 g/100 g gelatin concentration. Moreover, lower applied voltages correlate with higher z-potential and diffusion coefficient values in electrospun gelatin nanofiber dispersions (Okutan, Terzi, & Altay, 2014). In another study, gelatin from cold-water fish skin, was investigated (Gomes and coworkers employed various techniques, including glutaraldehyde (GTA) vapor, genipin, and dehydrothermal treatment, for crosslinking). Fish gelatin modified with GTA emerged as the most suitable substrate among these. Solvents such as water or concentrated acetic acid were efficaciously employed for electrospinning (Gomes et al., 2013).

1.7 CONCLUSION AND FUTURE PERSPECTIVE

In conclusion, integrating nanobiotechnology into food science has ushered in a new era of innovation and possibilities. The synthesis, design, and morphological properties of NPs, organic and inorganic, along with the application of nanoclays, nanocoatings, and electrospun nanofibers, have significantly expanded our understanding of food systems and revolutionized their manipulation. These advancements have enhanced food quality, safety, and shelf life while opening avenues for novel functional foods and personalized nutrition. The utilization of organic and inorganic NPs has paved the way for controlled delivery systems, encapsulation of bioactive compounds, and improved sensory attributes in foods.

The future of nanobiotechnological techniques in food science is promising and brimming with potential. Further research could focus on tailoring nanoparticle properties for specific applications, optimizing processing methods to ensure scalability, and investigating the long-term effects of nanoparticle consumption on human health and the environment. Developing sustainable and biocompatible nanomaterials could address current concerns and broaden the acceptance of nanotechnology in the food industry. As nanobiotechnological techniques evolve, collaborations between scientists, engineers, and regulatory bodies will play a pivotal role in establishing robust guidelines and standards for their safe and effective implementation. Additionally, advancements in characterization tools and techniques will facilitate deeper insights into nanoparticle behavior within complex food matrices.

Integrating nanobiotechnological techniques into food science presents exciting possibilities for enhancing food quality and safety and prompts us to navigate ethical, safety, and regulatory considerations conscientiously. By harnessing the full potential of these techniques, we can reshape the landscape of the food industry and ultimately contribute to healthier, more sustainable, and technologically advanced food systems for the benefit of all.

BIBLIOGRAPHY

Aguilera-Miguel, A., López-Gonzalez, E., Sadtler, V., Durand, A., Marchal, P., Castel, C., & Choplin, L. (2018). Hydrophobically modified dextrans as stabilizers for O/W highly concentrated emulsions. Comparison with commercial non-ionic polymeric stabilizers. *Colloids and Surfaces A: Physicochemical and Engineering Aspects, 550*, 155–166. doi: https://doi.org/10.1016/j.colsurfa.2018.04.022.

Aqel, A., Abou El-Nour, K. M., Ammar, R. A., & Al-Warthan, A. (2012). Carbon nanotubes, science and technology: Part (I) structure, synthesis and characterisation. *Arabian Journal of Chemistry, 5*(1), 1–23.

Astefanei, A., Núñez, O., & Galceran, M. T. (2015). Characterisation and determination of fullerenes: A critical review. *Analytica Chimica Acta, 882*, 1–21. doi: https://doi.org/10.1016/j.aca.2015.03.025.

Baksh, H., Buledi, J. A., Khand, N. H., Solangi, A. R., Mallah, A., Sherazi, S. T., & Abro, M. I. (2020). Ultra-selective determination of carbofuran by electrochemical sensor based on nickel oxide nanoparticles stabilized by ionic liquid. *Monatshefte für Chemie-Chemical Monthly, 151,* 1689–1696.

Barikloo, H., & Ahmadi, E. (2018). Shelf life extension of strawberry by temperatures conditioning, chitosan coating, modified atmosphere, and clay and silica nanocomposite packaging. *Scientia Horticulturae, 240,* 496–508. doi: https://doi.org/10.1016/j.scienta.2018.06.012.

Buledi, J. A., Amin, S., Haider, S. I., Bhanger, M. I., & Solangi, A. R. (2021). A review on detection of heavy metals from aqueous media using nanomaterial-based sensors. *Environmental Science and Pollution Research, 28,* 58994–59002.

Buledi, J. A., Hyder, A., Khand, N. H., Memon, S. A., Batool, M., & Solangi, A. R. (2023). Potential mitigation of dyes through mxene composites. In: *Handbook of Functionalized Nanostructured MXenes: Synthetic Strategies and Applications from Energy to Environment Sustainability* (pp. 283–300): Springer.

Buledi, J. A., Mahar, N., Mallah, A., Solangi, A. R., Palabiyik, I. M., Qambrani, N., ... Karimi-Maleh, H. (2022). Electrochemical quantification of mancozeb through tungsten oxide/reduced graphene oxide nanocomposite: a potential method for environmental remediation. *Food and Chemical Toxicology, 161,* 112843.

Buledi, J. A., & Solangi, A. R. (2023). Low-temperature aqueous growth of SDS-stabilized CuO electrocatalyst for ranitidine monitoring in pharmaceuticals. *Chemical Papers, 77,* 6377–6386. https://doi.org/10.1007/s11696-023-02946-6

Bumbudsanpharoke, N., & Ko, S. (2019). Nanoclays in food and beverage packaging. *Journal of Nanomaterials, 2019,* 8927167. doi: 10.1155/2019/8927167.

Costa, C., Conte, A., Buonocore, G. G., Lavorgna, M., & Del Nobile, M. A. (2012). Calcium-alginate coating loaded with silver-montmorillonite nanoparticles to prolong the shelf-life of fresh-cut carrots. *Food Research International, 48*(1), 164–169. doi: https://doi.org/10.1016/j.foodres.2012.03.001.

Dasgupta, N., Ranjan, S., Mundekkad, D., Ramalingam, C., Shanker, R., & Kumar, A. (2015). Nanotechnology in agro-food: from field to plate. *Food Research International, 69,* 381–400.

Deshmukh, R. K., Hakim, L., Akhila, K., Ramakanth, D., & Gaikwad, K. K. (2023). Nano clays and its composites for food packaging applications. *International Nano Letters, 13*(2), 131–153. doi: 10.1007/s40089-022-00388-8

Dos Santos, C. A., Ingle, A. P., & Rai, M. (2020). The emerging role of metallic nanoparticles in food. *Applied Microbiology and Biotechnology, 104,* 2373–2383.

Duncan, R., & Izzo, L. (2005). Dendrimer biocompatibility and toxicity. *Advanced Drug Delivery Reviews, 57*(15), 2215–2237.

Duncan, T. V. (2011). Applications of nanotechnology in food packaging and food safety: barrier materials, antimicrobials and sensors. *Journal of Colloid and Interface Science, 363*(1), 1–24.

Ebrahimi, Y., Peighambardoust, S. J., Peighambardoust, S. H., & Karkaj, S. Z. (2019). Development of antibacterial carboxymethyl cellulose-based nanobiocomposite films containing various metallic nanoparticles for food packaging applications. *Journal of Food Science, 84*(9), 2537–2548.

Elliott, J. A., Shibuta, Y., Amara, H., Bichara, C., & Neyts, E. C. (2013). Atomistic modelling of CVD synthesis of carbon nanotubes and graphene. *Nanoscale, 5*(15), 6662–6676.

Espitia, P. J. P., Otoni, C. G., & Soares, N. F. F. (2016). Chapter 34 - Zinc oxide nanoparticles for food packaging applications. In: J. Barros-Velázquez (Ed.), *Antimicrobial Food Packaging* (pp. 425–431). San Diego, CA: Academic Press.

Eustis, S., & El-Sayed, M. A. (2006). Why gold nanoparticles are more precious than pretty gold: Noble metal surface plasmon resonance and its enhancement of the radiative and nonradiative properties of nanocrystals of different shapes. *Chemical Society Reviews, 35*(3), 209–217. doi: 10.1039/b514191e.

Ezhilarasi, P., Karthik, P., Chhanwal, N., & Anandharamakrishnan, C. (2013). Nanoencapsulation techniques for food bioactive components: a review. *Food and Bioprocess Technology, 6,* 628–647.

Ezzati Nazhad Dolatabadi, J., Omidi, Y., & Losic, D. (2011). Carbon nanotubes as an advanced drug and gene delivery nanosystem. *Current Nanoscience, 7*(3), 297–314.

Faivre, D., & Bennet, M. (2016). Magnetic nanoparticles line up. *Nature, 535*(7611), 235–236. doi: 10.1038/535235a.

Fakhouri, F., Casari, A., Mariano, M., Yamashita, F., Mei, L. I., Soldi, V., & Martelli, S. (2014). Effect of a gelatin-based edible coating containing cellulose nanocrystals (CNC) on the quality and nutrient retention of fresh strawberries during storage. Paper presented at the *IOP Conference Series: Materials Science and Engineering,* 2nd International Conference on Structural Nano Composites (NANOSTRUC 2014) 20–21 May 2014, Madrid, Spain *64,* 012024. doi:10.1088/1757-899X/64/1/012024

Garavand, F., Cacciotti, I., Vahedikia, N., Rehman, A., Tarhan, Ö., Akbari-Alavijeh, S., ... Jafari, S. M. (2022). A comprehensive review on the nanocomposites loaded with chitosan nanoparticles for food packaging. *Critical Reviews in Food Science and Nutrition, 62*(5), 1383–1416. doi: 10.1080/10408398.2020.1843133.

Garcia, C. V., Shin, G. H., & Kim, J. T. (2018). Metal oxide-based nanocomposites in food packaging: Applications, migration, and regulations. *Trends in Food Science & Technology, 82*, 21–31.

Gomes, S. R., Rodrigues, G., Martins, G. G., Henriques, C. M. R., & Silva, J. C. (2013). In vitro evaluation of crosslinked electrospun fish gelatin scaffolds. *Materials Science and Engineering: C, 33*(3), 1219–1227. doi: https://doi.org/10.1016/j.msec.2012.12.014

Hasan, M., Chong, E. W. N., Jafarzadeh, S., Paridah, M. T., Gopakumar, D. A., Tajarudin, H. A., ... Abdul Khalil, H. P. S. (2019). Enhancement in the physico-mechanical functions of seaweed biopolymer film via embedding fillers for plasticulture application—a comparison with conventional biodegradable mulch film. *Polymers, 11*(2), 210. doi: 10.3390/polym11020210.

Hoseinnejad, M., Jafari, S. M., & Katouzian, I. (2018). Inorganic and metal nanoparticles and their antimicrobial activity in food packaging applications. *Critical Reviews in Microbiology, 44*(2), 161–181.

Hyder, A., Buledi, J. A., Nawaz, M., Rajpar, D. B., Orooji, Y., Yola, M. L., ... Solangi, A. R. (2022). Identification of heavy metal ions from aqueous environment through gold, silver and copper nanoparticles: An excellent colorimetric approach. *Environmental Research, 205*, 112475.

Ibrahim, K. S. (2013). Carbon nanotubes–Properties and applications: A review. *Carbon Letters, 14*(3), 131–144.

Jafarzadeh, S., Ariffin, F., Mahmud, S., Alias, A. K., Hosseini, S. F., & Ahmad, M. (2017). Improving the physical and protective functions of semolina films by embedding a blend nanofillers (ZnO-nr and nano kaolin). *Food Packaging and Shelf Life, 12*, 66 75. doi: https://doi.org/10.1016/j.fpsl.2017.03.001.

Jafarzadeh, S., & Jafari, S. M. (2021). Impact of metal nanoparticles on the mechanical, barrier, optical and thermal properties of biodegradable food packaging materials. *Critical Reviews in Food Science and Nutrition, 61*(16), 2640–2658.

Jafarzadeh, S., Mohammadi Nafchi, A., Salehabadi, A., Oladzad-abbasabadi, N., & Jafari, S. M. (2021). Application of bio-nanocomposite films and edible coatings for extending the shelf life of fresh fruits and vegetables. *Advances in Colloid and Interface Science, 291*, 102405. doi: https://doi.org/10.1016/j.cis.2021.102405.

Jeong, I., Lee, J., Vincent Joseph, K. L., Lee, H. I., Kim, J. K., Yoon, S., & Lee, J. (2014). Low-cost electrospun WC/C composite nanofiber as a powerful platinum-free counter electrode for dye sensitized solar cell. *Nano Energy, 9*, 392–400. doi: https://doi.org/10.1016/j.nanoen.2014.08.010.

Jiang, H., Fang, D., Hsiao, B. S., Chu, B., & Chen, W. (2004). Optimization and characterization of dextran membranes prepared by electrospinning. *Biomacromolecules, 5*(2), 326–333. doi: 10.1021/bm034345w.

Khan, I., Saeed, K., & Khan, I. (2019). Nanoparticles: Properties, applications and toxicities. *Arabian Journal of Chemistry, 12*(7), 908–931.

Khlebtsov, N. G., & Dykman, L. A. (2010). Optical properties and biomedical applications of plasmonic nanoparticles. *Journal of Quantitative Spectroscopy and Radiative Transfer, 111*(1), 1–35. doi: https://doi.org/10.1016/j.jqsrt.2009.07.012.

Lang, G., Neugirg, B. R., Kluge, D., Fery, A., & Scheibel, T. (2017). Mechanical testing of engineered spider silk filaments provides insights into molecular features on a mesoscale. *ACS Applied Materials & Interfaces, 9*(1), 892–900. doi: 10.1021/acsami.6b13093.

Le Guével, X. (2017). Overview of inorganic nanoparticles for food science applications. In: Axelos, M. & Van der Voorde, M. H. (Eds.), *Nanotechnology in Agriculture and Food Science* (pp. 197–208). Weinheim, Germany: Wiley-VCH Verlag GmbH & Co. KGaA. https://doi.org/10.1002/9783527697724.ch12

Li, J., Sun, Q., Sun, Y., Chen, B., Wu, X., & Le, T. (2019). Improvement of banana postharvest quality using a novel soybean protein isolate/cinnamaldehyde/zinc oxide bionanocomposite coating strategy. *Scientia Horticulturae, 258*, 108786. doi: https://doi.org/10.1016/j.scienta.2019.108786.

Li, W., Li, L., Cao, Y., Lan, T., Chen, H., & Qin, Y. (2017). Effects of PLA film incorporated with ZnO nanoparticle on the quality attributes of fresh-cut apple. *Nanomaterials, 7*(8), 207. doi:10.3390/nano7080207.

Liu, D., Li, C., Zhou, F., Zhang, T., Zhang, H., Li, X., ... Li, Y. (2015). Rapid synthesis of monodisperse Au nanospheres through a laser irradiation-induced shape conversion, self-assembly and their electromagnetic coupling SERS enhancement. *Scientific Reports, 5*(1), 7686. doi: 10.1038/srep07686.

Majeed, K., Jawaid, M., Hassan, A., Abu Bakar, A., Abdul Khalil, H. P. S., Salema, A. A., & Inuwa, I. (2013). Potential materials for food packaging from nanoclay/natural fibres filled hybrid composites. *Materials & Design* (1980–2015), *46*, 391–410. doi: https://doi.org/10.1016/j.matdes.2012.10.044.

Makaremi, M., Pasbakhsh, P., Cavallaro, G., Lazzara, G., Aw, Y. K., Lee, S. M., & Milioto, S. (2017). Effect of morphology and size of halloysite nanotubes on functional pectin bionanocomposites for food packaging applications. *ACS Applied Materials & Interfaces, 9*(20), 17476–17488. doi: 10.1021/acsami.7b04297.

Mo, D., Hu, L., Zeng, G., Chen, G., Wan, J., Yu, Z., ... Cheng, M. (2017). Cadmium-containing quantum dots: properties, applications, and toxicity. *Applied Microbiology and Biotechnology, 101*(7), 2713–2733.

Mogilevsky, G., Hartman, O., Emmons, E. D., Balboa, A., DeCoste, J. B., Schindler, B. J., ... Karwacki, C. J. (2014). Bottom-up synthesis of anatase nanoparticles with graphene domains. *ACS Applied Materials & Interfaces, 6*(13), 10638–10648. doi: 10.1021/am502322y.

Moraru, C. I., Panchapakesan, C. P., Huang, Q., Takhistov, P., Liu, S., & Kokini, J. L. (2003). Nanotechnology: a new frontier in food science understanding the special properties of materials of nanometer size will allow food scientists to design new, healthier, tastier, and safer foods. *Nanotechnology, 57*(12), 24–29.

Nawaz, M., Shaikh, H., Buledi, J. A., Solangi, A. R., Karaman, C., Erk, N., ... Camarada, M. B. (2024). Fabrication of ZnO-doped reduce graphene oxide-based electrochemical sensor for the determination of 2, 4, 6–trichlorophenol from aqueous environment. *Carbon Letters, 34*, 201–214. https://doi.org/10.1007/s42823-023-00562-8.

Nikolic, M. V., Vasiljevic, Z. Z., Auger, S., & Vidic, J. (2021). Metal oxide nanoparticles for safe active and intelligent food packaging. *Trends in Food Science & Technology, 116*, 655–668.

Okutan, N., Terzi, P., & Altay, F. (2014). Affecting parameters on electrospinning process and characterization of electrospun gelatin nanofibers. *Food Hydrocolloids, 39*, 19–26. doi: https://doi.org/10.1016/j.foodhyd.2013.12.022.

Othman, S. H., Abd Salam, N. R., Zainal, N., Kadir Basha, R., & Talib, R. A. (2014). Antimicrobial activity of TiO2 nanoparticle-coated film for potential food packaging applications. *International Journal of Photoenergy, 2014*, 945930. doi: 10.1155/2014/945930.

Pan, K., & Zhong, Q. (2016). Organic nanoparticles in foods: fabrication, characterization, and utilization. *Annual Review of Food Science and Technology, 7*, 245–266.

Pirozzi, A., Del Grosso, V., Ferrari, G., & Donsì, F. (2020). Edible coatings containing oregano essential oil nanoemulsion for improving postharvest quality and shelf life of tomatoes. *Foods, 9*(11), 1605. doi:10.3390/foods9111605.

Qu, W., Zuo, W., Li, N., Hou, Y., Song, Z., Gou, G., & Yang, J. (2017). Design of multifunctional liposome-quantum dot hybrid nanocarriers and their biomedical application. *Journal of Drug Targeting, 25*(8), 661–672.

Qureshi, M., Karthikeyan, S., Karthikeyan, P., Khan, P., Uprit, S., & Mishra, U. (2012). Application of nanotechnology in food and dairy processing: an overview. *Pakistan Journal of Food Science, 22*(1), 23–31.

Rai, M., Yadav, A., & Gade, A. (2009). Silver nanoparticles as a new generation of antimicrobials. *Biotechnology Advances, 27*(1), 76–83.

Reiss, G., & Hütten, A. (2005). Applications beyond data storage. *Nature Materials, 4*(10), 725–726. doi: 10.1038/nmat1494.

Resano-Goizueta, I., Ashokan, B. K., Trezza, T. A., & Padua, G. W. (2018). Effect of nano-fillers on tensile properties of biopolymer films. *Journal of Polymers and the Environment, 26*(9), 3817–3823. doi: 10.1007/s10924-018-1260-1.

Saeed, K., & Khan, I. (2014). Preparation and properties of single-walled carbon nanotubes/poly(butylene terephthalate) nanocomposites. *Iranian Polymer Journal, 23*(1), 53–58. doi: 10.1007/s13726-013-0199-2.

Saeed, K., & Khan, I. (2016). Preparation and characterization of single-walled carbon nanotube/nylon 6, 6 nanocomposites. *Instrumentation Science & Technology, 44*(4), 435–444.

Saravanan, S., Chawla, A., Vairamani, M., Sastry, T., Subramanian, K., & Selvamurugan, N. (2017). Scaffolds containing chitosan, gelatin and graphene oxide for bone tissue regeneration in vitro and in vivo. *International Journal of Biological Macromolecules, 104*, 1975–1985.

Sari, T., Mann, B., Kumar, R., Singh, R., Sharma, R., Bhardwaj, M., & Athira, S. (2015). Preparation and characterization of nanoemulsion encapsulating curcumin. *Food Hydrocolloids, 43*, 540–546.

Shah, S.-Z., Taqvi, I. H., Ameen, S., Mallah, A., Buledi, J. A., Khand, N. H., & Solangi, A. R. (2023). Plant based fabrication of CuO/NiO nanocomposite: a green approach for low-level quantification of vanillin in food samples. *Pure and Applied Chemistry, 95*(5), 501–511.

Shahidi, F., Arachchi, J. K. V., & Jeon, Y.-J. (1999). Food applications of chitin and chitosans. *Trends in Food Science & Technology, 10*(2), 37–51. doi: 10.1016/S0924-2244(99)00017-5.

Shi, S., Wang, W., Liu, L., Wu, S., Wei, Y., & Li, W. (2013). Effect of chitosan/nano-silica coating on the physicochemical characteristics of longan fruit under ambient temperature. *Journal of Food Engineering, 118*(1), 125–131. doi: 10.1016/j.jfoodeng.2013.03.029.

Swaroop, C., & Shukla, M. (2019). Development of blown polylactic acid-MgO nanocomposite films for food packaging. *Composites Part A: Applied Science and Manufacturing, 124*, 105482. doi: 1016/j.compositesa.2019.105482.

Taqvi, S. I. H., Solangi, A. R., Buledi, J. A., Khand, N. H., Junejo, B., Memon, A. F., ... Vasseghian, Y. (2022). Plant extract-based green fabrication of nickel ferrite (NiFe2O4) nanoparticles: an operative platform for non-enzymatic determination of pentachlorophenol. *Chemosphere, 294*, 133760.

Voon, H. C., Bhat, R., Easa, A. M., Liong, M. T., & Karim, A. A. (2012). Effect of addition of halloysite nanoclay and SiO2 nanoparticles on barrier and mechanical properties of bovine gelatin films. *Food and Bioprocess Technology, 5*(5), 1766–1774. doi: 10.1007/s11947-010-0461-y.

Wegner, K.D., Resch-Genger, U. (2024). The 2023 Nobel Prize in Chemistry: Quantum dots. Analytical and Bioanalytical Chemistry, 416, 3283–3293. https://doi.org/10.1007/s00216-024-05225-9

Wang Yihan, W. M. (2014). Nanofiber fabrication techniques and its applicability to chitosan. Progress *in Chemistry, 26*(11), 1821–1831.

Wu, W., He, Q., & Jiang, C. (2008). Magnetic iron oxide nanoparticles: Synthesis and surface functionalization strategies. *Nanoscale Research Letters, 3*(11), 397. doi: 10.1007/s11671-008-9174-9.

Xiao, Y.-Y., Gong, X.-L., Kang, Y., Jiang, Z.-C., Zhang, S., & Li, B.-J. (2016). Light-, pH-and thermal-responsive hydrogels with the triple-shape memory effect. *Chemical Communications, 52*(70), 10609–10612.

Xing, Y., Yang, H., Guo, X., Bi, X., Liu, X., Xu, Q., ... Zheng, Y. (2020). Effect of chitosan/Nano-TiO2 composite coatings on the postharvest quality and physicochemical characteristics of mango fruits. *Scientia Horticulturae, 263*, 109135. doi: 10.1016/j.scienta.2019.109135.

Xue, J., Xie, J., Liu, W., & Xia, Y. (2017). Electrospun nanofibers: New concepts, materials, and applications. *Accounts of Chemical Research, 50*(8), 1976–1987. doi: 10.1021/acs.accounts.7b00218.

Yang, Y. J., Choi, Y. S., Jung, D., Park, B. R., Hwang, W. B., Kim, H. W., & Cha, H. J. (2013). Production of a novel silk-like protein from sea anemone and fabrication of wet-spun and electrospun marine-derived silk fibers. *NPG Asia Materials, 5*(6), e50. doi: 10.1038/am.2013.19.

Yu, H., Park, J.-Y., Kwon, C. W., Hong, S.-C., Park, K.-M., & Chang, P.-S. (2018). An overview of nanotechnology in food science: Preparative methods, practical applications, and safety. *Journal of Chemistry, 2018,* 5427978.

Zhou, X., Pu, H., & Sun, D.-W. (2021). DNA functionalized metal and metal oxide nanoparticles: Principles and recent advances in food safety detection. *Critical Reviews in Food Science and Nutrition, 61*(14), 2277–2296.

2 Analysis of Macronutrients in Food

Iqra Zubair Awan

2.1 INTRODUCTION TO FOOD ANALYSIS

Food, the fuel for human life's existence, has always been a critical subject of interest for producers, developers, consumers, and health experts. Contemporary advancement in food science and technology demands the determination of food composition, physical characteristics, shelf life, nutrition, and impacts on human health. In recent years, the rapidly changing lifestyle of the public has altered their eating habits, demanding the production of newer food items. Therefore, it has become very challenging for scientists to monitor and abide by the national and international food regulations to ensure food quality and safety. All food commodities, from raw through end product, must qualify through the quality control check procedure prior to being taken to the market. The food analysis could include nutritional value, macronutrients, micronutrients, additives, pesticides, pathogens, allergens, and organic/inorganic contaminants. Consumers will likely choose food based on quality, nutritional profile, dietary patterns, and lifestyle. A significant concern for many is the relationship between health and diet. For instance, the demand for low-fat content urges scientists to develop food and dairy products with reduced/low or free fat to achieve long-term goals of health benefits like low risk of cardiac diseases and cancer (Powell-Wiley et al., 2021).

As carbohydrates, proteins, and fats constitute the principal components of food, their relative ratio in a particular food product might make them a top choice for one and inconsumable for another person. It would be accurate to discuss the food matrix at this point, which explains the nutritional and non-nutritional components of food and their molecular relationships, i.e., chemical bonds, to each other. The dietary components include vitamins, minerals, and other health-regulating constituents like antioxidants. In contrast, the non-nutritional are the added components that impart specific properties or texture to the food (Aguilera, 2018). Analytical methods are required for the characterization to address the food safety concerns and the accuracy of nutrient and non-nutrient components (a test of the low or fat-free product as in the above example). The method selection in food analysis to achieve the targeted application is linked to its:

- Purpose of analysis
- Characteristics of methods
- Characteristics of interests

For accurate analysis, it is inevitable to make a suitable method selection and precise sample preparation. The test must be performed vigilantly repeatedly, and careful data analysis must be done to omit any ambiguity before reporting the data. This chapter discusses the application of nanotechnology for analyzing macronutrients (carbohydrates, proteins, and lipids), food additives, and colorants.

2.2 CARBOHYDRATES

The energy booster foods in our diet are carbohydrates, which play a significant role in the biochemistry and physiology of the human body. Carbohydrates (often termed as "carbs," $[C_x(H_2O)_y]$) exist in the form of sugar molecules, starch, and fiber; which are then further classified into two

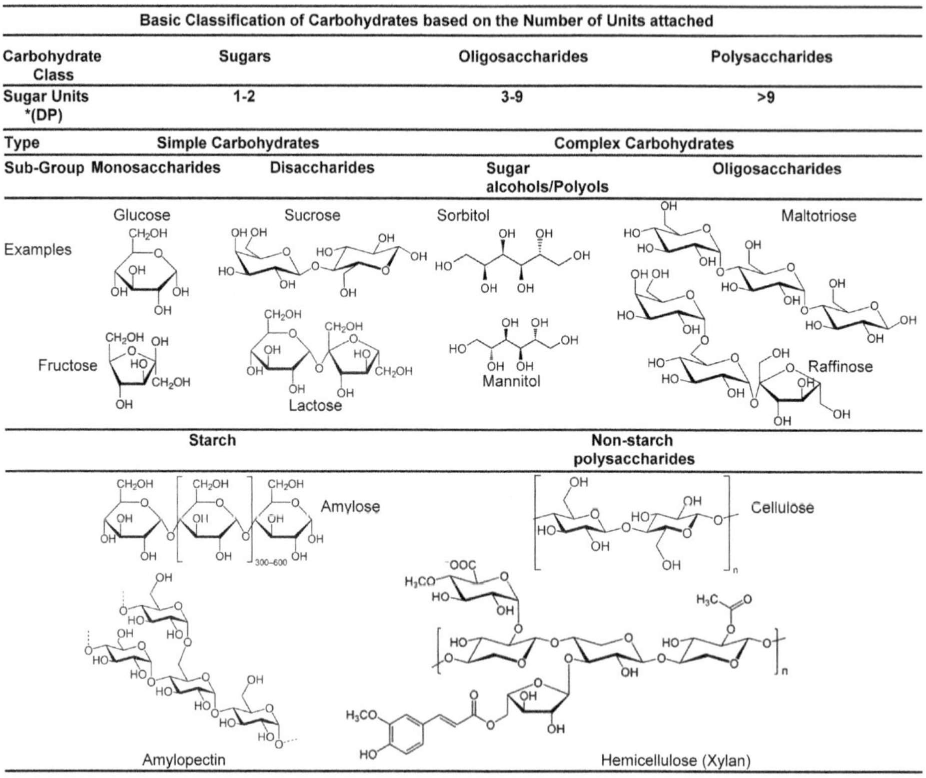

Basic Classification of Carbohydrates based on the Number of Units attached			
Carbohydrate Class	Sugars	Oligosaccharides	Polysaccharides
Sugar Units *(DP)	1-2	3-9	>9
Type	Simple Carbohydrates	Complex Carbohydrates	
Sub-Group	Monosaccharides	Disaccharides	Sugar alcohols/Polyols	Oligosaccharides

FIGURE 2.1 Classification and some examples of carbohydrates.

major groups: simple and complex carbohydrates (Figure 2.1). Simple carbohydrates are monosaccharides and disaccharides or, in other words, carbohydrates having short chains of sugar molecules that are easy to digest, such as glucose, maltose, fructose, lactose, and so on. On the other hand, complex carbohydrates are polysaccharides (having long chain of sugar molecules) that are either non-digestible (such as cellulose) or take longer time to digest (such as starch) (amylose and amylopectin). Fiber, also a type of complex carbohydrate, is of two types, i.e., soluble and insoluble. Soluble fiber dissolves in water to form a gel in the digestive tract which slows down digestion and helps lower blood glucose and cholesterol. On the contrary, insoluble fiber does not absorb water and just adds bulk to the stool and prevents constipation and irregular stools. During digestion, the enzymes break down carbohydrates into glucose, the body's preferred form of energy. This glucose is absorbed in the bloodstream, called blood sugar or glycemic response.

2.2.1 ROLE OF CARBOHYDRATES

Carbohydrates serve as energy stores, form structural and protective components like plasma membrane, intermediates in the biosynthesis of fats and proteins, aid in the regulation of nerve tissue, along with lipids and proteins form surface antigens, receptor molecules, vitamins, and antibiotics, the structural framework of RNA and DNA, play a role in intra and inter-cell communication participate in cholesterol and triglyceride metabolism, important constituent of connective tissues. Therefore, based on the 2,000 calorie count, it is recommended to derive 45%–65% (approximately 225–325 g) of your energy from carbohydrates (Trumbo et al., 2002).

2.2.2 FOOD SOURCES

Healthy or good carbohydrate sources are natural or minimally processed. Their natural nutritional profile is intact, promoting good health by delivering vitamins, minerals, fiber, and phytonutrients such as whole grains, green vegetables, fiber-rich fruits, and beans. Contrary to this, refined carbohydrates are passed through a manufacturing process that removes most of their nutritional value by cutting out fiber, minerals, vitamins, and antioxidants/polyphenols. Refined carbohydrates are present in manufactured food products in which flavor, color, and taste are the priority; therefore, they also contain added fat, salt, and sugar that load up the food with calories but poor nutrition. White bread, pastries, sodas, white flour, artificial sweets, and savory snacks are refined carbohydrate sources. These products contain carbohydrates, which are digested quickly, and you no longer feel hungry. Frequent meal consumption leads to weight gain and diseases linked to heart and diabetes.

A comparison of some examples of simple and complex carbohydrates showing carbohydrate content, dietary fiber, and their percentage Daily Value (% DV) and health effects have been indicated (Table 2.1). The total sugars in complex carbohydrates are far less than in refined carbs. Purified sucrose/dextrose added to boost food's flavor is refined sugar and added sugar.

Since carbohydrate sources have spread widely, being selective in the diet can offer prodigious health benefits. In food development and engineering, carbohydrates are added to various processed food forms to provide bulkiness, desirable texture, viscosity, stability to emulsions and foams, flavors, aromas, colorants, water-holding capacity, and feeling of fullness. As discussed earlier, carbohydrates are divided into healthy (good) and unhealthy (bad) categories based on the refinement

---

TABLE 2.1

Some Examples of Complex and Refined Carbs Foods and Their Health Effects

Type of Food	Carbohydrate Content, g^b (% DV)	Dietary Fibers, g^b (% DV)	Total Sugars, g^b (% DV)	Health Effects
Healthy Complex Carbs Food, 100 g				
Cooked Brown Rice	23.5 (9)	1.8 (6)	–	Decreases risk of cardiac disease
Sweet Potatoes	20.1 (7)	3 (11)	4.2 (8)	Prevents free-radical damage and promote gut functioning
Quinoa Cooked	21.3 (8)	2.8 (10)	0.9 (2)	Balances blood sugar and weight loss
Oat Bran	66.2 (24)	15.4 (55)	1.5 (3)	Blood sugar control, bowel function, and weight loss
Kidney beans	22.8 (8)	6.4 (23)	0.3 (1)	Healthy levels of blood sugar, slow-release carbs
Banana	22.8 (8)	2.6 (9)	12.2 (24)	Improves gut, kidney, and heart health
Wholegrain Penne Pasta	62.5 (23)	0.7 (38)	1.8 (4)	Improves digestive health
Unhealthy Refined Carbs Food, 100 g				
White Bread	49.4 (18)	2.7 (10)	5.7 (11)	Obesity, heart disease, and diabetes
Coca Cola	10.8 (4)	-	10.8 (22)	Obesity, diabetes, and hyperactivity in children
Cooked White Rice	28.2 (10)	0.4 (1)	0.1 (0)	High triglyceride levels, high blood pressure
Corn Flakes	84.1 (31)	3.3 (12)	9.5 (19)	Promotes insulin spikes and impairs glucose tolerance
Canned Orange Juice	11 (4)	0.3 (1)	8.8 (18)	Blood sugar spikes and weight gain
Cookies	~71.4 (26)[a]	~3.6 (13)	~50 (100)	Increased risk of developing chronic health issues
Brownie	~64.7 (24)	~2 (7)	~45.1 (90)	Tooth decay and unhealthy weight gain

[a] ~Approximate value may vary with each brand.

[b] The carbohydrate, protein content, and % DV have been retrieved from the USDA data source. (USDA, 2019) [U.S. Department of Agriculture. (2019). *Agricultural Research Service. FoodData Central.* fdc.nal.usda.gov.].

they have gone through. For this reason, carbohydrate analysis is essential from many viewpoints. The qualitative and quantitative analyses of carbohydrates allow us to determine the compositions of edibles and beverages and their ingredients (Chang, 2010). For the accurate transfer of information to the consumer, it is crucial to ensure that the ingredient chart and associated nutritional values are precise and corresponding. As mentioned on the label, quantitative determinations focus on the sequence of ingredients listed per their quantity. Such analysis allows a quality control check over food items sold in the supermarket. For instance, the Red Bull, a popular beverage among youngsters, is marketed as an energy drink with the claim that it gives an energy boost to improve physical and cognitive performance. However, not many studies stand in support of this idea (Alsunni, 2015). The ingredients of Red bull and other energy drinks contain approximately ~11 g/100 mL of carbohydrates. Now, one standard serving of Red Bull is 250 mL, which makes 26.4 g of sugar, 53% of total sugars, and 10% of carbohydrates as per the % Daily Value. Moreover, the quantitative analysis helps identify the ingredients' components (like added sugar and fiber content), their concentration, and calorie count. Hence, qualitative and quantitative analyses can help make a carbohydrate profile of any food item. Controlling the sugar content in food is necessary due to the health issues related to sugar (glucose, sucrose, fructose, etc.) consumption, such as obesity and diabetes (Alam et al., 2022). This part of this chapter focuses on carbohydrate analysis concerning nanobiotechnology. For example, the development of glucose sensors extensively started after the first generation enzymatic glucometer in 1962 (Clark & Lyons, 1962; Toghill & Compton, 2010). Contemporary studies have found a link between classical sugar analysis and novel materials.

2.2.3 Conventional Methods for Analysis of Carbohydrates in Food

Several methods evolved with time to analyze carbohydrates; however, they may be tailored as per the requirement of the specific component to be explored. The list of procedures below shows a transition from old to new analysis methods.

1. Colored reactions
   - Molisch Test: for the presence of carbohydrates (Devor, 1950).
   - Fehling's Test: for reducing sugars (Fehling, 1849).
   - Benedict's Test: presence of reducing sugars (Benedict, 1909).
   - Barfoed's Test: for monosaccharides (Nurprialdi et al., 2022).
   - Bial's Test: for pentoses (Fernell & King, 1953).
   - Mucic Acid: for galactose (Pigman et al., 1949).
   - Seliwanoff: to differentiate aldose and ketose sugars (Katoch, 2011; Sánchez-Viesca & Gómez, 2018)).
   - Iodine Test: for starch and glycogen (Cooke, 1935).
   - Others.
2. Enzymatic Methods (Bilitewski et al., 1993; Ernst & Arditti, 1972).
3. Physical Methods:
   - Polarimetry (Singleton et al., 2002).
   - Refractive Index (Magwaza & Opara, 2015).
   - Density (Finley, 1976).
   - Infrared (Wiercigroch et al., 2017).
   - Immunoassay (Kurbanov et al., 2021).
4. Chemical Reactions
   - Titrimetric Method: Lane-Eynon method - for quantification of reducing sugars (Lane & Eynon, 1923).
   - Gravimetric Method: Munson and Walker - for quantification of reducing sugars (Ekelund, 1946).
   - Colorimetric Method: Anthrone method - for quantification of total sugars in a sample (Ludwig & Goldberg, 1956), Phenol - Sulfuric Acid method (DuBois et al., 1956).

5. Electrochemical Methods (such as Cyclic Voltammetry).

Some of the techniques mentioned above are now outdated but often coupled with new methods for double check and more precise quantification. Among the modern methods, high performance liquid chromatography (HPLC) is at the top of the list, followed by other forms of chromatography like gas liquid chromatography (GLC) and thin layer chromatography (TLC). In chromatography, carbohydrates are separated based on their differential adsorption characteristics. Other techniques include electrochemical methods. Nevertheless, this chapter discusses the role of materials chemistry, precisely the nanoparticles, for qualitative and quantitative analyses of carbohydrates paired with these techniques.

2.2.3.1   Sample Preparation

Carbohydrates vary widely based on the number of attached sugar units, structure, and solubility properties. Therefore, it is impossible to provide a method that fits every type. However, Figure 2.2 shows a generalized flow diagram of the extraction of carbohydrates from raw material or a finished product. The sample preparation may vary according to the sample, ingredient, or component. As shown in Figure 2.2, the first step of sample preparation is drying: a weighed amount of the sample is placed in an oven under a vacuum below 55°C. Once dried, the material is finely ground, and lipids/fats are extracted using a mixture of chloroform and methanol. Polysaccharides and insoluble dietary fiber are separated, and mono or disaccharides are collected and analyzed. Apart from monosaccharides and oligosaccharides, other molecules, such as organic acids, vitamins, and pigments, might also be present in the extract/matrix and interfere with the analysis. These molecules must be separated before the carbohydrate analysis is carried out by passing them through a clarifying agent or ion-exchange resins.

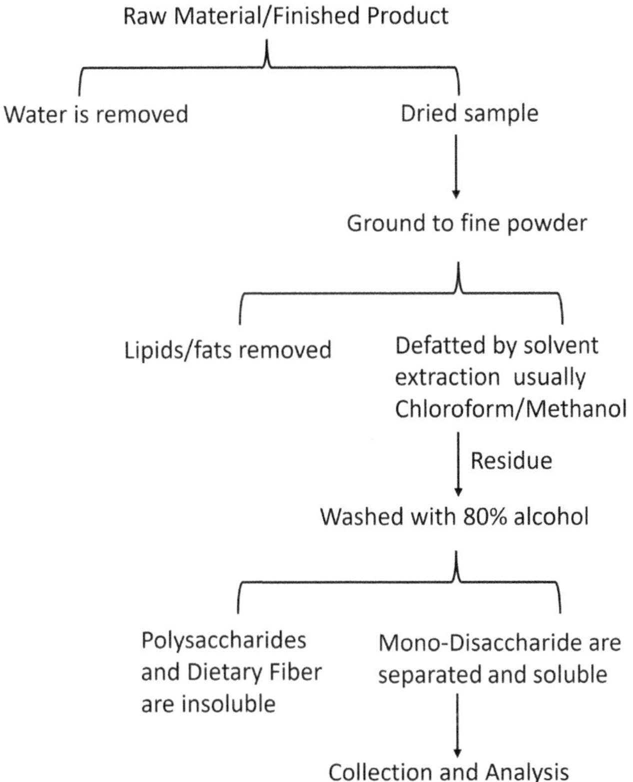

FIGURE 2.2   Schematic illustration of carbohydrate isolation from food samples.

2.2.3.2 Total Carbohydrates

The total carbohydrate content in food is measured by proximate analysis by the added difference of all other food components such as water, protein, fat, etc.

% total carbohydrates = 100 g – (protein + fat + moisture (i.e., total solids/water content) + ash) in g per 100 g of food (Sumar, 1994).

However, complex carbohydrates (polysaccharides such as starch) are calculated by the added difference of sugar and fiber.

2.3 THE ROLE OF NANOTECHNOLOGY IN FOOD SCIENCE AND ENGINEERING

Nanoparticles are typically defined as tiny materials (sizes between 1 and 100 nm) that show extraordinary properties, different from the bulk sample of the same material (Auffan et al., 2009). Nanoparticles have gained significant attention recently due to their extensive commercial applications in nearly all science fields. They are unique due to their high surface area-to-volume ratio, solubility, enhanced reactivity, strength, thermodynamics, and other physicochemical properties. The utilization of the properties of nano-sized materials with the activity of biological molecules is known as "nanobiotechnology". Combined with biotechnology and food science, nanotechnology creates impressive advancement opportunities in both fields.

Literature shows that biotechnology in food processing has no limitations. Starting from nano fertilizers to nanopesticide and nanobiosensors development, detection of toxic substances in foods, adulterant detection, and packaging, nanobiotechnology has it all (Chang et al., 2019; Hotessa Halake & Muda Haro, 2022). One of the main reasons for this versatility is the tunable properties of nanoparticles. Therefore, considering it a modern technology, this chapter aims to address recent advancements in biotechnology to determine macronutrients.

2.4 NANOBIOTECHNOLOGY IN CARBOHYDRATE ANALYSIS

2.4.1 Electrochemical Techniques

An electrochemical analysis deals with the qualitative and quantitative information obtained as an electrical stimulation based on the result of chemical reactivity in the sample upon a given voltage. Although several such types of techniques exist, they are mainly divided into five major categories: (i) potentiometry, (ii) voltammetry, (iii) coulometry, (iv) conductometric, and (v) electrometry. Among voltammetry, cyclic voltammetry is prominent in detecting sugars in samples. This technique applies a triangular potential waveform via potentiostat to an electroactive solution to measure the current in response to the redox reaction. The method involves using three electrodes: a working electrode, a reference electrode, and a counter electrode. In studies carried out for sugar detection, the working electrode is modified to be utilized as a catalyst for the reaction. As the catalytic effect varies with each material, so does the working electrode and redox efficiency of response. The analytical data obtained is interpreted as a voltammogram that plots the working electrode's produced current vs. potential. The electroactive materials for working electrodes are of many categories, such as metals (Ni, Cu, Pt, Au, Ag), non-metals (glassy carbon), alloys (Cu-Ni, Ag-Pt), single and mixed oxides (CuO, Fe_3O_4, NiO-ZnO), hydroxides ($Cu(OH)_2$ and composites (Au/TiO_2, Cu/CuO) are used. For instance, a common reaction studied in this context is the electrooxidation of glucose in the alkaline medium, as shown in Equation 2.1.

$$M(high\ oxidation\ state) + Glucose \rightarrow M(low\ oxidation\ state) + Glucose\ derivative \qquad (2.1)$$

The development of electrodes for macronutrient analysis is favored owing to its ease, speed, selectivity, and cost-effectiveness. A few examples are quoted in Table 2.2.

TABLE 2.2

Some Examples of Nanoparticles Used for Carbohydrate Determination

Nanoparticle	Analyte	LOD, μmol L^{-1}	Sensitivity, μA mM^{-1}cm^{-2}	Response Time (s)	Ref.
CNT/MoS$_2$/NiNPs	Glucose	0.197	1,212	3	Fall et al. (2023)
GLAD Ni and GLAD NiO on Ni foils		0.007	4,400	–	Singer et al. (2020)
Ni@PS nanocomposite		0.2	–	<3	Ensafi et al. (2017)
Ni/NCNs		0.070	210.56	1.6	Jia et al. (2021)
NiO-NPs		3.2	170.85	<3	Youcef et al. (2022)
Pt$_1$/Ni(OH)$_2$/NG		–	220.75	4.6	Long et al. (2022)
Co$_4$-N nanosheets		0.1	1,137	1.7	Liu et al. (2018)
(Ni-Co)$_3$S$_4$, rGO-PEDOT		0.503	938.4	<1.5	Meng et al. (2019)
CNTs/Pd NPs		20	70	–	Wang et al. (2014)
Ppy-CS-Fe$_3$O$_4$-NP/ ITO		234	12	<2	AL-Mokaram et al. (2016)
ZnO-CuO NRS/FTO		0.40	2,961.8	<3	Ahmad et al. (2017)
Co$_3$O$_4$ nanofibers		0.97	36.25	7	Ding et al. (2010)
Co$_3$O$_4$ NPs		0.8	27.33	5	Soomro et al. (2015)
Cu/FGS		1.05	7,254	–	Li et al. (2020)
GalOxNPs/CHIT	Galactose	0.05	7×10^{-3}	–	Sharma et al. (2019)
3D Co-Ni hydroxide nanostructures	Glucose	0.127	1,911.5		Li et al. (2019)
CuNP/SWCNT/ Nafion modified GCE	Glucose galactose arabinose xylose Fructose mannose	250 The given value is for glucose	256±3 The given value is for glucose	10 The given value is for glucose	Male et al. (2004)
PBTh-Pd particles/ ITO	Glucose	7	5,620	–	Lamiri et al. (2020)
Cu (NCA/NPC)		0.200	1,621	3	Chen et al. (2019)
Cu/Co/CoO/GC	Glucose Fructose	0.22	1,035 663	4	He et al. (2018)
Co/CoO/NGC	Glucose Fructose	1.01	310 183	4	
Cu/NGC	Glucose Fructose	0.35	580 320	3	
ZnO nanorods	Glucose	1,000	2.97	–	Chung et al. (2017)
CuO/ ZnO-DSDSHNM		0.358	1,536.80	1.60	Haghparas et al. (2021)
Cu$_2$O-Zn		0.13	441.2	<3	Manna et al. (2020)
Au@Ni		15.7	23.17	3	Gao et al. (2020)
Ag@TiO$_2$@ZIF-67		0.99	0.788	5	Arif et al. (2020)

(Continued)

TABLE 2.2 (*Continued*)

Some Examples of Nanoparticles Used for Carbohydrate Determination

Colorimetry

Nanoparticle	Analyte	LOD, mol L^{-1}	Wavelength, nm	Analysis Technique	Ref.
AgNPs	D-(+)-glucose	$8.7\pm0.5\times10^{-6}$	430	LSPR	Della Pelle et al. (2019)
	D-(+)-galactose	$8.7\pm0.3\times10^{-6}$			
	D-(-)-fructose				
	sucrose				
	D-(+)-raffinose				
	D-(+)-maltose				
	D-(+)-trehalose				
	D-mannitol				
	D-sorbitol				
	i-inositol				
	xylitol				
	D-(+)-xylose.				
Au-NPs	Maltose	4.7×10^{-6}	434	SPR	El-Shishtawy et al. (2023)
Au-NPs	Glucose	1.0×10^{-6}	520–580	LSPR	Palazzo et al. (2012)
AgNPs	D-(+)-glucose	4.0×10^{-9}	410	LSPR	Durmazel, Uzer, et al. (2019)
	D-(+)-galactose				
	D-(−)-fructose				
	D-(+)-mannose				
	D-(+)-maltose monohydrate				
	D-(+)-lactose monohydrate				
AgNPs	Glucose	$\leq 8.7\times10^{-6}$	430	LSPR	Scroccarello et al. (2019)
	Fructose				
	Sorbitol				
AuNPs	Chlorogenic acid	$\leq 3.3\times10^{-6}$	540		
	Epicatechin				
	Gallic acid				
AuNPs	Fructose	6.7×10^{-3}	525	LSPR	Brasiunas et al. (2021)
	Glucose	8.1×10^{-3}			
	Lactose	8.7×10^{-3}			
	mannose	1.1×10^{-4}			
GOx@GA-Fe nanozyme	Glucose	4.3×10^{-10}	652	UV	Zhang et al. (2021)

Mass Spectroscopy

Nanoparticle	Analyte	LOD, μmol L^{-1}	Technique Used	Ref.
Au NPs-nSi 20–50 (Si) 350×50 nm (Au thin film)	Glucose	0.010	SALDI/MS	Tsao and Yang (2015)
DNP	β-cyclodextrin	–	MALDI-TOF	Wu et al. (2013)
HgTe nanostructure	fructose	15	SALDI/MS	Wang et al. (2014)
	maltose	10		

Carbon nanotubes (CNTs) have acquired a special place as effective electrode substances in electrochemical sensing due to their strength and high electric and thermal conductivity. Fall et al. reported the performance of carbon nanotubes functionalized with molybdenum disulfide and nickel (CNT/MoS$_2$/NiNPs) nanocomposites with different CNT/MoS$_2$ ratios, prepared by hydrothermal technique for the direct sensing of glucose in lab-based and commercial energy drinks sold in Senegal. A nickel catalyst was used for glucose electrooxidation, and the electrocatalytic activity was studied by cyclic voltammetry and amperometry in alkaline media. High sensitivity above > 1,200 μA·mmol L^{-1}·cm^{-2} with a wide linear range with a wide linear range (0.05–0.65 mmol L^{-1}) and a Limit of Detection (LOD) of 0.197 μmol L^{-1} was reported in a short time (Fall et al., 2023). Likewise, Wang et al. investigated the glucose-sensing ability of Pd nanoparticles (5 nm) decorated on CNTs, synthesized using supercritical carbon dioxide (SCCO$_2$) and air — the results of 20 g/100 g Pd/SCOO$_2$ loading were compared with 20 g/100 g, Pd/air that indicated the former electrode to be 70 μA mmol^{-1}L^{-1}cm^{-2}, whereas the latter showed only 30 μA mmol^{-1}L^{-1}cm^{-2} — similarly, the detection limit of 20 g/100 g Pd/SCOO$_2$ was far lower than the electrode prepared by 20 g/100 g, Pd/air (Wang et al., 2014). Male and coworkers developed electrochemical sensors fabricated with SWCNT/Cu nanocomposite to improve the electroactivity and selectivity of glucose. The results were compared with the bare Cu disk electrode. Along with glucose (100 μmol L^{-1}), current responses of other mono and disaccharides were also investigated. Monosaccharides galactose (120%), arabinose (94%), xylose (109%), fructose (99%), and mannose (97%) produced similar results to glucose however, the disaccharides maltose (50%), lactose (56%), sucrose (39%), and trehalose (28%), as well as the trisaccharide raffinose (56%). Current signals were relatively low that indicated varying sensitivity of these analytes for the SWCNT/Cu fabricated electrode (Male et al., 2004). The supported and unsupported use of transition metals has also been widely studied for improving the efficiency of electrochemical non-enzymatic sensors. For instance, single atom Pt supported on Ni(OH)$_2$ nanoplates/nitrogen-doped graphene Pt$_1$/Ni(OH)$_2$/NG is an excellent recent approach used for glucose detection. By comparison of Ni(OH)$_2$, Ni(OH)$_2$/NG, and Pt$_1$/Ni(OH)$_2$/NG, the significance of Pt and Ni supported on nitrogen-doped graphene (NG) has been explained by Density Functional Theory (DFT), which indicated a strong interaction between single-atom Pt active centers and their adjacent Ni atoms, where NG helped improve the conductance (Long et al., 2022). Another work reported using copper and cobalt nanoparticles simultaneously on nitrogen-doped graphic carbon (NGC) to detect glucose and fructose in wine samples. The excellent performance of this sensor was attributed to a synergistic effect between the Cu, Co, and CoO nanoparticles coated by NGC Cu/Co/CoO/NGC. Furthermore, economical and disposable compact screen-printed carbon electrodes modified by Cu/Co/CoO/NGC were also synthesized, and the recovery results were compared with HPLC (He et al., 2018). Other works on nanoparticles can be studied here (Dong & Huang, et al., 2018; Dong & Song, et al., 2018; Dong et al., 2019; Gong et al., 2019; Naikoo et al., 2021; Shen et al., 2022).

2.4.2 COLORIMETRY

Colorimetry, a photometric technique, determines analyte concentration at a particular wavelength as a function of color changes in the solution. The working principle is based on Beer's Lambert Law, as the physical change results from the chemical reaction between the detection chemical and the analyte. For instance, in the following example, Durmazel and coworkers reported the non-enzymatic spectrophotometric analysis of reducing sugars (glucose, galactose, fructose, mannose, maltose, and lactose) by the in-situ formation of silver nanoparticles, inspired by the silver mirror reaction. The reducing sugars converted Ag(NH$_3$)$_2^{2+}$ to zero-valent silver nanoparticles in an alkaline medium in aqueous ammonia. Therefore, sugars were quantified as a function of the formation of silver nanoparticles in the mixture at absorption spectra measured at 410 nm. This method was also suggested to detect polyphenolic compounds in commercial fruit juice, ultra-high temperature (UHT) milk, and honey samples (Durmazel, Üzer, et al., 2019). Likewise, Pelle and colleagues also worked on the synthesis of silver nanoparticles by benefiting from the reducing properties of D-glucose, D-galactose, D-fructose, sucrose, D-raffinose D-maltose, D-trehalose, D-mannitol, D-sorbitol, i-inositol, xylitol, and D-xylose. The absorption studies were carried out by localized surface plasmon resonance (LSPR) absorption band of 430 nm. It was found that polyols and monosaccharides are more reactive than disaccharides. This method also determined sugars in six soft drinks and five apple extracts. Moreover, the results produced by the studied assay were comparable to values produced by ion-chromatography ($R=0.994$) (Della Pelle et al., 2019). Scroccarello and coworkers investigated the sugar content and antioxidant properties of 42 apple samples by forming silver and gold nanoparticles. After the reduction reaction, the produced AuNPs and AgNPs were found in red (540 nm) and yellow (430 nm) color suspension (Scroccarello et al., 2019). Zhang and coworkers worked on the simultaneous detection of glucose, sucrose, fructose, lactose, and bacterial contamination in apple juice with the help of gallic acid and glucose oxidase (origin from *Aspergillus niger*) that functioned as a nanozyme GOx@GA-Fe. The sugar concentration was analyzed by the extent of Tetramethylbenzidine (TMB) oxidation. The bacterial strains studied were *E. coli*, *S. aureus*, *S. enterica* subsp. *enterica* serotype Typhimurium, *L. monocytogenes*, and *E. sakazalii* (Zhang et al., 2021). Some other examples of colorimetric analysis of sugars are reported in Table 2.2.

2.4.3 MASS SPECTROMETRY (MS)

Mass spectrometry is an analytical technique that detects analyte(s) based on their mass/charge ratio of ions. Matrix-Assisted Laser Desorption Ionization (MALDI) is a type of mass spectrometry in which matrix molecules softly ionize the sample through laser, and the sample molecules are converted to their gas phase without being fragmented. This technique has also been used to analyze all forms of carbohydrates. The nanoparticle is used as a matrix to ionize the sugars and make the lower quantities easily detectable by increasing the sensitivity of detection of the respective analyte.

For example, Lin-Su and coworkers reported the analysis of ribose, glucose, cellobiose, and maltose using surface-assisted laser desorption/ionization mass spectrometry (SALDI-MS) in positive mode using AuNPs as matrices to cationize molecular ions from monosaccharides and disaccharides. The LOD for glucose, ribose, cellobiose, and maltose were 41, 82, 144, and 151 nmol L^{-1}, respectively (Su & Tseng, 2007). Tsao and coworkers studied the use of gold grafting on nanostructured silica for the detection of low contents of glucose. The glucose molecule deposited on the AuNPs-nSi surface was catalyzed to gluconic acid. The nanomaterial fabrication with electroless deposition parameters was used to quantify glucose (Tsao & Yang, 2015). Wu and coworkers (2013) incorporated diamond nanoparticles as the matrix in MALDI to improve the ionization of underivatized carbohydrates (dextran, a polysaccharide) for sensitive detection even in the presence of proteins. Likewise, mono monosaccharides and disaccharides were quantitated on five honey samples by SALDI-MS using HgTe nanostructures as the matrix. The method is simple, reproducible (Relative Standard Deviation, RSD < 15%), and rapid, detecting monosaccharides, disaccharides, and oligosaccharides in some honey samples' m/z range of 650–2,700 (Wang et al., 2014). Literature on the use of Mass spectrometry for carbohydrates can be reviewed here (Abdelhamid, 2019; Harvey, 2015, 2018, 2023; Kailemia et al., 2014; Pomastowski & Buszewski, 2019).

2.4.4 HIGH-PERFORMANCE LIQUID CHROMATOGRAPHY

High-Performance Liquid Chromatography (HPLC) is an analytical tool for separating, qualitative, and quantitatively analyzing mixtures based on their relative polarity with the column used. HPLC has ultraviolet (UV) and refractive index (RI) detectors. The technique is widely used for analyzing food mixtures, pharmaceutical chemistry, and detecting organic compounds. The mixture is separated, and quantification is compared to the standards. Xu and coworkers determined the concentration of fructose, glucose, and sucrose in sugarcane molasses by HPLC, separation based on an amino column, and a mobile phase based on acetonitrile and water. The mixture of sugars was isolated from sugarcane bagasse by solid phase extraction (SPE) (Xu et al., 2015). It is to be noted that several other methods were available for detecting carbohydrates that have not been discussed in detail in this chapter.

2.5 PROTEINS

Proteins are complex molecules of amino acid building blocks that account for about 14%–16% of the body mass, ~11 kg in men and 9 kg in women (Duda et al., 2019). An amino acid is a bi-functional organic molecule that contains both a carboxyl group [as well as an amine group ($-NH_2$)].

2.5.1 FUNCTION

Proteins are found throughout the body in muscles, skin, hair, nails, and virtually all body parts. Though not known precisely, scientists believe there should be approximately 20,000 different proteins in the human body that keep it running (Ponomarenko et al., 2016). Various forms of proteins are assigned responsible for transport function, signaling molecules, antibodies, structural support, channels and pumps, metabolism, hormones, enzymes, acid-base balance, fluid balance, etc (Morris et al., 2022). Proteins are considered to be very heavy and complicated biomolecules as their mass ranges from nearly 5,000 to a million Daltons. Structural units that make up proteins join to form short polymer chains called peptides or longer chains called either polypeptides or proteins. The process of making proteins is called translation and involves the step-by-step addition of amino acids to a growing protein chain by a ribozyme called a ribosome. Amino acids are of 20 types: carbon, hydrogen, nitrogen, oxygen, and sulfur elements. Nitrogen is the key element in the structure, making peptide bonds between C-N upon dehydration of two distinct amino acids (Figure 2.3). The amino acids, like the R group attached to the α carbon atom, differ. The nature of the R group determines the properties of proteins. The proportion of the amino acid varies

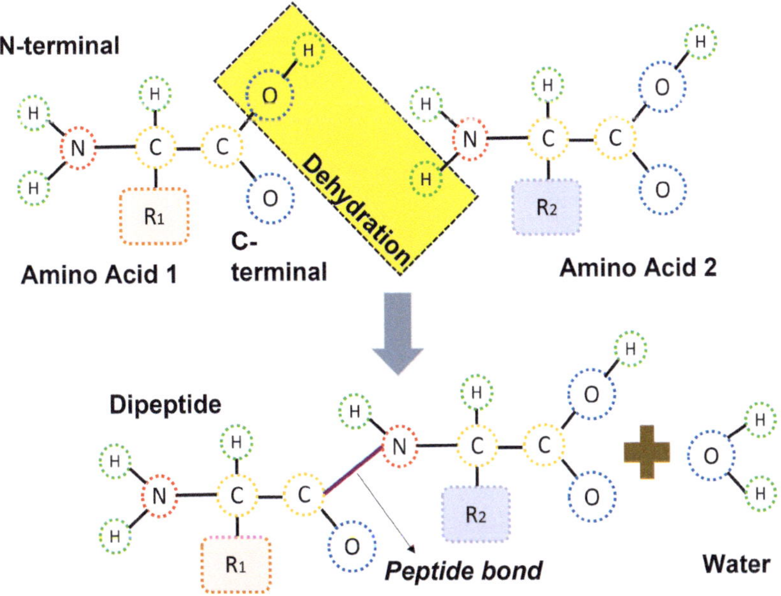

FIGURE 2.3 Dipeptide formation.

as characteristic protein; however, all proteins contain at least some amount of each amino acid except gelatin. Despite having the same protein content, food items might have a varying abundance of different amino acids.

2.5.2 Classification of Proteins

Though there are different criteria for classifying proteins, proteins are generally ranked based on their structure, composition, biological properties, and solubility properties.

2.5.2.1 Structural Classification

In the case of structural classification, proteins are categorized as primary, secondary, tertiary, and quaternary structures.

2.5.2.1.1 Primary Structure

A protein is regarded to have a primary structure when its amino acids are arranged in a linear sequence.

2.5.2.1.2 Secondary Structure

A secondary structure of protein results from repetitive folding of polypeptides due to interactions between atoms of the backbone. Common examples are: alpha helix and β-helix or pleated sheath, beta strands, beta turns, and random coil.

1. Alpha Helix:

    An alpha helix is a spiral arrangement of amino acid chain. In an α helix, the carbonyl group (C=O) of one amino acid forms a hydrogen bond with the amino group (NH_2) of the fourth amino acid in chain. This bonding results in pulling the polypeptide chain into a helical structure, looking like a curled structure. The R group of amino acids lies outwards, with the freedom to interact (Godbey, 2022).

2. β-Helix or Pleated Sheath:

    Beta strand is nearly a linear secondary structure however as a result of hydrogen bonding between two or more segments of polypeptide chains of two adjacent beta strands, a sheet-like structure is formed, called β-helix or pleated sheath. The H-bonds are formed between C=O group and NH_2 groups of the backbone, whereas the R groups lie above and below the sheet plane. The strands of β pleated sheet could be parallel or anti-parallel to each other (Godbey, 2022).

3. Beta Turns:

    Turns result in a change in the direction of protein chain for the sake of connection between two other elements of the secondary structure (Rose et al., 1985). For instance, beta turns, in which the direction of polypeptide alters in the space of amino acid residues labeled as i, $i + 1$, $i + 2$ and $i + 3$. Beta turns are formed as an H-bond formed between C=O of residue i and the N-H of residue $i + 3$ which means the first residue connecting with the fourth residue of the turn (Ahn et al., 2017).

4. Random Coil:

    Random coils, as the name indicates, are unlike the regular secondary structure and do not form by consistent hydrogen connection between the strands. They are located at two locations in proteins; terminal arms (both C and N terminus) and loops (the unstructured regions of regular secondary structure elements) (Smith et al., 1996).

2.5.2.1.2.1 Super-Secondary Structures   A supersecondary structure, also called the structural motif, is defined as a 3D-protein structure formed by combination of elements of a secondary structure. Supersecondary structures could be from alpha helices, Beta motifs, and combination motifs (Rudnev et al., 2021).

1. Helix Supersecondary Structures
    1. *Helix hairpin*

       A helix hairpin, also called alpha-alpha hairpin is formed by a connection of two anti-parallel helices linked by a loop of multiple residues, making a structure similar to hairpin. The longer the loops, the more the possibilities of conformations (Rudnev et al., 2021).
    2. *Helix corner*

       A helix corner, also termed as alpha-alpha corner forms by the connection of two alpha helices (approximately at the right angle to each other) via a short loop of hydrophobic residues (Rudnev et al., 2021).
    3. *Helix-loop-helix*

       The helix-loop-helix structure has two helices connected by a loop. These types of structures are usual in ligand binding (Jones, 2004).
    4. *Helix-turn-helix*

       Helix-turn-helix results in two helices linked by a short sequence, or turn of amino acids (Religa et al., 2007).
2. Beta Sheet Supersecondary Structures
    1. *Beta hairpin*

       Likewise, alpha hairpin, a beta hairpin is formed by link of two anti-parallel beta strands connected by a loop, making a structure similar to hairpin. Mostly, loops made by less than seven residues are seen, however, they may comprise 2–16 residues (DuPai et al., 2021; Efimov, 1991).
    2. *Beta corner*

       Beta corner is formed as a result of two antiparallel strands (at nearly 90°) to each other (Efimov, 1991).
    3. *Greek key motif*

       Greek key motif is formed as a combination of four sequentially adjacent beta strands (up, down, up, down) that may not be geometrically aligned (Hutchinson & Thornton, 1993). Typically, the beta sheet is anti-parallel while the alternate strands lie in the same direction. Moreover, the first and last strands are next to each other and connected by a hydrogen bond.
3. Mixed Supersecondary Structures
    1. *Beta-alpha-beta motifs*

       This type of motif is formed by a combination of two beta strands linked by an alpha helix through connecting loops. With the parallel beta strands, alpha helix is nearly parallel to the strands as well. Such helices can be left-handed or right-handed (Liang et al., 2009).
    2. *Rossman fold*

       Rossman folds are formed by three beta strands and two helices in an alternate manner i.e. beta strand, helix, beta strand, helix, beta strand. Such motif has a tendency to reverse the direction of the peptide chain within a protein (Hanukoglu, 2015).
    3. *The zinc finger*

       Zinc finger is formed by the combination of two antiparallel beta strands and an alpha helix that runs perpendicular to the beta strands. This is called zinc finger, as a zinc atom is linked to polypeptide sequences in such a way that zinc atoms seem to be held in fingers (Cassandri et al., 2017).

2.5.2.1.3 Tertiary Structure

Tertiary structure of the protein is a comprehensive 3-D combination of α-helices and β-sheets that fold next to each other due to interactions between side groups of the amino acids and the surrounding environment.

2.5.2.1.4 *Quaternary Structure*

The quaternary structure of a protein is a combination of multiple protein chains or subunits in a closed-packed manner.

2.5.2.2 Functional Classification

In terms of functional classification, proteins can be divided into enzymatic (Pepsin), transport (hemoglobin), defense (lectins), structural (Keratin), regulatory (Insulin), and reserve/storage protein (Gliadin).

2.5.2.3 Constitution Classification

Simple, conjugate, and derived proteins are three different categories of constitutional classification.

- Simple proteins are divided into fibrous (fibrin) and globular (myoglobin).
- Conjugate proteins are divided into nucleoproteins (histone), glycoproteins (ceruloplasmin), mucoproteins (mucus), Chromoproteins (flavoproteins), lipoproteins (Chylomicrons), and metalloprotein (Transferrin).
- Derived proteins are divided into primary-derived proteins and secondary-derived proteins.

2.5.3 FOOD SOURCES

Foods recommended for high protein intake include eggs, lean meat, white meat, chicken, turkey, fish, nuts, and dairy products. Table 2.3 shows the selected high protein foods with % DV, top three abundant amino acids in the related food, and health benefits. Notably, the recommended daily intake (RDI) for a sedentary adult is $0.8\,g\,kg^{-1}$ of body weight. For instance, a person that weighs 75 kg (165 pounds), must consume 60 g of protein daily.

2.5.4 PROTEIN ANALYSIS

Protein analysis refers to the characterization and quantification techniques developed for protein in an entity. Protein analysis is crucial for understanding the role of proteins in biological processes, disease mechanisms, drug development, and various scientific and medical applications. Many methods have been used to measure the protein content. These methods determine nitrogen content, aromatic amino acids, peptide bonds, dye-binding capacity, ultraviolet absorptivity of proteins, and light-scattering properties. A few objectives of protein analysis of food are quantification of protein to study the amino acid composition, protein content that can be derived after consuming food, and the nutritional value of that protein.

2.5.4.1 Protein Quantification

Protein quantification refers to understanding the total protein content in terms of abundance or concentration in the given sample. Protein quantification is vital for comparing protein levels among various food items. Typical methods used in the quantification of protein are the Kjeldahl Method, Dumas (Nitrogen Combustion) Method, Infrared Spectroscopy, Biuret Method, Lowry Method, Dye-Binding Method, Bradford Dye-Binding Method, Bicinchoninic Acid Method, Ultraviolet 280 nm Absorption Method, HPLC, and radio-chemically labeled proteins (Chang, 2010; Hussain et al., 2019).

2.6 ROLE OF NANOTECHNOLOGY IN THE ANALYSIS OF PROTEINS

Nanotechnology is an emerging trend for studying proteomics forms of protein in an organism. It has revolutionized the identification of proteins and helped understand their interaction with the body. It has made possible the simultaneous study of hundreds of proteins in a mixture, making the process more economical and viable in nanoscale studies, which require less use of samples and reagents while increasing sensitivity.

TABLE 2.3

Selected High Protein Foods with % DV, Top Three Amino Acids and Health Benefits

Type of Food	Protein Content, g[a]	Percentage Daily Value, % DV[a]	Most Abundant Amino acids	Health Benefits
Meat, Fish, Poultry, 100 g				
Eggs (one large) 50 g	6.3	13	Valine Isoleucine Leucine	Source of good cholesterol
Egg Scrambled (2 eggs) 96 g	13.3	27	Valine Isoleucine Leucine	Maintains heart health
Chicken breast (Skinless)	22.5	45	Isoleucine Lysine Valine	Builds muscle mass Reduces appetite
Roasted turkey breast	30.1	60	Lysine Threonine Leucine	Supports thyroid function
Pork loin	27.3	55	Isoleucine Lysine Valine	source of zinc, selenium, and vitamins B_{12} and B_6
Ground lamb	24.8	50	Isoleucine Lysine Valine	promotes muscle growth & maintenance
Lean beef	27.8	56	Isoleucine Lysine Valine	Rich source of iron
Smoked pink salmon	18.3	37	Isoleucine Threonine Valine	Low risk of heart disease & type 2 diabetes
Smoked tuna	25.5	51	Isoleucine Lysine Valine	Reduces inflammation, manages heart rhythm & lowers triglycerides in blood fats
Nuts, 100 g				
Almonds (Raw)	6	12	Glutamic acid Glycine Arginine	Lower risk of heart diseases
Pistachios (Raw)	5.73	12	Valine Isoleucine Arginine	Promote weight loss Help control blood sugar
Cashews (Raw)	5.2	10%	Valine Isoleucine Tryptophan	Reduces risk of colon cancer
Dairy Products, 100 g				
Goat Milk	3.6	7	Isoleucine Valine Lysine	Boosts Immune System Improves bone & teeth Health
Parmesan Cheese	40	80	Valine Isoleucine Proline	Boosting metabolism Building muscle mass

(Continued)

TABLE 2.3 (*Continued*)

Selected High Protein Foods with % DV, Top Three Amino Acids and Health Benefits

Type of Food	Protein Content, g[a]	Percentage Daily Value, % DV[a]	Most Abundant Amino acids	Health Benefits
Cottage cheese	11.1	22	Isoleucine Valine Proline	bone health Aids in Digestion
Cream Cheese	6.2	12	Isoleucine Leucine Lysine	Boosts immunity
Buttermilk	3.2	6	Isoleucine Valine Leucine	Good source of vitamins A and D
Fruits and Vegetables, 100 g				
Avocado	4	8	Valine Isoleucine Lysine	Lower blood pressure
Bananas	1.1	2	Histidine Leucine Isoleucine	Prevents constipation
Broccoli	2.8	6	Valine Isoleucine Tryptophan	Improving skin health
Baked potatoes	2.1	4	Valine Isoleucine Aspartic acid	Fiber, weight loss
Legumes, 100 g				
Red kidney beans (boiled)	9.5	19	Isoleucine Leucine Valine	Lowers heart and fatty liver disease
Soybeans	18.2	36	Isoleucine Valine Leucine	Source of iron, phosphorus, and potassium
Chickpeas	8.8	18	Isoleucine Lysine Valine	Controls blood sugar, weight loss & gut health
Cereals and Pastas, 100 g				
Whole Wheat Pasta	6	12	Proline Glutamic acid Isoleucine	Improves your digestive health
Oatmeal	2.5	5	Tryptophan Valine Isoleucine	Prevents constipation
Corn Flakes	7.5	15[b]	Tryptophan Tyrosine Lysine	Source of folic acid It helps produce new blood cells

[a] The protein content and % DV have been retrieved from the USDA data source.

[b] ~Approximate value may vary with each brand.

A known amount of protein is mixed with nanoparticles in detecting proteins with nanoparticles. Subsequently, the residual protein is measured in the supernatant when the reaction has been carried out. This way, the difference between fed and residual protein is calculated, determining the nanoparticle's ability to take up protein. Several kinds of magnetic nanoparticles (MNPs), metal oxides (MO), and layered double hydroxides (LDH) exemplify the ability of nanoparticles to detect proteins when used with different techniques such as MALDI, HPLC, adsorption, and metal oxide affinity chromatography, and some of these examples are shown in Table 2.4. It is observed that gold nanoparticles are essential for protein analysis due to their optical properties, compatibility with a wide range of wavelengths, and chemical stability. GNPs, once they have captured proteins, show a deviation in the emission spectrum of deviated light because of the binding with protein. For instance, Martínez et al. reported a homogeneous aggregation immunoassay of measurement of soy proteins by using AuNPs and light scattering detection (Sanchez-Martinez et al., 2009). Similarly, Lu and his mates worked on AuNPs using uGA to analyze beta-lactoglobulin, lactoferrin, and lysozyme protein. They also addressed the problem of AuNP aggregation due to protein corona formation by changing the solution pH of the uGA assay. Also, they suggested the effect of pH plays a vital role in determining targeted protein (Lu et al., 2020). Likewise, Oviedo and colleagues reported a spectrophotometric method for nanoparticle protein content determination. The method involved using Coomassie blue dye that tends to bind with basic and aromatic amino acid residues, particularly arginine and lysine. Protein-loaded nanoparticles were mixed with the known quantity of Coomassie blue dye reagent. Later, the unconjugated protein was mixed with bovine serum albumin (BSA) for titration. The method was distinctive in that direct measurement of reacted dye was done rather than the typical method in which the final protein concentration is determined in the supernatant after centrifugation by the difference between conjugated and non-conjugated proteins. However, for this method, the cut-off limit for protein coating is lower than 1ppm (Oviedo et al., 2019). Kailasa and colleagues studied the ZrO_2-SiO_2 nanoparticles (NPs) and nanorods (NRs) studied the enrichment of leucine-enkephalin, methionine-enkephalin and thiopeptide peptides phosphopeptides and phosphoproteins by using MALDI-TOF. ZrO_2 NPs showed better affinity for phosphopeptides than nanorods (Kailasa & Wu, 2010). Similarly, other examples of protein analysis can be found here (Pourmadadi et al., 2022; Wen et al., 2022).

2.7 LIPIDS

Lipids are composed of repeating units of fatty acid ($CH_3(CH_2)nCOOH$) and are regarded as one of the macronutrients required for the proper physiological functioning of the body (Asokapandian et al., 2021). Chemically, lipids are the esters of glycerol and fatty acids or the triglycerides of fatty acids. Nevertheless, unlike other biomolecules, lipids are known by their definition of solubility. They are soluble in organic solvents (ether, chloroform, benzene, and acetone) but insoluble in water. However, this definition does not fit all forms of lipids since the existence of some forms that do dissolve in water, such as triglycerols, are insoluble/hydrophobic. Still, on the other hand, mono and di-glycerol are somewhat soluble in water and polar solvents. Short fatty acid chain molecules (C_1–C_4) have a higher solubility than long-chain fatty acids. Lipids are not polymers but rather small molecules. The terms lipids, fats, and oils are commonly used as substitutable. At room temperature, fats are solids, whereas oils are liquids.

2.7.1 CLASSIFICATION

Lipids are classified as follows:

1. Simple lipids, which are further classified as fats/oils and waxes,
2. Compound lipids are categorized as phospholipids, glycolipids, sphingolipids, lipoproteins, and sulfolipids.
3. Derived lipids-fatty acids, steroids, and sterols.

TABLE 2.4

Some Examples of Nanoparticles Used for Protein Analysis

Nanoparticle	Nanoparticle Size, nm	Protein Type	Method	Ref.
Azide-and alkyne-functionalized gold nanoparticles (AuNPs).	–	BSA (canine, duck, equine, feline & rabbit) Milk	CAP Biuret reagent	Zhu et al. (2012)
AuNPs	–	Soy Protein fruit juice Non-milk yogurt	light scattering 530 nm	Sanchez-Martinez et al. (2009)
HA: Eu^{3+}	50	BSA	Coomassie blue dye 595 nm	Oviedo et al. (2019)
$Zn_{1-.48}Cd_{1-.60}$ SxOy/ $H_8(Al_8Si_{40}O_{96})\cdot24H_2O$ LDH	8–12			
ZHCl-Asp	10			
ZHCl-Gly	10			
$Bi_4Ge_3O_{12}$	6–12			
Y_2O_3:Eu^{3+} (with silica shell)	50–80			
$Zn_{0.75}Al_{0.25}(OH)_2(NO_3)_{0.25}\cdot nH_2O$	20			
MWCNT–COOH	–			
(3-aminopropyl)-trimethoxysilane (APTMS) functionalized ZnO	110	Cytochrome c Catalase Laccase Versatile peroxidase BSA	light scattering 595 nm	
AuNPs		beta-lactoglobulin Lactoferrin lysozyme	uGA assay	Lu et al. (2020)
Ti^{4+}-immobilized magnetic composite microspheres	390	Phosphorylated protein from milk	Metal ion affinity chromatography	Ma et al. (2013)
Fe_3O_{4-} citric acid NP (MAG-200) Fe_3O_4,-Citric acid-Cu NPs (MAG-200–17)	40	β-lactoglobulin in cheese whey	Magnetic solid-phase extraction/ Adsorption	Nicolás et al. (2019)
$Cu–Al_2O_3$-g-C_3N_4–Pd	LOD=3.3 fg nL^{-1}; Linear detection range = 10 fg mL^{-1} up to 100 ng mL^{-1}	amyloid β-protein	Electroanalysis	Miao et al. (2019)
MoO_3@GO	4.61–12.25	β-Casein	Metal oxide affinity chromatography	Sun et al. (2017)

2.7.2 FUNCTION

Lipids serve as energy stores of the body. When excess carbohydrates are consumed, they are converted into lipids by an acetyl CoA in a process known as lipogeneses. Lipids act as a cushion against external shocks for the body organs, for example, during injury. The cell membrane is composed of phospholipids, cholesterol, and glycolipids. The cell membrane regulates transport functions, absorbing fat-soluble vitamins and hormone synthesis. They transmit nerve impulses, transport essential nutrients, and are body insulators. Some examples of lipids are cholesterol, steroids, phospholipids, and fat-soluble vitamins and waxes.

2.7.3 FOOD SOURCES

Food may contain various lipids, but the most important are triacylglycerols and phospholipids. Just like carbohydrates, it is advised to opt for healthy fat sources. There are two types of fats: saturated (no double bonds between carbon atoms) and unsaturated (at least one double bond in the fatty acid chain). Unsaturated fats are healthier fat sources because they help lower cholesterol levels and maintain heart health. Saturated fats sources: whole milk, butter, cheese, margarine, coconut oil, vegetable oil, meat, peanuts, fried foods, etc. Unsaturated fats sources: walnuts, flax, avocado, sunflower oil, soybean oil, fish oil, canola oil, red meat, etc. According to a report published by the British Heart Foundation, it is advised that men should consume no more than 30 g of saturated fat/per day, whereas women should consume 20 g of saturated fat/per day along with less than 5 g *trans*-fat/day, for both.

Consumption of more fat than required is linked to cardiac diseases, heart attacks, memory impairment, obesity, low energy, high cholesterol, and reduced vision (Dhaka et al., 2011; Duan et al., 2018; Sacks et al., 2017). In Table 2.5, a brief description of the food source (100g) and the fat derived from it with % daily value has been shown. Though oils are not consumed in such high amounts per day, 100 g has been chosen as a standard to understand the high and low-fat content in all forms of food.

TABLE 2.5
Selected Fat Sources Foods with % DV and Health Effects

Type of Food	Total Fat Content, g (% DV)[a]	Saturated Fat, g[a] (% DV)[a]	*Trans* Fat, g (% DV)[a]	Health Effects
		Food Source (100 g)		
Olive oil	100 (128)	13.8 (69)	–	High in antioxidants
Canola oil	100 (128)	7.4 (37)	0.4 (–)	Reduce the risk of heart disease
Low-Fat sour cream	14.1 (18)	8.7 (43)	–	Weight gain and increased risk of heart disease
Raw Cashews	43.9 (56)	7.8 (39)	0	reduces LDL cholesterol
Butter	81.1 (104)	51.4 (257)	3.3	Lowers risks of lung cancer and prostate cancer
Fried egg	14.8 (19)	4.3 (22)	0	Oxysterols are linked to heart diseases
Chia Seeds	30.7 (39)	3.3 (17)	0.1	High in fiber, low Blood pressure
Vegetable oil	100 (128)	13.7 (68)	–	Neurodegenerative disorders, heart diseases
Whole Milk	3.3 (4)	1.9g (9)	–	Source of minerals essential for body functioning
Margarine	80.5 (103)	16.7 (84)	–	It makes blood platelets stickier and causes heart disease.
Salmon	4.3 (6)	0.9 (5)	–	Omega-3 fatty acids, rich in vitamins
Avocado	14.7 (19)	2.1 (11)	–	Source of potassium, folate, and fiber
Roasted Turkey	7.4 (9)	2.2 (11)	–	Boosts Mental Health and well-being.
Peanut	49.2 (63)	6.3g (31%)	0	An essential source of vitamin B_3, Lowers LDL cholesterol.

[a] The fat content and % DV have been retrieved from the USDA data source.

2.8 ROLE OF NANOBIOTECHNOLOGY IN THE ANALYSIS OF LIPIDS

Analysis of lipids in food develops their total fat, saturated fat, and trans-fat profile. It is crucial to determine the allowed level of a particular food a person can consume daily according to the recommended daily intake. The objective of nanoparticle lipid analysis is the same as that of carbohydrates and proteins, such as determining nutritional profile health effects and detecting minimal quantities of fats in food. A few examples of nanoparticle lipid analysis have been presented in Table 2.6. Shahamirifard et al. studied a glassy carbon electrode modified by Hydroxyapatite-ZnO-Pd NPs for vitamin C analysis. Electrochemical impedance spectroscopy and cyclic voltammetry studies were carried out for electroanalysis. Having applied the potential difference of 0.23 V, HAP- ZnO–Pd NPs/CPE showed great catalytic activity for the oxidation of vitamin C in fruit juice and skin-brightening creams (Shahamirifard & Ghaedi, 2019). Similarly, Lee and colleagues utilized nanoparticles to adsorb proteins and lipids in heavy dairy cream. The ability of three different types, cellulose nanofibrils NPs, TiO_2 NPs, and polystyrene NPs, were understood by LC-MS/MS. The nanoparticles showed high selectivity for Triacylglycerol (TAG), Diacylglycerol (DAG), Phospholipases (PLs), and Lipopolysaccharides (LysoPL) (Lee et al., 2018). Lipid analysis is more complicated than carbohydrates and proteins due to multifunctional groups, such as complex lipids. Several kinds of hormones and vitamins studied on qualitative and quantifying lipids can be found here (Engel et al., 2022; Leopold et al., 2018; Yu et al., 2021).

TABLE 2.6
Some Examples of Nanoparticles Used for Protein Analysis

Nanoparticles	Analyte	Method Used	LOD	Linear Range	Ref.
Au NPs-APTS modified silica gel	progesterone testosterone	Thin Layer Chromatography	0.16 ng 0.13 ng	1–200 ng	Amoli-Diva and Pourghazi (2015)
Garlic (Allium sativum L.) capped Ag nanoparticles (G-Ag NPs)	Cholesterol	Electroanalysis	0.186 mmol L^{-1}	0.4–5.17 mmol L^{-1}	El-Naka et al. (2023)
Cellulose nanofibrils (CNF)	TAG DAG PLs LysoPL	LC-MS/MS	TAG DAG PLs LysoPL	59.3% 12.8% 24.7% 3.2%	Lee et al. (2018)
TiO_2 NPs			TAG DAG PLs LysoPL	51.4% 13.5% 24.4% 9.4%	
Polystyrene (PS) NPs			TAG DAG PLs LysoPL	58.7% 14.5% 25.7% 2.4%	
HAP-ZnO-Pd NPs/CPE	Vitamin C	Electroanalysis	19.4 nmol L^{-1}	0.12–55.36 μM	Shahamirifard and Ghaedi (2019)
BSA/Ab-VD$_2$/CD-CH/ITO	vitamin D_2		1.35 ng mL^{-1}	10–50 ng mL^{-1}	Sarkar et al. (2018)
Au/PAn/γ-Al$_2$O$_3$)	vitamin E		0.06 μmol L^{-1}	–	Parvin et al. (2018)

2.9 FOOD ADDITIVES

Food additives are artificially synthesized, or natural chemical substances added to food to impart properties like color, texture, aroma, flavor, and longer shelf-life. They are divided into colorants, stabilizers, thickening agents, preservatives, and flavors. Table 2.7 shows the common food additives, their function, and examples. Every day, billions consume food additives, but their flavors, aroma, and taste remain dominant in the worse health effects associated with them. In advanced countries, the concept of organic food has increased in recent years (Popa et al., 2019), but still, consumers prefer conventional food production. In this chapter, we will discuss how nanoparticles have helped food scientists determine the level of food additives used in culinary. Further literature on applications of nanomaterials for food safety and analysis can be studied here (Ghosh et al., 2022; Mohammad et al., 2022).

Table 2.8 shows the utilization of nanoparticles for qualitative and quantitative determination of food dyes.

TABLE 2.7

Some Examples and Functions of Food Additives Commonly Used

Food Additives	Function	Common Examples
Anti-caking agents	Refrain ingredients from becoming lumpy	Talc, Powdered cellulose, Mannitol
Emulsifiers	Stop fats from clotting together	Lecithin, Guar gum, Carrageenan
Thickeners and vegetable gums	Improves texture and consistency	Gelatin, Cornstarch, Xanthan Gum
Artificial Sweeteners	Intensify sweetness	Aspartame, Sucralose, Xylitol
Stabilizers and firming agents	Maintain even food dispersion	Aluminum Sulphate, Calcium Citrate, Ammonium Aluminum Sulphate
Colors	Increase the sweetness	Quinoline yellow, Patent Blue V
Raising agents	Increase the volume of food by gas production	baking soda, baking powder, cream of tartar
Preservatives	Hinders production of microbes and food spoilage	Sodium benzoate, Benzoic acid, Potassium sorbate
Foaming agents	Maintain uniform aeration of gases in foods	Sodium Alginate, Alginic acid, Gelatine
Mineral salts	Enhance texture and flavor	Sodium Acetate 3-hydrate, Calcium Lactate, Ferrous Gluconate
Glazing agent	It makes the appearance better and protects food	Beeswax, carnauba wax, vegetable oils, or animal fats
Bulking agents	To enhance food volume without significant changes to its available energy	Powdered Cellulose, polydextrose, inulin
Fortifying agents	To improve the nutritional profile of food	Probiotic Vitamins, minerals, and amino acids
Acidity regulators	For controlling food pH/ stability and also to affect the activity of enzymes	Lactic acid, Citric acid, Tartaric acid
Humectants	Control moisture; do not let the food dry	Glycerol, Mono-glycerides, Propylene Glycol
Flavorings	To enhance the natural flavor	Isoamyl acetate (banana-like odor), Methyl anthranilate (grape), Ethyl maltol (sugar cotton candy)
Antifoaming agents	Prevent foaming in foods.	Polyoxyethylene stearate, Polysorbate 65, Dimethylpolysiloxane Formulations

TABLE 2.8

Some Examples of Nanoparticles Used for the Determination of Food Additives

Nanoparticles	Used in Food as	Analyte	Method Used	Linear Range	LOD	Ref.
ZnO NPs/CPE		Sudan III	Electroanalysis	0.05–15.0 µmol L^{-1}	0.00254 µmol L^{-1}	Heydari et al. (2019)
Au/Ag alloy nanoparticles		Melamine	SERS	0.001–10 µmol L^{-1}	0.001 µmol L^{-1}	Nie et al. (2019)
Sulfur and nitrogen-doped carbon QDs	Colorants	Clenbuterol	Immunofluorescent	0.07-1.7 ng mL^{-1}	23 pg mL^{-1}	Yao et al. (2019)
Conjugated polymer nanoparticles (CPNs) and AuNPs		Melamine	Fluorescence resonance energy transfer (FRET)	0.1–1.5 µmol L^{-1}	0.0017 µmol L^{-1}	Zhang et al. (2018)
SWCNTs	Packaging	Bisphenol A	Field-effect transistor	4.38×10^{-12} to 4.38×10^{-5} mol L^{-1}	2.19×10^{-12} mol L^{-1}	Sánchez-Acevedo et al. (2009)
ZnO/CNTs			Electroanalysis	0.002–700 µmol L^{-1}	9.0 nmol L^{-1}	Najafi et al. (2014)
PdNPs/SWCNTs	Artificial Sweetener	Aspartame	Electroanalysis	1–120 µmol L^{-1}	0.001 µmol L^{-1}	Chen et al. (2021)
Pt-NPs@Chitosan/ZnTiO$_3$ NCs	Acidity regulator/ preservative	Lactic acid	Electroanalysis	0.30–2.40 mmol L^{-1}	22.36±1.12 µmol L^{-1}	Faisal et al. (2023)
Ag-NPs	preservative	Sodium benzoate	SERS	10–500 mg L^{-1}	0.56 mg L^{-1}	Xue et al. (2020)
PANI/NiO NPs	Stabilizer, Thickener, Gelling agent	Porcine gelatin	Quartz Crystal Microbalance	100–400 mg L^{-1}	51.2 mg L^{-1}	Kurniawan et al. (2022)
TiO$_2$-Ag/Polypyrrole nanocomposite/GCE	Fortifying agents	Tyramine	Electroanalysis	4.10×10^{-8} to 3.06×10^{-6} mol L^{-1}	2×10^{-8} mol L^{-1}	Erdogan et al. (2018)
AgNPs/GO/Screen-Printed silver electrode	Flavoring agent Thinker Stabilizer	Sodium ion (Na$^+$)	Electroanalysis	0–100 mmol L^{-1}	9.344 mmol L^{-1}	Traiwatcharanon et al. (2020)
Fe$_3$O$_4$	Flavoring agent	Ethyl maltol Maltol	HPLC	0.25–800 µg mL^{-1} 0.25–800 µg mL^{-1}	0.05 µg mL^{-1} 0.04 µg mL^{-1}	Zhang et al. (2017)

2.10 COLORANTS

Food coloring or color additives are dyes and pigments often added to food to enhance the sensory aspects. One of the objectives of color addition in food is to recover color loss by exposure to environmental conditions such as light, moisture, air, and temperature fluctuations, particularly during transport, for example, in fruits. Enhancing the colors, which give pleasing effects to the eyes, increases the chances of selling it. Therefore, sugar candies prepared for kids are often seen in bright, attractive colors (Garber et al., 2001; Spence, 2015). Like the detection of macronutrients and food additives, several methods have been suggested for detecting colorants in food. The rapid and relatively more straightforward analysis of food pigments is owing to their color, which can form a base of study by classical spectrophotometry, spectroscopy, and chromatography. Moreover, multiple other advanced methods have also been indicated in the literature, for instance, Fluorescence Resonance Energy Transfer (FRET), electrochemical analysis, Diffuse reflectance Fourier Transform Infrared (DRS-FTIR), and Surface-Enhanced Raman spectroscopy (SERS). A few examples of these latest techniques have been discussed in this chapter.

Balram and coworkers effectively detected sunset yellow (sulfonated azo-dye, orange to yellow, E 110) by a modified electrochemical sensor. They were prepared by Ni-Co nanoparticles dopped on graphene oxide sheets by hydrothermal synthesis. They were fabricated on screen-printed carbon electrodes to determine the electrocatalytic properties of the nanoparticles. Along with high sensitivity, low LOD, and linear range of 0.125–108.5 mmol L^{-1}, the experiment was reproducible and had excellent stability. Furthermore, the jelly, candy, ice cream, and soft drinks samples were also analyzed (Balram et al., 2022). Likewise, Karim-Nezhad and coworkers studied tartrazine (E102, Bright lemon yellow) analysis by the catalytic properties of nanocomposite of zinc oxide nanoparticles (ZnO NPs) and p-aminobenzenesulfonic acid (p-ABSA)-carbon paste modified electrode with the linear concentration range of 0349–5.44 mm. The nanocomposite was also used to study the concentration of tartrazine in soft drinks (Karim-Nezhad et al., 2017). Wang and coworkers investigated graphene-nanomeshes' effectiveness against Amaranth (E123, dark red solid). The porosity of graphene nanomeshes also played a role in dye adsorption and electrochemical detection. Therefore, a comprehensive linear range response of 5.0×10^{-9} to 1.0×10^{-6} mol L^{-1} was observed for this type of electrode modification (Wang et al., 2018). Dorraji reported excellent simultaneous detection of sunset yellow and tartrazine with differential pulse voltammetry by ZnO/Cysteic acid/GCE modified electrode along with results on commercial soft drinks, obtaining RSD of < 5% (Dorraji & Jalali, 2017). The redox behavior of Indigo carmine (Blue, E 132) was also quantified by the Selenium Dioxide nanoparticles in the presence of a cationic surfactant cetylpyridinium bromide (CPB) and glassy carbon electrode (GCE). Results of cyclic voltammetry, chronoamperometry, and electrochemical impedance spectroscopy (EIS) showed that the surfactant-modified electrode was 4.2 times better than GCE (Kavieva & Ziyatdinova, 2022). Allura Red (E129) was found to be analyzed by a nanocomposite of Poly (diallyldimethylammonium chloride) functionalized Graphene and Ni, giving a high electrochemical activity due to the synergistic effect of large surface area and better electron-transfer opportunity. Similar works are reported on removing various dyes in Table 2.9 and reference (Al-Hawary et al., 2023; Mahale et al., 2022; Zhang et al., 2020).

2.11 CONCLUSION

Through a thorough literature survey of the analysis of carbohydrates, proteins, lipids, food additives, and colorants, it can be concluded that nanobiotechnology has played a significant role in advancing food engineering and safety. It has made consumers and producers more conscious about food quality and nutritional value. It can be rightly said that with every passing day, the progress in nanobiotechnology can offer healthier, safer, high-quality, fortified food for the upcoming generations.

TABLE 2.9

Some Examples of Nanoparticles Used for Food Colorant Analysis

Nanoparticle	Food Color (Analyte)	[a]E-Number	Method Used	Linear Range	LOD	Ref.
Ni-Co$_3$O$_4$ NPs/GO	Sunset Yellow	E110	Electroanalysis	0.125–108.5 µmol L^{-1}	0.9 nmol L^{-1}	Balram et al. (2022)
Pp-ABSA/ZnO NPs-CPE	Tartrazine	E102		0.0349–5.44 µmol L^{-1}	0.08 µmol L^{-1}	Karim-Nezhad et al. (2017)
Graphene nanomeshes (GNM)	Amaranth	E123		5.0×10^{-3} to 1.0 µmol L^{-1}	7.0×10^{-4} µmol L^{-1}	Wang et al. (2018)
Graphene with 2-aminoethanethiol functionalized gold nanoparticles	Erythrosine	E127		3.75×10^{-2} to 150 µmol L^{-1}	4.77×10^{-3} µmol L^{-1}	Zhao et al. (2018)
ZnO/Cysteic acid/GCE	Sunset Yellow	E110		0.1–3.0 µmol L^{-1}	0.03 µmol L^{-1}	Dorraji and Jalali (2017)
	Tartrazine			0.07–1.86 µmol L^{-1}	0.01 µmol L^{-1}	
SeO$_2$ NP/cetyl pyridinium bromide (CPB)	Indigo Carmine	E132		0.025–1.0 µmol L^{-1}	4.3 nmol L^{-1}	Kavieva and Ziyatdinova (2022)
SeO$_2$ NP/GCE				1.0–10 µmol L^{-1}	14.3 nmol L^{-1}	
PDDA-Gr-Ni nanocomposite	Allura red	E129		0.05–10.0 µmol L^{-1}	8.0 nmol L^{-1}	Yu et al. (2016)
MIP/f-MWCNTs/GCE	Sunset yellow	E110		0.05–100 µmol L^{-1}	0.005 µmol L^{-1}	Arvand et al. (2017)
Ag NPs	Carmine	E120	SERS	0.01–1,000 µmol L^{-1}	0.01 µmol L^{-1}	Wu et al. (2017)
Au NPs-embedded MIHs	New red	–		1.64×10^{-6} to 1.64×10^{-4} mol L^{-1}	1.64×10^{-7} mol L^{-1}	Neng et al. (2019)
CsPbBr$_3$ QDs	Rhodamine 6G (R6G)	–	FRET	0–10 µg mL^{-1}	0.01 µg mL^{-1}	Chan et al. (2022)
Ni-Sn-oxide NSs/CPE	Erythrosine	E127	Electroanalysis	0.1–100 µmol L^{-1}	2.1×10^{-3} µmol L^{-1}	Arvand et al. (2018)
Poly(ionic liquid) immobilized magnetic nanoparticles	Allura red	E129	fluorescence spectrophotometry	0.01–0.50 µg mL^{-1}	0.002 µg mL^{-1}	Bakheet and Zhu (2017)
AgNPs	Tartrazine	E102	DRS-FTIR	1–160 ng mL^{-1}	2.44 ng mL^{-1}	Tiwari and Deb (2019)
PEDOT: PSS/AuNPs/CA/SPCE	Carmine	E120	Electroanalysis	9.0×10^{-3} to 3.9 µmol L^{-1}	6.05×10^{-3} µmol L^{-1}	Zhao et al. (2018)

[a] The E-number for each colorant is a code for all approved European additives.(Scotter, 2015; European Commision, 2024).

BIBLIOGRAPHY

Abdelhamid, H. N. (2019). Nanoparticle-based surface assisted laser desorption ionization mass spectrometry: a review. *Microchimica Acta, 186*, 1–35.

Aguilera, J. M. (2018). The food matrix: implications in processing, nutrition and health. *Critical Reviews in Food Science and Nutrition, 59*(22), 3612–3629. https://doi.org/10.1080/10408398.2018.1502743.

Ahmad, R., Tripathy, N., Ahn, M.-S., Bhat, K. S., Mahmoudi, T., Wang, Y., Yoo, J.-Y., Kwon, D.-W., Yang, H.-Y., & Hahn, Y.-B. (2017). Highly efficient non-enzymatic glucose sensor based on CuO modified vertically-grown ZnO nanorods on electrode. *Scientific Reports, 7*(1), 5715. https://doi.org/10.1038/s41598-017-06064-8.

Ahn, J., Kassees, K., Lee, T., Manandhar, B., & Yousif, A. (2017). Strategy and tactics for designing analogs: biochemical characterization of the large molecules. https://doi.org/10.1016/B978-0-12-409547-2.12413-8.

Al-Hawary, S. I. S., Bali, A. O., Askar, S., Lafta, H. A., Kadhim, Z. J., Kholdorov, B., Riadi, Y., Solanki, R., & Mustafa, Y. F. (2023). Recent advances in nanomaterials-based electrochemical and optical sensing approaches for detection of food dyes in food samples: a comprehensive overview. *Microchemical Journal*, 108540. https://doi.org/10.1016/j.microc.2023.108540.

AL- Mokaram, A. M. A. A., Yahya, R., Abdi, M. M., & Mahmud, H. N. M. E. (2016). One-step electrochemical deposition of Polypyrrole-Chitosan-Iron oxide nanocomposite films for non-enzymatic glucose biosensor. *Materials Letters, 183*, 90–93. https://doi.org/10.1016/j.matlet.2016.07.049.

Alam, Y. H., Kim, R., & Jang, C. (2022). Metabolism and health impacts of dietary sugars. *Journal of Lipid and Atherosclerosis, 11*(1), 20. https://doi.org/10.12997/jla.2022.11.1.20.

Alsunni, A. A. (2015). Energy drink consumption: beneficial and adverse health effects. *International Journal of Health Sciences, 9*(4), 468. https://www.ncbi.nlm.nih.gov/pubmed/26715927.

Amoli-Diva, M., & Pourghazi, K. (2015). Gold nanoparticles grafted modified silica gel as a new stationary phase for separation and determination of steroid hormones by thin layer chromatography. *Journal of Food and Drug Analysis, 23*(2), 279–286. https://doi.org/10.1016/j.jfda.2014.11.005.

Arif, D., Hussain, Z., Sohail, M., Liaqat, M. A., Khan, M. A., & Noor, T. (2020). A non-enzymatic electrochemical sensor for glucose detection based on Ag@ TiO2@ metal-organic framework (ZIF-67) nanocomposite. *Frontiers in Chemistry, 8*, 573510. https://doi.org/10.3389/fchem.2020.573510.

Arvand, M., Pourhabib, A., & Asadi, M. (2018). Template-based synthesis of uniform bimetallic nickel-tin oxide hollow nanospheres as a new sensing platform for detection of erythrosine in food products. *Sensors and Actuators B: Chemical, 255*, 1716–1725. https://doi.org/10.1016/j.snb.2017.08.189.

Arvand, M., Zamani, M., & Ardaki, M. S. (2017). Rapid electrochemical synthesis of molecularly imprinted polymers on functionalized multi-walled carbon nanotubes for selective recognition of sunset yellow in food samples. *Sensors and Actuators B: Chemical, 243*, 927–939. https://doi.org/10.1016/j.snb.2016.12.077.

Asokapandian, S., Sreelakshmi, S., & Rajamanickam, G. (2021). Lipids and oils: an overview. *Food Biopolymers: Structural, Functional and Nutraceutical Properties*, 389–411. https://doi.org/10.1007/978-3-030-27061-2_16.

Auffan, M., Rose, J., Bottero, J.-Y., Lowry, G. V., Jolivet, J.-P., & Wiesner, M. R. (2009). Towards a definition of inorganic nanoparticles from an environmental, health and safety perspective. *Nature Nanotechnology, 4*(10), 634–641. https://doi.org/10.1038/nnano.2009.242.

Bakheet, A. A. A., & Zhu, X. S. (2017). Poly (ionic liquid) immobilized magnetic nanoparticles as sorbent coupled with fluorescence spectrophotometry for separation/analysis of Allura red. *Journal of Molecular Liquids, 242*, 900–906. https://doi.org/10.1016/j.molliq.2017.07.097.

Balram, D., Lian, K.-Y., Sebastian, N., Al-Mubaddel, F. S., & Noman, M. T. (2022). Ultrasensitive detection of food colorant sunset yellow using nickel nanoparticles promoted lettuce-like spinel Co3O4 anchored GO nanosheets. *Food and Chemical Toxicology, 159*, 112725. https://doi.org/10.1016/j.fct.2021.112725.

Benedict, S. R. (1909). A reagent for the detection of reducing sugars. *Journal of Biological Chemistry, 5*(5), 485–487.

Bilitewski, U., Jäger, A., Rüger, P., & Weise, W. (1993). Enzyme electrodes for the determination of carbohydrates in food. *Sensors and Actuators B: Chemical, 15*(1–3), 113–118. https://doi.org/10.1016/0925-4005(93)85036-A.

Brasiunas, B., Popov, A., Ramanavicius, A., & Ramanaviciene, A. (2021). Gold nanoparticle based colorimetric sensing strategy for the determination of reducing sugars. *Food Chemistry, 351*, 129238. https://doi.org/10.1016/j.foodchem.2021.129238.

Cassandri, M., Smirnov, A., Novelli, F., Pitolli, C., Agostini, M., Malewicz, M., Melino, G., & Raschellà, G. (2017). Zinc-finger proteins in health and disease. *Cell Death Discovery, 3*(1), 1–12. https://doi.org/10.1038/cddiscovery.2017.71.

Chan, K. K., Yap, S. H. K., Giovanni, D., Sum, T. C., & Yong, K.-T. (2022). Water-stable perovskite quantum dots-based FRET nanosensor for the detection of Rhodamine 6G in water, food, and biological samples. *Microchemical Journal, 180*, 107624. https://doi.org/10.1016/j.microc.2022.107624.

Chang, P.-S., Umakoshi, H., & Kim, H. (2019). Nanotechnology for food engineering: biomembrane and nanocarriers. *Journal of Chemistry, 2019*, 1–3. https://doi.org/10.1155/2019/4629365.

Chang, S. K. C. (2010). Protein Analysis. In *Food Analysis* (pp. 133–146). Springer US. https://doi.org/10.1007/978-1-4419-1478-1_9.

Chen, H., Fan, G., Zhao, J., Qiu, M., Sun, P., Fu, Y., Han, D., & Cui, G. (2019). A portable micro glucose sensor based on copper-based nanocomposite structure. *New Journal of Chemistry, 43*(20), 7806–7813. https://doi.org/10.1039/C9NJ00888H.

Chen, Y., Wei, D., He, K., Li, H., & Sun, F. (2021). Electrochemical method for detection of aspartame in beverage using Pd modified single walled carbon nanotubes sensor (PdNPs/SWCNTs). *International Journal of Electrochemical Science, 16*(5), 210522. https://doi.org/10.20964/2021.05.56.

Chung, R.-J., Wang, A.-N., Liao, Q.-L., & Chuang, K.-Y. (2017). Non-enzymatic glucose sensor composed of carbon-coated nano-zinc oxide. *Nanomaterials, 7*(2), 36. https://doi.org/10.3390/nano7020036.

Clark Jr, L. C., & Lyons, C. (1962). Electrode systems for continuous monitoring in cardiovascular surgery. *Annals of the New York Academy of sciences, 102*(1), 29–45. https://doi.org/10.1111/j.1749-6632.1962.tb13623.x.

Cooke, F. (1935). Chemical and Physical Considerations of Carbohydrates.Chemistry Honors Papers. *52*. https://digitalcommons.ursinus.edu/chem_hon/52/

Della Pelle, F., Scroccarello, A., Scarano, S., & Compagnone, D. (2019). Silver nanoparticles-based plasmonic assay for the determination of sugar content in food matrices. *Analytica Chimica Acta, 1051*, 129–137. https://doi.org/10.1016/j.aca.2018.11.015.

Devor, A. W. (1950). Carbohydrate tests using sulfonated α-naphthol. *Journal of the American Chemical Society, 72*(5), 2008–2012. https://doi.org/10.1021/ja01161a037.

Dhaka, V., Gulia, N., Ahlawat, K. S., & Khatkar, B. S. (2011). Trans fats-sources, health risks and alternative approach-A review. *Journal of Food Science and Technology, 48*, 534–541.

Ding, Y., Wang, Y., Su, L., Bellagamba, M., Zhang, H., & Lei, Y. (2010). Electrospun Co3O4 nanofibers for sensitive and selective glucose detection. *Biosensors and Bioelectronics, 26*(2), 542–548. https://doi.org/10.1016/j.bios.2010.07.050.

Dong, Q., Huang, Y., Song, D., Wu, H., Cao, F., & Lei, Y. (2018). Dual functional rhodium oxide nanocorals enabled sensor for both non-enzymatic glucose and solid-state pH sensing. *Biosensors and Bioelectronics, 112*, 136–142. https://doi.org/10.1016/j.bios.2018.04.021.

Dong, Q., Song, D., Huang, Y., Xu, Z., Chapman, J. H., Willis, W. S., Li, B., & Lei, Y. (2018). High-temperature annealing enabled iridium oxide nanofibers for both non-enzymatic glucose and solid-state pH sensing. *Electrochimica Acta, 281*, 117–126. https://doi.org/10.1016/j.electacta.2018.04.205.

Dong, Q., Wang, X., Liu, H., Ryu, H., Zhao, J., Li, B., & Lei, Y. (2019). Heterogeneous iridium oxide/gold nanocluster for non-enzymatic glucose sensing and pH probing. *Engineered Science, 8*(4), 46–53. https://doi.org/10.30919/es8d512.

Dorraji, P. S., & Jalali, F. (2017). Electrochemical fabrication of a novel ZnO/cysteic acid nanocomposite modified electrode and its application to simultaneous determination of sunset yellow and tartrazine. *Food Chemistry, 227*, 73–77. https://doi.org/10.1016/j.foodchem.2017.01.071.

Duan, Y., Zeng, L., Zheng, C., Song, B., Li, F., Kong, X., & Xu, K. (2018). Inflammatory links between high fat diets and diseases. *Frontiers in Immunology, 9*, 2649. https://doi.org/10.3389/fimmu.2018.02649.

DuBois, M., Gilles, K. A., Hamilton, J. K., Rebers, P. t., & Smith, F. (1956). Colorimetric method for determination of sugars and related substances. *Analytical Chemistry, 28*(3), 350–356. https://doi.org/10.1021/ac60111a017.

Duda, K., Majerczak, J., Nieckarz, Z., Heymsfield, S. B., & Zoladz, J. A. (2019). Human body composition and muscle mass. In *Muscle and Exercise Physiology* (pp. 3–26). Elsevier. https://doi.org/10.1016/B978-0-12-814593-7.00001-3.

DuPai, C. D., Davies, B. W., & Wilke, C. O. (2021). A systematic analysis of the beta hairpin motif in the Protein Data Bank. *Protein Science, 30*(3), 613–623. https://doi.org/10.1002/pro.4020.

Durmazel, S., Üzer, A., Erbil, B., Sayın, B., & Apak, R. (2019). Silver nanoparticle formation-based colorimetric determination of reducing sugars in food extracts via Tollens' reagent. *ACS Omega, 4*(4), 7596–7604. https://doi.org/10.1021/acsomega.9b00761

Efimov, A. (1991). Structure of coiled β-β-hairpins and β-β-corners. *FEBS Letters, 284*(2), 288–292. https://doi.org/10.1016/0014-5793(91)80706-9.

Ekelund, S. (1946). Determining reducing sugar. A critical study of Munson and Walker general-method. *Acta agriculturae Suecana, 1*, 239–319.

El-Naka, M. A., El-Dissouky, A., Ali, G., Ebrahim, S., & Shokry, A. (2023). Garlic capped silver nanoparticles for rapid detection of cholesterol. *Talanta, 253*, 123908. https://doi.org/10.1016/j.talanta.2022.123908.

El-Shishtawy, R. M., Al Angari, Y. M., Alotaibi, M. M., & Almulaiky, Y. Q. (2023). Novel and facile colorimetric detection of reducing sugars in foods via in situ formed gelatin-capped silver nanoparticles. *Polymers, 15*(5), 1086. https://doi.org/10.3390/polym15051086.

Engel, K. M., Prabutzki, P., Leopold, J., Nimptsch, A., Lemmnitzer, K., Vos, D. N., Hopf, C., & Schiller, J. (2022). A new update of MALDI-TOF mass spectrometry in lipid research. *Progress in Lipid Research, 86*, 101145. https://doi.org/10.1016/j.plipres.2021.101145.

Ensafi, A. A., Ahmadi, N., & Rezaei, B. (2017). Nickel nanoparticles supported on porous silicon flour, application as a non-enzymatic electrochemical glucose sensor. *Sensors and Actuators B: Chemical, 239*, 807–815. https://doi.org/10.1016/j.snb.2016.08.088.

Erdogan, Z. O., Akin, I., & Kucukkolbasi, S. (2018). A new non-enzymatic sensor based on TiO2-Ag/polypyrrole for electrochemical detection of tyramine. *Synthetic Metals, 246*, 96–100. https://doi.org/10.1016/j.synthmet.2018.10.006.

Ernst, R., & Arditti, J. (1972). Enzymatic quantitative determination of hexoses, singly and in mixtures with their oligosaccharides. *New Phytologist, 71*(2), 307–315. https://doi.org/10.1111/j.1469-8137.1972.tb04077.x.

European Commission. (2024). Regulation (EC) No 1333/2008 of the European Parliament and of the Council of 16 December 2008 on food additives. *Official Journal of the European Union, 354*, 16–33. https://eur-lex.europa.eu/legal-content/EN/TXT/?uri=CELEX%3A02008R1333-20240423

Faisal, M., Alam, M., Ahmed, J., Asiri, A. M., Alsareii, S., Alruwais, R. S., Alqahtani, N. F., Rahman, M. M., & Harraz, F. A. (2023). Efficient electrochemical detection of L-lactic acid using platinum nanoparticle decorated Chitosan/ZnTiO3 nanocomposites. *Journal of Industrial and Engineering Chemistry, 118*, 362–371. https://doi.org/10.1016/j.jiec.2022.11.021.

Fall, B., Sall, D. D., Hémadi, M., Diaw, A. K. D., Fall, M., Randriamahazaka, H., & Thomas, S. (2023). Highly efficient non-enzymatic electrochemical glucose sensor based on carbon nanotubes functionalized by molybdenum disulfide and decorated with nickel nanoparticles (GCE/CNT/MoS2/NiNPs). *Sensors and Actuators Reports, 5*, 100136. https://doi.org/10.1016/j.snr.2022.100136.

Fehling, H. (1849). The quantitative determination of sugar and starch by means of copper sulfate. *Annals of Chemical Pharmacy, 72*, 106–113.

Fernell, W., & King, H. (1953). The simultaneous determination of pentose and hexose in mixtures of sugars. *Analyst, 78*(923), 80–83. https://doi.org/10.1039/AN9537800080.

Finley, J. (1976). Density separation of protein and carbohydrates in a nonaqueous solvent system. *Journal of Food Science, 41*(4), 882–885. https://doi.org/10.1111/j.1365-2621.1976.tb00744_41_4.x.

Gao, X., Du, X., Liu, D., Gao, H., Wang, P., & Yang, J. (2020). Core-shell gold-nickel nanostructures as highly selective and stable nonenzymatic glucose sensor for fermentation process. *Scientific Reports, 10*(1), 1365. https://doi.org/10.1038/s41598-020-58403-x.

Garber Jr, L. L., Hyatt, E. M., & Starr Jr, R. G. (2001). Placing food color experimentation into a valid consumer context. *Journal of Food Products Marketing, 7*(3), 3–24. https://doi.org/10.1300/J038v07n03_02.

Ghosh, T., Raj, G. B., & Dash, K. K. (2022). A comprehensive review on nanotechnology based sensors for monitoring quality and shelf life of food products. *Measurement: Food, 7*, 100049. https://doi.org/10.1016/j.meafoo.2022.100049.

Godbey, W. T. (2022). Proteins. In W. T. Godbey (Ed.), *Biotechnology and Its Applications* (2nd ed., pp. 47–72). Academic Press, Elsevier. https://doi.org/10.1016/B978-0-12-817726-6.00003-4.

Gong, X., Gu, Y., Zhang, F., Liu, Z., Li, Y., Chen, G., & Wang, B. (2019). High-performance non-enzymatic glucose sensors based on CoNiCu alloy nanotubes arrays prepared by electrodeposition. *Frontiers in Materials, 6*, 3. https://doi.org/10.3389/fmats.2019.00003.

Haghparas, Z., Kordrostami, Z., Sorouri, M., Rajabzadeh, M., & Khalifeh, R. (2021). Highly sensitive non-enzymatic electrochemical glucose sensor based on dumbbell-shaped double-shelled hollow nanoporous CuO/ZnO microstructures. *Scientific Reports, 11*(1), 344. https://doi.org/10.1038/s41598-020-79460-2.

Hanukoglu, I. (2015). Proteopedia: Rossmann fold: a beta-alpha-beta fold at dinucleotide binding sites. *Biochemistry and Molecular Biology Education, 43*(3), 206–209. https://doi.org/10.1002/bmb.20849.

Harvey, D. J. (2015). Analysis of carbohydrates and glycoconjugates by matrix-assisted laser desorption/ionization mass spectrometry: an update for 2009-2010. *Mass Spectrometry Reviews, 34*(3), 268–422. https://doi.org/10.1002/mas.21411.

Harvey, D. J. (2018). Analysis of carbohydrates and glycoconjugates by matrix-assisted laser desorption/ionization mass spectrometry: an update for 2013–2014. *Mass Spectrometry Reviews, 37*(4), 353–491. https://doi.org/10.1002/mas.21530.

Harvey, D. J. (2023). Analysis of carbohydrates and glycoconjugates by matrix-assisted laser desorption/ionization mass spectrometry: an update for 2017–2018. *Mass Spectrometry Reviews, 42*(1), 227–431. https://doi.org/10.1002/mas.21721.

He, J., Zhong, Y., Xu, Q., Sun, H., Zhou, W., & Shao, Z. (2018). Nitrogen-doped graphic carbon protected Cu/Co/CoO nanoparticles for ultrasensitive and stable non-enzymatic determination of glucose and fructose in wine. *Journal of the Electrochemical Society, 165*(13), B543. https://doi.org/10.1149/2.0151813jes.

Heydari, M., Ghoreishi, S. M., & Khoobi, A. (2019). Response surface modeling of electrochemical data for sensitive determination of Sudan III in food products at the surface of a nanocomposite modified electrode. *Food Analytical Methods, 12*, 1781–1790. https://doi.org/10.1007/s12161-019-01528-1.

Hotessa Halake, N., & Muda Haro, J. (2022). Role of nanobiotechnology towards agri-food system. *Journal of Nanotechnology, 2022*. https://doi.org/10.1155/2022/6108610.

Hussain, M. T., Forbes, N., & Perrie, Y. (2019). Comparative analysis of protein quantification methods for the rapid determination of protein loading in liposomal formulations. *Pharmaceutics, 11*(1), 39. https://doi.org/10.3390/pharmaceutics11010039.

Hutchinson, E. G., & Thornton, J. M. (1993). The Greek key motif: extraction, classification and analysis. *Protein Engineering, Design and Selection, 6*(3), 233–245. https://doi.org/10.1093/protein/6.3.233.

Jia, H., Shang, N., Feng, Y., Ye, H., Zhao, J., Wang, H., Wang, C., & Zhang, Y. (2021). Facile preparation of Ni nanoparticle embedded on mesoporous carbon nanorods for non-enzymatic glucose detection. *Journal of Colloid and Interface Science, 583*, 310–320. https://doi.org/10.1016/j.jcis.2020.09.051.

Jones, S. (2004). An overview of the basic helix-loop-helix proteins. *Genome Biology, 5*, 1–6. https://doi.org/10.1186/gb-2004-5-6-226.

Kailasa, S. K., & Wu, H.-F. (2010). Multifunctional ZrO2 nanoparticles and ZrO2-SiO2 nanorods for improved MALDI-MS analysis of cyclodextrins, peptides, and phosphoproteins. *Analytical and Bioanalytical Chemistry, 396*, 1115–1125.

Kailemia, M. J., Ruhaak, L. R., Lebrilla, C. B., & Amster, I. J. (2014). Oligosaccharide analysis by mass spectrometry: a review of recent developments. *Analytical Chemistry, 86*(1), 196–212. https://doi.org/10.1021/ac403969n.

Karim-Nezhad, G., Khorablou, Z., Zamani, M., Dorraji, P. S., & Alamgholiloo, M. (2017). Voltammetric sensor for tartrazine determination in soft drinks using poly (p-aminobenzenesulfonic acid)/zinc oxide nanoparticles in carbon paste electrode. *Journal of Food and Drug Analysis, 25*(2), 293–301. https://doi.org/10.1016/j.jfda.2016.10.002.

Katoch, R. (2011). Carbohydrate Estimations. In *Analytical Techniques in Biochemistry and Molecular Biology* (pp. 67–76). Springer, New York. https://doi.org/10.1007/978-1-4419-9785-2_5.

Kavieva, L., & Ziyatdinova, G. (2022). Voltammetric sensor based on SeO2 nanoparticles and surfactants for indigo carmine determination. *Sensors, 22*(9), 3224. https://doi.org/10.3390/s22093224.

Kurbanov, M., Mukhamadiev, B., Kalanova, D., Muzafarova, K., & Kurbanova, S. (2021). Immune-enzyme methods of food safety analysis. *IOP Conference Series: Earth and Environmental Science, 848*, 012185. doi:10.1088/1755-1315/848/1/012185

Kurniawan, F., Nugroho, A., Baskara, R. A., Candle, L., Pradini, D., Madurani, K. A., Sugiarso, R. D., & Juwono, H. (2022). Rapid analysis to distinguish porcine and bovine gelatin using PANI/NiO nanoparticles modified Quartz Crystal Microbalance (QCM) sensor. *Heliyon, 8*(5), e09401.https://doi.org/10.1016/j.heliyon.2022.e09401

Lamiri, L., Belgherbi, O., Dehchar, C., Laidoudi, S., Tounsi, A., Nessark, B., Habelhames, F., Hamam, A., & Gourari, B. (2020). Performance of polybithiophene-palladium particles modified electrode for non-enzymatic glucose detection. *Synthetic Metals, 266*, 116437. https://doi.org/10.1016/j.synthmet.2020.116437.

Lane, J., & Eynon, L. (1923). Methods for determination of reducing and nonreducing sugars. *Journal of Science, 42*, 32–37.

Lee, J. Y., Wang, H., Pyrgiotakis, G., DeLoid, G. M., Zhang, Z., Beltran-Huarac, J., Demokritou, P., & Zhong, W. (2018). Analysis of lipid adsorption on nanoparticles by nanoflow liquid chromatography-tandem mass spectrometry. *Analytical and Bioanalytical Chemistry, 410*, 6155–6164.

Leopold, J., Popkova, Y., Engel, K. M., & Schiller, J. (2018). Recent developments of useful MALDI matrices for the mass spectrometric characterization of lipids. *Biomolecules, 8*(4), 173. https://doi.org/10.3390/biom8040173.

Li, H., Zhang, L., Mao, Y., Wen, C., & Zhao, P. (2019). A simple electrochemical route to access amorphous Co-Ni hydroxide for non-enzymatic glucose sensing. *Nanoscale Research Letters, 14*, 1–12.

Li, J.-H., Tang, J.-X., Wei, L., He, S.-J., Ma, L.-Q., Shen, W.-C., Kang, F.-Y., & Huang, Z.-H. (2020). Preparation and performance of electrochemical glucose sensors based on copper nanoparticles loaded on flexible graphite sheet. *New Carbon Materials, 35*(4), 410–419. https://doi.org/10.1016/S1872-5805(20)60498-X.

Liang, H., Chen, H., Fan, K., Wei, P., Guo, X., Jin, C., Zeng, C., Tang, C., & Lai, L. (2009). De novo design of a βαβ motif. *Angewandte Chemie International Edition*, *48*(18), 3301–3303. https://doi.org/10.1002/anie.200805476.

Liu, T., Li, M., & Guo, L. (2018). Designing and facilely synthesizing a series of cobalt nitride (Co4N) nanocatalysts as non-enzymatic glucose sensors: a comparative study toward the influences of material structures on electrocatalytic activities. *Talanta*, *181*, 154–164. https://doi.org/10.1016/j.talanta.2017.12.082.

Long, B., Zhao, Y., Cao, P., Wei, W., Mo, Y., Liu, J., Sun, C.-J., Guo, X., Shan, C., & Zeng, M.-H. (2022). Single-atom pt boosting electrochemical nonenzymatic glucose sensing on ni (oh) 2/n-doped graphene. *Analytical Chemistry*, *94*(4), 1919–1924. https://doi.org/10.1021/acs.analchem.1c04912.

Lu, D., Zhang, D., Zhao, Q., Lu, X., & Shi, X. (2020). A critical factor for quantifying proteins in unmodified gold nanoparticles-based aptasensing: the eEffect of pH. *Chemosensors*, *8*(4), 98. https://doi.org/10.3390/chemosensors8040098.

Ludwig, T. G., & Goldberg, H. J. (1956). The anthrone method for the determination of carbohydrates in foods and in oral rinsing. *Journal of Dental Research*, *35*(1), 90–94. https://doi.org/10.1177/00220345560350012301.

Ma, W., Zhang, Y., Li, L., Zhang, Y., Yu, M., Guo, J., Lu, H., & Wang, C. (2013). Ti4+-immobilized magnetic composite microspheres for highly selective enrichment of phosphopeptides. *Advanced Functional Materials*, *23*(1), 107–115. https://doi.org/10.1002/adfm.201201364.

Magwaza, L. S., & Opara, U. L. (2015). Analytical methods for determination of sugars and sweetness of horticultural products-A review. *Scientia Horticulturae*, *184*, 179–192. https://doi.org/10.1016/j.scienta.2015.01.001.

Mahale, R. S., Shashanka, R., Vasanth, S., & Vinaykumar, R. (2022). Voltammetric determination of various food azo dyes using different modified carbon paste electrodes. https://doi.org/10.33263/BRIAC124.45574566.

Male, K. B., Hrapovic, S., Liu, Y., Wang, D., & Luong, J. H. (2004). Electrochemical detection of carbohydrates using copper nanoparticles and carbon nanotubes. *Analytica Chimica Acta*, *516*(1–2), 35–41. https://doi.org/10.1016/j.aca.2004.03.075.

Manna, A. K., Guha, P., Solanki, V. J., Srivastava, S., & Varma, S. (2020). Non-enzymatic glucose sensing with hybrid nanostructured Cu2O-ZnO prepared by single-step coelectrodeposition technique. *Journal of Solid State Electrochemistry*, *24*, 1647–1658. https://doi.org/10.1007/s10008-020-04635-w.

Meng, A., Yuan, X., Li, Z., Zhao, K., Sheng, L., & Li, Q. (2019). Direct growth of 3D porous (Ni-Co) 3S4 nanosheets arrays on rGO-PEDOT hybrid film for high performance non-enzymatic glucose sensing. *Sensors and Actuators B: Chemical*, *291*, 9–16. https://doi.org/10.1016/j.snb.2019.04.042.

Miao, J., Li, X., Li, Y., Dong, X., Zhao, G., Fang, J., Wei, Q., & Cao, W. (2019). Dual-signal sandwich electrochemical immunosensor for amyloid β-protein detection based on Cu-Al2O3-g-C3N4-Pd and UiO-66@PANI-MB. *Analytica Chimica Acta*, *1089*, 48–55. https://doi.org/10.1016/j.aca.2019.09.017.

Mohammad, Z. H., Ahmad, F., Ibrahim, S. A., & Zaidi, S. (2022). Application of nanotechnology in different aspects of the food industry. *Discover Food*, *2*(1), 12. https://doi.org/10.1007/s44187-022-00013-9.

Morris, R., Black, K. A., & Stollar, E. J. (2022) Uncovering protein function: from classification to complexes. *Essays in Biochemistry*, *66*(3), 255–285. https://doi.org/10.1042/EBC20200108.

Naikoo, G. A., Salim, H., Hassan, I. U., Awan, T., Arshad, F., Pedram, M. Z., Ahmed, W., & Qurashi, A. (2021). Recent advances in non-enzymatic glucose sensors based on metal and metal oxide nanostructures for diabetes management-a review. *Frontiers in Chemistry*, *9*, 748957. https://doi.org/10.3389/fchem.2021.748957.

Najafi, M., Khalilzadeh, M. A., & Karimi-Maleh, H. (2014). A new strategy for determination of bisphenol A in the presence of Sudan I using a ZnO/CNTs/ionic liquid paste electrode in food samples. *Food Chemistry*, *158*, 125–131. https://doi.org/10.1016/j.foodchem.2014.02.082.

Neng, J., Xu, K., Wang, Y., Jia, K., Zhang, Q., & Sun, P. (2019). Sensitive and selective detection of new red colorant based on surface-enhanced Raman spectroscopy using molecularly imprinted hydrogels. *Applied Sciences*, *9*(13), 2672. https://doi.org/10.3390/app9132672.

Nicolás, P., Ferreira, M. L., & Lassalle, V. (2019). Magnetic solid-phase extraction: a nanotechnological strategy for cheese whey protein recovery. *Journal of Food Engineering*, *263*, 380–387. https://doi.org/10.1016/j.jfoodeng.2019.07.020.

Nie, B., Luo, Y., Shi, J., Gao, L., & Duan, G. (2019). Bowl like Pore array made of hollow Au/Ag alloy nanoparticles for SERS detection of melamine in solid milk powder. *Sensors and Actuators B: Chemical*, *301*, 127087. https://doi.org/10.1016/j.snb.2019.127087.

Nurprialdi, B., Gani, V. O. T., Halda, S., Pratama, P. A., & Panjaitan, R. S. (2022). Qualitative and quantitative identification of carbohydrates in commercial yoghurt products. *Indonesian Journal of Pharmaceutical Research*, *2*(2), 11–21. https://doi.org/10.31869/ijpr.v2i2.4134.

Oviedo, M. J., Quester, K., Hirata, G. A., & Vazquez-Duhalt, R. (2019). Determination of conjugated protein on nanoparticles by an adaptation of the Coomassie blue dye method. *MethodsX, 6*, 2134–2140. https://doi.org/10.1016/j.mex.2019.09.015.

Palazzo, G., Facchini, L., & Mallardi, A. (2012). Colorimetric detection of sugars based on gold nanoparticle formation. *Sensors and Actuators B: Chemical, 161*(1), 366–371. https://doi.org/10.1016/j.snb.2011.10.046.

Parvin, M. H., Arjomandi, J., & Lee, J. Y. (2018). γ-Al2O3 nanoparticle catalyst mediated polyaniline gold electrode biosensor for vitamin E. *Catalysis Communications, 110*, 59–63. https://doi.org/10.1016/j.catcom.2018.03.009.

Pigman, W., Browning, B., McPherson, W., Calkins, C., & Leaf, R. (1949). Oxidation of D-galactose and cellulose with nitric acid, nitrous acid and nitrogen oxides. *Journal of the American Chemical Society, 71*(6), 2200–2204. https://doi.org/10.1021/ja01174a076.

Pomastowski, P., & Buszewski, B. (2019). Complementarity of matrix-and nanostructure-assisted laser desorption/ionization approaches. *Nanomaterials, 9*(2), 260. https://doi.org/10.3390/nano9020260.

Ponomarenko, E. A., Poverennaya, E. V., Ilgisonis, E. V., Pyatnitskiy, M. A., Kopylov, A. T., Zgoda, V. G., Lisitsa, A. V., & Archakov, A. I. (2016). The size of the human proteome: the width and depth. *International Journal of Analytical Chemistry, 2016*, 7436849. doi: 10.1155/2016/7436849

Popa, M. E., Mitelut, A. C., Popa, E. E., Stan, A., & Popa, V. I. (2019). Organic foods contribution to nutritional quality and value. *Trends in Food Science & Technology, 84*, 15–18. https://doi.org/10.1016/j.tifs.2018.01.003.

Pourmadadi, M., Rajabzadeh-Khosroshahi, M., Saeidi Tabar, F., Ajalli, N., Samadi, A., Yazdani, M., Yazdian, F., Rahdar, A., & Díez-Pascual, A. M. (2022). Two-dimensional graphitic carbon nitride (g-C3N4) nanosheets and their derivatives for diagnosis and detection applications. *Journal of Functional Biomaterials, 13*(4), 204. https://doi.org/10.3390/jfb13040204.

Powell-Wiley, T. M., Poirier, P., Burke, L. E., Després, J.-P., Gordon-Larsen, P., Lavie, C. J., Lear, S. A., Ndumele, C. E., Neeland, I. J., Sanders, P., & St-Onge, M.-P. (2021). Obesity and cardiovascular disease: a scientific statement from the American Heart Association. *Circulation, 143*(21), e984–e1010. https://doi.org/10.1161/CIR.0000000000000973.

Religa, T. L., Johnson, C. M., Vu, D. M., Brewer, S. H., Dyer, R. B., & Fersht, A. R. (2007). The helix-turn-helix motif as an ultrafast independently folding domain: the pathway of folding of Engrailed homeodomain. *Proceedings of the National Academy of Sciences, 104*(22), 9272–9277. https://doi.org/10.1073/pnas.0703434104.

Rose, G. D., Glerasch, L. M., & Smith, J. A. (1985). Turns in peptides and proteins. *Advances in Protein Chemistry, 37*, 1–109. https://doi.org/10.1016/S0065-3233(08)60063-7.

Rudnev, V. R., Kulikova, L. I., Nikolsky, K. S., Malsagova, K. A., Kopylov, A. T., & Kaysheva, A. L. (2021). Current approaches in supersecondary structures investigation. *International Journal of Molecular Sciences, 22*(21), 11879.

Sacks, F. M., Lichtenstein, A. H., Wu, J. H., Appel, L. J., Creager, M. A., Kris-Etherton, P. M., Miller, M., Rimm, E. B., Rudel, L. L., & Robinson, J. G. (2017). Dietary fats and cardiovascular disease: a presidential advisory from the American Heart Association. *Circulation, 136*(3), e1–e23. https://doi.org/10.1161/CIR.0000000000000510.

Sánchez-Acevedo, Z. C., Riu, J., & Rius, F. X. (2009). Fast picomolar selective detection of bisphenol A in water using a carbon nanotube field effect transistor functionalized with estrogen receptor-α. *Biosensors and Bioelectronics, 24*(9), 2842–2846. https://doi.org/10.1016/j.bios.2009.02.019

Sanchez-Martinez, M., Aguilar-Caballos, M., & Gómez-Hens, A. (2009). Homogeneous immunoassay for soy protein determination in food samples using gold nanoparticles as labels and light scattering detection. *Analytica Chimica Acta, 636*(1), 58–62. https://doi.org/10.1016/j.aca.2009.01.043.

Sánchez-Viesca, F., & Gómez, R. (2018). Reactivities involved in the Seliwanoff reaction. *Modern Chemistry, 6*(1), 1–5. https://doi.org/10.11648/j.mc.20180601.11.

Sarkar, T., Bohidar, H., & Solanki, P. R. (2018). Carbon dots-modified chitosan based electrochemical biosensing platform for detection of vitamin D. *International Journal of Biological Macromolecules, 109*, 687–697. https://doi.org/10.1016/j.ijbiomac.2017.12.122.

Scotter, M. (2015). Overview of EU regulations and safety assessment for food colours. In *Colour Additives for Foods and Beverages* (pp. 61–74). Elsevier. https://doi.org/10.1016/B978-1-78242-011-8.00003-9.

Scroccarello, A., Della Pelle, F., Neri, L., Pittia, P., & Compagnone, D. (2019). Silver and gold nanoparticles based colorimetric assays for the determination of sugars and polyphenols in apples. *Food Research International, 119*, 359–368. https://doi.org/10.1016/j.foodres.2019.02.006.

Shahamirifard, S. A., & Ghaedi, M. (2019). A new electrochemical sensor for simultaneous determination of arbutin and vitamin C based on hydroxyapatite-ZnO-Pd nanoparticles modified carbon paste electrode. *Biosensors and Bioelectronics*, *141*, 111474. https://doi.org/10.1016/j.bios.2019.111474.

Sharma, M., Yadav, P., & Sharma, M. (2019). Novel electrochemical sensing of galactose using GalOxNPs/CHIT modified pencil graphite electrode. *Carbohydrate Research*, *483*, 107749. https://doi.org/10.1016/j.carres.2019.107749.

Shen, L., Liang, Z., Chen, Z., Wu, C., Hu, X., Zhang, J., Jiang, Q., & Wang, Y. (2022). Reusable electrochemical non-enzymatic glucose sensors based on Au-inlaid nanocages. *Nano Research*, *15*(7), 6490–6499. https://doi.org/10.1007/s12274-022-4219-4.

Singer, N., Pillai, R. G., Johnson, A. I., Harris, K. D., & Jemere, A. B. (2020). Nanostructured nickel oxide electrodes for non-enzymatic electrochemical glucose sensing. *Microchimica Acta*, *187*, 1–10.

Singleton, V., Horn, J., Bucke, C., & Adlard, M. (2002). A new polarimetric method for the analysis of dextran and sucrose. *Journal of American Society of Sugarcane Technology*, *22*, 112–119.

Smith, L. J., Fiebig, K. M., Schwalbe, H., & Dobson, C. M. (1996). The concept of a random coil: residual structure in peptides and denatured proteins. *Folding and Design*, *1*(5), R95–R106.

Soomro, R. A., Nafady, A., Ibupoto, Z. H., Sherazi, S. T. H., Willander, M., & Abro, M. I. (2015). Development of sensitive non-enzymatic glucose sensor using complex nanostructures of cobalt oxide. *Materials Science in Semiconductor Processing*, *34*, 373–381. https://doi.org/10.1016/j.mssp.2015.02.055.

Spence, C. (2015). On the psychological impact of food colour. *Flavour*, *4*(1), 21. https://doi.org/10.1186/s13411-015-0031-3.

Su, C.-L., & Tseng, W.-L. (2007). Gold nanoparticles as assisted matrix for determining neutral small carbohydrates through laser desorption/ionization time-of-flight mass spectrometry. *Analytical Chemistry*, *79*(4), 1626–1633. https://doi.org/10.1021/ac061747w.

Sumar, S., Coultate, T. P., & Davies, J. (1994). Food and nutrition update:an analysis of macronutrients in the diet. *Nutrition & Food Science*, *94*(6), 31–35. https://doi.org/10.1108/00346659410069719.

Sun, H., Zhang, Q., Zhang, L., Zhang, W., & Zhang, L. (2017). Facile preparation of molybdenum (VI) oxide-modified graphene oxide nanocomposite for specific enrichment of phosphopeptides. *Journal of Chromatography A*, *1521*, 36–43. https://doi.org/10.1016/j.chroma.2017.08.029.

Tiwari, S., & Deb, M. K. (2019). Modified silver nanoparticles-enhanced single drop microextraction of tartrazine in food samples coupled with diffuse reflectance Fourier transform infrared spectroscopic analysis. *Analytical Methods*, *11*(28), 3552–3562. https://doi.org/10.1039/C9AY00713J.

Toghill, K. E., & Compton, R. G. (2010). Electrochemical non-enzymatic glucose sensors: a perspective and an evaluation. *International Journal of Electrochemical Science*, *5*(9), 1246–1301.

Traiwatcharanon, P., Siriwatcharapiboon, W., & Wongchoosuk, C. (2020). Electrochemical sodium ion sensor based on silver nanoparticles/graphene oxide nanocomposite for food application. *Chemosensors*, *8*(3), 58. https://doi.org/10.3390/chemosensors8030058.

Trumbo, P., Schlicker, S., Yates, A. A., & Poos, M. (2002). Dietary reference intakes for energy, carbohydrate, fiber, fat, fatty acids, cholesterol, protein and amino acids (Commentary). *Journal of the American Dietetic Association*, *102*(11), 1621–1631. https://doi.org/10.1016/s0002-8223(02)90346-9.

Tsao, C.-W., & Yang, Z.-J. (2015). High sensitivity and high detection specificity of gold-nanoparticle-grafted nanostructured silicon mass spectrometry for glucose analysis. *ACS Applied Materials & Interfaces*, *7*(40), 22630–22637. https://doi.org/10.1021/acsami.5b07395.

Usda, U. (2019). Department of agriculture agricultural research service. *Food Data Central*, *335*, 336.

Wang, C.-H., Lee, S.-W., Tseng, C.-J., Wu, J.-W., Hung, I.-M., Tseng, C.-M., & Chang, J.-K. (2014). Nanocrystalline Pd/carbon nanotube composites synthesized using supercritical fluid for superior glucose sensing performance. *Journal of Alloys and Compounds*, *615*, S496–S500. https://doi.org/10.1016/j.jallcom.2013.12.187.

Wang, C.-W., Chen, W.-T., & Chang, H.-T. (2014). Quantification of saccharides in honey samples through surface-assisted laser desorption/ionization mass spectrometry using HgTe nanostructures. *Journal of the American Society for Mass Spectrometry*, *25*(7), 1247–1252. https://doi.org/10.1007/s13361-014-0886-z.

Wang, M., Cui, M., Zhao, M., & Cao, H. (2018). Sensitive determination of Amaranth in foods using graphene nanomeshes. *Journal of Electroanalytical Chemistry*, *809*, 117–124. https://doi.org/10.1016/j.jelechem.2017.12.059.

Wen, M., Li, J., Zhong, W., Xu, J., Qu, S., Wei, H., & Shang, L. (2022). High-throughput colorimetric analysis of nanoparticle-protein interactions based on the enzyme-mimic properties of nanoparticles. *Analytical Chemistry*, *94*(24), 8783–8791. https://doi.org/10.1021/acs.analchem.2c01618.

Wiercigroch, E., Szafraniec, E., Czamara, K., Pacia, M. Z., Majzner, K., Kochan, K., Kaczor, A., Baranska, M., & Malek, K. (2017). Raman and infrared spectroscopy of carbohydrates: a review. *Spectrochimica Acta Part A: Molecular and Biomolecular Spectroscopy*, *185*, 317–335. https://doi.org/10.1016/j.saa.2017.05.045.

Wu, C.-L., Wang, C.-C., Lai, Y.-H., Lee, H., Lin, J.-D., Lee, Y. T., & Wang, Y.-S. (2013). Selective enhancement of carbohydrate ion abundances by diamond nanoparticles for mass spectrometric analysis. *Analytical Chemistry*, *85*(8), 3836–3841. https://doi.org/10.1021/ac3036469.

Wu, Y.-X., Liang, P., Dong, Q.-M., Bai, Y., Yu, Z., Huang, J., Zhong, Y., Dai, Y.-C., Ni, D., & Shu, H.-B. (2017). Design of a silver nanoparticle for sensitive surface enhanced Raman spectroscopy detection of carmine dye. *Food Chemistry*, *237*, 974–980. https://doi.org/10.1016/j.foodchem.2017.06.057.

Xu, W., Liang, L., & Zhu, M. (2015). Determination of sugars in molasses by HPLC following solid-phase extraction. *International Journal of Food Properties*, *18*(3), 547–557. https://doi.org/10.1080/10942912.2013.837064.

Xue, L., Chen, L., Dong, J., Cai, L., Wang, Y., & Chen, X. (2020). Dispersive liquid-liquid microextraction coupled with surface enhanced Raman scattering for the rapid detection of sodium benzoate. *Talanta*, *208*, 120360. https://doi.org/10.1016/j.talanta.2019.120360.

Yao, D., Liang, A., & Jiang, Z. (2019). A fluorometric clenbuterol immunoassay using sulfur and nitrogen doped carbon quantum dots. *Microchimica Acta*, *186*, 1–9.

Youcef, M., Hamza, B., Nora, H., Walid, B., Salima, M., Ahmed, B., Malika, F., Marc, S., Christian, B., & Wassila, D. (2022). A novel green synthesized NiO nanoparticles modified glassy carbon electrode for non-enzymatic glucose sensing. *Microchemical Journal*, *178*, 107332. https://doi.org/10.1016/j.microc.2022.107332.

Yu, J., Kang, Y., Zhang, H., Yang, F., Zhen, H., Zhu, X., Wu, T., & Du, Y. (2021). A polymer-based matrix for effective SALDI analysis of lipids. *Journal of the American Society for Mass Spectrometry*, *32*(5), 1189–1195. https://doi.org/10.1021/jasms.1c00010.

Yu, L., Shi, M., Yue, X., & Qu, L. (2016). Detection of allura red based on the composite of poly (diallyldimethylammonium chloride) functionalized graphene and nickel nanoparticles modified electrode. *Sensors and Actuators B: Chemical*, *225*, 398–404. https://doi.org/10.1016/j.snb.2015.11.061.

Zhang, C.-J., Gao, Z.-Y., Wang, Q.-B., Zhang, X., Yao, J.-S., Qiao, C.-D., & Liu, Q.-Z. (2018). Highly sensitive detection of melamine based on the fluorescence resonance energy transfer between conjugated polymer nanoparticles and gold nanoparticles. *Polymers*, *10*(8), 873. https://doi.org/10.3390/polym10080873.

Zhang, J., Hu, S., Du, Y., Cao, D., Wang, G., & Yuan, Z. (2020). Improved food additive analysis by ever-increasing nanotechnology. *Journal of Food and Drug Analysis*, *28*(4), 622. https://doi.org/10.38212/2224-6614.1152.

Zhang, Q., Wang, X., Kang, Y., Sun, H., Liang, Y., Liu, J., Su, Z., Dan, J., Luo, L., & Yue, T. (2021). Natural products self-assembled nanozyme for cascade detection of glucose and bacterial viability in food. *Foods*, *10*(11), 2596. https://doi.org/10.3390/foods10112596.

Zhang, X., Niu, J., Yang, Y., Qin, P., Tian, S., Zhu, J., & Lu, M. (2017). Fe3O4 nanoparticles as the adsorbent of magnetic solid-phase extraction for clean and preconcentration of maltol and ethyl maltol in food samples followed by HPLC analysis. *Journal of Liquid Chromatography & Related Technologies*, *40*(16), 832–838. https://doi.org/10.1080/10826076.2017.1373671.

Zhao, X., Ding, J., Bai, W., Wang, Y., Yan, Y., Cheng, Y., & Zhang, J. (2018). PEDOT: PSS/AuNPs/CA modified screen-printed carbon based disposable electrochemical sensor for sensitive and selective determination of carmine. *Journal of Electroanalytical Chemistry*, *824*, 14–21. https://doi.org/10.1016/j.jelechem.2018.07.030.

Zhao, X., Hu, W., Wang, Y., Zhu, L., Yang, L., Sha, Z., & Zhang, J. (2018). Decoration of graphene with 2-aminoethanethiol functionalized gold nanoparticles for molecular imprinted sensing of erythrosine. *Carbon*, *127*, 618–626. https://doi.org/10.1016/j.carbon.2017.11.041.

Zhu, K., Zhang, Y., He, S., Chen, W., Shen, J., Wang, Z., & Jiang, X. (2012). Quantification of proteins by functionalized gold nanoparticles using click chemistry. *Analytical Chemistry*, *84*(10), 4267–4270. https://doi.org/10.1021/ac3010567.

3 Analysis of Metal Content in Food

Application of Nanotechnology

Christopher Chiedozie Obi,
Chukwunonso Onyeka Aniagor,
Matthew Ndubuisi Abonyi,
Henry Chukwuka Oyeoka, and
Martin Emeka Ibenta

3.1 INTRODUCTION

Metals are naturally occurring elements of the earth's crust and are dispersed throughout the ecosystem in powdered form or leached into water bodies through wind and water-induced erosion (Hashem et al., 2021). However, the metals released through these nature-inducing processes are few compared to those from human activities (Hejna et al., 2018). When these elements are present in high environmental concentrations, they proliferate and move up the food web (Mohamed et al., 2021). "Heavy metals" refers to naturally occurring metals with an atomic number and elemental density more significant than 20 and 5 g cm^{-3}, respectively (Ali & Khan, 2018, 2019). Examples are iron (Fe), cobalt (Co), manganese (Mn), chromium (Cr), molybdenum (Mo), zinc (Zn), copper (Cu), and selenium (Se). Others are lead (Pb), cadmium (Cd), and mercury (Hg) (Hejna et al., 2018). Due to chemical characteristics and environmental behavior comparisons, metalloids like arsenic (As) often fall under heavy metals (Bati et al., 2017).

Metals are of two categories: essential and non-essential metals. Essential metals (such as Co, Cu, Fe, Mn, Mo, Se, Ni, and Zn) are required to sustain different physiological and biochemical processes in living organisms (Islam et al., 2023; Narayanan & Ma, 2023). They are also referred to as trace elements, microelements, or micronutrients. On the contrary, non-essential metals (commonly referred to as toxic metals, including Pb, Cd, F, and Hg) have no established biological usefulness. Even at low concentrations, non-essential metals pose severe toxicity and can cause organ damage to living organisms. Hence, they are contaminants (Aniagor et al., 2021a, b). Essential and non-essential metals will further be discussed in this chapter.

There is growing concern about the quality and safety of food due to heavy metal contamination. Consuming tainted meat, fish, and vegetables exposes humans to metallic elements (Nuapia et al., 2018). As an example, Gebeyehu and Bayissa (2020) recorded that the vegetables consumed in the Mojo area of Ethiopia contained Pb (3.63–7.56 mg kg^{-1}), Cr (1.49–4.63 mg kg^{-1}), Cd (0.56–1.56 mg kg^{-1}), As (1.93–5.73 mg kg^{-1}), and Hg (3.43–4.23 mg kg^{-1}). The target hazard quotient (THQ), based on the estimated daily intake (EDI) of the metallic elements, was more significant than 1, indicating a severe health risk. This could impact the physiology and metabolomics of humans and has been linked to conditions including mental retardation, cancer, and suppression of the immune response (Manwani et al., 2023). Conversely, the concentrations of Cd, Hg, Cu, Pb, As, Cr, and Ni in 1,066 meat samples from Zhejiang, China, were analyzed by Han and coworkers (2022) and were reported to be 0.002, 0.0038, 0.801, 0.029, 0.018, 0.061, and 0.055 mg kg^{-1}, respectively. The consumption of the samples reportedly posed a relatively minimal risk to human health, according to the exposure analysis. Levent and coworkers (2020) analyzed the levels of heavy metals in two species of fish, the red mullet and whiting found in the Southern Black Sea and found that the average values of Zn, Cu, Pb Hg, and Cd were 6.4, 0.29, 0.05, 0.036, and 0.017 mg kg^{-1} for red mullet and 9.05, 0.195, 0.07, 0.024, and 0.013 mg kg^{-1} for whiting, respectively. These levels were lower than the permissible limits by the UK, EU, and Turkish regulators with Hazard Index (HI < 1), indicating no significant health risk to consumers. The toxic effects of heavy metals, whether essential or non-essential depend on several factors such as exposure route, dose, and chemical species; in addition to the gender, age, genetics, and nutritional state of the person exposed (Tchounwou et al., 2012). Acute poisoning occurs when high dosages are consumed over a short time, but chronic poisoning, also known as bioaccumulation, occurs when low amounts are consumed slowly over an extended time (Hejna et al., 2018).

Several factors, such as processing, postharvest handling, cooking methods, and prevailing conditions of the environment during the growth of food, affect the content of heavy metals in foods (Morgan, 1999). For instance, certain crops grown in polluted soils or environments have higher metal content, whereas post-harvest processing procedures like washing can eliminate or reduce metal contamination. The amount of metals in some foods can be reduced during cooking, while some can take in metals if the water used for cooking is contaminated. Metals utilized in food processing devices or materials for packaging food may contaminate food. Also, food preparation and storage in the kitchen may result in contamination (Morgan, 1999).

The efficacy of analytical tools for chemical examination has significantly increased over the last few decades. The conventional techniques used to analyze metals in food samples include atomic fluorescence spectrometry (AFS), graphite furnace atomic absorption spectrometry (GFAAS), atomic absorption spectrometry (AAS), flame atomic absorption spectrometry (FAAS), inductively coupled plasma mass spectrometry (ICP-MS), and inductively coupled plasma atomic emission spectrometry (ICP-AES) (Wang et al., 2020). While these techniques are capable of precisely determining the number of heavy metals in food, their drawbacks, such as intricate sample pretreatment procedures, a lengthy detection period, high equipment and operating costs, limit their applicability to laboratory analysis and render them challenging for field applications (Huang et al., 2012). Hence, quick detection techniques for heavy metal contamination need to be developed.

Nanotechnology has increased interest because it has opened up tremendous new possibilities and unique applications in several sectors, such as the food industry. Nanomaterials exhibit unique physical and chemical characteristics due to their exceedingly small size. Hence, applying nanosensors (chemical nanosensors (usually referred to as nanosensors) and nano biosensors) as a rapid detection technique for heavy metal contamination in food has become popular. Nanosensors are any chemical, biological, or surgical sensory site that relays information about nanoparticles to the macroscopic environment (Foster, 2005). Nanosensors are thus a modified form of chemical sensor or biosensor that uses nanomaterial in a piece of analytical tool. Chemical sensors and biosensors differ because the former uses chemical reagents and bioreagents (Kuswandi et al., 2017). Nanosensors can detect toxic metals in food at very low levels by using a variety of nanomaterials.

Analyzing metals in food samples is pertinent for the protection of human health. This chapter will discuss metals' sources and transfer pathways (trophic transfers) in the food chain and nanotechnology techniques for analyzing metals in food. Bioaccumulation and analysis of metal concentration in human nutrition will also be discussed. This chapter will also provide insight into the health implications of heavy metal content in the food chain and the strategies for mitigating toxic metal contamination.

3.2　ESSENTIAL AND NON-ESSENTIAL METALS

Essential metals are those that living organisms need for various biological and physiological processes in the body (Ali & Khan, 2019; Hejna et al., 2018). Mn, Fe, Co, Cu, Zn, Cr, Mo, and Se are examples of essential heavy metals (Ali & Khan, 2019; Elmorsi et al., 2019). They are generally present in trace concentrations (in the order of µg or ng per L or kg), and their nutritional requirements are usually low, albeit with varying bioavailability (Hambidge, 2003). Essential metals are commonly incorporated into animal feed as dietary additions to boost health and optimize output (Hejna et al., 2018). Excessive levels of crucial heavy metals might be hazardous; their concentrations in habitats must not go above the upper permitted limits (UPLs) or they will be regarded as pollutants. UPLs differ based on the kind and use of the water for aquatic habitat and the contaminant's adverse synergistic effects on living organisms (Elmorsi et al., 2019; Sauliutė & Svecevičius, 2015). The close margin between essentiality and toxicity of essential heavy metals such as Mn, Fe, Cu, and Zn has piqued the interest of environmental and ecotoxicological researchers (Ali & Khan, 2019; Yousafzai & Shakoori, 2007). Long-term Cu exposure harms human health, and Chronic Cu poisoning can result in liver damage and severe neurological issues (Afrin et al., 2015; Ali & Khan, 2019; Rossi et al., 2014).

Non-essential metals, on the other hand, have no recognized biological purpose in the body (Ali & Khan, 2019). The four most dangerous heavy metals and metalloids for the environment, namely Cd, Pb, Hg, and As, are widely classified as non-essential metals (Ali & Khan, 2019). Therefore, they have no known biological activities and are deemed pollutants and undesirable chemicals even in animal feeds. Furthermore, As, Cd, Pb, and Hg, previously identified as a public health risk, have a severe toxic effect given that they can cause damage to organs even at low exposure levels (Hejna et al., 2018).

3.3 SOURCES OF METALS IN FOOD

Typically, heavy metals can find their way into agricultural soil via various anthropogenic activities like frequent and disproportionate application of metal-containing chemicals on soil such as pesticides and fertilizers (Mukhles et al., 2022), irrigation using contaminated ground/surface water (Veluprabakaran & Kavitha, 2023) or untreated industrial wastewater (AL-Huqail et al., 2022), atmospheric accumulations caused by industrial and automobile emissions (Liu et al., 2023), smelting processes (Luo et al., 2023). However, the most prevalent human activities in the subject matter are industrial effluents and intensive agricultural activities. Metals such as Pb Cu, Cr, and As were reportedly caused by the discharge of untreated industrial effluent and the irrigation of agrarian farmlands with water from contaminated sources (Islam et al., 2015; Kormoker et al., 2021; Shah & Shaheen, 2007). At the same time, Ni and Cd appeared to be linked to lithogenic sources as well as to human activities such as the frequent use of inorganic fertilizers and industrial effluents and emissions (Atafar et al., 2010; Islam et al., 2015; Kormoker et al., 2021).

Contamination may occur during plant or animal growth, depending on the environmental conditions. The number of metals accumulated in the atmosphere can influence metal concentration in plant crops. It is possible for contamination to happen either directly on the plant itself or on the soil and then be ingested by the plant. Pesticides and fertilizers used in agricultural processes are other plausible contamination sources. Further, higher quantities of toxic elements from animal sources may result from their use in the formulation of animal feeds (Granados-Chinchilla et al., 2015; Morgan, 1999). To this effect, the detection of toxic metals (such as Cr, Cd, Pb, and Hg) in different animal mineral feeds has been reported; specifically, Pd, Cd, and Cr have been detected in cattle feeds (Sigarini et al., 2017), Hg, Pb, and Al in cat and dog feeds (Zafalon et al., 2021), and Cd, Pd, and Cr in poultry feeds (Hossain et al., 2022). In the course of processing, storage, and marketing, foods may also become contaminated. Market-sold foods tend to accumulate heavy metals (Nuapia et al., 2018). The consumption of grains, pulses, vegetables, fruits, and other products cultivated on soil contaminated with toxic metals exposes humans to these metals daily. When a high level of heavy metals is found in food consumed by humans, the metals can harm a variety of biochemical processes, which can result in cardiovascular, nervous, and neurological issues as well as problems with the kidney, brain, liver, bones, and nervous system (Ekhator et al., 2017).

Foods may become contaminated with heavy metals as a result of postharvest handling. Following harvest, food is stored at the farm, transported to the processing facility, and stored there. All of these reflect phases in the food production process where further contamination can take place. Several procedures that take place at the processing facility itself have the potential to contaminate food items with metal. Contamination of the product could result from exposure to processing machinery. Another possible source of heavy metals in food is the water used for washing and rinsing at the plant and added as a component or an additive to the food product. Additives are frequently found in processed foods and are employed for some reasons in the finished product. Another possible source of toxic metals in food is trace metals in these additives. The final factor that could contaminate the food in the processing stage is the material used to package it for sale. Cans made of tin or aluminum both have the potential to contaminate food with metal. Lead contamination used to be primarily caused by the lead solder that was used to seal cans. Because of this, lead solder was banned from being used in food products in many countries, including the US (Morgan, 1999).

Cooking or food preparation in the house is another possible source of metals in food. Food may be exposed to unclean or contaminated kitchen counters, utensils, and pots, among other contaminated surfaces. Materials the utensils and kitchenware are made of may also be a factor. Another source of food ingestion of metals may be the use of contaminated water for cooking or food preparations. Metal contamination may also result from inappropriate household storage conditions, like keeping acidic foods in metal instead of plastic containers.

3.4 METAL TROPHIC TRANSFERS

The trophic transfer of heavy metals in the food web is a crucial subject in environmental research. It describes how contaminants move from a particular trophic level to another in the food web (Figure 3.1). Heavy metals are transmitted from the abiotic environment (water, sediments, and soils) to living creatures and accumulate in the biota, contaminating the food chain (Ali & Khan, 2019).

Heavy metal transfer from soil to plant is a critical pathway for harmful heavy metals from the environment into the food chain. Earth is one of the largest reservoirs of heavy metals and impacts food safety at the start of the food chain (Mao et al., 2019). It is the primary route humans are exposed to metal pollution (Cui et al., 2004). Irrigating agricultural areas with sewage wastewater causes heavy metal deposition in soils, which transfers to food crops such as cereals, vegetables, and milk (Singh et al., 2010). Mao et al. (2019) investigated the geochemical characteristics of heavy metals in soil and rice plants in the Yangtze River Delta. The study's findings showed that heavy metal levels in soil dropped as soil pH increased, whereas rice shoots collected heavy metals quickly under reduced soil pH conditions. Heavy metals are sometimes elevated at successive trophic levels of food chains, resulting in the biomagnification of these compounds in the food chain (Ali & Khan, 2019; Green et al., 2010). Rice farming is particularly of concern when using water contaminated with heavy metals for irrigation because the growth process requires more water from seed germination to harvest. This was first hinted at by Arunakumara et al. (2013) and re-echoed by Ali and Khan (2019). Detecting heavy metals in the soil irrigated with sewage and industrial effluents in the Dera Ghazi Khan area of Pakistan inspired Attah and coworkers (Atta et al., 2023) to analyze heavy metals in different plant species cultivated in the area. Effective soil-to-plant metal transfer was more evident in tomato, red corn, spinach, wheat, and brinjal than in other species. The Pb, Cr, Mn, and Ni samples exceeded the WHO/FAO recommended safe limits (the report (Atta et al., 2023) can be consulted for details). The health risk index (HRI) analysis gave 2.9, 3.1, 3.8, 4.3, 4.5, 4.7, 5.9, 6.4, 6.4, for luffa, cabbage, white corn, apple gourd, red corn, tomato, brinjal, wheat, and spinach, respectively; indicating that a potential heavy metal contamination risk in the order: spinach > wheat > brinjal > tomato > red corn > apple gourd > white corn > cabbage > luffa. The studies revealed that the intake of these plant foods cultivated on wastewater-irrigated land poses a substantial health risk to consumers. Vegetable contamination with toxic metals cannot be overlooked because they are crucial to human nutrition. Comparatively, vegetables cultivated on wastewater-irrigated soils reportedly had higher daily metal consumption levels than vegetables cultivated on control soils (Jan et al., 2010).

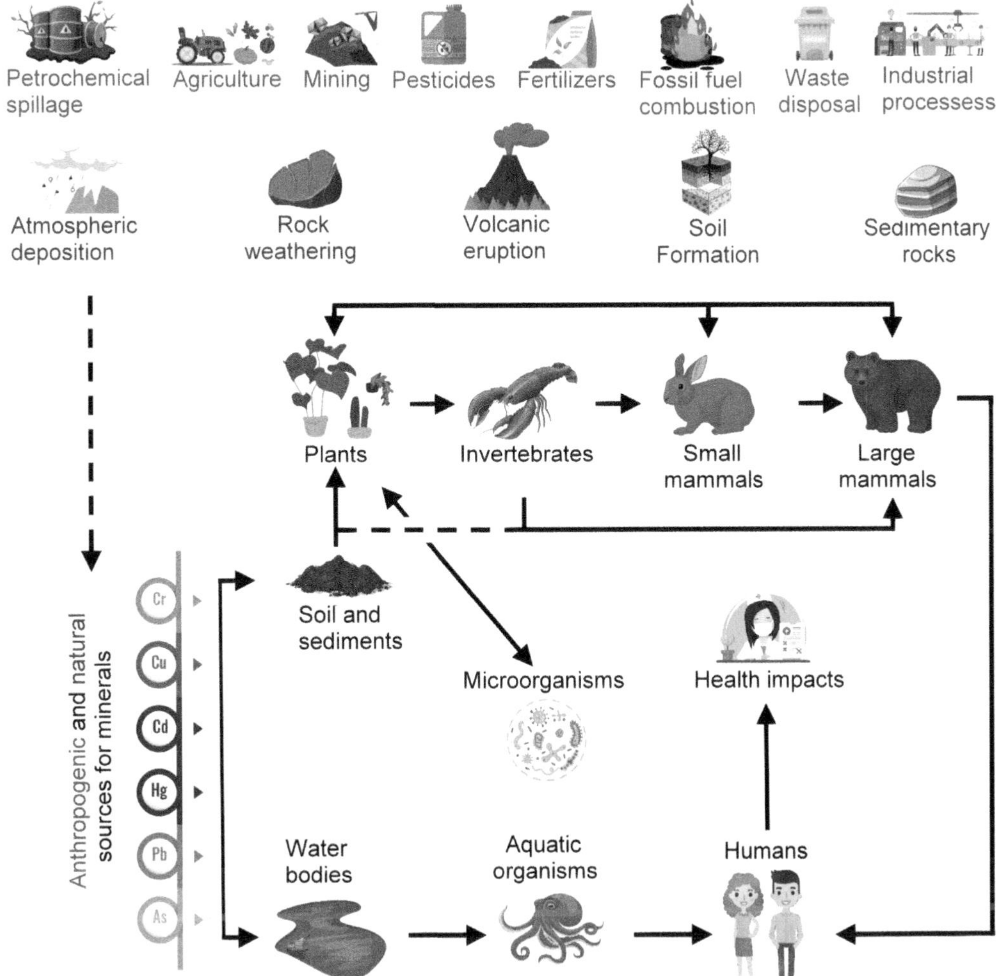

FIGURE 3.1   Metal transfer pathways in the food chain.

Another point of entry of metals into plant foods is the ariel transfers (Ali & Khan, 2019). This involves the atmospheric deposition of heavy metals on plant leaves and penetration of the plant system via foliar transfer. Another significant transfer channel is the uptake of toxic metals by tobacco plants and then onto humans who smoke tobacco. An effective route to Cd exposure is smoking tobacco, which was revealed in a US-based study highlighting that smoking significantly correlates with cancer-related mortality (Kim et al., 2023). The movement of metals from soil to humans via milk consumption is another vital transfer route. When cows, for instance, feed on grasses grown on metal-contaminated soil, consuming the milk derived from such cows can lead to the transfer of metals to humans. This has been demonstrated (Sai Chaithanya et al., 2022; Yan et al., 2022).

Similarly, heavy metals are transferred to aquatic habitats from a variety of sources, both human and natural, including waste from industries, home effluents, leaching from waste dumps, atmospheric sources, and other metal-based businesses, as well as E-Waste (Sonone et al., 2020). Heavy metal toxicity is mainly responsible for the degeneration of aquaculture populations, creating physical abnormalities in creatures and contaminating the aquatic environment. These poisonous heavy metals cause a variety of fish ailments. Because people consume fish, it has an indirect impact on them. Hence, the infiltration of heavy metals into water bodies and aquatic ecosystems significantly impacts the food chain. Such toxic elements rapidly accumulate in living organisms and are difficult to digest or eliminate (Baby et al., 2010). They are eaten and stored in biological organisms or transported to a greater trophic level via wetland food chains (Green et al., 2010; Xie et al., 2010).

3.5 BIOACCUMULATION AND ANALYSIS OF METALS IN FOOD

Bioaccumulation and biomagnification of heavy metals in food chains seriously harm wildlife and human health (Ali & Khan, 2019). Bioaccumulation is the buildup or accumulation of contaminants like heavy metals in living organisms. In comparison, biomagnification is the term that refers to the rise in the concentration of heavy metals along the food web. Metals can bioaccumulate in food of both plant and animal origins. The following subsections will discuss bioaccumulation in the food web and highlight the concentrations of heavy metals in different food materials of plant and animal origin.

3.5.1 BIOACCUMULATION OF METALS AND ANALYSIS OF THEIR CONCENTRATION IN FOODS OF PLANT ORIGIN

Edible plants such as vegetables are essential components of the human diet; their contamination with heavy metals can harm human health. Heavy metals tend to build up in the tissues of food plants. When heavy metals infiltrate our environment, plants may bioaccumulate them. Vegetables, both leafy and nonleafy, are effective heavy metal accumulators. Though water-to-plant, air-to-plant, and soil-to-plant interactions can lead to the contamination of plants by heavy metals, soil-to-plant interaction is the primary source of metal bioaccumulation in plants (Khan et al., 2015). Metals can infiltrate the soil through different routes, such as using phosphate-based fertilizers and metal-containing pesticides in agricultural land, water runoff from industrial facilities and roads, atmospheric deposition, untreated sewage sludge on lands, etc. However, the metal's mobility and soil availability influence the likelihood that a metal will make it into a food chain. When metals enter the soil, their cations are attached to negatively charged soil particles like organic matter and clay. When they break away from the soil particles and penetrate the soil solution, their bioavailability is enabled, and they become susceptible to accumulate in organisms that inhabit the soil (Gall et al., 2015). The primary process by which humans are exposed to soil contamination through the food web occurs via the bioaccumulation and translocation of metallic elements in plants. The response of a common medicinal plant (*Mentha piperita*) under soil contamination with Pb, As Cd, and Ni and their soil to plant transfer and translocation in the roots, stems, and leaves were studied by Dinu et al. (2021) and compared with a reference sample with no metal addition. The metal concentration analysis revealed that Cd, Ni, and Pb did accumulate in various parts of the plant, except for As. The metal buildup in the organs of the plant followed the order of roots > stems > leaves. There is a close correlation between the heavy metals accumulated in the soil and those in the food crops (Khan et al., 2015). As the concentration of heavy metals in the surrounding soil increases, so does the average amount that plants accumulate (Chaves et al., 2011).

The potential for a hyper-accumulation of heavy metals is thought to exist in leafy vegetables such as lettuce (Ramos et al., 2002). Green leafy vegetables can accumulate heavy metals in their tissues without showing any signs of toxicity. As reported by Monteiro and coworkers (Monteiro et al., 2007), the levels of heavy metals in the roots and shoots of lettuce increased with an increase in the exposure time. High metal concentrations can be found in the vegetative and non-vegetative parts of plants cultivated in soil irrigated with metal-contaminated wastewater. Additionally, the accumulation of heavy metals in plants is also influenced by individual species of plant and diverse cultivars within a species (Khan et al., 2015). Heavy metal buildup and translocation from soil to plant are of great concern because bioaccumulation in humans is ensured when humans consume edible plants.

To study the impact of phosphorus industry effluents on food plants, Guo et al. (2023) assessed the heavy metal contents of vegetables cultivated around a phosphorus production plant in the Guizhou Province of China. Results revealed that the mean levels of metals in vegetables were Cd (0.382 mg kg^{-1}), Hg (0.015 mg kg^{-1}), Pb (0.227 mg kg^{-1}), As (0.728 mg kg^{-1}), Cr (0.850 mg kg^{-1}), Zn (6.438 mg kg^{-1}), Mn (27.227 mg kg^{-1}), and Co (0.525 mg kg^{-1}). Significant variation was observed in metal accumulation in different vegetable species. Chayote, chard, white radish, potato, crown daisy, eggplant, sweet potato, pumpkin, and carrot were chosen as the vegetables with the lowest levels of metal accumulation. All vegetables, except chayote, have health risks that are both carcinogenic and noncarcinogenic; eating sweet potato leaves bears the most health risk.

Amongst the different vegetables, fruits, and other crops (such as bananas, carrots, beans, jackfruit, green chili, potato, mango, onions, and tomato) studied for their heavy metal levels in Bangladesh, Shaheen et al. (2016) discovered that the concentration of Pb in mango and Cd in tomato were higher than the WHO/FAO permissible limits (see Table 3.1 for details). The concentration of Pb in mango was discovered to be six times the acceptable safe limit. The THQs for all the metals analyzed were 1, indicating no health risks for the adult demography. However, the intake of vegetables resulted in total THQs of Mn and Cu greater than 1, indicating substantial health hazards. HI was determined to be less than 1 for fruit consumption but greater than 1 for the consumption of vegetables, meaning that only vegetable consumption has the potential for adverse health impacts. A similar study by Laboni et al. (2023) in an industrial district of Narsingdi in Bangladesh analyzed 72 samples of spinach, watercress, red amaranth, eggplant, cauliflower, and alligator weed. It determined that Ni, Cd, Pb, and Cr were present in most samples examined. About 1.39%, 44.44%, and 79.17% of the samples had concentrations above the FAO/WHO permissible limit for Ni, Cd, and Pb, respectively.

The concentrations of Co, Pb Cd, As, Cr, Ni, and Cu in several commonly consumed leguminous foods (tarragon, bay laurel, dill, Syrian mesquite, vine leaves, thymes, arugula, basil, common purslane, and parsley) in the city of Erbil, Iraq were analyzed (Abuzed Sadee & Jameel Ali, 2023). The heavy metals concentration in the legumes ranged as follows: Co (0.10–0.32), Pb (0.61–1.65), Cd (0.110–0.612), As (0.15–0.63), Cr (0.45–1.76), Ni (0.29–1.27), and Cu (0.64–1.92), all measured in µg g^{-1}. The Hazard quotient ranged from 0.03–0.09, 0.83–2.23, 0.60–3.32, 2.71–11.38, 0.81–3.18, 0.08–0.34, and 0.09–0.26, for Co, Pb Cd, As, Cr, Ni, and Cu, respectively. The study revealed that the Erbil city dwellers are at no carcinogenic health risk. The presence of metals (Ni, Cu, Zn, and Fe) was equally detected in all the commonly consumed vegetables (yard long bean, fern leaves, spiny gourd pointed gourd, jute leaves, and green arum leaves) analyzed in the Assam region of India (Boro et al., 2023). Though essential, the content levels were reported to be of food safety concern. Fe content was highest in most leaves, followed by Zn, Cu, and Ni. In a different study, Qi et al. (2023) created a database of metals in Chinese vegetable soils via a comprehensive review of relevant literature and gathering field samples. The average levels of metals in parts of the vegetables that are edible were Hg (0.007 mg kg^{-1}), Pb (0.137 mg kg^{-1}), Cd (0.093 mg kg^{-1}), Cr (0.118 mg kg^{-1}), As (0.024 mg kg^{-1}), Zn (3.272 mg kg^{-1}), and Cu (0.622 mg kg^{-1}). The five toxic metals had exceedance rates in this increasing order: As (0.21%), Cr (4.03%), Hg (11.5%), Cd (12.9%), and Pb (18.5%). Leafy vegetables exhibited a high accumulation of Cd, while root vegetables exhibited a high Pb accumulation. Solanaceous and leguminous vegetables generally demonstrated lesser bioaccumulation of metals. According to the health risk data, the non-carcinogenic risk associated with the consumption of individual vegetables was within the permissible limit, with children being more at health risk than adults.

The analysis of the metal levels contained in the rice and other grains (beans, peas, lentils, oats, barley, and wheat) cultivated in the US, Italy, India, and Thailand revealed that the average concentrations of Pb, Cd, and As in the white rice produced in the US were 2.8, 6.5, and 131 µg/ kg, respectively (TatahMentan et al., 2020). White rice cultivated in Italy, Thailand, and India had higher average heavy metal concentrations of Pb (3.6 µg kg^{-1}), Cd (8.4 µg kg^{-1}), and As (155 µg kg^{-1}) than the US white rice. The average Pb, Cd, and As values in the US brown rice were 4.5, 17.4, and

217 µg kg^{-1}, respectively. None of the grains surpassed the codex standards for Cd and Pb in cereals, pulses, and polished rice. The brown rice exceeded the codex standard for As. Other grains had higher levels of essential metals than brown and white rice. The summary of some reported works on the concentration of the non-essential heavy metals of most concern in different foods of plant origin is presented in Table 3.1.

3.5.2 Bioaccumulation of Metals and Analysis of Their Concentration in Foods of Animal Origin

Heavy metals are accumulated by animals of both terrestrial and aquatic species from the food they consume and their abiotic environment. Marine species are susceptible to bioaccumulation because their bodies cannot get rid of the contaminants as quickly as they are absorbed from the water in their environment. Consuming contaminated aquatic species or contact with contaminants in the food we eat, the air we breathe, or the water we drink can lead to human bioaccumulation. Heavy metals can remain in our bodies for long since they do not biodegrade.

From the tiniest microorganisms to humans, bioaccumulation occurs throughout the food web. Microorganisms are consumed by zooplankton, fish consume zooplankton, and humans eat fish. Heavy metals bioaccumulate in every living organism at every stage of this process, and by the time they reach humans, we have consumed substantial amounts of the heavy metals. Physiology, food, and behavior are among the factors affecting the accumulation of heavy metals in animals (Ali & Khan, 2019).

Insects continually come into contact with toxic metals through their food material, growth habitats, and processing operations (Haque et al., 2021; Malematja et al., 2023). When an insect consumes contaminated substrates like plant byproducts with pesticide or fertilizer residues or animal excrement with veterinary medicine residues, it bioaccumulates metals in its tissue (Muhammad et al., 2022). Pesticides deployed to keep insects away from farming areas may harm livestock, especially if the feeder insects were collected from such an environment. This is still a significant issue in some developing nations where chickens and other poultry are fed with locusts and grasshoppers collected from the vicinity of such farms (Malematja et al., 2023; Mézes, 2018). Insect species such as termites, super worms, locusts, yellow mealworms, and grasshoppers, have been reported to contain varying concentrations of Cd, Pb, Fe, As, Cu, Hg, and Zn, the type and concentration of the metal in the substrate being responsible for the variations in their content levels (Lange & Nakamura, 2021; Malematja et al., 2023; Mlček et al., 2017; Van der Fels-Klerx et al., 2018).

Two commercially available edible aquatic insects (*Cybister tripunctatus* and *Dytiscus marginalis*) were analyzed for the presence of Pb Cu, As, Zn, Co, Fe, Cr, Se, and Ca (Aydoğan, 2023). The aquatic insects were found to contain toxic metals (As and Pb), but their concentrations were below the toxicity levels to human health. Contrarily, when heavy metal concentrations in edible insects like locusts, termites, and grasshoppers were examined by Muhammad and coworkers (Muhammad et al., 2022), they discovered that these insects contained heavy metals in concentrations that are above the WHO-recommended limits.

Going beyond lower organisms, the levels of metals (Al, As, Cr, Cd, Cu, Ni, Pb, and Zn) detected in the meat (beef, pork, and mutton) consumed in the industrial areas of Ningxia, China, was reported by Wang et al. (2023) (see Table 3.2). The decreasing order of the metal content in the meat samples analyzed was Zn> Cu >Cr> As >Pb> Al >Ni/Cd. The assessment of the health risk status showed that the HI was generally < 1, signifying that residents' consumption of meat samples poses no significant risk to their health. Similarly, the contents of essential and nonessential metals (Cd, Pb, As, Cr, Hg, Mn, Zn, and Cu (Table 3.2)) in the tissues of both male and female wild boar and their effect on food safety was studied by Lénárt and coworkers (2023). In all the tissue samples from both sexes, the Cd, As, and Hg levels were less than the detection threshold (Cd< 0.05 mg kg^{-1}, As, Hg: < 0.5 mg kg^{-1}). The Pb levels in the muscles of the males and females were found to be (0.22 ± 0.06) and (0.36 ± 0.16) mg kg^{-1}, respectively, suggesting a considerable variation ($p = 0.0184$)

TABLE 3.1

Concentration of Heavy Metals in Foods of Plant Origin from Different Countries

Country	Food Crop	Metals					References
		Cd	As	Cr	Pb	Hg	
Ethiopia	Tomato (mg kg^{-1})	0.56±0.05[a,c]	1.93±0.50[a]	1.49±0.01[ab]	3.63±0.11[ac]	3.43±0.05[b]	Gebeyehu & Bayissa (2020)
	Cabbage (mg kg^{-1})	1.56±0.05[ac]	5.73±0.37[a]	4.63±0.20[ab]	7.56±0.23[ac]	4.23±0.28[b]	
China	Cereals (millet rice, corn, flour); Mean, (mg kg^{-1})	0.0128	0.0158	0.2564	0.0453	NP	Wang et al. (2023)
	Beans (tofu, cowpeas); Mean, (mg kg^{-1})	0.0045	0.0000	0.7536	0.4636	NP	
	Potatoes (sweet potato, potato); Mean, (mg kg^{-1})	0.0000	0.1475	0.0322	0.1798	NP	
	solanaceous fruits (cucumber, pepper, bitter gourd, eggplant); Mean, (mg kg^{-1})	0.0185	0.0965	0.2728	0.1659	NP	
	Vegetables (cabbage, spinach, cauliflower); Mean, (mg kg^{-1})	0.0277	0.0042	0.5642	0.1603	NP	
	Fruits (watermelons, grapes, pears, apples); Mean, (mg kg^{-1})	0.0173	0.0473	0.0673	0.1496	NP	
US	White rice (µg kg^{-1})	1.7–71	65–202	NP	0.2–32	NP	TatahMentan et al. (2020)
Italy/Thailand/ India	White rice (µg kg^{-1})	3.1–27	58–183	NP	2–96	NP	
Iraq	Legumes (µg kg^{-1})	0.110–0.612	0.15–0.63	0.45–1.76	0.61–1.65	NP	Abuzed Sadee & Jameel Ali (2023)
Pakistan	Brinjal (mg kg^{-1})	0.18	0.001	4.61[a]	0.33[a]	NP	Atta et al. (2023)
	Tomato (mg kg^{-1})	0.12	0.0002	3.02[s]	0.15	NP	
	Luffa (mg kg^{-1})	0	0	1.22	0.004	NP	
	Apple Gourd (mg kg^{-1})	0.11	0	3.7[a]	0.23	NP	
	Cabbage (mg kg^{-1})	0.13	0	3[a]	0.09	NP	
	Spinach (mg kg^{-1})	0.14	0	4.17[a]	0.09	NP	
	Red Corn (mg kg^{-1})	0.12	0	2.4[a]	0.34[a]	NP	
	White Corn (mg kg^{-1})	0	0	1.34	0.22	NP	
	Wheat (mg kg^{-1})	0.14	0.0002	2.61[a]	0.17	NP	

(Continued)

TABLE 3.1 (*Continued*)

Concentration of Heavy Metals in Foods of Plant Origin from Different Countries

Country	Food Crop	Metals					References
		Cd	As	Cr	Pb	Hg	
Bangladesh	Banana (mg kg⁻¹)	NP	NP	0.317±0.012	0.003±0.003	NP	Shaheen et al. (2016)
	Jackfruit (mg kg⁻¹)	0.037±0.0	0.007±0.0	0.863±0.080	0.017±0.015	NP	
	Mango (mg kg⁻¹)	0.005±0.0	0.013±0.0	0.893±0.149	0.642±0.556[a]	NP	
	Brinjal (mg kg⁻¹)	0.041±0.032	0.006±0.001	0.497±0.029	0.011±0.011	NP	
	Bean (mg kg⁻¹)	0.018±0.007	0.008±0.001	1.110±0.054	0.057±0.050	NP	
	Carrot (mg kg⁻¹)	0.023±0.003	0.006±0.001	0.296±0.021	0.029±0.025	NP	
	Green chili (mg kg⁻¹)	0.023±0.011	0.004±0.001	0.650±0.039	0.006±0.005	NP	
	Onion (mg kg⁻¹)	0.023±0.017	0.008±0.002	0.542±0.131	0.027±0.023	NP	
	Potato (mg kg⁻¹)	0.013±0.007	0.006±0.001	0.528±0.051	0.007±0.006	NP	
	Tomato (mg kg⁻¹)	0.056±0.004[a]	0.006±0.002	0.795±0.059	0.005±0.004	NP	
Iran	Wheat (mg kg⁻¹)	2.640	0.094	0.016	0.044	NP	Ghanati et al. (2019)
Poland	Flour (mg kg⁻¹)	NP	NP	0.043	0.067	NP	Bielecka et al. (2022)
Malaysia	Grain (mg kg⁻¹)	0.032	0.082	0.001	0.013	NP	Zulkafflee et al. (2022)
Burkina Faso	Rice (mg kg⁻¹)	NP	1.98	0.08	0.15	NP	Bazié et al. (2022)
Croatia	Beans (mg kg⁻¹)	9.2	< dl	0.04	0.25	NP	Stančić et al. 2(016)
	White potato (mg kg⁻¹)	8.2	< dl	0.23	0.36	NP	
	Red potato (mg kg⁻¹)	5.9	< dl	0.28	0.17	NP	
Colombia	Carrot (mg kg⁻¹)	6.80	8.80	0.15	6.96	NP	Lizarazo et al. (2020)
Ethiopia	Cabbage (mg kg⁻¹)	15.96	4.50	1.00	6.70	NP	Bayissa & Gebeyehu (2021)
India	Apple (mg kg⁻¹)	NP	NP	0.002	0.068	NP	Mawari et al. (2022)
Poland	Strawberry (mg kg⁻¹)	NP	NP	0.050	0.10	NP	Rusin et al. (2021)
	Raspberry (mg kg⁻¹)	NP	NP	0.050	0.10	NP	

Note: The source article provides references to the permissible limit standards.

[a] Above FAO/WHO permissible limit.

[b] Above EU permissible limit.

[c] Above China EPA permissible limit.

NP: Not provided.

between the sexes. The intake of the tissues of the wild boars evaluated could be considered toxicologically undesirable as food, primarily because of the Pb levels, and as a result, constitute a risk to regular consumers of this kind of meat from wildlife in the Transdanubia region of Hungary.

Canned meat and chicken (canned corned beef, canned beef, and chicken luncheon) sold in the Egyptian market were assessed by Morshdy et al. (2023) for the presence of Cd, Pb, As, Al, Hg, and Sn. The estimated HQ (hazard quotient) and HI for the assessed heavy metals in daily consumption showed that canned beef and chicken lunches had values less than 1.0, indicating consumer safety. In contrast, canned corned beef had a value of 1.021, indicating possible health risks. The details of the individual concentrations can be seen in Table 3.2. The metals (Cd, Co, Pb, Cr, Se, As, Ni, Zn, and Cu) contained in the meat products of sheep, turkey, beef, and ostriches collected from the Kurdistan region of Iran were equally assessed by Raeeszadeh et al. (2022). The levels of other metals in the examined meats, excluding cobalt, exceeded the maximum permissible limits. Although cadmium's THQ in beef and sheep meats exceeded their maximum allowable limits, its cancer target risk (TR) was within acceptable limits.

The analysis of liver, kidney, and muscle samples of sheep collected from different parts of Australia (MacLachlan et al., 2016) showed that among the metals examined (Pb, As, Se, Co, Cu, Mo, and Zn), Pb, As, and Hg levels in meat and organs were considerably low, while Cd levels in the liver and kidneys were relatively high occasionally (see Table 3.2 for the mean concentrations of the metals). Chicken liver samples collected from Erbil City, Iraq, were analyzed for Cd, Co, Cr, Cu, Mn, Ni, Pb, Hg, Zn, and Se contents (Ali et al., 2020). For the trace elements, the average values (mg kg^{-1}) were Co (0.060±0.027), Cr (0.060±0.05), Cu (2.050±0.34), Mn (1.850±0.47), Ni (0.150±0.17), and Zn (33.535±5.24). Copper values exceeded the WHO's maximum limit. The nonessential metals had average amounts (mg kg^{-1}) of Cd (0.070±0.037), Pb (0.2780±0.10), Hg (0.110±0.083), and Se (2.010±0.454). The levels of Pb and Hg were also found to be above the WHO limits. In a different study, two common fish species (demersal and pelagic) obtained from the Galapagos Islands of Ecuador revealed higher concentrations of Mg, Cr, Mo, and K in the demersal species and high levels of Zn in the pelagic species (Franco-Fuentes et al., 2021). Consequently, the authors advised that men and women should not consume more than 86 and 73 g of *M. olfax* (a demersal species) per week, respectively, due to the possibility of Cd toxicity. *M. olfax* (sailfin grouper) is the most common finfish caught using artisanal handlines in the Galápagos. The dried flesh is shipped to Ecuador's mainland for use in a traditional Easter meal called fanesca. The fishery is active all year long, but it reaches its peak between October and April when Easter-related demand is at its highest.

Additionally, a study by Badr and Arafa (2023) made a comparison of the metal contents of the same kind of food products (Chicken Shawarma, Hawaweshi, Beef Shawarma) sold in the streets of the Gharbia region of Egypt and a laboratory-produced version of them. The results revealed that the average values in the examined street samples were significantly different from the laboratory products, attributed to adding low-cost and substandard components at varied percentages to maximize their earnings. The concentrations of Cd and Pb in the foods sold by street vendors were above the EOS (Egyptian Organization of Standardization)-permitted limits (Badr & Arafa, 2023). Further, the content of metals such as Cd, Pb, As, Cu, Zn, Ni, and Fe in different kinds of meats (chicken, pork, camel, birds, cow, sheep, and seafood (crab and fish) reviewed by Khanverdiluo et al. (2023) revealed that Camel meat had the highest levels of Cd (1.965 mg kg^{-1}) and As (1.503 mg kg^{-1}). Generally, the camel's grazing practices expose them to tainted water and food sources, and their digestive systems are different from those of other ruminants, making them more tolerant of harmful dietary elements such as heavy metals (Abdelrahman et al., 2023; Ullah et al., 2023). This could account for the elevated levels of Cd and other heavy metals that bioaccumulated in their tissues and organs.

The highest levels of trace elements (Zn, Ni, Cu, and Fe) were recorded in seafood, cow, and bird meat. The continental distribution showed that the Eastern Mediterranean and African regions had the highest Pb, Cd, Fe, Ni, and As levels. The risk analysis showed that different countries had

TABLE 3.2

Concentration of Heavy Metals in Foods of Animal Origin

Country	Food Crop	Metals (mg kg⁻¹)					References
		Cd	As	Cr	Pb	Hg	
China	Meat (beef, pork, and mutton); (Mean values)	0.0000	0.2183	0.4719	0.0780	NP	Wang et al. (2023)
Iran	Sheep	2.79	0.23	1.66	11.79	NP	Raeeszadeh et al. (2022)
	Beef	4.31	0.2	1.83	7.01	NP	
Kuwait	Mutton	0.301	349	362	0.482	NP	Abd-Elghany et al. (2020)
Australia	Sheep muscle (Mean values)	0.0035	0.010	NP	0.007	0.0025	MacLachlan et al. (2016)
	Sheep liver (Mean values)	0.280	0.010	NP	0.040	0.0034	
	Sheep kidney (Mean values)	0.853	0.011	NP	0.057	0.0061	
China	Fresh meat	0.002	0.018	0.061	0.029	NP	Han et al. (2022)
India	Meat	0.004	NP	0.21	NP	NP	Dubey et al. (2016)
Italy	Baked ham	0.01	NP	0.15	0.22	NP	Barone et al. (2021)
	Raw ham	0.01	NP	0.19	0.30	NP	
	Mortadella	0.02	NP	0.20	0.34	NP	
Hungary	Wild boar (male muscle)	< 0.05	< 0.5	0.13±0.06	0.22±0.06	< 0.5	Lénárt et al. (2023)
	Wild boar (female muscle)	< 0.05	< 0.5	0.14±0.08	0.36±0.16	< 0.5	
	Wild boar (male fat)	< 0.05	< 0.5	0.08±0.07	< 0.2	< 0.5	
	Wild boar (female fat)	< 0.05	< 0.5	0.08±0.07	< 0.2	< 0.5	
Egypt	Canned beef (Mean values)	0.073±0.006	0.24±0.01	NP	0.35±0.04	0.075±0.003	Morshdy et al. (2023)
	Canned corned beef (Mean values)	0.081±0.007	0.28±0.02	NP	0.39±0.04	0.086±0.003	
	Canned chicken luncheon (Mean values)	0.071±0.007	0.27±0.02	NP	0.21±0.02	4.51±0.55	
Iraq	Chicken liver (Mean values)	0.070±0.037	Np	Np	0.2780±0.10	0.110±0.083	Ali et al. (2020)
Hungary	Muscle tissue of roe deer (male)	<LOD	0.33±0.18	Np	0.42±0.22	1.07±0.30	Lehel et al. (2016)
	Muscle tissue of roe deer (female)	<LOD	0.23±0.23	Np	0.55±0.20	0.61±0.38	

LOD, limit of detection

varied risk patterns. See Khanverdiluo et al. (2023) for more details on that report. Table 3.2 summarizes the concentration of metals in animal foods from different countries.

Summarily, finding a solution to the issue of heavy metal bioaccumulation is a challenging task. Scientists and policymakers still have a lot of work to do to safeguard the biotic environment from heavy metal contamination and to create effective techniques for identifying the presence of heavy metals at unsafe levels.

3.6 NANOTECHNOLOGY TECHNIQUES FOR THE ANALYSIS OF METALS IN FOOD

The application of nanotechnology in studying metal content in food cuts across the analytical methods and materials. The most common nanotechnology-based techniques in which different categories of nanosensors are used to analyze metals include electrochemical, colorimetric, and fluorescent analyses.

3.6.1 ELECTROCHEMICAL ANALYSIS BASED ON NANOSENSORS AND BIONANOSENSORS

Recently, interest in using nanomaterials to create electrochemical nanosensors has increased. Freestanding nanoelectrode gadgets or microelectrodes doped with nanomaterials exhibit improved physical and chemical properties, making these devices desirable for enhancing existing electroanalytical applications (Barry & O'Riordan, 2016). The nanomaterials-based sensors (nanosensors), be it chemical nanosensors (chemical sensors) and nano biosensors (biosensors), can both be used online and incorporated into prevailing production and distribution lines, or they can be used offline as quickly, easily, portable, and disposable sensors for the detection of food contaminants. Nanosensors are, therefore, quicker, cheaper, and exhibit more sensitivity than traditional and instrumental methods (Kuswandi et al., 2017). Metal nanoparticles, metal oxide nanoparticles, metal nanoclusters, carbon and metal quantum dots, carbon nanotubes (CNTs), graphene, and nanocomposites are examples of nanoscale materials that can enhance the sensitivity of nanosensors via signal amplification and incorporate multiple transduction principles, including improved catalytic activity, electrochemical, Raman, superparamagnetic, and optical properties into the nanosensors (Srivastava et al., 2018).

Advanced nano-electrocatalysts for the construction of electrochemical sensors have been produced using carbon-based nanomaterials such as graphene, graphene oxides (GO), and reduced graphene oxide (rGO), all of which can adsorb heavy metal ions (Gong et al., 2021).

Lu and coworkers developed a novel nanocomposite-based electrochemical analytical technique using 3D graphene aerogel and MOF nanocomposite to determine the trace levels of different metal ions in vegetables (spinach) (Lu et al., 2018). In addition to providing backbone support to the MOF, the graphene also improves the conductivity of the nanocomposite by speeding up electron transmission in the matrix. Approximate values for the measured potentials of the different metal ions were 0.84, 0.62, 0.28, and 0.08 V for Cd(II), Pb(II), Cu(II), and Hg(II), respectively. Their limits of detection were 1.5, 2, 7, and 20 nmol L^{-1} for Pb(II), Hg(II), Cu(II), and Cd(II), respectively. Mitra et al. (2019) equally demonstrated the applicability of a 3D graphene-based electrochemical sensor for detecting Cd(II) ion accumulated in bacterial and rice tissues. Electrostatically induced phytic acid (Dai et al., 2016) and an on-site bismuth film (Kiliç & Kizil, 2019) coating on graphene-based electrodes, respectively, were found to exhibit enhanced sensitivity to simultaneous Pb(II) and Cd(II) detection. The modification of a gold (Au) electrode with improved charge transfer and mass transport performance, which facilitated the sensitive and selective detection of metal ions, was accomplished using a nanocomposite of GO coupled with a bipyridine ligand (Gumpu et al., 2017).

Carbon nanotubes (CNTs), cylindrical macromolecules with hexagonally organized carbon atoms, have some notable attributes, including an ordered structure, high electrical conductivity, superior mechanical strength, and high object ratio (Gong et al., 2021). Due to their narrow band gap, single-walled CNTs (SWCNTs) are particularly sensitive to the local redox or dielectric milieu. They also exhibit intense redox reactions with Hg^{2+} (Gong et al., 2021). Rapid detection of Cd(II) in meat using an electrochemical sensor with a metal-organic framework modified by multi-walled CNTs (MWCNTs) was reported by Wang et al. (2021). The outstanding performance of the nanocomposite electrochemical nanosensor was attributed to the synergistic interaction between MWCNTs with exceptional conductivity and the MOF with a unique octahedral structure and large surface area. The developed sensor demonstrated an excellent linear correlation spanning 0.5–170 μg L^{-1} at optimal testing conditions, with a recorded detection limit of 0.2 μg L^{-1}. A nanocomposite made of oxidized MWCNTs and porous nanosheets of graphitic carbon nitride (p-g-C_3N_4-NSs) was fabricated for the detection of heavy metal ions (Cd(II), Hg(II), Pb(II) and Zn(II)) in various spiked food samples (Ramalingam et al., 2019). When simultaneously detecting the heavy metal ions, the nanosensor demonstrated high sensitivity and selectivity, generally at –1.16, –0.5, –0.78, and +0.35 V (against Ag/AgCl) for Zn(II), Pb(II), Cd(II), and Hg(II), respectively. The detection limits varied from 8 to 60 ng L^{-1}. In another study, trace levels of Pb(II) and Cd(II) ions in organic vegetables from parts of Luzon, Philippines, were detected using a tri-nanocomposite of MWCNTs, silver nanoparticles (AgNPs), and Nafion adapted glass-like carbon electrodes (Nafion-GCE) (Palisoc et al., 2018). The electrode treated with 2 mg of MWCNTs and 1 mg of AgNPs was discovered to have the most favorable analytical response for detecting Pb(II) and Cd(II) ions. For Pb(II) ion, the electrode showed linearity between 0.493 and 157.2 μg L^{-1}, while for Cd(II) ion, linearity was exhibited from 1.864 to 155.1 μg L^{-1}. The limits of detection were 0.216 and 0.481 μg L^{-1} for Pb(II) and Cd(II) ions, respectively.

Since its invention in 1967, the electrochemical biosensor has been utilized extensively in food safety testing, medicine, animal husbandry, fermentation engineering, and environmental testing (Wang et al., 2020). It is an analytical technique for the electrochemical signal detection of bio-molecules. An electrochemical aptasensor (biosensor) with excellent sensitivity was developed by Yuan and coworkers for the simultaneous detection of trace levels of Pb(II) and Cd(II) ions in vegetables and fruits (Yuan et al., 2022). Au-S bonds were used to immobilize the two-stranded DNA and aptamers on the electrode. The linear response of the electrochemical aptasensor Cd(II) and Pb(II) ions ranged from 0.1 to 1,000 nmol L^{-1} at optimum conditions, and the detection limits were 89.31 and 16.44 pmol L^{-1} for Cd(II) and Pb(II), respectively. In addition, outstanding stability and reproducibility were demonstrated by the aptasensor. When nanomaterials are incorporated with biosensors, it can aid the improvement of the detection sensitivity of the sensors as in silica mesoporous/nanospheres (Cheng et al., 2015), silver nanoparticles (AgNPs) (Jeong et al., 2023) and rGO (Promphet et al., 2015).

3.6.2 Colorimetric Analysis Based on Nanoparticles-Based Sensors

Colorimetric techniques are popular because they are simple and cost-effective. There is typically no need for complex or expensive extra instruments because detection using these sensors can usually be accomplished simply by seeing the color changes with the naked eye (Alberti et al., 2021). Additionally, they can be utilized for point-of-care testing and field analysis (Wang et al., 2019). Colorimetric tools that are quick and affordable have proven successful in biosensing, environmental monitoring, and medical diagnosis. The rapid growth of these sensors' development in recent years demonstrates the scientific community's intense interest in this area. Several substrates with chromogenic properties and color tags have been investigated, especially those with excellent sensitivity, selectivity, cost-effectiveness, and suitability for real-world applications. Among these, sensors based on nanoparticles and dyes have received much attention (Alberti et al., 2020; Zhang et al., 2018). Due to the relatively low extinction coefficients associated with dyes and the limited precision in detecting minute color alterations with the naked eye, the dye-based approach is considered not excellently sensitive. In addition, it is frequently challenging to identify specific components in a mixture without using chemometric techniques (Kangas et al., 2017).

Nanoparticle-based sensors have recently received much interest due to nanomaterials' outstanding attributes, such as conductivity, biocompatibility, and catalytic activity (Alberti et al., 2021; Pham et al., 2018). The drawbacks of conventional organic dye-based colorimetric detectors can be overcome using nanomaterials. Several studies have been conducted recently on the applications of nanomaterial-based colorimetric sensors that may easily be used by the naked eye to detect trace concentrations of specific analytes (Lim & Gao, 2016; Paterson & De La Rica, 2015). The nanoparticles of noble metals (such as gold and silver) exhibit more significant extinction coefficients than dyes. They are endowed with different physicochemical and optical attributes, making it possible to sense target analytes by changes in color that can be seen with the naked eye (Alberti et al., 2021; Ma et al., 2019).

Srisukjaroen and coworkers (2022) developed a rapid and highly selective colorimetric chemosensor functionalized with gold nanoparticles (AuNPs) for the detection of Pb(II) ions in meat samples.

The Pb(II) detection process was completed when a color change (colorless to pink) was observed in the detection zone within 60 seconds. Pb(II) concentrations increased linearly with the chemosensor strip's color intensity. With the developed novel sensor, the detection limit was low (0.12 mg L^{-1}). The AuNPs-modified paper test strip achieved a nine-fold greater color response than the unmodified paper test strip. Additionally, the Pb(II) detection in meat samples using the paper test-based strip was consistent with the ICP-MS technique. In another study, Pang (2020) developed a colorimetric technique for the detection of Ca(II) ions based on the Ca(II) ion-induced aggregation of AuNPs and the corresponding evidential change in color from red to blue-purple. The detection limit for Ca(II) was 0.6 μmol L^{-1} and could be detected at the concentration range of 2.5–37.5 μmol L^{-1}. A colorimetric detection method based on silver nanoparticles (AgNPs) wrapped with chalcone carboxylic acid was developed and tested for Cd(II) detection (Dong et al., 2017). Particle aggregation brought about by introducing Cd(II) ions resulted in a visible yellow-to-orange color change. The developed technique makes it possible to detect Cd(II) ions selectively, with other metal ions showing no significant interference. The instrumental and visual detection limits were 0.13 and 0.23 μmol L^{-1}, respectively. The results and those from the standard ICP-AES were reported to be in good agreement.

Fish is a commonly consumed food globally. However, various contaminants, including heavy metals, can contaminate fish. Vonnie et al. (2022) developed a selective and sensitive colorimetric metal detection technique using tapioca starch and AuNP-based film (TS-AuNP) for the detection of trace concentration of Cd in fish. The detection mechanism was based on the formation of aggregates between the Cd and AuNP, with a corresponding color change from red to purple-grey. The quantification limit, as recorded, was 40.04 mmol L^{-1}. In a different study, N,N'-bis(2-hydroxyethyl) dithiooxamide modified AuNP was used in the development of a paper-based microfluidic analytical technique for the detection of Hg(II) (Shariati & Khayatian, 2020). The color of the adapted AuNPs changed in the test zone immediately after Hg(II) was added, and this change could be seen with the naked eye or with a digital camera. The linear range of Hg(II) at optimum experimental conditions was 0.020–8.0 μmol L^{-1}, with a detection limit of 12.5 nmol L^{-1} in aqueous media. In real-life samples like salmon fish, the novel technique gave 0.025–2.0 μmol L^{-1} and 15 nmol L^{-1} as linear ranges of Hg(II) and detection limit, respectively. The results obtained using this method were reported to agree with those of the established ASTM standard procedure. A similar microfluidic colorimetric sensing device based on $Fe_3O_4^-$ urease nanoprobe demonstrated its effectiveness in analyzing the presence of Pb(II), Cd(II), and Hg(II) in muscle tissue and fish gill (Swain et al., 2020). The quantification limit was 0.1, 0.1, and 0.5 ng L^{-1} for Pb(II), Cd(II), and Hg(II), respectively, while the limit of detection was recorded as 0.1 ng L^{-1}. With the microfluidic system and transparent microwell plates, visible results could be seen with the naked eye.

3.6.3 Fluorescent Analysis Based on Fluorescent Nanomaterials

Fluorescence-based analytical techniques have become popular in sensing technologies due to the ease and simplicity of optical signal transduction, the regular availability of fluorescent tags featuring a variety of spectrum characteristics, and swift advances in optical spectroscopy (Sharma et al., 2021). The reporter molecules in fluorescence-based analysis that are most frequently utilized are organic dyes. However, the application of organic dyes in biosensors is constrained by their reduced photostability, restricted quantum yield, and poor dye-to-dye tagging ratio (Resch-Genger et al., 2008).

The rise of nanotechnology in material science brought about a disruptive change, driving a transition from traditional fluorescent tags to nanoparticles and nano quenchers-based fluorescence (Charbgoo et al., 2016). Fluorescent nanomaterials-based detectors for analyzing heavy metal ions have been constructed with outstanding sensitivity. Nanomaterials exhibit exceptional electrical, optical, and catalytic capabilities with superior surface reactivity and significant adsorption capacity (Lian et al., 2020). Fluorescent nanomaterials can perform either or both of the two distinct roles in the fluorescent nanosensor, comprising a fluorescent component that produces observable signals and a receptor component for identifying specific ions. Much research has gone into discovering and manufacturing several kinds of nanomaterials for application in fluorescence-based biosensors in the future (Sharma et al., 2021). Nanomaterials such as quantum dots (QDs), metal nanomaterials, carbon-based nanomaterials, and metal oxide nanomaterials are commonly used as a substitute to or performance enhancer for fluorescent molecules-based fluorophores (Lian et al., 2020).

Since heavy metals represent one of the significant threats to food safety, their accumulation in food products and agricultural materials has caused great concern for people around the world and must be promptly resolved. Analysis based on Fluorescence resonance energy transfer (FRET) has recently become one of the most promising analytical tools for detecting and quantifying the contamination of heavy metals in food products and agriculture materials. These methods demonstrate a straightforward, quick, economical, stable, accurate, real-time, selective, and sensitive method (Shen et al., 2022).

Carbon dots (CDs) were found to be effective as selective sensors for the analysis of Hg(II), Cr(III), Pb(II), Cr(IV), Cu(II), and Ni(II) ions (Chaudhary et al., 2020; Chini et al., 2019; Khan et al., 2019; Liu et al., 2019; Singh et al., 2020). Most of the publications focused on detecting heavy metal ions using CDs that only have one fluorescence band exclusive to one metal cation. A novel fluorimetric method was developed based on the quenching impact of metal ions on the fluorescent emission of o-phenanthroline and 8-hydroxyquinoline for the detection of trace concentration of Pb(II) in Argentine honey (Talio et al., 2019). The calibration graph was linear in the 0.105–51.8 μg L^{-1} range, with R^2 higher than 0.998. The limits of quantification and detection at optimum conditions were 0.105 μg L^{-1} and 0.035 μg L^{-1}, respectively. The reliability of the technique was evaluated using the ICP-MS analysis of comparable samples. The proposed approach produced acceptable results when used to determine the presence of trace levels of lead in honey. It also demonstrated great sensitivity, sufficient selectivity, and adequate tolerance to external ions. Talio et al. (2017) studied the quantification of Cd(II) and Ni(II) ions in different tobacco samples by molecular fluorescence. Fluorophore eosin and carbon nanotubes were used via chemofiltration on the nylon membrane. Ni(II) ion was preferentially held on the solid matrix, while Cd(II) was retained in the filtrate solution. A spectrofluorimetric detection of both metal ions was performed on the solid matrix and the filtered solution with 0.019 and 0.041 μg L^{-1} limits of detection, respectively (Table 3.3).

3.7 HEALTH IMPLICATIONS OF HEAVY METAL CONTENT IN THE FOOD WEB

Most heavy metals in soil can build up in crops, which can then spread to other media via the food chain. In the crop-soil interface, notably in crucial global staple crops like wheat and maize, numerous heavy metals' bioconcentration factor (BCF) has been established (Rai et al., 2019; Wang et al., 2017a, b). According to numerous studies, the buildup of heavy metals in the human body is the cause of a wide range of severe health problems, including developmental abnormalities, kidney diseases, neuromuscular defects, cancer, and mental illness (Ahmad et al., 2018; Rai et al., 2019;

TABLE 3.3

Representative Studies of Nanotechnology-Based Techniques and Nanosensors for the Analysis of Metals in Food

Analytical Technique	Food Sample	Nanomaterial	Analyte	Linear Range	Limit of Detection	References
Electrochemical	Meat	UiO-66-NH$_2$ MOF and MWCNTs	Cd(II)	0.5–170 µg L^{-1}	0.2 µg L^{-1}	Wang et al. (2021)
Electrochemical	Spinach	Graphene aerogel-MOF	Cd(II)	0.01–1.5 µmol L^{-1}	9 nmol L^{-1}	Gao et al. (2019)
			Pb(II)	0.001–2 µmol L^{-1}	1 nmol L^{-1}	
			Cu(II)	0.01–1.6 µmol L^{-1}	8 nmol L^{-1}	
			Hg(II)	0.001–2.2 µmol L^{-1}	0.9 nmol L^{-1}	
Electrochemical	Spiked foods	MWCNTs and (p-g-C$_3$N$_4$-NSs)	Cd(II)		8–60 ng L^{-1}	Ramalingam et al. (2019)
			Hg(II)			
			Pb(II)			
			Zn(II)			
Electrochemical	Vegetable (spinach)	3D graphene aerogel and UiO-66-NH MOF	Cd(II)		20 nmol L^{-1}	Lu et al. (2018)
			Pb(II)		1.5 nmol L^{-1}	
			Cu(II)		7 nmol L^{-1}	
			Hg(II)		2 nmol L^{-1}	
Electrochemical	Organic vegetables	MWCNTs/AgNPs/Nafion-GCE	Pb(II)	0.493–157.2 µg L^{-1}	0.216 µg L^{-1}	Palisoc et al. (2018)
			Cd(II)	1.864–155.1 µg L^{-1}	0.481 µg L^{-1}	
Electrochemical	Rice	3D graphene	Cd(II)			Mitra et al. (2019)
Electrochemical	Fruits and vegetables	Aptasensor (Double-stranded DNA)	Cd(II)	0.1–1,000 nmol L^{-1}	89.31 pmol L^{-1}	Yuan et al. (2022)
			Pb(II)		16.44 pmol L^{-1}	
Colourimetric	Meat	AuNPs-modified chemosensor	Pb(II)		0.12 mg L^{-1}	Srisukjaroen et al. (2022)
Colourimetric	Salmon fish	Modified AuNP	Hg(II)	0.025–2.0 µM	15 nmol L^{-1}	Shariati & Khayatian (2020)
Colourimetric	Muscle tissue and fish gill	Fe$_3$O$_4$–urease nanoprobe	Pb(II)		0.1 ng L^{-1}	Swain et al. (2020)
			Cd(II)			
			Hg(II)			

Talbi et al., 2018). The buildup of heavy metals in the human body has toxic, carcinogenic effects, and neurotoxic (Ashot et al., 2020; Pipoyan et al., 2018; Pipoyan et al., 2023b).

Heavy metals can build up in human bones or fatty tissues through nutritional consumption, depleting vital nutrients and weakening immune systems. In addition, it has been suggested that several heavy metals, such as Al, Cd, Mn, and Pb, may contribute to intrauterine growth retardation (El-Kady & Abdel-Wahhab, 2018; Rai et al., 2019). Humans require a small amount of the mineral Cr, which we must get through food (Genchi et al., 2021). Supplemental Cr is beneficial for those with diabetes or low glucose tolerance. Cr increases the body's ability to break down fat, which helps people weigh less. Cr helps maintain healthy blood sugar levels, enhances mental skills, and breaks down carbohydrates and lipids (Chen et al., 2022). Gastrointestinal and hypoglycemia distress (hypoglycemia) may come from consuming too much Cr in meals. High Cr supplementation can harm the kidneys, liver, and nerves, which can cause irregular heartbeats (Broadhurst & Domenico, 2006).

Also, the increasing levels of Pb toxicity have been proven by numerous research (Ashraf et al., 2017; Iyer et al., 2015). In addition to seriously contaminating the environment, excessive Pb exposure can substantially negatively affect human health. A toxicological review of Pb has revealed that blood Pb levels as low as 5 g L^{-1} can cause developmental issues like impaired cognitive function, stunted growth, behavioral disorder, and diminished hearing. D-aminolaevulinic dehydratase, one of the key enzymes responsible for the biosynthesis of heme, is also inhibited by Pb (Ihedioha et al., 2017; Nkwunonwo et al., 2020). Pb concentrates primarily in bones in humans and animals, hindering the expected maturity of erythroid elements in the bone marrow. Pb also directly interferes with calcium metabolism and tampers with calciferol metabolism (Al-Saleh et al., 2017; Plant et al., 2000; Zhou et al., 2016). Children have experienced these effects at blood Pb concentrations ranging from 12 to 120 g L^{-1}. Pb causes sub-encephalopathic neurological and behavioral consequences because it harms the peripheral and central nervous systems. A guideline value obtained on this basis would also be protective against carcinogenic effects, according to human studies that show deleterious neurotoxic consequences other than cancer may occur at very low Pb concentrations (Nkwunonwo et al., 2020).

Ni is an essential element in animals. However, numerous secondary consequences, including cardiovascular, allergies, renal conditions, lung and nasal cancer, and pulmonary fibrosis, can be brought on by Ni exposure. Oxidative stress and mitochondrial dysfunction are necessary for Ni-induced toxicity, even though the exact molecular mechanisms behind this toxicity are unknown (Genchi et al., 2020; Ugulu, 2015). Additionally, exposure to an environment heavily polluted with Ni can have several degenerative repercussions (Khan et al., 2023; Zambelli & Ciurli, 2013). Humans exposed to Ni over an extended period may develop pulmonary cancer, lung fibrosis, and renal and cardiovascular disease (Begum et al., 2022; Pipoyan et al., 2023a). However, animal research has shown that several Ni compounds, including nickel oxide, nickel subsulfide, nickel sulfate, and nickel chloride, can potentially cause cancer (Oller et al., 2023). Additionally, it was noted that nickel nitrite exposure has significant adverse effects on the blood and several organs, including the liver, heart, and kidney, indicating that nickel nitrate is a dangerous contaminant (Al-Rikaby, 2021).

In addition to causing vomiting, hemodynamic instability, and nervous system injury, exposure to soluble inorganic arsenic can result in rapid death. A significant dose of arsenic can cause this fatal outcome (Mahlknecht et al., 2023). The lowest amounts of arsenic exposure cause vomiting, long-term neurological damage, and malignancies in organs other than the kidneys, skin, lungs, and bladder (Bhowmick et al., 2018; Palma-Lara et al., 2020; Shakoor et al., 2017). Diabetes and fertility are only marginally affected by nickel exposure, but high blood pressure, heart attacks, and other circulatory illnesses are significantly affected. Vegetables, fruits, shellfish, seafood, and other food and water sources can all be contaminated with arsenic, which can result in poisoning. Arsenic exposure over an extended period can lead to anomalies in the sensory and motor nerves and skin changes, including redness and edema. Additionally, liver and kidney function may be impacted (Zaw & Emett, 2002).

The accumulation of Cd in the human body, particularly in the kidneys, can harm the kidneys. According to studies by Luo and coworkers (2017) and Shahriar and coworkers (2020), chronic exposure to Cd metal can cause kidney issues, anemia, anosmia (loss of sense of smell), emphysema, renal problems, cardiovascular diseases, and hypertension. Itai-itai illness, which is extremely painful and results in bone loss and embrittlement (Nkwunonwo et al., 2020), seems Cd-related. High-density lipoprotein concentrations and the immune system can be hampered by excessive Zn levels in the body (Zhou et al., 2016). Like animals, excessive Cu consumption in humans can result in liver damage and other issues with the stomach (Rahman et al., 2014; Zhou et al., 2016). According to Liu and coworkers (2013), heavy metals in soil, such as Cr, Cu, and Zn, can result in non-carcinogenic risks to human health, such as liver disease, headaches, and neurologic issues.

3.8 STRATEGIES FOR THE MITIGATION OF TOXIC METALS CONTAMINATION

Heavy metal contamination is a global concern, with standard treatment procedures failing to provide a cost-effective and environmentally benign solution (Rai et al., 2019). For example, Cd is generally recognized as a hazardous environmental contaminant that endangers the well-being of humans (Nair et al., 2013). According to Schaefer and coworkers (2020), a critical step in decreasing heavy metals like Cd in the diet is to limit or avoid early uptake by plants used as food or feed crops. The researchers further recommended stopping Cd glazes and coatings on pots and pans (Cd-plated utensils) and galvanized items in food processing and minimizing the use of Cd-containing stabilizers in plastics as strategies for mitigating excess metals like Cd in food.

Reduced heavy metal concentrations in food items can be reduced by lowering their soil concentrations, particularly in soil solutions (Grabowska, 2011). Gaur and coworkers (2021) reported that the redox reaction in microbial and phytoremediation enables the valence of certain metals to be shifted, making them less hazardous, and that enzymes released by bacteria and the entire microbial cell are regarded as an eco-efficient biocatalyst for heavy metal removal from contaminated locations. Furthermore, microorganisms secrete urease enzyme, which breaks down urea into ammonium ions and carbonates, such that during the biocarbonation process, carbonates build an insoluble combination with heavy metals which has proven to be effective in lowering the content of heavy metals from polluted soil (Abdel-Gawwad et al., 2020). Singh and coworkers (2019) noted that exopolysaccharides (EPS) are negatively charged and attract positively charged heavy metal ions, forming a complex. Hence, EPS has evolved as an outstanding heavy metal scavenger with various lucrative advantages such as low cost, sustainability, and eco-friendliness. The efficacy of incorporating straw into polluted soil has also been examined by Tlustos and coworkers (1998), and reductions in Cd and Zn concentrations in spinach of 40% and 25%, respectively, were reported.

Monitoring and control are critical components of heavy metal mitigation efforts. Heavy metals in pesticides and fertilizers should be monitored to ensure they are not contained in hazardous concentrations. More care should be taken in food processing, storage, and marketing to mitigate against practices that expose food materials to heavy metals. The risk of contamination resulting from processing equipment can be reduced if the processing equipment and materials are carefully selected. To ensure food safety, metal-free water should be used in food processing plants and food preparation in the home. The use of metal-containing additives in food processing should be minimized. The packaging materials should not be made of metals. Home contamination can also be minimized by ensuring that kitchen counters, utensils, and pots, among other possible contamination surfaces, are always kept clean and sanitized. Storage materials that can expose food to metals, such as keeping acidic foods in metal containers, should be discouraged. To mitigate against excessive buildup in the food chain, it is crucial to regularly assess the presence of heavy metals in foods consumed by the population of different localities.

3.9 CONCLUSION

Food is one of the significant sources of human exposure to metals. Metals can bioaccumulate in foods of both plant and animal origins. While essential metals are utilized by living organisms for different physiological and biochemical activities, non-essential metals have no usefulness and are regarded as contaminants. Even vital metals can contaminate if contained beyond the body's required concentration. The contamination of food with metals can have disastrous effects on human health, including the causation of developmental abnormalities, kidney diseases, neuromuscular defects, cancer, mental illness, etc. When heavy metals infiltrate our environment, plants may bioaccumulate them. Most heavy metals in soil can build up in crops, which can then spread to other media via the food chain. The use of phosphate-based fertilizers and metal-containing pesticides in agricultural land, water runoff from industrial facilities and roads, atmospheric deposition, use of untreated sewage sludge on lands, and so on are some pathways through which metals penetrate the soil.

Similarly, both terrestrial and aquatic species accumulate heavy metals through the food they consume and their abiotic environment. Humans consume contaminated plant or animal food products and are exposed to heavy metal contamination. Hence, various national and international regulatory bodies have set standards for maximum permissible limits of the content of toxic metals in different food products. Close monitoring of foods consumed by humans is paramount to mitigating heavy metal contamination of foods.

BIBLIOGRAPHY

Abd-Elghany, S. M., Mohammed, M. A., Abdelkhalek, A., Saad, F. S. S., & Sallam, K. I. (2020). Health risk assessment of exposure to heavy metals from sheep meat and offal in Kuwait. *Journal of Food Protection*, *83*(3), 503–510.

Abdel-Gawwad, H. A., Hussein, H. S., & Mohammed, M. S. (2020). Bio-removal of Pb, Cu, and Ni from solutions as nano-carbonates using a plant-derived urease enzyme-urea mixture. *Environmental Science and Pollution Research*, *27*, 30741–30754.

Abdelrahman, M. M., Alhidary, I. A., Matar, A. M., Alobre, M. M., Ayadi, M., & Aljumaah, R. S. (2023). Heavy metals levels in soil, water and feed and relation to slaughtered camels' tissues (Camelus dromedarius) from five districts in Saudi Arabia during spring. *Life*, *13*(3), 732.

Abuzed Sadee, B., & Jameel Ali, R. (2023). Determination of heavy metals in edible vegetables and a human health risk assessment. *Environmental Nanotechnology, Monitoring & Management*, *19*, 100761. https://doi.org/10.1016/j.enmm.2022.100761.

Afrin, R., Mia, M. Y., Ahsan, M. A., & Akbor, A. (2015). Concentration of heavy metals in available fish species (Bain, Mastacembelus armatus; Taki, Channa punctatus and Bele, Glossogobius giuris) in the Turag river, Bangladesh: Heavy metal concentration in Bangladesh fish. *Biological Sciences-PJSIR*, *58*(2), 104–110.

Ahmad, T., Park, J., Keel, S., Yun, J., Lee, U., Kim, Y., & Lee, S.-S. (2018). Behavior of heavy metals in air pollution control devices of 2,400 kg/h municipal solid waste incinerator. *Korean Journal of Chemical Engineering*, *35*, 1823–1828.

AL-Huqail, A. A., Kumar, P., Eid, E. M., Adelodun, B., Abou Fayssal, S., Singh, J., Arya, A. K., Goala, M., Kumar, V., & Širić, I. (2022). Risk assessment of heavy metals contamination in soil and two rice (Oryza sativa L.) varieties irrigated with paper mill effluent. *Agriculture*, *12*(11), 1864.

Al-Rikaby, A. A. (2021). Hematologic evaluation and histopathological alteration of nickel nitrate exposure in male rabbits. *Annals of the Romanian Society for Cell Biology*, *25*(4), 1307–1319.

Al-Saleh, I., Al-Rouqi, R., Elkhatib, R., Abduljabbar, M., & Al-Rajudi, T. (2017). Risk assessment of environmental exposure to heavy metals in mothers and their respective infants. *International Journal of Hygiene and Environmental Health*, *220*(8), 1252–1278.

Alberti, G., Zanoni, C., Magnaghi, L. R., & Biesuz, R. (2020). Disposable and low-cost colorimetric sensors for environmental analysis. *International Journal of Environmental Research and Public Health*, *17*(22), 8331.

Alberti, G., Zanoni, C., Magnaghi, L. R., & Biesuz, R. (2021). Gold and silver nanoparticle-based colorimetric sensors: New trends and applications. *Chemosensors*, *9*(11), 305.

Ali, H., & Khan, E. (2018). What are heavy metals? Long-standing controversy over the scientific use of the term 'heavy metals'–Proposal of a comprehensive definition. *Toxicological & Environmental Chemistry*, *100*(1), 6–19.

Ali, H., & Khan, E. (2019). Trophic transfer, bioaccumulation, and biomagnification of non-essential hazardous heavy metals and metalloids in food chains/webs: Concepts and implications for wildlife and human health. *Human and Ecological Risk Assessment: An International Journal*, *25*(6), 1353–1376.

Ali, H. S., Almashhadany, D. A., & Khalid, H. S. (2020). Determination of heavy metals and selenium content in chicken liver at Erbil city, Iraq. *Italian Journal of Food Safety*, *9*(3), 8659. doi: 10.4081/ijfs.2020.8659.

Aniagor, C., Abdel-Halim, E., & Hashem, A. (2021a). Evaluation of the aqueous Fe(II) ion sorption capacity of functionalized microcrystalline cellulose. *Journal of Environmental Chemical Engineering*, *9*(4), 105703.

Aniagor, C. O., Afifi, M., & Hashem, A. (2021b). Heavy metal adsorptive application of hydrolyzed corn starch. *Journal of Polymer Research*, *28*(11), 1–10.

Arunakumara, K., Walpola, B. C., & Yoon, M.-H. (2013). Current status of heavy metal contamination in Asia's rice lands. *Reviews in Environmental Science and Bio/technology*, *12*, 355–377.

Ashot, D. P., Sergey, A. H., Radik, M. B., Arthur, S. S., & Mantovani, A. (2020). Risk assessment of dietary exposure to potentially toxic trace elements in emerging countries: A pilot study on intake via flour-based products in Yerevan, Armenia. *Food and Chemical Toxicology*, *146*, 111768.

Ashraf, U., Hussain, S., Anjum, S. A., Abbas, F., Tanveer, M., Noor, M. A., & Tang, X. (2017). Alterations in growth, oxidative damage, and metal uptake of five aromatic rice cultivars under lead toxicity. *Plant Physiology and Biochemistry*, *115*, 461–471.

Atafar, Z., Mesdaghinia, A., Nouri, J., Homaee, M., Yunesian, M., Ahmadimoghaddam, M., & Mahvi, A. H. (2010). Effect of fertilizer application on soil heavy metal concentration. *Environmental Monitoring and Assessment, 160*, 83–89.

Atta, M. I., Zehra, S. S., Dai, D.-Q., Ali, H., Naveed, K., Ali, I., Sarwar, M., Ali, B., Iqbal, R., & Bawazeer, S. (2023). Amassing of heavy metals in soils, vegetables and crop plants irrigated with wastewater: Health risk assessment of heavy metals in Dera Ghazi Khan, Punjab, Pakistan. *Frontiers in Plant Science, 13*, 1080635.

Aydoğan, Z. (2023). Determination of selected elements in two commercially available edible aquatic insects (Coleoptera) and their worldwide updated list. *Environmental Monitoring and Assessment, 195*(1), 249.

Baby, J., Raj, J. S., Biby, E. T., Sankarganesh, P., Jeevitha, M., Ajisha, S., & Rajan, S. S. (2010). Toxic effect of heavy metals on aquatic environment. *International Journal of Biological and Chemical Sciences, 4*(4), 939–952. doi: 10.4314/ijbcs.v4i4.62976

Badr, M. R., & Arafa, S. (2023). Assessment of street vended foods in the Gharbia governorate vs laboratory foods: Approximate chemical composition and heavy metals. *Journal of Sustainable Agricultural Sciences, 49*(1), 31–38.

Barone, G., Storelli, A., Quaglia, N. C., Garofalo, R., Meleleo, D., Busco, A., & Storelli, M. M. (2021). Trace metals in pork meat products marketed in Italy: Occurrence and health risk characterization. *Biological Trace Element Research, 199*, 2826–2836.

Barry, S., & O'Riordan, A. (2016). Electrochemical nanosensors: Advances and applications. *Reports in Electrochemistry, 6*, 1–14.https://doi.org/10.2147/RIE.S80550

Bati, K., Mogobe, O., & Masamba, W. R. (2017). Concentrations of some trace elements in vegetables sold at Maun market, Botswana. *Journal of Food Research, 6*(1), 1927–0887.

Bayissa, L. D., & Gebeyehu, H. R. (2021). Vegetables contamination by heavy metals and associated health risk to the population in Koka area of Central Ethiopia. *PLoS One, 16*(7), e0254236.

Bazié, B. S. R., Compaoré, M. K. A., Bandé, M., Kpoda, S. D., Méda, N.-S.-B. R., Kangambega, T. M. O., Ilboudo, I., Sandwidi, B. Y., Nikiema, F., & Yakoro, A. (2022). Evaluation of metallic trace elements contents in some major raw foodstuffs in Burkina Faso and health risk assessment. *Scientific Reports, 12*(1), 4460.

Begum, W., Rai, S., Banerjee, S., Bhattacharjee, S., Mondal, M. H., Bhattarai, A., & Saha, B. (2022). A comprehensive review on the sources, essentiality and toxicological profile of nickel. *RSC Advances, 12*(15), 9139–9153.

Bhowmick, S., Pramanik, S., Singh, P., Mondal, P., Chatterjee, D., & Nriagu, J. (2018). Arsenic in groundwater of West Bengal, India: A review of human health risks and assessment of possible intervention options. *Science of the Total Environment, 612*, 148–169.

Bielecka, J., Markiewicz-Żukowska, R., Puścion-Jakubik, A., Grabia, M., Nowakowski, P., Soroczyńska, J., & Socha, K. (2022). Gluten-free cereals and pseudocereals as a potential source of exposure to toxic elements among polish residents. *Nutrients, 14*(11), 2342.

Boro, B., Basumatary, S., & Das, B. (2023). Metal contents of some selected vegetables grown in Bodoland territorial region of Assam, India. *Current Chemistry Letters, 12*(3), 471–476.

Broadhurst, C. L., & Domenico, P. (2006). Clinical studies on chromium picolinate supplementation in diabetes mellitus–A review. *Diabetes Technology & Therapeutics, 8*(6), 677–687.

Charbgoo, F., Soltani, F., Taghdisi, S. M., Abnous, K., & Ramezani, M. (2016). Nanoparticles application in high sensitive aptasensor design. *TrAC Trends in Analytical Chemistry, 85*, 85–97.

Chaudhary, N., Gupta, P. K., Eremin, S., & Solanki, P. R. (2020). One-step green approach to synthesize highly fluorescent carbon quantum dots from banana juice for selective detection of copper ions. *Journal of Environmental Chemical Engineering, 8*(3), 103720.

Chaves, L. H. G., Estrela, M. A., & de Souza, R. S. (2011). Effect on plant growth and heavy metal accumulation by sunflower. *Journal of Phytology, 3*(12), 04–09.

Chen, F., Ma, J., Akhtar, S., Khan, Z. I., Ahmad, K., Ashfaq, A., Nawaz, H., & Nadeem, M. (2022). Assessment of chromium toxicity and potential health implications of agriculturally diversely irrigated food crops in the semi-arid regions of South Asia. *Agricultural Water Management, 272*, 107833.

Cheng, Z., Li, G., & Liu, M. (2015). Metal-enhanced fluorescence effect of Ag and Au nanoparticles modified with rhodamine derivative in detecting Hg^{2+}. *Sensors and Actuators B: Chemical, 212*, 495–504.

Chini, M. K., Kumar, V., Javed, A., & Satapathi, S. (2019). Graphene quantum dots and carbon nano dots for the FRET based detection of heavy metal ions. *Nano-Structures & Nano-Objects, 19*, 100347.

Cui, Y.-J., Zhu, Y.-G., Zhai, R.-H., Chen, D.-Y., Huang, Y.-Z., Qiu, Y., & Liang, J.-Z. (2004). Transfer of metals from soil to vegetables in an area near a smelter in Nanning, China. *Environment International, 30*(6), 785–791.

Dai, H., Wang, N., Wang, D., Ma, H., & Lin, M. (2016). An electrochemical sensor based on phytic acid functionalized polypyrrole/graphene oxide nanocomposites for simultaneous determination of Cd (II) and Pb (II). *Chemical Engineering Journal, 299*, 150–155.

Dinu, C., Gheorghe, S., Tenea, A. G., Stoica, C., Vasile, G.-G., Popescu, R. L., Serban, E. A., & Pascu, L. F. (2021). Toxic metals (As, Cd, Ni, Pb) impact in the most common medicinal plant (Mentha piperita). *International Journal of Environmental Research and Public Health, 18*(8), 3904.

Dong, Y., Ding, L., Jin, X., & Zhu, N. (2017). Silver nanoparticles capped with chalcon carboxylic acid as a probe for colorimetric determination of cadmium (II). *Microchimica Acta, 184*, 3357–3362.

Dubey, V., Agnihotri, M., & Shukla, A. (2016). Determination of heavy metals in selected meat and meat products from meat market of Singrauli. *IOSR-JAC, 9*, 52–54.

Ekhator, O., Udowelle, N., Igbiri, S., Asomugha, R., Igweze, Z., & Orisakwe, O. (2017). Safety evaluation of potential toxic metals exposure from street foods consumed in mid-west Nigeria. *Journal of Environmental and Public Health, 2017*, 8458057.

El-Kady, A. A., & Abdel-Wahhab, M. A. (2018). Occurrence of trace metals in foodstuffs and their health impact. *Trends in Food Science & Technology, 75*, 36–45.

Elmorsi, R. R., Abou-El-Sherbini, K. S., Mostafa, G. A.-H., & Hamed, M. A. (2019). Distribution of essential heavy metals in the aquatic ecosystem of Lake Manzala, Egypt. *Heliyon, 5*(8), e02276.https://doi.org/10.1016/j.heliyon.2019.e02276

Foster, L. E. (2005). *Nanotechnology: Science, Innovation, and Opportunity*. Prentice Hall PTR.

Franco-Fuentes, E., Moity, N., Ramírez-González, J., Andrade-Vera, S., Hardisson, A., González-Weller, D., Paz, S., Rubio, C., & Gutiérrez, Á. J. (2021). Metals in commercial fish in the Galapagos Marine Reserve: Contribution to food security and toxic risk assessment. *Journal of Environmental Management, 286*, 112188. https://doi.org/10.1016/j.jenvman.2021.112188.

Gall, J. E., Boyd, R. S., & Rajakaruna, N. (2015). Transfer of heavy metals through terrestrial food webs: A review. *Environmental Monitoring and Assessment, 187*, 1–21.

Gao, N., He, C., Ma, M., Cai, Z., Zhou, Y., Chang, G., Wang, X., & He, Y. (2019). Electrochemical co-deposition synthesis of Au-ZrO$_2$-graphene nanocomposite for a nonenzymatic methyl parathion sensor. *Analytica Chimica Acta, 1072*, 25–34.

Gaur, V. K., Sharma, P., Gaur, P., Varjani, S., Ngo, H. H., Guo, W., Chaturvedi, P., & Singhania, R. R. (2021). Sustainable mitigation of heavy metals from effluents: Toxicity and fate with recent technological advancements. *Bioengineered, 12*(1), 7297–7313.

Gebeyehu, H. R., & Bayissa, L. D. (2020). Levels of heavy metals in soil and vegetables and associated health risks in Mojo area, Ethiopia. *PLoS One, 15*(1), e0227883.

Genchi, G., Carocci, A., Lauria, G., Sinicropi, M. S., & Catalano, A. (2020). Nickel: Human health and environmental toxicology. *International Journal of Environmental Research and Public Health, 17*(3), 679.

Genchi, G., Lauria, G., Catalano, A., Carocci, A., & Sinicropi, M. S. (2021). The double face of metals: The intriguing case of chromium. *Applied Sciences, 11*(2), 638.

Ghanati, K., Zayeri, F., & Hosseini, H. (2019). Potential health risk assessment of different heavy metals in wheat products. *Iranian Journal of Pharmaceutical Research: IJPR, 18*(4), 2093.

Gong, Z., Chan, H. T., Chen, Q., & Chen, H. (2021). Application of nanotechnology in analysis and removal of heavy metals in food and water resources. *Nanomaterials, 11*(7), 1792.

Grabowska, I. (2011). Reduction of heavy metals transfer into food. *Polish Journal of Environmental Studies, 20*(3), 635–642.

Granados-Chinchilla, F., Mena, S. P., & Arias, L. M. (2015). Inorganic contaminants and composition analysis of commercial feed grade mineral compounds available in Costa Rica. *International Journal of Food Contamination, 2*(1), 1–14.

Green, I., Diaz, A., & Tibbett, M. (2010). Factors affecting the concentration in seven-spotted ladybirds (Coccinella septempunctata L.) of Cd and Zn transferred through the food chain. *Environmental Pollution, 158*(1), 135–141.

Gumpu, M. B., Veerapandian, M., Krishnan, U. M., & Rayappan, J. B. B. (2017). Simultaneous electrochemical detection of Cd (II), Pb (II), As (III) and Hg (II) ions using ruthenium (II)-textured graphene oxide nanocomposite. *Talanta, 162*, 574–582.

Guo, Z., Dai, H., Wang, M., & Pan, S. (2023). Health risk assessment of heavy metal exposure through vegetable consumption around a phosphorus chemical plant in the Kaiyang karst area, southwestern China. *Environmental Science and Pollution Research, 30*(13), 35617–35634.

Hambidge, M. (2003). Biomarkers of trace mineral intake and status. *The Journal of Nutrition, 133*(3), 948S–955S.

Han, J. L., Pan, X. D., & Chen, Q. (2022). Distribution and safety assessment of heavy metals in fresh meat from Zhejiang, China. *Scientific Reports, 12*(1), 3241.

Haque, M. M., Hossain, N., Jolly, Y. N., & Tareq, S. M. (2021). Probabilistic health risk assessment of toxic metals in chickens from the largest production areas of Dhaka, Bangladesh. *Environmental Science and Pollution Research, 28*(37), 51329–51341.

Hashem, A., Aniagor, C., Taha, G. M., & Fikry, M. (2021). Utilization of low-cost sugarcane waste for the adsorption of aqueous Pb (II): Kinetics and isotherm studies. *Current Research in Green and Sustainable Chemistry, 4*, 100056.

Hejna, M., Gottardo, D., Baldi, A., Dell'Orto, V., Cheli, F., Zaninelli, M., & Rossi, L. (2018). Nutritional ecology of heavy metals. *Animal, 12*(10), 2156–2170.

Hossain, M. M., Hannan, A. S. M. A., Kamal, M. M., Hossain, M. A., & Quraishi, S. B. (2022). Appraisal and validation of a method used for detecting heavy metals in poultry feed in Bangladesh. *Veterinary World, 15*(9), 2217.

Huang, M.-R., Guo-Li, G., Feng-Ying, S., & Xin-Gui, L. (2012). Development of potentiometric lead ion sensors based on ionophores bearing oxygen/sulfur-containing functional groups. *Chinese Journal of Analytical Chemistry, 40*(1), 50–58.

Ihedioha, J., Ukoha, P., & Ekere, N. (2017). Ecological and human health risk assessment of heavy metal contamination in soil of a municipal solid waste dump in Uyo, Nigeria. *Environmental Geochemistry and Health, 39*, 497–515.

Islam, M. R., Akash, S., Jony, M. H., Alam, M. N., Nowrin, F. T., Rahman, M. M., Rauf, A., & Thiruvengadam, M. (2023). Exploring the potential function of trace elements in human health: A therapeutic perspective. *Molecular and Cellular Biochemistry, 478*, 1–31.

Islam, M. S., Ahmed, M. K., Habibullah-Al-Mamun, M., & Raknuzzaman, M. (2015). The concentration, source and potential human health risk of heavy metals in the commonly consumed foods in Bangladesh. *Ecotoxicology and Environmental Safety, 122*, 462–469.

Iyer, S., Sengupta, C., & Velumani, A. (2015). Lead toxicity: An overview of prevalence in Indians. *Clinica Chimica Acta, 451*, 161–164.

Jan, F. A., Ishaq, M., Khan, S., Ihsanullah, I., Ahmad, I., & Shakirullah, M. (2010). A comparative study of human health risks via consumption of food crops grown on wastewater irrigated soil (Peshawar) and relatively clean water irrigated soil (lower Dir). *Journal of Hazardous Materials, 179*(1–3), 612–621.

Jeong, S., Yang, S., Lee, Y. J., & Lee, S. H. (2023). Laser-induced graphene incorporated with silver nanoparticles applied for heavy metal multi-detection. *Journal of Materials Chemistry A, 11*(25), 13409–13418.

Kangas, M. J., Burks, R. M., Atwater, J., Lukowicz, R. M., Williams, P., & Holmes, A. E. (2017). Colorimetric sensor arrays for the detection and identification of chemical weapons and explosives. *Critical Reviews in Analytical Chemistry, 47*(2), 138–153.

Khan, A., Khan, S., Khan, M. A., Qamar, Z., & Waqas, M. (2015). The uptake and bioaccumulation of heavy metals by food plants, their effects on plants nutrients, and associated health risk: A review. *Environmental Science and Pollution Research, 22*, 13772–13799.

Khan, Z. I., Ahmad, K., Ahmad, T., Zafar, A., Alrefaei, A. F., Ashfaq, A., Akhtar, S., Mahpara, S., Mehmood, N., & Ugulu, I. (2023). Evaluation of nickel toxicity and potential health implications of agriculturally diversely irrigated wheat crop varieties. *Arabian Journal of Chemistry, 16*(8), 104934.

Khan, Z. M., Rahman, R. S., Islam, S., & Zulfequar, M. (2019). Hydrothermal treatment of red lentils for the synthesis of fluorescent carbon quantum dots and its application for sensing Fe3+. *Optical Materials, 91*, 386–395.

Khanverdiluo, S., Talebi-Ghane, E., Ranjbar, A., & Mehri, F. (2023). Content of potentially toxic elements (PTEs) in various animal meats: A meta-analysis study, systematic review, and health risk assessment. *Environmental Science and Pollution Research, 30*(6), 14050–14061.

Kiliç, H. D., & Kizil, H. (2019). Simultaneous analysis of Pb^{2+} and Cd^{2+} at graphene/bismuth nanocomposite film-modified pencil graphite electrode using square wave anodic stripping voltammetry. *Analytical and Bioanalytical Chemistry, 411*, 8113–8121.

Kim, J., Song, H., Lee, J., Kim, Y. J., Chung, H. S., Yu, J. M., Jang, G., Park, R., Chung, W., & Oh, C.-M. (2023). Smoking and passive smoking increases mortality through mediation effect of cadmium exposure in the United States. *Scientific Reports, 13*(1), 3878.

Kormoker, T., Proshad, R., Islam, M. S., Shamsuzzoha, M., Akter, A., & Tusher, T. R. (2021). Concentrations, source apportionment and potential health risk of toxic metals in foodstuffs of Bangladesh. *Toxin Reviews, 40*(4), 1447–1460.

Kuswandi, B., Futra, D., & Heng, L. (2017). Nanosensors for the detection of food contaminants. In *Nanotechnology Applications in Food* (pp. 307–333). Elsevier.

Laboni, F. A., Ahmed, M. W., Kaium, A., Alam, M. K., Parven, A., Jubayer, M. F., Rahman, M. A., Meftaul, I. M., & Khan, M. S. I. (2023). Heavy metals in widely consumed vegetables grown in industrial areas of Bangladesh: A potential human health hazard. *Biological Trace Element Research, 201*(2), 995–1005.

Lange, K. W., & Nakamura, Y. (2021). Edible insects as future food: Chances and challenges. *Journal of Future Foods, 1*(1), 38–46.

Lehel, J., Laczay, P., Gyurcsó, A., Jánoska, F., Majoros, S., Lányi, K., & Marosán, M. (2016). Toxic heavy metals in the muscle of roe deer (Capreolus capreolus)-food toxicological significance. *Environmental Science and Pollution Research, 23*, 4465–4472.

Lénárt, Z., Bartha, A., Abonyi-Tóth, Z., & Lehel, J. (2023). Monitoring of metal content in the tissues of wild boar (Sus scrofa) and its food safety aspect. *Environmental Science and Pollution Research, 30*(6), 15899–15910.

Levent, B., Ayşah, Ö., Elif, A., & Fatih, Ş. (2020). Health risk assessment: Heavy metals in fish from the southern Black Sea. *Foods and Raw Materials, 8*(1), 115–124.

Lian, J., Xu, Q., Wang, Y., & Meng, F. (2020). Recent developments in fluorescent materials for heavy metal ions analysis from the perspective of forensic chemistry. *Frontiers in Chemistry, 8*, 593291.

Lim, W. Q., & Gao, Z. (2016). Plasmonic nanoparticles in biomedicine. *Nano Today, 11*(2), 168–188.

Liu, P., Wu, Q., Hu, W., Tian, K., Huang, B., & Zhao, Y. (2023). Effects of atmospheric deposition on heavy metals accumulation in agricultural soils: Evidence from field monitoring and Pb isotope analysis. *Environmental Pollution, 330*, 121740.

Liu, X., Song, Q., Tang, Y., Li, W., Xu, J., Wu, J., Wang, F., & Brookes, P. C. (2013). Human health risk assessment of heavy metals in soil-vegetable system: A multi-medium analysis. *Science of the Total Environment, 463*, 530–540.

Liu, Z., Jin, W., Wang, F., Li, T., Nie, J., Xiao, W., Zhang, Q., & Zhang, Y. (2019). Ratiometric fluorescent sensing of Pb^{2+} and Hg^{2+} with two types of carbon dot nanohybrids synthesized from the same biomass. *Sensors and Actuators B: Chemical, 296*, 126698.

Lizarazo, M. F., Herrera, C. D., Celis, C. A., Pombo, L. M., Teherán, A. A., Pineros, L. G., Forero, S. P., Velandia, J. R., Díaz, F. E., & Andrade, W. A. (2020). Contamination of staple crops by heavy metals in Sibaté, Colombia. *Heliyon, 6*(7), e04212. doi: 10.1016/j.heliyon.2020.e04212

Lu, M., Deng, Y., Luo, Y., Lv, J., Li, T., Xu, J., Chen, S.-W., & Wang, J. (2018). Graphene aerogel-metal-organic framework-based electrochemical method for simultaneous detection of multiple heavy-metal ions. *Analytical Chemistry, 91*(1), 888–895.

Luo, H.-F., Zhang, J.-Y., Jia, W.-J., Ji, F.-M., Yan, Q., Xu, Q., Ke, S., & Ke, J.-S. (2017). Analyzing the role of soil and rice cadmium pollution on human renal dysfunction by correlation and path analysis. *Environmental Science and Pollution Research, 24*, 2047–2054.

Luo, X., Wu, C., Lin, Y., Li, W., Deng, M., Tan, J., & Xue, S. (2023). Soil heavy metal pollution from Pb/Zn smelting regions in China and the remediation potential of biomineralization. *Journal of Environmental Sciences, 125*, 662–677.

Ma, X., He, S., Qiu, B., Luo, F., Guo, L., & Lin, Z. (2019). Noble metal nanoparticle-based multicolor immunoassays: An approach toward visual quantification of the analytes with the naked eye. *ACS Sensors, 4*(4), 782–791.

MacLachlan, D. J., Budd, K., Connolly, J., Derrick, J., Penrose, L., & Tobin, T. (2016). Arsenic, cadmium, cobalt, copper, lead, mercury, molybdenum, selenium and zinc concentrations in liver, kidney and muscle in Australian sheep. *Journal of Food Composition and Analysis, 50*, 97–107. https://doi.org/10.1016/j.jfca.2016.05.015.

Mahlknecht, J., Aguilar-Barajas, I., Farias, P., Knappett, P. S., Torres-Martínez, J. A., Hoogesteger, J., Lara, R. H., Ramírez-Mendoza, R. A., & Mora, A. (2023). Hydrochemical controls on arsenic contamination and its health risks in the Comarca Lagunera region (Mexico): Implications of the scientific evidence for public health policy. *Science of the Total Environment, 857*, 159347.

Malematja, E., Manyelo, T. G., Sebola, N. A., Kolobe, S. D., & Mabelebele, M. (2023). The accumulation of heavy metals in feeder insects and their impact on animal production. *Science of the Total Environment, 885*, 163716. https://doi.org/10.1016/j.scitotenv.2023.163716.

Manwani, S., Devi, P., Singh, T., Yadav, C. S., Awasthi, K. K., Bhoot, N., & Awasthi, G. (2023). Heavy metals in vegetables: A review of status, human health concerns, and management options. *Environmental Science and Pollution Research, 30*(28), 71940–71956.

Mao, C., Song, Y., Chen, L., Ji, J., Li, J., Yuan, X., Yang, Z., Ayoko, G. A., Frost, R. L., & Theiss, F. (2019). Human health risks of heavy metals in paddy rice based on transfer characteristics of heavy metals from soil to rice. *Catena, 175*, 339–348.

Mawari, G., Kumar, N., Sarkar, S., Daga, M. K., Singh, M. M., Joshi, T. K., & Khan, N. A. (2022). Heavy metal accumulation in fruits and vegetables and human health risk assessment: Findings from Maharashtra, India. *Environmental Health Insights*, *16*, 11786302221119151.

Mézes, M. (2018). Food safety aspect of insects: A review. *Acta Alimentaria*, 47 (4), 513–522. In.

Mitra, S., Purkait, T., Pramanik, K., Maiti, T. K., & Dey, R. S. (2019). Three-dimensional graphene for electrochemical detection of Cadmium in Klebsiella michiganensis to study the influence of Cadmium uptake in rice plant. *Materials Science and Engineering: C*, *103*, 109802. https://doi.org/10.1016/j.msec.2019.109802.

Mlček, J., Adámek, M., Adámková, A., Borkovcová, M., Bednářová, M., & Skácel, J. (2017). Detection of selected heavy metals and micronutrients in edible insect and their dependency on the feed using XRF spectrometry. *Potravinarstvo Slovak Journal of Food Sciences*, *11*(1), 725–730.

Mohamed, L. A., Aniagor, C. O., & Hashem, A. (2021). Isotherms and kinetic modelling of mycoremediation of hexavalent chromium contaminated wastewater. *Cleaner Engineering and Technology*, *4*, 100192.

Monteiro, M., Santos, C., Mann, R. M., Soares, A. M., & Lopes, T. (2007). Evaluation of cadmium genotoxicity in Lactuca sativa L. using nuclear microsatellites. *Environmental and Experimental Botany*, *60*(3), 421–427.

Morgan, J. N. (1999). Effects of processing on heavy metal content of foods. In: Jackson, L.S., Knize, M.G., Morgan, J.N. (eds) *Impact of Processing on Food Safety*. Advances in Experimental Medicine and Biology, vol 459, (pp. 195-211). Springer, Boston, MA. https://doi.org/10.1007/978-1-4615-4853-9_13

Morshdy, A., Yousef, R., Thariwat, A., & Hussein, M. (2023). Risks assessment of toxic metals in canned meat and chicken. *Food Research*, *7*(1), 151–157.

Muhammad, A., Auwal, Y., & Usman, A. (2022). Determination of heavy metals (Co, Cu, Cd, Fe, Pb, Zn) in some edible insects and fingerlings in dutsin-ma town. *Fudma Journal of Sciences*, *6*(4), 6–11.

Mukhles, M. B., Rahman, M. M., Rana, M. R., Huda, N., Ferdous, J., Rahman, F., Rafi, M. H., & Biswas, S. K. (2022). Effect of pesticides on nitrification activity and its interaction with chemical fertilizer and manure in long-term paddy soils. *Chemosphere*, *304*, 135379.

Nair, A. R., DeGheselle, O., Smeets, K., Van Kerkhove, E., & Cuypers, A. (2013). Cadmium-induced pathologies: Where is the oxidative balance lost (or not)? *International Journal of Molecular Sciences*, *14*(3), 6116–6143.

Narayanan, M., & Ma, Y. (2023). Metal tolerance mechanisms in plants and microbe-mediated bioremediation. *Environmental Research*, *222*, 115413.

Nkwunonwo, U. C., Odika, P. O., & Onyia, N. I. (2020). A review of the health implications of heavy metals in food chain in Nigeria. *The Scientific World Journal*, *2020*, 6594109.

Nuapia, Y., Chimuka, L., & Cukrowska, E. (2018). Assessment of heavy metals in raw food samples from open markets in two African cities. *Chemosphere*, *196*, 339–346.

Oller, A. R., Buxton, S., March, T. H., & Benson, J. M. (2023). Comparative pulmonary and genotoxic responses to inhaled nickel subsulfide and nickel sulfate in F344 rats. *Journal of Applied Toxicology*, *43*(5), 734–751.

Palisoc, S. T., Natividad, M. T., De Jesus, N., & Carlos, J. (2018). Highly sensitive AgNP/MWCNT/Nafion modified GCE-based sensor for the determination of heavy metals in organic and non-organic vegetables. *Scientific Reports*, *8*(1), 17445.

Palma-Lara, I., Martínez-Castillo, M., Quintana-Pérez, J., Arellano-Mendoza, M., Tamay-Cach, F., Valenzuela-Limón, O., García-Montalvo, E., & Hernández-Zavala, A. (2020). Arsenic exposure: A public health problem leading to several cancers. *Regulatory Toxicology and Pharmacology*, *110*, 104539.

Pang, S. (2020). A novel colorimetric assay for calcium ion and calmodulin detection based on gold nanoparticles. *Inorganic and Nano-Metal Chemistry*, *51*(5), 673–682.

Paterson, S., & De La Rica, R. (2015). Solution-based nanosensors for in-field detection with the naked eye. *Analyst*, *140*(10), 3308–3317.

Pham, X.-H., Hahm, E., Kim, T. H., Kim, H.-M., Lee, S. H., Lee, Y.-S., Jeong, D. H., & Jun, B.-H. (2018). Enzyme-catalyzed Ag growth on Au nanoparticle-assembled structure for highly sensitive colorimetric immunoassay. *Scientific Reports*, *8*(1), 6290.

Pipoyan, D., Beglaryan, M., Costantini, L., Molinari, R., & Merendino, N. (2018). Risk assessment of population exposure to toxic trace elements via consumption of vegetables and fruits grown in some mining areas of Armenia. *Human and Ecological Risk Assessment: An International Journal*, *24*(2), 317–330.

Pipoyan, D., Stepanyan, S., Beglaryan, M., & Mantovani, A. (2023a). Risk characterization of the Armenian population to nickel: Application of deterministic and probabilistic approaches to a total diet study in Yerevan City. *Biological Trace Element Research*, *201*(6), 2721–2732.

Pipoyan, D., Stepanyan, S., Beglaryan, M., Stepanyan, S., Mendelsohn, R., & Deziel, N. C. (2023b). Health risks of heavy metals in food and their economic burden in Armenia. *Environment International, 172*, 107794.

Plant, J., Smith, D., Smith, B., & Williams, L. (2000). Environmental geochemistry at the global scale. *Journal of the Geological Society, 157*(4), 837–849.

Promphet, N., Rattanarat, P., Rangkupan, R., Chailapakul, O., & Rodthongkum, N. (2015). An electrochemical sensor based on graphene/polyaniline/polystyrene nanoporous fibers modified electrode for simultaneous determination of lead and cadmium. *Sensors and Actuators B: Chemical, 207*, 526–534.

Qi, H., Zhuang, J., Zhuang, Z., Wang, Q., Wan, Y.-N., & Li, H.-F. (2023). Enrichment characteristics of heavy metals and health risk in different vegetables. *Huan Jing ke Xue= Huanjing Kexue, 44*(6), 3600–3608.

Raeeszadeh, M., Gravandi, H., & Akbari, A. (2022). Determination of some heavy metals levels in the meat of animal species (sheep, beef, turkey, and ostrich) and carcinogenic health risk assessment in Kurdistan province in the west of Iran. *Environmental Science and Pollution Research, 29*(41), 62248–62258.

Rahman, M. A., Rahman, M. M., Reichman, S. M., Lim, R. P., & Naidu, R. (2014). Heavy metals in Australian grown and imported rice and vegetables on sale in Australia: Health hazard. *Ecotoxicology and Environmental Safety, 100*, 53–60.

Rai, P. K., Lee, S. S., Zhang, M., Tsang, Y. F., & Kim, K.-H. (2019). Heavy metals in food crops: Health risks, fate, mechanisms, and management. *Environment International, 125*, 365–385.

Ramalingam, M., Ponnusamy, V. K., & Sangilimuthu, S. N. (2019). A nanocomposite consisting of porous graphitic carbon nitride nanosheets and oxidized multiwalled carbon nanotubes for simultaneous stripping voltammetric determination of cadmium (II), mercury (II), lead (II) and zinc (II). *Microchimica Acta, 186*, 1–10.

Ramos, I., Esteban, E., Lucena, J. J., & Gárate, A. N. (2002). Cadmium uptake and subcellular distribution in plants of Lactuca sp. Cd-Mn interaction. *Plant Science, 162*(5), 761–767.

Resch-Genger, U., Grabolle, M., Cavaliere-Jaricot, S., Nitschke, R., & Nann, T. (2008). Quantum dots versus organic dyes as fluorescent labels. *Nature Methods, 5*(9), 763–775.

Rossi, L., Eleonora, F., Boglioni, M., Giromini, C., Rebucci, R., & Baldi, A. (2014). Effect of zinc oxide and zinc chloride on human and swine intestinal epithelial cell lines. *International Journal of Health, Animal Science and Food Safety, 1*(2), 1–7.

Rusin, M., Domagalska, J., Rogala, D., Razzaghi, M., & Szymala, I. (2021). Concentration of cadmium and lead in vegetables and fruits. *Scientific Reports, 11*(1), 11913.

Sai Chaithanya, M., Bhaskar, D., & Vidya, R. (2022). Metal transfer and related human health risk assessment through milk from cattle grazing at an industrial discharge area. *Food Additives & Contaminants: Part A, 39*(2), 295–310.

Sauliutė, G., & Svecevičius, G. (2015). Heavy metal interactions during accumulation via direct route in fish: A review. *Zoology and Ecology, 25*(1), 77–86.

Schaefer, H. R., Dennis, S., & Fitzpatrick, S. (2020). Cadmium: Mitigation strategies to reduce dietary exposure. *Journal of Food Science, 85*(2), 260–267.

Shah, M. H., & Shaheen, N. (2007). Annual TSP and trace metal distribution in the urban atmosphere of Islamabad in comparison with mega-cities of the world. *Human and Ecological Risk Assessment, 13*(4), 884–899.

Shaheen, N., Irfan, N. M., Khan, I. N., Islam, S., Islam, M. S., & Ahmed, M. K. (2016). Presence of heavy metals in fruits and vegetables: Health risk implications in Bangladesh. *Chemosphere, 152*, 431–438.

Shahriar, S., Rahman, M. M., & Naidu, R. (2020). Geographical variation of cadmium in commercial rice brands in Bangladesh: Human health risk assessment. *Science of the Total Environment, 716*, 137049.

Shakoor, M. B., Nawaz, R., Hussain, F., Raza, M., Ali, S., Rizwan, M., Oh, S.-E., & Ahmad, S. (2017). Human health implications, risk assessment and remediation of As-contaminated water: A critical review. *Science of the Total Environment, 601*, 756–769.

Shariati, S., & Khayatian, G. (2020). Microfluidic paper-based analytical device using gold nanoparticles modified with N, N′-bis (2-hydroxyethyl) dithiooxamide for detection of Hg (ii) in air, fish and water samples. *New Journal of Chemistry, 44*(43), 18662–18667.

Sharma, A., Majdinasab, M., Khan, R., Li, Z., Hayat, A., & Marty, J. L. (2021). Nanomaterials in fluorescence-based biosensors: Defining key roles. *Nano-Structures & Nano-Objects, 27*, 100774. https://doi.org/10.1016/j.nanoso.2021.100774.

Shen, Y., Nie, C., Wei, Y., Zheng, Z., Xu, Z.-L., & Xiang, P. (2022). FRET-based innovative assays for precise detection of the residual heavy metals in food and agriculture-related matrices. *Coordination Chemistry Reviews, 469*, 214676. https://doi.org/10.1016/j.ccr.2022.214676.

Sigarini, K. D. S., de Oliveira, A. P., Martins, D. L., Brasil, A. S., de Oliveira, K. C., & Villa, R. D. (2017). Determination of the lead, cadmium, and chromium concentration in mineral feeds and supplements for cattle produced in the Mato Grosso State, Brazil. *Biological Trace Element Research, 177*, 209–214.

Singh, A., Sharma, R. K., Agrawal, M., & Marshall, F. M. (2010). Health risk assessment of heavy metals via dietary intake of foodstuffs from the wastewater irrigated site of a dry tropical area of India. *Food and Chemical Toxicology, 48*(2), 611–619.

Singh, J., Kaur, S., Lee, J., Mehta, A., Kumar, S., Kim, K.-H., Basu, S., & Rawat, M. (2020). Highly fluorescent carbon dots derived from Mangifera indica leaves for selective detection of metal ions. *Science of The Total Environment, 720*, 137604.

Singh, S., Kant, C., Yadav, R. K., Reddy, Y. P., & Abraham, G. (2019). Cyanobacterial exopolysaccharides: Composition, biosynthesis, and biotechnological applications. In *Cyanobacteria* (pp. 347–358). Elsevier.

Sonone, S. S., Jadhav, S., Sankhla, M. S., & Kumar, R. (2020). Water contamination by heavy metals and their toxic effect on aquaculture and human health through food chain. *Letters in Applied NanoBioScience, 10*(2), 2148–2166.

Srisukjaroen, R., Wechakorn, K., & Teepoo, S. (2022). A smartphone based-paper test strip chemosensor coupled with gold nanoparticles for the Pb2+ detection in highly contaminated meat samples. *Microchemical Journal, 179*, 107438. https://doi.org/10.1016/j.microc.2022.107438.

Srivastava, A. K., Dev, A., & Karmakar, S. (2018). Nanosensors and nanobiosensors in food and agriculture. *Environmental Chemistry Letters, 16*, 161–182.

Stančić, Z., Vujević, D., Gomaz, A., Bogdan, S., & Vincek, D. (2016). Detection of heavy metals in common vegetables at Varaždin City Market, Croatia. *Arhiv za Higijenu Rada I Toksikologiju, 67*(4), 340–350.

Swain, K. K., Balasubramaniam, R., & Bhand, S. (2020). A portable microfluidic device-based Fe3O4-urease nanoprobe-enhanced colorimetric sensor for the detection of heavy metals in fish tissue. *Preparative Biochemistry & Biotechnology, 50*(10), 1000–1013.

Talbi, A., Kerchich, Y., Kerbachi, R., & Boughedaoui, M. (2018). Assessment of annual air pollution levels with PM1, PM2.5, PM10 and associated heavy metals in Algiers, Algeria. *Environmental Pollution, 232*, 252–263.

Talio, M. C., Alesso, M., Acosta, M., Wills, V. S., & Fernández, L. P. (2017). Sequential determination of nickel and cadmium in tobacco, molasses and refill solutions for e-cigarettes samples by molecular fluorescence. *Talanta, 174*, 221–227.

Talio, M. C., Muñoz, V., Acosta, M., & Fernández, L. P. (2019). Determination of lead traces in honey using a fluorimetric method. *Food Chemistry, 298*, 125049. https://doi.org/10.1016/j.foodchem.2019.125049.

TatahMentan, M., Nyachoti, S., Scott, L., Phan, N., Okwori, F. O., Felemban, N., & Godebo, T. R. (2020). Toxic and essential elements in rice and other grains from the United States and other countries. *International Journal of Environmental Research and Public Health, 17*(21), 8128.

Tchounwou, P. B., Yedjou, C. G., Patlolla, A. K., & Sutton, D. J. (2012). Heavy metal toxicity and the environment. *Experientia supplementum, 101*, 133–164.

Tlustos, P., Pavlikova, D., Balik, J., Szakova, J., Hanc, A., & Balikova, M. (1998). The accumulation of arsenic and cadmium in plants and their distribution. *Rostlinna Vyroba, 44*, 463–470.

Ugulu, I. (2015). Determination of heavy metal accumulation in plant samples by spectrometric techniques in Turkey. *Applied Spectroscopy Reviews, 50*(2), 113–151.

Ullah, S., Ennab, W., Wei, Q., Wang, C., Quddus, A., Mustafa, S., Hadi, T., Mao, D., & Shi, F. (2023). Impact of cadmium and lead exposure on camel testicular function: Environmental contamination and reproductive health. *Animals, 13*(14), 2302.

Van der Fels-Klerx, H., Camenzuli, L., Belluco, S., Meijer, N., & Ricci, A. (2018). Food safety issues related to uses of insects for feeds and foods. *Comprehensive Reviews in Food Science and Food Safety, 17*(5), 1172–1183.

Veluprabakaran, V., & Kavitha, M. (2023). Evaluation of heavy metals in ground and surface water in Ranipet, India utilizing HPI model. *Environmental Monitoring and Assessment, 195*(7), 1–11.

Vonnie, J. M., Rovina, K., Erna, K. H., Mantihal, S., Huda, N., & Wahab, R. A. (2022). Development of a colorimetric sensor based on tapioca starch and gold nanoparticles for the detection of cadmium residues in fish products. *Journal of Food & Nutrition Research, 61*(4), 315–322.

Wang, H., Rao, H., Luo, M., Xue, X., Xue, Z., & Lu, X. (2019). Noble metal nanoparticles growth-based colorimetric strategies: From monocolorimetric to multicolorimetric sensors. *Coordination Chemistry Reviews, 398*, 113003.

Wang, L., Peng, X., Fu, H., Huang, C., Li, Y., & Liu, Z. (2020). Recent advances in the development of electrochemical aptasensors for detection of heavy metals in food. *Biosensors and Bioelectronics, 147*, 111777.

Wang, N., Xue, X.-M., Juhasz, A. L., Chang, Z.-Z., & Li, H.-B. (2017a). Biochar increases arsenic release from an anaerobic paddy soil due to enhanced microbial reduction of iron and arsenic. *Environmental Pollution, 220*, 514–522.

Wang, S., Wu, W., Liu, F., Liao, R., & Hu, Y. (2017b). Accumulation of heavy metals in soil-crop systems: A review for wheat and corn. *Environmental Science and Pollution Research, 24*, 15209–15225.

Wang, X., Xu, Y., Li, Y., Li, Y., Li, Z., Zhang, W., Zou, X., Shi, J., Huang, X., Liu, C., & Li, W. (2021). Rapid detection of cadmium ions in meat by a multi-walled carbon nanotubes enhanced metal-organic framework modified electrochemical sensor. *Food Chemistry, 357*, 129762. https://doi.org/10.1016/j.foodchem.2021.129762.

Wang, Y., Cao, D., Qin, J., Zhao, S., Lin, J., Zhang, X., Wang, J., & Zhu, M. (2023). Deterministic and probabilistic health risk assessment of toxic metals in the daily diets of residents in industrial regions of Northern Ningxia, China. *Biological Trace Element Research, 201*, 1–15.

Xie, L., Funk, D. H., & Buchwalter, D. B. (2010). Trophic transfer of Cd from natural periphyton to the grazing mayfly Centroptilum triangulifer in a life cycle test. *Environmental Pollution, 158*(1), 272–277.

Yan, M., Niu, C., Li, X., Wang, F., Jiang, S., Li, K., & Yao, Z. (2022). Heavy metal levels in milk and dairy products and health risk assessment: A systematic review of studies in China. *Science of the Total Environment, 851*, 158161.

Yousafzai, A. M., & Shakoori, A. (2007). Heavy metal bioaccumulation in the muscles of Mahaseer, Tor putitora, as an evidence of heavy metal pollution in River Kabul, Pakistan. *Pakistan Journal of Zoology, 39*(1), 1.

Yuan, M., Qian, S., Cao, H., Yu, J., Ye, T., Wu, X., Chen, L., & Xu, F. (2022). An ultra-sensitive electrochemical aptasensor for simultaneous quantitative detection of Pb2+ and Cd2+ in fruit and vegetable. *Food Chemistry, 382*, 132173. https://doi.org/https://doi.org/10.1016/j.foodchem.2022.132173.

Zafalon, R. V. A., Pedreira, R. S., Vendramini, T. H. A., Rentas, M. F., Pedrinelli, V., Rodrigues, R. B. A., Risolia, L. W., Perini, M. P., Amaral, A. R., & de Carvalho Balieiro, J. C. (2021). Toxic element levels in ingredients and commercial pet foods. *Scientific Reports, 11*(1), 21007.

Zambelli, B., & Ciurli, S. (2013). Nickel and human health. *Metal Ions in Life Sciences, 13*, 321–357.

Zaw, M., & Emett, M. T. (2002). Arsenic removal from water using advanced oxidation processes. *Toxicology Letters, 133*(1), 113–118.

Zhang, Z., Wang, H., Chen, Z., Wang, X., Choo, J., & Chen, L. (2018). Plasmonic colorimetric sensors based on etching and growth of noble metal nanoparticles: Strategies and applications. *Biosensors and Bioelectronics, 114*, 52–65.

Zhou, H., Yang, W.-T., Zhou, X., Liu, L., Gu, J.-F., Wang, W.-L., Zou, J.-L., Tian, T., Peng, P.-Q., & Liao, B.-H. (2016). Accumulation of heavy metals in vegetable species planted in contaminated soils and the health risk assessment. *International Journal of Environmental Research and Public Health, 13*(3), 289.

Zulkafflee, N. S., Mohd Redzuan, N. A., Nematbakhsh, S., Selamat, J., Ismail, M. R., Praveena, S. M., Yee Lee, S., & Abdull Razis, A. F. (2022). Heavy metal contamination in Oryza sativa L. at the eastern region of Malaysia and its risk assessment. *International Journal of Environmental Research and Public Health, 19*(2), 739.

4 Analysis of Pesticides

*Nuran Gokdere, Ummugulsum Polat,
and Ismail Murat Palabiyik*

4.1 INTRODUCTION

Pesticides are synthetic or natural chemical substances used to protect plants, vegetables, and fruits used for various purposes from harmful organisms, pests, rodents, and invasive plants. These chemicals, essential in food production, storage, and use, are also involved in achieving sustainable agriculture goals. Pesticides, which can be classified according to their various properties, areas of use, and the pests they affect, are compounds that can bioaccumulate and are harmful to living life and environmental ecology. Pesticides taken into the human body through the respiratory tract, skin, and eyes can cause nervous system damage, cancer, and genetic disorders. Pesticide use worldwide is limited to specific concentrations, and various authorities have determined maximum residue levels.

Some studies have found that commercial and homemade washing agents do not altogether remove pesticide residues from foods. In a study by Yang and coworkers, they looked at surface pesticide residues on apples using different washing solvents such as sodium bicarbonate, tap water, and Clorox bleach, and effective removal was only achieved by sodium bicarbonate. Related to this phenomenon, if food is exposed to too much pesticide before it is placed on the market, analysis of these compounds is fundamental (Yang et al., 2017).

Nowadays, pesticides, which can be obtained from multiple matrices such as food, plants, and biological samples, using different extraction techniques, are determined by separation techniques electrochemical and spectroscopic methods. In addition to the aim of the analysis being simple, fast response, and low cost, selectivity and reaching a low detection limit are very important regarding residue analysis. Biosensors are one of the tools that can be used for these purposes. Nanomaterial-based biosensors are widely used to determine pesticide residues due to their low detection limit, good repeatability and reproducibility, and significant response amplification. These sensors, which facilitate or eliminate sample pre-preparation techniques, can be classified according to the aptamer, nanoparticle, graphene, carbon nanotube, enzyme, or hybrid structures they produce from or according to the analysis method. Analyzes of the developed nanobiosensors from actual samples, predominantly plants, vegetables, and fruits, are included in the scientific literature.

4.2 PESTICIDES

4.2.1 PESTICIDES DEFINITION

Pesticides are chemicals that protect food from microorganisms and pests harmful to the health and development of living things and remove or destroy them during food production, consumption, and storage stages (Bajwa & Sandhu, 2014). According to the World Health Organization, pesticides are chemical compounds used to kill pests such as insects, rodents, fungi, and wild plants. Since pesticides harm organisms, they must be used safely and disposed of appropriately. Pesticides should be applied correctly and in recommended quantities under appropriate environmental conditions. When pesticides are used correctly, they protect plants from diseases, pests, and insects, ensure that the product does not deteriorate for a long time, and increase earnings (World Health Organization, 1999; Chwatko et al., 2019).

DOI: 10.1201/9781003514039-4

Pesticides have an essential role in improving the productivity and quality of agricultural products and controlling plant diseases spread by insects. Pesticides are also used against diseases such as malaria, yellow fever, plague, typhoid, and typhus, protecting people from these (Tudi et al., 2021).

Pesticides are generally used against a particular organism. The ideal situation is that pesticides poison only the targeted organism and do not harm others. Substances with high selectivity kill unwanted creatures at a specific concentration while not affecting other animals and plants much. The extent to which different species are affected by these chemicals largely depends on the concentrations and type of exposure (e.g., bees; Zioga et al., 2023). However, complete selectivity is not possible. Commonly used pesticides and their effects on human health are listed in Table 4.1.

Since pesticides affect human health, cause environmental toxicity, and lack complete selectivity, new pesticide designs have begun to be studied in recent years. One of these is contemporary nematode management. Its rate of spread in the soil is relatively slow, and it needs to be used a lot to be effective. Excessive use causes contamination. Nanoparticle technology prevents premature degradation caused by biology and photolysis and directs the substance to the target. Synthetic nanoparticles were first studied, but they had problems such as high costs and stability to biodegradation. Plant virus-based nanocarriers were reviewed by Caparco et al. (2023). These nanocarriers are very stable in the soil, and contamination is reduced due to these properties. These nanocarriers have been shown to have better soil mobility and pesticide distribution than synthetic nanoparticles (Caparco et al., 2023).

4.2.2 Pesticide Classification

Pesticides are classified according to their appearance, physical structure, formulation, disease groups and pests they are effective on, biological effects, contents, toxicity levels, and the techniques used.

Pesticides are classified as follows according to their formulation forms:

1. Powder drugs (Dust)
2. Wettable powder drugs (WP)
3. Emulsion concentrate drugs (EC or EM)
4. Solution-concentrated Drugs (SC)
5. Water-soluble powder drugs (SP)
6. Summer and winter oils
7. Granules (G)
8. Pellets
9. Tablets
10. Powdered seed preparations
11. Liquid seed medicine
12. Aerosols
13. Toxic baits
14. Formulations in capsule form
15. Flowing concentrates
16. Dry fluids.

Pesticides are divided into two according to their usage techniques:

1. Directly used drugs (powder drugs, ULV formulations, granules, and some nematocytes),
2. Medicines used by diluting with water or organic solvent.

The classification of pesticides according to their physical state is as follows:

TABLE 4.1

Effects and Toxicity of Commonly Used Pesticides on Human Health

Name	Effects	Toxicity	References
Paraoxon	Anticholinesterase Glutamate release Diminished GABA Uptake Oxidative damage Neurodegeneration	Neural and Excitotoxic damage on brain tissues	Farizatto and Bahr (2017)
Methyl parathion	Bind to acetylcholinesterase Overstimulation of the cholinergic system	Carcinogenic, immunosuppressive Liver Cardiovascular Muscular system toxicity	Garcia et al. (2003)
Malathion	Inhibition of acetylcholine esterase	Carcinogenic Neurotoxic Increasing oxidative stress and inflammatory conditions Hepatotoxic Nephrotoxic Reproductive and hepatotoxicity	Badr (2020)
Dichlorvos	Choline esterase inhibition	Neurotoxic Mutagenic Carcinogenic Increasing oxidative stress Genotoxic	Binukumar and Gill (2010)
Carbaryl	Inhibition of acetylcholine esterase	Neurotoxic Reproductive system toxicity Genotoxic Carcinogenic	Balcker et al. (2010)
Chlorpyrifos	Inhibition of acetylcholine esterase	Mutagenic Carcinogenic Teratogenic Neurotoxic	Eaton et al. (2008)
Acetamiprid	Agonist at nicotinic acetylcholine receptor	Increasing oxidative stress Genotoxicity Developmental toxicity Neurotoxic Hepatotoxic Immunotoxic	Phogat et al. (2022)
Atrazine	binding to the plastoquinone-binding protein	Endocrine disruption Inducing oxidative stress Genotoxic	Chang et al.(2022)
Paraquat	Producing reactive oxygen species Inhibition of photosynthesis	Hepatorenal toxicity Cardiovascular toxicity	Kim and Kim (2020)

1. Solid formulations (Powder, WP granule, and so on)
2. Liquid (Liquid) formulations (EC, oils solutions, and so on)
3. Gaseous (Cyan, CS_2, CH_3Br)
4. Used as feed (Sodium arsenite, BHC + bran).

Pesticides are classified according to their target pest organisms and given unique names according to their effects. In the naming, first, the name of the pest it kills is added, and then the Latin word "cide," which means kill or killer, is added.

Pesticides are classified as follows according to the pest group they are used in:

1. Insecticides (Insecticide)
2. Those that kill fungi (Fungicide)
3. Those that stop the activity of fungi (Fungistatic)
4. Those that kill weeds (Herbicides)
5. Those who kill spiders (Acaricide)
6. Those that kill bacteria (Bactericide)
7. Those that kill aphids (Aficid)
8. Those that kill rodents (Rodenticide)
9. Those that kill nematodes (Nematocyte)
10. Those who kill snails (Molluscide)

Due to their toxic nature, pesticides adversely affect humans and their environment, directly or indirectly, apart from the targeted pest or disease. These toxic substances enter the human body in three ways:

a. Oral
b. Absorption
c. Inhalation

Chemical classification is more complex than others. Pesticides generally used or produced recently are organic and may be of synthetic or plant origin. In addition, inorganic compounds can be used as pesticides.

4.2.3 PESTICIDE ANALYSIS

The use of pesticides is necessary for the production of large quantities of crops and the prevention of essential vector diseases. However, pesticides cause acute and chronic toxicity, carcinogenicity, teratogenicity in non-target organisms, environmental and food contamination, and pest resistance. Therefore, although the use of pesticides cannot be prevented entirely, continuous monitoring of human, animal, and environmental exposure is necessary for taking precautions. In addition, it is legally required to determine and control the maximum permissible residue limits in plant and animal foods that are predicted not to cause harmful effects in humans. For example, pesticide residues are found in milk because animals are fed with feed containing pesticides, pesticides are applied to their bodies, and pesticides are used in the areas where milk is processed. Dairy operators face a number of management problems during the production season including pests that can cause human disease (Ledoux, 2011). For these reasons, sensitive, reliable, and reproducible analytical methods are needed to detect pesticides or their metabolites in environmental, food, and biological samples (Yavuz & Aksoy, 2016).

The European Commission has determined maximum residue levels (MRLs) for quantifying high-risk pesticide residues in food and nutrients humans consume. There is a need for newly developed selective analytical methods to detect human exposure to organic pollutants and pesticides periodically and to determine MRLs by European Union authorities (Baranowska et al., 2006).

The chromatographic techniques used in pesticide residue analysis are Gas Chromatography (GC), Liquid Chromatography (LC), LC-MS which combines LC and GC with mass spectrometric detection (MS), GC-MS and reversed-phase liquid chromatography (RPLC) equipped with a UV or Fluorescence detector suitable for polar pesticide analysis (Hogendoorn & Van Zoonen, 2000 and Atapattu & Johnson, 2020)

GC-MS analyzes low molecular weight, highly volatile, and temperature-stable compounds and compounds chemically altered by derivatization. LC-MS/MS systems provide information on the chemical structure of pesticides, reduce the matrix effect on the analysis result, require less sample clean-up, thus saving time, and can lower detection limits (Sobhanzadeh et al., 2009).

Electron capture detections (ECD), flame photometric detections (FPD), nitrogen-phosphorus detections (NPD), phosphorus detections, and diode array detections (DAD) were used for pesticide analysis as qualitative and quantitative (Ledoux, 2011).

4.2.3.1 Extraction Methods Used in Pesticide Analysis

The extraction process in pesticide residue analysis is challenging as it separates different groups of pesticides simultaneously with a single method. Pesticides have other chemical and physical properties and behave differently in the matrix. Various extraction methods have been developed to facilitate multiple pesticide residue analyses and increase extraction efficiency. Additionally, the detection limit is reduced by using nanomaterials such as nanoparticles and carbon nanotubes. In this way, the sensitivity of the analysis increases. As researchers have turned to green chemistry in recent years, solid-phase extraction and solid-phase microextraction methods, which use less solvent, have increased. Extraction methods used in pesticide analyses are shown in Table 4.2.

TABLE 4.2

Extraction Methods and Applications Used in Pesticide Analysis

Extraction Method	Example of Application of Extraction Methods	Reference
Liquid-Liquid Extraction	Pesticide residue analysis was performed using fabric phase sorptive extraction and liquid-liquid extraction. A very low-cost analysis has been developed since the sorbent can be used for up to 12 studies.	Aghdam et al. (2023)
Supercritical fluid extraction (SFE)	Flutriafol enantiomers were analyzed using supercritical fluid chromatography-tandem mass spectrometry. The developed method was applied to tomato, cucumber, apple, grape, and soil samples, and good results were obtained.	Tao et al. (2014)
QuEChERS Method	A total of 122 pesticide residue analyses were developed using the Fe_3O_4-pDA and QuEChERS methods. The developed procedure analyzed pesticide residue on 100 fruit samples, including peach, pear, grape, and watermelon.	Qi et al. (2023)
Accelerated Solvent Extraction (ASE)	Using ASE and gel permeation chromatography, a method was developed for clotrimazole residue in pork, chicken, beef, fish, beef kidney, and chicken liver.	Han et al. (2021)
Matrix solid-phase dispersion (MSPD)	A method was developed for 30 vegetable pesticide residues using Fe_3O_4 and MSPD. The developed method has high specificity, sensitivity, accuracy, and precision.	Liu et al. (2021)
Microwave-assisted extraction (MAE)	A pesticide residue analysis method was developed using MAE and LC-MS/MS. The analysis showed good linearity, acceptable accuracy, and acceptable interassay precision.	Kalogiouri et al. (2021)
Solid Phase Extraction (SPE)	A method was developed for detecting organophosphate pesticides from fruit juice using MSPE and $NiFe_2O_4$@PDA@Mg/Al-LDH. The developed method exhibited high accuracy and sensitivity.	Du et al. (2019)
Solid Phase Microextraction Method (SPME)	A method was developed for detecting triazole pesticides in water samples using GO/APS and SPME. The developed method can determine low triazole pesticide concentrations.	Piryaei et al. (2023)
Stir-Bar Sorptive Extraction (SBSE)	SBSE kit combined with HPLC-UV has been developed for organophosphate detection. The designed kit was successfully applied for pesticide analysis on actual water samples. The method has low detection limits, high extraction efficiency, and a wide linear range.	Gorji et al. (2019)

Fe_3O_4-pDA: poly-dopamine-modified magnetic nanomaterial; Fe_3O_4: iron oxide nanoparticles; GO/APS: Graphene-Oxide/Acrylonitrile Polystyrene; $NiFe_2O_4$@PDA@Mg/Al-LDH: $NiFe_2O_4$ nanoparticles coated with polydopamine and Mg/Al layered double hydroxides.

4.3 APPLICATIONS OF NANOBIOMATERIALS IN PESTICIDE ANALYSIS

Nanotechnology is the science that studies materials in the nanorange, usually from 1 to 100 nm. Nanomaterials have optical, magnetic, and electrical properties that have made them attractive (Raghu et al., 2020). Coating the surface of the nanomaterial with a biomaterial such as DNA, protein, aptamer, or antibody enables the development of a more specific sensor for the substance to be analyzed. Aptamers are single-stranded DNA or RNA oligonucleotides selected *in vitro* using exponential enrichment. They have several advantages, such as faster synthesis, stability, and convenient reinforcement. It has a higher affinity than other biomaterials. Carbon nanotubes, nanoparticles, quantum dots, nanowires, nanoflowers, and nanorods are frequently used nanomaterials for nanobiosensors. Nanobiosensors are widely used in biomedical, diagnostic, and environmental applications (Malik et al., 2013; Li et al., 2019).

In the last decade, nanobiosensors have been used in pesticide analysis. Because nanobiosensors can provide analysis even if the substance to be analyzed is in very small amounts. Compared to traditional methods, nanobiosensors offer a more cost-effective and faster analysis. Nanobiosensors are used in many transducers. The most commonly used are electrochemical and colorimetric transducers (Mirres et al., 2022).

4.3.1 ELECTROCHEMICAL METHODS

4.3.1.1 Carbon Nanotubes

Carbon nanotubes (CNTs) are structures formed by coiling a single graphene sheet of carbon atoms to form a cylinder (Büyük, 2017). It is widely used in analysis due to its outstanding mechanical strength, high surface area, good electrical conductivity, electrochemical stability, and high thermal conductivity (Sireesha, 2018). Electrochemical nanobiosensors using carbon nanotubes as nanomaterial are listed in Table 4.3.

The study planned to develop a biosensor with high selectivity, simple application, and fast response to determine organophosphorus pesticides (Han, 2018). In this study, it has been shown that amino acid ionic liquid increases the dispersion and biocompatibility of carbon nanotubes in water. Thus, this composite structure has good conductivity and electrocatalytic activity. Compared to literature methods, the biosensor developed in this way has been shown to have 2–8 times greater sensitivity (LOD 3 fmol L^{-1}). The developed biosensor showed excellent selectivity since it can be measured at negative potentials and has high specificity. It was used to determine pesticide residues at femtomolar levels in actual samples. In this way, it can detect paraoxon at very low concentrations (Han et al., 2018).

In another study, Xu coworkers (2018) created a regenerative electrochemical aptasensor based on a nanocomposite of CuO NFs and carboxyl-functional single-walled carbon nanotubes (c-SWCNTs) for the detection of chlorpyrifos. They used CuO NFs to expand the surface area and c-SWCNTs to improve the ability to immobilize amine probes and hybridize between aptamers and amine probes. Differential pulse voltammetry (DPV) was used for analysis. As a result of the investigation, a linear response was observed between 0.1 and 150 ng mL^{-1}, and the LOD was 70 pg mL^{-1}. It has a very low detection limit for chlorpyrifos. The developed nanobiosensor has been successfully applied to detect chlorpyrifos in apple and celery sprouts. In addition, the developed nanobiosensor showed proper stability after five regenerations using urea (Xu et al., 2018).

In a study developed for the electrochemical determination of malathion (Kaur, 2016), an electrode was used as the composite structure consisting of conducting polymer (CP) - PEDOT and MWCNTs, which were electrochemically deposited onto fluorine-doped tin oxide. The –COOH group in MWCNTs is related to the covalent immobilization of the acetylcholine esterase enzyme. In the study, pH, buffer solution, acetylcholine esterase, and substrate concentration are the experimental factors examined for optimization. Under optimized conditions, the linear working range was calculated as 1 fmol L^{-1} to 1 µmol L^{-1}, and the detection limit was 1 fmol L^{-1}. Regeneration

TABLE 4.3

Electrochemical Analyses Using Carbon Nanotubes as Nanomaterial

Analyte	Nanobiosensor	Linear Range	Limit of Detection (LOD)	Recovery (%)	Application in Food	Reference
Paraoxon	f-MWCNT/poly(SNS-NH2)/AChE	0.005–0.1 and 0.1–10 µg L^{-1}	2.46 ng L^{-1}	–	Water	Kesik et al. (2014)
Parathion		0.001–0.01 and 0.01–7 µg L^{-1}	0.542 ng L^{-1}		Tap water	
Chlorfenvinphos		0.005–0.1 and 0.1–12.5 µg L^{-1}	4.90 ng L^{-1}			
Paraoxon	Nafion/M-Cell/CNT@AAIL/ GCE	10–800 fmol L^{-1} and 0.8–5 pmol L^{-1}	3 fmol L^{-1}	97.5–104.0	Tap water Spinach juice s	Han et al. (2018)
Malathion	Au/VNSWCNTs/	1×10^{-5}–1.0 ppb	1.96×10^{-6} ppb	95–105	Cabbage water	Xu et al. (2019)
Chlorpyrifos	AuNPs/AChE		2.06×10^{-6} ppb		Tap water Purified water	
Methyl parathion			3.04×10^{-6} ppb		River water Lake water	
Chlorpyrifos	Aptamer/Fc@WMCNTs /OMC/GCE	1–105 ng mL^{-1}	0.33 ng mL^{-1}	98.5–107.2	Lettuce Leek Pakchoi	Jiao et al. (2016)
Chlorpyrifos	Aptamer/AMP/CuO NFs-SWCNTs/Nafion/GCE	0.1–150 ng mL^{-1}	70 pg mL^{-1}	95.58±4.32– 106.72±2.66	Apple Celery Cabbage	Xu et al. (2018)
Chlorpyrifos	AChE/MWCNTs/IL/ SPE	0.05–1×10^{5} µg L^{-1}	0.05 µg mL^{-1}	87–101	Rape Cabbage Lettuce	Chen et al. (2017)
Parathion Malathion	AChE/DAR/ PB-SWCNTs/ DAR/P-ABSA/ GCE	1.88×10^{-13} g L^{-1} 3.11×10^{-13} g L^{-1}	1×10^{-6}–1×10^{-12} g L^{-1}	96–105	Chopped vegetables	Jiang et al. (2019)
Malathion	AChE/PEDOT-MWCNTs/ FTO	1 fM–1 µM	1 fmol L^{-1}	96–98	Lettuce	Kaur et al. (2016)

(Continued)

TABLE 4.3 (*Continued*)

Electrochemical Analyses Using Carbon Nanotubes as Nanomaterial

Analyte	Nanobiosensor	Linear Range	Limit of Detection (LOD)	Recovery (%)	Application in Food	Reference
Dimeotate	SPCE | CNTs/ZrO$_2$/PB/ Nafion | GMP-AChE	1.0×10^{-3}ng mL^{-1}	5.6×10^{-4}ng mL^{-1}	88–105	Chinese Cabbage	Gan et al. (2010)
Pirimicarb	AChE/PB-MWCNTs/SPEs	1.0×10^{-6}–1.0g L^{-1}	5.32×10^{-8}g L^{-1}	–	River water	Chai et al. (2013)
Carbaryl	GC/MWCNT/PANI/ AChE	9.9–49.6 µmol L^{-1}	1.4 µmol L^{-1}	97.3–103.4	Apple	Cesarino et al. (2012)
Methomyl	GC/MWCNT/PANI/ AChE	9.9–49.6 µmol L^{-1}	0.95 µmol L^{-1}	94.6–101.2 and 95.4–101.9	Cabbage Broccoli	Cesarino et al. (2012)
Carbaryl	[(PDDA–SWCNTs)/AchE]$_5$/ GCE	1×10^{-6}–1×10^{-11}g L^{-1}	1×10^{-12}g L^{-1}	99.8±2.7	Water	Firdoz et al. (2010)
Paraoxon Carbaryl	MWCNT–(PEI/DNA)$_2$/OPH/ AChE	0.5–40 µmol L^{-1} 10–80 µmol L^{-1}	0.5 µmol L^{-1} 10 µmol L^{-1}	–	Apple	Zhang et al. (2015)

AAIL: mino acid ionic liquid; AChE: acetylcholinesterase; AMP: Amino-modified capture probe; AuNPs: gold nanoparticles; CuO NFs: copper oxide nanoflowers; DAR: diazo-resin; f-MWCNT: multi-walled carbon nanotube; Fc@WMCNTs: ferrocene hybrid multiwalled carbon nanotubes; FTO: fluorine doped tin oxide; GCE: glassy carbon electrode; GMP: magnetic nanoparticulate; IL: ionic liquid; M-Cell: fused cell; OMC: mesoporous carbon; PANI:polyaniline; PB: Prussian blue; PDDA:poly(diallyldimethylammonium chloride); PEDOT: Poly(3,4-ethylenedioxythiophene); PEI: polyethyleneimine; poly(SNS-NH2): poly(4-(2,5-di(thiophen-2-yl)-1H-pyrrol-1-yl)benzenamine); SPCE:screen-printed carbon electrodes; VNSWCNTs: nitro- gen-doped single-walled carbon nanotubes; ZrO$_2$: Zirconia nanoparticle.

of the inhibited acetylcholinesterase enzyme was achieved with 99% accuracy by 2-PAM. The developed electrode was used to determine this organophosphate pesticide in lettuce samples with an accuracy of 96%–98%. In addition, this study has the lowest detection limit compared to the other studies mentioned (Kaur et al., 2016).

4.3.1.2 Nanoparticles

Nanoparticles can be produced from materials with different chemical properties, most commonly metals, metal oxides, non-oxide ceramics, silicates, organics, polymers, carbon, and biomolecules. They are also found in various shapes, such as spheres, cylinders, and tubes. There are two nanoparticle synthesis methods: bottom-up and top-down (Kamacı, 2019). Each approach has its characterization and application. Electrochemical nanobiosensors using nanoparticles as nanomaterials are listed in Table 4.4.

Ma and coworkers (2018) developed a nanobiosensor with a core-shell structure to determine malathion, chlorpyrifos, and methyl parathion. This composite structure contains an N-doped carbon shell and Pt and Pd metals. To form this structure, the reduction of K_2PtCl_4, H_2PtCl_6, and Na_2PdCl_4, the self-polymerization of dopamine as a reducer, and the incorporation of Pluronic F127 using one-pot application are required. As a result of the characterization studies, it is mentioned in the survey that Pd-Pt nanoparticles have a diameter of approximately 15 nm and are enclosed in N-doped carbon shells with a thickness of about 35 nm. The linear working ranges in molarity for these three organophosphorus pesticides were between 10^{-5} and 10^{-14}. Meanwhile, the detection limits were calculated as (or below) 10^{-14}. It was shown in the study that the developed electrode determined these pesticides from cabbage samples with acceptable accuracy and had high selectivity, reproducibility, and stability (Ma et al., 2018).

In a study published in 2020 (Hu, 2020), the acetylcholine esterase–chitosan/titanium dioxide structure was developed to determine organophosphorus pesticides, which was then modified to the glass carbon surface. With this electrode, the working range for dichlorvos was determined as 1.13–22.6 nmol L^{-1}, and the detection limit was 0.23 nmol L^{-1}. Interference studies were also carried out, and the new and stable electrode developed was applied with reasonable accuracy to determine this pesticide from cabbage juice. It has been mentioned that screen-printed biosensors can increase stability and sensitivity, and alternatives to titanium dioxide can be used in future studies (Hu et al., 2020).

Unlike other literature studies, Zhao et al. (2021) developed a new nanogold/mercaptomethamidophos electrochemical biosensor to analyze 11 organophosphorus pesticides simultaneously. The nanobiosensor was fabricated by combining strong Au-S bonds and the interaction between AChE and mercaptometamidofos. This biosensor exhibited a more comprehensive linear range of 0.019–0.077 and 0.9905–0.9992 ng mL^{-1} and a lower LOD of 0.1–1,500 ng mL^{-1}. The nanobiosensor detected trichlorfon and dichlorvos in real apple and cabbage samples. The average recovery was 81.45%–116.81%, the RSD was 10%, and the developed nanobiosensor can effectively detect organophosphorus pesticides (Zhao et al., 2021).

4.3.1.3 Nanoflower

Nanoflowers are advanced classes of nanoparticles with a structure similar to plant flowers. Although nanoflowers have a small structure, they have several leaves with a large surface area surrounding this structure, and thanks to these features, they are used in analysis and drug delivery systems (Güven & Kar, 2022).

In a study published in 2019, Bao and his colleagues developed an electrochemical acetylcholinesterase biosensor based on graphene-copper oxide nanoflower nanocomposites (3DG-CuO NFs) to detect malathion. 3DG-CuO NFs increase the surface area and biosensor performance in the designed biosensor. As a result of optimization studies, the designed biosensor exhibited a linear range between 3 pmol L^{-1} and 46.665 nmol L^{-1} with an LOD of 0.92 pmol L^{-1}. The developed biosensor was used to detect malathion in two different tap waters and showed recovery between 94% and 106% (Bao et al., 2019).

TABLE 4.4

Electrochemical Analyses Using Nanoparticles as Nanomaterials

Analyte	Nanobiosensor	Linear Range	Limit of Detection (LOD)	Recovery, %	Application in Food	Reference
Paraoxon	TMB-MnNS-ATCh	0.1–$20\,ng\,mL^{-1}$	$0.025\,ng\,mL^{-1}$	99.72–102.41	Pakchoi	Wu et al. (2021)
Malathion Chlorpyrifos Methyl parathion	AChE/PtPd@NCS/GCE	1×10^{-14}–$1\times10^{-10}\,mol\,L^{-1}$ and 1×10^{-9}–$1\times10^{-5}\,mol\,L^{-1}$ 1×10^{-13}–$1\times10^{-6}\,mol\,L^{-1}$ 1×10^{-14}–$1\times10^{-11}\,mol\,L^{-1}$ and 1×10^{-10}–$1\times10^{-5}\,mol\,L^{-1}$	$7.9\times10^{-15}M$ $7.1\times10^{-14}M$ $8.6\times10^{-15}M$	93.21–104.3	Cabbage	Ma et al. (2018)
Methyl parathion	AChE/Nafion/ AuNPs/rGO/ GCE	1×10^{-10}–$1\times10^{-6}\,g\,L^{-1}$	$2.78\times10^{-11}\,g\,L^{-1}$	94–106	Tap water Mineral water Chinese cabbage	Dong et al. (2019)
Methyl paraoxon Methyl parathion Ethyl paraoxon	TiO$_2$@DA@S/H/E nanoenzyme	500–$100\,\mu mol\,L^{-1}$	$0.2\,\mu mol\,L^{-1}$	86.38–98.19	Lactuca sativa L.	Qui et al. (2019)
Malathion Trichlorfon	poly(FBThF)/Ag-rGO-NH$_2$/AChE/GCE	0.099–$9.9\,\mu g\,L^{-1}$ and 0.0206–$2.06\,\mu g\,L^{-1}$.	0.032–$0.001\,\mu g\,L^{-1}$	95–108	Apple Tap water	Zhang et al. (2019)
Malathion	AChE/MHCS/GCE AChE/Fe$_3$O$_4$@MHCS/GCE	0.01–$100.0\,mg\,L^{-1}$ and 100–$600\,mg\,L^{-1}$ 0.01–$50.0\,mg\,L^{-1}$ and 50–$600\,mg\,L^{-1}$	$0.0148\,mg\,L^{-1}$ $0.0182\,mg\,L^{-1}$	97.80–104.10	Pear	Luo et al. (2018)
Parathion	Au@Pt nanozyme	0.01–$40.0\,\mu g\,kg^{-1}$	2.13×10–$3\,\mu g\,kg^{-1}$	73.12–116.29	Rice Pear Apple Cabbage	Chen et al. (2020)

(Continued)

TABLE 4.4 (*Continued*)

Electrochemical Analyses Using Nanoparticles as Nanomaterials

Analyte	Nanobiosensor	Linear Range	Limit of Detection (LOD)	Recovery, %	Application in Food	Reference
Methyl parathion	Au–Ag@BSA/GCE	0.02–8.0 and 8–200 mmol L^{-1}	8.2 nmol L^{-1}	95.60–104.00	Soil Apple Cabbage Spinach Lettuce Wastewater River water Tap water	Rahmani et al. (2018)
Dichlorvos	AChE/[BSmim]HSO$_4$-AuNPs-porous carbon/BDD	1×10^{-10}–1×10^{-6} gL^{-1}	6.61×10^{-11} g L^{-1}	80.8–93.1	Lettuce	Wei & Wang (2015)
Dichlorvos	AChE-CS/TiO$_2$/GCE	1.13–22.6 µmol L^{-1}	0.23 nmol L^{-1}	97.4–110.0	Cabbage	Hu et al. (2020)
Chlorpyrifos	AChE/Fe$_3$O$_4$/GR/SPE	0.05–100.0 µg L^{-1}	0.02 µg L^{-1}	91.14–108.3	Cabbage Spinach	Wang et al. (2016)
Chlorpyrifos	MCH/Aptamer/Fc/Au/Mo$_2$C/Mo$_2$N/ GCE	0–400 ng mL^{-1}	0.036 ng mL^{-1}	94.7–116.7	Apple Pakchoi	Lin et al. (2021)
Chlorpyrifos	HRP-AuNP-BSA-CPF	1×10^{-3}–1×10^{-1} ng mL^{-1}	0.070 pg mL^{-1}	85–110	Chinese Cabbage Lettuce	Hou et al. (2020)

[BSmim]HSO$_4$: honeycomb-like hierarchically ion liquids; ATCh: acetylthiocholine; BSA: bovine serum albumin; CPF: chlorpyrifos; CS: chitosan; DA:dopamine; FBThF:4, 7-di (furan-2-yl) benzo thiadiazole; HRP: horseradish peroxidase; MCH: 6-mercapto-1- hexanol; MHCS: mesoporous hollow carbon spheres; MnNS: manganese dioxide nanosheets; Mo2N/Mo$_2$C: Molybdenum nanoparticle; PtPd@NCS: Pt-Pd nanoparticles encapsulated in N-doped carbon shells; S/H/E: serine /histamine /glutamic acid; TiO$_2$:titanium nanoparticle; TMB: tetramethylbenzidine.

4.3.1.4 Nanowire

Nanowires have cylindrical single-crystalline shapes with a diameter of 10–100 nm and a length of several microns. Although they generally have a cylindrical solid tube shape, they can sometimes have a pentagonal symmetry or spiral shape (Demiç, 2011).

In a study reported by Turan and coworkers (2016), they used a conductive polymer (poly(5,6-bis (octyloxy)-4,7-di(thieno[3][3,2-b]thiophen-2-yl)benzo[c][1,2,5]oxoadiazole) (PTTBO) for paraoxon detection. An amperometric butyrylcholinesterase (BChE) biosensor was developed based on silver nanowires and graphite electrodes. The conductive polymer used in the biosensor increased the stability, and silver nanowires increased the biosensor performance. The designed biosensor showed a binaural range of 0.5–8 µmol L^{-1} and 10–120 µmol L^{-1}, with a low LOD of 212 µmol L^{-1} for butyryl thiocholine iodide. Also, the proposed biosensor was used for paraoxon detection, and an LOD of 0.9 µg L^{-1} (3.2×10^{-9} mol L^{-1}) was obtained. In addition, the biosensor was used for paraoxon detection from real tap water and milk samples. Recovery results were obtained between 96.3% and 104% (Turan et al., 2016).

In another study, Li and coworkers (2020) developed a biosensor to determine malathion rapidly. The targeted biosensor was formed by attaching acetylcholinesterase to the glassy carbon electrode by modification of copper nanowires (CuNWs) and graphene oxide-tetraethylenepentamine (rGO-TEPA). CuNWs—rGO-TEPA structure was used to increase the chargeability of acetylcholinesterase to the electrode and improve the electrode's conductivity. Malathion significantly reduces the current by inhibiting acetylcholinesterase activity. Using this sensor, the linear working range was calculated as 1–20 ng mL^{-1}, and the detection limit was calculated as 0.39 ng mL^{-1}, and it was applied to cabbage and carrot samples (Li et al., 2020).

4.3.1.5 Nanorods

One-dimensional nanomaterials are synthesized using metals such as Au, Ag, Zn, Ti, etc. It has a rigid structure. Nanorods facilitate electron transfer. Thanks to these features, it is used in analysis. Its optical and photonic properties also offer various uses (Makvandi et al., 2021; Naidu et al., 2021).

In a study published in 2019, Cui and his colleagues developed a sensitive electrochemical acetylcholinesterase biosensor in which Au nanorods (AuNRs)@mesoporous SiO$_2$ (MS) core-shell nanoparticles were doped into the CS/TiO$_2$-CS matrix for the analysis of organophosphorus pesticides. The developed nanobiosensor was characterized using scanning electron microscopy and electrochemical techniques, and doping conditions were optimized. In this way, an increase in electroconductivity and bioelectrocatalytic activity was observed. The linear working range of the method was determined as 0.018–13.6 µmol L^{-1}. LOD values were 1.3 and 5.3 nmol L^{-1} for fenthion and dichlorvos, respectively. Accuracy and repeatability studies were performed on fruit juice samples. As a result, it has been stated that this developed nanobiosensor has high stability and sensitivity for analyzing these pesticides, and analyses are performed with good precision and accuracy (Cui et al., 2019).

4.3.2 Optical Methods

4.3.2.1 Carbon Nanotubes

In an aptamer-based study, a turn-on sensor was developed for monitoring and quantifying acetamiprid (Lin, 2016). The structure of this sensor consists of an acetamiprid-aptamer-modified ZnS:Mn probe. The synthesized aptamer sensor has low toxicity and excellent fluorescence properties. The probe's fluorescence is turned off by adding carbon nanotubes to the medium. When acetamiprid is added to the medium, the fluorescence is turned on as the probe-aptamer will specifically bind with this substance. The developed sensor calculated the linear working range as 0–150 nmol L^{-1}, and the detection limit was 0.7 nmol L^{-1}. The method has been successfully applied to determine acetamiprid from river water and cabbage leaves (Lin et al., 2016).

4.3.2.2 Nanoparticles

Nanoparticles are frequently used in optical nanobiosensors. The gold nanoparticle is the most used. Optical nanobiosensors using nanoparticles as nanomaterials are listed in Table 4.5.

Bala et al. developed a high-sensitivity, low-cost, and fast colorimetric method for determining malathion from various samples, using aptamer, cationic peptide, and silver nanoparticles for biosensor purposes. Malathion binds to the hexapeptide in the medium with high affinity. When an oligonucleotide is added as an aptamer, the aptamer combines with malathion, releasing the peptide. The released peptide from complex aggregates with the silver nanoparticles and the solution color changes from yellow to orange. The intensity of this color change was measured at 390 and 520 nm, depending on the pesticide concentration. The study calculated the % recovery and relative standard deviation values as 89%–120% and 2.98%–4.78%, respectively. The developed biosensor has been successfully applied by the spike method determination of malathion from lake water, tap water, and apple samples (Bala et al., 2018).

4.3.2.3 Nanoclusters

Transition metal nanoclusters have small particle sizes. In this way, it has very active surface atoms. Its size is less than 2 nm. It is frequently used to design fluorescent probes due to its molecule-like properties (Kaboggoza, 2019).

In a study Sharma (2021) in which bimetallic structures were used to determine pesticides, an enzyme-based fluorimetric measurement technique was developed for ethyl parathion's simple, inexpensive, and highly sensitive determination. The structure in this study is bimetallic BSA@ AuAg nanoclusters (NC). The fluorescence, which decreases with the addition of Cu^{2+} to the medium, is switched on again with the help of thiocholine, the catalytic product produced as a result of the enzymatic reaction of acetylcholine using the acetylcholinesterase enzyme. Ethyl parathion added to the medium suppresses the formation of thiocholine by inhibiting the acetylcholine esterase enzyme. It reduces the thiocholine – Cu complex formation, thus reversing the quenched property of bimetallic nanoclusters. The detection limit was calculated as 2.40 pmol L^{-1} in this study. In the study, it was mentioned that the bimetallic structure showed better optoelectronic and physical properties. This method, which has specificity studies carried out using other pesticides, has been successfully applied to analyze this pesticide in tap water, spring water, and orange and apple juice (Sharma et al., 2021).

4.3.2.4 Quantum Dots

Quantum dots (QD) are crystalline nanomaterials, generally carbon and graphene-based, with a size between 1 and 10 nm. Optical properties may vary depending on size. It can glow in different colors (Sağlam, 2017).

In a study published in 2020, Korram and coworkers developed a glutathione-capped CdTe quantum dot-based fluorescent sensor to determine organophosphate pesticides. The basis of the study is that GSH-capped CdTe QDs are sensitive to H_2O_2 produced as a result of the enzymatic reaction of acetylcholine esterase and choline oxidase, and as a result of this interaction, fluorescence turn-off formation of CdTe QDs-GSH-capped. Turn-on fluorescence occurs in the presence of organophosphorus at 520 nm. Detection limits obtained using this sensor are 2.30×10^{-10} to 7.53×10^{-16} mol L^{-1} for pesticides such as paraoxon, dichlorvos, malathion, and triazophos. This limit is lower than other methods reported in the literature, especially for paraoxon. The developed biosensor was also used as a fluorescence nanoprobe for the enzyme activity of acetylcholinesterase, and it was mentioned in the study that it could be used for the analysis of these pesticides in food, water, and environmental samples (Korram et al., 2020).

In another study using quantum dots, Sun and coworkers (2011) developed an optical nanobiosensor for detecting monocrotophos by layer-by-layer bonding of enzymes and nanoparticles using chitosan. Acetylcholinesterase and CdTe were used as recognition elements, and CdTe was used as

TABLE 4.5

Optical Analyzes Using Nanoparticles as Nanomaterials

Analyte	Nanobiosensor	Linear Range	Limit of Detection (LOD)	Recovery, %	Application in Food	Reference
Methyl parathion	Fluorimetric/DNA–Tb–dured–AgNPs	$0.03–5.0\,mg\,L^{-1}$	$0.034\,mg\,L^{-1}$	97.0–102.0	Apple Tap water	Wang and Liu (2021)
Methyl parathion	TRY-PRO-AuNPs/RF-QDs	$0.04–400.0\,ng\,mL^{-1}$	$0.018\,ng\,mL^{-1}$	95–110	Tap water Milk Rice	Yan et al. (2015)
Acetamiprid	Colorimetric/ACA-CV-GNPs	$3.0\times10^{-8}–4.0\times10^{-6}\,mol\,L^{-1}$	$1.76\times10^{-8}\,mol\,L^{-1}$	98.45–104.5	Tea	Li et al. (2019)
Malathion	Colorimetric/Aptamer -AuNP$_s$- KKKRRR	$0.01–0.75\,nmol\,L^{-1}$	$0.5\,pmol\,L^{-1}$	89–120	Lake water Tap water Apple	Bala et al. (2018)
Acetamiprid	Fluorimetric/ ABA- (dsDNA- AuNPs)+(DNA-UCNPs)	$0.025–1.0\,\mu mol\,L^{-1}$	$0.36\,nmol\,L^{-1}$	96.3–99.7	Celery leaves Chinese green tea	Yang et al. (2019)
Carbofuran, Oxamyl Methomyl Carbaryl	FH–A–AChE–GNPs	$2–10\,nmol\,L^{-1}$ $10–100\,nmol\,L^{-1}$ $100–500\,nmol\,L^{-1}$ $200–1,000\,nmol\,L^{-1}$	$2\,nmol\,L^{-1}$ $21\,nmol\,L^{-1}$ $113\,nmol\,L^{-1}$ $236\,nmol\,L^{-1}$	93–105	Orange Tomato Prunes Cabbage	Kestwal et al. (2015)

A: agarose; ABA: acetamiprid-binding aptamer; ACA: acetamiprid aptamer; CV: crystal violet; GNPs: aptamer decorated gold nanoparticles; FH: fenugreek hydrogel; KKKRRR: hexapeptide; PRO: protamine; RF: ratiometric fluorescent; Tb: terbium; TRY: trypsin; UCNPs: upconversion nanoparticles.

a pH indicator. In the absence of monocrotophos, acetylcholine has been hydrolyzed to acetic acid and choline, and there has been a pH drop that CdTe could detect due to the increased amount of acid in the environment. In the presence of monocrotophos, the effect of acetylcholine decreased, and there was a change in fluorescence intensity. The developed nanobiosensor exhibited a linear range of 8.95×10^{-8} to 6.09×10^{-7} mol L^{-1} with a LOD of 3.20×10^{-8} mol L^{-1} (Sun et al., 2011).

4.4 CONCLUSION

This study mentions nanobiosensors with different structures and analytical technique applications used in pesticide analysis. When the literature is examined, it can be seen that the aim is to create sensors with low detection limits, selective, fast, and good accuracy and precision. It is seen that by increasing the analysis surfaces, especially by using hybrid structures, a higher response signal can be obtained, and thus, lower concentrations of pesticide residues can be determined. Aptamers and enzymes, which are biological molecules used in the structure of nanobiosensors and can be attached to the sensor surface by different methods, appear to increase selectivity and specificity.

When literature studies are examined, pesticide analyses are generally performed on raw fruits and vegetables, water, and fruit juices using nanobiomaterials. However, the presence of pesticides in animal foods should also be considered. It is known that animals can consume food and water containing pesticides, and when people consume animal foods such as meat and milk, they can take pesticides into their bodies. When the studies are examined, it is seen that the literature is limited on the detection of pesticides in animal products using nanobiomaterials.

GC and HPLC are among the current analytical techniques for detecting pesticides. GC and HPLC techniques are sensitive but require complex procedures and expensive equipment. Before this analysis, complex sample pretreatment techniques are needed, which is time-consuming. Another disadvantage is that the derivatization technique should be used in GC analysis for non-volatile compounds. Electrochemical and optical analyses using nanobiomaterials have lower detection limits, simple sample preparation methods, cheaper equipment, and no need for any derivatization related to compound structure compared to current analytical techniques.

In the future, producing more selective and lower-limit nanobiosensors combined with varying systems of detector and analytical techniques that can be used in the analysis of pesticides, which have an essential place in sustainable food policies, is among the goals. In addition, it may be aimed to increase the turn of these sensors into chips, especially for on-site analysis.

ACKNOWLEDGMENT

Nuran Gokdere thanks the financial support from the Technological Research Council of Turkey (TUBITAK) under the BIDEB/2211-A Ph.D.

BIBLIOGRAPHY

Aghdam, M. B., Farajzadeh, M. A., & Mogaddam, M. R. A. (2023). Fabric phase sorptive extraction combined with dispersive liquid-liquid extraction for the extraction of some pesticide residues from fruit juices using partially carbonized cotton textile followed by gas chromatography-flame ionization detector. *Journal of Food Composition and Analysis, 124*, 105625. https://doi.org/10.1007/s10311-021-01353-1

Arduini, F., Cinti, S., Scognamiglio, V., & Moscone, D. (2016). Nanomaterials in electrochemical biosensors for pesticide detection: advances and challenges in food analysis. *Microchimica Acta, 183*, 2063–2083. https://doi.org/10.1007/s00604-016-1858-8.

Atapattu, S. N., & Johnson, K. R. (2020). Pesticide analysis in cannabis products. *Journal of Chromatography A, 1612*, 460656. https://doi.org/10.1016/j.chroma.2019.460656. https://doi.org/10.1016/j.chroma.2019.460656.

Badr, A. M. (2020). Organophosphate toxicity: Updates of malathion potential toxic effects in mammals and potential treatments. *Environmental Science and Pollution Research, 27*(21), 26036–26057. https://doi.org/10.1007/s11356-020-08937-4.

Bajwa, U., & Sandhu, K. S. (2014). Effect of handling and processing on pesticide residues in food-a review. *Journal of Food Science and Technology, 51*, 201–220. https://doi.org/10.1007/s13197-011-0499-5.

Bala, R., Mittal, S., Sharma, R. K., & Wangoo, N. (2018). A supersensitive silver nanoprobe based apta-sensor for low cost detection of malathion residues in water and food samples. *Spectrochimica Acta Part A: Molecular and Biomolecular Spectroscopy, 196*, 268–273. https://doi.org/10.1016/j.saa.2018.02.007.

Bao, J., Huang, T., Wang, Z., Yang, H., Geng, X., Xu, G., ... & Hou, C. (2019). 3D graphene/copper oxide nano-flowers based acetylcholinesterase biosensor for sensitive detection of organophosphate pesticides. *Sensors and Actuators B: Chemical, 279*, 95–101. https://doi.org/10.1016/j.snb.2018.09.118.

Baranowska, I., Barchańska, H., & Pacak, E. (2006). Procedures of trophic chain samples preparation for determination of triazines by HPLC and metals by ICP-AES methods. *Environmental Pollution, 143*(2), 206–211. https://doi.org/10.1016/j.envpol.2005.11.039.

Binukumar, B. K., & Gill, K. D. (2010). Cellular and molecular mechanisms of dichlorvos neurotoxicity: cholinergic, noncholinergic, cell signaling, gene expression and therapeutic aspects. https://nopr.niscpr.res.in/handle/123456789/9737.

Blacker, A. M., Lunchick, C., Lasserre-Bigot, D., Payraudeau, V., & Krolski, M. E. (2010). Toxicological profile of carbaryl. In *Hayes' Handbook of Pesticide Toxicology* (pp. 1607–1617). Academic Press. https://doi.org/10.1016/B978-0-12-374367-1.00074-4.

Büyük, M. (2017). Karbon nanotüp üretiminde metal katalizörün taban malzemeye tutunma etkileri (Master's thesis, Fen Bilimleri Enstitüsü).

Caparco, A. A., González-Gamboa, I., Hays, S. S., Pokorski, J. K., & Steinmetz, N. F. (2023). Delivery of nematicides using TMGMV-derived spherical nanoparticles. *Nano Letters, 23*(12), 5785–5793. https://doi.org/10.1021/acs.nanolett.3c01684.

Cesarino, I., Moraes, F. C., Lanza, M. R., & Machado, S. A. (2012). Electrochemical detection of carbamate pesticides in fruit and vegetables with a biosensor based on acetylcholinesterase immobilised on a composite of polyaniline-carbon nanotubes. *Food Chemistry, 135*(3), 873–879. https://doi.org/10.1016/j.foodchem.2012.04.147.

Chai, Y., Niu, X., Chen, C., Zhao, H., & Lan, M. (2013). Carbamate insecticide sensing based on acetylcholinesterase/Prussian blue-multi-walled carbon nanotubes/screen-printed electrodes. *Analytical Letters, 46*(5), 803–817. https://doi.org/10.1080/00032719.2012.733899.

Chang, J., Fang, W., Chen, L., Zhang, P., Zhang, G., Zhang, H., ... & Ma, W. (2022). Toxicological effects, environmental behaviors and remediation technologies of herbicide atrazine in soil and sediment: A comprehensive review. *Chemosphere, 307*, 136006. https://doi.org/10.1016/j.chemosphere.2022.136006.

Chen, D., Fu, J., Liu, Z., Guo, Y., Sun, X., Wang, X., & Wang, Z. (2017). A Simple acetylcholinesterase biosensor based on ionic liquid/multiwalled carbon nanotubes-modified screen-printed electrode for rapid detecting chlorpyrifos. *International Journal of Electrochemical Science, 12*(10), 9465–9477. https://doi.org/10.20964/2017.10.12.

Chen, G., Jin, M., Ma, J., Yan, M., Cui, X., Wang, Y., ... Wang, J. (2020). Competitive bio-barcode immunoassay for highly sensitive detection of parathion based on bimetallic nanozyme catalysis. *Journal of Agricultural and Food Chemistry, 68*(2), 660–668. https://doi.org/10.1021/acs.jafc.9b06125.

Chwatko, G., Krawczyk, M., Iciek, M., Kamińska, A., Bilska-Wilkosz, A., Marcykiewicz, B., & Głowacki, R. (2019). Determination of lipoic acid in human plasma by high-performance liquid chromatography with ultraviolet detection. *Arabian Journal of Chemistry, 12*(8), 4878–4886. https://doi.org/10.1016/j.arabjc.2016.10.006.

Cui, H. F., Zhang, T. T., Lv, Q. Y., Song, X., Zhai, X. J., & Wang, G. G. (2019). An acetylcholinesterase biosensor based on doping Au nanorod@ SiO_2 nanoparticles into TiO_2-chitosan hydrogel for detection of organophosphate pesticides. *Biosensors and Bioelectronics, 141*, 111452. https://doi.org/10.1016/j.bios.2019.111452.

Demiç, Ö. (2011). *III-V bileşik yarıiletkenlerle nanotel üretilmesi ve bazı karakteristikleri* (Master's thesis, Gaziosmanpaşa Üniversitesi, Fen Bilimleri Enstitüsü).

Dong, P., Jiang, B., & Zheng, J. (2019). A novel acetylcholinesterase biosensor based on gold nanoparticles obtained by electroless plating on three-dimensional graphene for detecting organophosphorus pesticides in water and vegetable samples. *Analytical Methods, 11*(18), 2428–2434. https://doi.org/10.1039/C9AY00549H.

Du, L., Wang, X., Liu, T., Li, J., Wang, J., Gao, M., & Wang, H. (2019). Magnetic solid-phase extraction of organophosphorus pesticides from fruit juices using $NiFe_2O_4$@ polydopamine@ Mg/Al-layered double hydroxides nanocomposites as an adsorbent. *Microchemical Journal, 150*, 104128. https://doi.org/10.1016/j.microc.2019.104128.

Eaton, D. L., Daroff, R. B., Autrup, H., Bridges, J., Buffler, P., Costa, L. G., … Spencer, P. S. (2008). Review of the toxicology of chlorpyrifos with an emphasis on human exposure and neurodevelopment. *Critical Reviews in Toxicology*, *38*(sup2), 1–125. https://doi.org/10.1080/10408440802272158.

Farizatto, K. L., & Bahr, B. A. (2017). Paraoxon: An anticholinesterase that triggers an excitotoxic cascade of oxidative stress, adhesion responses, and synaptic compromise. *European Scientific Journal*, *13*, 29. https://doi.org/10.19044/esj.2017.c1p4.

Firdoz, S., Ma, F., Yue, X., Dai, Z., Kumar, A., & Jiang, B. (2010). A novel amperometric biosensor based on single walled carbon nanotubes with acetylcholine esterase for the detection of carbaryl pesticide in water. *Talanta*, *83*(1), 269–273. https://doi.org/10.1016/j.talanta.2010.09.028.

Gan, N., Yang, X., Xie, D., Wu, Y., & Wen, W. (2010). A disposable organophosphorus pesticides enzyme biosensor based on magnetic composite nanoparticles modified screen printed carbon electrode. *Sensors*, *10*(1), 625–638. https://doi.org/10.3390/s100100625.

Garcia, S., Abu-Qare, A., Meeker-O'Connell, W., Borton, A., & Abou-Donia, M. (2003). Methyl parathion: A review of health effects. *Journal of Toxicology and Environmental Health Part B: Critical Reviews*, *6*(2), 185–210. https://doi.org/10.1080/10937400306471.

Gorji, S., Biparva, P., Bahram, M., & Nematzadeh, G. (2019). Stir bar sorptive extraction kit for determination of pesticides in water samples with chemometric data processing. *Microchemical Journal*, *148*, 313–321. https://doi.org/10.1016/j.microc.2019.04.056.

Güven, O. C., & Kar, M. (2022). *Bakır (Cu) nanoçiçeklerinin yeşil sentez yöntemi kullanılarak sentezlenmesi, karakterizasyonu, fotokatalitik, antimikrobiyal ve antioksidan aktivitesinin belirlenmesi* (Master's thesis, Nevşehir Hacı Bektaş Veli Üniversitesi).

Han, C., Hu, B., Jin, N., Jin, J., Yu, Z., Huang, C., & Shen, Y. (2021). Accelerated solvent extraction-gel permeation chromatography-gas chromatography-tandem mass spectrometry to rapidly detect clotrimazole residue in animal-derived food. *LWT*, *144*, 111248. https://doi.org/10.1016/j.lwt.2021.111248.

Han, L., Chen, D., & Li, F. (2018). Rational integration of biomineralization, microbial surface display, and carbon nanocomposites: Ultrasensitive and selective biosensor for traces of pesticides. *Advanced Materials Interfaces*, *5*(24), 1801332. https://doi.org/10.1002/admi.201801332.

Hogendoorn, E., & van Zoonen, P. (2000). Recent and future developments of liquid chromatography in pesticide trace analysis. *Journal of Chromatography A*, *892*(1–2), 435–453. https://doi.org/10.1016/S0021-9673(00)00151-5.

Hou, L., Zhang, X., Kong, M., Jiang, G., Sun, Y., Mo, W., … Zhao, S. (2020). A competitive immunoassay for electrochemical impedimetric determination of chlorpyrifos using a nanogold-modified glassy carbon electrode based on enzymatic biocatalytic precipitation. *Microchimica Acta*, *187*, 1–9. https://doi.org/10.1007/s00604-020-4175-1.

Hu, H., Wang, B., Li, Y., Wang, P., & Yang, L. (2020). Acetylcholinesterase sensor with patterned structure for detecting organophosphorus pesticides based on titanium dioxide sol-gel carrier. *Electroanalysis*, *32*(8), 1834–1842. https://doi.org/10.1002/elan.202060027.

Jiang, B., Lu, M., & Xu, M. (2019). Amperometric sensing of organophosphorus pesticides based on covalently attached multilayer assemblies of diazo-resin, Prussian blue single-walled carbon nanotubes, and acetyl cholinesterase. *Revue Roumaine de Chimie*, *64*(9), 763–774.

Jiao, Y., Jia, H., Guo, Y., Zhang, H., Wang, Z., Sun, X., Zhao, J. (2016). An ultrasensitive aptasensor for chlorpyrifos based on ordered mesoporous carbon/ferrocene hybrid multiwalled carbon nanotubes. *RSC Advances*, *6*(63), 58541–58548. https://doi.org/10.1039/C6RA07735H.

Kaboggoza, Herbert Cirrus (2019). *Peroksil radikal süpürme etkinliği ölçümü için altın nanoküme esaslı yöntem geliştirilmesi* (Master's thesis. İstanbul Universitesi Fen Bilimleri Enstitüsü).

Kalogiouri, N. P., Papadakis, E. N., Maggalou, M. G., Karaoglanidis, G. S., Samanidou, V. F., & Menkissoglu-Spiroudi, U. (2021). Development of a microwave-assisted extraction protocol for the simultaneous determination of mycotoxins and pesticide residues in apples by LC-MS/MS. *Applied Sciences*, *11*(22), 10931. https://doi.org/10.3390/app112210931.

Kamacı, Y. (2019). *Etkili yanma yöntemiyle metal oksit nanopartiküllerin sentezi* (Master's thesis, Bartın Üniversitesi, Fen Bilimleri Enstitüsü).

Kaur, N., Thakur, H., & Prabhakar, N. (2016). Conducting polymer and multiwalled carbon nanotubes nanocomposites based amperometric biosensor for detection of organophosphate. *Journal of Electroanalytical Chemistry*, *775*, 121–128. https://doi.org/10.1016/j.jelechem.2016.05.037.

Kesik, M., Kanik, F. E., Turan, J., Kolb, M., Timur, S., Bahadır, M., & Toppare, L. (2014). An acetylcholinesterase biosensor based on a conducting polymer using multiwalled carbon nanotubes for amperometric detection of organophosphorous pesticides. *Sensors and Actuators B: Chemical*, *205*, 39–49. https://doi.org/10.1016/j.snb.2014.08.058.

Kestwal, R. M., Bagal-Kestwal, D., & Chiang, B. H. (2015). Fenugreek hydrogel-agarose composite entrapped gold nanoparticles for acetylcholinesterase based biosensor for carbamates detection. *Analytica Chimica Acta, 886,* 143–150. https://doi.org/10.1016/j.aca.2015.06.004.

Kim, J. W., & Kim, D. S. (2020). Paraquat: Toxicology and impacts of its ban on human health and agriculture. *Weed Science, 68*(3), 208–213. https://doi.org/10.1017/wsc.2019.70.

Korram, J., Dewangan, L., Karbhal, I., Nagwanshi, R., Vaishanav, S. K., Ghosh, K. K., & Satnami, M. L. (2020). CdTe QD-based inhibition and reactivation assay of acetylcholinesterase for the detection of organophosphorus pesticides. *RSC Advances, 10*(41), 24190–24202. https://doi.org/10.1039/D0RA03055D.

LeDoux, M. (2011). Analytical methods applied to the determination of pesticide residues in foods of animal origin. A review of the past two decades. *Journal of Chromatography A, 1218*(8), 1021–1036. https://doi.org/10.1016/j.chroma.2010.12.097.

Li, H., Hu, W., Hassan, M. M., Zhang, Z., & Chen, Q. (2019). A facile and sensitive SERS-based biosensor for colormetric detection of acetamiprid in green tea based on unmodified gold nanoparticles. *Journal of Food Measurement and Characterization, 13,* 259–268. https://doi.org/10.1007/s11694-018-9940-z.

Li, S., Qu, L. M., Wang, J. F., Ran, X. Q., & Niu, X. (2020). Acetylcholinesterase based rGO-TEPA-Copper nanowires biosensor for detecting malathion. *International Journal of Electrochemical Science, 15*(1), 505–514. https://doi.org/10.20964/2020.01.75.

Lin, B., Yu, Y., Li, R., Cao, Y., & Guo, M. (2016). Turn-on sensor for quantification and imaging of acetamiprid residues based on quantum dots functionalized with aptamer. *Sensors and Actuators B: Chemical, 229,* 100–109. https://doi.org/10.1016/j.snb.2016.01.114.

Lin, Z., Liu, X., Li, Y., Li, C., Yang, L., Ma, K., … Huang, H. (2021). Electrochemical aptasensor based on Mo$_2$C/Mo$_2$N and gold nanoparticles for determination of chlorpyrifos. *Microchimica Acta, 188*(5), 170. https://doi.org/10.1007/s00604-021-04830-0.

Liu, J., Ji, C., Liu, X., Li, X., Wu, H., & Zeng, D. (2021). Fe$_3$O$_4$ nanoparticles as matrix solid-phase dispersion extraction adsorbents for the analysis of thirty pesticides in vegetables by ultrahigh-performance liquid chromatography-tandem mass spectrometry. *Journal of Chromatography B, 1165,* 122532. https://doi.org/10.1016/j.jchromb.2021.122532.

Luo, R., Feng, Z., Shen, G., Xiu, Y., Zhou, Y., Niu, X., & Wang, H. (2018). Acetylcholinesterase biosensor based on mesoporous hollow carbon spheres/core-shell magnetic nanoparticles-modified electrode for the detection of organophosphorus pesticides. *Sensors, 18*(12), 4429. https://doi.org/10.3390/s18124429.

Ma, L., Zhou, L., He, Y., Wang, L., Huang, Z., Jiang, Y., & Gao, J. (2018). Hierarchical nanocomposites with an N-doped carbon shell and bimetal core: Novel enzyme nanocarriers for electrochemical pesticide detection. *Biosensors and Bioelectronics, 121,* 166–173. https://doi.org/10.1016/j.bios.2018.08.038.

Makvandi, P., Zarepour, A., Zheng, X., Agarwal, T., Ghomi, M., Sartorius, R., … Mattoli, V. (2021). Non-spherical nanostructures in nanomedicine: From noble metal nanorods to transition metal dichalcogenide nanosheets. *Applied Materials Today, 24,* 101107. https://doi.org/10.1016/j.apmt.2021.101107.

Malik, P., Katyal, V., Malik, V., Asatkar, A., Inwati, G., & Mukherjee, T. K. (2013). Nanobiosensors: concepts and variations. International Scholarly Research Notices, *2013,* 327435. https://doi.org/10.1155/2013/327435.

Mirres, A. C. D. M., Silva, B. E. P. D. M. D., Tessaro, L., Galvan, D., Andrade, J. C. D., Aquino, A., … & Conte-Junior, C. A. (2022). Recent advances in nanomaterial-based biosensors for pesticide detection in foods. *Biosensors, 12*(8), 572. https://doi.org/10.3390/bios12080572

Naidu, K. C. B., Kumar, N. S., Banerjee, P., & Reddy, B. V. S. (2021). A review on the origin of nanofibers/nanorods structures and applications. *Journal of Materials Science: Materials in Medicine, 32*(6), 68. https://doi.org/10.1007/s10856-021-06541-7.

Phogat, A., Singh, J., Kumar, V., & Malik, V. (2022). Toxicity of the acetamiprid insecticide for mammals: A review. *Environmental Chemistry Letters, 20,* 1–26. https://doi.org/10.1007/s10311-021-01353-1.

Piryaei, M., Mahdi Abolghasemi, M., Zahedi, E., & Torabbeigi, M. (2023). Nonporous grapheneoxide coated by acrylonitrile polystyrene as new adsorbent in dispersed solid phase microextraction for estimating pesticides in aqueous samples. *ChemistrySelect, 8*(27), e202300123. https://doi.org/10.1002/slct.202300123.

Qi, P., Wang, J., Liu, Z., Zhao, H., Wang, Z., Di, S., & Wang, X. (2023). Fabrication of poly-dopamine-modified magnetic nanomaterial and development of integrated QuEChERS method for 122 pesticides residue analysis in fruits. *Journal of Chromatography A, 1708,* 464336. https://doi.org/10.1016/j.chroma.2023.464336.

Qiu, L., Lv, P., Zhao, C., Feng, X., Fang, G., Liu, J., & Wang, S. (2019). Electrochemical detection of organophosphorus pesticides based on amino acids conjugated nanoenzyme modified electrodes. *Sensors and Actuators B: Chemical, 286,* 386–393. https://doi.org/10.1016/j.snb.2019.02.007.

Raghu, H. V., Parkunan, T., & Kumar, N. (2020). Application of nanobiosensors for food safety monitoring. *Environmental Nanotechnology, 4,* 93–129. https://doi.org/10.1007/978-3-030-26668-4_3.

Rahmani, T., Hajian, A., Afkhami, A., & Bagheri, H. (2018). A novel and high performance enzyme-less sensing layer for electrochemical detection of methyl parathion based on BSA templated Au-Ag bimetallic nanoclusters. *New Journal of Chemistry, 42*(9), 7213–7222. https://doi.org/10.1039/C8NJ00425K.

Sağlam, M. E. (2017). *Suda çözünebilen kuantum noktaların sentezi ve sensör uygulamaları* (Master's thesis, Necmettin Erbakan Üniversitesi Fen Bilimleri Enstitüsü).

Sharma, D., Wangoo, N., & Sharma, R. K. (2021). Sensing platform for pico-molar level detection of ethyl parathion using Au-Ag nanoclusters based enzymatic strategy. *Talanta, 221*, 121267. https://doi.org/10.1016/j.talanta.2020.121267.

Sireesha, M., Jagadeesh Babu, V., Kranthi Kiran, A. S., & Ramakrishna, S. (2018). A review on carbon nanotubes in biosensor devices and their applications in medicine. *Nanocomposites, 4*(2), 36–57. https://doi.org/10.1080/20550324.2018.1478765.

Sobhanzadeh, E., Bakar, N. K. A., Abas, M. R., & Nemati, K. (2009). Sample preparation methods for pesticides analysis in food matrices and environmental samples by chromatography-based techniques: A review. *Malaysian Journal of Fundamental and Applied Sciences, 5*(2), 923–950. https://doi.org/10.11113/mjfas.v5n2.294.

Sun, X., Liu, B., & Xia, K. (2011). A sensitive and regenerable biosensor for organophosphate pesticide based on self-assembled multilayer film with CdTe as fluorescence probe. *Luminescence, 26*(6), 616–621. https://doi.org/10.1002/bio.1284.

Tao, Y., Dong, F., Xu, J., Liu, X., Cheng, Y., Liu, N., … Zheng, Y. (2014). Green and sensitive supercritical fluid chromatographic-tandem mass spectrometric method for the separation and determination of flutriafol enantiomers in vegetables, fruits, and soil. *Journal of Agricultural and Food Chemistry, 62*(47), 11457–11464. https://doi.org/10.1021/jf504324t.

Tudi, M., Daniel Ruan, H., Wang, L., Lyu, J., Sadler, R., Connell, D., … Phung, D. T. (2021). Agriculture development, pesticide application and its impact on the environment. *International Journal of Environmental Research and Public Health, 18*(3), 1112. https://doi.org/10.3390/ijerph18031112.

Turan, J., Kesik, M., Soylemez, S., Goker, S., Coskun, S., Unalan, H. E., & Toppare, L. (2016). An effective surface design based on a conjugated polymer and silver nanowires for the detection of paraoxon in tap water and milk. *Sensors and Actuators B: Chemical, 228*, 278–286. https://doi.org/10.1016/j.snb.2016.01.034.

Wang, H., Zhao, G., Chen, D., Wang, Z., & Liu, G. (2016). A sensitive acetylcholinesterase biosensor based on screen printed electrode modified with Fe3O4 nanoparticle and graphene for chlorpyrifos determination. *International Journal of Electrochemical Science, 11*, 10906–10918. https://doi.org/10.20964/2016.12.90.

Wang, T., & Liu, Y. (2021). A lanthanide-based ratiometric fluorescent biosensor for the enzyme-free detection of organophosphorus pesticides. *Analytical Methods, 13*(17), 2005–2010. https://doi.org/10.1039/D1AY00345C.

Wei, M., & Wang, J. (2015). A novel acetylcholinesterase biosensor based on ionic liquids-AuNPs-porous carbon composite matrix for detection of organophosphate pesticides. *Sensors and Actuators B: Chemical, 211*, 290–296. https://doi.org/10.1016/j.snb.2015.01.112.

World Health Organization. (1999). *Pesticide residues in food: 1998: toxicological evaluations* (No. WHO/PCS/99.18). World Health Organization.

Wu, J., Yang, Q., Li, Q., Li, H., & Li, F. (2021). Two-dimensional MnO_2 nanozyme-mediated homogeneous electrochemical detection of organophosphate pesticides without the interference of H_2O_2 and color. *Analytical Chemistry, 93*(8), 4084–4091. https://doi.org/10.1021/acs.analchem.0c05257.

Xu, G., Huo, D., Hou, C., Zhao, Y., Bao, J., Yang, M., & Fa, H. (2018). A regenerative and selective electrochemical aptasensor based on copper oxide nanoflowers-single walled carbon nanotubes nanocomposite for chlorpyrifos detection. *Talanta, 178*, 1046–1052. https://doi.org/10.1016/j.talanta.2017.08.086.

Xu, M., Jiang, S., Jiang, B., & Zheng, J. (2019). Organophosphorus pesticides detection using acetylcholinesterase biosensor based on gold nanoparticles constructed by electroless plating on vertical nitrogen-doped single-walled carbon nanotubes. *International Journal of Environmental Analytical Chemistry, 99*(10), 913–927. https://doi.org/10.1080/03067319.2019.1616714.

Yan, X., Li, H., Han, X., & Su, X. (2015). A ratiometric fluorescent quantum dots based biosensor for organophosphorus pesticides detection by inner-filter effect. *Biosensors and Bioelectronics, 74*, 277–283. https://doi.org/10.1016/j.bios.2015.06.020.

Yang, G., He, Y., Zhao, J., Chen, S., & Yuan, R. (2021). Ratiometric electrochemiluminescence biosensor based on Ir nanorods and CdS quantum dots for the detection of organophosphorus pesticides. *Sensors and Actuators B: Chemical, 341*, 130008. https://doi.org/10.1016/j.snb.2021.130008.

Yang, L., Sun, H., Wang, X., Yao, W., Zhang, W., & Jiang, L. (2019). An aptamer based aggregation assay for the neonicotinoid insecticide acetamiprid using fluorescent upconversion nanoparticles and DNA functionalized gold nanoparticles. *Microchimica Acta, 186*, 1–11. https://doi.org/10.1007/s00604-019-3422-9.

Yang, T., Doherty, J., Zhao, B., Kinchla, A. J., Clark, J. M., & He, L. (2017). Effectiveness of commercial and homemade washing agents in removing pesticide residues on and in apples. *Journal of Agricultural and Food Chemistry, 65*(44), 9744–9752.

Yavuz, O., & Aksoy, A. (2016). Pestisit analizlerinde kullanılan metotlar. *Turkiye Klinikleri Veterinary Sciences-Pharmacology and Toxicology-Special Topics, 2*, 89–100.

Zhang, P., Sun, T., Rong, S., Zeng, D., Yu, H., Zhang, Z., … Pan, H. (2019). A sensitive amperometric AChE-biosensor for organophosphate pesticides detection based on conjugated polymer and Ag-rGO-NH$_2$ nanocomposite. *Bioelectrochemistry, 127*, 163–170. https://doi.org/10.1016/j.bioelechem.2019.02.003.

Zhang, Y., Arugula, M. A., Wales, M., Wild, J., & Simonian, A. L. (2015). A novel layer-by-layer assembled multi-enzyme/CNT biosensor for discriminative detection between organophosphorus and non-organophosphrus pesticides. *Biosensors and Bioelectronics, 67*, 287–295. https://doi.org/10.1016/j.bios.2014.08.036.

Zhao, G., Zhou, B., Wang, X., Shen, J., & Zhao, B. (2021). Detection of organophosphorus pesticides by nano-gold/mercaptomethamidophos multi-residue electrochemical biosensor. *Food Chemistry, 354*, 129511. https://doi.org/10.1016/j.foodchem.2021.129511.

Zheng, Q., Yu, Y., Fan, K., Ji, F., Wu, J., & Ying, Y. (2016). A nano-silver enzyme electrode for organophosphorus pesticide detection. *Analytical and Bioanalytical Chemistry, 408*, 5819–5827. https://doi.org/10.1007/s00216-016-9694-6.

Zioga, E., White, B., & Stout J. C. (2023). Honey bees and bumble bees may be exposed to pesticides differently when foraging on agricultural areas. *Science of the Total Environment, 896*, 166214. https://doi.org/10.1016/j.scitotenv.2023.166214.

5 Analysis of Organic Contaminants

*Shazia Chohan, Sanam Iram Soomro,
Najma Memon, and Hüseyin Kara*

5.1 INTRODUCTION

Organic contaminants constitute a diverse grouping of chemical compounds, generating increasing concern within various environmental and industrial spheres. These compounds, predominantly carbon atoms, exhibit molecular structures and can originate from natural sources and human activities. The rising interest in organic contaminants stems from their potential to adversely impact ecosystems, human well-being, and the broader environmental milieu. These substances are ubiquitously distributed across various environments, encompassing aquatic systems, terrestrial substrates, atmospheric matrices, and the food chain. The environment gets contaminated via mechanisms such as industrial effluents, agricultural runoff, wastewater discharges, and the utilization of pesticides and herbicides. The spectrum of organic contaminants encompasses a wide range of farm chemicals, industrial agents, pharmaceuticals, and personal care products. Comprehending organic contaminants' presence, behavior, and ramifications is paramount in safeguarding the environment and human health. Effective screening, evaluation, and mitigation strategies are imperative to mitigate these compounds' potential risks, ensuring ecosystems' sustenance and communities' well-being (Ridgway, Lalljie, and Smith 2007). Consequently, this chapter presents the outlook of organic contaminant detection using bionanotechnological.

Organic contaminant detection has assumed an increasingly pivotal role across diverse domains, including environmental monitoring, food safety, and healthcare. In response to the burgeoning demand for quick, highly sensitive, and selective detection techniques, nano-biosensors have emerged as an innovative technology to revolutionize the landscape of organic contaminant detection. Analyzing organic contaminants within food matrices presents a multifaceted challenge, necessitating the discernment of specific contaminants originating from distinct stages of food production, processing, or packaging (Srivastava, Dev, and Karmakar 2018). Herein, we have targeted three organic contaminants in food (nitrosamine, acrylamide, and melamine), for which a good deal of literature related to nanotechnological sensing for food products has emerged. Therefore, the forthcoming sections will talk about the nanobiosensors, nitrosamine, acrylamide, and melamine as food contaminants, followed by literature specific to these contaminants using nanobiosensors.

Nanobiosensors harness the distinctive characteristics of nanomaterials and biological molecules to create highly efficient detection tools, bringing a paradigm shift in organic contaminant identification. These biosensors rely on transduction mechanisms critical for converting bioanalyte interactions into discernible electrical signals (X. Huang, Zhu, and Kianfar 2021). The judicious selection of nanomaterials assumes paramount importance in this context, given their remarkable surface area-to-volume ratios and electromechanical attributes, rendering them exceptionally well-suited for this purpose. The fundamental architecture of a biosensor comprises three essential constituents: a bioreceptor, a transducer, and a detector. The bioreceptor is a scaffold for the targeted material and can be derived from diverse biologically responsive substances. The transducer assumes the crucial role of converting the

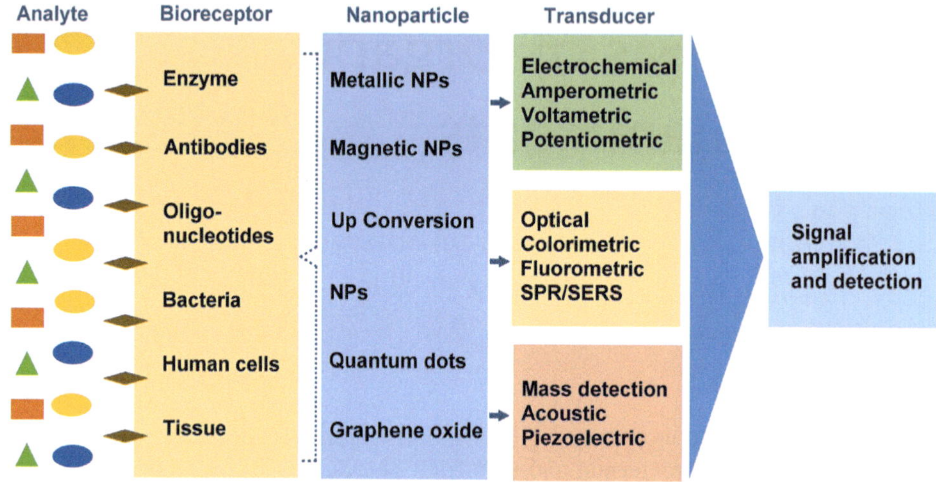

FIGURE 5.1 Configuration of a typical nanobiosensor showing various bioreceptors, nanoparticles types, transducers, and detectors.

interaction between the bioanalyte and bioreceptor into an electrical signal, thereby facilitating the translation of biochemical responses into electrical data (Bogue 2008). The detector system receives and amplifies this electrical signal, making it amenable to comprehensive analysis. An essential requirement for the efficacy of nanobiosensors pertains to the availability of immobilization techniques, which enhance the efficiency of bioreceptor–analyte interactions. Immobilization confers cost-effectiveness to the biological sensing process, albeit subject to modulation by environmental factors such as temperature, pH, and physicochemical fluctuations. Nanobiosensors are used to prepare nanomaterials, typically ranging in size from 1 to 100 nm, characterized by distinct physicochemical properties that set them apart from macro-scale materials. They are pivotal in advancing biosensor technology and have contributed to developing nanoelectromechanical systems (NEMS) (Bogue 2009). Various nanomaterials, such as nanotubes, nanowires, nanorods, nanoparticles, and nanocrystalline thin films, have been extensively studied for their unique electronic and mechanical attributes. Figure 5.1 depicts a nanobiosensor comprising analytes, bioreceptors, and nanostructure-integrated transducers that collectively enable the detection and transmission of biological responses to the detector.

These devices find diverse applications, including amperometric devices for glucose detection, quantum dots for detecting binding events, and bioconjugated nanomaterials for precise biomolecular identification. Colloid nanoparticles can be combined with antibodies for immunosensing and immunolabeling purposes, while metal-based nanoparticles excel in electronic and optical applications. The enhancement of biosensor signals has been achieved by employing various nanomaterials, including magnetic nanoparticles, zinc oxide-based nanostructures, and carbon nanotubes (Mun'delanji et al. 2015).

5.1.1 *N*-NITROSAMINE

5.1.1.1 Overview

The initial observation and publication of the chemical interaction between nitrous acid and primary aromatic amines can be attributed to Peter Griess in 1864. Following this discovery, further research was conducted by prominent scientists such as Baeyer, Caro, and Otto Witt during the 1870s. Otto Witt, in particular, introduced the term "nitrosamine" in his 1878 publication, defining it as "any substituted ammonia containing the univalent nitrosyl group, 'NO,' (Witt 1878) directly linked to the ammoniacal nitrogen." Notably, in 1956, there was documented evidence of developing malignant primary hepatic tumors in animal test subjects exposed to NDMA (*N*-Nitrosodimethylamine) (Barnes and Magee 1954).

Nitrosamines, constituting a significant group of genotoxic chemical carcinogens, are pervasive in the human diet and various environmental media. Their capacity to induce cancer in experimental animals, including higher primates, closely resembling human malignancies, was established. Extensive testing and epidemiological data indicate a connection between certain human cancers, such as esophageal, stomach, and nasopharyngeal cancers, and *N*-nitroso compounds in foods (Tricker and Preussmann 1991). Over the decades, research has accumulated substantial knowledge about *N*-nitroso compounds (NOC), beginning in the early 20th century, revealing their potent carcinogenic properties. In 1956, Barnes and Magee were among the first to describe the potent carcinogenicity of *N*-Nitroso-dimethylamine (NDMA) (Magee and Barnes 1956).

N-nitrosamines, highly carcinogenic, are frequently encountered as contaminants in food. The report by Magee and Barnes in 1956 regarding the hepatotoxic and carcinogenic properties of *N*-nitrosodimethylamine (NDMA) stemmed from observations of liver disease in minks fed nitrite-treated herring meal. Subsequent investigations of over 300 compounds unveiled their mutagenic and carcinogenic characteristics in laboratory and real-world settings. The International Agency for Research on Cancer (IARC) categorizes various *N*-nitrosamines based on their carcinogenic potential to humans. At the same time, the U.S. Environmental Protection Agency (USEPA) lists several N-nitrosamines as contaminants (Lahiri et al. 1988; Przybyla et al. 2022; Hecht 1997; Schwarzenegger, Adams, and Denton 2006).

Considering the significance of these contaminants, the analysis and regulation of N-nitrosamine content using state-of-the-art technology, such as bionanosensors, have become critically important (Korostynska, Mason, and Al-Shamma 2012). The emergence of bionanosensors represents a compelling approach that integrates biorecognition elements with nanotechnology to establish a robust and precise platform. This integration offers selectivity in detecting *N*-nitrosamines, overcoming the limitations of conventional methods, and introducing a reliable approach. By leveraging the inherent specificity of biomolecules such as antibodies or aptamers, immobilized on nanomaterial surfaces, bionanosensors provide a tailored interface for capturing *N*-nitrosamines. The analysis of *N*-nitrosamines through bionanosensors marks a significant advancement in analytical methodology, capitalizing on the synergistic combination of biorecognition elements and nanotechnology. Bionanosensors offer a customized and biologically responsive platform for the selective detection of *N*-nitrosamine compounds, thereby circumventing the constraints of conventional methods. Using biomolecules as recognition elements, such as antibodies or aptamers, immobilized on nanomaterial surfaces, these sensors provide a sensitive and specific interface for capturing N-nitrosamines. These interactions generate measurable signals, including electrochemical, optical, or piezoelectric responses, enabling rapid and quantitative analysis. The convergence of biorecognition and nanoscale engineering equips bionanosensors with exceptional sensitivity, facilitating *N*-nitrosamine detection at deficient concentrations and proving pivotal in addressing health and environmental concerns associated with these carcinogenic agents.

5.1.1.2   Structure and Mechanism of N-Nitrosamine Formation

A category of compounds known as *N*-nitroso compounds (NOCs), specifically nitrosamines, encompass a joint functional group >N-N=O. These compounds can be divided into two primary groups: *N*-nitrosamines and *N*-nitrosamides. Their chemical stability characterizes *N*-nitrosamines, while *N*-nitrosamides exhibit lower stability and produce electrophilic alkylating or arylating species depending on their structural and physicochemical characteristics.

The formation of NOCs can occur under various circumstances when nitrosating agents come into contact with *N*-nitrosatable amino compounds. In aqueous acidic environments, amines, typically

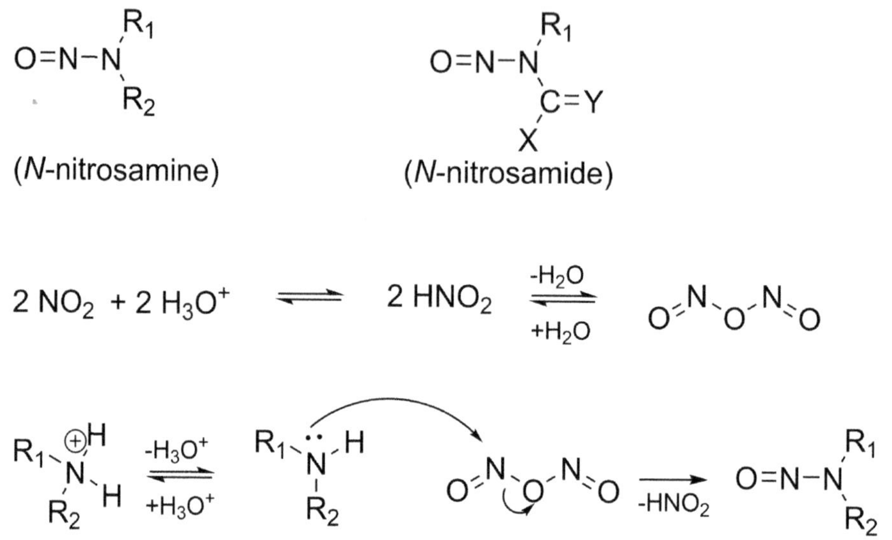

SCHEME 5.1   Structure of N-nitrosamine, N-nitrosoamide, and reaction mechanism of N-nitrosoamine formation.

secondary amines, react with nitrous acid to generate nitrosating species, which can be dinitrogen trioxide (N_2O_3), tetroxide (N_2O_4), or nitrous acidium (Nitrosonium) ion NO_2-H_2O. The nitrosation rate is influenced by the pK_a value of a given amine (Mirvish 1975). In aldehydes, particularly formaldehyde, nitrosation is significantly catalyzed and can occur even under neutral or basic pH conditions (Keeper and Roller 1973). This reaction has been observed in specific environmental settings, including certain workplaces, where amines, aldehydes, and nitrosating agents coexist. Early on, acid-catalyzed nitrosation has been proposed as a potential mechanism for in vivo. N-nitroso compound formation in the stomach (Druckrey et al. 1961). This in-vivo formation is particularly relevant for weakly basic amines. Tertiary amines can also produce nitrosamines through dealkylating nitrosation, although typically at much slower reaction rates compared to the formation of nitrosamines from secondary amines. Nitrosation of primary amines in aqueous solutions leads to unstable diazonium intermediates, which subsequently react with water to form alcohols (Smith and Loeppky 1967; Loeppky et al. 1994).

N-nitrosamines are considered important pollutants because they are found almost everywhere in various environmental settings, even in small amounts measured in ng kg^{-1} and ng L^{-1}. This group of hydrophilic substances consists of over 300 different types, and they all share a common feature: a nitroso group connected to an amine (López-Rodríguez et al. 2020).

5.1.1.3   *N*-Nitrosoamine Present in Food

Nitrosamines, a group of chemical compounds, are intricately linked to multiple stages of the food supply chain, including manufacturing, storage, cooking, and packaging. These compounds are generated due to the interaction between N-nitrosatable amines and nitrosating agents, and they are prevalent in our food due to both natural processes and human activities (Sen 1988; Bartsch and Spiegelhalder 1996; Pfundstein et al. 1991). The conversion of nitrate and nitrite, commonly found in food, into nitrogen oxides (NO_x) with the assistance of microbial processes plays a significant role in their presence. Technical procedures like drying, smoking, and kilning, combined with specific handling and storage conditions can substantially contribute to the formation of nitrosamines. Notable examples of these compounds include NDMA (N-Nitroso-dimethylamine), NPYR (N-Nitrosopyrrolidine), NPIP (N-nitrosopiperidine), and NTHZ (N-nitrosothiazolidine). Additionally, N-nitrosated amino acids such as sarcosine (NSAR), 3-hydroxyproline and proline (NPRO), thiazolidine-4-carboxylic acid (NTCA), oxazolidine-4-carboxylic acid (NOCA) and *N*-nitroso-2-methyl-thiazolidine-4-carboxylic acid (NMTCA) as well as its oxazolidine analog *N*-nitroso-2-methyl-oxazolidine-4-carboxylic acid (NMOCA) are present in foods. Still, their potential to cause cancer varies (Tricker and Kubacki 1992). NPRO and NTCA are frequently found in food, but *N*-nitrosated amino acids, in general, are not linked to mutations or cancer, apart from NSAR, which has relatively low carcinogenicity.

Meat and meat-based products often contain nitrosamines due to various processing methods such as smoking, curing, and salting. For example, in the curing process, specialized microorganisms catalyze the conversion of nitrate and nitrite into nitrogen oxide (NO). This process extends the shelf life of meat products and enhances their color and flavor, contributing to the characteristic red hue of cured meats. Regarding regulation, the European Food Safety Authority (EFSA) conducted a comprehensive reassessment of acceptable intake levels for nitrate and nitrite in 2017 to address concerns about nitrosamine formation (Committee et al. 2017). Literature from the early 1980s highlighted the prevalence of volatile *N*-nitrosamines in meat products, leading to subsequent efforts to reduce their levels (Spiegelhalder, Eisenbrand, and Preussmann 1980a, b). Strategies such as lowering nitrite content and using nitrosation inhibitors were implemented, resulting in significant reductions.

Interestingly, processing and thermal treatments can increase the concentration of specific *N*-nitrosamines. For instance, smoke-treated products have been found to contain *N*-nitroso thiazolidine (NTHZ) at concentrations up to 10 mg kg^{-1}. In comparison, *N*-nitroso-oxazolidine (NOCA) was detected in smoked mutton at 40–70 mg kg^{-1} (Additives et al. 2017a, b; Helgason et al. 1984; Ellen, Egmond, and Sahertian 1986). Further analysis conducted by the Bavarian Food Surveillance Office in 2014 demonstrated the presence of nitrosamines in smoked bacon samples, with only 15 out of 40 samples showing no detectable levels beyond the detection limit (Zhu et al. 2015). NDMA, a prevalent volatile *N*-nitrosamine, was found in various fish products, with concentrations ranging from 0.5 to 8.0 mg kg^{-1}, influenced by diverse preparation methods and conservation practices (Tricker et al. 1991). Importantly, differences were noted between Asian and Western fish products, with the former having higher NDMA concentrations (Bartsch and Spiegelhalder 1996). A Finland study found NDMA in smoked fish, with levels up to 28.5 mg kg^{-1}. EU fish products generally contain 0.2–3.0 mg kg^{-1} of NDMA. At the same time, Japan and China occasionally report higher values, up to 5.9 mg kg^{-1} in Japan and over 130 mg kg^{-1} in certain Chinese fish products (Penttilä, Räsänen, and Kimppa 1990; Habermeyer et al. 2015). Beer, a widely consumed beverage, has also been scrutinized for nitrosamine contamination, particularly NDMA. The heating processes involved in malt preparation, especially direct firing, contribute to the presence of NDMA. In a study of 158 beer samples, 70% contained detectable levels of NDMA, with concentrations averaging 2.7 mg kg^{-1} and reaching a maximum of 68 mg kg^{-1}. The presence of nitrosamines in food is a complex issue influenced by various factors throughout the food production and processing chain. The wide range of concentrations observed in different products underscores the complexity of managing and reducing nitrosamine levels in our food supply (Habermeyer and Eisenbrand 2019).

5.1.1.4   Human Exposure to *N*-Nitrosamines

N-nitrosamines from food in various countries, as reported by Habermeyer et al. in 2009 (Habermeyer and Eisenbrand 2009), it was found that NDMA is the most prevalent N-nitrosamine in food, contributing significantly to human exposure. Mitigation efforts implemented since the 1980s, such as reducing nitrite content in curing salt, adding nitrosation inhibitors like ascorbic acid and tocopherols, and improving kilning processes for barley malt production, have led to a decrease in mean exposure to NOC from 1.1 mg·person^{-1}·day^{-1} in Germany in 1980 to approximately 0.2–0.3 mg kg^{-1} in 1990 (Hinz et al. 2023; Tricker and Kubacki 1992). It's important to note that humans can also be exposed to NOC in specific workplace environments besides dietary intake. However, this aspect is not discussed in detail here and may require consideration in individual risk assessments. For most industrialized countries, the estimated dietary intake of volatile *N*-nitrosamines is around < 0.1–0.6 mg person^{-1} or < 1.7–27 ng kg^{-1} body weight (Bartsch and Spiegelhalder 1996;

Österdahl 1988; Vermeer et al. 1998). The primary sources of exposure are beer, processed meat products, and fish.

In contrast, when assessing non-volatile *N*-nitroso compounds using NeNO-group specific analysis to detect "apparent total *N*-nitroso compounds" (ATNC), daily exposure levels were found to be much higher, ranging from approximately 10–100 mg person^{-1}. Beer and bacon consumption were identified as significant contributors to this exposure. Notably, while many non-volatile N-nitroso compounds found in food, like most N-nitroso amino acids, are not considered biologically active, there is still a need for chemical identification and biological characterization of the full range of non-volatile NOC. It is estimated that only about 10% of the total ATNC content has been characterized. Another significant exposure pathway is the endogenous formation of NOC. While most nitrosated amino acids are not carcinogenic or mutagenic and are excreted mainly in urine, they can serve as biomarkers for endogenous nitrosation. Assessing the health relevance and cancer risk associated with the endogenous formation of carcinogenic NOC remains challenging. Factors like inflammatory diseases, infections, and other conditions can increase endogenous nitrosation, potentially elevating the risk of carcinogenic NOC formation. The validity of NPRO as a biomarker for endogenous NOC formation beyond nitrosation of amino acids or peptides is still under investigation, emphasizing the need to develop more predictive biomarkers to assess overall human risk from endogenous NOC formation in the future (Tricker and Kubacki 1992; Challis 1996; Meah, Harrison, and Davies 1994).

5.1.1.5 Methods for Detection of *N*-Nitrosamine

The established daily intake limit for nitrite has been imposed as a precautionary measure due to concerns regarding its potential conversion into *N*-nitrosamines, which elevate the risk of cancer and methemoglobinemia. Ensuring food safety requires developing a highly sensitive, selective, rapid, and precise method for quantifying nitrite content. Traditional approaches for nitrite detection, such as the Griess assay, necessitate the time-consuming preprocessing of food samples. In contrast, modern analytical techniques such as high-performance liquid chromatography, gas chromatography, ion chromatography, mass spectrometry, and potentiometry demand sizable equipment, stable receptors, and trained personnel for implementation. The need for a practical and efficient nitrite detection method in food safety protocols remains paramount due to the potential health risks associated with excessive nitrite consumption. Recently, novel nanobiosensors have emerged for detecting *N*-nitrosamines, achieving low detection limits across various sample types. These nanobiosensors employ highly sensitive, selective, and reproducible techniques. A variety of methods have been devised to detect *N*-nitrosamines effectively. In this context, we elucidate distinct approaches for detecting *N*-nitrosamines (Parr and Joseph 2019; Bhattacharya, Samanta, and Chakraborty 2021).

5.1.1.6 DNA-Based Electrochemical Sensors

The burgeoning field of DNA-based electrochemical sensors has garnered considerable attention owing to their exceptional sensitivity and facile detection modalities. Nevertheless, the prerequisite immobilization of DNA on the electrode surface is imperative to ensure precision. Nanostructured materials such as graphene, carbon nanotubes, and gold nanoparticles have emerged as promising candidates, offering a plethora of binding sites for DNA tethering. Carbon dots (CDs), characterized by their biocompatibility and eco-friendliness, have emerged as compelling candidates for sensor applications. This study endeavors to engineer an electrochemical biosensor employing CDs to discern N-nitrosamine, a potent mutagenic compound. Leveraging naturally sourced CDs, this approach underscores environmental sustainability, as these materials exhibit minimal toxicity and impose no ecological hazards. CDs' inherent multiple binding sites represent a novel advancement in biosensor design, marking a seminal achievement in the field, particularly for detecting carcinogenic and mutagenic compounds such as NDMA and NDEA. Notably, the biosensor's unparalleled selectivity and extraordinary sensitivity demonstrated its efficacy across a broad detection range from 1×10^{-8} to 8×10^{-7} mol L^{-1} (Majumdar, Thakur, and Chowdhury 2020).

5.1.1.7 2-Dimensional Material Nanbiosensors

Electrochemical sensors have garnered significant attention among researchers due to their advantages, including swift detection capabilities, an extensive linear detection range, low detection thresholds, and high durability. Two-dimensional materials like MoS_2 and graphene are widely employed in electrochemical detection owing to their abundant active sites and substantial specific surface area. Electrochemical detection enhancement can be achieved by incorporating metal nanoparticles (G. Zhang, Liu, et al. 2016). Haldorai et al. pioneered the development of electrochemical nitrite sensors utilizing spindle-shaped Co_3O_4 and rGO nanocomposites, achieving commendable reproducibility, stability, sensitivity, and selectivity (Haldorai et al. 2016). Furthermore, Zhang and coworkers harnessed Fe_3O_4/MoS_2 nanocomposite structures on MoS_2 nanosheets for nitrite detection, offering a wide detection range and low detection limits (Y. Zhang, Chen, et al. 2016).

5.1.1.8 Fluorescence-Based Nanobiosensors

Fluorescence techniques have found extensive applications in hazard detection, particularly in ultra-sensitive and highly accurate analysis of compounds like *N*-nitrosamines. Among these methods, fluorescence-based approaches using quantum dots hold significant promise, owing to their simplicity, rapid response kinetics, cost-effectiveness, superior sensitivity, and selectivity. Carbon dots (CDs) have emerged as an attractive candidate for fluorescence-based sensing due to their desirable attributes, including facile synthesis, notable photoluminescence, exceptional photostability, biocompatibility, and water solubility. The advanced *N*-nitrosodimethylamine (NDMA) detection methods featuring Lanthanide-doped upconversion nanoparticles (UCNPs) offer considerable advantages, including stable fluorescence signals, heightened sensitivity, and diminished autofluorescence, thus minimizing background interference. Functional cobalt (III) tetraphenylporphyrin ($Co(tpp)ClO_4$) stands out as an ideal receptor in fluorescence sensing, exhibiting remarkable specificity towards analyte binding. An innovative fluorescence nanosensor is designed by incorporating functional cobalt porphyrin materials for the sensitive detection of NDMA in pickled meat, achieving detection at $\mu g\ kg^{-1}$ levels. The synthesized $Co(tpp)ClO_4$ preferentially interacts with NDMA, leading to fluorescence quenching attributed to dipolar resonance contributions. This underlying principle is extended to enable quantitative NDMA detection, presenting a novel approach for rapid and sensitive determination of *N*-nitrosamines by harnessing UCNPs and functional porphyrin materials. The fluorescence quenching arises from the strong interaction between $Co(tpp)ClO_4$ and NDMA, resulting in a redshift in the UV-Vis spectrum. This nanosensor exhibits exceptional selectivity and a low limit of detection (LOD) of $0.18\ ng\ mL^{-1}$ for NDMA, characterized by rapid response kinetics, high sensitivity, and a wide dynamic detection range, positioning it as a promising candidate for enhancing food safety control measures (Ouyang et al. 2023; Ouyang et al. 2021).

A fluorescent biosensor for real-time NDMA monitoring at the point of use was developed using zein film and upconversion nanoparticles (UCNPs) functionalized with aptamers. The morphology, crystal structure, and surface groups of UCNPs were characterized using instrumental techniques such as TEM, XRD, and FT-IR to enhance and test the favorable adherence of UCNPs to zein film and their specificity for detection.

This biosensor's initial production involved synthesizing hydrophilic UCNPs, which were combined with a zein film and bonded to an aptamer. Additional DNA molecules with complimentary ends that have had dabcyl at 5' end were used to join the zein film. The UCNPs and dabcyl undergo fluorescence resonance energy transfer (FRET), which underlies the biosensor's functions. The particular aptamer specifically caught NDMA, and its presence caused the DNA structure to be disturbed. Since this disruption prevented the FRET mechanism, the zein film emit a detectable fluorescence signal. Under ideal conditions, the zein film-based biosensor showed remarkable performance for on-site and portable NDMA detection, with an astonishing lower detection limit as low as $0.017\ ng\ mL^{-1}$ (Ouyang et al.).

5.1.2 Acrylamide

5.1.2.1 Overview

In April 2002, researchers from the Swedish National Food Administration (SNFA) and the University of Stockholm reported acrylamide formation in foods rich in starch or subjected to high-temperature cooking (Lofstedt 2003). Subsequent findings have corroborated these results, with numerous countries, including Norway, Switzerland, the UK, and the USA, independently confirming the presence of acrylamide. This chemical compound has been identified as a potential instigator of a broad spectrum of adverse effects, encompassing human neurotoxicity and carcinogenicity in animals. The hypothesis that food heating could be a significant source of acrylamide exposure in humans has been proposed, though no acrylamide has been detected in boiled food products. Two distinct regulatory limits have been established for acrylamide: one pertaining to drinking water and the other concerning the migration of acrylamide from packaging materials into food. The latter is stipulated to be undetectable within a detection limit (LOD) of 10 µg of acrylamide in 1 kg of food.

In comparison, an estimated daily intake of approximately 10 µg is contingent on dietary habits. This situation has raised concerns among food producers and regulatory authorities. Valuable insights into acrylamide and its toxicological characteristics are elucidated in reports issued by the Scientific Committee on Food (SCF) in 2002 and the Heatox Report in 2007 (Keramat, LeBail, Prost, and Jafari 2011). Extensive investigations into acrylamide's carcinogenicity have been undertaken in experimental animals, prompted by reports of unusual cancers among factory workers exposed to vinyl chloride. Studies conducted in the 1980s on mice and rats revealed multiple tumors upon systemic administration of acrylamide. Concurrently, epidemiological assessments of cancer risk among workers exposed to acrylamide were initiated, encompassing cohorts from both monomer production and polymerization industries. These evaluations yielded no consistent evidence of acrylamide exposure exerting a discernible impact on cancer incidence. Within the body, acrylamide undergoes biotransformation to its epoxide counterpart, glycidamide, which has been shown to exhibit genotoxicity in various *in vitro* and *in vivo* assessments. The International Agency for Research on Cancer (IARC) evaluated acrylamide in 1994, classifying it as "probably carcinogenic to humans" based on positive findings in bioassays involving mice and rats (Loomis, Huang, and Chen 2014). However, this assessment primarily considered acrylamide as a synthetic industrial chemical, with significant human exposures presumed to be limited to occupational settings. Epidemiological studies to ascertain potential cancer risks stemming from acrylamide in human diets are still emerging in the scientific literature. The highest levels of acrylamide are produced during the frying of potatoes, the roasting of coffee and cocoa beans, the baking of bread and cakes, the thermal processing of cereals, and the roasting of meat. Recognizing the toxicological threat posed by acrylamide intake via food, the Joint Expert Committee of FAO/WHO on Food Additives (JECFA) has emphasized the importance of strategies to reduce acrylamide content in thermally processed foods (Capuano and Fogliano 2011; Abt et al. 2019). Advanced methods for quantifying acrylamide are imperative to gauge human exposure to this detrimental compound accurately. The necessity for quantifying acrylamide and its adducts to enhance epidemiological investigations was underscored during a meeting of the European Food Safety Authority (EFSA) 2008 (Basaran and Faiz 2022). Techniques such as HPLC-MS/MS and immunoenzymatic assays are particularly well-suited for the quantitative determination of adducts formed by acrylamide and its metabolite glycidamide with hemoglobin or DNA, as well as for measuring metabolites excreted in the urine, such as mercapturic acid, which can serve as biomarkers of acrylamide exposure. For acrylamide quantification in foods, it is essential that procedures yield reliable quantitative data and are cost-effective, convenient, and swift (Sharp 2003). The collection and preparation of samples prior to instrumental analysis must be executed with precision to ensure accurate and reproducible outcomes. The European Committee of Normalization (Comité Européen de Normalization, CEN) is anticipated to establish an official method for acrylamide quantification in foods by the end of 2013 (Halford and Curtis 2019).

5.1.2.2 Structure and Formation of Acrylamide in Food

Acrylamide, or acrylic amide, is an organic compound represented by $CH_2=CHC(O)NH_2$. It is a colorless and odorless solid that exhibits excellent solubility in water and various organic solvents. From a chemical standpoint, acrylamide is a primary amide ($CONH_2$) that has undergone vinyl substitution. Its primary industrial utility lies in serving as a precursor for polyacrylamides, widely utilized as versatile water-soluble substances with applications as thickening agents and facilitators of flocculation (Keramat, LeBail, Prost, and Soltanizadeh 2011; Lingnert et al. 2002).

Most research efforts have concentrated on elucidating the mechanisms responsible for generating acrylamide in heat-treated foods. Nevertheless, due to incomplete comprehension of these mechanisms, it remains challenging to devise an effective strategy for minimizing its occurrence across diverse heat processing techniques by regulating pivotal steps in food processing. The quantification of acrylamide content in foods pertains to the residual amount of acrylamide that persists after its formation and subsequent degradation.

The Maillard reaction, a multifaceted sequence of reactions occurring during the thermal treatment of food, plays a pivotal role in imparting the brown hue, crustiness, and distinctive flavor observed in baked, fried, and toasted edibles. This intricate process involves the convergence of reducing sugars, such as glucose and fructose, with amino acids, predominantly asparagine, leading to *N*-glycosides forming. These reactions subsequently give rise to Amadori rearrangement products, which undergo diverse transformations, ultimately producing melanoidin and responsible for the generation of brown color, crust texture, and appealing flavor profile characteristics of baked, fried, and toasted foods.

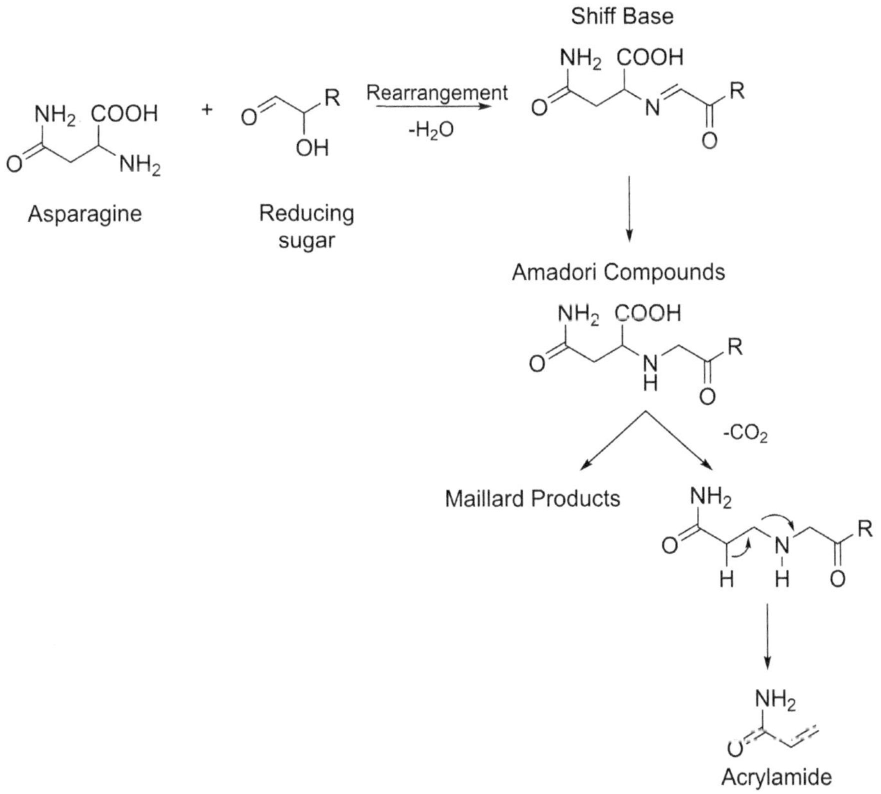

SCHEME 5.2 Structure and mechanism of acrylamide formation in food.

Scientific investigations have established that reducing sugars is the rate-limiting factor in the case of potatoes, whereas asparagine plays this role in cereal products. Recent studies have also identified the compound 3-aminopropionamide in cocoa beans, coffee, and cereal products, contributing to the Maillard reaction (Mottram, Wedzicha, and Dodson 2002). Notably, the response is predominantly a surface-driven phenomenon, meaning that acrylamide in bread is primarily concentrated in the crust with minimal presence in the crumb. In the case of potato crisps, the elevated acrylamide content can be attributed to the crisps' structure, which essentially comprises two thin surfaces with minimal material between them. Overall, the essential prerequisites for acrylamide formation in heat-processed foods involve the availability of free asparagine, free reducing sugars, high temperatures exceeding 120°C, and low moisture conditions at the food's surface (Friedman 2003; Stadler et al. 2002).

The Maillard reaction also applies to the formation of advanced glycation end products (AGEs) in food, which are complex compounds that develop when we cook or process food. When these compounds accumulate in your body, they can lead to health issues and contribute to chronic diseases.

AGEs are produced during the reaction due to the progressing Maillard reaction. These AGEs can exist in various forms, including as free molecules, bonded to peptide chains, or attached to proteins. They are further classified into two categories based on their molecular weight: low molecular weight AGEs and high molecular weight AGEs. Examples of well-known free-form AGEs include $N\varepsilon$-(carboxymethyl)-lysine (CML), pyrraline, and pentosidine. On the other hand, peptide and protein-bound AGEs constitute a diverse group of peptides and proteins that have undergone modification through glycation with distinct structural changes. These AGE-altered compounds can potentially influence the biological properties of native proteins. This influence extends to their ability to induce pathological effects, impact immune responses, and affect metabolic processes. More than 40 AGEs have been identified and characterized in various studies.

Among these, $N\varepsilon$-(carboxyethyl)-lysine (CEL) is a homolog of the well-known AGE, $N\varepsilon$-(carboxymethyl)-lysine (CML), and is formed through the reaction of methylglyoxal (MG) with lysine. Pyrraline, a pyrrole derivative containing the $N\varepsilon$-amino lysine group, is another prevalent

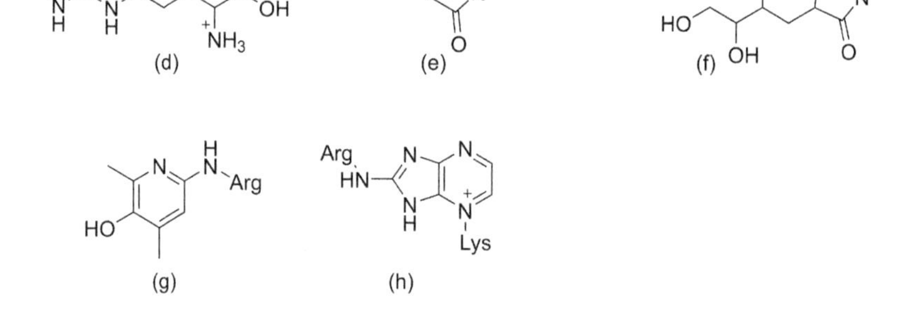

SCHEME 5.3 This scheme shows the structure of the free form of dAGEs (a) $N\varepsilon$-(carboxymethyl)-lysine (CML), (b) $N\varepsilon$-(carboxyethyl)-lysine (CEL), (c) pyraline, (d) methylglyoxal-derived hydroimidazolones (MG-H^1), (e) Glyoxal-derived hydroimidazolone (G-H1), (f) 3- deoxyglucosone-derived dihydroxyimidazoline (3DG-H^1), (g) Argpyrimidine, (h) Pentosidine.

AGE found in food items. Additionally, arginine's guanidino group serves as a significant site for the formation of methylglyoxal-derived hydroimidazolones, glyoxal-derived hydroimidazolone, and 3-deoxyglucosone-derived dihydroxyimidazoline. In parallel, arginine can react with methylglyoxal to form argpyrimidine. Bound-form AGEs have been elucidated with various molecular structures involving saccharides and amines. These reactions, known as Maillard reactions (MR), can result in irreversible crosslinking and non-crosslinking modifications of protein structures when an amino group is linked to a protein. Non-crosslinking changes, as exemplified by CML, CEL, and pyrraline, involve the glycation of a single lysine residue. In contrast, crosslinking modifications connect an AGE-modified residue to another amino acid chain group within or among polypeptide chains. Pentosidine is an example of an AGE crosslinker formed by the reaction between lysine and arginine residues.

Lysine dimers, such as GOLD, MOLD, and DOLD, are generated from reactions between two lysine side-chains and two molecules of glucose (G), methylglyoxal (MG), and 3-deoxyglucosone (3DG). It's worth noting that crosslinking structures between two arginine residues have not yet been fully elucidated. Additionally, while some of these AGEs have been identified through various research, it's important to highlight that "nine out of ten" AGEs were detected *in vivo* or in systems that resemble biological conditions. This underscores the potential for discovering new AGEs in processed foods that contain Maillard reaction products (MRPs)—explained diagrammatically in schemes 5.3 and 5.4 above (Zhang, 2020 #133).

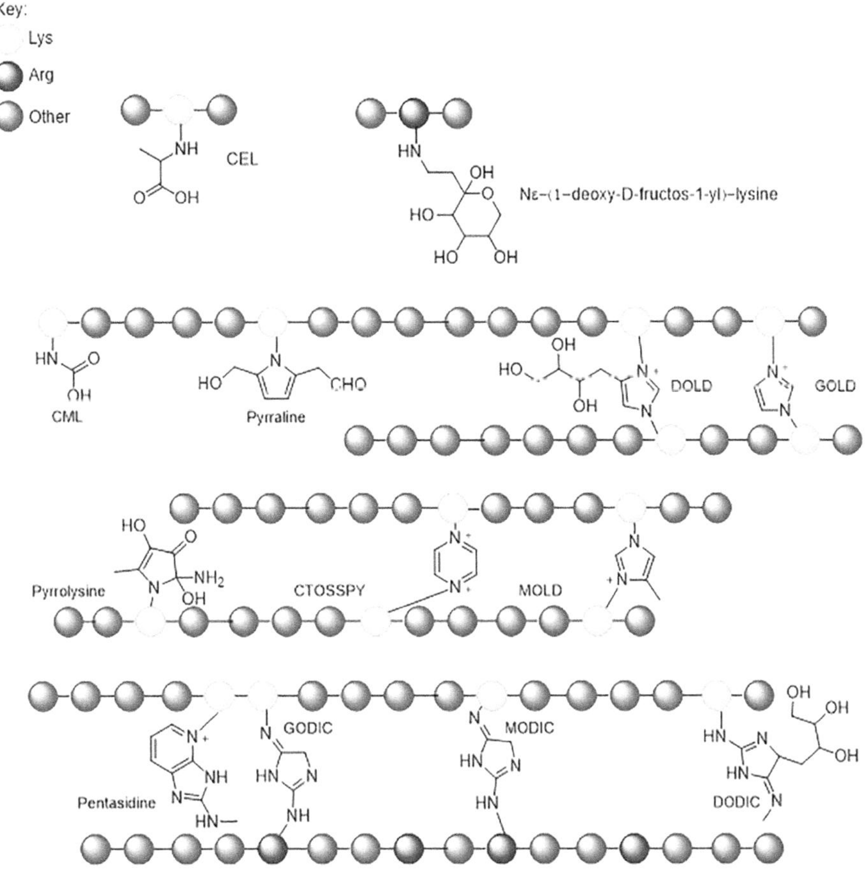

SCHEME 5.4 This scheme describes the peptide/protein-bound AGEs in food.

5.1.2.3 Exposure to AA in Human

Acrylamide (AA) is encountered through oral, dermal, and inhalation routes. Smokers face a higher AA exposure than nonsmokers, and non-food industrial applications such as tobacco smoke also contribute to AA exposure. The amalgamation of sources contributing to AA exposure encompasses diet, smoking, drinking water, and occupational factors. The International Agency for Cancer Research has classified acrylamide as a substance with various adverse effects on health, including neurotoxicity, carcinogenicity, and genotoxicity. These harmful properties are primarily linked to its transformation into glycidamide, a highly mutagenic by-product that can induce (Çebi 2024) enzyme mutations during normal physiological processes.

Furthermore, acrylamide can react with specific chemical groups (-SH, -OH, and -NH$_2$) in DNA, damaging the DNA structure. Acrylamide is also recognized as a mutagen that affects germ cells. According to the Joint Expert Committee on Food Additives, acrylamide (AA) is a potential human carcinogen predominantly found in heat-processed foods, specifically fried, deep-fried, or baked foods. The daily intake of AA varies depending on countries and dietary patterns, but the average mean intake stands at approximately 0.4 mg kg^{-1} of body weight per day (bw d^{-1}) (Dybing et al. 2005). High-level consumers might have an average intake of 1.0 mg kg^{-1}·bw d^{-1}. According to the World Health Organization (WHO), AA exhibits no reliably identifiable threshold of effects. This means that exposure to low doses may result in a symptom-silent period during which the harmful effects may not be clinically evident. Still, morphological and biochemical changes may be present. The tolerable daily intake for neurotoxicity due to AA is estimated to be 40 mg·kg^{-1}·d^{-1}, while for cancer, it is estimated at 2.6 and 16 mg·kg^{-1}·d^{-1} based on AA or glycidamide, respectively. A study conducted in Lebanon discovered that daily AA consumption from potato and corn chips exceeded the WHO's risk intake by 7–40 times, although it remained below the neurotoxic risk threshold. This has raised concerns regarding cancer risk for the Lebanese population and highlights the need for further epidemiological studies and ongoing monitoring of AA levels in food products. Another study revealed that caffeinated beverages contained an average of 29,176 mg kg^{-1} of AA, surpassing the WHO's risk intake limits for carcinogenicity and neurotoxicity. The findings underscore the necessity for regulating the caffeinated product industry in Lebanon by implementing legislation and standard protocols for product preparation to limit AA content and safeguard consumers (Timmermann et al. 2021). Food packaging containing polyacrylamide may indirectly expose individuals to residual AA monomers. The diet is recognized as the primary source of AA exposure for the general nonsmoking population, with approximately 38% of caloric intake coming from food sources known to contain AA (Rifai and Saleh 2020). Additionally, scientific research has demonstrated that exposure to acrylamide through dietary sources or environmental factors can lead to cardiac developmental toxicity (CDT), as observed in experiments involving zebrafish embryos. When acrylamide is introduced into the system after fertilization, it has been shown to result in the abnormal shrinking of the heart and the distortion of its typical development (Abt et al. 2019).

5.1.2.4 Detection Approaches for Acrylamide

The chromatographic methods offer high sensitivity, selectivity, stability, and reproducibility. Still, these techniques are costly for a single analysis, time-consuming, and do not provide real-time results for acrylamide analysis in food products. Furthermore, ELISA (enzyme-linked immunosorbent assay) also necessitates supplementary robust confirmation methods for better analytical results (Abt et al. 2019). Recently, biosensing methods have been developed to offer simplicity, sensitivity, specificity, and rapid response measurements as viable alternatives. Biosensors are quantitative or semi-quantitative analytical devices that incorporate living recognition entities such as enzymes, antibodies, phages, aptamers, or single-stranded DNA, coupled with suitable physicochemical, optical, thermometric, piezoelectric, and magnetic transducers (Bahadır and Sezgintürk 2015). In essence, biosensors are chemical sensors in which the biological recognition system utilizes a biochemical mechanism to interface with optoelectronic systems.

5.1.2.5   Electrochemical Nanobiosensors for Detection of Acrylamide

Enzymatic or protein-based nanoparticles, including Hb, myoglobin, and cytochrome C, harbor four heme groups within their redox centers. Among these, Hb emerges as a more pragmatic choice for constructing acrylamide biosensors, owing to its commercial availability, cost-effectiveness, stability, and structural attributes. These factors facilitate the formation of Hb-acrylamide adducts, rendering native Hb-based biosensors more specific and economically viable for acrylamide detection applications. Various forms of protein-based biosensors have been devised for the detection of acrylamide. These biosensors adopt diverse configurations, including Hb/sol-gel film-modified carbon paste electrodes, Hb/gold nanoparticles (AuNPs)/carboxylated multiwalled carbon nanotubes (cMWCNT)/glassy carbon electrodes, Poly (maleic anhydride-alt-butyl vinyl ether (AM41)-polyethylene glycol (PEG)/Hb/citric acid nanoparticles (NPs), Hb/silver nanoparticles (AgNPs)-modified boron-doped diamond electrodes, and dimethyl octadecyl-ammonium bromide (DDAB)/Hb/glassy carbon electrodes. These biosensors have found applications across diverse fields, such as electrochemical bionanonsors, which work on the principle of amperometric and voltammetric detection of acrylamide, which is discussed in the following sections (Pundir, Yadav, and Chhillar 2019).

Identifying acrylamide in processed food items has been rendered feasible by advancing electrochemical techniques for acrylamide detection. These methods involve the measurement of parameters such as current generation, voltage, and impedance in the context of electron transfer resistance (R_{CT}), which is proportionate to the concentration of the analyte. These electrochemical biosensors are crafted by conjugating the catalytic core of hemoglobin (Hb) with acrylamide, forming an adduct during redox reactions (Abt et al. 2019).

The researcher developed the electrochemical biosensor based on single-stranded DNA (ssDNA) immobilized on a glassy carbon electrode (GE) using Au-S bonding, a novel electrochemical biosensor for the detection of acrylamide (AA). The association between ssDNA and AA was studied using differential pulse voltammetry (DPV) and UV-Vis absorption spectrometry. This ssDNA/GE biosensor immediately responded to AA, long-term stability, outstanding reliability, and high sensitivity. The detection threshold for AA under ideal circumstances was 8.1 nmol L^{-1} with a detection limit of 0.4–10.0 μmol L^{-1}. The biosensor had shown potential for monitoring AA pollution in environmental samples when it was successfully used to detect AA in tap water and potato chips. Further research found that a two-electron, two-proton transfer method was used to electrochemically oxidize the AA-ssDNA adduct on the GE surface, forming a one-to-one combination between ssDNA and AA (S. Huang et al. 2016).

Various voltammetric and amperometric biosensors have been specifically tailored to detect acrylamide. Voltammetric biosensors have harnessed configurations such as Hb/DDAB/carbon paste electrodes and Hb/single-walled carbon nanotube (SWCNT)/glassy carbon electrodes. On the other hand, amperometric biosensors have been founded upon setups like Hb/gold nanoparticles (AuNPs), Hb/cMWCNT/copper nanoparticles (CuNPs)/polyaniline (PANI)/pencil graphite electrodes, and Hb/cMWCNT-Fe$_3$O$_4$NP/chitosan (CHIT)/gold electrodes. Nonetheless, these methodologies are not without their constraints, which include the intricate fabrication of working electrodes and less-than-optimal analytical performance.

Immobilization of Hemoglobin NPs on a Gold Electrode: By covalently immobilizing HbNPs onto an Au electrode (HbNPs/AuE), an enhanced amperometric acrylamide biosensor was created and characterized by atomic force microscopy (AFM), transmission electron microscopy (TEM), UV-vis spectroscopy, Fourier transformation infrared (FTIR) spectroscopy, and X-ray diffraction (XRD) and used for the detection of acrylamide in biscuit, crisps and fried products. The electrochemical reaction occurs at the electrode surface, causing a drop in current. As a result, the electroactivity of Hb is altered because it decreases the flow of electrons from the active Au electrode that HbNPs immobilize. The response of the HbNPs/AuE biosensor was evaluated by monitoring the reduction in current during cyclic voltammetry, which resulted from the formation of

Hb-acrylamide adducts during the electrochemical reaction on the functional Au electrode surface. The HbNPs/AuE biosensor demonstrated a working potential range of −0.750 to +0.500 V, with the maximum current observed at 0.26 V. Optimization studies revealed an optimal pH of 5.0 In comparison to prior Hb-based biosensors, the HbNPs-based acrylamide biosensor demonstrated improved analytical performance in terms of a lower detection limit (0.10 nmol L^{-1}), working range of (0.05–100 mmol L^{-1}), and a more extended storage stability of 120 days (Yadav, Chhillar, and Pundir 2018).

Haemoglobin Immobilization Modified with Boron-Doped Diamond: The development of a biosensor for acrylamide (AA) using a modified boron-doped diamond electrode (Pt-BDD) with platinum (Pt) and hemoglobin (Hb) was carried out in recent studies. Characterization of the prepared electrode was carried out with the help of scanning electron microscopy with energy dispersive spectroscopy (SEM-EDS) and X-ray photoelectron spectroscopy (XPS). The Hb solution was dropped onto the Pt-BDD surface, immobilizing Hb. The globin structure of hemoglobin comprises electroactive heme sites. The PtNPs-modified Pt-BDD surface exhibited exceptional stability. The analysis of Hb using CV produced an oxidation peak of the quasi-reversible reaction caused by Hb-Fe^{3+}/Hb-Fe^{2+} as an active site in metal-based sensors, which includes gold and platinum (Umam, Saepudin, and Ivandini, 2017; Wulandari et al. 2019). Furthermore, the Hb-Pt-modified BDD (Hb-Pt-BDD) demonstrated a linear response in cyclic voltammetry (CV) within an acetate-buffered saline solution (0.2 mol L^{-1}, pH 4.8) over an AA concentration range of 0.01–1 nmol L^{-1}. The limit of detection (LOD) and limit of quantification (LOQ) were in the range of 0.0085 nmol L^{-1} and 0.026 nmol L^{-1}, respectively. The Hb-Pt-BDD electrode they developed exhibits high stability, excellent sensitivity, and reusability since it can remove Hb adducts without affecting the Pt on the BDD surface (Wulandari et al. 2019).

Single-walled carbon nanotubes (SWCNTs) and hemoglobin (Hb) were deposited onto glassy carbon electrodes to detect acrylamide in food directly. These electrodes work on the same principle of amperometric bionanosensors mentioned in the previous section. The innovative electrochemical biosensor in which Haemoglobin has been used as a biosensor's analytical active component has been developed. Four electroactive iron (III) hemes are prosthetic groups in hemoglobin, a redox-active protein. The electroactivity is caused by the reversible transformation of Hb-Fe^{3+}(III) to Hb-Fe^{2+}(II). These nano-biosensors operate on the principle of electron exchange between a recognition element and an electrode. The direct transmission of electrons by hemoglobin is complex, so electron mediators such as carbon paste electrodes modified with hemoglobin were used. Adduct hemoglobin-acrylamide is created by reacting both species, i.e., Hb-NH$_2$ groups. Hence, the electrode works on the voltammetry principle and osteryoung square-wave voltammogram (OSWV). Increasing the adduct concentration leads to decreased current, which acts as the signal for detecting acrylamide. The method was very sensitive and selective, with a detection limit of 1.2 × 10^{-10} mol L^{-1} (Stobiecka, Radecka, and Radecki 2007).

Immobilization of Hemoglobin (Hb) on Nanocomposites: The electrode's construction (Hb/cMWCNT/Fe$_3$O$_4$/CHIT/Au) of carboxylated multi-walled carbon nanotubes (cMWCNT) and iron oxide nanoparticles (Fe$_3$O$_4$NPs) electrodeposited onto Au electrode through chitosan (CHIT) film. This electrode was the enhancement of a gold electrode-based amperometric acrylamide biosensor that was meticulously characterized using scanning electron microscopy (SEM), Fourier-transform infrared (FTIR) spectroscopy, electrochemical impedance spectroscopy (EIS), and differential pulse voltammetry at different stages of development. This bionanosensor was modified amperometric, worked on the principle of cyclic voltammetry, and was used to evolve acrylamide in potato crisps. The enhanced biosensor primarily relied on the interaction between acrylamide and Hb, impacting the redox reactions of Hb, mainly the current generated during its reversible conversion between Fe^{2+} and Fe^{3+}. This biosensor exhibited optimal performance under conditions of pH 5.0, a temperature of 30°C, an 8-second response time, and a broad concentration ranging from 3–90 nmol L^{-1}, boasting an impressive detection limit of 0.02 nmol L^{-1} and a sensitivity of 36.9 µA/nmol L^{-1}/cm^2 (Batra et al. 2013).

A composite biosensor of Hb-DDAB/PtAuPd NPs/Ch-IL/MWCNTsIL/GCE was developed to achieve highly sensitive amperometric detection of acrylamide in various thermally processed foods. The biosensor involved the immobilization of Hb dimethyldioctadecylammonium bromide (Hb-DDAB), platinum-gold-palladium alloy nanoparticles (PtAuPd NPs), and chitosan-1-ethyl-3-methylimidazolium bis(trifluoromethylsulfonyl)imide (Ch-IL) onto a glassy carbon electrode (GCE) using multiwalled carbon nanotubes-IL (MWCNTs-IL). The biosensor operated by forming an adduct through the reaction of acrylamide with the α-NH_2 group of N-terminal valine in Hb, reducing the peak current of Hb-Fe^{3+}. The immobilized GCE was thoroughly characterized using various techniques, including electrochemical impedance spectroscopy (EIS), cyclic voltammetry (CV), energy dispersive X-ray spectroscopy (EDS), and scanning electron microscopy (SEM). Under optimal conditions, the biosensor could detect acrylamide using square wave voltammetry (SWV) within two linear concentration ranges: 0.03–39.0 nmol L^{-1} and 39.0–150.0 nmol L^{-1}. Notably, it achieved an impressive detection limit (LOD) of 0.01 nmol L^{-1}. The biosensor displayed exceptional selectivity, even in high concentrations of common interfering compounds. Recent studies have further validated this biosensor's stability, sensitivity, reproducibility, and rapid response, with a response time of less than 8 seconds. Consequently, the biosensor proved highly effective for detecting acrylamide in potato chips, yielding results comparable to the standard gas chromatography-mass spectrometry (GC-MS) method (Varmira et al. 2018).

A potentiometric biosensor for acrylamide quantification was engineered utilizing intact bacterial cells from Pseudomonas aeruginosa that were immobilized on a biorecognition element featuring an amidase enzyme. This enzyme facilitated the breakdown of acrylamide, releasing ammonium ions (NH_4^+) and organic acid. The biosensor was constructed by immobilizing these cells onto various membrane types in the presence of glutaraldehyde and an ammonium ion-selective electrode. It demonstrated a linear response within the concentration range of 0.1–4.0×10^{-3} mol L^{-1}, achieved a minimum response time of 55 seconds, and exhibited a storage stability of 54 days. Selectivity tests revealed that the biosensor cross-reacted with acetamide and formamide but not phenylacetamide, p-nitrophenylacetamide, and acetanilide (Silva et al. 2009).

The biosensor proved effective for quantifying acrylamide in industrial effluents, boasting an average substrate recovery rate of 93.3%. The use of polymeric membranes for disk preparation contributed to its affordability, as it relied on whole cells as a source of amidase activity. However, the biosensor's stability was compromised prematurely, leading to the development of a "sandwich" configuration utilizing two membrane disks. This innovative approach allowed the cells to persist between the membranes, preventing premature loss and ensuring prolonged functionality of the biosensor. Further investigation of the biosensor's performance revealed a characteristic response at 120 mV, a Nernstian slope of 48 mV decade^{-1}, a half-life of 27 days, and a 58.99 mV·mmol L^{-1} sensitivity for acrylamide. An innovative marker-free electrochemical sensor based on cells was developed to assess the potential harmful effects of acrylamide on pheochromocytoma cells. This sensor, known for its simplicity and high sensitivity, underwent surface modification involving the immobilization of gold nanoparticles (AuNPs) and subsequent electrochemical reduction with graphene oxide (GO). Comprehensive characterization of the modified AuNPs/GO electrode was performed using cyclic voltammetry (CV), electrochemical impedance spectroscopy (EIS), and differential pulse voltammetry (DPV). Incorporating reduced GO significantly enhanced the electron-transfer rate between the cell and the electrode surface, while AuNPs preserved the cells' bioactivity. The biosensor demonstrated a unique correlation with the logarithmic value of cell numbers within the range of 1.6×10^{-4} to 1.6×10^{-7} cells mL^{-1}, yielding an impressive relative standard deviation (RSD) value of 1.68%. Additionally, the DPV signal at a cell adsorption concentration of 1.6×10^{-7} cells mL^{-1} exhibited a decrease corresponding to the acrylamide concentration in the 0.1–5 mmol L^{-1} range, with an outstanding detection limit of 0.04 mmol L^{-1}. Morphological analysis, based on scanning electron microscopy (SEM) and 3-(4,5-dimethylthiazol-2-yl)-2,5-diphenyltetrazolium bromide, validated the findings from the electrochemical investigation (Pundir, Yadav, and Chhillar 2019; Batra and Pundir 2024).

5.1.3 MELAMINE

5.1.3.1 Overview

Melamine is an organic molecule and a base with a molecular formula of $C_3H_6N_6$. It is also known as 2,4,6-triamino-1,3,5-triazine. It was then produced by heating ammonia-containing carbonate, thiourea, cyanamide, or dicyandiamide (Li et al. 2015). Melamine may be decomposed into the metabolites ammeline, ammelide, and cyanuric acid, which have been used to adulterate foods or milk to boost their perceived protein level (Bann and Miller 1958). Since it has a significant amount of nitrogen, 66% by mass, introducing 1% melamine may cause the crude protein levels to increase over 4%. Melamine was unlawfully added to raw milk and dairy products to artificially increase the apparent crude protein content because traditional testing techniques like the Kjeldahl or Dumas tests cannot distinguish nitrogen sources (Kobayashi et al. 2010). Melamine and its metabolites are thought to be assimilated in the gastrointestinal system and accumulate as stones inside the kidney. The WHO recommended a new acceptable daily consumption of $0.2\,\text{mg}\,\text{kg}^{-1}$ of melamine based on weight in 2008.

5.1.3.2 Health Issue Regarding Consumption of Melamine

In recent times, melamine's existence has been identified in various food products, including milk-containing products, dried whole eggs, uncooked chicken eggs, non-dairy creamer, and ammonium bicarbonate. Melamine contamination has had severe consequences, particularly causing the formation of nephrolith resembling stones and sand in children's kidneys, leading to global fatalities attributed to melamine adulteration, associated with inflammatory responses and urinary bladder hyperplasia. The European Food Safety Authority (EFSA) 2008 established $0.5\,\text{mg}\,\text{kg}^{-1}$ of total body weight as the tolerated daily intake (TDI) to account for potential health risks from melamine exposure over a lifetime. This TDI considers short-term exposure scenarios, like repeated consumption of melamine-tainted products. Due to the kidney's vulnerability to high melamine levels, the EFSA applied the $0.5\,\text{mg}\,\text{kg}^{-1}$ TDI in a 2007 contamination case. Recent findings indicate that melamine and its analogs might have been responsible for fatalities linked to contaminated food consumption. Because of its propensity to form crystals, melamine is associated with kidney damage and has links to carcinogenicity via urolithiasis. Urolithiasis becomes more prevalent when melamine is mixed with similar substances like cyanuric acid.

Melamine induces kidney-related toxicity due to crystal formation and is linked to carcinogenicity via urolithiasis. Urolithiasis becomes more prevalent when melamine is combined with analogs like cyanuric acid (Puschner et al. 2007; Buur et al. 2008). In general, melamine-related human death rates are minimal. Based on analytical test sensitivity, the US FDA has established a safe level of 50 melamine parts billion^{-1} (ppb) (Tyan et al. 2009).

5.1.3.3 Detection Methods for Melamine

Sensor technology has been widely employed for detecting foodstuff residues for around 16 years. Chemical sensors and biosensors, in particular, have progressed quickly and attracted much attention. Chemical sensors serve as devices that convert chemical data, encompassing the concentration of a specific component in an illustration or the overall composition analysis, into a signal that can be analyzed effectively. Typically, these sensors consist of two fundamental elements linked in a series: a chemical identification system, a receptor, and a physicochemical transducer (Thévenot et al. 2001). A biosensor combines a biological component with a physicochemical detector to act as a tool for detecting a particular analyte. A biorecognition element, a biotransducer element, and an electrical system outfitted with a signal amplifier, processor, and display are typically its three primary parts. Based on their unique bio-recognition mechanisms, biosensors may be categorized into whole cells, DNA, tissues, organelles, microorganisms, and enzymatic and non-enzymatic receptor biosensors. The detected signals in these biosensors are typically transformed into electrical, thermal, or optical outputs for further analysis and interpretation (Farré, Brix, and Barceló 2005).

Sensors and biosensors for detecting melamine have been divided into various categories based on the recognition principle of the analyte. Resent sensors used for the detection of melamine are discussed here.

5.1.3.4 Sensors for the Detection of Melamine

Optical and electrochemical sensors have been implemented significantly to detect melamine in foods and environmental samples. Visual transduction methods are employed for chemical information spanning from the concentration of analyte and coupling kinetics towards molecular structure and microscopic imaging. Optical sensors with photonic properties are used in several ways for signal transduction. These consist of polarization, refractive index, fluorescence intensity, chemiluminescence, transmission, absorbance, and reactivity, as highlighted in the reference (Luchansky and Bailey 2012). Hence, colorimetric, fluorescence, and chemiluminescence optical sensors are used for melamine analysis. The colorimetric sensor employs a stable cyanuric acid-melamine complex facilitated by three complementary hydrogen bonds. This interaction, relying on diaminopyridine and diimide moieties, showcases reversible, specific, and cooperative properties ideal for molecular self-assembly. This method utilizes nanomaterials like AuNPs, AgNPs, Quantum Dots (QDs), and magnetic nanoparticles for melamine identification. Melamine's presence induces color changes in AuNPs (Ling and Huang 2010) due to hydrogen-bonding recognition, directly indicating its presence. Fluorescence sensors (Duan and Yang, 2024) employ diverse materials like organic dyes (Chen et al. 2013), quantum dots, and metal nanoclusters for their simplicity and sensitivity. Quantum dots (Gill, Zayats, and Willner 2008), with their quantum mechanical properties and unique optical characteristics, have been notably applied in fluorescence sensing. Chemiluminescence sensors utilize chemical reactions to emit light, with advantages like sensitivity and simplicity, often integrated into flow injection analysis systems. Photoluminescence techniques use mercapto-propionic acid-capped cadmium selenide quantum dots (m-CdSe QDs) as fluorescent labels for detecting melamine in milk, employing their structural and functional properties (N. Li et al. 2014). Electrochemical sensors are portable, simple and have moderate cost and offer benefits like sensitivity, selectivity, speed, and applicability to various analytical fields. Various electrochemical methods, including square wave voltammetry (SWV), differential pulse voltammetry (DPV), and cyclic voltammetry (CV), are useful for finding electroactive chemicals. Additionally, electrochemiluminescent (ECL) assays are favorable due to their combination of electrochemistry's simplicity with the chemiluminescence method's sensitivity and wide linear range.

5.1.3.5 Bionanosensors for Detection of Melamine

Biosensors have been widely employed as complementary approaches to standard analytical methods. These biosensors include enzymes, immune systems, heat, and piezoelectric. The various types of biosensors rely on nanomaterials to produce signals. Researchers have extensively explored and recognized distinct biosensors based on nanomaterial that detects food-related substances. Metal, carbon, quantum dots, and upconversion nanomaterials have been synthesized. Biosensors, such as metal-organic and covalent organic frameworks, recognize, transmit, and amplify signals. Nevertheless, challenges have arisen when utilizing individual or composite chemical materials for evaluating sensitive and precise detection (Huo et al. 2021).

Significant advancements in the field of optical biosensors have been accomplished over decades. They are leading to their swift evolution and widespread utilization across various critical areas such as food safety, security, life sciences, environmental monitoring, and medicine. Due to the progress in automated DNA-synthesis technology, DNA is recognized as a nanomaterial form. Additionally, DNA can be used to guide the assembly of various metal nanoparticle shapes. The initial instance of a functional nucleic acid-based biosensor observed the idea of 'molecular beacons' (Tyagi and Kramer 1996). These innovative DNA nanostructures capable of switching open up novel biosensing avenues, particularly in nucleic acid detection. Various bionanosensors, such as immunosensors, optical, electrochemical, aptamer-based, calorimetric, and fluorescence-based biosensors, have been reported.

5.1.3.6 Immunosensors

Immunoassay is an analytical technique primarily reliant on the specific binding affinity between an antibody and its corresponding antigen. By employing a suitable transducer, one can convert the level of binding into easily readable output signals. In creating biosensors utilizing an immunoassay as their biorecognition element, monoclonal, polyclonal, or recombinant antibody varieties are commonly employed (Farré, Kantiani, and Barceló 2007). These immunosensors find widespread applications across pharmaceutical analysis, toxicological examination, bioanalysis, clinical chemistry, and environmental assessment and can potentially analyze challenging matrices without requiring extensive pre-treatment. Various types of immunosensors have been fabricated for food contaminants such as melamine.

An inhibitory immunoassay-based optical biosensor method has been reported to examine liquid milk and baby formula samples. The study focused on a polyclonal melamine antibody made with a structural imitation known as 6-hydrazino-1,3,5-triazine-2,4-diamine (HTD). This method does away with the need for occurring derivatization. Instead, the carrier protein was given a bifunctional cross-linker before interacting with the hapten. After that, the produced antibody was added to an SPR optical biosensor immunoassay. The test's sensitivity was calculated as an IC50 and found to be 67.9 ng mL^{-1} in a buffer solution. There was no cross-reactivity between the antibody and by-products of the manufacture of melamine. Nevertheless, a notable cross-reactivity was found through the insecticide cyromazine, whose metabolite melamine is. After applying a sample matrix to the assay, a 0.5 mg mL^{-1} detection threshold was established for infant formula and liquid milk samples (Fodey et al. 2011).

Multiplexed Planar Waveguide Fluorescence Immunosensor (MPWFI): These immunoassays followed an indirect competitive format, where the target analytes competed with corresponding antibody-bioreceptor binding sites on the chip's antigen-coated surface. Each round of measurement took 20 minutes, comprising of regeneration. After the incubation of 4 minutes, the mixture flowed on the chip. Any analyte-antibody-Cy5.5 molecules are free after 5 minutes of the reaction interacting with the immobilized BSA-analyte conjugates within the chip. The sensor responses in each cycle decreased inversely with analyte concentration, indicating higher concentrations resulted in weaker antibody binding and vice versa. Minimal fluorescent signals from sites without binding BSA-analyte conjugates suggested free analyte-Ab-Cy5.5, and non-specific adsorption did not contribute to fluorescence. The immunosensor demonstrated ppb-level sensitivity in buffer and diluted milk samples, outperforming ELISA, SPR, and HPLC methods in detecting melamine and aflatoxin M1, except for LC/ESI-MS/MS due to complexity. The stability and reusability of the immunosensor were addressed by covalently attaching BSA-analyte conjugates for regeneration without compromising functionality. The sensor remained stable and reusable across multiple assays. Recovery rates were 85%–103%, compared to standard deviations between 1.3% and 6.5%, validating the method's accuracy and reliability for different milk products. However, two analytes can be detected simultaneously, so double-channel standard curves and suitable logistic correlation ($R^2 > 0.99$) were obtained. The detection limit was about 0.045 and 13.37 ng mL^{-1}, with corresponding working ranges of 0.073–0.400 ng mL^{-1} and 26.38–270.0 ng mL^{-1} (H. Guo et al. 2016).

5.1.3.7 Aptamer-Based Biosensors

A synthetic aptamer is a single-stranded nucleic acid or peptide molecule exhibiting high specificity and strong affinity for a specific target. Typically, it undergoes evaluation through a combinatorial technique known as the systematic evolution of ligands by exponential enrichment (SELEX) (Citartan et al. 2012). Aptamers have many benefits over antibodies since they can be changed rapidly, are appropriately stored, and are spontaneously synthesized. Due to their inherent advantages, they are frequently utilized as recognition components in developing biosensors. They are instrumental in detecting various substances, including metal ions, chemical compounds, peptides, proteins, and even organisms (Song et al. 2009). Extensive research has been conducted on establishing hydrogen bonds between thymine and 2,6-diamino-5-methylpyrimidin-4-1, a compound structurally akin to melamine.

Additionally, the ability of pyrimidine to form such bonds with purines has been thoroughly investigated. Utilizing these hydrogen bonds, single-stranded oligonucleotides containing thymine are linked to melamine. As a result, several aptamer-based biosensors for detecting melamine have been developed recently (Thadke et al. 2018).

Aptamer-Based Gold Nanoparticles: Advancements in nanobiosensor technology for detecting melamine include the utilization of aptamer-based gold nanoparticles, serving as a colorimetric biosensor. As Huang et al. detailed, this approach encompasses label-free and labeled gold nanoparticles. In this context, citrate ions with negative charges envelop the gold nanoparticles, ensuring their stability by preventing aggregation in aqueous solutions—however, these same citrate ions aggregate when there is a lot of salt present. Aptamers, specifically poly-T10, exhibit a robust affinity for gold nanoparticles, reinforcing their stability against NaCl-induced aggregation. When melamine is introduced, the aptamers competitively bond with melamine due to their stronger association, diminishing the salt tolerance of gold nanoparticles. This reduction in tolerance leads to the eventual aggregation of the gold nanoparticles. In this scenario, gold nanoparticles labeled with single-stranded DNA (ssDNA) are employed to detect melamine. The method depends on the interaction between thymine and melamine. Initially, selected ssDNA molecules are attached to the gold nanoparticle surface. The Gold nanoparticles with DNA functions form when the matching oligonucleotides hybridize, resulting in observable color changes. Melamine prompts this hybridization process. Upon melamine introduction, the functionalized gold nanoparticles congregate upon oligonucleotide-melamine binding, eliciting a shift in color between red and blue. The mechanism is shown in Figure 5.1. Melamine was measured at concentrations as minimal as 41.7 and 46.5 nmol L^{-1} (H. Huang et al. 2011) (Figure 5.2).

Fluorescence-Based Aptamer: A highly sensitive and specific melamine detection system in milk was developed using a fluorescent aptasensor that relies on the unique fluorescence characteristics of Tb^{3+} ions when interacting with G-rich single-stranded DNA. Specifically, a modified cDNA strand with six guanine bases (G6) hybridizes with the aptamer of melamine, forming double-stranded DNA. When melamine is absent, the structure exhibits weak fluorescence sensitization to Tb^{3+} ions, resulting in a low background signal. Nevertheless, the presence of melamine forms a unique "T-melamine-T" structure with the aptamer through hydrogen bonds (NH···O and NH···N), preventing the aptamer from pairing with the cDNA, causing the cDNA to remain single-stranded and allowing Tb^{3+} ions to emit strong fluorescence, serving as a clear indicator of melamine presence. Without the assistance of experts, the whole detection procedure, including sample preparation, can finished in 2 hours.

Furthermore, the aptasensor was cost-effective, with low expenses for Tb^{3+} and oligonucleotide synthesis. Utilized a standard fluorescence spectrophotometer keeps testing costs minimal. The method was proven simple, rapid, and reliable, successfully detected melamine in milk samples with linear ranges between 1.0 and 10 µg mL^{-1}, and the detection limit was 0.02 µg mL^{-1}, respectively. Finally, this aptasensor employed label-free fluorescent and recognition probes, working through a "mix-and-detect" process without immobilization or separation steps (C. Yang et al. 2022).

5.1.3.8 Electrochemical Biosensor

An electrochemical bionanosensor comprises a biosensor that utilizes an electrochemical transducer regarded as a chemically modified electrode (CME) since a biological layer was applied to a semiconducting, electronic, and ionic conducting material. Due to their quick reaction time, inexpensiveness, simplicity of usage, and great sensitivity, electrochemical biosensors offer a variety of benefits over other types of sensors. Electrochemical accumulation generates the concentrations

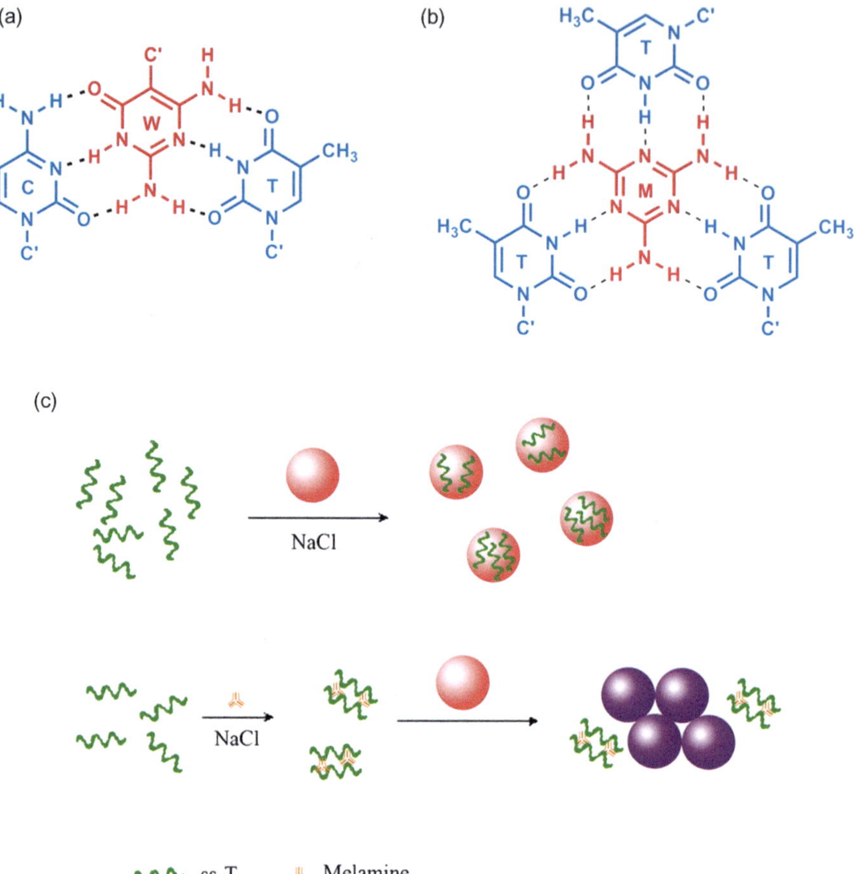

FIGURE 5.2 (a) Mentions the formation of a Janus Wedge base triplet resulting from the interaction between a target residue's W and C faces. (b) It describes the structure of the three hydrogen bonds recognized by melamine and thymine. (c) It outlines the operation of a melamine detection system that utilizes an AuNP optical sensor without needing label analysis of melamine Reproduced with permission (H. Huang et al. 2011).

necessary for examination by anodic stripping voltammetry and flameless atomic absorption spectrometry. A cyclic voltammetric method of melamine accumulation on a glassy carbon electrode (GCE) and chemical connection of horseradish peroxidase (HRP) with the melamine accumulating via the linkage using glutaraldehyde were presented by authors as the two phases for the sensor manufacturing process Guaiacol is oxidized by the coupled HRP, producing an amber-colored end product. By calculating the colored product's absorbance intensities, melamine was quantitatively analyzed. This method established a sizable range of linearity for melamine detection, ranging from 1.0×10^{-11} to 1.0×10^{-8} mol L^{-1} in concentration (Xu, Wei, Du, Li, Ji, Hu, et al. 2013).

The method combined the electrochemical accumulation process employing enzyme colorimetric assay (EA-ECA). Melamine's electrochemical behavior using unmodified glassy carbon electrodes (GCEs) was examined using cyclic voltammetry. Two distinct melamine concentrations and their cyclic voltammograms were investigated in a pH 3.0 solution. No oxidation peak was seen during the positive potential scan for a solution containing 1.0×10^{-8} mol L^{-1} melamine. However, oxidation peak currents increased at potentials higher than 1.2 V compared to reference electrodes such as standard calomel electrodes.

On the contrary, an anodic peak at approximately 1.5 V during the first positive scan developed when the solution had 1.0×10^{-4} mol L^{-1} melamine, suggesting the electro-oxidation of melamine. Several samples related to infant formula powders and fish feed were analyzed for melamine to evaluate the reliability of the EA-ECA technique. The results discovered the EA-ECA method demonstrated satisfactory recoveries, falling within the range of 94.5%–102.8%, which closely matched the added melamine quantities, which suggested that the EA-ECA approach is suitable for detecting the presence of melamine in authentic samples, mainly when melamine concentrations fall within the ranging from 1.0×10^{-8} to 5.0×10^{-10} mol L^{-1} (Xu, Wei, Du, Li, Ji, and Hu 2013).

5.1.3.9 Optical Biosensors

Optical biosensors based on optoelectronic and microelectronic methods to detect food allergens. In these biosensors, alterations in surface properties are measured, and the binding of an analyte with a sensor chip is responsible for creating a target-receptor complex. These sensors offer superior analysis for intricate food compositions and necessitate minimal sample processing (Nishi et al. 2015). Among non-surface plasmon resonance (SPR) methods, those utilizing DNA arrays can identify numerous food contaminants within a single assay, demonstrating significant efficacy (Watson 2004). DNA from contaminated samples can be amplified using polymerase chain reaction (PCR) to enhance sensitivity and precision. The amplified DNA molecules are then detected using colorimetric or fluorescent enzyme-labeled DNA hybridization assays. Optical signals can be observed with the naked eye or using sensor electrodes within point-of-care (POC) configurations (Alves et al. 2016). Surface plasmon resonance (SPR) biosensors employ surface plasmon waves, which are electromagnetic waves, to identify alterations that occur when the target analyte interacts with a biorecognition element on the sensor. As the SPR biosensor encounters changes, it triggers modifications in the refractive index. This alteration is then utilized to gauge or observe the reaction (Perumal and Hashim 2014).

A colorimetric sensor was created utilizing AuNPs treated with 1-(2-mercaptoethyl)-1,3,5-triazinane-2,4,6-trione (MTT) for detecting melamine in milk and dairy products. When the MTT-stabilized AuNPs were exposed to melamine, the hydrogen bonding between the two substances caused the AuNPs to aggregate. This caused the color of the AuNPs to change from red wine to purple, following the shift in the UV-vis spectra (Ai, Liu, and Lu 2009).

Dopamine-Stabilized Silver-Based Bionanosensors: Another researcher utilized a colorimetric output to identify melamine called dopamine-stabilized silver nanoparticles (AgNPs). Melamine and dopamine were designed to bind together by Michael's addition and Schiff base reactions, aggregating AgNPs and causing a colorimetric response that changed from yellow to red (Ma et al. 2011).

Fluorescence Probe Biosensors: Fluorescence techniques have revealed their efficacy as potent optical methods due to their great sensitivity, dependable equipment, simplicity of use, and capacity to evaluate different fluorescence properties. Many studies in this area utilize fluorescence resonance energy transfer (FRET) spectroscopy, which occurs when the donor and the acceptor's emission and absorption spectra overlap. Many melamine sensing systems leveraging FRET are built on the foundation of fluorescence generated by organic dyes (L. Guo et al. 2011; Tang, Du, and Su 2013). However, a significant drawback of most organic fluorophores is their susceptibility to photobleaching and poor stability under ambient conditions, resulting in varying fluorescence signals during analysis. Specific nanomaterials, like quantum dots (QDs), have been employed as signaling probes to overcome these restrictions. However, the use of QDs can be encumbered by complexities in their preparation process and ongoing debates about their relatively higher toxicity. Various types of nanobiosensors have been introduced based on fluorescence methods which include carbon dots and gold nanoparticles (AuNPs), gold nanoparticle incapacitate Sodium Yttrium fluoride (NaYF$_4$), rhodamine B and gold nanoparticles, porous MOF confined CsPbBr$_3$ quantum dots, CsPbBr$_3$ NCs with BaSO$_4$ and the AuNPs, Melamine–Thymine Recognition by label-free fluorescence detection respectively.

Carbon Dots and Gold Nanoparticles (AuNps): Carbon dots (C-dots), a burgeoning class of carbon nanomaterials, have incited considerable study endeavors across several fields. Numerous methods may produce these dots inexpensively (Buur et al. 2008). C-dots, acting as energy donors, were juxtaposed with AuNPs, which served as energy acceptors. Compared to conventional quantum dots (QDs) and organic dyes, photoluminescent C-dots exhibit remarkable attributes, including significant chemical inertness, impressive aqueous solubility, simple functionalization, photobleach resistance, and low toxicity. In a recent study, the effectiveness of Förster resonance energy transfer (FRET) between gold nanoparticles (AuNPs) and C-dots was reduced when melamine was present, leading to the development of a fluorescence sensor for melamine. When C-dots are added to an AuNPs solution, the close closeness of the C-dots to the surface of the AuNPs causes a quenching of fluorescence. Melamine with amino groups can outperform C-dots and restore fluorescence (Lv et al. 2018).

Nanobiosensor with Gold Nanoparticle Incapacitate Sodium Yttrium fluoride ($NaYF_4$): An additional nanobiosensor tailored for melamine detection in milk. This innovation involved merging gold nanoparticles with upconversion nanoparticles. Uncommon earth metals lanthanides (Ytterbium-Yb and Erbium-Er) were employed to incapacitate Sodium Yttrium fluoride ($NaYF_4$), creating upconversion nanoparticles distinguished by their unique fluorescence characteristics. The fluorescence of $NaYF_4$:Yb^{3+} Er^{3+} nanoparticles was suppressed due to their interaction with accompanying gold nanoparticles. However, the addition of melamine caused the gold nanoparticles to aggregate, causing them to detach from the surface of the upconversion nanoparticles. This event reinstated the fluorescence of the upconversion nanoparticles. The nanobiosensors exhibited a linear range from 32.0 to 500 nmol L^{-1}, boasting a detection limit of 18.0 nmol L^{-1} at a pH of 7.0. The entire process required an incubation period of 12 minutes, and the sensors displayed a sensitivity rating of 0.968 when applied to untreated milk samples (Wu et al. 2015).

Rhodamine B and Gold Nanoparticles: The study investigated the effect of melamine on absorption and fluorescence spectra of RB (a dye), gold nanoparticles, and RB-AuNPs associations. Mixing RB with melamine didn't cause any changes in their spectra, indicating no interaction. Combining RB with AuNPs led to a slight shift in the nanoparticles' absorption due to RB adsorption on their surface through electrostatic interactions. When melamine was added to RB-AuNP assemblies, the original absorption peak weakened, and a new peak emerged due to AuNP aggregation driven by melamine concentration. RB had minimal impact on this aggregation process. Both AuNPs and RB-AuNPs aggregated with melamine. A melamine detection method was proposed in dairy products, utilizing FRET among AuNPs and RB. FRET dampened RB's fluorescence when melamine was absent, but melamine disrupted this interaction, causing RB's fluorescence to "turn on." The method offered higher sensitivity compared to traditional colorimetric assays for melamine detection. The process involved FRET quenching RB's fluorescence in the presence of AuNPs and its recovery when melamine disrupted the FRET interaction. After optimization, the melamine detection method's performance was assessed. As melamine concentration increased, RB's fluorescence, previously reduced by FRET, recovered. The limit of detection (LOD) was 0.18 mg L^{-1}, and precision was confirmed for multiple measurements of 50 mg L^{-1} melamine; the relative standard deviation (RSD) was 2.1%. This method emphasized the consistent fluorescence response of RB-AuNPs to melamine (Cao et al. 2014).

A Porous MOF Confined $CsPbBr_3$ Quantum Dots: In recent years, the Perovskite quantum dots (PeQDs) nanocomposite-based method, enhancing sensitivity and stability for melamine detection, has been introduced. The researchers created a novel $CsPbBr_3$/HZIF-8 nanocomposite via a two-step in-situ growth process, incorporating luminescent $CsPbBr_3$ into a porous ZIF-8 metal-organic framework. The composite displayed robust green emission and remarkable resistance to moisture and UV light, even in aqueous settings, and served as a successful luminescent probe for detecting melamine in actual samples, capitalizing on MOF's porous structure for efficient analyte detection. The study established linear detection relationships for melamine within concentration ranges of 3–500 nmol L^{-1}, respectively, with detection limits of approximately 42.64 nmol L^{-1}. The CsPbBr3 PeQDs were evenly dispersed throughout the HZIF-8 MOF matrix and displayed strong green fluorescence at 510 nm with an FWHM value of 25 nm in sunshine. The nanocomposite demonstrated improved resistance to moisture and UV radiation. The $CsPbBr_3$/HZIF-8 composite was used as an on-off-on luminous probe for melamine detection in actual samples. The target analytes may be identified quickly due to the MOF's porous nature, and the nanosensor was discovered to be extremely sensitive to melamine detection (Ahmed et al. 2023).

Melamine–Thymine Recognition by Label-Free Fluorescence Detection: The suggested approach for melamine detection uses a simple fluorescence technique without labels. It employs polythymine DNA that can fold into a double-stranded structure in the presence of melamine. The interaction between melamine and polythymine DNA prevents degradation. SYBR Green I dye binds to the double-stranded DNA, leading to intense fluorescence, and was sensitive for assessing different melamine concentrations. The fluorescence signal intensity increased with melamine concentration, even at $6\,\mu\mathrm{mol}\ \mathrm{L}^{-1}$, below the maximum safe level ($7.9\,\mu\mathrm{mol}\ \mathrm{L}^{-1}$) set by the FDA for baby milk products. The fluorescence signal change exhibited a linear correlation in the melamine concentration of $10–200\,\mu\mathrm{mol}\ \mathrm{L}^{-1}$, with a calculated detection limit (LOD) of $1.58\,\mu\mathrm{mol}\ \mathrm{L}^{-1}$, below the FDA's limit for non-infant-formula products ($19.8\,\mu\mathrm{mol}\ \mathrm{L}^{-1}$). This approach demonstrates promise for melamine detection, leveraging melamine-thymine recognition and offering potential applications in milk quality assessment (Yang et al. 2018).

5.1.4 3-Monochloropropane-1,2-diol (3-MCPD)

5.1.4.1 Overview

3-Monochloropropane-1,2-diol (3-MCPD) is classified within the chloropropanols group, a category of chemical compounds characterized by their alcohol structure consisting of a three-carbon backbone accompanied by one or two chlorine atoms. During the 1980s, Professor Velı́šek and his colleagues initially demonstrated that chloropropanols, including chloroesters, could be generated as byproducts during the production of hydrolyzed vegetable proteins (HVP). This synthesis process involved subjecting protein by-products, such as soybean and rapeseed meal, to hydrolysis using hydrochloric acid (HCl). Acid-HVP found widespread use in manufacturing savory ingredients and culinary products like soy sauce. As research progressed into the 1990s and early 2000s, multiple surveys unveiled the presence of chloropropanols in acid HVP and various non-HVP food products, encompassing bread, bakery goods, cheese, fish, and meats. Concerns arose due to the relatively elevated levels of 3-MCPD in acid HVP, prompting manufacturers in the 1990s to explore alternative methods to reduce its prevalence. These efforts focused on altering processing conditions or transitioning to fully fermented ingredients. These endeavors resulted in a substantial reduction of 3-MCPD content in commercial savory products and soy sauces. Although initial reports regarding 3-MCPD esters (3-monochloropropane-1,2-diol esters) surfaced in the 1980s, their significance became more pronounced approximately two decades later when elevated concentrations were identified in refined vegetable oils/fats and food products. Subsequent comprehensive research endeavors prioritized the development of analytical techniques and an improved understanding of the formation mechanisms of MCPDEs (3-monochloropropane-1,2-diol esters) and glycidyl esters (GEs) in refined vegetable fats and oils. GEs came under scrutiny during investigations aimed at reconciling disparities among various analytical methods employed for MCPDE determination (Stadler 2015; Hamlet et al. 2002).

5.1.4.2 Structure and Formation of 3-Monochloropropane-1,2-diol in Food

The concentration of the precursor raw material before processing plays a significant role in forming 3-MCPDEs. Since 1991, researchers have been investigating cation and free radical-mediated molecular processes. 3-MCPD fatty acid esters are generated when monoacylglycerols (MAG), diacylglycerols (DAG), and triacylglycerols (TAG) react with chloride ions at elevated temperatures. Notably, the formation of 3-MCPDE is approximately 2–5 times more pronounced when derived from DAG precursors than MAG precursors. While partial acylglycerols (DAGs and MAGs) are identified as precursors in refined oils, their origin from TAGs remains debatable. These precursors and specific processing conditions can intensify the production of 3-MCPDE contaminants.

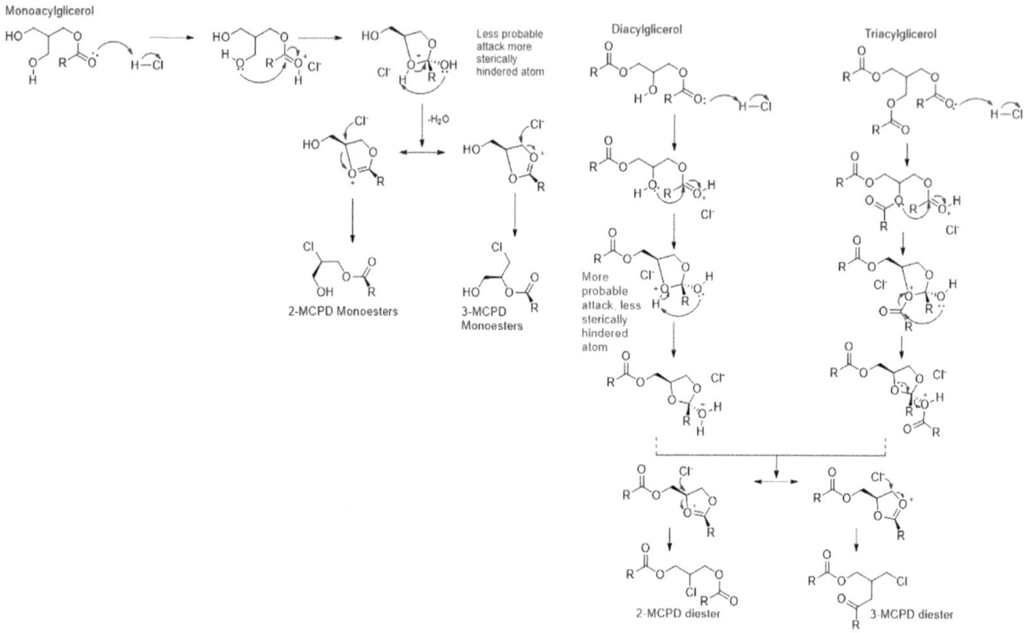

SCHEME 5.5 Mechanism of formation for MCPDE.

The formation of 3-MCPDE may be associated with refining fats or oils, particularly during the deodorization process, which involves steam distillation at temperatures ranging from 200°C to 270°C to eliminate undesirable compounds. This process can also lead to the production of trans-isomers, the thermal breakdown of triglycerides into free fatty acids (FFA), and the loss of sterols and tocopherols. Importantly, 3-MCPDE is not detectable in virgin or unrefined oils but can be generated through enzymatic release, acid hydrolysis, and thermal treatment (Syed Putra et al. 2023).

5.1.4.3 Exposure of 3-Monochloropropane-1,2-diol in Humans

A recent European Food Safety Authority (EFSA) report overviews 3-MCPD esters in various food items and includes an initial exposure assessment. In their evaluation, EFSA incorporated data spanning the years 2009 to 2011, considering advancements in data collection and analytical methodologies specific to 3-MCPD during that period. Exposure levels across all demographic groups varied from less than 1–$1.5\,mg\,kg^{-1}\,bw\,d^{-1}$, with 95% of exposure falling below 2–$3\,mg\,kg^{-1}\,bw\,d^{-1}$. Various studies have reported levels of 3-MCPD esters in different food commodities, revealing a wide range of values across distinct food groups and products. For instance, in cereals, values ranged from <0.01 to $1.4\,mg\,kg^{-1}$, while in coffee, they ranged from <0.1 to $0.4\,mg\,kg^{-1}$. Dairy products exhibited reported values spanning from non-detectable to $1.3\,mg\,kg^{-1}$. In the case of infant and baby food, values ranged from <0.01 to $1\,mg\,kg^{-1}$. For malt and beer, the observed range was 0.004–$0.7\,mg\,kg^{-1}$. Meat and meat products displayed values from non-detectable to $6.4\,mg\,kg^{-1}$, whereas smoked and pickled fish meat fell within the 0.3–$1.1\,mg\,kg^{-1}$ range.

Most studies focusing on ester-bound 3-MCPD have centered on vegetable oils and fats, cereals, and potato products. Fewer investigations have been conducted on other food categories, such as meat and dairy products. A recent study conducted by Schallschmidt and coworkers (2012) explored the formation of free 3-MCPD in meat and the influence of various grilling conditions (electric, gas, charcoal grilling, addition of salt, use of oil or emulsion marinade). The findings indicated significant variations in 3-MCPD levels depending on the pre-treatment and grilling conditions, with charcoal and gas-grilled steaks exhibiting higher levels than those prepared using an electric grill.

The highest contamination was observed in a steak pre-treated with an oily marinade and grilled on a charcoal grill with a closed lid. The analysis also examined the contribution of specific food groups to the overall mean exposure in various population groups. It was found that 'Margarine and similar products' and 'Vegetable fats and oils (excluding walnut oil)' were the primary contributors to total dietary exposure across the population groups. They were followed by 'Bread and rolls,' 'Fine bakery snacks,' and 'Preserved meat (smoked).

The EFSA reports and surveys show that, for most population groups (60 out of 64), the daily intake of 3-MCPD in their diets was generally below 1 μg kg^{-1} of body weight. In specific surveys involving 'Toddlers' and 'Other children,' the average dietary exposure ranged from 1 to 1.5 μg kg^{-1} bw day^{-1}. Among high consumers, around half of the population groups (30 out of 64, or 47%) had a 95th percentile (P95) dietary exposure below 1 μg kg^{-1} bw day^{-1}. For 26 population groups, the P95 exposure fell between 1 and 2 μg kg^{-1} bw day^{-1}. Additionally, eight population groups had P95 exposure levels ranging from 2 to 3 μg kg^{-1} bw day^{-1}. These groups included three from surveys of children, Adults, and Elderly (EFSA, 2018).

5.1.4.4 Toxicity Level of 3-Monochloropropane-1,2-diol

The International Agency for Research on Cancer (IARC) has categorized 3-MCPD as potentially carcinogenic to humans (Group 2B) and set a tolerable daily intake (TDI) of 0.002 mg kg^{-1} bw. There is the consideration that a benchmark dose approach might lead to a higher TDI of around 0.007 mg kg^{-1} bw. Limited data exists on the toxicity of 2-MCPD, but its mechanism of action is expected to be similar to 3-MCPD. A significant concern in assessing the risk of chloroesters exposure is the fate of MCPDEs and GEs in the gastrointestinal tract. Early studies suggested incomplete hydrolysis of these compounds, but later research confirmed the quantitative release of 3-MCPD and glycidol in the human intestine, potentially impacting dietary exposures. Glycidol is of higher concern due to its genotoxic mode of action, classified as probably carcinogenic to humans (Group 2A) by IARC. EFSA is conducting a risk assessment of chloroesters and GEs in foods, with the report expected in 2016. However, data gaps exist for composite foods due to non-validated analysis methods and potential overestimation of MCPD levels, partly because of poor fat extraction and the presence of glycidol (Syed Putra et al. 2023; Fattore et al. 2023).

5.1.4.5 Detection Methods of 3-Monochloropropane-1,2-diol

5.1.4.5.1 Molecularly Imprinted Polymer Bases Nanobiosensors

Nanoporous Gold Capped with Molecularly Imprinted Polymer (p-ATP): Molecularly imprinted polymer (MIP) stands out as a promising technique for detecting 3-MCPD in biosensors. MIP is fashioned through a process in which functional monomers are co-polymerized with template molecules. This results in the creation of cavities that precisely match the template molecules' shape, size, and functional groups. This unique feature enables MIP to identify and mitigate interference from other substances in a manner reminiscent of a "lock and key" mechanism, similar to how natural antibody-antigen and enzyme-substrate systems function. Various MIP sensors have been developed to selectively detect 3-MCPD, including electrochemical, impedimetric, and fluorescence sensing methods. Among these, the electrochemical approach, which relies on the redox reaction of the analyte within an electrochemical platform, offers distinct advantages such as cost-effectiveness, rapid response times, simplified operation, and affordable instrumentation. In the context of this sensor, an innovative electrochemical sensor was devised by combining nanoporous gold (NPG) with MIP for the swift detection of 3-MCPD. NPG was chosen for its uniform porous structure, enhanced conductivity, and exceptional catalytic properties, which hold significant potential for improving the sensitivity of electrochemical sensing. The sensor's construction involved the application of an NPG film onto the surface of a glass carbon electrode (GCE). Subsequently, the MIP membrane was formed on the NPG surface through a self-assembly process facilitated by electropolymerization. This approach augmented the sensor's electrocatalytic capabilities and furnished ample surface area for creating a stable MIP membrane. To validate the practical utility of the MIP/NPG/GCE sensor, it was employed to detect 3-MCPD. The preparation of the MIP involved following electrochemically grown film; a self-assembled monolayer was initially deposited onto the NPG/GCE electrode. The process consisted of these steps: (i) self-assembly of the functional monomer (p-ATP) onto the NPG/GCE electrode, (ii) adsorption of the template molecules (3-MCPD), (iii) electropolymerization, and (iv) subsequent removal of the template molecules.

The outcomes of investigations revealed that the MIP/NPG/GCE sensor displayed exceptional conductivity and exhibited sensitive electrochemical responses to 3-MCPD. These attributes were attributed to the synergistic impact of NPG's remarkable electrocatalytic properties and porous structure, as well as the outstanding specificity of the MIP membrane. Consequently, the optimized MIP/NPG/GCE sensor delivered exceptional detection capabilities, featuring a significantly broader linear detection range spanning from 10^{-16} to 10^{-7} mol L^{-1}, along with a notably low detection limit of 3.5×10^{-17} mol L^{-1} (S/N = 3). Additionally, the sensor demonstrated remarkable specificity, robustness against interference, and reproducibility. When applied to actual samples, the developed sensor yielded highly satisfactory results, with recovery rates falling from 105.90% to 107.52%. The constructed electrochemical sensor offered various advantages, including straightforward preparation, exceptional sensitivity, a broad linear detection range, and strong specificity. Furthermore, this approach holds the potential for future extension to detect other types of analytes (Cheng et al. 2022).

Molecularly Imprinted Overoxidized Polypyrrole MIP(oPPy) on Graphene Oxide (GO): Molecular Imprinting Technology (MIT) is employed for the precise detection of a wide range of substances, offering exceptional selectivity, affordability, and stability. Polypyrrole (PPy), a conjugated polymer, is frequently chosen in MIT due to its straightforward preparation and desirable attributes. Over-oxidized polypyrrole (oPPy) emerges as a promising polymer for MIT-based sensors. It can be synthesized through electrochemical means, creating specific binding sites tailored to target molecules. To augment MIT's efficacy, nanostructured materials like magnetic nanoparticles (MNPs), graphene oxide (GO), and carbon nanotubes are utilized to modify the surface of electrodes, enhancing sensitivity. GO is particularly advantageous due to its distinctive traits, such as a substantial surface area and adjustability through oxygen functionalization. Combining GO with MIT yields sensors characterized by superior selectivity and sensitivity. This research introduces an innovative sensor system that utilizes MIP(oPPy)-GO-modified disposable graphite surfaces to perform impedimetric analysis of 3-MCPD, a substance not previously studied. This sensor leverages the strengths of Electrochemical Impedance Spectroscopy (EIS), MIT, and single-use graphite electrodes to enable a straightforward, cost-effective, rapid, and exceptionally selective assessment of 3-MCPD. The sensor's performance is enhanced through meticulous optimization of factors like electropolymerization conditions, extraction, and rebinding durations, resulting in an extensive measurement range and a low detection threshold. The MIP(oPPy)-GO-based impedimetric sensor developed in this study exhibits outstanding capabilities in detecting 3-MCPD. Its practical applicability is demonstrated in the analysis of soy sauce samples. This sensor platform harbors significant potential for detecting various analytes within real-world matrices in forthcoming applications.

Sensor surfaces for 3-MCPD detection were prepared in two steps. First, Platinum Glassy Electrodes (PGE) were soaked in a Graphene Oxide (GO) solution and left to dry. Then, pyrrole was electropolymerized with or without 3-MCPD molecules in a pH 4.0 buffer. This created the sensor surface known as MIP(oPPy)-GO/PGE (before extraction). A stock solution of 3-MCPD was prepared and added to the pyrrole solution for making MIP surfaces. The research aimed to create a disposable impedimetric assay using MIP(oPPy)-GO for precisely detecting 3-MCPD, a contaminant often found in food processing, at concentrations as low as nanomolar levels. This is of paramount importance due to the potential health hazards associated with it. Remarkably, a remarkably low detection limit of 1.82 nmol L^{-1} was achieved without the necessity for enzymes, crosslinking agents, initiators, or mediators. The GO and MIP(oPPy) combination presented an economically efficient approach for detecting 3-MCPD. Additionally, the study examined potential interfering substances, including chloropropanol analogs, and found that they did not significantly affect the detection process (Yaman et al. 2021).

Molecularly Imprinted Polymers/Carbon Dots-Grafted Paper Sensor: Molecular imprinting is a promising approach to minimize interference from other substances in a sample and enhance the selective detection of specific target molecules. Molecularly imprinted polymers (MIPs) can be precisely tailored to match the size and shape of the template molecules, resulting in exceptional selectivity, especially when compared to closely related compounds. Previously, an indirect electrochemical sensor for detecting 3-MCPD was created using a MIP/nano Au electrode. However, its performance was affected by various factors, such as electrode polishing, temperature, humidity, anti-fouling properties, and interference from active molecules. Additionally, there have been attempts to develop an anion-imprinted polymer grafted paper-based sensor for detecting Cd(II) ions, but sensitivity remained a challenge. The ideal future approach would involve combining a paper-based device with MIPs/fluorescence composites, offering both high selectivity for target molecules and minimal interference from the sample matrix while retaining the practical handling characteristics of a typical paper device. Although some efforts have been made to apply fluorescence monomers or encapsulate fluorescent nanoparticles in MIP-based fluorescence sensors, these synthesis processes are often intricate. Carbon dots (CD), a type of fluorescence material, have gained popularity in rapid detection sensors due to their distinct emission spectra and vigorous signal intensity. In this context, carbon dots with amino groups were attached to filter paper containing carboxyl groups through a negative pressure and heating process. The MIP film was then synthesized using 3-MCPD as a template and methacrylic acid (MAA) as a monomer. The interactions between amino and carboxyl groups facilitated this direct formation on the surface. The resulting paper-based sensor was tested to detect 3-MCPD selectively and sensitively. The study utilized a one-step synthesis method to create a sensor based on the electrostatic interaction between amino groups (CDs) and carboxyl groups (MAA). This process involved treating filter paper with 5 M H_2O_2 for 3 hours, followed by thorough washing with ultrapure water. Amino group-containing CDs were passed through the filter paper ten times under negative pressure at 80°C. The resulting sensor was fabricated using this one-pot synthesis approach and subsequently cleaned with acetonitrile/acetic acid and methanol to remove any templates and unreacted components.

A non-imprinted composite (NIPs/CDs grafted paper) was also prepared using the same procedure but without adding 3-MCPD as a template molecule. A sensor utilizing MIPs/CDs grafted report was prepared and employed to detect 3-MCPD. In this process, CDs were directly imprinted onto the filter paper, capitalizing on electrostatic attraction to provide substantial fluorescence intensity. An MIP film was also synthesized on the CDs-coated paper, enabling the selective extraction of 3-MCPD. The maximum adsorption capacity reached 68.97 mg per gram, and the pseudo-second-order model proved effective for the selective adsorption of 3-MCPD. Under optimized conditions, the sensor demonstrated a linear detection range from 1 to 150 ng mL^{-1}, with an impressively low detection limit of 0.6 ng mL^{-1}. Following the guidelines of the Chinese National Standard (GB/T 2717-2003), the MIPs/CDs grafted paper-based sensor can be employed to detect 3-MCPD. Furthermore, this approach offers several appealing advantages, including high sensitivity, swift response, user-friendly operation, cost-effectiveness, and portability. It is worth noting that this sensor concept can be adapted for detecting other target molecules by designing new MIP films (Fang et al. 2019).

5.2 CONCLUSION

The increased consumption of thermally processed food items poses a significant hazard to all living organisms. Among the myriad food contaminants, one notable toxic substance is organic food, produced during different reactions and commonly found in baked goods, fried, oven-cooked, and cooked at high temperatures. It is responsible for various potential health risks, including carcinogenicity, neurotoxicity, reproductive toxicity, and cardiac toxicity. Different conventional detection methods have been employed to identify food organic contaminants, and biosensing methods are considered superior due to their simplicity, sensitivity, selectivity, speed, accuracy, and cost-effectiveness.

Based on our literature survey, we found only a few reports on methods using biosensors to detect nitrosamine. Two prominent methods for detecting nitrosamine in the literature are constructed using carbon dots as nanoparticles, whereas DNA and zien film are bio-components. One method uses an electrochemical sensing platform, while the other employs fluorometry. Besides, indirect biosensing methods of nitrosamine using nitrites as marker molecules were also found, utilizing two-dimensional materials like MoS_2 and graphene, or Co_3O_4 and rGO nanocomposites and Fe_3O_4/MoS_2 nanocomposite structures.

Biosensors for the detection of acrylamide have been widely reported with Hb/sol-gel film-modified carbon paste, gold nanoparticles (AuNPs)/carboxylated multiwalled carbon nanotubes (cMWCNT)/glassy carbon, poly [maleic anhydride-alt-butyl vinyl ether (AM41)-polyethylene glycol], citric acid nanoparticles (NPs), silver nanoparticles (AgNPs)-modified boron-doped diamond, and dimethyldioctadecyl-ammonium bromide (DDAB) glassy carbon systems with various electrochemical detection modes such as amperometry, voltammetry, and potentiometry are reported with biological species such as DNA, enzyme, protein hemoglobin (Hb) respectively.

A considerable number of biosensors for melamine detection have also been reported, such as optical, electrochemical, immunosensors, and aptamers using various nanoparticles that are bio-modified with DNA, nucleic acid, antigen, and peptide molecules like hemoglobin. Nanoparticles such as gold nanoparticles (AuNPs) treated with 1-(2-mercaptoethyl)-1,35-triazinane-2,4,6-trione (MTT), Sodium Yttrium fluoride ($NaYF_4$), rhodamine B and $CsPbBr_3$ NCs/$BaSO_4$ has been used for the detection of melamine in foods.

This review finds a good amount of work on biosensing selected organic contaminants in foods. There is a need to develop more biosensors for nitrosamine in food matrices, and increased efforts are also needed to commercialize the biosensors for all three contaminants. For that purpose, electrochemical or paper methods would be suitable for in-field or on-site analysis of foods.

Thus, Future research efforts should concentrate on developing miniaturizing laboratory models of food organic contaminants using nanobiosensors that could lead to the development of portable models for field use, offering insights for a more comprehensive understanding of food organic contaminants biochemistry, detection, and safety, and promoting progress in reducing potential toxic effects.

ACKNOWLEDGMENTS

We would like to sincerely thank the Scientific and Technological Research Council of Turkey (TUBITAK) for their generous support through the 2221-TUBITAK Scholarship program under the number 1059B212200690.

BIBLIOGRAPHY

Abt, Eileen, Lauren Posnick Robin, Sara McGrath, Jannavi Srinivasan, Michael DiNovi, Yoko Adachi, and Stuart Chirtel. 2019. "Acrylamide levels and dietary exposure from foods in the United States, an update based on 2011-2015 data." *Food Additives & Contaminants: Part A* 36 (10): 1475–1490.

Additives, EFSA Panel on Food, Nutrient Sources added to Food, Alicja Mortensen, Fernando Aguilar, Riccardo Crebelli, Alessandro Di Domenico, Birgit Dusemund, Maria Jose Frutos, Pierre Galtier, David Gott, and Ursula Gundert-Remy. 2017a. "Re-evaluation of potassium nitrite (E 249) and sodium nitrite (E 250) as food additives." *EFSA Journal* 15 (6): e04786.

Additives, EFSA Panel on Food, Nutrient Sources added to Food, Alicja Mortensen, Fernando Aguilar, Riccardo Crebelli, Alessandro Di Domenico, Birgit Dusemund, Maria Jose Frutos, Pierre Galtier, David Gott, and Ursula Gundert-Remy. 2017b. "Re-evaluation of sodium nitrate (E 251) and potassium nitrate (E 252) as food additives." *EFSA Journal* 15 (6): e04787.

Ahmed, Shahnaz, Suman Lahkar, Simanta Doley, Dambarudhar Mohanta, Swapan Kumar Dolui. 2023. "A hierarchically porous MOF confined CsPbBr3 quantum dots: Fluorescence switching probe for detecting Cu (II) and melamine in food samples." *Journal of Photochemistry, and Photobiology A: Chemistry* 443: 114821.

Ai, Kelong, Yanlan Liu, and Lehui Lu. 2009. "Hydrogen-bonding recognition-induced color change of gold nanoparticles for visual detection of melamine in raw milk and infant formula." *Journal of the American Chemical Society* 131 (27): 9496–9497.

Alves, Rita C, M Fátima Barroso, María Begoña González-García, M Beatriz PP Oliveira, Cristina Delerue-Matos. 2016. "New trends in food allergens detection: Toward biosensing strategies." *Critical Reviews in Food Science, and Nutrition* 56 (14): 2304–2319.

Bahadır, Elif Burcu, and Mustafa Kemal Sezgintürk. 2015. "Applications of commercial biosensors in clinical, food, environmental, and biothreat/biowarfare analyses." *Analytical Biochemistry* 478: 107–120.

Bann, Bernard, and Samuel A Miller. 1958. "Melamine and derivatives of melamine." *Chemical Reviews* 58 (1): 131–172.

Barnes, John M., Peter N. Magee. 1954. "Some toxic properties of dimethylnitrosamine." *British Journal of Industrial Medicine* 11 (3): 167.

Herbert D. Bartsch, Berthold Spiegelhalder 1996. "Environmental exposure to N-nitroso compounds (NNOC) and precursors: An overview." *European Journal of Cancer Prevention* 5: 11–17.

Basaran, Burhan, and Ozlem Faiz. 2022. "Determining the levels of acrylamide in some traditional foods unique to Turkey and risk assessment." *Iranian Journal of Pharmaceutical Research: IJPR* 21 (1): e123948.

Batra, Bhawna, Suman Lata, CS Pundir. 2013. "Construction of an improved amperometric acrylamide biosensor based on hemoglobin immobilized onto carboxylated multi-walled carbon nanotubes/iron oxide nanoparticles/chitosan composite film." *Bioprocess and Biosystems Engineering* 36: 1591–1599.

Batra, Bhawna, and Chandra S Pundir. 2024. "Detection of acrylamide by biosensors." In *Acrylamide in Food*, 581–590. Elsevier.

Bhattacharya, Sagarika, Subhra Samanta, and Biswarup Chakraborty. 2021. *Nitric Oxide Sensing*. CRC Press.

Bogue, Robert. 2008. "Nanosensors: A review of recent progress." *Sensor Review* 28 (1): 12–17.

Bogue, Robert. 2009. "Nanosensors: A review of recent research." *Sensor Review* 29 (4): 310–315.

Buur, Jennifer L, Ronald E Baynes, Jim E Riviere. 2008. "Estimating meat withdrawal times in pigs exposed to melamine contaminated feed using a physiologically based pharmacokinetic model." *Regulatory Toxicology, and Pharmacology* 51 (3): 324–331.

Cao, Xianyi, Fei Shen, Minwei Zhang, Jiajia Guo, Yeli Luo, Jingyue Xu, Ying Li, Chunyan Sun. 2014. "Highly sensitive detection of melamine based on fluorescence resonance energy transfer between rhodamine B and gold nanoparticles." *Dyes, and Pigments* 111: 99–107.

Capuano, Edoardo, and Vincenzo Fogliano. 2011. "Acrylamide and 5-hydroxymethylfurfural (HMF): A review on metabolism, toxicity, occurrence in food and mitigation strategies." *LWT-Food Science and Technology* 44 (4): 793–810.

Çebi, Ayşegül. 2024. "Acrylamide intake, its effects on tissues and cancer." In *Acrylamide in Food*, 65–93. Elsevier.

Challis, Brian C. Challis 1996. "Environmental exposures to N-nitroso compounds and precursors: General review of methods and current status." *European Journal of Cancer Prevention* 5: 19–26.

Chen, Gengwen, Fengling Song, Xiaoqing Xiong, Xiaojun Peng. 2013. "Fluorescent nanosensors based on fluorescence resonance energy transfer (FRET)." *Industrial, and Engineering Chemistry Research* 52 (33): 11228–11245.

Cheng, Weiwei, Qiaoyun Zhang, Di Wu, Yuling Yang, Yan Zhang, and Xiaozhi Tang. 2022. "A facile electrochemical method for rapid determination of 3-chloropropane-1,2-diol in soy sauce based on nanoporous gold capped with molecularly imprinted polymer." *Food Control* 134: 108750.

Citartan, Marimuthu, Subash CB Gopinath, Junji Tominaga, Soo-Choon Tan, Thean-Hock Tang. 2012. "Assays for aptamer-based platforms." *Biosensors, and Bioelectronics* 34 (1): 1–11.

Committee, EFSA Scientific, Anthony Hardy, Diane Benford, Thorhallur Halldorsson, Michael John Jeger, Katrine Helle Knutsen, Simon More, Alicja Mortensen, Hanspeter Naegeli, and Hubert Noteborn. 2017. "Update: Use of the benchmark dose approach in risk assessment." *EFSA Journal* 15 (1): e04658.

Hermann Druckrey, Rudolf A. Preussmann, Dietrich F. Schmähl, and M Müller. 1961. "Erzeugung von Magenkrebs durch Nitrosamide an Ratten." *Naturwissenschaften* 48 (6): 165–165.

Duan, Ning Duan, Shaoxiang Yang 2024. "Research Progress on Multifunctional Fluorescent Probes for Biological Imaging, Food and Environmental Detection" *Critical Reviews in Analytical Chemistry* 54 (4): 775-817. doi: 10.1080/10408347.2022.2098670.

Dybing, Erik, Farmer, Peter B.F. Melvin E. Andersen, Sam P.D. Lalljie, Detlef J.G. Müller, Stephen S. Olin, Barbara J. Petersen, Josef Rudolf Schlatter, Gabriele Scholz, Joseph A. Scimeca, Nadia Slimani, Margareta Å. Törnqvist, Sandra Tuijtelaars, Philippe J.P. Verger. 2005. "Human exposure and internal dose assessments of acrylamide in food." *Food and Chemical Toxicology* 43 (3): 365–410.

Ellen, Geert, Els Egmond, and Emile T Sahertian. 1986. "N-nitrosamines and residual nitrite in cured meats from the Dutch market." *Zeitschrift fur Lebensmittel-Untersuchung und-Forschung* 182 (1): 14–18.

European Food Safety Authority (EFSA). 2018. "Update of the risk assessment on 3-monochloropropanediol and its fatty acid esters." *EFSA Journal* 16 (1): 5083.

Farré, Marinella, Rikke Brix, and Damià Barceló. 2005. "Screening water for pollutants using biological techniques under European Union funding during the last 10 years." *TrAC Trends in Analytical Chemistry* 24 (6): 532–545.

Farré, Marinella, Lina Kantiani, and Damià Barceló. 2007. "Advances in immunochemical technologies for analysis of organic pollutants in the environment." *TrAC Trends in Analytical Chemistry* 26 (11): 1100–1112.

Fang, Min, Lv Zhou, Hu Zhang, Liang Liu, Zhi-Yong Gong. 2019. "A molecularly imprinted polymers/ carbon dots-grafted paper sensor for 3-monochloropropane-1,2-diol determination." *Food Chemistry* 274: 156–161.

Fattore, Elena, Alessia Lanno, Alberto Danieli, Simone Stefano, Alice Passoni, Alessandra Roncaglioni, Renzo Bagnati, and Enrico Davoli. 2023. "Toxicology of 3-monochloropropane-1,2-diol and its esters: a narrative review." *Archives in Toxicology* 97 (5): 1247–1265.

Fodey, Terence L, Colin S Thompson, Imelda M Traynor, Simon A Haughey, D Glenn Kennedy, and Steven RH Crooks. 2011. "Development of an optical biosensor based immunoassay to screen infant formula milk samples for adulteration with melamine." *Analytical Chemistry* 83 (12): 5012–5016.

Friedman, Mendel. 2003. "Chemistry, biochemistry, and safety of acrylamide. A review." *Journal of Agricultural and Food Chemistry* 51 (16): 4504–4526.

Gill, Ron, Maya Zayats, and Itamar Willner. 2008. "Semiconductor quantum dots for bioanalysis." *Angewandte Chemie International Edition* 47 (40): 7602–7625.

Guo, Hongli, Xiaohong Zhou, Yan Zhang, Baodong Song, Jingxuan Zhang, and Hanchang Shi. 2016. "Highly sensitive and simultaneous detection of melamine and aflatoxin M1 in milk products by multiplexed planar waveguide fluorescence immunosensor (MPWFI)." *Food Chemistry* 197: 359–366.

Guo, Liangqia, Jianhai Zhong, Jinmei Wu, FengFu Fu, Guonan Chen, Yongxuan Chen, Xiaoyan Zheng, and Song Lin. 2011. "Sensitive turn-on fluorescent detection of melamine based on fluorescence resonance energy transfer." *Analyst* 136 (8): 1659–1663.

Habermeyer, Michael, and Gerhard Eisenbrand. 2009. "N-nitrosamines, including N-nitrosoaminoacids and potential further nonvolatiles." In *Process-Induced Food Toxicants: Occurrence, Formation, Mitigation, and Health Risks*, 365–386. Wiley.

Habermeyer, Michael, and Gerhard Eisenbrand. 2019. N-Nitroso Compounds in Foods. In: Varelis, P., Melton, L., Shahidi, F. *Encyclopedia of Food Chemistry*, 593–602. Elsevier. doi:10.1016/ b978-0-08-100596-5.21824-6

Habermeyer, Michael, Angelika Roth, Sabine Guth, Patrick Diel, Karl-Heinz Engel, Bernd Epe, Peter Fürst, Volker Heinz, Hans-Ulrich Humpf, and Hans-Georg Joost. 2015. "Nitrate and nitrite in the diet: How to assess their benefit and risk for human health." *Molecular Nutrition & Food Research* 59 (1): 106–128.

Haldorai, Yuvaraj, Jun Yeong Kim, AT Ezhil Vilian, Nam Su Heo, Yun Suk Huh, Young-Kyu Han. 2016. "An enzyme-free electrochemical sensor based on reduced graphene oxide/Co_3O_4 nanospindle composite for sensitive detection of nitrite." *Sensors, and Actuators B: Chemical* 227: 92–99.

Halford, Nigel G, and Tanya Curtis. 2019. *Acrylamide in Food*. World Scientific.

Colin G. Hamlet, Peter A. Sadd, Colin Crews, Jan Velíšek, Denise E. Baxter. 2002 "Occurrence of 3- chloropropane-1,2-diol (3-chloro-1,2-propanediol and related compounds in foods: a review." *Food Additives and Contaminants* 19: 619–631.

Hecht, Stephen S. 1997. "Approaches to cancer prevention based on an understanding of N-nitrosamine carcinogenesis." *Proceedings of the Society for Experimental Biology and Medicine* 216 (2): 181–191.

Stanley W. B. Ewen, John M. Stowers. No hits for Helgason T, JR Outram or JR Pollock. 1984. "N-Nitrosamines in smoked meats and their relation to diabetes." *IARC Scientific Publications* (57): 911–920.

Hinz, Jana, Tessema F Mekonnen, Jonas Bergrath, Savanna Sewell, Yannic Schneck, Michaela Wirtz, and Ursula Telgheder. 2023. "Analysis of nine Nitrosamines relevant to occupational safety by ion mobility spectroscopy and preliminary gas chromatographic separation." *Journal of Chromatography Open*: 100102.

Huang, Hui, Li Li, Guohua Zhou, Zhihong Liu, Qiao Ma, Yuqi Feng, Guoping Zeng, Philip Tinnefeld, and Zhike He. 2011. "Visual detection of melamine in milk samples based on label-free and labeled gold nanoparticles." *Talanta* 85 (2): 1013–1019.

Huang, Shan, Shuangyan Lu, Chusheng Huang, Jiarong Sheng, Lixia Zhang, Wei Su, and Qi Xiao. 2016. "An electrochemical biosensor based on single-stranded DNA modified gold electrode for acrylamide determination." *Sensors and Actuators B: Chemical* 224: 22–30. https://doi.org/10.1016/j.snb.2015.10.008. https://www.sciencedirect.com/science/article/pii/S0925400515304755.

Huang, Xiaoping, Yufang Zhu, and Ehsan Kianfar. 2021. "Nano biosensors: Properties, applications and electrochemical techniques." *Journal of Materials Research and Technology* 12: 1649–1672.

Huo, Bingyang, Yuling Hu, Zhixian Gao, and Gongke Li. 2021. "Recent advances on functional nucleic acid-based biosensors for detection of food contaminants." *Talanta* 222: 121565.

Keeper, Larry K, and Peter P Roller. 1973. "N-nitrosation by nitrite ion in neutral and basic medium." *Science* 181 (4106): 1245–1247.

Keramat, Javad, Alain LeBail, Carole Prost, and Maryam Jafari. 2011. "Acrylamide in baking products: a review article." *Food and Bioprocess Technology* 4: 530–543.

Keramat, Javad, Alain LeBail, Carole Prost, and Nafiseh Soltanizadeh. 2011. "Acrylamide in foods: chemistry and analysis. A review." *Food and Bioprocess Technology* 4: 340–363.

Kobayashi, Takahiro, Atsushi Okada, Yasuhiro Fujii, Kazuhiro Niimi, Shuzo Hamamoto, Takahiro Yasui, Keiichi Tozawa, and Kenjiro Kohri. 2010. "The mechanism of renal stone formation and renal failure induced by administration of melamine and cyanuric acid." *Urological Research* 38: 117–125.

Olga Korostynska, Andrew Lawrence Mason and Abdullrahman A. Al-shamma. 2012. "Monitoring of nitrates and phosphates in wastewater: current technologies and further challenges." *International Journal on Smart Sensing and Intelligent Systems* 5 (1): 149.

Vijay Lahiri, Pankaj Khanna, Bimla Elhence, Kalpana Singh, and PK Wahal. 1988. "Nitrosamine in leather dust extracts." *British Journal of Industrial Medicine* 45 (9): 647.

Li, Na, Danqing Liu, Hua Cui. 2014. "Metal-nanoparticle-involved chemiluminescence and its applications in bioassays." *Analytical, and Bioanalytical Chemistry* 406: 5561–5571.

Li, Ying, Jingyue Xu, and Chunyan Sun. 2015. "Chemical sensors and biosensors for the detection of melamine." *Rsc Advances* 5 (2): 1125–1147.

Ling, Jian, and Cheng Zhi Huang. 2010. "Energy transfer with gold nanoparticles for analytical applications in the fields of biochemical and pharmaceutical sciences." *Analytical Methods* 2 (10): 1439–1447.

Lingnert, Hans, Spiros Grivas, Margaretha Jägerstad, Kerstin Skog, Margareta Törnqvist, and Per Åman. 2002. "Acrylamide in food: mechanisms of formation and influencing factors during heating of foods." *Scandinavian Journal of Nutrition* 46 (4): 159–172.

Loeppky, Richard N, Yen T Bao, Jaeyoung Bae, Li Yu, and Graziella Shevlin. 1994. "Blocking nitrosamine formation: Understanding the chemistry of rapid nitrosation." ACS Publications.

Lofstedt, Ragnar E. 2003. "Science communication and the Swedish acrylamide "alarm"." *Journal of Health Communication* 8 (5): 407–32. https://doi.org/10.1080/713852123.

Loomis, Dana, Wei Huang, and Guosheng Chen. 2014. "The International Agency for Research on Cancer (IARC) evaluation of the carcinogenicity of outdoor air pollution: Focus on China." *Chinese Journal of Cancer* 33 (4): 189.

López-Rodríguez, Rocío, James A McManus, Natasha S Murphy, Martin A Ott, and Michael J Burns. 2020. "Pathways for N-nitroso compound formation: Secondary amines and beyond." *Organic Process Research & Development* 24 (9): 1558–1585.

Luchansky, Matthew S, and Ryan C Bailey. 2012. "High-Q optical sensors for chemical and biological analysis." *Analytical Chemistry* 84 (2): 793–821.

Lv, Man, Yang Liu, Jinhui Geng, Xiaohong Kou, Zhihong Xin, Dayong Yang. 2018. "Engineering nanomaterials-based biosensors for food safety detection." *Biosensors* 106: 122–128.

Ma, Yurong, Hongyun Niu, Xiaole Zhang, and Yaqi Cai. 2011. "One-step synthesis of silver/dopamine nanoparticles and visual detection of melamine in raw milk." *Analyst* 136 (20): 4192–4196.

Magee, Peter N, and JM Barnes. 1956. "The production of malignant primary hepatic tumours in the rat by feeding dimethylnitrosamine." *British Journal of Cancer* 10 (1): 114.

Majumdar, Sristi, Debajit Thakur, and Devasish Chowdhury. 2020. "DNA carbon-nanodots based electrochemical biosensor for detection of mutagenic nitrosamines." *ACS Applied Bio Materials* 3 (3). 1796–1803.

Meah, MN, N Harrison, and A Davies. 1994. "Nitrate and nitrite in foods and the diet." *Food Additives & Contaminants* 11 (4): 519–532.

Mirvish, Sidney S. 1975. "Formation of N-nitroso compounds: chemistry, kinetics, and in vivo occurrence." *Toxicology and Applied Pharmacology* 31 (3): 325–351.

Mottram, Donald S, Bronislaw L Wedzicha, and Andrew T Dodson. 2002. "Acrylamide is formed in the Maillard reaction." *Nature* 419 (6906): 448–449.

Mun'delanji, C Vestergaard, Kagan Kerman, I- Ming Hsing, and Eiichi Tamiya. 2015. *Nanobiosensors and Nanobioanalyses.* Springer.

Nishi, Kentaro, Shin-Ichiro Isobe, Yun Zhu, and Ryoiti Kiyama. 2015. "Fluorescence-based bioassays for the detection and evaluation of food materials." *Sensors* 15 (10): 25831–25867.

Österdahl, Bengt Göran. 1988. "Volatile nitrosamines in foods on the Swedish market and estimation of their daily intake." *Food Additives & Contaminants* 5 (4): 587–595.

Ouyang, Qin, Yanna Rong, Baoning Wang, Waqas Ahmad, Shuangshuang Liu, and Quansheng Chen. "Transforming On-Site Detection of N-Nitrosodimethylamine: A Novel Fluorescent Biosensor Utilizing Zein Film and Upconversion Nanoparticles." *Available at SSRN 4478145.*

Ouyang, Qin, Li Wang, Waqas Ahmad, Yawen Rong, Huanhuan Li, Yuqian Hu, and Quansheng Chen. 2021. "A highly sensitive detection of carbendazim pesticide in food based on the upconversion-MnO$_2$ luminescent resonance energy transfer biosensor." *Food Chemistry* 349: 129157.

Ouyang, Qin, Mingming Zhang, Baoning Wang, Waqas Ahmad, and Quansheng Chen. 2023. "Development of a novel upconversion fluorescence nanosensor based on metalloporphyrin element for sensitive detection of N-nitrosodimethylamine." *Sensors and Actuators B: Chemical* 393: 134260.

Parr, Maria Kristina, and Jan F Joseph. 2019. "NDMA impurity in valsartan and other pharmaceutical products: Analytical methods for the determination of N-nitrosamines." *Journal of Pharmaceutical and Biomedical Analysis* 164: 536–549.

Penttilä, Pirjo-Liisa, Leena Räsänen, and Sinikka Kimppa. 1990. "Nitrate, nitrite, and N-nitroso compounds in Finnish foods and the estimation of the dietary intakes." *Zeitschrift fur Lebensmittel-Untersuchung und-Forschung* 190 (4): 336–340.

Perumal, Veeradasan, and Uda Hashim. 2014. "Advances in biosensors: Principle, architecture and applications." *Journal of Applied Biomedicine* 12 (1): 1–15.

Pfundstein, B, AR Tricker, E Theobald, B Spiegelhalder, and R Preussmann. 1991. "Mean daily intake of primary and secondary amines from foods and beverages in West Germany in 1989–1990." *Food and Chemical Toxicology* 29 (11): 733–739.

Przybyla, Jennifer, Heather Carlson-Lynch, Nickolette Roney, Mario Citra, and Claire Heit. 2022. "Toxicological profile for n-nitrosodimethylamine (NDMA): Draft for public comment." Centers for Disease Control and Prevention. https://stacks.cdc.gov/view/cdc/115424

Pundir, Chandra S, Neelam Yadav, and Anil Kumar Chhillar. 2019. "Occurrence, synthesis, toxicity and detection methods for acrylamide determination in processed foods with special reference to biosensors: A review." *Trends in Food Science & Technology* 85: 211–225.

Puschner, Birgit, Robert H Poppenga, Linda J Lowenstine, Michael S Filigenzi, and Patricia A Pesavento. 2007. "Assessment of melamine and cyanuric acid toxicity in cats." *Journal of Veterinary Diagnostic Investigation* 19 (6): 616–624.

Ridgway, Kathy, Sam PD Lalljie, and Roger M Smith. 2007. "Sample preparation techniques for the determination of trace residues and contaminants in foods." *Journal of Chromatography A* 1153 (1–2): 36–53.

Rifai, Lubna, and Fatima A Saleh. 2020. "A review on acrylamide in food: Occurrence, toxicity, and mitigation strategies." *International Journal of Toxicology* 39 (2): 93–102.

Schallschmidt, Kristin, Alexander Hitzel, Margarete Pöhlmann, Fredi Schwägele, Karl Speer, and Wolfgang Jira. 2012. "Determination of 3-MCPD in grilled meat using pressurized liquid extraction and gas chromatography-high resolution mass spectrometry." *Journal fuer Verbraucherschutz und Lebensmittelsicherheit* 7 (3): 203–210.

Schwarzenegger, Arnold, Linda S Adams, and Joan E Denton. 2006. "*N*-nitrosodimethylamine." Public Health Goals for Chemicals in Drinking Water. https://oehha.ca.gov/media/downloads/water/chemicals/phg/122206ndmaphg.pdf

Sen, Nrisinha P. 1988. "Migration and formation of N-nitrosamines from food contact materials." *ACS Symposium Series* 365: 146–158. doi: 10.1021/bk-1988-0365.ch012

Sharp, David. 2003. "Acrylamide in food." *The Lancet* 361 (9355): 361–362.

Silva, Nelson, Dulce Gil, Amin Karmali, and Manuel Matos. 2009. "Biosensor for acrylamide based on an ion-selective electrode using whole cells of Pseudomonas aeruginosa containing amidase activity." *Biocatalysis and Biotransformation* 27 (2): 143–151.

Smith, Peter AS, and Richard N Loeppky. 1967. "Nitrosative cleavage of tertiary amines." *Journal of the American Chemical Society* 89 (5): 1147–1157.

Song, Yujun, Chao Zhao, Jinsong Ren, and Xiaogang Qu. 2009. "Rapid and ultra-sensitive detection of AMP using a fluorescent and magnetic nano-silica sandwich complex." *Chemical Communications* (15): 1975–1977.

Spiegelhalder, Berthold, Eisenbrand, Gerhard, Preussmann, Rudolf A. 1980a. "Occurrence of volatile nitrosamines in food: a survey of the West German market." *IARC Scientific Publications* (31): 467–479.

Spiegelhalder, Berthold, Eisenbrand, Gerhard, Preussmann, Rudolf A. 1980b. "Volatile nitrosamines in food." *Oncology* 37 (4): 211–216.

Srivastava, Anup K, Atul Dev, and Surajit Karmakar. 2018. "Nanosensors and nanobiosensors in food and agriculture." *Environmental Chemistry Letters* 16: 161–182.

Stadler, Richard H, Imre Blank, Natalia Varga, Fabien Robert, Jörg Hau, Philippe A Guy, Marie-Claude Robert, and Sonja Riediker. 2002. "Acrylamide from Maillard reaction products." *Nature* 419 (6906): 449–450.

Stadler, Richard. 2015. "Monochloropropane-1,2-diol esters (MCPDEs) and glycidyl esters (GEs): An update" *Current Opinion in Food Science* 6: 12–18.

Stobiecka, Agata, Hanna Radecka, and Jerzy Radecki. 2007. "Novel voltammetric biosensor for determining acrylamide in food samples." *Biosensors and Bioelectronics* 22 (9): 2165–2170. https://doi.org/10.1016/j.bios.2006.10.008. https://www.sciencedirect.com/science/article/pii/S0956566306004969.

Syed Putra, Sharifah Shahira, Wan Jefrey Basirun, Amal Elgharbawy, Maan Hayyan, Waleed Al Abdulmonem, Abdullah Aljohani and Adeeb Hayyan Basirun. 2023. "3-Monochloropropane-1,2-diol (3-MCPD): a review on properties, occurrence, mechanism of formation, toxicity, analytical approach and mitigation strategy." *Food Measure* 17: 3592–3615.

Tang, Guangchao, Liping Du, and Xingguang Su. 2013. "Detection of melamine based on the fluorescence resonance energy transfer between CdTe QDs and Rhodamine B." *Food Chemistry* 141 (4): 4060–4065.

Thadke, Shivaji A, J Dinithi R Perera, VM Hridya, Kirti Bhatt, Ashif Y Shaikh, Wei-Che Hsieh, Mengshen Chen, Chakicherla Gayathri, Roberto R Gil, Gordon S Rule. 2018. "Design of bivalent nucleic acid ligands for recognition of RNA-repeated expansion associated with Huntington's disease." *Biochemistry* 57 (14): 2094–2108.

Thévenot, Daniel R, Klara Toth, Richard A Durst, and George S Wilson. 2001. "Electrochemical biosensors: recommended definitions and classification." *Analytical Letters* 34 (5): 635–659.

Timmermann, Clara Amalie Gade, Signe Sonne Mølck, Manik Kadawathagedara, Anne Ahrendt Bjerregaard, Margareta Törnqvist, Anne Lise Brantsæter, and Marie Pedersen. 2021. "A review of dietary intake of acrylamide in humans." *Toxics* 9 (7): 155.

Anthony R. Tricker, Stanislaw J. Kubacki. 1992. "Review of the occurrence and formation of non-volatile N-nitroso compounds in foods." *Food Additives & Contaminants* 9 (1): 39–69.

Anthony R. Tricker, Beate Pfundstein, E. Theobald, Rudolf A. Preussmann, Berthold Spiegelhalder. 1991. "Mean daily intake of volatile N-nitrosamines from foods and beverages in West Germany in 1989–1990." *Food and Chemical Toxicology* 29 (11): 729–732.

Anthony R. Tricker, Rudolf A. Preussmann. 1991. "Carcinogenic N-nitrosamines in the diet: Occurrence, formation, mechanisms and carcinogenic potential." *Mutation Research/Genetic Toxicology* 259 (3–4): 277–289.

Tyagi, Sanjay, and Fred Russell Kramer. 1996. "Molecular beacons: probes that fluoresce upon hybridization." *Nature Biotechnology* 14 (3): 303–308.

Tyan, Yu-Chang, Ming-Hui Yang, Shiang-Bin Jong, Chih-Kuang Wang, Jentaie Shiea. 2009. "Melamine contamination." *Analytical and Bioanalytical Chemistry* 395: 729–735.

Khoirul Umam, Endang Saepudin, Tribidasari Anggraningrum Ivandini. 2017. "Preparation of hemoglobin-modified boron-doped diamond for acrylamide biosensors." *IOP Conference Series: Materials Science and Engineering*. IOP Publishing.

Varmira, Kambiz, Omid Abdi, Mohammad-Bagher Gholivand, Hector C. Goicoechea, and Ali R. Jalalvand. 2018. "Intellectual modifying a bare glassy carbon electrode to fabricate a novel and ultrasensitive electrochemical biosensor: Application to determination of acrylamide in food samples." *Talanta* 176: 509–517. https://doi.org/10.1016/j.talanta.2017.08.069. https://www.sciencedirect.com/science/article/pii/S0039914017308974.

Ingrid T. M. Vermeer, Jan W. Dallinga, Jos C. S. Kleinjans, Jan M. S. van Maanen. 1998. "Volatile N-nitrosamine formation after intake of nitrate at the ADI level in combination with an amine-rich diet." *Environmental Health Perspectives* 106 (8): 459–463.

Watson, David. 2004. *Pesticide, Veterinary and Other Residues in Food.* Sawston, UK: Woodhead Publishing.

Witt, Otto N. 1878. "XXIII.-On aromatic nitrosamines." *Journal of the Chemical Society, Transactions* 33: 202–211.

Wu, Qiongqiong, Qian Long, Haitao Li, Youyu Zhang, and Shouzhuo Yao. 2015. "An upconversion fluorescence resonance energy transfer nanosensor for one step detection of melamine in raw milk." *Talanta* 136: 47–53.

Wulandari, Retno, Tribidasari Anggraningrum Ivandini, Endang Saepudin, Yasuaki Einaga. 2019. "Modification of boron-doped diamond electrodes with platinum to increase the stability and sensitivity of haemoglobin-based acrylamide sensors." *Sensors and Materials* 31: 1105–1117.

Xu, Qin, Huan Ping Wei, Shi Du, Hong Bo Li, Zhen Ping Ji, and Xiao Ya Hu. 2013. "Detection of subnanomolar melamine based on electrochemical accumulation coupled with enzyme colorimetric assay." *Journal of Agricultural and Food Chemistry* 61 (8): 1810–1817. https://doi.org/10.1021/jf304034e. https://doi.org/10.1021/jf304034e.

Yadav, Neelam, Anil Kumar Chhillar, and Chandra S Pundir. 2018. "Preparation, characterization and application of haemoglobin nanoparticles for detection of acrylamide in processed foods." *International Journal of Biological Macromolecules* 107: 1000–1013.

Yang, Chuanyu, Caiyi Du, Ruifang Su, Junyang Wang, Ying Li, Xinyue Ma, Zhihong Li, and Chunyan Sun. 2022. "A signal-on fluorescent aptasensor by sensitized Tb3+ luminescence for detection of melamine in milk." *Talanta* 236: 122842.

Yaman, Yesim Tugce, Gulcin Bolat, Turkan Busra Saygin, Serdar Abaci. 2021. "Molecularly imprinted label-free sensor platform for impedimetric detection of 3-monochloropropane-1,2-diol." *Sensors and Actuators B: Chemical* 328: 128986.

Yang, Hualin, Jiujun Wang, Qinghua Wu, Yun Wang, Li Li, and Baomiao Ding. 2018. "Simple and label-free fluorescent detection of melamine based on melamine-thymine recognition." *Sensors* 18 (9): 2968.

Zhang, Gong, Huijuan Liu, Jiuhui Qu, and Jinghong Li. 2016. "Two-dimensional layered MoS 2: rational design, properties and electrochemical applications." *Energy & Environmental Science* 9 (4): 1190–1209.

Zhang, Ya, Peng Chen, Fangfang Wen, Bo Yuan, and Honggui Wang. 2016. "Fe3O4 nanospheres on MoS2 nanoflake: Electrocatalysis and detection of Cr (VI) and nitrite." *Journal of Electroanalytical Chemistry* 761: 14–20.

Zhong, Wenwan. 2009. "Nanomaterials in fluorescence-based biosensing." *Analytical and Bioanalytical Chemistry* 394: 47–59.

Zhu, Yingchun, Kunsheng Zhang, Lizhen Ma, Nairui Huo, Hua Yang, and Jiaomin Hao. 2015. "Sensory, physicochemical, and microbiological changes in vacuum packed channel catfish (Clarias lazera) patties during controlled freezing-point storage." *Food Science and Biotechnology* 24: 1249–1256.

6 Analysis of Inorganic Ions

Jamil A. Buledi, Amber R. Solangi,
Muhammad Nawaz, and Ali Hyder

6.1 INTRODUCTION

Nanomaterials have ushered in a new era of advancement in the field of analytical chemistry, particularly in the detection and quantification of inorganic ions like nitrates (NO_3^-), sulfates (SO_4^{2-}), and phosphates (PO_4^{3-}). These ions play critical roles in various natural and human-induced processes, making their accurate assessment crucial for evaluating water quality, soil fertility, and pollution levels. While conventional methods for detecting these ions have been widely used, they often suffer from sensitivity, precision, and speed limitations. The emergence of nanomaterials has opened up exciting possibilities for enhancing ion detection and quantification, capitalizing on their unique properties (Amali et al., 2021; Sushila Singh, Sangwan, Sharma, Devi, & Moond, 2021; Taqvi et al., 2022). Nanomaterials encompass diverse structures, including metallic nanoparticles, nanotubes, and nanocomposites, each offering distinct advantages for improving the accuracy and precision of inorganic ion analysis. One key advantage of nanomaterials is their size-dependent behavior, which can lead to enhanced ion adsorption and selective recognition (Buledi, Amin, Haider, Bhanger, & Solangi, 2021; Buledi, Mahar, et al., 2022; Hyder et al., 2022). For instance, nanoparticles' large surface area allows for efficient binding of target ions, resulting in heightened sensitivity in detection methods (Amin et al., 2021; Khand et al., 2021). Researchers have explored various types of nanomaterials, such as graphene-based substances, metal nanoparticles, and nanowires, to develop innovative ion sensors and probes (Buledi, Shah, Mallah, & Solangi, 2022; Singh, Hasan, Sharma, & Narang, 2022; Wang, Yin, Zeng, & Zhu, 2019; Yu et al., 2015). In the domain of nitrate analysis, nanomaterials have exhibited significant promise. Nitrate contamination in food, especially water bodies, poses considerable health and environmental risks. Traditional colorimetric methods often lack the requisite sensitivity for accurate quantification, especially at low concentrations.

In contrast, nanomaterial-based sensors offer the potential for real-time and ultra-sensitive nitrate detection. Modified carbon nanotubes, for example, have demonstrated the capacity to enhance nitrate adsorption and enable rapid electrochemical detection. These advancements facilitate on-site monitoring and early detection of nitrate pollution in food sources (Mahmud et al., 2020; Tyagi, Rawtani, Khatri, & Tharmavaram, 2018). Similarly, sulfate analysis benefits from nanomaterials' distinct attributes. Sulfate ions are widespread across industries, demanding precise quantification for regulatory compliance and quality control. Nanomaterial-enhanced analytical methods have streamlined sulfate ion separation and preconcentration from intricate matrices, improving detection thresholds. Magnetic nanoparticles coated with specific ligands, constituting nanocomposite materials, have proven effective in selectively capturing sulfate ions from food samples. This approach enhances analytical performance and simplifies the sample preparation (Copley & Barton, 1994; Fernando, Ilankoon, Syed, & Yellishetty, 2018). Nanomaterial-based techniques have demonstrated the ability to amplify phosphate signals in phosphate ion monitoring, which is crucial for effective agricultural practices and preventing over-fertilization. This enables swift and accurate quantification. Researchers have advanced phosphate detection using nanoporous materials, functionalized nanoparticles, and nanoscale catalysts through fluorescence, colorimetry, and

DOI: 10.1201/9781003514039-6

electrochemistry techniques. These innovations contribute significantly to sustainable agriculture and environmental preservation (Ratkovski et al., 2020; Razanajatovo et al., 2021).

Moreover, this chapter focuses on the different nanomaterials used as sensing tools to analyze nitrates, phosphates, and sulfate ions in food samples.

6.2 NITRATE ION ANALYSIS

In most approaches, like colorimetric techniques, the detection of nitrate commonly involves an indirect process where it is first converted to nitrite. This conversion is achieved by leveraging enzymes or a range of metal catalysts like copper, zinc, mercury, or cadmium. However, these reactions require significant time and are not well-suited for on-site applications. The advancement of direct measurement techniques for nitrate is unquestionably essential. These methods offer simplicity, cost-effectiveness, and the potential for real-time analysis. Embracing such direct measurement approaches would greatly aid in managing nitrate levels and mitigating its environmental impact (Nishan et al., 2023).

6.3 NANOMATERIAL-BASED ELECTROCHEMICAL SENSOR FOR DETECTION OF NITRATES

Commercial ion-selective electrodes (ISEs) designed for nitrate detection typically exhibit responses based on the Hofmeister series, which ranks ions by their affinity for the electrode surface. In this series, perchlorate (ClO_4^-), thiocyanate (SCN^-), nitrate (NO_3^-), bicarbonate (HCO_3^-), acetate (CH_3COO^-), sulfate (SO_4^{2-}), and hydrogen phosphate (HPO_4^{2-}) ions are arranged in order of decreasing affinity (Cattrall, 1997). Given the high concentration of chloride ions in seawater, achieving accurate nitrate detection necessitates a selectivity coefficient for nitrate over chloride in the range of 105–107 (Cattrall, 1997). Notably, Watt and coworkers (Watt, Zakharov, Haley, & Johnson, 2013) introduced a receptor that exhibited a strong nitrate-π interaction, displaying a binding trend of $NO_3^- > Cl^- > Br^- > I^-$. This receptor was built upon a bis(arylethynyl)pyridine framework.

Ionophores play a crucial role in facilitating specific ionic transport across membranes. They form reversible complexes with analyte ions, consisting of a hydrophilic pocket for ion transport and a hydrophobic exterior for membrane incorporation. (Crespo, 2017; Mistlberger, Crespo, & Bakker, 2014). While nitrate ISE determination in water samples poses fewer challenges due to minute background ion concentrations, accurate quantification in marine waters is more complex. A potentiometric sensor based on a plasticized PVC membrane has been shown to provide reliable data in wastewater analysis. However, its relatively short membrane lifetime limits its use in continuous long-term monitoring (Hassan, Sayour, & Al-Mehrezi, 2007). Mazloum and coworkers (Ardakani, Salavati-Niasari, & Jamshidpoor, 2004) achieved an impressive nitrate over chloride selectivity of ~4,000 using bis(2-hydroxyacetophenone)ethylenediamine vanadium(IV) as an ionophore. Hassan and coworkers (Hassan et al., 2007) improved potentiometric performance by incorporating multi-walled carbon nanotubes (MWCNTs), resulting in nanomolar sensitivity and reduced potential drift. Le Goff and coworkers (Le Goff, Braven, Ebdon, & Scholefield, 2002) demonstrated enhanced durability and reduced detection limits (~0.34 μmol L^{-1}) using membranes immobilizing triallyl leucine betaine chloride on a plasticized polystyrene-block-polybutadiene-block-polystyrene copolymer.

Wang and coworkers (2018) achieved excellent selectivity for nitrate detection (10–1,000 mg L^{-1} or kg^{-1}) by incorporating a NO_3^- ionophore film on graphene microelectrode arrays. Traditional ISEs can be influenced by variations in natural organic matter and surfactant concentrations, prompting the development of polypyrrole membrane-based potentiometric sensors that exhibit linear responses to nitrate over a wide concentration range. These sensors have shown a promising correlation with laboratory results and improved durability through chemical immobilization of the ionophore (Bendikov, Kim, & Harmon, 2005).

Ion-selective field-effect transistors (ISFETs) offer the potential for miniaturization and seamless integration into electronic circuits alongside charge-sensitive membranes. These transistors present a promising solid-state alternative to traditional ion-selective electrodes (ISEs) (Bi et al., 2006). An innovative design by Myers and coworkers introduced a highly sensitive nitrate sensor based on an AlGaN/GaN transistor. This sensor employed a PVC-based membrane and operated without a reference electrode (Myers et al., 2013). The AlGaN/GaN interface's spontaneous formation of a two-dimensional electron gas (2DEG) layer eliminated the requirement for a reference electrode, known for its long-term stability issues.

Additionally, AlGaN/GaN exhibited enhanced chemical stability, electron transduction, and mobility compared to silicon (Asadnia et al., 2017). Figure 6.1 depicts the device, accompanied by

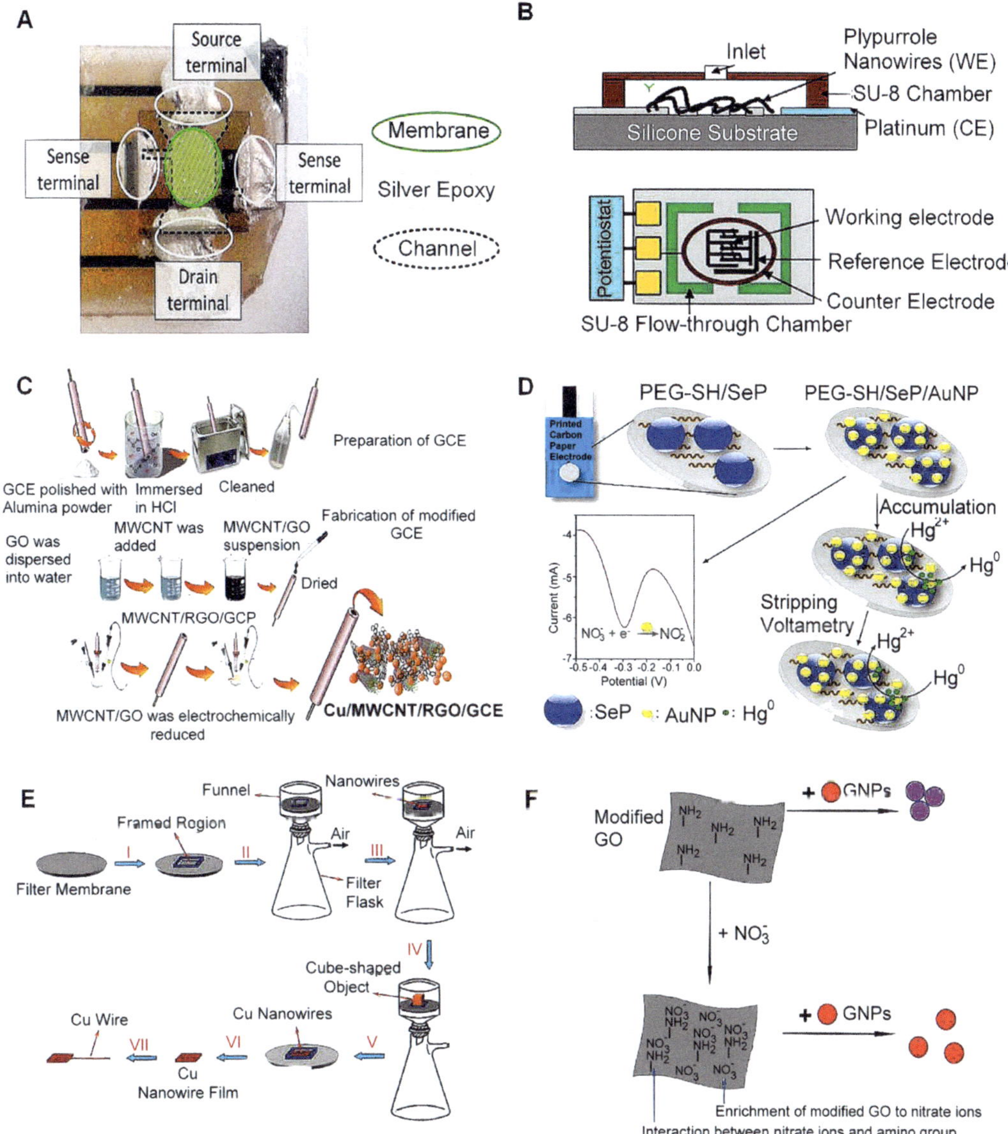

FIGURE 6.1 Nitrate detection chemical sensors in aquatic medium, (a) A nitrate ion detection device employs an AlGaN/GaN heterostructure. The sensor layout shows an overlay illustration. (b) A microfluidic sensor specific to nitrate utilizes doped-polypyrrole nanowires. The example includes a flow-through electrochemical test cell. (c) Electrode fabrication involves depositing Cu/MWCNT/rGO onto a glassy carbon electrode (GCE). The procedure comprises GCE polishing, cleaning, dispersion of GO and MWCNT, drop-casting onto GCE, drying, electrochemical reduction, and immersion in a CuSO₄·5H₂O and H₂SO₄ solution. Copyright © 2023 with permission from Elsevier (Mahmud et al., 2020).

a schematic of the sensor layout. Notably, the sensor's sensitivity without a PVC coating was significantly weaker than that of the PVC-coated nitrate-sensitive membrane. However, the challenge of PVC membrane adhesion to the ALGaN layer could hinder long-term device performance. To address this, an alternative approach could involve depositing chalcogenides on the GaN cap layer replacing the PVC membrane (Gao & Bao, 2020).

In a distinct method, Cuartero and coworkers achieved direct potentiometric nitrate determination in seawater through coupling with an electrochemical desalination module. This module reduced chloride concentration by three orders of magnitude, mitigating a significant interference related to ionophore-based membranes. The nitrate-selective electrode in this setup utilized lipophilic functionalized multi-wall carbon nanotubes (f-MWCNTs) as an ion-to-electron transducer, achieving a limit of detection of 0.5 μmol L^{-1} (Cuartero, Crespo, & Bakker, 2015). Another innovation involved using graphene oxide (rGO) nanosheets, a well-known field-effect transistor (FET) material, to enhance nitrate sensor sensitivity. Surface decoration of rGO sheets in the FET device with benzyl triethylammonium chloride (TEBAC) led to improved selectivity, rapid response, and sensitivity. Despite these advancements, challenges related to sensor adhesion persisted (Chen et al., 2018).

While potentiometric methods have demonstrated success in nitrate quantification, challenges in achieving selectivity within complex sample matrices like seawater persist. Pankratova et al. developed an array of potentiometric sensors using polymeric membrane materials to measure various ions in lake water. Their integrated system allowed continuous monitoring of nutrients by pumping lake samples from specific depths and calibrating the sensor array at regular intervals (Pankratova, Cuartero, Cherubini, Crespo, & Bakker, 2017). The same group achieved *in situ* nitrate and nitrite measurements in seawater using a submersible probe containing MWCNT-coated glassy carbon electrodes. Coupling desalination, acidification units, and membrane electrodes enabled nitrate and nitrite detection with a rapid response time and long-term stability. Miniaturized sensors within a flow cell reduced nutrient detection limits to 0.6 μmol L^{-1} through inline desalination and acidification (Cuartero et al., 2015).

6.4 NANOMATERIAL MODIFIED VOLTAMMETRIC SENSORS FOR NITRATE DETECTION

It has been the focus of extensive research, spanning almost two decades, to develop various nitrate sensors for applications in food samples. Aravamudhan and Bhansali introduced a nitrate-selective sensor based on doped-polypyrrole nanowires (PPy-NWs). This involved synthesizing PPy-NWs through controlled oxidation of pyrrole monomers with FeCl$_3$ within nanoporous alumina templates. The resulting PPy-NWs were effectively utilized as a working electrode for electrochemical measurements of nitrate concentrations (Aravamudhan & Bhansali, 2008). A highly sensitive amperometric nanobiosensor was developed by Gokhale and coworkers, involving the immobilization of nitrate reductase on a poly(3,4-ethylenedioxythiophene) (PEDOT) copolymer within the pores of a polycarbonate membrane. This nanosensor demonstrated swift fabrication, rapid response times, low background noise, and a low detection limit of around 0.16 mg L^{-1}. Its linear response range was 200–1,100 μg L^{-1}. Notably, the binding efficiency of nitrate reductase in the PEDOT matrix was notably higher than for interfering ions, signifying elevated sensitivity and selectivity towards nitrate ions. However, the sensor's performance decreased after storage at 4°C due to the instability of the PEDOT micro-emulsion (Gokhale, Lu, Weerasiri, Yu, & Lee, 2015).

Bagheri and coworkers developed an innovative electrochemical sensor using Cu nanoparticles on a composite of multi-wall carbon nanotubes and reduced graphene oxide nanosheets fixed onto a glassy carbon electrode for simultaneous nitrite and nitrate ion determination. This configuration yielded enhanced sensitivity and selectivity, demonstrating selectivity over 200 times compared to common interfering anions at nanomolar concentrations (Bagheri, Hajian, Rezaei, & Shirzadmehr, 2017).

6.5 DIRECT DETECTION OF NITRATES IONS USING NANOMATERIALS

In a research conducted in 2013 by Mora and coworkers, they explored the application of cysta-mine-functionalized AuNPs for directly detecting nitrate in water samples, eliminating the need for secondary reactions (Mura et al., 2015). A robust S-Au bond facilitated the attachment of cys-teamine to the AuNPs' surface. The exposed $-NH_2$ groups of cysteamine interacted with nitrate ions, inducing the cross-linking (aggregation) of the nanoparticles. This process shifted visible color from red to purple/blue. Nevertheless, the practicality of the direct nanoplasmonic approach was constrained due to the relatively sluggish reactivity of nitrate ions. As a result, alternative strategies for nitrate detection were explored, focusing on indirect methods that involve nitrate reduction. This can be achieved through various means, including using enzymes and metals for catalyzing nitrate reduction. These approaches offer promising avenues for enhancing the sensitivity and selectivity of nitrate detection techniques.

6.6 INDIRECT ANALYSIS OF NITRATES IONS

In an indirect approach, the detection of nitrate involves a secondary step of reduction to nitrite, facilitated either by enzymes or metals. Enzymatic reduction entails the conversion of nitrate to nitrite through the action of nitrate reductase enzymes. Despite its simplicity, environmental friend-liness, rapidity, and heightened sensitivity (Hooda, Sachdeva, & Chauhan, 2016), this technique is associated with considerable expenses. Moreover, challenges such as limited nitrate reductase activity, susceptibility to heat, poor electrical conductivity, and subsequent instability within laboratory conditions are noteworthy drawbacks. While a diverse array of organisms, spanning plants, animals, and microorganisms, could potentially yield nitrate reductases, only a restricted resource pool has been harnessed for large-scale purification, with fungal nitrate reductase, par-ticularly Aspergillus niger, being favored due to its potent catalytic prowess (Hooda et al., 2016). Within enzyme-based methodologies, two variants, namely the soluble and immobilized forms, are prevalent. Immobilized enzymes exhibit enhanced cost-effectiveness and practicality owing to superior stability and reusability (Hooda et al., 2016). Recently, significant attention has gravi-tated toward employing AuNPs (gold nanoparticles) as platforms for enzyme immobilization due to their compatibility with biological systems, electrical conductance, and robust enzyme loading capabilities (Sachdeva & Hooda, 2014). Nevertheless, the intricate separation of nanobioconjugated mixtures suspended in solutions has prompted the exploration of alternative strategies, with one such approach involving the tethering of nanoparticles onto insoluble supports. Epoxy supports have gained recognition for this purpose, courtesy of the augmented stability they confer to the matrix. Consequently, the utilization of immobilized nitrate reductase-based methodologies, bolstering the stability of nitrate reductase, stands poised to facilitate the advancement of nanosensors as uncom-plicated, sensitive, and economically viable technologies for nitrate level determination (Sachdeva & Hooda, 2014).

In an alternative approach for nitrate reduction, nitrate is converted into nitrite utilizing spe-cific metals such as zinc, cadmium, copper, and vanadium (Jaikang, Wangkarn, Paengnakorn, & Grudpan, 2019; Mir, 2007). While this method is cost-effective, it is not without its limitations. Empirical observations have highlighted that certain metals can reduce nitrite into other com-pounds, including ammonia. Consequently, the accuracy of nitrate estimation is compromised in the presence of nitrite (Mir, 2007). This methodology entails initially analyzing nitrate and nitrite samples for nitrite content. Subsequently, the nitrite-containing portion of the sample undergoes urea treatment to eliminate nitrite. The resulting urea-treated sample can then be subjected to metal reduction, enabling a precise measurement of nitrate concentration.

Furthermore, despite their reduction capabilities, each of these metals is not without challenges. For instance, while proficient in reducing nitrate with high efficiency, cadmium poses environmen-tal and health risks due to its classification as a toxic heavy metal. Proper disposal of cadmium is problematic, further compounding its drawbacks (Patton & Kryskalla, 2011). Zinc NPs have been explored as an alternative reducing agent, exhibiting lower toxicity effects. However, it is less efficient in the reduction process. Research findings have indicated that enhancing salinity and ionic strength can enhance the nitrate-to-nitrite reduction process through the cadmium-zinc column (Jaikang et al., 2019).

6.7 NANOMATERIAL-BASED ELECTROCHEMICAL SENSORS FOR PHOSPHATE DETECTION IN FOOD SAMPLES

Monitoring phosphate levels using bio-electroanalytical biosensors has been extensively studied, as highlighted by Villalba and coworkers (2009). Numerous potentiometric sensors have been explored for detecting phosphate in aquatic environments. Due to their exceptional selectivity and sensitivity characteristics, a focus has been placed on phosphate-selective ionophores embedded within polymer matrices. Notably, Ganjali and coworkers made pioneering contributions in this field. They initially reported on a sensor utilizing a vanadyl salen (VS) ionophore and PVC membrane for potentiometric determination of HPO_4^{2-}. While the sensor demonstrated reasonable selectivity for HPO_4^{2-} over chloride, its sensitivity and detection limit were suboptimal (Ganjali, Norouzi, Hatambeygi, & Salavati-Niasari, 2006). Subsequent improvements were achieved by Ganjali et al., who enhanced the detection limit and selectivity coefficient using different ionophores. They utilized bis(2-hydroxyanil) acetylacetonedioxomolybdenum (VI) to improve sensitivity and selectivity.

Moreover, oxo-molybdenum methyl-salen (MS) ionophore enabled even greater sensitivity and selectivity for HPO_4^{2-} detection (Ganjali, Norouzi, Ghomi, & Salavati-Niasari, 2006). Other researchers, such as Ashok et al., compared the sensing performance of different ionophores, revealing varying levels of sensitivity and selectivity (Kumar, Mehtab, Singh, Aggarwal, & Singh, 2008). Topcu et al. proposed a potentiometric sensor based on a chitosan-smectite nanocomposite, exhibiting a wide linear detection range and high selectivity for HPO_4^{2-} over interfering ions (Topcu et al., 2018).

Alizadeh and Atayi (2018) utilized hydrogen phosphate ion-imprinted polymer nanoparticles to enhance sensing capability. Ahmad et al. (2016) developed a field-effect transistor (FET) biosensor with enhanced sensitivity through ZnO nanorods. Li, Jiang, Yu, and Yang (2016) focused on molybdenum-based ion selective electrodes, showing good selectivity but some interference to HPO_4^{2-} which leads to a falling signal of HPO_4^{2-} in the actual matrix. Zhou and coworkers (Zhao et al., 2019) recently introduced an rGO-FET platform with high sensitivity and selectivity for phosphate ion detection in water samples. Using rGO nanosheets and Al_2O_3 coating contributed to improved immobilization and stability.

6.8 NANOMATERIAL MODIFIED AMPROMTERIC SENSORS FOR PHOSPHATE DETECTION

The researchers have been focusing on developing amperometric sensor systems to monitor phosphate levels in highly saline waters. An example of successful detection comes from Quintana and colleagues, who demonstrated amperometric phosphate detection in seawater down to 0.3 μmol L^{-1} levels. This method utilizes a phosphomolybdate complex that is reduced at a carbon paste electrode maintained at +0.3 V compared to an Ag/AgCl reference electrode, as displayed in Figure 6.2 (a, b, c, and d). The sensor system addressed the challenge of chloride interference and exhibited a linear response within a concentration range of 1–20 μmol L^{-1}. Another approach was taken by Talarico et al. (2015), who designed a sensor using a carbon black nanoparticle-modified screen-printed electrode. This modification significantly improved the electrochemical response, making the sensor suitable for actual water samples due to its simplicity and cost-effectiveness.

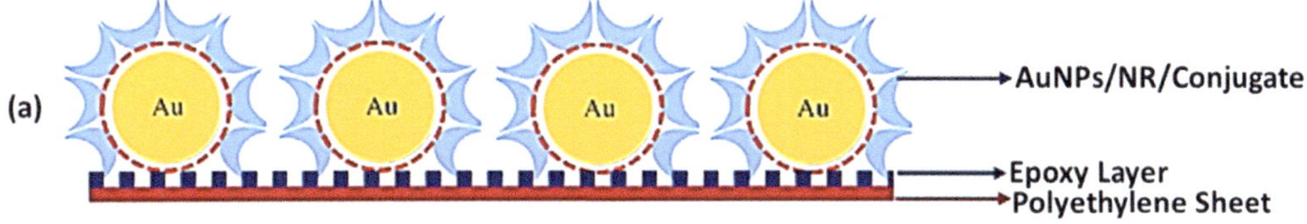

FIGURE 6.2 (a) Diagram of the setup for phosphate detection using a carbon paste microelectrode placed within a glass capillary tube. (b) Illustration depicting the experimental arrangement utilized for the continuous analysis of phosphate. (c) Layout of a microfluidic chip designed for phosphate analysis. The chip includes inlets for sample and reagent introduction, a serpentine channel for reactions, a designated viewing area, and a waste outlet. A separate image offers a cross-sectional perspective of the microfluidic chip holder and the integrated detection components. (d) Schematic representations showcasing the reflection-mode fluorescent imaging technique employed for a paper-based phosphate sensor. Copyright © 2023 with permission from Elsevier (Mahmud et al., 2020).

Cinti and coworkers (Cinti, Talarico, Palleschi, Moscone, & Arduini, 2016) introduced a reagentless paper-based sensor for phosphate using a three-step manufacturing process. This sensor relied on forming a phosphomolybdic complex for both colorimetric and voltammetric methods, allowing versatile electrochemical detection.

In a different development, Song and coworkers (Y. Song, Bian, Li, Tong, & Xia, 2016) created a chemically modified electrode for phosphate detection through sequential injection analysis (SIA). Notably, this sensor utilized oxalic acid to counter silicate interference, a common issue in sodium molybdate-based methods for phosphate quantification.

6.9 ANALYSIS OF SULFATE IONS

In analytical chemistry, the fusion of nanotechnology has significantly amplified the advancement of sophisticated sensing platforms. These innovations find compelling applications in sulfate ion (SO_4^{2-}) detection and quantification. Sulfate ions, crucial anionic species with natural and human-induced origins, have garnered substantial attention due to their profound implications in environmental surveillance, industrial processes, and health-related concerns. The amalgamation of nanomaterials into sensor design, exploiting their distinctive physicochemical properties, shows immense potential for achieving heightened sensitivity and selectivity in sulfate ion detection. This discourse explores pioneering strides in nanomaterial-driven sulfate ion sensors, providing a comprehensive overview of the array of nanomaterials utilized, the strategies employed for their functionalization, and the mechanisms for signal transduction. Moreover, this discussion also highlights prospects and challenges within this captivating domain (Badmus, Oyehan, & Saleh, 2021; Safari, Rahimi, Lah, Gholami, & Khur, 2020).

Nanomaterials offer a pathway to fine-tune sensor performance at the molecular scale, owing to their exceptional surface-to-volume ratio and quantum effects. Carbon-based nanomaterials, exemplified by carbon nanotubes (CNTs) and graphene, have been pivotal in sulfate ion sensing due to their outstanding electrical, optical, and electrochemical attributes (Buledi, Solangi, et al., 2022; Pato et al., 2021). Conjugating these nanomaterials with specific receptors such as crown ethers, cyclodextrins, or metal-organic frameworks enhances their selectivity and sensitivity towards sulfate ions. Similarly, metal and metal oxide nanomaterials, including gold nanoparticles (AuNPs), silver nanoparticles (AgNPs), and zinc oxide nanoparticles (ZnO NPs), have been harnessed for sulfate ion detection in food samples by virtue of their plasmonic, catalytic, and semiconducting characteristics (Buledi, Batool, et al., 2023; Khatoon et al., 2022; Qambrani et al., 2022). These nanomaterials serve as versatile platforms for designing sensing interfaces that interact with sulfate ions, leading to discernible alterations in their physicochemical traits (Laborda et al., 2016; Xu et al., 2020; Yusof et al., 2022). The process of functionalizing nanomaterials with selective receptors or recognition elements emerges as a pivotal step in augmenting the specificity and sensitivity of sulfate ion sensors. Molecularly imprinted polymers (MIPs), an engineered class of receptors, have emerged as potent tools for impressing the spatial arrangement of target sulfate ions. Coating nanomaterial surfaces with MIPs yields specific binding sites that facilitate the highly selective capture of sulfate ions, even in intricate samples.

Additionally, aptamers—single-stranded DNA or RNA sequences—can be immobilized onto nanomaterial surfaces, creating affinity-based sensors. These aptamer-functionalized nanomaterials exhibit remarkable specificity, enabling real-time monitoring of sulfate ion levels in complex environments (Buledi, Hyder, et al., 2023; Duan, Ma, Lian, & Zheng, 2014; Semenova & Silina, 2019; Shah et al., 2023). Regarding signal transduction mechanisms, nanomaterial-based sulfate ion sensors encompass a diverse array of strategies to convert the recognition event into a measurable signal. Electrochemical sensors, for instance, exploit alterations in electrical conductivity, capacitance, or impedance upon sulfate ion interaction with nanomaterial surfaces (Xia et al., 2010). This approach has been effectively harnessed with modified CNTs, graphene, and metal/metal oxide nanoparticles, enabling label-free, real-time sulfate ion detection.

Additionally, optical sensors leverage nanomaterials' plasmonic, fluorescence, or colorimetric properties for signal generation. Functionalized AuNPs or quantum dots, for instance, manifest changes in their visual traits upon interacting with sulfate ions, facilitating facile and sensitive detection through spectroscopic measurements. Furthermore, piezoelectric sensors, capitalizing on the mechanical deformation of nanomaterials upon sulfate ion binding, provide an alternative approach for transducing the recognition event into a quantifiable signal (Ng, Koneswaran, & Narayanaswamy, 2016).

While the strides in nanomaterial-based sulfate ion sensors are undoubtedly noteworthy, several challenges and prospects warrant consideration. One key challenge involves potential interference from other coexisting ions or molecules in complex samples, possibly leading to inaccurate results (Ali et al., 2023; Song, Wu, Xu, Wu, & Yao, 2022). Efforts to enhance selectivity through inventive receptor design and refined signal processing techniques are pivotal to mitigating this concern (Hosseini et al., 2019). Moreover, the practical deployment of these sensors, especially in real-world scenarios like environmental monitoring or clinical diagnostics, necessitates rigorous validation and standardized procedures. Collaborative endeavors involving researchers, regulatory bodies, and industry stakeholders are imperative to bridge this gap (Ji, Li, Zhang, Wang, & Ozaki, 2021).

6.10 MISCELLANEOUS IONS

There is considerably less information regarding these ions and their nanobiotechnological detection systems compared to the above ions described. However, there are a couple of primers for fluoride and halide ions that are suggested to the reader (Guo et al., 2017; Abkhalimov et al., 2018).

Most nanotechnological applications for these ions are for freshwater (Table 6.1). As time has advanced, authors have seen the necessity of implementing such swift and on-site measurement techniques in more complex matrices such as foods (Table 6.1). Noteworthy, iron oxide nanoparticles have also been used to determine and monitor food pH values as a quality characteristic (Meng et al., 2016). The availability of different measurement systems for these ions is important as even nanoparticle application in food has a profound impact on sensory properties (Moncada et al., 2015).

TABLE 6.1

Examples of Nanobiotechnology in Halide Analysis

Nanoparticle Used	Matrix	Measurement Principle	Detection Limit, mg L^{-1}	Determination Range, mg L^{-1}	Reference
Bromide					
Silver triangular nanoplates	Mineral water, river and sea water, algae, shrimps, fish, bread, and pharmaceuticals	Colorimetric paper strips	0.020	0.05–1.5	Gorbunova et al. (2020)
Chloride					
Silver triangular nanoplates	Pharmaceutical formulations, river water, and tomato juice	Conversion of Cl$^-$ to Cl$_2$, dynamic gas extraction, colorimetric detection-modified paper test strips	0.04	0.1–1.5	Gorbunova et al. (2018)
Epoxy-silver nanocomposite	Seawater and a commercial electrolyte solution	Colorimetric (green–brown in the presence of Cl$^-$)	497	7.10×10^2– 1.42×10^4	Franco et al. (2022)
Silver itaconate nanocomposites	Water (remediation)	Colorimetric paper strips	0.053	0.15–2.40	Kolesnikova et al. (2022)
Ag nanocomposites in carbon	Water and canned fish	Colorimetric paper strips	0.036	0.15–0.24	Gorbunova et al. (2023)
Fluoride					
Ag doped CdS/ZnS core/shell	Water	L-cysteine fluorimetric probe	0.100	0.19–22.80	Sankar Boxi and Paria, 2016)
Al$_2$O$_3$-Chitosan (*Ziziphus jujuba* Mill. seed meal extract)	Water	Colorimetric	Not reported	Up to 2	Yadav et al. (2020)
Amino phenylboronic acid-zirconium oxide	Water and toothpaste	Voltammetric	1.9×10^{-5}	5.7–190	Akhtar et al. (2022)
ZnO nanoparticles and iron chloride	Toothpaste	Colorimetric	Not reported	Up to 1,450	Watjanavarreerat et al. (2022)
Tannic acid-functionalized silver	Water	Colorimetric	0.004	Not reported	Bezuneh et al. (2023)
4-Quinonimine functionalized gold nanoparticles	Water	Surface Plasmon absorption band at 535 nm. Decoloration purple-red at 880 nm	0.003	Up to 6.43	Kundu and Kar (2023)
Iodide					
Mercapto-functionalized thymine gold	Aqueous solutions	Colorimetric (blue to red)	0.0013	0.0025–0.08	Chen et al. (2013)
Chitosan oligosaccharide lactate-capped silver	Tap, river water, kelp (*Laminaria japonica* Aresch)	Colorimetric	0.002	0.028	Maruthupandi et al. (2019)
Gold nanostars probes	River water, table salt, seaweed, complex vitamin tablet	Surface plasmon resonance	6.45×10^{-4}	Not reported	Zhou et al. (2021)

6.11 CONCLUSION AND FUTURE PERSPECTIVE

In summary, exploring nanomaterials for analyzing inorganic ions, particularly nitrates, sulfates, and phosphates, within food samples presents a promising and innovative avenue. By incorporating nanomaterials into analytical techniques, ion detection's sensitivity, selectivity, and efficiency have improved substantially, contributing to food safety and quality assessment advancements. This chapter underscores the potential of nanomaterial-centric approaches in accurately quantifying these ions, even at minute concentrations, which holds great significance due to their implications for human health and the environment. The findings from these investigations emphasize the need for ongoing exploration and progress in this field. As nanotechnology evolves, there exists considerable opportunity for refining and expanding nanomaterial-based analytical methodologies. Researchers can delve into creating novel nanomaterials endowed with enhanced characteristics, such as heightened stability, increased surface area, and superior affinity for target ions. Furthermore, combining nanomaterials with emerging analytical technologies like microfluidics and biosensors could pave the way for swifter and more sensitive ion detection techniques.

In the foreseeable future, applying nanomaterials to scrutinize inorganic ions in food samples could have far-reaching implications for food safety regulations and quality management. Regulatory entities and industries could gain from adopting these sophisticated approaches to ensure the precise evaluation of ion levels in diverse food items. Furthermore, integrating nanomaterial-based sensors into compact and user-friendly devices could empower consumers to make knowledgeable decisions about their dietary preferences, thereby fostering healthier eating habits. Nevertheless, it is imperative to acknowledge potential obstacles and ethical considerations linked to nanotechnology. As advancements are made in this field, a comprehensive assessment of the environmental impact and potential risks of using nanomaterials in food analysis becomes essential. Continuous research is warranted to address these apprehensions and to ensure the safe and responsible implementation of nanomaterial-centric methodologies.

BIBLIOGRAPHY

Abkhalimov, E. V., Timofeev A. A., & Ershov B. G. (2018). Electrochemical mechanism of silver nanoprisms transformation in aqueous solutions containing the halide ions. *Journal of Nanoparticle Research, 20*, 26.

Ahmad, A., Zaini, N. M., Rozi, N., Abd Karim, N. H., Hasbullah, S. A., Heng, L. Y., & Hanifah, S. A. (2016). A fluorescence phosphate sensor based on poly (glycidyl methacrylate) microspheres with aluminium-morin. *Malaysian Journal of Analytical Sciences, 20*(4), 704–712.

Akhtar, K., Baig J. A., Kazi, T. G., Sirajuddin, Afridi, H. I., Talpur, F. N., Solangi I. B., & Samaijo, S. (2022). Novel fluoride selective voltammetric sensing method by amino phenylboronic acid-zirconium oxide nanoparticles modified gold electrode. *Microchemical Journal, 174*, 107073.

Ali, H., Ali, A., Buledi, J. A., Memon, A. A., Solangi, A., Yang, J., & Thebo, K. H. (2023). MXene-based nanocomposites: An emerging aspirant for removal of antibiotics, dyes, and heavy metal ions. *Materials Chemistry Frontiers, 7*, 5519–5544.

Alizadeh, T., & Atayi, K. (2018). Synthesis of nano-sized hydrogen phosphate-imprinted polymer in acetonitrile/water mixture and its use as a recognition element of hydrogen phosphate selective all-solid state potentiometric electrode. *Journal of Molecular Recognition, 31*(2), e2678.

Amali, R. K. A., Lim, H. N., Ibrahim, I., Huang, N. M., Zainal, Z., & Ahmad, S. A. A. (2021). Significance of nanomaterials in electrochemical sensors for nitrate detection: A review. *Trends in Environmental Analytical Chemistry, 31*, e00135. https://doi.org/10.1016/j.teac.2021.e00135.

Amin, S., Solangi, A. R., Hassan, D., Hussain, N., Ahmed, J., & Baksh, H. (2021). Recent trends in development of nanomaterials based green analytical methods for environmental remediation. *Current Analytical Chemistry, 17*(4), 438–448.

Aravamudhan, S., & Bhansali, S. (2008). Development of micro-fluidic nitrate-selective sensor based on doped-polypyrrole nanowires. *Sensors and Actuators B: Chemical, 132*(2), 623–630.

Ardakani, M. M., Salavati-Niasari, M., & Jamshidpoor, M. (2004). Selective nitrate poly (vinylchloride) membrane electrode based on bis (2-hydroxyacetophenone) ethylenediimine vanadyl (IV). *Sensors and Actuators B: Chemical, 101*(3), 302–307.

Asadnia, M., Myers, M., Umana-Membreno, G. A., Sanders, T. M., Mishra, U. K., Nener, B. D., . . . Parish, G. (2017). Ca2+ detection utilising AlGaN/GaN transistors with ion-selective polymer membranes. *Analytica Chimica Acta, 987*, 105–110.

Badmus, S. O., Oyehan, T. A., & Saleh, T. A. (2021). Synthesis of a novel polymer-assisted AlNiMn nanomaterial for efficient removal of sulfate ions from contaminated water. *Journal of Polymers and the Environment, 29*, 2840–2854.

Bagheri, H., Hajian, A., Rezaei, M., & Shirzadmehr, A. (2017). Composite of Cu metal nanoparticles-multiwall carbon nanotubes-reduced graphene oxide as a novel and high performance platform of the electrochemical sensor for simultaneous determination of nitrite and nitrate. *Journal of Hazardous Materials, 324*, 762–772.

Bendikov, T. A., Kim, J., & Harmon, T. C. (2005). Development and environmental application of a nitrate selective microsensor based on doped polypyrrole films. *Sensors and Actuators B: Chemical, 106*(2), 512–517.

Bezuneh, T. T., Ofgea, N. M., Tessema, S. S., Bushira F. A. (2023). Tannic acid-functionalized silver nanoparticles as colorimetric probe for the simultaneous and sensitive detection of aluminum(III) and fluoride ions. *ACS Omega, 8*(40), 37293–37301. https://doi.org/10.1021/acsomega.3c05092.

Bi, H., Zhong, W., Meng, S., Kong, J., Yang, P., & Liu, B. (2006). Construction of a biomimetic surface on microfluidic chips for biofouling resistance. *Analytical Chemistry, 78*(10), 3399–3405.

Boxi, S. S., & Paria, S. (2016). Fluorometric selective detection of fluoride ions in aqueous media using Ag doped CdS/ZnS core/shell nanoparticles. *Dalton Transactions, 45*, 811–819.

Buledi, J. A., Amin, S., Haider, S. I., Bhanger, M. I., & Solangi, A. R. (2021). A review on detection of heavy metals from aqueous media using nanomaterial-based sensors. *Environmental Science and Pollution Research, 28*, 58994–59002.

Buledi, J. A., Batool, M., Ameen, S., Solangi, A. R., Mallah, A., Palaybik, I. M., ... Ali, S. (2023). Polyethylene glycol functionalized CuO/rGO nanocomposite based electrochemical sensor for ultra-sensitive electro-oxidation of bromoxynil in vegetables. *Chemical Papers, 77*(11), 7215–7223.

Buledi, J. A., Hyder, A., Khand, N. H., Memon, S. A., Batool, M., & Solangi, A. R. (2023). Potential mitigation of dyes through mxene composites. In Rizwna, K., Khan, A., Ahmed Asiri, A. M. (Eds.), *Handbook of Functionalized Nanostructured MXenes: Synthetic Strategies and Applications from Energy to Environment Sustainability* (pp. 283–300). Springer.

Buledi, J. A., Mahar, N., Mallah, A., Solangi, A. R., Palabiyik, I. M., Qambrani, N., ... Karimi-Maleh, H. (2022). Electrochemical quantification of mancozeb through tungsten oxide/reduced graphene oxide nanocomposite: A potential method for environmental remediation. *Food and Chemical Toxicology, 161*, 112843.

Buledi, J. A., Shah, Z.-U.-H., Mallah, A., & Solangi, A. R. (2022). Current perspective and developments in electrochemical sensors modified with nanomaterials for environmental and pharmaceutical analysis. *Current Analytical Chemistry, 18*(1), 102–115.

Buledi, J. A., Solangi, A. R., Hyder, A., Khand, N. H., Memon, S. A., Mallah, A., ... Behzadpour, M. (2022). Selective oxidation of amaranth dye in soft drinks through tin oxide decorated reduced graphene oxide nanocomposite based electrochemical sensor. *Food and Chemical Toxicology, 165*, 113177.

Cattrall, R. W. (1997). Chemical sensors. Issue 52 of Oxford chemistry primers,). Oxford, New York, Oxford University Press. ISSN 1367-109X

Chen, L., Lu, W., Wang, X., & Chen, L. (2013). A highly selective and sensitive colorimetric sensor for iodide detection based on anti-aggregation of gold nanoparticles. *Sensors and Actuators B: Chemical, 182*, 482–488.

Chen, X., Pu, H., Fu, Z., Sui, X., Chang, J., Chen, J., & Mao, S. (2018). Real-time and selective detection of nitrates in water using graphene-based field-effect transistor sensors. *Environmental Science: Nano, 5*(8), 1990–1999.

Cinti, S., Talarico, D., Palleschi, G., Moscone, D., & Arduini, F. (2016). Novel reagentless paper-based screen-printed electrochemical sensor to detect phosphate. *Analytica Chimica Acta, 919*, 78–84.

Copley, R. R., & Barton, G. J. (1994). A structural analysis of phosphate and sulphate binding sites in proteins: Estimation of propensities for binding and conservation of phosphate binding sites. *Journal of Molecular Biology, 242*(4), 321–329.

Crespo, G. A. (2017). Recent advances in ion-selective membrane electrodes for in situ environmental water analysis. *Electrochimica Acta, 245*, 1023–1034.

Cuartero, M., Crespo, G. A., & Bakker, E. (2015). Tandem electrochemical desalination-potentiometric nitrate sensing for seawater analysis. *Analytical Chemistry, 87*(16), 8084–8089.

Duan, X., Ma, J., Lian, J., & Zheng, W. (2014). The art of using ionic liquids in the synthesis of inorganic nanomaterials. *CrystEngComm, 16*(13), 2550–2559.

Fernando, W. A. M., Ilankoon, I., Syed, T. H., & Yellishetty, M. (2018). Challenges and opportunities in the removal of sulphate ions in contaminated mine water: A review. *Minerals Engineering, 117*, 74–90.

Franco, A., Velásquez-Ordoñez, C., Ojeda-Martínez, M., Ojeda-Martínez, M., Barrera-Calva, E., & Rentería-Tapia, V. (2022). Paper-based colorimetric sensor for detection of chloride anions in water using an epoxy-silver nanocomposite. *Journal of Nanoparticle Research, 24*, 54. https://doi.org/10.1007/s11051-022-05435-1

Ganjali, M. R., Norouzi, P., Ghomi, M., & Salavati-Niasari, M. (2006). Highly selective and sensitive mono-hydrogen phosphate membrane sensor based on molybdenum acetylacetonate. *Analytica Chimica Acta, 567*(2), 196–201.

Ganjali, M. R., Norouzi, P., Hatambeygi, N., & Salavati-Niasari, M. (2006). Anion recognition: Fabrication of a highly selective and sensitive HPO$_4$ (2-) PVC sensor based on a oxo-molybdenum methyl-salen. *Journal of the Brazilian Chemical Society, 17*, 859–865.

Gao, S., & Bao, X. (2020). Chalcogenide taper and its nonlinear effects and sensing applications. *Iscience, 23*(1), 100802.

Gokhale, A. A., Lu, J., Weerasiri, R. R., Yu, J., & Lee, I. (2015). Amperometric detection and quantification of nitrate ions using a highly sensitive nanostructured membrane electrocodeposited biosensor array. *Electroanalysis, 27*(5), 1127–1137.

Gorbunova, M. O., Garshina M. S., Kulyaginova, M. S., Apyari, V. V., Furletov, A. A., Garshev A. V., Dmitrienkoc, S. G., & Zolotovce Y. A. (2020). A dynamic gas extraction-assisted paper-based method for colorimetric determination of bromides. *Analytical Methods, 12*, 587–594.

Gorbunova, M. O., Shevchenko A. V., Apyari V. V., Furletov, A. A., Volkov, P. A., Garshev A. V., & Dmitrienko S. G. (2018). Selective determination of chloride ions using silver triangular nanoplates and dynamic gas extraction. *Sensors and Actuators B: Chemical, 256*, 699–705.

Gorbunova, M. O., Uflyand, I. E., Zhinzhilo, V. A., Zarubina, A. O., Kolesnikova, T. S., Spirin M. G., & Dzhardimalieva, G. I. (2023). Preparation of Rreactive Iindicator Ppapers Bbased on Ssilver-Ccontaining Nnanocomposites for the Aanalysis of Cchloride Iions. *Micromachines, 14*(9), 1682.

Guo, Y., Li, J., Chai, S., & Yao, J. (2017). Nanomaterials for the optical detection of fluoride. *Nanoscale, 9*(45), 17667–17680.

Hassan, S. S., Sayour, H., & Al-Mehrezi, S. S. (2007). A novel planar miniaturized potentiometric sensor for flow injection analysis of nitrates in wastewaters, fertilizers and pharmaceuticals. *Analytica Chimica Acta, 581*(1), 13–18.

Hooda, V., Sachdeva, V., & Chauhan, N. (2016). Nitrate quantification: Recent insights into enzyme-based methods. *Reviews in Analytical Chemistry, 35*(3), 99–114.

Hosseini, S., Usefi, M. B., Habibi, M., Parvizian, F., Van der Bruggen, B., Ahmadi, A., & Nemati, M. (2019). Fabrication of mixed matrix anion exchange membrane decorated with polyaniline nanoparticles to chloride and sulfate ions removal from water. *Ionics, 25*, 6135–6145.

Hyder, A., Buledi, J. A., Nawaz, M., Rajpar, D. B., Orooji, Y., Yola, M. L., … Solangi, A. R. (2022). Identification of heavy metal ions from aqueous environment through gold, silver and copper nanoparticles: An excellent colorimetric approach. *Environmental Research, 205*, 112475.

Jaikang, P., Wangkarn, S., Paengnakorn, P., & Grudpan, K. (2019). Microliter operation for determination of nitrate-nitrogen via simple zinc reduction and color formation in a Well Plate with a Smartphone. *Analytical Sciences, 35*(4), 421–425.

Ji, W., Li, L., Zhang, Y., Wang, X., & Ozaki, Y. (2021). Recent advances in surface-enhanced Raman scattering-based sensors for the detection of inorganic ions: Sensing mechanism and beyond. *Journal of Raman Spectroscopy, 52*(2), 468–481.

Khand, N. H., Solangi, A. R., Ameen, S., Fatima, A., Buledi, J. A., Mallah, A., … Orooji, Y. (2021). A new electrochemical method for the detection of quercetin in onion, honey and green tea using Co$_3$O$_4$ modified GCE. *Journal of Food Measurement and Characterization, 15*(4), 3720–3730.

Khatoon, A., Syed, J. A., Buledi, J. A., Shakeel, S., Mallah, A., Solangi, A. R., … Shah, M. R. (2022). Bio-green fabrication of bell pepper mediated silver nanoparticles: An efficient material for electrochemical sensing of arbutin in cosmetics. *Journal of the Iranian Chemical Society, 19*(8), 3659–3672.

Kolesnikova, T. S., Zarubina, A. O., Gorbunova, M. O., Zhinzhilo, V. A., Dzhardimalieva, G. I., & Uflyand, I. E. (2022). Silver itaconate as single-source precursor of nanocomposites for the analysis of chloride ions. *Materials, 15*(23), 8376.

Kumar, A., Mehtab, S., Singh, U. P., Aggarwal, V., & Singh, J. (2008). Tripodal cadmium complex and macrocyclic ligand based sensors for phosphate ion determination in environmental samples. *Electroanalysis: An International Journal Devoted to Fundamental and Practical Aspects of Electroanalysis, 20*(11), 1186–1193.

Kundu, S., & Kar, P. (2023). Selective colorimetric sensing of fluoride ion in water by 4-quinonimine functionalized gold nanoparticles. *Journal of Cluster Science, 34*(6), 2799–2809. https://doi.org/10.1007/s10876-023-02427-6.

Laborda, F., Bolea, E., Cepriá, G., Gómez, M. T., Jiménez, M. S., Pérez-Arantegui, J., & Castillo, J. R. (2016). Detection, characterization and quantification of inorganic engineered nanomaterials: A review of techniques and methodological approaches for the analysis of complex samples. *Analytica Chimica Acta, 904*, 10–32.

Le Goff, T., Braven, J., Ebdon, L., & Scholefield, D. (2002). High-performance nitrate-selective electrodes containing immobilized amino acid betaines as sensors. *Analytical Chemistry, 74*(11), 2596–2602.

Li, Y., Jiang, T., Yu, X., & Yang, H. (2016). Phosphate sensor using molybdenum. *Journal of the Electrochemical Society, 163*(9), B479.

Mahmud, M. P., Ejeian, F., Azadi, S., Myers, M., Pejcic, B., Abbassi, R., ... Asadnia, M. (2020). Recent progress in sensing nitrate, nitrite, phosphate, and ammonium in aquatic environment. *Chemosphere, 259*, 127492.

Maruthupandi, M., Chandhru, M.,Rani S. K., & Vasimalai, N. (2019). Highly selective detection of iodide in biological, food, and environmental samples using polymer-capped silver nanoparticles: Preparation of a paper-based testing kit for on-site monitoring. *ACS Omega, 4*(7), 11372–11379.

Meng, X., Ryu J., Kim, B., & Ko, S. (2016). Application of iron oxide as a pH-dependent indicator for improving the nutritional quality. *Clinical Nutrition Research, 5*(3), 172–179.

Mir, S. A. (2007). An improved zinc reduction method for direct determination of nitrate in presence of nitrite. *Asian Journal of Chemistry, 19*(7), 5703.

Mistlberger, G., Crespo, G. A., & Bakker, E. (2014). Ionophore-based optical sensors. *Annual Review of Analytical Chemistry, 7*, 483–512.

Moncada, M., Astete, C., Sabliov, C., Olson, D., Boeneke, C., & Aryana, K. J. (2015). Nano spray-dried sodium chloride and its effects on the microbiological and sensory characteristics of surface-salted cheese crackers. *Journal of Dairy Science, 98*(9), 5946–54.

Mura, S., Greppi, G., Roggero, P. P., Musu, E., Pittalis, D., Carletti, A., ... Irudayaraj, J. (2015). Functionalized gold nanoparticles for the detection of nitrates in water. *International Journal of Environmental Science and Technology, 12*, 1021–1028.

Myers, M., Khir, F. L. M., Podolska, A., Umana-Membreno, G. A., Nener, B., Baker, M., & Parish, G. (2013). Nitrate ion detection using AlGaN/GaN heterostructure-based devices without a reference electrode. *Sensors and Actuators B: Chemical, 181*, 301–305.

Ng, S. M., Koneswaran, M., & Narayanaswamy, R. (2016). A review on fluorescent inorganic nanoparticles for optical sensing applications. *RSC Advances, 6*(26), 21624–21661.

Nishan, U., Rehman, S., Ullah, R., Bari, A., Afridi, S., Shah, M., ... Khan, N. (2023). Fabrication of a colorimetric sensor using acetic acid-capped drug-mediated copper oxide nanoparticles for nitrite biosensing in processed food. *Frontiers in Materials, 10*, 1169945.

Pankratova, N., Cuartero, M., Cherubini, T., Crespo, G. A., & Bakker, E. (2017). Inline acidification for potentiometric sensing of nitrite in natural waters. *Analytical Chemistry, 89*(1), 571 575.

Palo, A. H., Balouch, A., Alveroglu, E., Buledi, J. A., Lal, S., & Mal, D. (2021). A practical non-enzymatic, ultra-sensitive molybdenum oxide (MoO3) electrochemical nanosensor for hydroquinone. *Journal of the Electrochemical Society, 168*(5), 056503.

Patton, C. J., & Kryskalla, J. R. (2011). Colorimetric determination of nitrate plus nitrite in water by enzymatic reduction, automated discrete analyzer methods. *US Geological Survey Techniques and Methods* book 5, chap. B8, 34 p.

Qambrani, N., Buledi, J. A., Khand, N. H., Solangi, A. R., Ameen, S., Jalbani, N. S., ... Shojaei, M. (2022). Facile synthesis of NiO/ZnO nanocomposite as an effective platform for electrochemical determination of carbamazepine. *Chemosphere, 303*, 135270.

Ratkovski, G. P., do Nascimento, K. T., Pedro, G. C., Ratkovski, D. R., Gorza, F. D., da Silva, R. J., ... de Melo, C. P. (2020). Spinel cobalt ferrite nanoparticles for sensing phosphate ions in aqueous media and biological samples. *Langmuir, 36*(11), 2920–2929.

Razanajatovo, M. R., Gao, W., Song, Y., Zhao, X., Sun, Q., & Zhang, Q. (2021). Selective adsorption of phosphate in water using lanthanum-based nanomaterials: A critical review. *Chinese Chemical Letters, 32*(9), 2637–2647.

Sachdeva, V., & Hooda, V. (2014). A new immobilization and sensing platform for nitrate quantification. *Talanta, 124*, 52–59.

Safari, M., Rahimi, A., Lah, R. M., Gholami, R., & Khur, W. S. (2020). Sustaining sulfate ions throughout smart water flooding by nanoparticle based scale inhibitors. *Journal of Molecular Liquids, 310*, 113250.

Semenova, D., & Silina, Y. E. (2019). The role of nanoanalytics in the development of organic-inorganic nano-hybrids-seeing nanomaterials as they are. *Nanomaterials, 9*(12), 1673.

Sepahvand, M., Ghasemi, F., & Hosseini, H. M. S. (2021). Plasmonic nanoparticles for colorimetric detection of nitrite and nitrate. *Food and Chemical Toxicology, 149*, 112025.

Shah, S.-Z., Taqvi, I. H., Ameen, S., Mallah, A., Buledi, J. A., Khand, N. H., & Solangi, A. R. (2023). Plant based fabrication of CuO/NiO nanocomposite: A green approach for low-level quantification of vanillin in food samples. *Pure and Applied Chemistry, 95*(5), 501–511.

Singh, S., Hasan, M. R., Sharma, P., & Narang, J. (2022). Graphene nanomaterials: The wondering material from synthesis to applications. *Sensors International, 3*, 100190. https://doi.org/10.1016/j.sintl.2022.100190.

Singh, S., Sangwan, S., Sharma, P., Devi, P., & Moond, M. (2021). Nanotechnology for sustainable agriculture: An emerging perspective. *Journal of Nanoscience and Nanotechnology, 21*(6), 3453–3465.

Song, W., Wu, Z., Xu, X., Wu, H., & Yao, Y. (2022). Nitrogen-doped carbon nanosheets with Fe-based nanoparticles for highly efficient degradation of antibiotics and sulfate ion enhancement effect. *Chemosphere, 294*, 133704.

Song, Y., Bian, C., Li, Y., Tong, J., & Xia, S. (2016). Electrochemical determination of phosphate in freshwater free of silicate interference. *Journal of Biomedical Science, 5*, 1–7.

Talarico, D., Cinti, S., Arduini, F., Amine, A., Moscone, D., & Palleschi, G. (2015). Phosphate detection through a cost-effective carbon black nanoparticle-modified screen-printed electrode embedded in a continuous flow system. *Environmental Science & Technology, 49*(13), 7934–7939.

Taqvi, S. I. H., Solangi, A. R., Buledi, J. A., Khand, N. H., Junejo, B., Memon, A. F., … Vasseghian, Y. (2022). Plant extract-based green fabrication of nickel ferrite ($NiFe_2O_4$) nanoparticles: An operative platform for non-enzymatic determination of pentachlorophenol. *Chemosphere, 294*, 133760.

Topcu, C., Caglar, B., Onder, A., Coldur, F., Caglar, S., Guner, E. K., … Tabak, A. (2018). Structural characterization of chitosan-smectite nanocomposite and its application in the development of a novel potentiometric monohydrogen phosphate-selective sensor. *Materials Research Bulletin, 98*, 288–299.

Tyagi, S., Rawtani, D., Khatri, N., & Tharmavaram, M. (2018). Strategies for nitrate removal from aqueous environment using nanotechnology: A review. *Journal of Water Process Engineering, 21*, 84–95.

Wang, X., Tan, W., Ji, H., Liu, F., Wu, D., Ma, J., & Kong, Y. (2018). Facile electrosynthesis of nickel hexacyanoferrate/poly (2, 6–diaminopyridine) hybrids as highly sensitive nitrite sensor. *Sensors and Actuators B: Chemical, 264*, 240–248.

Wang, X., Yin, R., Zeng, L., & Zhu, M. (2019). A review of graphene-based nanomaterials for removal of antibiotics from aqueous environments. *Environmental Pollution, 253*, 100–110. https://doi.org/10.1016/j.envpol.2019.06.067.

Watjanavarreerat, W., Steier, L., & Locharoenrat, K. (2022). Use of zinc oxide nanoparticles for detection of fluoride in toothpaste gel. *Journal of Environmental Science and Health. Part A, Toxic/Hazardous Substances & Environmental Engineering, 57*(9), 789–796.

Watt, M. M., Zakharov, L. N., Haley, M. M., & Johnson, D. W. (2013). Selective nitrate binding in competitive hydrogen bonding solvents: Do anion-π interactions facilitate nitrate selectivity? *Angewandte Chemie International Edition, 52*(39), 10275–10280.

Xia, F., Zuo, X., Yang, R., Xiao, Y., Kang, D., Vallée-Bélisle, A., … Heeger, A. J. (2010). Colorimetric detection of DNA, small molecules, proteins, and ions using unmodified gold nanoparticles and conjugated polyelectrolytes. *Proceedings of the National Academy of Sciences, 107*(24), 10837–10841.

Xu, X., Niu, X., Li, X., Li, Z., Du, D., & Lin, Y. (2020). Nanomaterial-based sensors and biosensors for enhanced inorganic arsenic detection: A functional perspective. *Sensors and Actuators B: Chemical, 315*, 128100.

Yadav, M., Kumari, S., & Khan, S. (2020). Development of affordable nanoparticle-chitosan based colorimetric detector for fluoride ions. *Sustainable Chemistry and Pharmacy, 18*, 100344.

Yu, L., Zhang, Q., Xu, Q., Jin, D., Jin, G., Li, K., & Hu, X. (2015). Electrochemical detection of nitrate in PM2.5 with a copper-modified carbon fiber micro-disk electrode. *Talanta, 143*, 245–253. https://doi.org/10.1016/j.talanta.2015.04.049.

Yusof, N., Hashim, S., Ghoshal, S., Azlan, M., Zaid, M., Boukhris, I., & Kebaili, I. (2022). Spectrographic analysis of zinc-sulfate-magnesium-phosphate glass containing neodymium ions: Impact of silver-gold nanoparticles plasmonic coupling. *Journal of Luminescence, 242*, 118571.

Zhao, S., Tong, J., Li, Y., Sun, J., Bian, C., & Xia, S. (2019). Palladium-gold modified ultramicro interdigital array electrode chip for nitrate detection in neutral water. *Micromachines, 10*(4), 223.

Zhou, R., Huang, X., An, Q., Xu, W., Liu, Y., Xu, D., Lin, Q., Wang, S., & Zhang, J. (2021). A convenient and sensitive colorimetric iodide assay based on directly inducing morphological transformation of gold nanostars. *Journal of Food and Drug Analysis, 29*, 144e152.

7 Nanofertilizers and Nanopesticides for Sustainable Agriculture, Food Security and Environmental Quality

Nadir H. Khand, Bindia Junejo, Amber R. Solangi, Tariq Aziz, and Zia-ul Hassan

7.1 INTRODUCTION

Sustainable agriculture is a practice that meets our current food needs while safeguarding resources for future generations (Velten, Leventon, Jager, & Newig, 2015). It emphasizes eco-friendly methods that minimize environmental harm, boost soil health, and foster biodiversity. In contrast, conventional crop management often relies heavily on intensive chemical inputs in monoculture farming, resulting in resource depletion (Tahat, Alananbeh, Othman, & Leskovar, 2020). Conventional agriculture faces several challenges, including soil degradation, water pollution due to excessive fertilizer use, enhanced pest resistance to pesticides, and biodiversity loss. Sustainable agriculture, on the other hand, prioritizes practices such as crop rotation, organic farming, reduced chemical usage, and efficient water management (Gomiero, 2018). The primary goal of sustainable agriculture is to tackle these challenges by promoting long-term viability, resilience, and a harmonious relationship between agriculture and the environment (Pretty, 2018). It's not only about securing food production but also about conserving natural ecosystems and mitigating the impacts of climate change. Moving to the realm of nanotechnology, this cutting-edge field offers immense promise for agriculture. It enables precise monitoring of soil conditions through nanosensors (Johnson, Sajeev, & Nair, 2021), allowing farmers to make data-driven decisions about irrigation and fertilization. In modern agriculture, chemical fertilizers like urea (containing 46% nitrogen, N) and phosphorus (P) containing Diammonium phosphate—commonly called DAP (including 18% N and 46% P_2O_5), are the key players in the human quest to enhance crop yields (Sharma & Singhvi, 2017). Simultaneously, chemical pesticides, including insecticides, e.g., chlorpyrifos and neonicotinoids; herbicides, e.g., glyphosate; and fungicides, e.g., chlorothalonil, are extensively being successfully used to combat pests and diseases (Rani et al., 2021).

Nonetheless, their widespread and non-judicious use has resulted in creating potential environmental consequences, such as the contamination of soil and water bodies, eutrophication, the degradation of soil, greenhouse gas emissions, acidification, unintended harm to non-target organisms, human health concerns, loss of biodiversity, and worries about food safety and security (Feng et al., 2021; Fu et al., 2022; Yadav & Devi, 2017). To address these persistent challenges, nanotechnology has appeared to be one of the most practicable solutions of modern times. One notable application is nanofertilizers, which are developed using nanoparticles (NPs) to enhance the precise delivery of nutrients to plants, enhance nutrient-use efficiency, and mitigate the adverse impacts associated with traditional agricultural practices (Pramanik et al., 2020). Resistance development nanopesticides target specific pests, reducing the overall need for chemical usage and helping combat resistance issues (Liu et al., 2021). Moreover, nanotechnology enhances nutrient delivery by utilizing NPs that improve plant nutrient uptake while minimizing runoff, contributing to sustainability and efficiency (An et al., 2022). Nanomaterials also play a role in soil remediation by effectively removing pollutants and revitalizing soil health. Smart delivery systems (Kumar et al., 2019) utilizing NPs transport nutrients effectively to plant cells, optimizing their effectiveness and ensuring resource-efficient agriculture. While nanotechnology holds great promise, it is essential to stress the necessity of careful regulation and monitoring to address potential environmental and health concerns associated with these advanced materials. We have reviewed the role of nanotechnology in sustainable agriculture and discussed the mechanisms, the innovations, and the environmental implications related to nanofertilizers and nanopesticides. The ethical considerations and regulatory challenges accompanying this cutting-edge technology have also been discussed.

7.2 NANOFERTILIZERS

Fertilizers are substances that may be natural or synthetic and used to supply essential plant nutrients through soil application to cope with their deficiency. Fertilizers are, therefore, necessary for the average growth and development of plants and indispensable to achieve optimum crop yields. The nutrient level of soil gets decreased by the time of harvest because of crop removal; these nutrients must be replaced by the addition of fertilizers (synthetic or natural) to sustain soil fertility. Therefore, fertilizers are considered as vital constituents of the present-day agriculture. About half of the current global food production relies on synthetic fertilizers. Hence, it has been advocated that "Fertilizers feed the World." Though chemical fertilizers are the key contributor to sustainable crop production for the global population, their extensive use can bring serious challenges to the present and future generations in the shape of polluted air, water, and degraded soils, besides greenhouse gas emissions and loss of precious biodiversity. Synthetic fertilizers are found hazardous not only for humans but for plants as well. The low nutrient uptake efficiency and the leaching of synthetic fertilizers are the main reasons for environmental pollution (Dimkpa et al., 2019; Kumar, Kumar, & Prakash, 2019; Seleiman et al., 2020). Hence, keeping these hazardous effects of synthetic fertilizers in mind, it is highly indispensable to take the initiatives to balance chemical fertilizers by integrating them with organic or nanofertilizers to enhance crop production without compromising the environmental quality.

The NPs containing micro and macronutrients, which may be supplied to the crops efficiently, are called nanofertilizers (Shang et al., 2019). Nanofertilizers have unique physicochemical properties, making them more advantageous than traditional chemical fertilizers. Particle dimensions (< 100 nm) enable the Nanofertilizers to enter the plant vascular system through soil or foliar application. These Nanofertilizers possess high surface area due to their ultra-small size (Hussain, Bilal, & Iqbal, 2022), resulting in increased absorption and retention capacity compared to commercial fertilizers. A gradual release of nutrients is provided to the crops as required, refraining from undesirable consequences (Siddiqi & Husen, 2017). The foliar application of Nanofertilizers can cause a 10%–25% increase in the yields of cereal grains as compared to commercial fertilizers (Babu et al., 2022; Delfani, Baradarn Firouzabadi, Farrokhi, & Makarian, 2014; Liu & Lal, 2015; Rathore et al., 2019).

When explicitly applied, the nanofertilizers are synthesized by combining nutrients with nano dimensions singly or with various adsorbent materials, which can supply more than one nutrient to the crops. Different encapsulation methods are available for Nanofertilizers, e.g., the encapsulation of nutrients with nanomaterials like nanotubes or nanoporous materials, nutrient coating with a thin polymer layer, and nutrient delivery through nanoscale formulations. Due to enormous challenges regarding the declining crop yields and the deficiency of multi-nutrients faced by present-day agriculture, agriculture scientists need to exploit more efficient crop production strategies involving nanofertilizers (Chhipa, 2017).

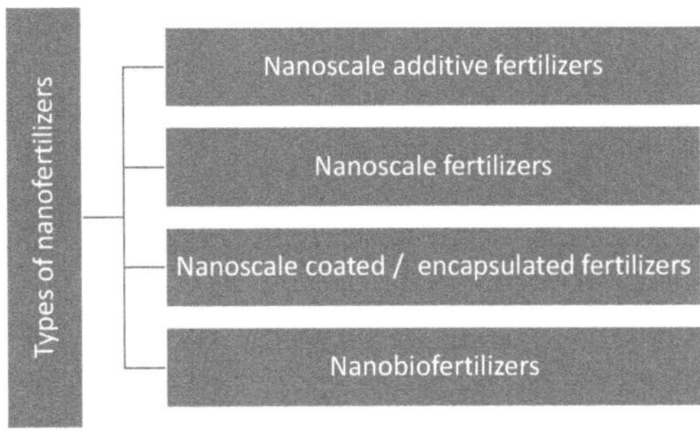

FIGURE 7.1 Main types of nanofertilizers.

7.2.1 Types of Nanofertilizers

Nanofertilizers are increasingly utilized in modern agriculture, precisely tailored to match the specific nutritional needs of crops and minimize nutrient losses. These nanofertilizers can be categorized into four main types, as depicted in Figure 7.1 (Mejias et al., 2021).

7.2.1.1 Nanoscale Additive Fertilizers

A traditional fertilizer to which nanoscale additives or other nanomaterials have been supplemented to enhance their performance.

7.2.1.2 Nanoscale Fertilizers

Comprising NPs containing essential nutrients, nanoscale fertilizers offer a targeted nutrient delivery system.

7.2.1.3 Nanoscale Coating Fertilizers

Traditional fertilizers are coated or loaded with NPs in this category, improving their nutrient release and utilization.

7.2.1.4 Nanobiofertilizers

Biofertilizers (fertilizers containing single or multiple nutrient-solubilizing microorganisms) are encapsulated with specific NPs to enhance the bioavailability of deficient nutrients from the soils.

The production of nanofertilizers primarily involves nutrient encapsulation within nanomaterials. These nanomaterials are generated using physical or chemical methods (top- and bottom-down, respectively). Nutrient encapsulation is achieved within the nanoporous materials and is coated by thin polymer film particles or incorporated into emulsions with nanoscale dimensions. Porous nanomaterials are crucial in minimizing N loss by regulating nutrient release according to plant demand and enhancing nutrient uptake. For example:

- **Graphene Oxide Films:** Carbon-based nanomaterials, e.g., potassium nitrate, extend the release of nutrients to reduce their leaching losses and enhance their effectiveness.
- **Ammonium-Charged Zeolites:** These materials enhance phosphate solubility, enhancing P availability and crop uptake.
- **Nanocalcite ($CaCO_3$–40 g/100 g) with Nano SiO_2 (4 g/100 g), MgO (1 g/100 g), and Fe_2O_3 (1 g/100 g):** These materials are highly beneficial for plant nutrition as they enhance the uptake of various nutrients, e.g., phosphorus, calcium, magnesium, iron, zinc, and manganese (Kahlel, Ghidan, Al-Antary, Alshomali, & Asoufi, 2020).

Furthermore, encapsulating beneficial microorganisms, such as bacteria or fungi, in nanofertilizers results in developing specific purpose nanobiofertilizers, which in turn enhance plant growth through increasing the bioavailability of essential elements in the rhizosphere, e.g., nitrogen, phosphorus, and potassium.

7.2.2 Classification of Nanofertilizers

Nanofertilizers are further classified based on their mode of action, e.g., nanofertilizers that release relatively slowly to meet the nutritional requirement throughout the life span of a plant, nanofertilizers that hit specific plant-part specific nanofertilizers, nanofertilizers that enhance the growth of plants and those that mainly control water and nutrient losses. These classes of nanofertilizers are illustrated in Figure 7.2.

7.2.2.1 Slow-Release Nanofertilizers

These materials are very competent compared to conventional fertilizers to address the challenges related to nutrient leaching losses and low nutrient-use efficiency. These fertilizers exert positive environmental impacts since nutrient release and uptake are controlled. The control-release nanofertilizers encapsulate the nutrients within nanoscale materials consisting of polymers, inorganic substances, and lipids. These carriers influence nutrient release depending on pH, temperature, stimuli-responsive mechanisms (enzymes-mediated biodegradation), and moisture. These materials offer several advantages contributing to sustainable agriculture. These control-release nanofertilizers have a well-established reputation for elevating crop productivity. Research findings underscore their capacity to enhance crop yields, optimize nutrient utilization, and promote overall plant vitality. These remarkable fertilizers hold promise for various crop species, e.g., rice, wheat, corn, and soybeans, and demonstrate encouraging outcomes regarding crop yields and nutrient utilization. To fully unlock the potential of these innovative fertilizers in meeting the growing global food production demand, it is imperative to maintain an unwavering commitment to ongoing research and development initiatives (Babu et al., 2022; DeRosa, Monreal, Schnitzer, Walsh, & Sultan, 2010; Ghormade, Deshpande, & Paknikar, 2011; Kah, Kookana, Gogos, & Bucheli, 2018; Liu & Lal, 2015).

7.2.2.2 Plant-Part Specific Nanofertilizers

Nanoaptamers represent an innovative breakthrough in fertilizer delivery systems, ushering in a transformative era for agricultural practices. These are engineered to specifically target molecules in the soil, facilitating the direct delivery of nutrients or other essential compounds to plants. Nanoaptamers can bind with plant hormones and enzymes, ensuring these vital components' efficient and effective conveyance. Scientific reports suggest that nanoaptamers can enhance the

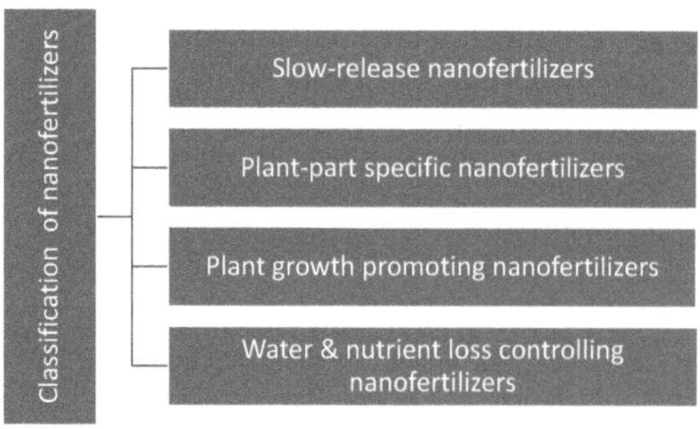

FIGURE 7.2 Classification of nanofertilizers.

uptake of plant nutrients from the soil (Kaushal & Wani, 2017; Rameshaiah, Pallavi, & Shabnam, 2015). Aptamers, which can be constructed from oligonucleotides or peptides, play a pivotal role in modifying the surface of nanofertilizers. This modification results in slow nutrient bioavailability encapsulated within the nanostructure, triggered by rhizosphere signals. Nanoaptamers achieve the delivery of fertilizers by attaching fertilizer components to NPs, such as gold NPs or liposomes. This innovative approach preserves the integrity of the aptamer, safeguarding it from degradation, and facilitates its precise delivery to designated plant cells. Subsequently, the aptamer binds to specific receptors on the plant cell surface and permits the cellular entry of NPs and the availability of fertilizer nutrients. In practical terms, nanoaptamers serve as regulators, establishing a vital link between soil microbes and rhizosphere and ensuring complete control over the required fertilizer dose administration (Majeed, Ramli, Mansor, & Man, 2015; Naz, Shukrullah, & Ghaffar, 2021).

Furthermore, the landscape of target-based nanofertilizers encompasses a diverse array of formulations, including urea with nano-coating, iron oxide NPs, nano-hydroxyapatite, C-based nano-nutrient carriers, with nanoemulsions and nano-encapsulated micronutrients, besides clay-based nanofertilizers (Liu & Lal, 2015; Raliya, Saharan, Dimkpa, & Biswas, 2017). Such novel formulations facilitate the controlled release of nutrients and augment nutrient uptake, promoting higher crop yields and coping with dire environmental consequences (Kottegoda, Munaweera, Madusanka, & Karunaratne, 2011)

7.2.2.3 Plant Growth-Promoting Nanofertilizers

Certain nanofertilizers, notably carbon nanotubes, can stimulate plant growth through their interaction with plant root systems and their role in promoting hormone synthesis. CNTs enrich the soil by increasing carbon content and enhancing nutrient availability. These cylindrical structures, composed of rolled-up sheets of C atoms, acquire unique traits that render them highly effective as fertilizers. CNTs exhibit the capacity to absorb and release nutrients, leading to improved soil structure, increased water retention, and enhanced plant growth. Remarkably, unlike the adverse effects often associated with high-dose fertilizer applications, low concentrations of C nanotubes have shown positive effects on the germination of seeds, development of roots, and transport of water without any phytotoxicity concerns (Mondal, Basu, Das, & Nandy, 2011). CNTs seamlessly integrate with the soil as fertilizer, gradually releasing essential plant nutrients. As a soil amendment, CNTs are directly incorporated into the soil, augmenting its ability to retain nutrients and moisture. Their ability to penetrate deep into the soil ensures a sustained supply of nourishment to plants over more extended periods. This controlled nutrient release minimizes over-fertilization risk, promoting healthy plant growth and environmental sustainability (Mondal et al., 2011).

7.2.2.4 Water and Nutrient Loss-Controlling Nanofertilizers

Nanofertilizers incorporate NPs with the capability to regulate the release of fertilizers into the soil, offering farmers the advantage of using less fertilizer while maintaining crop yields. Various strategies are being explored to design nanofertilizers that effectively manage nutrient release and mitigate water loss. One technique involves encapsulating nanofertilizers within a porous matrix, enabling a gradual nutrient release over time (Okey-Onyesolu et al., 2021). Another method entails modifying the surface of nanofertilizers to enhance their hydrophilicity, thereby increasing their water-holding capacity and reducing water loss due to evaporation (Sivarethinamohan & Sujatha, 2021).

Examples of nanofertilizer types that effectively control nutrient and water loss in soil include the coating of urea with the NPs of B, Ca, Cu, Fe, K, Mg, Mo, S, Zn, and ammonium sulfate, and so on (Ahmed & Fahmy, 2017; Jakhar et al., 2022). Additionally, nanobeads and nanoemulsions represent notable nanofertilizers designed to efficiently manage soil water and nutrient retention (Mali, Raj, & Trivedi, 2020).

7.2.3 Mechanisms of Nutrient Uptake by Plants

Nanofertilizers penetrate plants via their roots or through upper parts, i.e., cuticle and stomata. While the exact mechanism of NP uptake and translocation by plant cells remains unclear, existing reports underscore the significance of NPs' size, shape, and interaction with cell walls in this process. The cell wall's size exclusion limit, typically 5–20 nm, serves as a barrier and limits plant cell entrance of larger particles (Wang, Lombi, Zhao, & Kopittke, 2016). However, the surface functionalization of NPs can potentially enlarge pores. Other potential routes for NP entry include ion channels, endocytosis, complex formation with membrane transporters, or interaction with root exudates (Nair et al., 2010). Nanocarriers play a crucial role in shielding soil filtration of encapsulated nutrients, maintaining their balance with the rhizosphere. The encapsulated nutrients can enter the rhizosphere via various forces, viz. H-bonds, viscous or molecular forces, and surface tension expand their spatial distribution (Cai et al., 2014).

For the foliar supply, NPs can enter through stomatal openings or cuticles (Nair et al., 2010). The initial barrier restricting NPs is the leaf cuticle with < 5 nm size. NPs entering through stomatal pores can traverse the vascular system via apoplastic or symplastic pathways. NPs with 10–50 nm size usually prefer the symplastic path, while larger particles (50–200 nm) favor the apoplastic route. Several factors influence the efficacy of NPs, including leaf morphology and chemical composition, sunlight exposure, temperature, humidity, and photosphere microenvironment (Talan et al., 2018; Varshney, Pandey, & Chitikineni, 2018; Yang et al., 2018).

7.2.4 Benefits and Challenges of Nanofertilizers

Nanofertilizers represent a promising innovation in agriculture, offering several potential benefits while presenting specific challenges that must be addressed.

Benefits: One of the primary benefits of nanofertilizers is their capacity to enhance nutrient availability and plant uptake efficiency. Traditional chemical fertilizers can suffer from inefficient nutrient release and leaching into the soil, leading to nutrient wastage and environmental pollution. Nanofertilizers, however, are engineered at the nanoscale to release nutrients gradually and precisely, aligning nutrient availability with plant needs. This results in improved nutrient uptake by plants, promoting healthier and more robust crop growth. Furthermore, nanofertilizers can be designed to carry multiple nutrients in a single formulation, simplifying nutrient management practices for farmers (Bhardwaj et al., 2022; Zulfiqar, Navarro, Ashraf, Akram, & Munné-Bosch, 2019).

Nanofertilizers also have the potential to reduce the environmental impact associated with conventional fertilizers. The controlled release of nutrients minimizes excess runoff and leaching into groundwater, significantly contributing to water pollution and eutrophication in aquatic ecosystems. By improving nutrient use efficiency and reducing nutrient losses, nanofertilizers help safeguard water quality and protect natural habitats. Additionally, their precise nutrient delivery can reduce overall fertilizer usage, further decreasing the environmental footprint of agriculture (Bhardwaj et al., 2022; Zulfiqar et al., 2019). The process of achieving the goal of sustainable agriculture and a healthy environment through nanofertilizer use is depicted in Figure 7.3.

Challenges: While nanofertilizers offer exciting benefits, they also come with specific challenges that require careful consideration. The nanoparticles' small size and unique properties raise concerns about their potential toxicity to plants, soil microorganisms, and the environment (Gupta & Xie, 2018; Oberdörster et al., 2004).

It is crucial to thoroughly evaluate the safety of nanofertilizers to ensure they do not adversely affect agricultural ecosystems. Additionally, understanding the long-term effects of nanofertilizers is another critical challenge. This includes assessing their impact on soil health, microbial communities, and ecosystem dynamics over extended periods (Nowack & Bucheli, 2007; Shoults-Wilson et al., 2011). Long-term studies are essential to identify any potential accumulative or delayed effects. Regulatory frameworks for nanofertilizers are also still evolving, and there is a need

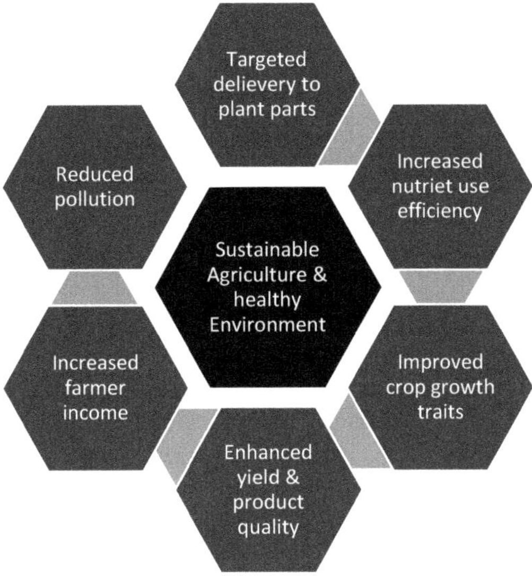

FIGURE 7.3 Achieving sustainable agriculture and healthy environment goal through nanofertilizer.

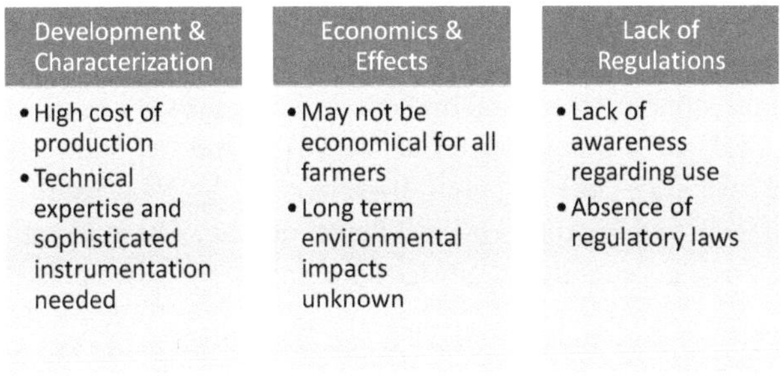

Development & Characterization	Economics & Effects	Lack of Regulations
• High cost of production • Technical expertise and sophisticated instrumentation needed	• May not be economical for all farmers • Long term environmental impacts unknown	• Lack of awareness regarding use • Absence of regulatory laws

FIGURE 7.4 Challenges in using nanofertilizers.

for standardized testing protocols and safety assessments specific to these nanoscale materials. Regulatory agencies must adapt to nanoparticles' unique properties and behaviors to ensure their safe use in agriculture.

Nanofertilizers have the potential to revolutionize nutrient management in agriculture by enhancing nutrient availability, reducing environmental impact, and improving overall crop productivity. However, addressing toxicity-related challenges, long-term effects, and regulatory oversight is essential to ensure the responsible adoption of nanofertilizers and their sustainable integration into modern agriculture. The significant difficulties in nanofertilizer use are summarized in Figure 7.4.

7.2.5 SIGNIFICANT APPLICATIONS OF NANOFERTILIZERS

In a report, Iqbal and coworkers (2022) have developed a protocol leveraging nanotechnology-based elicitors to facilitate *in vitro* shoot multiplication and callus induction in various mung bean varieties to enhance their phytochemical content. This research employed shoot and nodal tips of three mungbean genotypes as explants for the multiplication of shoots. The explants were grown (both in-vivo and in vitro) using Murashige & Skoog (MS) media, with different doses of benzyl amino purine (BAP) and indole butyric acid (IBA) as separate treatments.

Concurrently, calli were cultured on MS medium supplied with ZnO and CuO NPs as nano-elicitors intended to stimulate the production of glycosides and phenolic compounds. The findings revealed that in vitro explants exhibited superior shoot growth and leaves per explant as against in-vivo explants. Moreover, shoot tips demonstrated a more favorable response to in-vitro culture against nodal explants. This study suggests that applying NPs to elicit secondary metabolites from in-vitro cells of mungbean is a potential strategy (Iqbal et al., 2022).

Similarly, Geremew and their team (2023) evaluated the effect of ZnO-NPs developed from pecan tree leaves [*Carya illinoinensis* (Wangenh.) K. Koch] on mustard plant growth. In the face of challenges posed by climate change and land degradation to food production sustainability, the utilization of advanced nanotechnology plays a pivotal role. The research sought to establish a foundation for producing nanofertilizers and nanocomposite materials with broader agricultural applications and the potential for improving human nutrition quality. The focus was creating nanomaterials based on essential minerals like zinc, with zinc oxide nanoparticles (ZnO-NPs) as the primary target. The study aimed to explore how these ZnO-NPs, produced using *Carya illinoinensis* leaf extract, influenced various aspects of mustard plant development, including physiology, nutritional content, and antioxidant properties. The envisioned applications of ZnO-NPs encompassed promoting plant growth and augmenting crop yields through a novel soil supplement. Furthermore, the biofortification of *Brassica juncea* (L.) Coss plants with ZnO-NPs enhanced the crop's nutritional value and potentially amplified its therapeutic benefits (Geremew et al., 2023).

Ma and coworkers conducted a study revealing the beneficial effects of silver nanoparticles (Ag-NPs) on the quality and longevity of cut tree peony flowers. In their investigation, Ag-NPs were synthesized using *Eucommia ulmoides* Oliv. leaf extract, demonstrating their ability to inhibit bacterial growth effectively. Treating tree peony flowers with Ag-NPs resulted in notable improvements, including increased flower size, weight, and water balance. Moreover, the treated petals exhibited lower oxidative stress markers and higher concentrations of antioxidant enzymes. Notably, Ag-NPs also played a role in reducing bacterial growth within the stem vessels. This innovative technique involved pre-treating cut tree peony flowers with Ag-NPs, resulting in enhanced water uptake, prolonged vase life, and improved overall flower quality. The application of nanotechnology in the cultivation of cut tree peonies holds substantial promise for advancing the cut flower industry (Ma et al., 2023).

7.3 NANOPESTICIDES

Pesticides have long been an essential tool in agriculture, helping to safeguard crops from the relentless onslaught of pests and diseases. Pesticides, found in both chemically synthesized forms and as microorganisms, play a crucial role in agriculture for managing, eliminating, targeting, or providing resistance against pests, diseases, and parasites (Stocka, Tankiewicz, Biziuk, & Namieśnik, 2011; Yadav & Devi, 2017). They exhibit a diverse composition, incorporating organic and inorganic constituents, and can be classified according to their distinct chemical structures (Ogawa, Tokunaga, Kobayashi, Hirai, & Shibata, 2020). This classification encompasses organochlorines, organophosphates, carbamates, formamidines, thiocyanates, organotins, dinitrophenols, synthetic pyrethroids, and antibiotics (Amin et al., 2021; Vo, Raju, & Supreeth, 2020). These chemical formulations are designed to eliminate or deter the presence of unwanted organisms, thereby ensuring bountiful harvests and food security. However, the widespread use of conventional pesticides has raised significant concerns due to their adverse environmental and health effects (Sarkar, Gil, Keeley, & Jansen, 2021). One of the primary drawbacks of conventional pesticides is their detrimental effect on the environment. These chemicals can persist in soil and water, contaminating ecosystems. Runoff from treated fields can pollute nearby water bodies, endangering aquatic life and disrupting delicate ecosystems (Aktar, Sengupta, & Chowdhury, 2009). Conventional pesticides often lack specificity and can harm the intended pests and beneficial insects, birds, and other organisms. This indiscriminate approach can disrupt the balance of natural ecosystems, leading to unintended consequences (Mahmood et al., 2016). In addition, residues from conventional pesticides can accumulate in food products, posing health risks to consumers. They can lead to various health problems, including cancer, developmental issues, hormone disruption, etc. (Owens, Feldman, & Kepner, 2010).

Similarly, pesticide exposure can have severe health implications for farmworkers and nearby communities. Pesticides have been considered potential mutagenic agents due to their components capable of inducing changes in DNA. As per the assessment of the World Health Organization (WHO), approximately 1 million individuals experience acute poisoning as a result of exposure to pesticides, and a 0.4%–1.9% mortality rate is recorded every year (Eddleston, 2020; Thundiyil, Stober, Besbelli, & Pronczuk, 2008). However, the long-term use of conventional pesticides also develops resistance to pests over time, rendering them less effective (Georghiou, 2012; Liang, Tang, Cheke, & Wu, 2015; Onstad & Knolhoff, 2023). This phenomenon necessitates using higher quantities and more potent chemicals, exacerbating the environmental and health concerns associated with pesticides. Considering these drawbacks, there is a growing need for safer and more sustainable alternatives to conventional pesticides. This has led to the emergence of a promising solution: nanopesticides. Nanopesticides, a cutting-edge innovation in agriculture, aims to address many of the shortcomings of traditional pesticides (Behl et al., 2022). By harnessing the power of nanotechnology, these novel formulations offer precise and targeted pest control while minimizing the adverse effects on the environment and human health (Madhuban, Rajesh, & Arunava, 2012).

Nanopesticides are characterized by their small particle size, typically 1–100 nm. This reduced size allows for a high surface area-to-volume ratio, enhancing their reactivity and efficacy (Anandhi, Saminathan, Yasotha, Saravanan, & Rajanbabu, 2020). Nanopesticides offer precision in pest control. Their small size enables them to penetrate the cuticles of insects or the cell walls of pathogens more effectively, delivering the active ingredients directly to the target organisms (Dangi & Verma, 2021). The nanopesticides exhibit enhanced bioavailability and minimize environmental contamination due to their improved adhesion to target surfaces. This reduces the risk of chemical runoff and pollution of soil and water bodies (Xu et al., 2022). Compared to traditional formulations, nanopesticides often require lower quantities of active ingredients while maintaining the same level of effectiveness (Jampílek & Kráľová, 2017). This reduction in pesticide load can contribute to safer and more sustainable agricultural practices. Nanopesticide also offers the controlled release of the active ingredients and can be tailored to address specific pest challenges. They can incorporate multiple active ingredients to optimize the performance of selective pests (Huang et al., 2018). Their unique interaction mechanism can help reduce the development of pesticide resistance in target pests, extending their effectiveness over time. In comparison, nanopesticides hold promise for reducing risks to human health and the environment due to lower pesticide usage.

As agriculture faces the challenges of pest resistance, environmental conservation, and food security, nanopesticides have emerged as a transformative solution. These formulations offer the potential to enhance crop protection, reduce the ecological footprint of agriculture, and contribute to more sustainable and resilient food production systems. Nanopesticides provide a solution by capitalizing on three distinctive attributes: enhancing solubility, enabling controlled or targeted release, and safeguarding against premature degradation (Kah, Beulke, Tiede, & Hofmann, 2013). Nanopesticides encompass various categories and formulations (Figure 7.5), each tailored to specific pest control needs, including nanoparticle-based pesticides (Zhao et al., 2022), nanoemulsions (Heydari et al., 2020), nanocapsules (Shi et al., 2022), nanocarriers (Jiang et al., 2022), nano-pesticidal sprays (Zhao, Huang, Adeleye, & Keller, 2017), nanopowders (Polischuk, Churilov, Borychev, Byshov, & Nazarova, 2018), and polymeric nanoparticles (Vemula & Reddy, 2023), each of these nanopesticide formulations are designed to address specific agricultural challenges while minimizing environmental impact and enhancing pest control precision.

7.3.1 Widely Used Nanopesticides in Crop Management

In the quest to safeguard plants from insects and pests, integrating NPs holds significant promise for developing innovative pesticides, insecticides, and insect repellents. A diverse array of nanoparticle categories has found practical applications in formulating these pest-control solutions. The choice of which NPs to employ hinges on their intended action method or the specific properties they bring

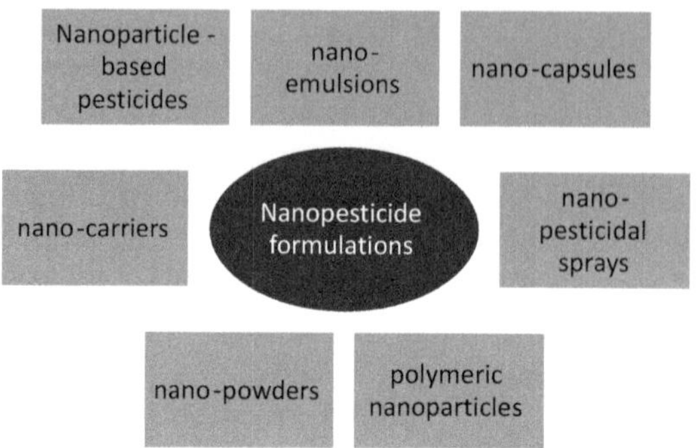

FIGURE 7.5 Major nanopesticide formulations/categories.

to the table. Notably, frequently utilized NPs in pesticide formulations encompass silver, gold, copper, titanium, zinc, silica, aluminum, chitin NPs, nano-clay, multi-walled carbon nanotubes (CNTs), graphene NPs, and more.

Zhong, Ximing, and their team crafted nanocarriers by functionalizing Zein with mesoporous silica to deliver avermectin pesticides. The synthesized nanopesticide exhibited dual responsiveness properties to pH and enzyme. The nanocarriers displayed a remarkable pesticide loading capacity, exceeding 280 mg g^{-1}. Furthermore, the fabricated nanocarriers contributed to a substantial 55% reduction in the pesticide's degradation rate when exposed to ultraviolet irradiation. Additionally, the nanopesticide offered notable advantages, enhancing wetting and adhesion properties when applied to cucumber foliar surfaces. The synthesized nanopesticide demonstrated high efficacy comparable to pure pesticide (Zhong et al., 2022). Likewise, Al Thabiani Aziz and coworkers harnessed *Artemisia herba-alba* Asso, commonly known as the wormwood plant, as a sustainable source for producing environmentally friendly nano insecticides targeted at mosquito vectors. Using plant extract, they synthesized Ag-NPs and conducted experiments on Anopheles, *Aedes*, and *Culex* mosquitoes from Indian and Saudi Arabian strains. The results were impressive, as the Artemisia herba-alba-synthesized Ag-NPs displayed potent larvicidal toxicity against these mosquito species. Additionally, the researchers optimized the relative LC$_{50}$ (lethal concentration for 50% mortality) values of the Ag-NPs for both Indian and Saudi Arabian mosquito strains (Alshehri et al., 2018). In a research study, novel nanopesticides were developed using smart, degradable metal-organic framework (MOF) nanocarriers gated with Fe^{III}-tannic acid networks. These nanopesticides exhibited seven stimuli-responsive behaviors, enabling the controlled release of encapsulated substances in response to environmental cues relevant to crop settings. The study demonstrated their efficacy in enhancing pesticide adhesion and retention on hydrophobic foliage. Additionally, these nanopesticides showed high fungicidal activity against pathogenic fungi and were safe for wheat seed germination and growth, suggesting their potential for improving agricultural production quality (Dong et al., 2021). Nahid Sarlak and coworkers developed MWCNT-graft-poly(citric acid) (MWCNT-g-PCA) hybrid materials by polymerizing citric acid onto oxidized multiwall CNTs. These materials, featuring conjugated citric acid branches, demonstrated water solubility and the ability to trap water-soluble chemicals and metal ions. The researchers successfully encapsulated pesticides (EP) like zineb and mancozeb in a polycitric acid shell, and their study revealed that the CNT-g-PCA-EP hybrid material exhibited enhanced toxicity against *Alternaria alternata* fungi compared to bulk pesticides. The *A. alternata* is a well-known species of fungus belonging to the genus '*Alternaria*'. Its presence is frequently reported in a variety of ecosystems, viz. soil, air, and plant surfaces. This species is typically believed to cause diseases in numerous plant species, including fruits, vegetables, and ornamental plants. It has been the main research focus for its mycotoxin production, e.g., 'alternariol' and 'alternariol monomethyl ether'—which possess the potential to contaminate food and pose health risks when consumed in bulk amounts (Sarlak, Taherifar, & Salehi, 2014).

In a study, Fowsiya et al. developed an eco-friendly nanoemulsion using green-synthesized NPs and phytochemicals derived from *Carissa edulis* (Forssk.) Vahl focuses on creating a green insecticide. They successfully isolated the bioactive caryophyllene and formulated a green nanoemulsion using Ag_2O and ZnO NPs. This nanoemulsion exhibited high insecticidal efficacy against damaging crop pests like *C. medinalis*, *S. incrtulas*, and *S. mauritia*, making it a promising eco-friendly insecticidal product (Fowsiya, Muthusamy, Alfarhan, & Madhumitha, 2023). Stepping to a cleaner and more sustainable nanopesticide for agriculture, Nguyen Hoang Ly et al. focused on developing nanocomposites (UiO-66@ZnO/Biochar) to encapsulate the pesticide carbendazim (CBZ) for precise and environmentally friendly spraying in agriculture. The nanocomposite exhibited high CBZ adsorption capacities, significantly extending the pesticide's photodegradation half-life. It also effectively inhibited fungal growth (Ly et al., 2023). Ying Ding and coworkers carried out another efficient and eco-friendly approach to aphid control in sustainable agriculture. The researchers devised a multifunctional nanopesticide delivery system to enhance agricultural pesticide efficiency. They encapsulated the neonicotinoid insecticide acetamiprid (Ace) in a mesoporous silica nanocarrier with redox-responsive properties, resulting in Ace@MSN-SS-C10. This nanopesticide exhibited controlled release triggered by glutathione (GSH). It demonstrated excellent wettability, adhesion, stability, and safety. Greenhouse experiments revealed that foliar spraying with 1.5 mg of Ace@MSN-SS-C10 per plant significantly reduced aphid populations on *Vicia faba* L. while maintaining Ace residue levels below international safety limits (Ding et al., 2023). Similarly, Wei Gan and the research team designed a pH-responsive fluorescent nanopesticide with the goal of combatting pine wilt disease caused by the pine wood nematode. Their nanopesticide, THI@PAMAM@MSNs, efficiently enclosed thiamethoxam (THI) within mesoporous silica nanoparticles (MSNs) and polyamidoamine (PAMAM) dendrimers, achieving an impressive THI loading rate of 21.8%. THI@PAMAM@MSNs also demonstrated superior formulation stability, wettability, and adhesion compared to the commercially available THI@WG. Notably, in experimental assessments, THI@PAMAM@MSNs exhibited enhanced efficacy in transporting and controlling pine wood nematodes and their vector insects (Gan et al., 2023).

The studies discussed above underscore the remarkable contribution of nanopesticides in advancing sustainable agriculture and crop management. These innovative formulations, developed by different researchers, offer multifunctional solutions to address critical agricultural challenges. Nanopesticides provide precise, controlled delivery of agrochemicals, minimizing environmental impact and reducing pesticide residues in crops. These findings illuminate the potential of nanopesticides as a pivotal tool in sustainable agriculture, offering efficient, eco-friendly alternatives that contribute to the goal of safer and more productive crop management.

7.3.2 Environmental and Safety Considerations Regarding Nanopesticides

Nanopesticides, hailed for their potential to revolutionize agriculture through precision pest control and reduced environmental impact, also have environmental and safety concerns that demand careful consideration. These concerns arise from the unique characteristics of nanopesticides, primarily their extremely small particle size, typically ranging from 1 to 100 nm. One significant environmental concern is the persistence of nanopesticides in the environment. Due to their small size, these particles can endure in soil and water for extended periods, raising questions about potential long-term ecological consequences and contamination risks. The other critical issue is the ecotoxicity of nanopesticides, especially their unintended effects on non-target organisms. Beneficial insects, pollinators, aquatic life, soil microorganisms, and birds can all be inadvertently exposed to these NPs. The small size and increased mobility of nanopesticides mean that non-target organisms may face a broader exposure range, potentially disrupting ecosystems and food chains.

Moreover, nanopesticides might enhance environmental mobility, heightening contaminant transport risk through soil and water. This could lead to off-site contamination and ecosystem disturbance, affecting terrestrial and aquatic ecosystems. Effective resistance management strategies are another concern. As with conventional pesticides, pests may resist nanopesticides over time, rendering them less effective. Sustainable resistance management practices must be implemented to maintain the long-term efficacy of these novel formulations.

Regarding human health and safety, potential risks associated with nanopesticide residues in food and water require careful investigation. Furthermore, occupational safety is a concern for farmworkers and pesticide applicators, who may face increased risks when handling nanopesticides. Adequate safety measures, including personal protective equipment and training, must be in place to safeguard workers. Nanopesticides can emit NPs into the air, raising the possibility of inhalation exposure. Proper application techniques and equipment are essential to minimize this risk. Finally, the end-of-life management of nanopesticides and their containers is crucial to prevent environmental contamination. Proper disposal and recycling methods must be established and adhered to (Kannan et al., 2023; Li, Xu, Kah, Lin, & Filser, 2019; Pandey, Giri, Kumar, Mishra, & Raja Rishi, 2018; Xu et al., 2022).

Balancing the potential benefits of nanopesticides, such as precision and reduced environmental impact, with these multifaceted concerns requires a comprehensive approach involving robust research, stringent regulation, and adopting responsible agricultural practices. Continuous assessment and monitoring are essential to ensure the safe and sustainable integration of nanopesticides into modern agriculture while minimizing their impact on the environment and non-target organisms.

7.4 REGULATIONS, RISK ASSESSMENT, AND ETHICAL CONSIDERATIONS

7.4.1 OVERVIEW OF REGULATORY FRAMEWORKS FOR EVALUATING NANOAGROCHEMICALS' SAFETY

Introducing nanoagrochemicals into agriculture necessitates a robust regulatory framework to ensure their safety and responsible use. Regulatory bodies worldwide are actively adapting and developing guidelines specific to nano-based products. These frameworks include product characterization, toxicity assessment, environmental fate, labeling requirements, and risk management strategies. These regulations are designed to safeguard human health, protect the environment, and maintain the sustainability of agriculture while facilitating innovation in the field (Amenta et al., 2015; Kumari et al., 2023).

7.4.2 CHALLENGES IN RISK ASSESSMENT DUE TO NANOPARTICLE COMPLEXITY

Assessing the safety of nanoagrochemicals presents unique challenges due to the complexity of NPs. Nanoscale materials may exhibit novel behaviors and interactions compared to their bulk counterparts. This complexity can complicate traditional risk assessment methods often designed for larger-scale chemicals. Challenges include understanding the bioavailability and bioaccumulation of NPs in organisms, predicting their behavior in different environmental conditions, and discerning potential long-term effects (Iavicoli, Leso, Beezhold, Shvedova, & pharmacology, 2017; Singh & Gurjar, 2022). Addressing these challenges requires interdisciplinary collaboration among scientists, regulators, and industry stakeholders to develop comprehensive risk assessment strategies.

7.4.3 IMPORTANCE OF STANDARDIZED TESTING PROTOCOLS FOR NANO-BASED PRODUCTS

Standardized testing protocols are paramount in ensuring the safety and efficacy of nano-based products in agriculture. These protocols provide a systematic and reproducible framework for evaluating product performance and safety. Given the unique properties of NPs, standardized testing is crucial for accurately assessing their behavior, toxicity, and environmental impact. Consistency in testing protocols facilitates regulatory compliance and enables meaningful comparisons between different nanoagrochemicals. Based on sound scientific principles and risk assessment criteria, developing standardized protocols contributes to transparent, reliable, and responsible innovation in nano-agriculture (Iavicoli et al., 2017).

7.4.4 ETHICAL CONSIDERATIONS: BALANCING BENEFITS AND POTENTIAL HARMS

Adopting nanotechnology in agriculture raises essential ethical considerations beyond scientific and regulatory domains. Balancing the benefits of increased crop yields, reduced environmental impact, and improved food security with potential harms is paramount. Ethical dilemmas include equitable access to nano-agriculture technologies, intellectual property rights, and possible unintended consequences. Moreover, ethical concerns extend to these technologies' social, economic, and cultural impacts, including their influence on farming practices, livelihoods, and rural communities (Bhattacharya, Roy, Bhattacharya, Prasad, & Mandal, 2023). To navigate these complexities, stakeholders must engage in informed and inclusive dialogues that consider the broader implications of nanotechnology in agriculture and prioritize sustainability, equity, and responsible innovation.

7.5 PROSPECTS AND CHALLENGES

The integration of nanotechnology in agriculture offers both exciting possibilities and formidable challenges. Nanobiotechnology is poised to lead the way in precision agriculture by customizing nutrient delivery, pest management, and crop monitoring on a nanoscale, potentially optimizing resources, reducing waste, and increasing yields. Prospects include the development of nanosensors for real-time data collection, enabling precise decision-making in farming practices. Eco-friendly nanopesticides are emerging as alternatives to traditional chemical pesticides, with ongoing research to enhance their effectiveness and ensure long-term safety for the environment and human health. Robust regulatory frameworks addressing safety and environmental concerns are crucial to integrating nanobiotechnology into agriculture successfully. Ethical considerations, such as equitable access and misuse prevention, must also be addressed in future policymaking. Transferring knowledge and technology to farmers, particularly in resource-limited regions, presents a significant challenge, necessitating educational programs and extension services to bridge the gap. Continued research into the long-term environmental impact of nanofertilizers and pesticides remains essential. Assessing effects on soil microbiota, water quality, and non-target species will be critical in maintaining ecosystem integrity. Economic viability, especially for smallholder farmers, will be crucial to widespread adoption, requiring cost-effective production methods and inclusive pricing structures. Given the global nature of agriculture, international collaboration will be pivotal. Sharing research findings, best practices, and regulatory experiences will accelerate the responsible integration of nanotechnology into agriculture globally.

In conclusion, the future of nanobiotechnology in sustainable agriculture holds great promise, but it also demands careful consideration of its implications and challenges. The dynamic interplay between technological innovation, policy development, and ethical practices will shape how nanobiotechnology contributes to the sustainability and resilience of our food systems in the decades to come. Researchers, policymakers, and agricultural stakeholders must work collaboratively to harness its full potential while safeguarding the environment and ensuring equitable access to its benefits.

7.6 CONCLUDING REMARKS AND FUTURE THRUST

In conclusion, integrating nanotechnology into agriculture represents a paradigm shift with immense potential for addressing critical challenges in modern crop management. While chemical fertilizers and pesticides have undeniably boosted agricultural productivity, they have come at a cost to environmental quality and sustainability. The extensive use of these chemicals has led to water pollution, the emergence of pesticide-resistant pests, and harm to non-target organisms. However, the advent of nanotechnology has ushered in a new era of precision agriculture. With their controlled release mechanisms and targeted pest control, nanopesticides offer a path toward more effective and environmentally friendly pest management. Likewise, nanofertilizers, with their nanoscale nutrient delivery systems, enhance nutrient efficiency while reducing the environmental impact of excess nutrient runoff. The case studies and research highlighted in this chapter provide tangible evidence of the efficacy and safety of nanopesticides and nanofertilizers. These innovative approaches have demonstrated their potential to improve crop yields, reduce environmental contamination, and foster sustainable agriculture practices. As we move forward, it is essential to continue research and development in the field of nanobiotechnology for agriculture. Rigorous testing, regulatory oversight, and responsible adoption are crucial to ensure these nanoscale solutions' long-term safety and sustainability. In the face of growing global food demand and mounting environmental challenges, nanotechnology offers hope.

BIBLIOGRAPHY

Ahmed, S., & Fahmy, A. (2017). Environmental Impact for applications of neem cake coated urea and nano iron foliar on rationalization of chemical nitrogen fertilizers and wheat yield. *Journal of Soil Sciences and Agricultural Engineering, 8*(11), 613–620.

Aktar, M. W., Sengupta, D., & Chowdhury, A. (2009). Impact of pesticides use in agriculture: Their benefits and hazards. *Interdisciplinary Toxicology, 2*(1), 1.

Alshehri, M. A., Panneerselvam, C., Murugan, K., Trivedi, S., Mahyoub, J. A., & Maggi, F. (2018). The desert wormwood (Artemisia herba-alba)-From Arabian folk medicine to a source of green and effective nanoinsecticides against mosquito vectors. *Journal of Photochemistry and Photobiology B: Biology, 180*, 225–234.

Amenta, V., Aschberger, K., Arena, M., Bouwmeester, H., Moniz, F. B., & Brandhoff, P. (2015). Regulatory aspects of nanotechnology in the agri/feed/food sector in EU and non-EU countries. *Pharmacology, 73*(1), 463–476.

Amin, S., Solangi, A. R., Hassan, D., Hussain, N., Ahmed, J., & Baksh, H. (2021). Recent trends in development of nanomaterials based green analytical methods for environmental remediation. *Current Analytical Chemistry, 17*(4), 438–448.

An, C., Sun, C., Li, N., Huang, B., Jiang, J., Shen, Y., ... Wang, C. (2022). Nanomaterials and nanotechnology for the delivery of agrochemicals: Strategies towards sustainable agriculture. *Journal of Nanobiotechnology, 20*(1), 1–19.

Anandhi, S., Saminathan, V., Yasotha, P., Saravanan, P., & Rajanbabu, V. (2020). Nano-pesticides in pest management. *Journal of Entomology and Zoology Studies, 8*, 685–690.

Babu, S., Singh, R., Yadav, D., Rathore, S. S., Raj, R., Avasthe, R., ... Yadav, B. (2022). Nanofertilizers for agricultural and environmental sustainability. *Chemosphere, 292*, 133451.

Behl, T., Kaur, I., Sehgal, A., Singh, S., Sharma, N., Bhatia, S., ... Bungau, S. (2022). The dichotomy of nanotechnology as the cutting edge of agriculture: Nano-farming as an asset versus nanotoxicity. *Chemosphere, 288*, 132533.

Bhardwaj, A. K., Arya, G., Kumar, R., Hamed, L., Pirasteh-Anosheh, H., Jasrotia, P., ... Singh, G. P. (2022). Switching to nanonutrients for sustaining agroecosystems and environment: The challenges and benefits in moving up from ionic to particle feeding. *Journal of Nanobiotechnology, 20*(1), 19.

Bhattacharya, B., Roy, P., Bhattacharya, S., Prasad, B., & Mandal, A. K. (2023). Nanotechnology and sustainable development: Overcoming the obstacles by adopting ethical practices for future farming. In *Engineered Nanomaterials for Sustainable Agricultural Production, Soil Improvement and Stress Management*, Husen, A. (Ed.) (pp. 431–445): Elsevier.

Cai, D., Wu, Z., Jiang, J., Wu, Y., Feng, H., Brown, I. G., ... Yu, Z. (2014). Controlling nitrogen migration through micro-nano networks. *Scientific Reports, 4*(1), 3665.

Chhipa, H. (2017). Nanofertilizers and nanopesticides for agriculture. *Environmental Chemistry Letters, 15*, 15–22.

Dangi, K., & Verma, A. K. (2021). Efficient & eco-friendly smart nano-pesticides: Emerging prospects for agriculture. *Materials Today: Proceedings, 45*, 3819–3824.

Delfani, M., Baradarn Firouzabadi, M., Farrokhi, N., & Makarian, H. (2014). Some physiological responses of black-eyed pea to iron and magnesium nanofertilizers. *Communications in Soil Science and Plant Analysis, 45*(4), 530–540.

DeRosa, M. C., Monreal, C., Schnitzer, M., Walsh, R., & Sultan, Y. (2010). Nanotechnology in fertilizers. *Nature Nanotechnology, 5*(2), 91.

Dimkpa, C. O., Singh, U., Bindraban, P. S., Adisa, I. O., Elmer, W. H., Gardea-Torresdey, J. L., & White, J. C. (2019). Addition-omission of zinc, copper, and boron nano and bulk oxide particles demonstrate element and size-specific response of soybean to micronutrients exposure. *Science of the Total Environment, 665*, 606–616.

Ding, Y., Xiao, Z., Chen, F., Yue, L., Wang, C., Fan, N., ... Wang, Z. (2023). A mesoporous silica nanocarrier pesticide delivery system for loading acetamiprid: Effectively manage aphids and reduce plant pesticide residue. *Science of the Total Environment, 863*, 160900.

Dong, J., Chen, W., Feng, J., Liu, X., Xu, Y., Wang, C., Yang, W., & Du, X. (2021). Facile, smart, and degradable metal-organic framework nanopesticides gated with FeIII-tannic acid networks in response to seven biological and environmental stimuli. *ACS Applied Materials & Interfaces, 13*(16), 19507–19520.

Eddleston, M. (2020). Poisoning by pesticides. *Medicine, 48*(3), 214–217.

Feng, J., Ma, Y., Chen, Z., Liu, Q., Yang, J., Gao, Y., ... Yang, W. (2021). Development and characterization of pyriproxyfen-loaded nanoemulsion for housefly control: Improving activity, reducing toxicity, and protecting ecological environment. *ACS Sustainable Chemistry & Engineering, 9*(14), 4988–4999.

Fowsiya, J., Muthusamy, K., Alfarhan, A., & Madhumitha, G. (2023). Promising insecticidal effect of C. edulis phytochemical loaded nanoemulsion using Ag2O and ZnO NPs: A synergistic combination by ultra-sonication against crop damaging insects. *South African Journal of Botany, 157*, 566–578.

Fu, H., Tan, P., Wang, R., Li, S., Liu, H., Yang, Y., & Wu, Z. (2022). Advances in organophosphorus pesticides pollution: Current status and challenges in ecotoxicological, sustainable agriculture, and degradation strategies. *Journal of Hazardous Materials, 424*, 127494.

Gan, W., Kong, X., Fang, J., Shi, X., Zhang, S., Li, Y., ... Zhang, F. (2023). A pH-responsive fluorescent nanopesticide for selective delivery and visualization in pine wood nematode control. *Chemical Engineering Journal, 463*, 142353.

Georghiou, G. P. (2012). *Pest Resistance to Pesticides*. Springer Science & Business Media.

Geremew, A., Carson, L., Woldesenbet, S., Wang, H., Reeves, S., Brooks Jr, N., ... Peace, E. (2023). Effect of zinc oxide nanoparticles synthesized from Carya illinoinensis leaf extract on growth and antioxidant properties of mustard (Brassica juncea). *Frontiers in Plant Science, 14*, 1108186.

Ghormade, V., Deshpande, M. V., & Paknikar, K. M. (2011). Perspectives for nanobiotechnology enabled protection and nutrition of plants. *Biotechnology Advances, 29*(6), 792–803.

Gomiero, T. (2018). Food quality assessment in organic vs. conventional agricultural produce: Findings and issues. *Applied Soil Ecology, 123*, 714–728.

Gupta, R., & Xie, H. (2018). Nanoparticles in daily life: Applications, toxicity and regulations. *Journal of Environmental Pathology, Toxicology and Oncology, 37*(3), 209–230.

Heydari, M., Amirjani, A., Bagheri, M., Sharifian, I., & Sabahi, Q. (2020). Eco-friendly pesticide based on peppermint oil nanoemulsion: Preparation, physicochemical properties, and its aphicidal activity against cotton aphid. *Environmental Science and Pollution Research, 27*, 6667–6679.

Huang, B., Chen, F., Shen, Y., Qian, K., Wang, Y., Sun, C., ... Zeng, Z. (2018). Advances in targeted pesticides with environmentally responsive controlled release by nanotechnology. *Nanomaterials, 8*(2), 102.

Hussain, N., Bilal, M., & Iqbal, H. M. N. (2022). Carbon-based nanomaterials with multipurpose attributes for water treatment: Greening the 21st-century nanostructure materials deployment. *Biomaterials and Poymer Horizon, 1* (1), 48–58.

Iavicoli, I., Leso, V., Beezhold, D. H., & Shvedova, A. A. (2017). Nanotechnology in agriculture: Opportunities, toxicological implications, and occupational risks. *Toxicology and Applied Pharmacology, 329*, 96–111.

Iqbal, Z., Javad, S., Naz, S., Shah, A. A., Shah, A. N., Paray, B. A., ... Abdelsalam, N. R. (2022). Elicitation of the in vitro cultures of selected varieties of Vigna radiata L. with zinc oxide and copper oxide nanoparticles for enhanced phytochemicals production. *Frontiers in Plant Science, 13*, 908532.

Jakhar, A. M., Aziz, I., Kaleri, A. R., Hasnain, M., Haider, G., Ma, J., & Abideen, Z. (2022). Nano-fertilizers: A sustainable technology for improving crop nutrition and food security. *NanoImpact, 27*, 100411.

Jampílek, J., & Kráľová, K. (2017). Nanopesticides: Preparation, targeting, and controlled release. In *New Pesticides and Soil Sensors*, Grumezescu, A. M. (Ed.) (pp. 81–127): Elsevier.

Jiang, Q., Xie, Y., Peng, M., Wang, Z., Li, T., Yin, M., ... Yan, S. (2022). A nanocarrier pesticide delivery system with promising benefits in the case of dinotefuran: Strikingly enhanced bioactivity and reduced pesticide residue. *Environmental Science: Nano, 9*(3), 988–999.

Johnson, M. S., Sajeev, S., & Nair, R. S. (2021). *Role of Nanosensors in agriculture.* Paper presented at the 2021 *International Conference on Computational Intelligence and Knowledge Economy (ICCIKE).* Dubai, United Arab Emirates, pp. 58–63, doi: 10.1109/ICCIKE51210.2021.9410709.

Kah, M., Beulke, S., Tiede, K., & Hofmann, T. (2013). Nanopesticides: State of knowledge, environmental fate, and exposure modeling. *Critical Reviews in Environmental Science and Technology, 43*(16), 1823–1867.

Kah, M., Kookana, R. S., Gogos, A., & Bucheli, T. D. (2018). A critical evaluation of nanopesticides and nanofertilizers against their conventional analogues. *Nature Nanotechnology, 13*(8), 677–684.

Kahlel, A., Ghidan, A., Al-Antary, T., Alshomali, I., & Asoufi, H. (2020). Effects of nanotechnology liquid fertilizers on certain vegetative growth of broad bean (Vicia faba L.). *Fresenius Environmental Bulletin, 29*(6), 4763–4768.

Kannan, M., Bojan, N., Swaminathan, J., Zicarelli, G., Hemalatha, D., & Zhang, Y. (2023). Nanopesticides in agricultural pest management and their environmental risks: A review. *International Journal of Environmental Science and Technology, 20*(9), 10507–10532.

Kaushal, M., & Wani, S. P. (2017). Nanosensors: Frontiers in precision agriculture. In: Prasad, R., Kumar, M., Kumar, V. (eds) *Nanotechnology: An Agricultural Paradigm,* 279–291. Springer, Singapore. https://doi.org/10.1007/978-981-10-4573-8_13.

Kottegoda, N., Munaweera, I., Madusanka, N., & Karunaratne, V. (2011). A green slow-release fertilizer composition based on urea-modified hydroxyapatite nanoparticles encapsulated wood. *Current Science,* 73–78.

Kumar, R., Kumar, R., & Prakash, O. (2019). Chapter-5 The impact of chemical fertilizers on our environment and ecosystem. In *Research Trends in Environmental Sciences* (pp.69–86). Chief Editor: Dr. Poonam Sharma. ISBN: 978-93-90217-13-7. Publisher: AkiNik Publications.

Kumar, S., Nehra, M., Dilbaghi, N., Marrazza, G., Hassan, A. A., & Kim, K.-H. (2019). Nano-based smart pesticide formulations: Emerging opportunities for agriculture. *Journal of Controlled Release, 294,* 131–153.

Kumari, R., Suman, K., Karmakar, S., Lakra, S. G., Saurav, G. K., & Mahto, B. K. (2023). Regulation and safety measures for nanotechnology-based agri-products. *Frontiers in Genome Editing, 5,* 1200987.

Li, L., Xu, Z., Kah, M., Lin, D., & Filser, J. (2019). *Nanopesticides: A Comprehensive Assessment of Environmental Risk Is Needed before Widespread Agricultural Application.* ACS Publications.

Liang, J., Tang, S., Cheke, R. A., & Wu, J. (2015). Models for determining how many natural enemies to release inoculatively in combinations of biological and chemical control with pesticide resistance. *Journal of Mathematical Analysis and Applications, 422*(2), 1479–1503.

Liu, R., & Lal, R. (2015). Potentials of engineered nanoparticles as fertilizers for increasing agronomic productions. *Science of the Total Environment, 514,* 131–139.

Liu, X., Cao, A., Yan, D., Ouyang, C., Wang, Q., & Li, Y. (2021). Overview of mechanisms and uses of biopesticides. *International Journal of Pest Management, 67*(1), 65–72.

Ly, N. H., Nguyen, N. B., Tran, H. N., Hoang, T. T. H., Joo, S.-W., Vasseghian, Y., … Klemeš, J. (2023). Metal-organic framework nanopesticide carrier for accurate pesticide delivery and decrement of groundwater pollution. *Journal of Cleaner Production, 402,* 136809.

Ma, Z., Zhang, K., Guo, W., Yu, W., Wang, J., & Li, J. (2023). Green synthesis of silver nanoparticles using Eucommia ulmoides leaf extract for inhibiting stem end bacteria in cut tree peony flowers. *Frontiers in Plant Science, 14,* 1176359.

Madhuban, G., Rajesh, K., & Arunava, G. (2012). Nano-pesticides-A recent approach for pest control. *The Journal of Plant Protection Sciences, 4*(2), 1–7.

Mahmood, I., Imadi, S. R., Shazadi, K., Gul, A., & Hakeem, K. R. (2016). Effects of pesticides on environment. *Plant, Soil and Microbes: Volume 1: Implications in Crop Science,* 253–269.

Majeed, Z., Ramli, N. K., Mansor, N., & Man, Z. (2015). A comprehensive review on biodegradable polymers and their blends used in controlled-release fertilizer processes. *Reviews in Chemical Engineering, 31*(1), 69–95.

Mali, S. C., Raj, S., & Trivedi, R. (2020). Nanotechnology a novel approach to enhance crop productivity. *Biochemistry and Biophysics Reports, 24,* 100821.

Mejias, J. H., Salazar, F., Amaro, L. P., Hube, S., Rodriguez, M. & Alfaro, M. (2021). Nanofertilizers: A Cutting-Edge Approach to Increase Nitrogen Use Efficiency in Grasslands. *Frontiers in Environmental Science, 9,* 635114.

Mondal, A., Basu, R., Das, S., & Nandy, P. (2011). Beneficial role of carbon nanotubes on mustard plant growth: An agricultural prospect. *Journal of Nanoparticle Research, 13,* 4519–4528.

Nair, R., Varghese, S. H., Nair, B. G., Maekawa, T., Yoshida, Y., & Kumar, D. S. (2010). Nanoparticulate material delivery to plants. *Plant Science, 179*(3), 154–163.

Naz, M. Y., Shukrullah, S., & Ghaffar, A. (2021). Sensors detecting controlled fertilizer release. In *Controlled Release Fertilizers for Sustainable Agriculture*, Lewu, F. B., Volova, T., Thomas, S., & Rakhimol, K. R. (Eds.) pp. 131–153: Elsevier.

Nowack, B., & Bucheli, T. D. (2007). Occurrence, behavior and effects of nanoparticles in the environment. *Environmental Pollution, 150*(1), 5–22.

Oberdörster, G., Sharp, Z., Atudorei, V., Elder, A., Gelein, R., Kreyling, W., & Cox, C. (2004). Translocation of inhaled ultrafine particles to the brain. *Inhalation Toxicology, 16*(6–7), 437–445.

Ogawa, Y., Tokunaga, E., Kobayashi, O., Hirai, K., & Shibata, N. (2020). Current contributions of organofluorine compounds to the agrochemical industry. *Iscience, 23*(9).

Okey-Onyesolu, C. F., Hassanisaadi, M., Bilal, M., Barani, M., Rahdar, A., Iqbal, J., & Kyzas, G. Z. J. C. (2021). Nanomaterials as nanofertilizers and nanopesticides: An overview. *ChemistrySelect, 6*(33), 8645–8663.

Onstad, D. W., & Knolhoff, L. M. (2023). Major issues in insect resistance management. In *Insect Resistance Management*, Onstad, D. W., & Knolhoff, L. M. (Eds.) (pp. 1–29): Elsevier.

Owens, K., Feldman, J., & Kepner, J. (2010). Wide range of diseases linked to pesticides. *Pesticides and You, 30*(2), 13–21.

Pandey, S., Giri, K., Kumar, R., Mishra, G., & Raja Rishi, R. (2018). Nanopesticides: Opportunities in crop protection and associated environmental risks. *Proceedings of the National Academy of Sciences, India Section B: Biological Sciences, 88*, 1287–1308.

Polischuk, S., Churilov, G., Borychev, S., Byshov, N., & Nazarova, A. (2018). Nanopowders of cuprum, cobalt and their oxides used in the intensive technology for growing cucumbers. *International Journal of Nanobiotechnology, 15*(4–5), 352–369.

Pramanik, P., Krishnan, P., Maity, A., Mridha, N., Mukherjee, A., & Rai, V. (2020). Application of nanotechnology in agriculture. *Environmental Nanotechnology, 4*, 317–348.

Pretty, J. (2018). Intensification for redesigned and sustainable agricultural systems. *Science, 362*(6417), eaav0294.

Raliya, R., Saharan, V., Dimkpa, C., & Biswas, P. (2017). Nanofertilizer for precision and sustainable agriculture: Current state and future perspectives. *Journal of Agricultural and Food Chemistry, 66*(26), 6487–6503.

Rameshaiah, G., Pallavi, J., & Shabnam, S. (2015). Nano fertilizers and nano sensors-an attempt for developing smart agriculture. *International Journal of Engineering Research and General Science, 3*(1), 314–320.

Rani, L., Thapa, K., Kanojia, N., Sharma, N., Singh, S., Grewal, A. S., … Kaushal, J. (2021). An extensive review on the consequences of chemical pesticides on human health and environment. *Journal of Cleaner Production, 283*, 124657.

Rathore, S., Shekhawat, K. A., Singh, R., Updhyay, P., Shekhawat, R., & Premi, O. P. (2019). Effect of nano-particles on growth, productivity, profitability of Indian mustard (Brassica juncea) under semi-arid conditions. *Indian Journal of Agricultural Sciences, 89*(7), 1145-1150.

Sarkar, S., Gil, J. D. B., Keeley, J., & Jansen, K. (2021). *The Use of Pesticides in Developing Countries and Their Impact on Health and the Right to F*ood: European Union.

Sarlak, N., Taherifar, A., & Salehi, F. (2014). Synthesis of nanopesticides by encapsulating pesticide nanoparticles using functionalized carbon nanotubes and application of new nanocomposite for plant disease treatment. *Journal of Agricultural and Food Chemistry, 62*(21), 4833–4838.

Seleiman, M. F., Almutairi, K. F., Alotaibi, M., Shami, A., Alhammad, B. A., & Battaglia, M. L. (2020). Nano-fertilization as an emerging fertilization technique: Why can modern agriculture benefit from its use? *Plants, 10*(1), 2.

Shang, Y., Hasan, M. K., Ahammed, G. J., Li, M., Yin, H., & Zhou, J. (2019). Applications of nanotechnology in plant growth and crop protection: A review. *Molecules, 24*(14), 2558.

Sharma, N., & Singhvi, R. (2017). Effects of chemical fertilizers and pesticides on human health and environment: A review. *International Journal of Agriculture, Environment and Biotechnology, 10*(6), 675–680.

Shi, L., Yan, W., Sun, L., Hou, C., Wei, N., Chen, Z., & Feng, J. (2022). Preparation and characterization of emamectin benzoate nanocapsules based on the dual role of polydopamine. *Pest Management Science, 78*(10), 4407–4416.

Shoults-Wilson, W. A., Reinsch, B. C., Tsyusko, O. V., Bertsch, P. M., Lowry, G. V., & Unrine, J. M. (2011). Role of particle size and soil type in toxicity of silver nanoparticles to earthworms. *Soil Science Society of America Journal, 75*(2), 365–377.

Siddiqi, K. S., & Husen, A. (2017). Plant response to engineered metal oxide nanoparticles. *Nanoscale Research Letters, 12*, 1–18.

Singh, D., & Gurjar, B. R. (2022). Nanotechnology for agricultural applications: Facts, issues, knowledge gaps, and challenges in environmental risk assessment. *Journal of Environmental Management, 322*, 116033.

Sivarethinamohan, R., & Sujatha, S. (2021). Unlocking the potentials of using nanotechnology to stabilize agriculture and food production. *AIP Conference Proceedings*, 2327, 020022. https://doi.org/10.1063/5.0039418

Stocka, J., Tankiewicz, M., Biziuk, M., & Namieśnik, J. (2011). Green aspects of techniques for the determination of currently used pesticides in environmental samples. *International Journal of Molecular Sciences*, *12*(11), 7785–7805.

Tahat, M. M., Alananbeh, K. M., Othman, Y. A., & Leskovar, D. I. (2020). Soil health and sustainable agriculture. *Sustainability*, *12*(12), 4859.

Talan, A., Mishra, A., Eremin, S. A., Narang, J., Kumar, A., & Gandhi, S. (2018). Ultrasensitive electrochemical immuno-sensing platform based on gold nanoparticles triggering chlorpyrifos detection in fruits and vegetables. *Biosensors and Bioelectronics*, *105*, 14–21.

Thundiyil, J. G., Stober, J., Besbelli, N., & Pronczuk, J. (2008). Acute pesticide poisoning: A proposed classification tool. *Bulletin of the World Health Organization*, *86*, 205–209.

Varshney, R. K., Pandey, M. K., & Chitikineni, A. (2018). Plant genetics and molecular biology: An introduction. *Advances in Biochemical Engineering/Biotechnology*, *164*, 1–9.

Velten, S., Leventon, J., Jager, N., & Newig, J. (2015). What is sustainable agriculture? A systematic review. *Sustainability*, *7*(6), 7833–7865.

Vemula, M., & Reddy, A. V. (2023). Polymeric nanoparticles as effective delivery systems in agriculture sustainability. *Nanotechnology for Environmental Engineering*, 8(3), 805–814.

Vo, A., Raju, N., & Supreeth, M. (2020). Effect of bacteria–Azotobacter species. *Journal of Environmental Chemistry and Toxicology*, *4*(2), 1.

Wang, P., Lombi, E., Zhao, F.-J., & Kopittke, P. M. (2016). Nanotechnology: A new opportunity in plant sciences. *Trends in Plant Science*, *21*(8), 699–712.

Xu, Z., Tang, T., Lin, Q., Yu, J., Zhang, C., Zhao, X., … Li, L. (2022). Environmental risks and the potential benefits of nanopesticides: A review. *Environmental Chemistry Letters*, *20*(3), 2097–2108.

Yadav, I. C., & Devi, N. L. (2017). Pesticides classification and its impact on human and environment. *Environmental Science and Engineering*, *6*, 140–158.

Yang, J., Jiang, F., Ma, C., Rui, Y., Rui, M., Adeel, M. … Xing B. (2018). Alteration of crop yield and quality of wheat upon exposure to silver nanoparticles in a life cycle study. *Journal of Agricultural and Food Chemistry*, *66*(11), 2589–2597.

Zhao, L., Huang, Y., Adeleye, A. S., & Keller, A. A. (2017). Metabolomics reveals Cu (OH) 2 nanopesticide-activated anti-oxidative pathways and decreased beneficial antioxidants in spinach leaves. *Environmental Science & Technology*, *51*(17), 10184–10194.

Zhao, W., Liu, Y., Zhang, P., Zhou, P., Wu, Z., Lou, B., … Lynch, I. (2022). Engineered Zn-based nano-pesticides as an opportunity for treatment of phytopathogens in agriculture. *NanoImpact*, *28*, 100420.

Zhong, X., Wen, H., Zeng, R., Deng, H., Su, G., & Zhou, H., (2022). Zein-functionalized mesoporous silica as nanocarriers for nanopesticides with pH/enzyme dual responsive properties. *Industrial Crops and Products*, *188*, 115716.

Zulfiqar, F., Navarro, M., Ashraf, M., Akram, N. A., & Munné-Bosch, S. (2019). Nanofertilizer use for sustainable agriculture: Advantages and limitations. *Plant Science*, *289*, 110270.

8 Post-Harvest Management of Food Crops and Agro-waste Utilization in a Developing Economy

A Review

Matthew N. Abonyi, Chukwunonso O. Aniagor,
Christopher C. Obi, and Emmanuel C. Nwadike

8.1 INTRODUCTION

In the context of a rapidly growing global population, the sustainable management of agricultural resources has emerged as a pivotal concern for developing economies. Post-harvest losses and agro-waste generation pose significant challenges to food security, economic growth, and environmental sustainability (Balana et al., 2022; Ibrahim et al., 2022). Innovative solutions become crucial as these countries strive to meet an expanding population's demands while minimizing waste and maximizing resource utilization. In recent years, integrating nano-biotechnology into post-harvest and agro-waste management has shown promising potential to address these pressing issues (Pooja & Jyoti, 2022; Jinsong et al., 2021).

Nano-biotechnology, a multidisciplinary field at the intersection of nanotechnology and biotechnology, has gained momentum as a transformative tool to revolutionize various industries, including agriculture. Its application in post-harvest and agro-waste management opens new avenues for enhancing shelf life, reducing losses, and converting waste into valuable resources. In this review, we explore the current state of post-harvest and agro-waste management in developing economics and shed light on the potential of nano-biotechnology as a game-changer for sustainable development. In developing economies, post-harvest losses significantly impede food security and economic growth. Lack of modern storage facilities, transportation inefficiencies, and inadequate handling practices contribute to substantial losses of fresh produce before reaching the market (Aulakh et al., 2013; Muganyizi et al., 2023).

Furthermore, these losses have far-reaching consequences, including increased food prices, reduced income for farmers, and heightened environmental impact due to wastage (Ogundele, 2022; Whitmee et al., 2015). Efforts to combat post-harvest losses have been ongoing, but conventional methods often fail to address the root causes effectively. However, nano-biotechnology offers novel and promising solutions to extend the shelf life of perishable produce and reduce losses during storage and transportation. Nanomaterials, such as nano-composites and nano-coatings, have shown remarkable potential in enhancing the barrier properties of packaging materials, thereby preserving the freshness and quality of agricultural products (Odetayo et al., 2022). The generation of agrarian waste is another pressing concern in developing economies. Crop residues, food processing by-products, and other organic waste streams are often underutilized, leading to environmental pollution and loss of potential value (Pardeep et al., 2018; Gürkan et al., 2021; Varghese et al., 2023). Traditional waste management methods, including open burning and landfilling, contribute to

DOI: 10.1201/9781003514039-8

greenhouse gas emissions and pose health hazards (Yu et al., 2019; Zhang et al., 2021a). Integrating nano-biotechnology into agro-waste management offers a transformative approach to convert these waste streams into value-added products. Nanocatalysts and nano biocatalysts have shown great promise in enhancing the efficiency of biogas production from agricultural waste through anaerobic digestion (Taha et al., 2020; Misson et al., 2015).

Additionally, nanomaterials can be employed to develop bio-based nanocomposites using agro-waste as a precursor, opening up opportunities for sustainable packaging materials. While the potential of nano-biotechnology is immense, its successful implementation in developing economies requires overcoming various challenges. One significant challenge lies in the cost-effective and scalable production of nano-based solutions. Access to advanced technology and expertise is often limited, so fostering international collaborations and knowledge transfer is essential.

Addressing concerns related to nanomaterial safety and regulations is imperative for their broad acceptance and use. Thorough research into potential environmental impacts and human health effects of nano-biotechnological interventions is essential to ensure their responsible application. Despite these challenges, the potential of nano-biotechnology in post-harvest and agro-waste management is compelling. Generating value-added products from waste streams can stimulate fresh economic opportunities and foster circular economy principles. This approach, alongside reduced post-harvest losses and improved waste management, contributes to heightened food availability, thus enhancing food security in developing economies.

Consequently, post-harvest and agro-waste management warrant immediate attention in developing economies. Integrating nano-biotechnology in these sectors offers a promising avenue toward comprehensive sustainable development, concurrently addressing food security, economic growth, and environmental sustainability. By effectively harnessing the potential of nanomaterials and biotechnological processes, developing economies hold the capacity to reshape agricultural practices, forging a future where waste is minimized, and resources are optimally utilized.

8.2 POST-HARVEST CHALLENGES IN A DEVELOPING ECONOMY

8.2.1 FOOD LOSS AND WASTAGE

Food loss and wastage during post-harvest handling, storage, and transportation are pressing challenges that impact food security and economic viability, particularly in developing economies. Post-harvest losses in some selected food crops are schematically shown in Figure 8.1.

FIGURE 8.1 Images of spoilt food crops; the perfect example of post-harvest losses in (a) tomato (b) apples (c) raw vegetables and (d) corn.

According to UNEP (2021), if food loss and waste were a country, it would be the third biggest source of greenhouse gas emissions. Despite significant progress in agricultural production, a substantial percentage of food is lost or wasted before reaching consumers. Table 8.1 presents the magnitude of food loss per capita income in the selected countries around the globe. This review delves into the extent of food loss at different post-harvest stages, identifies contributing factors, and explores its profound implications on food availability and economic losses for farmers. Food loss occurs at various stages of the post-harvest supply chain, from harvesting to consumption. According to the Food and Agriculture Organization (FAO) report by Gustavsson and coworkers (2011), approximately one-third of the world's food production is estimated to be lost or wasted. Developing economies are particularly vulnerable to higher losses due to inadequate infrastructure, inefficient handling practices, and a lack of modern technologies (Kader, 2005). Parfitt and coworkers (2010) stressed that significant losses occur during post-harvest handling, storage, and transportation.

Multiple factors contribute to food loss during post-harvest operations. Inadequate storage facilities, lack of temperature-controlled transportation, and poor handling techniques lead to physical losses due to spoilage, rot, and pest infestations (Hodges et al., 2011). Climatic conditions and variations in humidity further exacerbate losses. Additionally, post-harvest losses vary by commodity; perishable fruits and vegetables are more susceptible to spoilage, while cereals and grains are prone to damage caused by pests and fungi (Thapa & Kumar, 2019). The loss and wastage of food during post-harvest stages reduce the overall availability of food for consumption.

In contrast, the already produced and available food is damaged or spoiled before reaching consumers. This shortfall in food availability negatively affects food security, particularly in regions where populations are highly dependent on locally produced agricultural products (FAO, 2022). Reduction in food availability leads to higher food prices, making it difficult for vulnerable populations to access nutritious and affordable food (Lipinski et al., 2013). Furthermore, food loss during post-harvest phases results in substantial economic losses for farmers. Pingali and coworkers (2019) found that farmers in Sub-Saharan Africa reported significant income reductions due to food loss, adversely affecting their livelihoods. Economic losses occur due to increased production costs,

---

TABLE 8.1

Global Food Loss Estimate *per capita*

S/N	Country	Study Area	Estimated Food Loss Wastage (kg capital^{-1})	References
1	Nigeria	Sapele	189	Orhorhoro et al. (2017)
2	South Africa	Nationwide	134	Ramukhwatho (2016)
3	Brazil	Nationwide	60	Araujo et al. (2018)
4	China	Urban China	150	Zhang et al. (2020)
5	Columbia	Bogota	70	JICA (2013)
6	India	Rajam, Andhra Pradesh	58	Ramakrishna (2016)
7	Kenya	Nairobi	99	Takeuchi (2019)
8	Japan	Nationwide		Food Industry Policy Office (2017)
9	Mexico	Nationwide	94	Kemper et al. (2019)
10	Zambia	Ndola	78	Edema et al. (2012)
11	Viet Nam	Da Nang	64	Vetter-Gindele et al. (2019)
12	Ghana	Nation wide	84	Miezah et al. (2015)
13	Israel	Nationwide	105	Leket Israel (2019)
14	Saudi Arabia	Nationwide	105	SAGO (2019)
15	Lebanon	Beirut	105	Chalak et al. (2019)

which are not compensated by the reduced revenue from the damaged produce (Hossain & Kabir, 2020). Additionally, post-harvest losses can limit market access for farmers, reducing their potential to earn higher profits (Aryeetey et al., 2017).

A multi-faceted approach is essential to mitigate food loss during post-harvest handling, storage, and transportation. Investments in infrastructure, such as modern storage facilities and transportation systems, can help reduce physical losses (Miao et al., 2019). Also, adopting improved handling and processing techniques, including proper packaging and cold chain technologies, can extend the shelf life of perishable products (Opara & Pathare, 2014). Moreover, promoting awareness among farmers and stakeholders about the importance of post-harvest management and providing training in best practices can play a crucial role in reducing losses (FAO, 2023).

In the complex web of global food systems, the issue of food loss stands as a formidable barrier to the realization of food security and the eradication of hunger. Food loss, which occurs at various stages from production to consumption, results in the squandering of valuable resources, both natural and human, contributing to a world where an estimated 9.9% of the global population, or approximately 768 million people, suffer from chronic undernourishment in 2020 (FAO, 2019, 2022). This is an unsettling paradox in a world where millions suffer from chronic hunger.

The statistics surrounding food loss are staggering. The Food and Agriculture Organization (FAO) has reported that roughly one-third of all food produced for human consumption is lost or wasted annually, totaling approximately 1.3 billion metric tons (FAO, 2019). This not only represents a substantial economic cost, estimated at $940 billion annually, but also equates to the food that could feed nearly 2 billion people, more than twice the population of India (FAO, 2019; WRI, 2021; WFP, 2019).

Hence, food loss and wastage during post-harvest handling, storage, and transportation are critical challenges affecting food availability and contributing to economic losses for farmers. Addressing these issues requires collaborative efforts among governments, NGOs, research institutions, and private sectors. Implementing efficient post-harvest management practices, adopting innovative technologies, and formulating supportive policies can help minimize food loss, enhance food availability, and create a more sustainable and secure food system.

8.2.2 Post-Harvest Loss Mitigation Technologies

Technological interventions, such as modified atmosphere storage, cold storage, and hermetic storage, can significantly reduce post-harvest losses. Modified atmosphere storage maintains specific gas concentrations, slowing ripening and reducing spoilage (Jha et al., 2019). Cold storage preserves perishable produce at low temperatures, extending shelf life and reducing wastage (Sharma et al., 2020). Hermetic storage prevents insect infestations, providing a chemical-free solution to control pests (Ogendo et al., 2019). Some of the post-harvest mitigation technologies in use are elucidated in subsequent subheadings.

8.2.2.1 Solar Drying Technology

Solar drying technology represents a sustainable and environmentally friendly solution for preserving fruits, vegetables, and grains in the face of pressing challenges related to food loss and security. This method harnesses renewable energy from the sun to effectively reduce moisture content in agricultural produce, inhibiting microbial growth and extending shelf life (Müller et al., 2019). Importantly, it offers a compelling example of how cutting-edge nanobiotechnology can be integrated into sustainable agricultural practices. Solar dryers, powered by abundant solar energy, are particularly significant in regions with limited access to electricity, empowering farmers to process and store their produce efficiently (Müller et al., 2019). The solar drying process involves the exposure of agricultural commodities to direct sunlight, where solar radiation heats the crop, causing moisture to evaporate and escape into the surrounding air. To ensure efficient drying, proper ventilation replaces humid air with drier air (Van Hoang et al., 2021).

One of the most significant advantages of solar drying is its remarkable reduction of post-harvest losses. This method effectively curtails the growth of mold, fungi, and bacteria, which thrive in high-moisture environments (Alzate-Florez & Romero-Gómez, 2019). By reducing moisture content, the risk of spoilage and rot is significantly minimized, thereby improving the shelf life of agricultural produce (Koyuncu et al., 2020). Moreover, solar drying aligns with sustainable and energy-efficient principles, relying exclusively on solar energy, thus reducing the reliance on fossil fuels and conventional electricity sources (Kaur & Singh, 2018). The cost-effectiveness of solar drying is particularly notable in regions blessed with abundant sunlight, offering an affordable solution for small-scale farmers and local communities. Additionally, this method preserves the nutritional quality of produce, ensuring that vital nutrients are retained during the drying process, which makes it an ideal choice for drying a wide range of agricultural products, including fruits, vegetables, grains, and herbs (Toghyani et al., 2018).

Recent research has further underscored the potential of solar drying, as demonstrated by the study of Ssemwanga and colleagues (2020). This study explored the effects of traditional and improved solar drying methods on the sensory quality and nutritional composition of dried mangoes and pineapples. Notably, an innovative, improved solar drying (ISD) method, enhanced with solar concentrator plates and specialized greenhouse covers, outperformed conventional solar drying (CSD), preserving sensory quality and nutritional content to a superior degree. While solar drying presents a promising pathway to reducing post-harvest losses and enhancing food security, it is crucial to acknowledge its weather-dependent nature. Cloudy or rainy days can hinder drying, potentially leading to uneven drying or mold formation. Moreover, the drying time may be prolonged during the monsoon season or in regions with limited sunlight. Thus, ongoing research and innovation, potentially integrating nanobiotechnology, are essential to addressing these challenges and further optimizing the effectiveness of this sustainable and promising post-harvest mitigation technology.

8.2.2.2 Cold Chain Infrastructure

In perishable goods management, establishing a cold chain infrastructure is pivotal, and nano-biotechnology is increasingly playing a transformative role in this critical process. This infrastructure encompasses a comprehensive network of refrigerated storage, transportation, and distribution systems, all aimed at maintaining the pristine quality of perishable goods throughout their journey from farm to market (Daryanto et al., 2021). Cold chain facilities are indispensable guardians against the adverse effects of temperature fluctuations and accelerated ripening. By ensuring that produce remains under controlled temperature conditions, they play a pivotal role in ensuring consumers receive these goods in the best possible condition. Moreover, they significantly contribute to post-harvest management, a critical facet of the agricultural supply chain responsible for preserving the quality and extending the shelf life of perishable food products, ranging from fruits and vegetables to dairy products and seafood. The cold chain infrastructure becomes an essential shield for temperature-sensitive agricultural produce, effectively slowing down both physiological and microbial deterioration. This proactive approach minimizes post-harvest losses and curtails food wastage. By meticulously maintaining the optimal temperature and humidity levels, the cold chain ensures that products arrive at their final destinations fresh, safe, and in prime condition. This enhances consumer satisfaction and bolsters these products' marketability (Sodhi et al., 2020).

The cold chain infrastructure consists of several interlinked components, each contributing to the seamless flow of perishable products along their journey from farm to fork. These components encompass pre-cooling facilities for rapid field heat removal. These cold storage units provide temperature-controlled environments, refrigerated transport trucks and containers for safe transit, and insulation and temperature-resistant packaging materials to maintain the desired temperature during storage and transportation. However, implementing an efficient cold chain infrastructure presents formidable challenges, particularly in developing economies. Chief among these hurdles is the substantial initial investment required. Building a comprehensive cold chain infrastructure demands concrete upfront financial commitments, which can be daunting for small-scale farmers and processors. Energy consumption for refrigeration and cooling forms a significant portion of operational costs, and in regions with erratic energy availability, reliability becomes a concern.

Furthermore, in remote or rural areas, inadequate road infrastructure, limited transportation facilities, and insufficient storage capacity can disrupt the smooth operation of the cold chain. Maintaining and operating this infrastructure necessitates a skilled and trained workforce, a resource that may be scarce in certain regions (Yao et al., 2018). Recent research among Kenyan smallholder mango farmers illustrates the potential of nano-biotechnology-enhanced post-harvest practices to bolster the cold chain and extend shelf life (Amwoka et al., 2021). This study achieved significantly lower fruit pulp temperatures and extended shelf life by 18 days by implementing straightforward post-harvest techniques and uncomplicated storage technologies. These findings underscore the promise of basic yet effective approaches in preserving quality and prolonging storage duration, thus enhancing post-harvest management and benefiting mango marketing and processing. Despite the challenges in implementation, it is concluded that the positive impact of cold chain infrastructure on reducing post-harvest losses, ensuring food safety, and enhancing market access makes it an indispensable component of the agricultural supply chain. To fully realize its benefits, governments, private sectors, and stakeholders must collaborate and invest in building and maintaining a robust and efficient cold chain infrastructure.

8.2.2.3 Time-Temperature Indicator (TTI) Technology

In the ever-evolving landscape of intelligent packaging, nano-biotechnology has emerged as a transformative force, significantly enhancing the functionality of Time-Temperature Indicators (TTIs). These unassuming labels, often compact and self-adhesive, now wield the power of nanoscale science to detect and respond to unfavorable environmental conditions, marking a paradigm shift in product quality assurance (Severini et al., 2018). Nano TTI labels, strategically affixed onto shipping containers or individual packages, undergo an irreversible transformation, primarily manifested as a precise color change triggered by subtle fluctuations in temperature and time (Severini et al., 2018). Once viewed as passive, these labels have evolved into dynamic guardians of product quality, serving dual roles as early warning systems and freshness indicators.

Throughout the supply chain, nano TTI labels act as sentinels, vigilantly safeguarding product integrity. From bulk shipments to individual products, they stand as vigilant protectors against temperature excursions that could compromise perishable goods' quality and safety (DiazPerez, 2014). The real-time color transformation within the TTI label becomes an unmistakable signal of breached temperature thresholds, enabling immediate intervention and informed decision-making.

Beyond their role as guardians, nano TTI labels have emerged as sophisticated tools for predicting the remaining shelf life of perishable items. By meticulously assimilating the cumulative impact of time and temperature fluctuations at the nanoscale, these indicators provide nuanced estimations of product freshness and viability (Severini et al., 2018). This invaluable information empowers retailers and consumers to make informed choices, significantly minimizing food waste and elevating the overall consumer experience. However, it is essential to acknowledge that integrating nanobiotechnology into packaging introduces an inevitable cost component. As with any innovation, adopting nano TTI labels necessitates a compelling cost-benefit analysis (Ghaani & Ghomi, 2018). Ensuring that the advantages derived from nanotechnology substantially outweigh the supplementary expenses incurred in its integration becomes imperative. This cost-effectiveness consideration is pivotal in guiding packaging advancements toward practicality, relevance, and sustainability. With their ability to provide real-time insights and predictive estimations, Nano TTI labels redefine product quality assurance standards. Their visual indications of temperature variations and their capability to extend shelf life offer tangible benefits for the industry and consumers. Yet, the ultimate success of these technologies hinges on their ability to demonstrate a clear value proposition that justifies the additional investment. In an era of evolving packaging solutions, nano TTI labels testify to how nanobiotechnology can elevate product quality assurance and enhance supply chain efficiency.

8.2.2.4 Digital Platform and Mobile Phone-Driven Innovations

Mobile applications and digital platforms offer innovative solutions to bridge information gaps for farmers. These technologies provide real-time market prices, weather forecasts, and best post-harvest practices (Makate et al., 2016). Access to such information empowers farmers to make informed decisions, reduce wastage, and negotiate better prices in the market. Post-harvest management is critical in ensuring food security and reducing food losses in the agricultural supply chain. Mobile-based technologies have emerged as powerful tools to enhance post-harvest processes, enabling farmers, producers, and other stakeholders to make informed decisions, improve efficiency, and reduce wastage.

Mobile-based technologies, such as mobile applications (apps), short message service (SMS), and sensor-based systems, have revolutionized communication and information dissemination in agriculture. These technologies provide real-time access to critical data, weather forecasts, market prices, and best practices, enabling stakeholders to respond promptly to challenges and opportunities in post-harvest processes. They bridge the information gap between farmers and relevant support services, empowering them with the knowledge to make better decisions and optimize their post-harvest operations (Mittal et al., 2018). Mobile-based technologies have enormous potential in promoting sustainable agriculture and responsible post-harvest management. Reducing food losses and improving market access contribute to more efficient resource utilization and economic growth in the agricultural sector. Moreover, by empowering farmers with information and knowledge, these technologies can support adopting sustainable practices, such as reduced pesticide use and better resource management.

Despite the numerous benefits of mobile-based technologies, they face multiple challenges, such as the digital divide. Limited smartphone access and internet connectivity in rural areas can also hinder small-scale farmers' adoption of mobile-based technologies. In addition, many mobile apps and services are available in specific languages, which may not cater to the diverse linguistic needs of farmers. The mobile apps must also be user-friendly and intuitive, considering users' varying digital literacy levels (Ajayi et al., 2020). Mobile-based technologies have emerged as powerful tools in post-harvest management, providing real-time information and empowering stakeholders to make informed decisions. Despite some challenges, their potential for sustainable agriculture and reduced food losses make them invaluable assets in the quest for global food security and responsible resource management.

8.2.3 Agro-waste Management

Agro-waste, generated during post-harvest operations, poses significant environmental, public health, and agricultural sustainability challenges. Agricultural waste also rises as agrarian production increases to meet the growing demand for food. This review examines the scale of agricultural waste generated, its environmental impact, and the pressing need to adopt sustainable utilization and disposal strategies to mitigate its adverse effects. Post-harvest operations generate substantial agricultural waste, including harvesting, processing, and storage. Crop residues, such as stems, leaves, and peels, are left behind after primary food products are extracted or harvested.

Additionally, rejected or damaged produce, discarded packaging materials, and by-products from agro-industrial processes contribute to the overall waste (Muthoni et al., 2018). Traditional and inefficient post-harvest practices in developing countries exacerbate waste generation (Bhargava et al., 2021). The accumulation of this agricultural waste can have adverse environmental consequences. Improper disposal practices, such as open burning or landfilling, release harmful greenhouse gases, including methane and carbon dioxide, contributing to climate change (Rathod et al., 2020). Decaying organic matter in landfills produces leachate, polluting groundwater and surface water bodies (Dalai & Singh, 2017). Moreover, agro-waste can act as the breeding ground for pests and pathogens, leading to potential disease outbreaks in plants and animals (Mukhtar et al., 2020).

Sustainable agricultural waste management is crucial to minimize its environmental impact and promote resource efficiency. Several strategies can be employed to ensure responsible utilization and disposal of agro-waste, as shown in Figure 8.2.

8.2.3.1 Composting

Composting is a widely recognized and environmentally friendly method of managing agro-waste, involving the controlled decomposition of organic materials into nutrient-rich humus. This natural process relies on the activity of microorganisms, such as bacteria, fungi, and other decomposers, in the presence of oxygen. Composting offers numerous benefits for sustainable agriculture and waste management, making it an essential component of responsible agro-waste utilization. In this review-like explanation, we will delve into the critical aspects of composting, its benefits, and the importance of further research and collaboration in its application, supported by relevant references. Composting begins with collecting organic materials, such as crop residues, pruning waste, food scraps, and animal manure, which are then subjected to controlled decomposition.

Various composting techniques include traditional heap, vermicomposting, and aerated static pile composting. Each method offers different advantages in terms of the speed of decomposition, quality of compost produced, and resource requirements (Wu et al., 2020). Proper management of the composting process, including maintaining the right balance of carbon-to-nitrogen ratio, adequate moisture levels, and oxygen availability, is crucial for successful composting (Gutiérrez-Miceli et al., 2018). Compost is a valuable organic amendment that enhances soil fertility and structure. It provides essential nutrients, such as nitrogen, phosphorus, and potassium, in a slow-release form, promoting plant growth and reducing the need for synthetic fertilizers (Wang et al., 2021). The improved soil structure from compost application enhances water retention, aeration, and drainage, reducing soil erosion and nutrient leaching (Wang et al., 2021). Compost also improves the soil's ability to suppress certain plant diseases and pests through the activity of beneficial microorganisms (Kumar et al., 2021b). Composting sequesters carbon, contributing to climate change mitigation by reducing greenhouse gas emissions (Chen et al., 2017).

Effective agro-waste management is critical to reduce environmental pollution and minimize the strain on landfill sites. Composting is crucial in diverting agro-waste from landfills, where it would otherwise produce methane, a potent greenhouse gas (Meng et al., 2017). By converting agro-waste into compost, valuable resources are recovered and reused, supporting a circular

FIGURE 8.2 Agro-waste conversion to valuable products, for example, (a) compost, (b) biogas, (c) bioenergy, and (d) animal feed.

economy in agriculture (Chen et al., 2020). Composting reduces waste disposal costs and provides farmers with a cost-effective, sustainable source of organic fertilizer, improving soil health and crop productivity (Kumar et al., 2019). Composting municipal solid waste into organic fertilizer emerges as a sustainable approach for agro-ecosystem land upkeep, recuperating nutrients, and minimizing environmental harm. A study by Jodar and coworkers (2017) evaluates compost quality generated from municipal solid waste, scrutinizing mineral nutrient contents like potassium, calcium, magnesium, and more. The composting process was tracked using temperature as a parameter. The results reveal a comparable mineral nutrient evolution in municipal solid waste mixtures, culminating in a mature compost suitable for agricultural application. This composting technique offers a viable solution for managing solid waste while furnishing agriculture with nutrient-rich organic fertilizer.

8.2.3.2 Biogas Production

Biogas production, when harnessed through the lens of nanobiotechnology, emerges as an ingenious and sustainable solution, effectively addressing multiple pressing challenges of our era. It capitalizes on the power of anaerobic digestion to transform agro-waste into a potent renewable energy source, primarily composed of methane and carbon dioxide. This process revolutionizes waste management, energy generation, and environmental sustainability, marking a significant stride in our quest for a greener, more sustainable future. Biogas production leverages the inherent capabilities of microorganisms operating in an oxygen-free environment to metabolize complex organic matter found in agricultural residues, crop leftovers, animal manure, and kitchen waste (Karthikeyan et al., 2018). This "microbial alchemy" occurs within sealed biogas digesters or reactors, where agro-waste is meticulously mixed with water, creating an ideal habitat for these anaerobic microorganisms. As these microorganisms diligently work their magic, the organic matter is transformed into biogas, an invaluable byproduct with many applications (Nurjanah et al., 2018). Biogas, in its versatility, emerges as a clean, renewable fuel. It powers cooking stoves, heating systems, electricity generators, and more, ushering in a sustainable energy era. Simultaneously, it tackles the growing problem of agro-waste accumulation, effectively curbing the environmental impact of waste disposal and stemming the tide of greenhouse gas emissions (Rekha et al., 2020). But the story of biogas production doesn't end here. It extends its influence to the very roots of agriculture. The residue left behind after biogas extraction, known as digestate, assumes the role of a nutrient-rich organic fertilizer. It breathes new life into the soil when thoughtfully applied to agricultural fields. It enhances fertility, augments water retention capacity, and stimulates crop growth, a testament to its ecological and agrarian benefits (Ojha et al., 2019).

However, the fusion of nanobiotechnology into the biogas production narrative makes this tale even more remarkable. Nanotechnology introduces a realm of possibilities where the efficiency of anaerobic digestion can be further enhanced, the yield of biogas optimized, and the digestate's nutrient content fine-tuned. Nanoscale materials, carefully engineered and tailored, can serve as catalysts, accelerating the microbial digestion process. As highlighted by Khan and coworkers (2022), nano-materials have significantly transformed the domains of anaerobic digestion, methane production, and the management of waste-activated sludge. They currently represent a focal point of interest within the research community. Nanobiotechnology can aid in monitoring and controlling various parameters critical to biogas production, making the process more efficient and environmentally friendly (Zhang et al., 2021b). Integrating nanobiotechnology into biogas production represents a monumental leap in sustainable energy practices. It enhances efficiency, reduces environmental impact, and elevates the agricultural sector's resilience. This amalgamation of cutting-edge science and sustainable approach is our beacon, guiding us toward a future where renewable energy generation coexists harmoniously with waste management and environmental stewardship.

Anaerobic digestion naturally breaks down organic matter without oxygen, yielding biogas, mainly methane. The resulting gas serves as cooking, power, or fuel. According to research by Abdelwahab and Fodah (2022), applying nanoparticles (NPs) offers several advantages in anaerobic digestion processes. These NPs serve as crucial nutrient sources, facilitating the production of essential enzymes and co-enzymes, thereby stimulating the activity of anaerobic microorganisms when maintained at optimal concentrations (e.g., 100 mg L^{-1} for Fe NPs, 2 mg L^{-1} for Ni NPs, and 1 mg L^{-1} for Co NPs). Interestingly, exceeding the concentration of 100 mg L^{-1} for Fe NPs is more effective in reducing H$_2$S production than increasing CH$_4$ production. Conversely, concentrations exceeding 2 mg L^{-1} for Ni NPs and 1 mg L^{-1} for Co NPs tend to decrease CH$_4$ production. Effluents enriched with Fe and Ni NPs exhibit superior fertilizer properties than those with Co NPs alone. Furthermore, the combination of Fe, Ni, and Co NPs demonstrates greater efficiency in enhancing CH$_4$ production than using individual NPs. In their study, Abdullah and coworkers (2014) optimized biogas production from rice waste via wet digestion to address escalating food waste. Experimental and simulated results were compared, showing 14.4 kg-mol h^{-1} gas production (experiment) and 19.82 kg-mol h^{-1} (simulation) for a 0.05 kg-mol h^{-1} starch loading rate. Biogas composition was 69% methane and 29% CO$_2$, thus highlighting biogas's potential for sustainable agro-waste management, transforming waste into valuable energy resources.

Biogas production offers economic benefits to farmers and communities by providing a source of renewable energy, reducing waste management costs, and creating opportunities for income generation through the sale of surplus biogas or digestate (Molino et al., 2021). Adopting biogas production for agro-waste management requires further research and development to optimize the process, increase biogas yields, and enhance system efficiency. Research should also focus on exploring the potential of co-digestion, where multiple organic materials are mixed, to improve biogas production and address waste management challenges more effectively (Hafifi et al., 2018). Policy support and incentives are essential to encourage the widespread adoption of biogas production systems, facilitating access to financing, technology, and knowledge sharing among farmers and rural communities (Kumar et al., 2021a).

8.2.3.3   Animal Feed

Nanotechnology, specifically nano-biotechnology, revolutionizes the utilization of agricultural waste, such as crop residues, as animal feed, offering a potent source of nutrition and reducing the strain on grazing lands (Samkol & Egbunike, 2019). In agro-waste management, nanotechnology assumes a pivotal role in transforming various agricultural residues and by-products into highly nutritious and balanced animal feed formulations. These feeds, tailored to the needs of livestock and poultry, efficiently convert agro-waste into valuable nutrients while mitigating waste, curbing environmental pollution, and bolstering the economic sustainability of agricultural practices.

Agro-waste encompasses diverse residues like crop remnants, straws, husks, peels, and pomace, often unsuitable for direct human consumption due to high fiber content, low digestibility, or anti-nutritional components. However, nanotechnology-enabled animal feeds extract the full potential of these materials, providing animals with essential energy, proteins, vitamins, minerals, and nutrients. Fine-tuned feed formulations must be fine-tuned to prevent nutrient imbalances affecting animal health and productivity (Madruga et al., 2021). Food packaging materials have diversified, incorporating cellulose nanocrystals derived from natural sources alongside biologically active plant extracts like eugenol, thymol, or carvacrol. Metal oxides produced through microbial nanotechnology processes are also utilized, along with biopolymers, to enhance food packaging applications (Salmiah et al., 2023). Nanotechnology's application in food and feed science, particularly in the form of natural nano antimicrobials like nano-propolis, is a recent but promising development. According to Pinar and coworkers (2018), nano-propolises, consisting of nano-sized (1–100 nm) particles of propolis, offer advantages in veterinary medicine, including improved health, performance, and reliable food production. These nanoparticles enhance propolis's effectiveness without altering its inherent properties, achieved through various size-modification techniques. Propolis itself carries benefits, including anti-inflammatory properties.

Prasetyo and Suhendrata (2021) researched the nutritional value of agricultural waste forages in the Grobogan Regency's agroecosystem, specifically for ruminants. This study examined how soil fertility and water availability in upland agroecosystems influenced the nutritional composition of tropical agricultural waste used in cattle, sheep, and goat forage. By utilizing rice straw, corn, soybeans, and their carrying capacities, they devised comprehensive feed formulas for ruminant livestock. These formulations encompassed critical nutritional components, including crude protein, fiber, fat, energy, and ash.

While harnessing agro-waste as feed presents significant benefits, it has its share of challenges. The availability and quality of agro-waste may fluctuate seasonally, necessitating proper storage and preservation techniques. Some agro-waste may contain anti-nutritional substances or toxins requiring suitable processing and treatment to ensure safe animal consumption. Furthermore, precise feed formulation remains crucial to meet the distinct nutritional requirements of various animal species and production stages (Ramos-Morales et al., 2019). Nano-biotechnology provides innovative avenues to address these challenges and optimize agro-waste utilization as high-quality animal feed.

8.2.3.4　Bioenergy Production

Conversion of agro-waste into bioenergy, such as biofuels or bioethanol, offers a sustainable alternative to fossil fuels and reduces dependence on non-renewable resources (Kumar et al., 2019). The conversion of agro-waste to bioenergy involves harnessing the energy stored in agricultural residues and by-products to produce renewable energy sources such as biofuels and biogas. This process is crucial for sustainable waste management and offers an environmentally friendly alternative to fossil fuels. Biofuels are liquid or gaseous fuels derived from biomass, including agro-waste. Two main types of biofuels are produced from agro-waste: bioethanol and biodiesel.

8.2.3.4.1　Bioethanol Production

Bioethanol is produced through a process called fermentation. Agro-waste with high sugar or starch content, such as sugarcane bagasse, corn stover, or rice straw, can be converted into bioethanol. The agro-waste is first subjected to pre-treatment and broken down to release the sugars or starches. Then, specific microorganisms, usually yeast, ferment these sugars into ethanol. The ethanol is then purified and concentrated as a biofuel for transportation and power generation (Minteer et al., 2020).

In recent years, nanotechnology, particularly the utilization of nanomaterials, has played a pivotal role in advancing bioenergy generation. According to Kalaimani and Wai (2022), nanomaterials can augment catalysts and modify feedstocks, improving bioenergy yield and efficiency. Moreover, it anticipates future applications, underscoring nanotechnology's ongoing significance in revolutionizing bioenergy generation processes and bolstering sustainable energy production, aligning with global efforts to combat climate change and enhance energy security. From the work done by Snehal and coworkers (2014), producing bioethanol using banana *pseudo* stem can be a viable way of utilizing the residue. The study explores the utilization of banana pseudo stem for bioethanol production via sugar release through diverse chemical and biological pretreatments. Co-culture fermentation with cellulolytic enzyme-producing strains, *A. ellipticus* and *A. fumigatus*, effectively breaks down holocellulose (a plant cell wall component consisting of cellulose and hemicellulose, essential in industries like paper manufacturing and biofuel production) in the banana pseudo stem, enhancing sugar yields. Alkali and microbial treatments yield a hydrolysate, further fermented by Saccharomyces cerevisiae NCIM 3570, resulting in 17.1 g L^{-1} ethanol after 72 hours with an 84% yield and 0.024 g L^{-1} productivity. This investigation underscores fungal pretreatment's significance for efficient cellulose saccharification and highlights the economic viability of using pretreated banana pseudo stem for ethanol production.

Wang et al. (2008) conducted a study involving simultaneous saccharification and fermentation (SSF) to produce ethanol from kitchen waste, employing both an open and closed fermentation approach. Their findings revealed that the open fermentation method, devoid of heat treatment, exhibited favorable outcomes due to preserving essential nutrients within the food waste. This approach yielded a peak ethanol concentration of 33.05 g L^{-1}. Through a process of simultaneous saccharification and fermentation (SSF) using a composite amylolytic enzyme complex containing amyloglucosidase, α-amylase, and protease, along with baker's yeast, S. cerevisiae (Hong & Yoon, 2011), food remnants were successfully transformed into ethanol. This method yielded 36 g L^{-1} ethanol from 100 g L^{-1} of food residues. Kim and coworkers (2008) investigated optimizing enzymatic saccharification and ethanol fermentation conditions for food waste. Their model anticipated that under optimal parameters, the highest feasible concentrations of reducing sugars and ethanol would be 117.0 and 57.6 g L^{-1}, respectively. Key variables influencing the generation of reducing sugars from food waste were pinpointed. The liquid phase of the food waste hydrolysate was harnessed for ethanol production through fermentation using *S. cerevisiae* H058 (Yan et al., 2011). Achieving an elevated reduction in sugar production of 164.8 g L^{-1} from food waste necessitated optimal conditions. Given the intricate composition of food waste, particularly challenging for ethanol-producing microorganisms like *S. cerevisiae*, a preliminary hydrolysis step to yield fermentable sugars becomes imperative.

8.2.3.4.2 Biodiesel Production

Biodiesel is produced from agro-waste oils and fats. The process, called transesterification, involves reacting the agro-waste oil or grease with alcohol, typically methanol, in the presence of a catalyst. This reaction converts the triglycerides in the oil into biodiesel and glycerol as a byproduct. Biodiesel can be used as a direct replacement for diesel fuel in diesel engines, providing a renewable and more environmentally friendly alternative (Biswas et al., 2018). On the other hand, biogas is produced through the anaerobic digestion of agro-waste without oxygen. Agro-waste, such as crop residues, animal manure, and food waste, is mixed with water to create a slurry. This slurry is placed in an airtight biogas digester, where microorganisms break down the organic matter and produce biogas. Biogas mainly consists of methane and carbon dioxide, and it can be used as a clean and renewable energy source for cooking, heating, and electricity generation (Sialve et al., 2009).

Erchamo et al. (2021) in their study investigated the enhanced production of biodiesel from waste cooking oil (WCO) using an innovative eggshell-derived calcium oxide (CaO) nano-catalyst in a mixed methanol–ethanol system. The catalyst's unique properties were improved through specific calcination and hydration-dehydration treatments. Factors affecting biodiesel yield were systematically analyzed, revealing optimal conditions yielding 94% biodiesel using a 1:12 oil-to-methanol molar ratio, 2.5 g/100 g catalyst loading, 60°C, and 120-minute reaction time. Further exploration showcased viable outcomes by adjusting methanol-ethanol ratios. The study underscores the potential of methanol-ethanol mixtures and CaO nano-catalysts for cost-effective, eco-friendly WCO conversion into biodiesel, aligning with fuel standards. Using agro-waste for bioenergy generation offers some advantages, including converting waste to energy. Agro-waste is often burned or left to decay in fields, leading to environmental pollution and greenhouse gas emissions. By converting agro-waste to bioenergy, these residues are utilized efficiently, reducing waste accumulation and its negative impact on the environment.

In conclusion, biofuels and biogas derived from agro-waste are considered renewable energy sources since they come from organic materials that can be replenished through agricultural practices. Using bioenergy helps reduce greenhouse gas emissions, particularly when compared to fossil fuels. Biofuels and biogas release carbon dioxide during combustion, offset by the carbon dioxide absorbed during the growth of the crops used to produce the bioenergy (Balat & Balat, 2009).

8.2.4 Nanotechnology in Post-Harvest Management

Nanotechnology represents a novel innovation that presents various postharvest management techniques by manipulating materials to a nanometric scale, typically ranging from 1 to 100 nm in at least one dimension. This reduction in particle size results in new materials exhibiting distinctive and enhanced properties compared to larger counterparts. The discovery of nanoparticles (NPs), with sizes in the nanoscale range, has become a remarkable advancement in nanotechnology, offering solutions to current global challenges. Nanoparticles possess unique characteristics, such as their size, significant surface area-to-volume ratio, and morphology, leading to new or improved properties. Their addition to edible coatings allows interfacial solid contact with the polymer, requiring only a minute quantity to achieve the desired effects. By incorporating nanoparticles into edible coatings, the shelf life of fruits can be significantly extended beyond what is achievable with pure polymer alone. Nanoparticles in these coatings improve their mechanical and barrier properties and enhance thermal stability compared to traditional methods. Various studies have demonstrated the potential of nanoparticles in postharvest applications (Magnuson et al., 2011; Parisi et al., 2015; Gad & Zag Zog, 2017; De Moura et al., 2009; Shankar & Rhim, 2015). Nanoparticles can be categorized into two main groups: organic NPs and inorganic NPs. The focus is primarily on inorganic nanoparticles due to their higher stability than organic counterparts, which are sensitive to heat. Inorganic nanoparticles encompass various materials, including metal or metal oxides such as gold (Au), silver (Ag), iron oxide (Fe_3O_4), titanium oxide (TiO_2), copper oxide (CuO), zinc oxide (ZnO), aluminum oxides, cerium dioxide hydroxides, calcium carbonate, and carbon-based materials (Bouwmeester et al., 2014; He & Hwang, 2016).

Numerous reports highlight nanotechnology as a highly effective technique for extending the shelf life of fruits (Bhusare & Kadam, 2020; Flores-López et al., 2016; Ijaz et al., 2020; Lloret et al., 2012; Ruffo Roberto et al., 2019). Some of the diverse applications of nanotechnology in post-harvest management include nano-particles for antimicrobial effect, nano-composites for enhanced barrier properties, nano-emulsion as natural preservatives, nanosensors for quality monitoring, nanocoatings for food preservation, nanoencapsulation for controlled release of bioactive compounds and nanomaterials for waste valorization (Figure 8.3).

8.2.4.1 Nanoparticles for Antimicrobial Effects

Nanoparticles (NPs), such as silver, zinc oxide, and copper nanoparticles, have strong antimicrobial properties (Khan et al., 2019). These nanomaterials exhibit a higher surface area-to-volume ratio, enabling better contact with microbial cells. They can penetrate the cell membrane, disrupting cellular functions and causing cell death. By incorporating antimicrobial nanoparticles into packaging materials or coatings, the growth of spoilage microorganisms can be suppressed, extending the shelf life of food crops (Bibby et al., 2016). NPs have emerged as a promising tool for combating microbial pathogens due to their unique physicochemical properties and surface reactivity. In recent years, extensive research has been conducted to explore the antimicrobial effects of nanoparticles against various microorganisms, including bacteria, fungi, and viruses. The applications of NPs in different fields, including medicine, agriculture, and food safety, are well documented. In therapy, nanoparticles have been explored as potential antimicrobial agents for wound dressings, implants, and drug delivery systems. Nanoparticles incorporated into wound dressings can prevent microbial colonization, reduce infection rates, and promote wound healing.

Additionally, nanoparticles can be used as carriers for antimicrobial drugs, improving their efficacy and reducing drug resistance (Hajipour et al., 2012). In agriculture, nanoparticles have demonstrated potential in controlling plant pathogens, such as fungi and bacteria. Nanoparticles can be applied as foliar sprays or incorporated into agricultural materials like seeds and fertilizers. Silver and copper nanoparticles have effectively controlled plant pathogens, reducing the need for chemical pesticides and promoting sustainable agriculture (Velázquez-Jiménez et al., 2019). Nanoparticles offer innovative solutions for enhancing food safety and extending the shelf life of

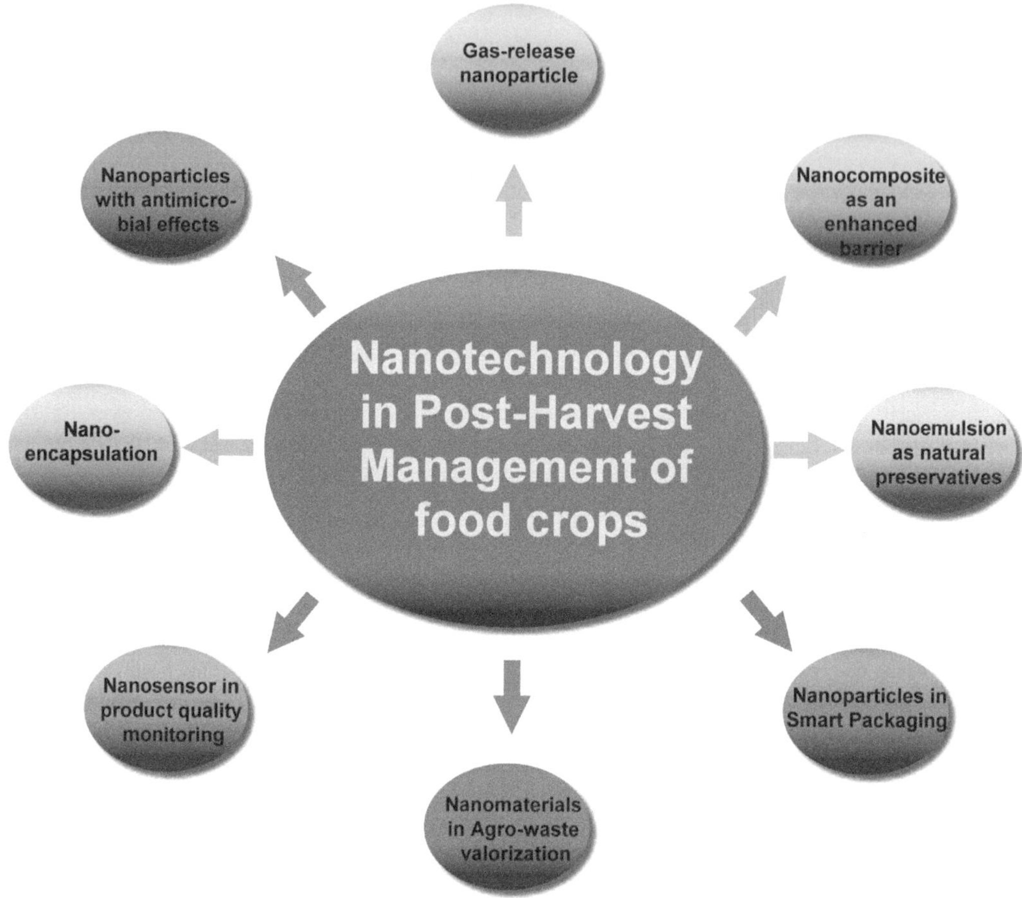

FIGURE 8.3 Applications of nano-materials in post-harvest management.

perishable food products. Silver nanoparticles have been used in food packaging materials to inhibit the growth of spoilage-causing bacteria and fungi, thereby reducing foodborne illnesses and food wastage. Additionally, nanoparticles can be applied to food surfaces as antimicrobial coatings, preserving the quality and safety of fresh produce (Jain et al., 2020).

8.2.4.2　Nanocomposites for Enhanced Barrier Properties

Nanocomposites typically consist of a nanoscale filler, such as montmorillonite nano clays, uniformly dispersed within a matrix material, often a polymer. The nanoparticles form a reinforcing network that restricts the movement of molecules and gases within the composite material. This improves barrier properties, including reduced gas permeability and controlled moisture transfer (Müller et al., 2020). The selective choice of nanoparticles and matrix materials allows tailoring the nanocomposite properties to meet specific packaging requirements. For instance, montmorillonite nanoclays can be incorporated into polymer films to reduce gas permeability and control moisture transfer, preventing oxidation and moisture-related spoilage (Azeredo, 2009). Improved Product Quality and Extended Shelf Life for Food Crops are Achieved Through Enhanced Gas Permeability in Food Packaging. Nanocomposites with montmorillonite nano clays have demonstrated enhanced gas barrier properties, particularly for oxygen and carbon dioxide. Reduced gas permeability minimizes the exchange of gases between the package and the external environment, preventing the oxidation of food components and the growth of spoilage microorganisms (Huang et al., 2016). Excessive moisture transfer can lead to undesirable effects on food quality, such as texture changes and mold growth. Nanocomposites with montmorillonite nano clays have been shown to control moisture diffusion, maintaining an optimal humidity level within the package. Significant for moisture-sensitive products, it guarantees the preservation of their texture, flavor, and overall quality during storage, as emphasized in a study by López de Dicastillo et al. in 2018.

Nanocomposite films containing montmorillonite and cellulose nanocrystals exhibit reduced gas and moisture permeability. Li and Liu (2021) highlighted the potential of these nanocomposites to extend the shelf life of perishable food products, catering to consumer demand for prolonged product freshness. Nano-sized fillers such as graphene oxide and nanocellulose are strategically dispersed within polymer matrices, improving oxygen and moisture barriers. Noteworthy research by Alboofetileh and coworkers (2020) underscores the potential of nanocomposites in retarding spoilage and maintaining quality in packaged foods. Incorporating nanomaterials such as montmorillonite and nanoparticles derived from agro-waste, like rice husk, has demonstrated reduced oxygen and water vapor permeability. Shankar and Rhim's work accentuates the eco-friendly potential of agro-based nanocomposites, offering sustainable alternatives to synthetic barriers and extending product shelf life (Shankar & Rhim, 2017). Bose and coworkers (2022) highlight the transformative potential of nanocomposites in developing the shelf life of packaged food products. These innovations signal a new era in food preservation technology, catering to the evolving needs of the food industry.

8.2.4.3   Nanoemulsions as Natural Preservatives

Nanoemulsions, which consist of tiny oil droplets dispersed in water with the help of surfactants, have gained attention as natural preservatives. Nanoemulsions are colloidal dispersions in which the droplet size ranges from 20 to 200 nm. They are created through various methods, including high-pressure homogenization, sonication, and spontaneous emulsification. The addition of surfactants stabilizes the nanodroplets, preventing their coalescence and leading to long-term stability (Liu et al., 2019). Essential oils and plant extracts with antimicrobial and antioxidant properties can be incorporated into nanoemulsions to inhibit microbial growth and delay oxidative processes (Jafari et al., 2020). Nanoemulsions offer an eco-friendly approach to enhancing the shelf life of food crops without the need for synthetic chemical preservatives. The small droplet size of nanoemulsions allows them to penetrate microorganisms' cell membranes effectively. The release of natural antimicrobial agents from the nanoemulsion disrupts the integrity of microbial cells, leading to their inhibition and preventing food spoilage (Gupta et al., 2019). Also, when loaded with natural antioxidants, Nano-emulsions scavenge free radicals and reactive oxygen species, which are responsible for the oxidative degradation of food components.

By controlling oxidative reactions, nanoemulsions preserve the quality and extend the shelf life of food products, ensuring they maintain their nutritional value, flavor, and appearance (Jafari et al., 2020). They improve the solubility and bioavailability of hydrophobic bioactive compounds, making it easier for them to interact with food components. This enhanced bioavailability leads to better dispersion and protection of bioactive agents, resulting in their more effective and uniform distribution in food matrices (Dehghannya et al., 2021). Nanoemulsions enable the controlled release of natural preservatives, ensuring their sustained efficacy over time. By encapsulating bioactive compounds in nanodroplets, a gradual release profile is achieved, enhancing the longevity of the preservative effects (Gupta et al., 2019). Nanoemulsions as natural preservatives align with the growing consumer demand for clean-label and sustainable food products. As a green preservation approach, nanoemulsions help reduce the reliance on synthetic chemical preservatives, contributing to environmentally friendly food practices (Salarbashi et al., 2018). Nanoemulsions represent a versatile and promising category of natural preservatives for enhancing food preservation and safety. Incorporating natural antimicrobial and antioxidant agents in nanoemulsions allows for targeted and controlled release, inhibiting microbial growth and oxidative reactions. Applying nanoemulsions as natural preservatives offers a green and sustainable approach to food preservation, meeting the increasing demand for clean-label products.

Jafari et al. (2020) worked on the fundamental aspects of nanoemulsions' application in food preservation. The work covers formulation techniques, properties, and various applications. Their extensive overview emphasizes nanoemulsions' pivotal role in enhancing food quality, bioavailability, and shelf life by improving stability and bioactive ingredient delivery. The authors' synthesis of research findings underscores nanoemulsions' potential as natural preservatives, aligning with the quest for sustainable food preservation solutions. McClements (2018) highlighted an overview of food emulsions, including nanoemulsions, highlighting principles, practices, and techniques. The focus extends to their potential as natural preservatives. Detailed methodologies for nanoemulsion formation, characterization, and application in food preservation were elucidated. This work is a valuable resource for researchers and practitioners, offering insights into developing effective natural preservative systems using nanoemulsions. Barros and Gonçalves (2021) explored the creation and characterization of nanoemulsions in various food contexts, focusing on their potential as natural preservatives. Their study delves into nanoemulsion preparation methods, encompassing high-energy and low-energy techniques. The authors delve into the influence of emulsifiers, stabilizers, and additives on nanoemulsion stability and their effectiveness in food preservation. Thakur and coworkers (2021) extensively evaluated nanoemulsions, covering formulation aspects and various applications, including natural food preservation. The methodology section outlines preparation methods such as high-pressure homogenization and phase inversion. The review emphasizes nanoemulsions' capacity to prolong perishable foods' shelf life. This work aids researchers in comprehending formulation complexities and harnessing nanoemulsions for natural preservation strategies.

This review guides researchers in understanding formulation intricacies and the potential of nanoemulsions for natural preservation strategies. Nanoemulsions exhibit potential in addressing challenges associated with food spoilage, owing to their enhanced stability, improved solubility, and controlled release properties. As advancements continue, personalized formulations can be tailored to target specific microorganisms, optimizing preservation efficacy. Integrating natural and bioactive compounds into nanoemulsions offers sustainable and clean-label preservation solutions. Nanotechnology-driven research will likely explore innovative delivery systems, such as stimuli-responsive nanoemulsions, enabling the on-demand release of preservatives. Continued collaboration between food scientists, technologists, and regulatory bodies will ensure the safe and effective utilization of nanoemulsions, elevating their role as efficient natural preservatives.

8.2.4.4 Nanomaterials for Intelligent Packaging

Smart packaging, an innovative approach in the field of food technology, integrates nanotechnology to create intelligent packaging materials capable of actively monitoring and controlling various environmental factors. Nanotechnology offers unique properties that enable the development of sensors and responsive materials for temperature, humidity, and gas exchange management. Nanotechnology involves manipulating and controlling materials at the nanoscale, granting novel properties and functionalities. Innovative packaging materials leverage nanotechnology-based sensors and responsive elements to detect and respond to environmental changes, ensuring the optimal preservation of perishable food crops during storage and transportation. Key nanomaterials in innovative packaging include nanoparticles, nanocomposites, and nano encapsulates. Nanosensors are at the core of intelligent packaging, allowing real-time monitoring of critical parameters. These sensors can detect changes in temperature, humidity, gas concentration (such as oxygen and ethylene), and other environmental factors that impact food freshness and safety. The information obtained from nanosensors helps maintain optimal storage conditions and allows timely actions to prevent quality degradation (Huang et al., 2018). Nanomaterials with responsive properties are designed to adapt to changing needs. For instance, pH-responsive nanoparticles can release antimicrobial agents in response to bacterial growth, safeguarding the food from contamination. Temperature-responsive nanomaterials can change permeability or mechanical properties based on temperature fluctuations, providing intelligent barriers for enhanced preservation (Meng et al., 2021). Smart packaging systems incorporating nanomaterials can be classified as active and intelligent packaging. Active packaging interacts with the food environment, utilizing nanomaterials to release antimicrobial agents, oxygen scavengers, or antioxidants.

On the other hand, intelligent packaging employs sensors and indicators to monitor food conditions and relay information to consumers or retailers (Zhu et al., 2018). Nanotechnology allows the development of smart tags and labels equipped with nanosensors, enabling product traceability and quality assurance throughout the supply chain. These smart labels can detect and record temperature excursions or other adverse conditions during transportation and storage, providing valuable data for food safety and regulatory compliance (Salarbashi et al., 2018).

The study by Torres-Giner et al. (2008) investigated smart packaging through bio nano-composites for cereal grain preservation. They focused on integrating thymol nanoparticles into poly(ethylene-co-vinyl alcohol) and poly(lactic acid) matrices. The results highlighted improved cereal grain preservation. This work underscores the potential of nanomaterial-incorporated packaging for shelf-life extension, though further research is needed to address long-term effects and potential migration concerns. Akhtar and coworkers (2013) investigated nanomaterials for intelligent packaging with antimicrobial properties. While focused on microbial interactions rather than packaging, it explores nanostructured surfaces to prevent biofilm growth. This work hints at a potential avenue for nanomaterial integration in packaging to combat microbial contamination. According to Zhang and coworkers (2018), active packaging technologies, particularly antimicrobial packaging, are among the critical research areas in food packaging. Various nanomaterials are highlighted for enhancing packaging properties. The review explores methods like layer-by-layer assembly and nanoparticle incorporation, showcasing nanomaterial potential for extending shelf life. However, practical scalability and regulatory considerations need more attention.

8.2.4.5 Nanomaterials for Agro-waste Valorization

Due to its large volume and potential negative impact, agro-waste, such as agricultural residues and by-products, represents a significant challenge for agriculture and the environment. However, agro-waste can be transformed into value-added products by applying nanotechnology-driven processes. Nanotechnology offers innovative solutions to convert agro-waste into nano fertilizers, nanocomposites, and bioactive compounds, providing sustainable alternatives to conventional practices and reducing waste generation. Conventional fertilizers are associated with nutrient losses and environmental pollution. Nanotechnology enables the development of nano fertilizers by incorporating nutrients or bioactive compounds from agro-waste into nanoscale carriers. For example, nanoparticles loaded with essential nutrients, such as nitrogen, phosphorus, and potassium, can improve nutrient uptake efficiency and reduce nutrient leaching (Taher & Abdollahi, 2018).

Additionally, the slow and controlled release of nutrients from nano fertilizers enhances nutrient utilization by plants, improving crop yields and reducing environmental impact. Agro-waste, such as lignocellulosic biomass, can be used as a reinforcing filler in nanocomposites to enhance their mechanical and thermal properties. For instance, nanocellulose derived from agricultural residues can be incorporated into polymer matrices to improve strength, flexibility, and barrier properties (Reis et al., 2020). Nanocomposites from agro-waste offer sustainable alternatives to conventional materials and reduce the reliance on non-renewable resources.

Nanotechnology presents an innovative and sustainable approach for agro-waste valorization, leading to the production of nano fertilizers, nanocomposites, and bioactive compounds from agricultural residues and by-products. These value-added products offer enhanced performance, reduced environmental impact, and economic benefits. Agricultural residues are converted into nano fertilizers, nanocomposites, and bioactive compounds through nanomaterial-enabled processes. This synergy addresses waste management challenges while promoting sustainable agriculture. Gupta and coworkers (2020) emphasize the potential of nanotechnology to revolutionize agro-waste utilization.

Similarly, Singh and coworkers (2019) underscore the improved nutrient availability and plant growth resulting from nano fertilizers. Rai and coworkers (2022) explored the integration of nano-technology for enhanced agro-waste valorization, showcasing the promise of sustainable practices aligned with nanotechnology principles. Collectively, these works reveal the transformative poten-tial of nanotechnology in agro-waste valorization.

8.2.5 Implementation Challenges and Safety Concerns for the Application of Nanotechnology

As nanotechnology continues to advance and find applications in various sectors, including agri-culture and food systems, there is a growing need for robust regulatory frameworks to ensure the safe and responsible development and use of nanomaterials. Nanotechnology holds tremendous potential in improving crop production, food quality, and safety. However, due to the unique prop-erties of nanomaterials, concerns about their potential toxicity and environmental impact have prompted authorities worldwide to establish safety assessments and guidelines. Nanotechnology introduces novel properties and behavior in materials, necessitating specific regulations to address potential risks associated with using nanomaterials in agriculture and food systems (Chaudhry et al., 2018). Existing rules may not adequately cover nanotechnology-related aspects, warranting the development of separate guidelines. Regulatory agencies emphasize the need for thorough safety assessments of nanomaterials for agricultural and food applications. Risk management strategies, such as exposure assessment and toxicity testing, are essential to evaluate the potential risks to humans, animals, and the environment (Levin et al., 2018). Regulatory frameworks should be adaptable to keep pace with technological advancements. Continuous review and updating of regulations enable the inclusion of emerging nanotechnology applications while ensuring safety standards (US FDA, 2018).

The safe and responsible development and use of nanotechnology in agriculture and food sys-tems require robust regulatory frameworks. These regulations are essential to clear definitions, standardized characterization, safety assessments, and international collaboration. A precaution-ary approach, adaptability, stakeholder involvement, and environmental impact assessments further enhance the effectiveness of nanotechnology-related guidelines. Public education and awareness campaigns foster transparency and trust in nanomaterials, paving the way for sustainable and inno-vative applications in the agricultural and food sectors.

8.3 FUTURE PERSPECTIVES AND PROSPECTS

The review of "Post-harvest management of food crops and associated agro-waste in a develop-ing economy" comprehensively assesses post-harvest challenges and potential solutions. It high-lights the critical role of efficient post-harvest management in ensuring food security and reducing agro-waste in developing economies. The prospects emanating from this study are promising. As developing economies advance technologically, integrating innovative approaches such as nano-technology, smart packaging, and mobile-based technologies can revolutionize post-harvest prac-tices. These technologies can mitigate storage, transportation, and distribution losses, enhancing food availability and quality.

Moreover, the study underscores the significance of collaboration between various stakehold-ers, including farmers, researchers, policymakers, and industry players. Effective policies, invest-ments, and infrastructure development are essential to create an enabling environment for improved post-harvest practices. By integrating modern technologies, developing economies can reduce food loss and generate value from agro-waste through composting, bioenergy production, and animal feed conversion.

8.4 CONCLUSION

The review emphasizes that post-harvest management is a multifaceted challenge that demands holistic solutions. Addressing food loss and agro-waste in developing economies requires a combination of technological innovation, policy support, and capacity building. Adopting advanced techniques such as nanotechnology, innovative packaging, and mobile-based technologies offers immense potential to transform the landscape of post-harvest practices. By enhancing food quality, extending shelf life, and minimizing waste, these technologies contribute to food security, economic growth, and environmental sustainability. However, successfully implementing these solutions necessitates a concerted effort from governments, researchers, industries, and local communities. The review calls for continued research, investment, and awareness-building to ensure the responsible and sustainable management of food crops and associated agro-waste in developing economies. Ultimately, this approach holds the promise of securing food supplies, reducing economic losses, and contributing to the overall well-being of populations in these regions.

BIBLIOGRAPHY

Abdelwahab, T.A.M., & Fodah, A.E.M. (2022). Utilization of nanoparticles for biogas production focusing on process stability and effluent quality. *SN Applied Sciences*, 4, 332.

Ajayi, A.E., Rufai, A.A., & Waziri, A.I. (2020). A review on mobile-based technologies for agriculture and rural development in developing countries. *Information Technology for Development*, 26(1), 1–20.

Akhtar, S., Husain, Q., & Watkinson, A.P. (2013). Antimicrobial susceptibility and biofilm formation of Staphylococcus aureus and methicillin-resistant Staphylococcus aureus isolated from household cockroaches. *Journal of Medical Entomology*, 50(4), 890–894.

Alboofetileh, M., Rezaei, M., & Shahidi, F. (2020). Nanocomposites in food packaging: A review. *Comprehensive Reviews in Food Science and Food Safety*, 19(5), 2658–2674.

Alzate-Florez, A.L., & Romero-Gómez, P. (2019). Solar drying of agricultural products: A review. *Food Research International*, 123, 217–248.

Amwoka, M.M., Ambuko, J.L., Jesang, H.M., & Owino, W.O. (2021). Effectiveness of selected cold chain management practices to extend shelf life of mango fruit. *Advances in Agriculture*, Article ID 8859144, https://doi.org/10.1155/2021/8859144

Araujo, G.P. de, Lourenço, C.E., Araújo, C.M.L. de, & Bastos, A. (2018). Intercâmbio Brasil-União Europeia sobre desperdício de alimentos: Relatório final, 40.

Aryeetey, E.O., Egyir, I.S., & Ayamdoo, E.A. (2017). Post-harvest losses along the maize value chain in Ghana. *Ghana Journal of Agricultural Science*, 51(1), 39–47.

Aulakh, J., Regmi, A., Fulton, J.R., & Alexander, C.E. (2013). Estimating postharvest food losses: Developing a consistent global estimation framework. *AgEcon Search*, 19, 45–53.

Azeredo, H.M.C. (2009). Nanocomposites for food packaging applications. *Food Research International*, 42(9), 1240–1253.

Baky, M.A.H., Khan, M.N.H., Kader, M.F., & Chowdhury, H.A. (2014). *Proceedings of the ASME 2014 8th International Conference on Energy Sustainability ES2014*, June 30–July 2, Boston, MA.

Balana, B., Aghadi, C.N., & Ogunniyi, A.I. (2022). Improving livelihoods through post-harvest-loss management: Evidence from Nigeria. *Food Security*, 14(1), 249–265.

Balat, M., & Balat, H. (2009). Recent trends in global production and utilization of bio-ethanol fuel. *Applied Energy*, 86(11), 2273–2282.

Barros, J., & Gonçalves, B. (2021). Nanoemulsions for food applications: Development and characterization. *Food and Bioprocess Technology*, 5, 854–867.

Bhargava, R., Pathak, V., & Rai, A. (2021). Agro-waste in developing countries: A review. *Renewable and Sustainable Energy Reviews*, 137, 110602.

Bhusare, S., & Kadam, S. (2020). Applications of nanotechnology in fruits and vegetables. *Food and Agriculture Spectrum Journal*, 1(3), 90–95.

Bibby, K., Viau, E., & Peplies, J. (2016). Distribution of silver nanoparticles in wastewater treatment plants and its potential significance in the environment. *Journal of Hazardous Materials*, 304, 221–228.

Biswas, B.K., Rahman, M.S., & Uddin, M.A. (2018). Biodiesel production from waste cooking oil and its performance analysis in a CI engine. *Heliyon*, 4(12), e01058.

Bose, S., Alkahtani, J., & Kim, J. (2022). Nanocomposites for food packaging: A review. *Journal of Food Science and Technology*, 59(1), 138–155.

Bouwmeester, H., Brandhoff, P., Marvin, H.J., Weigel, S., & Peters, R.J. (2014). State of the safety assessment and current use of nanomaterials in food and food production. *Trends in Food Science & Technology*, 40(2), 200–210.

Chalak, A., Abiad, M.G., Diab, M., & Nasreddine, L. (2019). The determinants of household food waste generation and its associated caloric and nutrient losses: The case of Lebanon. *PLoS One*, 14(12), e0225789.

Chaudhry, Q., Scotter, M., Blackburn, J., Ross, B., & Boxall, A. (2018). Applications and implications of nanotechnologies for the food sector. *Food Additives & Contaminants: Part A*, 25(3), 241–258.

Chen, J., Li, Y., Zhang, Y., Wu, J., & Xu, J. (2017). Greenhouse gas emission from co-composting of sewage sludge and agricultural waste. *Bioresource Technology*, 244(Part 1), 536–541.

Chen, X., Sun, Y., Yang, J., & Jiang, Y. (2020). Circular bioeconomy: A sustainable solution to biomass waste valorization. *Bioresource Technology*, 312, 123624.

Dalai, R., & Singh, R. (2017). Landfill leachate pollution and treatment-A review. *International Journal of Environmental Science and Technology*, 14(5), 1039–1058.

Daryanto, A., Wang, L., & Jacinthe, P.A. (2021). Cold chain infrastructure and post-harvest losses of vegetables: A global analysis. *Food Policy*, 101, 102060.

Dehghannya, J., Hamedi, M., Tajik, H., Khamiri, M., & Aliakbarlu, J. (2021). Nanoemulsion-based delivery systems for natural antimicrobials in food preservation. *Food Research International*, 145, 110406.

de Moura, M.R., Aouada, F.A., Avena-Bustillos, R.J., McHugh, T.H., Krochta, J.M., & Mattoso, L.H. (2009). Improved barrier and mechanical properties of novel hydroxypropyl methylcellulose edible films with chitosan/tripolyphosphate nanoparticles. *Journal of Food Engineering*, 92(4), 448–453.

DiazPerez, J.C. (2014). Current and potential applications of active and intelligent packaging for preservation of fresh fruits and vegetables. In *Postharvest Biology and Technology of Horticultural Crops: Principles and Practices for Quality Maintenance*, 243–266. Boca Raton, FL CRC Press.

Edema, M.O., Sichamba, V., & Ntengwe, F.W. (2012). Solid waste management – Case study of Ndola, Zambia. *International Journal of Plant, Animal and Environmental Sciences*, 2(3), 248–255.

Erchamo, Y.S., Mamo, T.T., Workneh, G.A., Mekonnen, Y.S. (2021). Improved biodiesel production from waste cooking oil with mixed methanol-ethanol using enhanced eggshell-derived CaO nano-catalyst. *Scientific Reports*, 11, 6708.

FAO. (2019). *Moving Forward on Food Loss and Waste Reduction*. Rome: The State of Food and Agriculture.

FAO. (2022). *The State of Food Security and Nutrition in the World 2022*. Rome: Food and Agriculture Organization of the United Nations.

FAO. (2023). *The FAO Action Plan on Food Loss and Waste Reduction*. Rome: Food and Agriculture Organization of the United Nations.

Flores-López, M.L., Cerqueira, M.A., de Rodríguez, D.J., & Vicente, A.A. (2016). Perspectives on utilization of edible coatings and nano-laminate coatings for extension of postharvest storage of fruits and vegetables. *Food Engineering Reviews*, 8(3), 292–305.

Food Industry Policy Office. (2017). *Reducing Food Loss and Waste & Promoting Recycling*. Food and Agriculture Organization of the United Nations, FAO. (2015). Global Initiative on Food Loss and Waste Reduction. Recovered from: https://openknowledge.fao.org/server/api/core/bitstreams/57f76ed9-6f19-4872-98b4-6e1c3e796213/content. Last Accessed: September 7th, 2024.

Gad, M., & Zag Zog, O. (2017). Mixing xanthan gum and chitosan nanoparticles to form new coating for maintain storage life and quality of Elmamoura guava fruits. *International Journal of Current Microbiology and Applied Sciences*, 6(11), 1582–1591.

Ghaani, M., & Ghomi, H. (2018). Use of smart packaging systems for monitoring of food quality changes. *Food Research International*, 108, 278–294.

Gupta, P., Mishra, A.K., & Sharma, A.K. (2019). Nanoemulsion-based delivery systems for encapsulation of food bioactive components: Nutritional, structural, and bioaccessibility aspects. *Journal of Food Science and Technology*, 56(8), 3269–3285.

Gupta, V.K. Flores-Contreras, E.A., González-González, R.B., Pizaña-Aranda, J.J.P., Parra-Arroyo, L., Rodríguez-Aguayo, A.A., Iñiguez-Moreno, M., González-Meza, G.M., Araújo, R.G., Ramírez-Gamboa, D., Parra-Saldívar, R., & Melchor-Martínez, E.M. (2024). Agricultural waste as a sustainable source for nanoparticle synthesis and their antimicrobial properties for food preservation. *Frontiers in Nanotechnology*, 6, 1346069. (2020). Nanotechnology-enabled agro-waste valorization: A sustainable solution for agricultural residues. *Science of the Total Environment*, 734, 139296.

Gürkan, A.K.G., Mahtem, M., & Ako, K. (2021). Utilization of agricultural wastes for sustainable development, *Black Sea Journal of Agriculture*, 4(4), 146–152.

Gustavsson, J., Cederberg, C., Sonesson, U., Van Otterdijk, R., & Meybeck, A. (2011). *Global Food Losses and Food Waste: Extent, Causes and Prevention*. Rome: Food and Agriculture Organization (FAO) of the United Nations.

Gutiérrez-Miceli, F.A., Santiago-Borraz, J., Molina-Ramírez, C., Nafate, C.C., Abud-Archila, M., Llaven-Maquedaa, H., & Dendooven, L. (2018). Vermicompost as a soil supplement to improve growth, yield and fruit quality of tomato (Solanum lycopersicum). *European Journal of Soil Biology*, 86, 72–80.

Hafifi, M.H.M., Daud, W.R.W., Sillanpää, M., Aroua, M.K., & Sulaiman, N.M.N. (2018). A review of co-digestion achievements in the anaerobic digestion of food waste for enhanced methane production. *Journal of Cleaner Production*, 184, 42–57.

Hajipour, M.J., Fromm, K.M., Akbar Ashkarran, A., Jimenez de Aberasturi, D., de Larramendi, I.R., Rojo, T., & Mahmoudi, M. (2012). Antibacterial properties of nanoparticles. *Trends in Biotechnology*, 30(10), 499–511.

He, X., & Hwang, H.M. (2016). Nanotechnology in food science: Functionality, applicability, and safety assessment. *Journal of Food and Drug Analysis*, 24(4), 671–681.

Hodges, R.J., Buzby, J.C., & Bennett, B. (2011). Postharvest losses and waste in developed and less developed countries: Opportunities to improve resource use. *The Journal of Agricultural Science*, 149(11), 37–45.

Hong, Y.S., & Yoon, H.H. (2011). Ethanol production from food residues. *Biomass and Bioenergy*, 35(7), 3271–3275.

Hossain, M., & Kabir, R. (2020). Economic analysis of food loss in developing economies: A case study of Bangladesh. *Cogent Food & Agriculture*, 6(1), 1761–1860.

Huang, Y., Wu, D., Chen, C., Chen, S., Zhu, Q., & Jin, T. (2016). A review of carbon nanomaterials' synthesis via the chemical vapor deposition method: Towards the low-cost production of nanocomposites. *Journal of Materials Chemistry A*, 4(15), 5489–5515.

Huang, Y., Zhang, X., Luo, Y., & Liu, X. (2018). Smart packaging based on nano-sensors for food quality and safety control. *Trends in Food Science & Technology*, 72, 1–6.

Ibrahim, H.I., Ibrahim, H.Y., Adeola, S.S., & Ojoko, E.A. (2022). Post harvest loss and food security: A case study of major food crops in Katsina state, Nigeria. *FUDMA Journal of Agriculture and Agricultural Technology*, 8(1), 393–403.

Ijaz, M., Zafar, M., Afsheen, S., & Iqbal, T. (2020). A review on Ag-nanostructures for enhancement in shelf time of fruits. *Journal of Inorganic and Organometallic Polymers and Materials*, 30(5), 1475–1482.

Jafari, S.M., McClements, D.J., & Şahin, S. (2020). Nanoemulsions as delivery systems for bioactive compounds: Impact of droplet size on their properties. *Food Chemistry*, 314, 126181.

Jain, P., Pandey, M.K., Sharma, G., Malviya, R., Agrawal, G.K., & Rajam, M.V. (2020). Nanotechnology in agro-food: From field to plate. *Food Research International*, 69, 381–400.

Jha, S.N., Tomar, R., & Choudhury, P.P. (2019). Modified atmosphere packaging of fruits and vegetables for extension of shelf-life – A review. *Food Packaging and Shelf Life*, 21, 100343.

JICA. (2013). *Project on Master Plan Study for Integrated Solid Waste Management in Bogota, D.C.* Chiyoda: Japan International Cooperation Agency, 2.

Jinsong, Z., Zhanting, Z., Zheng, C., Yu, W., Xiaojun, W., Bin, W., & Wenhua, G. (2021). Cellulose nanofibrils manufactured by various methods with application as paper strength additives. *Scientific Reports*, 11, 11918.

Jodar, J.R., Ramos, N., Carreira, J.A., Pacheco, R., & Fernández-Hernández, A. (2017). Quality assessment of compost prepared with municipal solid waste. *Open Engineering*, 7(1), 221–227.

Kader, A.A. (2005). Increasing food availability by reducing postharvest losses of fresh produce. *Acta Horticulturae*, 682, 2169–2176.

Kalaimani, M., & Wai, S.C. (2022). Perspectives on nanomaterials and nanotechnology for sustainable bioenergy generation. *Materials*, 15, 7769.

Karthikeyan, O.P., Visvanathan, C., & Verstraete, W. (2018). Biogas production from biomass and waste resources for energy generation. *Environmental Technology & Innovation*, 10, 309–327.

Kaur, R., & Singh, D. (2018). Solar drying: An eco-friendly technology for food preservation. *Journal of Renewable Energy and Sustainable Development*, 4(1), 15–21.

Kemper, K., Voegele, J., Hickey, V., Ahuja, P.S., Poveda, R., Edmeades, S., Kneller, C., Swannell, R., Gillick, S., Corallo, A., Aguilar, G., Alencastro, S., Felix, E., & Sebastian, A. (2019). Mexico Conceptual Framework for a National Strategy on Food Loss and Waste, 68.

Khan, M.S., Adil, S.F., Al-Warthan, A., & Siddiqui, M.R. (2019). Biogenic synthesis of metal nanoparticles and their potential applications. *Journal of Nanomaterials*, 1, 187–204.

Khan, S.Z., Zaidi, A.A., Naseer, M.N., & AlMohamadi, H. (2022). Nanomaterials for biogas augmentation towards renewable and sustainable energy production: A critical review. *Frontiers in Bioengineering and Biotechnology*, 10, 868454.

Kim, J.K., Oh, B.R., Shin, H.J., Eom, C.Y., & Kim, S.W. (2008). Statistical optimization of enzymatic saccharification and ethanol fermentation using food waste. *Process Biochemistry*, 43(11), 1308–1312.

Koyuncu, T., Çalik, G., & Arslan, D. (2020). Assessment of the effect of solar drying method on drying rate and dried material quality. *Food Science and Technology International*, 26(4), 321–333.

Kumar, A., Maunder, M., & Singh, G. (2021a). Bio gas as a renewable energy source: A review on recent developments and future perspectives. *Renewable and Sustainable Energy Reviews*, 137, 110574.

Kumar, A., & Yadav, D.N. (2019). Recent developments in pretreatment technologies on lignocellulosic waste to bioenergy: A review. *Bioresource Technology*, 279, 350–361.

Kumar, S., Dhaliwal, S.S., Kaur, R., & Gandhi, K.S. (2021b). Agricultural waste compost: A potent tool for sustainable soil fertility and plant growth enhancement. *Science of the Total Environment*, 753, 141984.

Kumar, S., Singh, S., & Mohan, D. (2019). Utilization of agro-industrial wastes for composting: Microbial diversity and effect of compost application on maize (Zea mays L.) crop. *Journal of Environmental Management*, 233, 688–698.

Leket Israel. (2019). Food waste and rescue in Israel: The economic. *Social and Environmental Impact*.

Levin, M., Langer, E., Rahman, S., & Brown, P. (2018). The regulation of nanotechnology. *International Journal of Nanotechnology*, 15(1/2/3), 39–61.

Li, J., & Liu, D. (2021). Recent advances in food packaging based on nanomaterials. *Advances in Food and Nutrition Research*, 96, 157–186.

Lipinski, B., Craig, H., James, L., Lisa, K., Richard, W., & Tim, S. (2013). *Reducing Food Loss and Waste*. World Resource Institute, 1–40.

Liu, X., Chen, Z., & Pan, L. (2019). Recent developments in nanoemulsion-based delivery systems. *Current Opinion in Food Science*, 26, 17–25.

Lloret, E., Picouet, P., & Fernández, A. (2012). Matrix effects on the antimicrobial capacity of silver-based nanocomposite absorbing materials. *LWT-Food Science and Technology*, 49(2), 333–338.

López de Dicastillo, C., Canet, W., Debeaufort, F., & Vilariño, J.M. (2018). Active edible films as carriers of antibacterial agents to improve food safety. *Food and Bioprocess Technology*, 11(5), 872–885.

Madruga, M.S., Figueiredo, A.M., Bezerra, T.K.A., Queiroga, R.C.R.E., Albuquerque, J.F., & Rocha, M.V.P. (2021). Nutritive value and in vitro fermentation kinetics of tropical agro-industrial by-products and their potential use in ruminant diets. *Animal Feed Science and Technology*, 276, 114899.

Magnuson, B.A., Jonaitis, T.S., & Card, J.W. (2011). A brief review of the occurrence, use, and safety of food-related nanomaterials. *Journal of Food Science*, 76(6), R126–R133.

Makate, C., Makate, M., Mango, N., & Mango, L. (2016). Mobile phone short message service (SMS) for value added agricultural extension delivery. *World Development*, 87, 21–30.

McClements, D.J. (2018). *Food Emulsions: Principles, Practices, and Techniques*. Boca Raton, FL: CRC Press.

Meng, J. Meng, Q., Liu, H., Zhang, H., Xu, Z., Lichtfouse, E., Yun, Y. (2022). Anaerobic digestion and recycling of kitchen waste: a review. *Environmental Chemistry Letters*, 20, 1745–1762. (2017). Recycling of kitchen waste to produce a novel compost of high quality through the two-step approach: Acidification plus alkalization. *Journal of the Science of Food and Agriculture*, 97(1), 90–97.

Meng, X., Nair, A.R., & Arora, A. (2021). Responsive nano-composites in food packaging: Recent advances, challenges, and opportunities. *Trends in Food Science & Technology*, 108, 120–132.

Miao, H., Wu, J., & Ding, C. (2019). Food loss and food waste in China: A review of food policies for industrial ecology. *Journal of Cleaner Production*, 231, 1294–1309.

Miezah, K., Obiri-Danso, K., Kádár, Z., Fei-Baffoe, B., & Mensah, M.Y. (2015). Municipal solid waste characterization and quantification as a measure towards effective waste management in Ghana. *Waste Management*, 46, 15–27.

Minteer, S.D., Carraher, J.M., & Verpoorte, R. (2020). Ethanol from biomass: A review. *Journal of the American Chemical Society*, 142(40), 17023–17032.

Misson, M., Zhang, H., & Jin, B. (2015). Nanobiocatalyst advancements and bioprocessing applications. *Journal of the Royal Society Interface*, 12, 20140891.

Mittal, N., Gupta, A.K., Parashar, P., Sood, N., & Agarwal, A. (2018). A review on role of nanotechnology in waste management. *Resource-Efficient Technologies*, 4(4), 409–422.

Molino, A., Zabaniotou, A., & Skoulou, V. (2021). Co-digestion of agro-industrial waste streams in Greece: An evaluation of technical and economic feasibility. *Journal of Cleaner Production*, 295, 126422.

Muganyizi, J., Bisheko, A., & Rejikumar. (2023). Major barriers to adoption of improved postharvest technologies among smallholder farmers in sub-Saharan Africa and South Asia: A systematic literature review. *World Development Sustainability*, 2, 100070.

Mukhtar, M., Ali, S., & Ali, S. (2020). Agro-waste management practices and environmental pollution: A review. *Environmental Science and Pollution Research*, 27(30), 37653–37669.

Müller, C., Hensel, O., Schönborn, A., Rohn, S., & Kroh, L.W. (2019). Drying of fruits and vegetables by hybrid solar-biomass dryers: A review. *Food and Bioproducts Processing*, 116, 241–260.

Müller, R.H., Keck, C.M., & Müller, B.W. (2020). Surface modification, surface characterization and nanoparticle coating of solid lipid nanoparticles (SLN) for controlled drug delivery. *Die Pharmazie-An International Journal of Pharmaceutical Sciences*, 53(9), 584–588.

Muthoni, J., Mutwiwa, U.N., & Kilonzo, J. (2018). Management of postharvest losses of fresh produce in developing countries: A review. *African Journal of Horticultural Science*, 15, 63–82.

Nurjanah, N., Saman, N., & Wahyudi, S. (2018). Biogas production from agro-waste: A review. *Journal of Advanced Research in Fluid Mechanics and Thermal Sciences*, 45(2), 216–226.

Odetayo, T., Tesfay, S., & Ngobese, N.Z. (2022). Nanotechnology-enhanced edible coating application on climacteric fruits. *Food Science & Nutrition*, 10, 2149–2167.

Ogendo, J.O., Makokha, A.O., Onyango, C.A., & Owino, W.O. (2019). Hermetic storage technology and cowpea postharvest storage losses: A review. *Food Security*, 11(6), 1283–1293.

Ogundele, F. (2022). Post harvest losses and food security in Nigeria: An empirical review. *African Journal of Agriculture and Food Science*, 5(3), 77–89. doi: 10.52589/AJAFS-C0442Z7J

Ojha, K.P., Park, K.J., & Han, S.K. (2019). Anaerobic digestion as an effective method for the management of food waste: A mini-review. *Journal of Environmental Management*, 237, 378–385.

Opara, L.U., & Pathare, P.B. (2014). Progress in postharvest handling and processing of fruits and vegetables. In *Emerging Technologies for Food Processing*, 2nd ed., 323–344. Cambridge, MA: Academic Press.

Orhorhoro, E.K., Ebunilo, P.O., & Sadjere, G.E. (2017). Determination and quantification of household solid waste generation for planning suitable sustainable waste management in Nigeria. *International Journal of Emerging Engineering Research and Technology*, 5(8), 10.

Pardeep, K.S., Surekha, D., & Joginder, S.D. (2018). Agro-industrial wastes and their utilization using solid-state fermentation: A review. *Bioresources and Bioprocessing*, 5(1), 1–15.

Parfitt, J., Barthel, M., & Macnaughton, S. (2010). Food waste within food supply chains: Quantification and potential for change to 2050. *Philosophical Transactions of the Royal Society B: Biological Sciences*, 365(1554), 3065–3081.

Parisi, C., Vigani, M., & Rodríguez-Cerezo, E. (2015). Agricultural nanotechnologies: What are the current possibilities? *Nano Today*, 10(2), 124–127.

Pinar, T.S., Ismail, S., Burcu, G.B., Seda, I.M., & Abdelfattah, Z.M.S. (2018). Nanotechnology and nano-propolis in animal production and health: An overview. *Italian Journal of Animal Science*, 17(4), 921–930.

Pingali, P., Alinovi, L., Sutton, J., & Anderson, C.L. (2019). Food loss in Sub-Saharan Africa: What do farmers say? *Global Food Security*, 23, 192–201.

Pooja, G., & Jyoti M. (2022). Application of agro-waste-mediated silica nanoparticles to sustainable agriculture. *Bioresources and Bioprocessing*, 9(9), 1–12.

Prasetyo, A., & Suhendrata, T. (2021). The nutritional value of agricultural waste forages fed to ruminants from agroecosystem in Grobogan Regency. *IOP Conference Series: Earth and Environmental Science*, 653, 012136

Rai, M. Schlavi, D., Di Lorenzo, V., Francesconi, S., Giovagnoli, S., Camaioni, E., & Balestra, G.M. (2022). Waste valorization by nanotechnology approaches for sustainable crop protection: a mini review. *IOP Conference Series: Materials Science and Engineering*, 1265, 012009. (2022). Nanotechnology-enabled agro-waste valorization: An emerging avenue for sustainable agriculture. *Current Opinion in Environmental Science & Health*, 25, 100274.

Ramakrishna, V. (2016). Municipal solid waste quantification, characterization and management in Rajam. *The International Journal of Engineering and Science*, 5(2), 8.

Ramos-Morales, E., Arco-Pérez, A., Martínez-Fernández, A., Lachica, M., Molina-Alcaide, E., Yáñez-Ruiz, D.R., & Martín-García, A.I. (2019). Nutrient composition, in vitro digestibility, and gas production of agro-industrial by-products for ruminant feeding. *Journal of Agricultural and Food Chemistry*, 67(19), 5481–5490.

Ramukhwatho, F.R. (2016). *An Assessment of the Household Food Wastage in a Developing Country: A Case Study of Five Areas in the City of Tshwane Metropolitan Municipality, Guateng Province*. South Africa: University of South Africa.

Rathore, P., & Sarmah, S.P. (2020). Economic, environmental and social optimization of solid waste management in the context of circular economy. *Computers & Industrial Engineering*, 145, 106510.

Reis, K.C., Rodrigues, F.H., Ferreira, A.G.M., & Lona, L.M.F. (2020). Nanocomposites based on biodegradable polymers and lignocellulosic fibers: A review. *Polymer Testing*, 83, 106316.

Rekha, P.D., Lai, Q., Savitha, K.S., & Aravind, J. (2020). Biogas from kitchen waste: A mini-review on challenges and prospects. *Environmental Technology & Innovation*, 18, 100753.

Ruffo Roberto, S., Youssef, K., Hashim, A.F., & Ippolito, A. (2019). Nanomaterials as alternative control means against postharvest diseases in fruit crops. *Nanomaterials*, 9(12), 1752.

SAGO. (2019). *Saudi FLW Baseline: Food Loss & Waste Index in Kingdom of Saudi Arabia*. Riyadh: Saudi Grains Organization.

Salarbashi, D., Barzegar, M., Sahari, M.A., & Azizi, M.H. (2018). Development of nanoliposome based on natural emulsifiers: Application in encapsulation of thymol and carvacrol for enhanced antimicrobial activity in food preservation. *Food Control*, 85, 38–45.

Salmiah, J.M.R., Sarina, M.R., Wan Nazwanie, W.A., & Nur Atiqah, N. (2023). Chapter 16- Potentialities of nano-biotechnology for creating preservatives for packaging livestock feed products. In *Nanobiotechnology for the Livestock Industry, Animal Health and Nutrition*, 303–327. doi: 10.1016/B978-0-323-98387-7.00013-6

Samkol, P., & Egbunike, G.N. (2019). Agro-industrial by-products and crop residues utilization as animal feed: A review. *Cogent Food & Agriculture*, 5(1), 1680260.

Severini, C., Derossi, A., Fiore, A.G., De Pilli, T., & Derossi, E. (2018). Smart packaging for food applications: A review. *Foods*, 7(11), 70.

Shankar, S., & Rhim, J.W. (2015). Amino acid mediated synthesis of silver nanoparticles and preparation of antimicrobial agar/silver nanoparticles composite films. *Carbohydrate Polymers*, 130, 353–363.

Shankar, S., & Rhim, J.W. (2017). Preparation and characterization of agro-nanocomposite films for food packaging applications. *Journal of Agricultural and Food Chemistry*, 65(4), 827–835.

Sharma, S., Singh, K.K., & Dileep Kumar, D. (2020). Post-harvest cold storage technologies for fruits and vegetables: A review. *Agricultural Reviews*, 41(1), 24–31.

Sialve, B., Bernet, N., & Bernard, O. (2009). Anaerobic digestion of microalgae as a necessary step to make microalgal biodiesel sustainable. *Biotechnology Advances*, 27(4), 409–416.

Singh, S. Preethi, B., Karmegam, N., Manikandan, S., Vickram, S., Subbaiya, R., Rajeshkumar, S., Gomadurai, C., & Govarthanan, M. (2024). Nanotechnology-powered innovations for agricultural and food waste valorization: A critical appraisal in the context of circular economy implementation in developing nations. *Process Safety and Environmental Protection*, 184, 477-491. (2019). Nanotechnology-driven agro-waste valorization: A sustainable approach for enhancing agricultural productivity. *Journal of Cleaner Production*, 230, 224–235.

Snehal, I., Sanket, J.J., & Akshaya, G. (2014). Production of bioethanol using agricultural waste: Banana pseudo stem. *Brazilian Journal of Microbiology*, 45(3), 855–892.

Sodhi, M.S., Soni, A., & Walia, S. (2020). Review on impact of cold chain management on post-harvest losses. *Journal of the Saudi Society of Agricultural Sciences*, 19(3), 220–225.

Ssemwanga, M., Makule, E., & Kayondo, S. (2020). The effect of traditional and improved solar drying methods on the sensory quality and nutritional composition of fruits: A case of mangoes and pineapples. *Heliyon*, 6, e04163.

Taha, Abdelfattah, M.A., Mahendra, K.M., Pradeepta, K.S., & Debaraj, B. (2020). Application of nanoparticles for biogas production: Current status and perspectives. *Energy Sources, Part A: Recovery, Utilization, and Environmental Effects*. https://doi.org/10.1080/15567036.2020.1767730

Taher, S., & Abdollahi, M. (2018). Nanotechnology-based fertilizers: A new paradigm for agriculture. *Environmental Science and Pollution Research*, 25(36), 35325–35336.

Takeuchi, N. (2019). Linkages with SDG 11.6.1 on MSW and composition analysis [Unpublished UN Habitat presentation].

Thakur, R., Ankit, K., Pundir, R., Sharma, R., Guleria, S., & Rathee, P. (2021). Nanoemulsions: A comprehensive review on various formulation aspects and application. *Journal of Microencapsulation*, 38(7), 707–729.

Thapa, G., & Kumar, S. (2019). Quantitative estimation of post-harvest losses of fruits and vegetables in India. *Journal of Food Science and Technology*, 56(11), 5135–5143.

Toghyani, S., Rahimi, M., Samani, A.H., & Tabatabaeefar, A. (2018). Performance analysis of a mixed-mode solar dryer for drying agricultural products: An experimental and computational study. *Renewable Energy*, 116, 177–186.

Torres-Giner, S., Giménez, E., Lagaron, J.M., & Gavara, R. (2008). Development of antimicrobial packaging films based on bionanocomposites of poly(ethylene-co-vinyl alcohol), poly(lactic acid), and thymol for the preservation of cereal grains. *Journal of Agricultural and Food Chemistry*, 56(12), 4690–4699.

United Nation Environmental Programme. (2021). Food Waste Index Report 2021, 1–100.

US FDA (United States Food and Drug Administration). (2018). Nanotechnology and food. Retrieved from https://www.fda.gov/food/nanotechnology-products-nano-what/nanotechnology-and-food

Van Hoang, N., Huan, N.M., & Tam, N.M. (2021). The development of solar drying technology in agricultural products in Vietnam. In *Renewable Energy and Sustainable Development*, 423–432. Singapore: Springer.

Varghese, S.A., Pulikkalparambil, H., Promhuad, K., Srisa, A., Laorenza, Y., Jarupan, L., Nampitch, T., Chonhenchob, V., & Harnkarnsujarit, N. (2023). Renovation of agro-waste for sustainable food packaging: A review. *Polymers*, 15, 648.

Velázquez-Jiménez, L.M., Flores-Rojas, G.G., Martínez-Gutiérrez, H., García-Pérez, B.E., Guzmán-Partida, A.M., & Lira-Saldivar, R.H. (2019). Nanoparticles as alternative to control plant bacterial diseases. *Agriculture*, 9(2), 34.

Vetter-Gindele, J., Braun, A., Warth, G., Bui, T.T.Q., Bachofer, F., & Eltrop, L. (2019). Assessment of household solid waste generation and composition by building type in Da Nang, Vietnam. *Resources*, 8(4), 171.

Wang, Q., Ma, H., Xu, W., Gong, L., Zhang, W., & Zou, D. (2008). Ethanol production from kitchen garbage using response surface methodology. *Biochemical Engineering Journal*, 39(3), 604–610.

Wang, Y., Wu, J., Zhao, L., & Qiao, H. (2021). Effects of different compost amendment rates on the yield, nutrient accumulation and allocation, and fruit quality of pepper plants. *Journal of Soils and Sediments*, 21(1), 75–87.

Whitmee, S., Haines, A., Beyrer, C., Boltz, F., Capon, A. G., Ferreira de Souza Dias, B., Ezeh, A., Frumkin, H., Gong, P., Head, P., Horton, R., Mace, G. M., Marten, R., Myers, S. S., Nishtar, S., Osofky, S. A., Pattanayak, S., Pongsiri, M., Romanelli, C., Soucat, A., Vega, J., & Yach, D. (2015). Safeguarding human health in the Anthropocene epoch: Report of The Rockefeller Foundation–Lancet Commission on planetary health. *The Lancet*, 38, 10007.

World Food Programme (WFP). (2019). Food waste: The facts. Retrieved from https://www.wfp.org/publications/food-waste-facts

World Resources Institute (WRI). (2021). *The Business Case for Reducing Food Loss and Waste: Restaurants.* Washington, DC: World Resources Institute (WRI)

Wu, W., Zhao, Y., & Gao, Q. (2020). Characteristics of leachate from vermicomposting of kitchen waste and cow dung and its impact on seed germination. *Environmental Science and Pollution Research*, 27(3), 2930–2941.

Yan, S., Li, J., Chen, X., Wu, J., Wang, P., & Ye, J. (2011). Enzymatical hydrolysis of food waste and ethano production from the hydrolysate. *Renewable Energy*, 36(4), 1259–1265.

Yao, L., Munir, S., Zhang, J., & Zou, X. (2018). Cold chain logistics in developing countries: A case study from China. *Food Control*, 90, 375–382.

Yu, Q., Liu, R., Li, K., & Ma, R. (2019). A review of crop straw pretreatment methods for biogas production by anaerobic digestion in China. *Renewable and Sustainable Energy Reviews*, 107, 51–586.

Zhang, H., Liu, G., Xue, L., Zuo, J., Chen, T., Vuppaladadiyam, A., & Duan, H. (2020). Anaerobic digestion based waste-to-energy technologies can halve the climate impact of China's fast-growing food waste by 2040. *Journal of Cleaner Production*, 277, 123490.

Zhang, M., Aiping, S., Muhammad, A., Lihua, Y., & Muhammad, A. (2021a). Comprehensive review on agricultural waste utilization and high-temperature fermentation and composting. *Biomass Conversion and Biorefinery*, 13, 5445–5468.

Zhang, S., Song, Z., Liu, Y., Qu, W., Yang, G., & Xu, R. (2021b). Understanding methane formation during the anaerobic digestion of agricultural straw under thermophilic conditions. *Waste Management*, 127, 329–336.

Zhang, Y., Han, J.H., & Han, Y. (2018). Active packaging technologies with an emphasis on antimicrobial packaging and its applications. *Journal of Food Protection*, 81(4), 670–687.

Zhu, L., Zheng, Y., He, X., Wang, H., Zhang, C., Xiao, Z., & Zeng, X. (2018). Active and intelligent packaging for food: A technological overview. *Food Reviews International*, 34(1), 62–84.

Adenike A. Akinsemolu and Helen Onyeaka

9.1 INTRODUCTION

Water has often been termed the 'elixir of life,' a phrase that underscores its significance as a basic necessity and a lifeline underpinning the tapestry of human existence. Water's centrality to life and culture is evident across ancient civilizations, from the banks of the Nile to the fertile basins of the Indus and Yellow Rivers (National Research Council & Safe Drinking Water Committee, 1977). Yet, as we find ourselves in the 21st century, the dynamics of this relationship have evolved. Now, more than ever, water quality intersects with critical issues of food safety, public hygiene, and global food supply (Vilakazi et al., 2019).

Historical records and narratives from ancient civilizations such as Mesopotamia and Egypt reveal that even in early human settlements, water sources were not just places to quench thirst but were also considered sacred, underlining their importance in daily life and rituals (Thompson, 2016). Fast forward to modern times, and water has transitioned from being a mere element of survival to a pivotal component in industrial processes, healthcare, agriculture, and, significantly, food production (Vilakazi et al., 2019; Ringler et al., 2023). The food we consume is, in many ways, a reflection of water that nourishes it. Whether it is in the cultivation of grains, the processing of raw materials, or the preparation of dishes, water quality is paramount. Just as poor soil quality can adversely impact the crops grown in it, subpar water can taint the food it interacts with, leading to a cascade of consequences for public health and safety (Okorogbona et al., 2018; Bhagwat, 2019). For instance, the outbreak of *E. coli* infections linked to contaminated water used in food production, reported in the US in the early 2000s, is a testament to these interdependencies (Doyle et al., 2006; Jarvie, 2014).

Agriculture, the most significant consumer of global freshwater resources, exemplifies the critical role of water in the food chain. A report revealed that almost 70% of global freshwater withdrawals are attributed to irrigation (Borsato et al., 2020). This statistic underscores the sheer volume of water needed to sustain global food systems and the magnitude of potential risk should this water be of inferior quality. Yet, water's role is not confined solely to agriculture. As urban populations burgeon, the demand for processed foods has escalated, further highlighting the interplay between water quality and food safety (FAO, 2017). With industrialization, the exposure of water sources to pollutants has grown, posing challenges in maintaining water quality. Chemical runoffs, heavy metals, and other contaminants have made their way into water sources, which are subsequently used in various stages of food production, presenting a pressing concern for both industry stakeholders and consumers (Lebelo et al., 2021; Linderhof et al., 2021).

Considering this intricate relationship, it's not just about water serving as a resource; it's about ensuring that every drop that integrates into the food system is optimal. Today, with advancements in science and technology, the means to remediate, decontaminate, and purify water has expanded exponentially. However, as with all scientific progress, it is incumbent upon us to assess the efficacy, sustainability, and implications of these methods (Nagar & Pradeep, 2020; Martínez-Huitle et al., 2023).

DOI: 10.1201/9781003514039-9

As we delve deeper into this topic, this chapter aims to unravel various facets of water's role in food production, its significance in public health, and the innovations to ensure its purity. In a world where resource constraints are becoming glaringly evident, understanding and addressing the nuances of water quality becomes not just a matter of academic interest but an imperative for sustainable survival.

9.2 GLOBAL WATER QUALITY ISSUES AND THEIR RELATIONSHIP TO FOOD PRODUCTION SYSTEMS

The ubiquity of water in human civilization is unquestionable, functioning as a veritable engine behind numerous sectors of the global economy. In comparison, the direct health implications of clean drinking water are frequently underscored, and the subtler, albeit equally consequential, repercussions of water quality on food production systems are not always in the spotlight. Understanding the intricacies of this relationship is crucial to ensuring food security and maintaining public health standards.

Agriculture, the cornerstone of food production, relies extensively on water resources. However, it's not just about quantity; water quality is paramount in determining crop yields, livestock health, and food products' overall safety (Umar et al., 2014; Malakar et al., 2019). The ramifications can be alarming when the water utilized in agricultural processes is contaminated with pollutants—ranging from heavy metals to microbial pathogens. This can result in diminished crop yields, compromised livestock health, and potential transmission of waterborne diseases through the food supply chain (Bhagwat, 2019).

Globally, various factors contribute to water quality challenges that intersect with food production systems:

a. **Industrial Runoff:** Rapid industrialization in many parts of the world has led to increased discharge of untreated industrial effluents into freshwater sources. These effluents, laden with toxins and heavy metals, can contaminate the water used in food production, thereby introducing these pollutants into the food chain (Afrad et al., 2020; Hussaini et al., 2023).

b. **Agricultural Chemicals:** Pesticides, herbicides, and synthetic fertilizers, integral to modern farming practices, often find their way into water sources due to runoff. When this contaminated water is used for irrigation, residues of these chemicals can be retained in the crops, posing health risks to consumers and the ecosystem (Tudi et al., 2021).

c. **Microbial Contamination:** Water sources can be breeding grounds for pathogens when lacking or ineffective sanitation infrastructure. Using such water in agriculture can lead to crops and livestock becoming carriers of these pathogens, leading to potential outbreaks of waterborne diseases (FAO & IWMI, 2017)

d. **Salinization:** In areas prone to excessive groundwater extraction, the intrusion of saline water into freshwater aquifers can become a significant concern. Elevated salinity levels can hamper crop growth and affect yields, particularly in regions reliant on groundwater for irrigation (Wang et al., 2020).

The linkage between water quality and food production is not limited to primary agricultural practices. As the global demand for processed foods rises, water needs in food processing plants escalate. Depending on the nature of the products, these plants might use water for cleaning raw materials, cooking, and even as an ingredient in some products. Contaminated water in such facilities can not only result in unsafe products. Still, the large-scale distribution of processed foods can also be a precursor to widespread foodborne disease outbreaks (Rather et al., 2017).

Moreover, the pertinence of water quality extends to fisheries and aquaculture, sectors witnessing burgeoning growth due to increasing global demand for seafood. Contaminated water bodies can lead to diseased fish stocks, making them unfit for consumption and jeopardizing a critical protein source for millions (Bashir et al., 2020). In addition to threatening food production by affecting the availability of safe seafood, contamination threatens other food production systems at different production levels. In dairy farming, for instance, the contamination of water resources affects the output and quality of pasture crops, affecting the industry's productivity (KC et al., 2020). Further, the industry's animals need significant amounts of clean drinking water. Together, the two impacts of contaminated water reduce milk and meat yields. Water contamination affects other levels of different food chains, playing distinct but universally critical roles in food production through agriculture and aquaculture, food storage and transportation, processing, the retail and provisioning of food, and its consumption (Linderhof et al., 2021).

To further magnify the significance of water quality in food systems, consider the socioeconomic implications. Farmers facing diminished yields due to contaminated irrigation water often face financial uncertainties. Moreover, contamination episodes can trigger food recalls, causing significant economic losses to businesses and undermining consumer confidence in food safety standards (Sarkar et al., 2021; WHO, 2021).

The nexus between water quality and food production systems is intricate and multidimensional. It's a confluence where environmental science, public health, and socioeconomics intersect. While the challenges posed by water quality issues are undoubtedly formidable, they underscore the necessity for proactive measures, robust policy frameworks, and technological innovations. As the global population surges, ensuring water purity in food production systems will be pivotal in safeguarding food security, public health, and economic stability.

9.3 INTERLINKING FOOD SECURITY, PUBLIC HEALTH, AND WATER MANAGEMENT

As societies grapple with increasing global challenges, there's a palpable intersection between food security, public health, and water management that cannot be overlooked. Delving deeper into the intricacies of these components illuminates their interdependent nature and the need for integrated solutions.

9.3.1 THE FOUNDATION OF FOOD SECURITY: WATER'S CENTRAL ROLE

When we speak of food security, we talk about a world where everyone has a fair shot at adequate and nutritious meals. However, this ideal is profoundly tethered to the quality and availability of water. Consider agriculture, the primary provider of our meals. Its thirst for water is insatiable, with vast swathes of our freshwater resources diverted to fields and farms (Abunnour et al., 2016; Yin et al., 2022). It's straightforward: the water that nourishes the crops eventually nourishes us. Should this water carry contaminants or pathogens, it could lead to crop diseases, reduced yields, and health risks for consumers, threatening the stability of food supply chains.

9.3.2 THE IMPERATIVE OF WATER IN PUBLIC HEALTH

But water's tale doesn't end in the fields. It's the lifeblood of communities, critical for sanitation, personal hygiene, and quenching thirst. Every drop of polluted water carries with it the specter of disease. To illustrate, waterborne diseases remain one of the most significant challenges to global health, exacting a heavy toll on lives lost, especially among the young (Ramírez-Castillo et al., 2015; Manetu & Karanja, 2021). Clean water's pivotal role in underpinning public health is non-negotiable.

9.3.3 The Synergy and its Consequences

To truly grasp the interconnectedness of food security and public health, one must appreciate the omnipresence of water. It's a cascading effect: an interruption in water supply could jeopardize agriculture, leading to diminished yields. The ripple effect? Escalating food prices and, consequently, nutritious food slipping out of reach for many, particularly those precariously perched on the economic ladder. Concurrently, water scarcity complicates sanitation, opening the floodgates to health complications.

Furthermore, we find ourselves in a world where the climate doesn't always play by predictable rules. Droughts, erratic rainfall, and other climate-induced anomalies amplify the challenges, underscoring the urgent need for adept water management. The repercussions of neglecting this are clear: compromised food production, water quality, and public health.

What becomes clear is that our strategies need a cohesive vision. Efforts directed towards food security should be intrinsically aware of water management nuances. Similarly, public health initiatives should account for water and food security dynamics. The path forward may be championing sustainable agricultural practices, like drip irrigation or rainwater harvesting. Such methods promise sustained yields, even when water is scarce, bolstering food security. In tandem, they lessen the burden on freshwater resources, a boon for public health.

Additionally, forging collaborations spanning water managers, agricultural experts, and public health advocates can be a game-changer. Such convergences can pave the way for pioneering solutions that tackle challenges holistically rather than in isolation. It can catalyze the creating and effective implementation of policies addressing food, water, and health challenges.

Our world is a web of interdependencies, and compartmentalized solutions might not always fit the bill. Recognizing the deep ties binding food security, public health, and water management is a clarion call for holistic strategies. As we navigate the myriad challenges of our times, it's this integrated vision that might hold the key to a sustainable, healthy, and food-secure future for all (Wichelns, 2015; Linderhof et al., 2021).

9.4 ADVANCED AND TRADITIONAL WATER REMEDIATION TECHNIQUES: A COMPARATIVE ANALYSIS

The quest for clean water has been a perennial concern for humanity. With the rising water quality challenges and their repercussions on food production systems, there has been a discernible emphasis on developing and deploying traditional and advanced water remediation techniques. This segment provides a comparative analysis of these techniques, evaluating their efficacy, limitations, and advances.

9.4.1 Traditional Water Remediation Techniques

a. **Boiling:** Boiling is one of the oldest and simplest water disinfection methods. By heating water to its boiling point, most pathogens are rendered inactive.
   Advantages:
   - Effective against a majority of pathogens.
   - No chemical residues were left in the treated water.
   Limitations:
   - Not efficient against chemical pollutants.
   - Requires a significant energy input, especially when treating large quantities (Sharma & Bhattacharya, 2017).
b. **Filtration:** Filtration uses porous materials to remove particulate matter physically and some microorganisms from water.

Advantages:
- Effective for turbid water sources.
- It can be used in combination with other treatment techniques.

Limitations:
- Does not eliminate dissolved chemical pollutants.
- Filters require regular maintenance and replacement to ensure efficacy (Cescon & Jiang, 2020).

c. **Chlorination:** Adding chlorine to water neutralizes microbial pathogens, making the water safe for consumption.

Advantages:
- Effective against a wide range of pathogens.
- Residual chlorine provides prolonged protection against possible contamination.

Limitations:
- It can produce harmful by-products when chlorine reacts with organic material in water.
- Taste and odor concerns.
- Ineffective against certain protozoa and chemical pollutants (Hashmi et al., 2009).

9.4.2 ADVANCED WATER REMEDIATION TECHNIQUES

a. **Advanced Oxidation Processes (AOPs):** AOPs involve generating highly reactive radicals, especially hydroxyl radicals, which can degrade various pollutants.

Advantages:
- Effective against many organic and some inorganic pollutants.
- Can achieve complete mineralization of contaminants.

Limitations:
- Typically require external energy input (e.g., UV light).
- It can be more expensive than traditional methods on a large scale (Nicholas, 2019).

b. **Nanotechnology-Based Water Treatment:** Using nanoparticles, especially metal-oxide nanoparticles, pollutants can be adsorbed or broken down at the molecular level. The following table summarizes the general uses of nanoparticles in the treatment and decontamination of water in various industries (Table 9.1).

TABLE 9.1

Nanotechnology-Based Water Treatment Pathways

Nanoparticle	Nanotechnology-Based Pathways	Applications	Reference
Carbon nanotubes	Adsorption	Removal of heavy metals and ions in wastewater	Jain et al. (2021)
Glutaraldehyde cross-linked magnetic chitosan nanoparticles	Biosorption	Removal of industrial dyes from industrial effluent and pesticides from agricultural wastewater	Tarhan et al. (2019)
Nanometal oxides	Adsorption	Removal of dyes and heavy metals from wastewater from oil refineries	Jain et al. (2021)
Polymer nanocomposites	Adsorption	Degradation and removal of heavy metals from oilfield wastewater	Jain et al. (2021)
Metal-organic framework nanoparticles	Adsorption	Removal of pharmaceutical waste contaminants and their metabolites from wastewater	Saroa et al. (2023)
Titanium dioxide nanoparticles	Photocatalytic oxidation	Decomposition of organic pollutants in agricultural and domestic wastewater	Yaqoob et al. (2020)

Advantages:
- High efficiency in contaminant removal.
- Can target specific pollutants.
- Offers potential for reuse and recycling of nanoparticles.

  Limitations:
- Concerns about nanoparticle release into the environment.
- Cost implications, especially for large-scale operations.
- Requires expertise and advanced infrastructure (Ajith & Rajamani, 2021).

c. **Bio-inspired Filtration:** Drawing inspiration from biological systems, these filters are designed to mimic processes like those seen in plant xylem or cellular membranes.

  Advantages:
- High specificity and efficiency.
- Potential for biodegradability, reducing environmental impact.

  Limitations:
- It might be less robust than synthetic counterparts.
- Limited scalability for industrial applications (Singh et al., 2022).

d. **Electrochemical Purification:** Using electrodes, contaminants can be adsorbed, precipitated, or degraded electrochemically.

  Advantages:
- Effective against diverse contaminants, including heavy metals and some organic pollutants.
- Energy-efficient as compared to some other advanced techniques.

  Limitations:
- Electrode material and maintenance can be expensive.
- Incomplete degradation may lead to secondary pollutants (Chaplin, 2018).

9.4.3 COMPARATIVE ANALYSIS

While both traditional and advanced techniques have merits, the choice often depends on the context. For instance, boiling or chlorination might be preferred in resource-limited settings due to their simplicity and cost-effectiveness. On the other hand, for treating industrial wastewater laden with complex pollutants, AOPs or nanotechnology might be more suitable. Cost is a pivotal factor. Traditional methods, being time-tested, often have established infrastructure and are more accessible. Conversely, many advanced techniques, despite their higher efficiency, come with steeper initial setup costs and may require specialized expertise (Dores et al., 2012). Efficacy is another crucial consideration. While traditional methods, especially chlorination, are adept at neutralizing many pathogens, they falter against certain protozoa and chemical pollutants. Advanced techniques, especially AOPs, and nanotechnology-based treatments, fill these gaps by tackling a broader spectrum of contaminants (Adeleye et al., 2016; Cardoso et al., 2021) (Table 9.2).

Environmental impact is an emergent concern. While techniques like chlorination pose risks due to by-products, matters related to nanoparticle release or energy consumption in some advanced methods must also be acknowledged. Incorporating scalability and adaptability, traditional techniques, being widely used, have demonstrated scalability. In contrast, while advanced techniques show immense promise, scaling them up, especially in diverse global settings, remains challenging (Bera et al., 2022). Water remediation, be it traditional or advanced, serves the essential purpose of ensuring water quality. As we grapple with increasingly complex water quality challenges, innovating and judiciously deploying techniques best suited to specific contexts is imperative. An integrated approach, harnessing the strengths of both traditional and advanced methods, can pave the way for a future where clean water remains an accessible resource for all, supporting robust food systems and safeguarding public health.

Ultimately, while both traditional and modern approaches to decontaminating wastewater co-exist, the modern and more advanced approaches are more widely used to treat wastewater. Nanotechnology is particularly effective and commonly used in decontaminating industrial wastewater and wastewater from the pharmaceutical, agricultural, food processing, and paper processing industries and wastewater from households, hospitals, and cities. The most widely used nanoparticles in the remediation of wastewater from these industries include polymeric nanoparticles, metal nanoparticles, carbon-based nanomaterials, and biopolymers, which decontaminate wastewater through different pathways, including filtration, disinfection, biosorption, and various forms of pathological control (Jain et al., 2021).

9.5 ECONOMIC AND ENVIRONMENTAL IMPLICATIONS OF ADVANCED TECHNIQUES

In today's rapidly evolving world, technological advancements in water treatment are emerging as promising solutions to address the multifaceted challenges of ensuring clean and accessible water for all. However, while these advanced techniques open doors to potential remedies, they usher in economic and environmental considerations. Unraveling the web of these implications is pivotal to ascertaining the costs and benefits of adopting such techniques on a large scale.

9.5.1 ECONOMIC IMPLICATIONS

Capital Expenditure and Return on Investment: Advanced water treatment techniques can be capital-intensive, especially those involving sophisticated equipment or materials. Nanotechnology-based water treatment or bio-inspired filtration, for instance, may require specialized infrastructure, skilled labor, and periodic maintenance (Adeleye et al., 2016). However, the flip side offers a compelling narrative: in regions plagued with severe water contamination, the long-term health savings and improved productivity resulting from clean water access can significantly offset initial costs.

Operational Costs: While some advanced techniques may present higher upfront costs, their operating expenses might be competitive, if not lower, than traditional methods. For instance, specific electrochemical purification systems might be energy efficient in the long run, resulting in cost savings (Alkhadra et al., 2022).

Market Opportunities: As demands for clean water surge globally, technologies that promise effective water treatment solutions are carving a niche in the market. The commercialization of these technologies can catalyze economic growth, drive innovation, and create employment opportunities (Singh et al., 2018).

9.5.2 ENVIRONMENTAL IMPLICATIONS

Resource Consumption: The production and operation of advanced water treatment technologies might involve consuming materials and energy. These resources' sourcing, processing, and transportation entail an environmental footprint. It's essential to weigh the environmental costs against the benefits of cleaner water.

By-products and Waste: Techniques like advanced oxidation processes or specific nanotechnology applications might produce results. Some of these could be benign, while others might necessitate careful disposal to prevent environmental harm (Cardoso et al., 2021).

Sustainability and Lifecycle Assessment: A comprehensive understanding of the environmental impact of any advanced technique necessitates a lifecycle assessment. This considers every phase, from raw material extraction to the technology's eventual decommissioning. Such reviews offer insights into the technology's overall sustainability.

TABLE 9.2
Different Types of Nanoparticles and Their Application Area in Water Treatment and Remediation

Application Area	Nanoparticle Type	Example	Mechanism of Action	Reference
Pharmaceutical treatment	Magnetite (Fe_3O_4)	Removal of antibiotics (e.g., Ciprofloxacin) and NSAIDs (e.g., Ibuprofen) from wastewater	Adsoprtion and magnet-assisted removal	Tatarchuk et al. (2023); Lin and Lee (2020); Salem Attia et al. (2013)
Antibiotic-resistant bacteria elimination	Silver (AgNPs)	Treating wastewater contaminated with bacterial DNA conveying antibiotic-resistant genes	Antimicrobial activity, inhibiting bacterial growth	Ezeuko et al. (2022); Nõlvak et al. (2023)
Heavy Metals	Iron nanoparticles	Lead removal from wastewater	Adsorption and conversion to non-toxic forms	Nizamuddin et al. (2019); Mensah et al. (2021)
Bacteria removal	Carbon nanotubes	Filtration of *E. coli* from water sources	Physical filtration and bacterial adherence	Arora and Attri (2020)
Removal or organic pollutants	Silica nanoparticles	Removal of dyes from water	Degradation of dye molecules	Jadhav et al. (2019); Alswieleh (2022)
Oil spill cleanup	Zinc Ferrite ($ZnFe_2O_4$) magnetic nanoparticles	Removal of crude oil spills from contaminated water	Adsorption of hydrophobic substances	Amar et al. (2021)

Ecological Footprint Reduction: On a brighter note, several advanced water treatment technologies are tailored to minimize ecological footprints. Bio-inspired filtration, for instance, draws inspiration from natural processes, aiming to merge technological efficacy with ecological harmony (Wu et al., 2022; Singh et al., 2022).

In synthesis, the economic and environmental ramifications of advanced water treatment techniques are as multifaceted as the technologies themselves. While they beckon with the allure of potent solutions to our water woes, they also demand meticulous scrutiny. Adopting a holistic lens is imperative, balancing immediate needs with long-term consequences. We can chart a course toward sustainable water management solutions by acknowledging and addressing these intertwined implications.

9.6 POLICY, REGULATION, & THE PATH FORWARD

Navigating the landscape of water treatment in our modern era requires integrating advanced technology with sound policy and regulation. History has shown that effective policies can foster innovation while ensuring equitable access and public safety.

9.6.1 THE SIGNIFICANCE OF POLICY IN WATER TREATMENT

Policies establish overarching standards and objectives, particularly in sectors as vital as water treatment. Here are several ways they shape our approach to this fundamental resource:

1. **Equity in Access:** One of the longstanding challenges surrounding water has been guaranteeing equal access for all. For instance, the United Nations' Sustainable Development Goal 6 aims to ensure the availability and sustainable management of water and sanitation

for everyone by 2030. Such global commitments drive national and local initiatives to direct resources and innovation towards underserved regions (Mondejar Our World in Data team, 2023).

2. **Safety Standards:** As we introduce new water treatment technologies, policies set the benchmarks for water purity. The US Safe Drinking Water Act (SDWA), for instance, authorizes the Environmental Protection Agency (EPA) to set national health-based standards for drinking water to protect against both naturally occurring and manufactured contaminants (Tiemann, 2014).

3. **Economic Strategies:** With financial considerations ever-present, policies can provide monetary incentives for adopting advanced water treatment techniques. The European Union's Horizon 2020, for instance, had provisions that promoted research and innovation in water treatment sectors, enabling economic growth while ensuring environmental safety (Arsić et al., 2015).

9.6.2 The Necessity of Vigilance in Regulation

Effective regulation is a safeguard against potential risks and ensures that innovations are both practical and safe:

1. **Environmental Oversight:** While technological advancements are promising, they may also have unforeseen ecological implications. Similar to the EU's Water Framework Directive, regulatory frameworks emphasize holistic water management, protecting ecosystems while accounting for economic considerations (Grizzetti et al., 2016).

2. **Consumer Safety:** Newer treatment methodologies may introduce new risks. Regulations, like Australia's Guidelines for Fresh and Marine Water Quality, provide comprehensive guidelines for ensuring the safety of both the environment and consumers against potential contaminants and treatment residues (Authority, 1993).

3. **Economic Fairness:** With the burgeoning market around advanced water treatments, financial regulations can prevent monopolistic tendencies. Bodies such as the Competition and Markets Authority in the UK work to prevent anti-competitive activities, ensuring a fair marketplace for water treatment solutions (Bakker, 2005).

9.6.3 Charting the Path Forward: Merging Technology and Governance

The future holds promise, but harnessing it requires a harmonious blend of technological advancements and insightful governance:

1. **Collaborative R&D:** Joint ventures between governments, academia, and the private sector can amplify innovation. The US EPA's partnerships with various organizations to promote research and share knowledge on water treatment is a case in point (EPA, 2023).

2. **Public Awareness Initiatives:** Proactive campaigns can educate the populace about water quality and the significance of novel treatment techniques. The World Health Organization's efforts in raising awareness about water-related diseases serve as an exemplary model (World Health Organization, 2019).

3. **Emphasizing Infrastructure:** Infrastructure is the backbone of any policy implementation. The Asian Development Bank's investments in water infrastructure across Asia aim to bring state-of-the-art water treatment facilities to regions previously lacking in resources (Biswas & Seetharam, 2008).

4. **Iterative Policy Development:** The dynamic nature of water treatment means policies must evolve. Feedback-driven adjustments, like those seen in Canada's adaptive management approach to water resources, ensure policies stay relevant and effective (Gilbert, 2022).
5. **Global Collaborations:** Collective global efforts, such as the International Water Association's initiatives, harness worldwide expertise and resources, emphasizing collaborative solutions to universal water challenges (International Water Association, n.d.)

9.7   CONCLUSION

As underscored throughout this discourse, water is a basic necessity and a linchpin of sustainable development and human prosperity. Its intrinsic value transcends beyond quenching thirst, holding within its molecular bonds the story of civilizations, ecosystems, and industries. Beyond its role in quenching our thirst, water tells tales of human endeavor, resilience, and ingenuity. The confluence of water quality with food security and public health is emblematic of a resilient society. Our mission to guarantee its sanctity and ubiquity goes beyond engineering feats—it's a moral and profound commitment.

Recent decades have witnessed a crescendo in concerns about water quality, further fueled by burgeoning populations, rapid industrial growth, urban sprawls, and the multifaceted repercussions of climate change. These have imbued the longstanding quest for clean water with layers of intricacy. Responding to these challenges, time-honored and contemporary water treatment methods have emerged, each with distinctive strengths, constraints, and nuances. Human creativity and adaptability are evidenced in the vast spectrum of water purification techniques, ranging from the rudimentary act of boiling to the sophisticated realm of nanotechnology. Yet, as we teeter on the precipice of what might be deemed a water renaissance, it's imperative to grasp that technology in isolation isn't the magic bullet. Integrating innovations seamlessly with robust policies, stringent oversight, and a keen insight into socio-economic realities remains essential. The intricate balance of innovation with its attendant economic and environmental ramifications can't be overemphasized. Even as we laud modern techniques' solutions, we must remain vigilant in assessing their feasibility, cost-effectiveness, and long-term sustainability. And as our discourse on policy and regulation has underscored, the tapestry of water treatment is woven with threads far beyond mere scientific and technical strands—it's deeply entwined with governance, justice, and vision.

Navigating the path to universally accessible clean water, integrated seamlessly into sustainable food mechanisms, is multifaceted. It demands both immediacy in action and sustained, collaborative efforts over time. History bears witness to the rise and fall of societies mainly based on their stewardship of water resources. Yet, equipped with modern-day knowledge, cutting-edge technology, and a spirit of cooperation, our generation possesses a unique advantage in addressing this challenge.

In conclusion, water, often dubbed life's elixir, remains unparalleled in its significance. Our endeavors to ensure its purity and widespread availability aren't mere technical exercises—they epitomize our dedication to life, enduring sustainability, and a collective vision for our world. Our decisions and the steps we take today will influence water quality and, by extension, the heritage we bequeath to those who follow in our footsteps.

BIBLIOGRAPHY

Abunnour, M. A., Hashim, N. B. M., & Jaafar, M. B. (2016). Agricultural water demand, quality and crop suitability in Souk-Alkhamis Al-Khums, Libya. In *IOP Conference Series: Earth and Environmental Science* (Vol. 37, No. 1, p. 012045). IOP Publishing.

Adeleye, A. S., Conway, J. R., Garner, K., Huang, Y., Su, Y., & Keller, A. A. (2016). Engineered nanomaterials for water treatment and remediation: Costs, benefits, and applicability. *Chemical Engineering Journal*, 286, 640–662. https://doi.org/10.1016/j.cej.2015.10.105.

Afrad, M. S. I., Monir, M. B., Haque, M. E., Barau, A. A., & Haque, M. M. (2020). Impact of industrial effluent on water, soil and rice production in Bangladesh: A case of Turag River Bank. *Journal of Environmental Health Science and Engineering*, 18, 825–834. https://doi.org/10.1007%2Fs40201-020-00506-8.

Ajith, M. P., & Rajamani, P. (2021). Nanotechnology for water purification-current trends and challenges. *Journal of Nanotechnology and Nanomaterials*, 2(2), 88–91. https://doi.org/10.33696/Nanotechnol.2.025.

Alkhadra, M. A., Su, X., Suss, M. E., Tian, H., Guyes, E. N., Shocron, A. N., … Bazant, M. Z. (2022). Electrochemical methods for water purification, ion separations, and energy conversion. *Chemical Reviews*, 122(16), 13547–13635. https://doi.org/10.1021/acs.chemrev.1c00396.

Alswieleh, A. M. (2022). Efficient removal of dyes from aqueous solution by adsorption on L-arginine-modified mesoporous silica nanoparticles. *Processes*, 10(6), 1079. https://doi.org/10.3390/pr10061079.

Amar, I., Faraj, S., Alsalheen, M., Abdalsamed, I., Altohami, F., & Samba, M. A. (2021). Oil spill removal from water surfaces using zinc ferrite magnetic nanoparticles as a sorbent material. *Iraqi Journal of Science*, 62, 718–728. https://doi.org/10.24996/ijs.2021.62.3.2.

Arora, B., & Attri, P. (2020). Carbon Nanotubes (CNTs): A potential nanomaterial for water purification. *Journal of Composites Science*, 4(3), 135. https://doi.org/10.3390/jcs4030135.

Arsić, S. M., Arsić, M. Ž., Tomić, R. P., Kokanović, M. M., & Mihić, M. M. (2015). Key aspects of managing an innovation project in the EU framework programme horizon 2020. *Journal of Process Management and New Technologies*, 3(1), 101–107.

Bakker, K. (2005). Neoliberalizing nature? Market environmentalism in water supply in England and Wales. *Annals of the Association of American Geographers*, 95(3), 542–565. https://doi.org/10.1111/j.1467-8306.2005.00474.x.

Bashir, I., Lone, F. A., Bhat, R. A., Mir, S. A., Dar, Z. A., & Dar, S. A. (2020). Concerns and threats of contamination on aquatic ecosystems. *Bioremediation and Biotechnology: Sustainable Approaches to Pollution Degradation*, 1–26. https://doi.org/10.1007%2F978-3-030-35691-0_1.

Bera, S. P., Godhaniya, M., & Kothari, C. (2022). Emerging and advanced membrane technology for wastewater treatment: A review. *Journal of Basic Microbiology*, 62(3–4), 245–259. https://doi.org/10.1002/jobm.202100259.

Bhagwat, V. R. (2019). Safety of water used in food production. In *Food Safety and Human Health* (pp. 219–247). Academic Press. https://doi.org/10.1016/B978-0-12-816333-7.00009-6.

Biswas, A. K., & Seetharam, K. E. (2008). Achieving water security for Asia: Asian water development outlook, 2007. *International Journal of Water Resources Development*, 24(1), 145–176. https://doi.org/10.1080/07900620701760556.

Borsato, E., Rosa, L., Marinello, F., Tarolli, P., & D'Odorico, P. (2020). Weak and strong sustainability of irrigation: A framework for irrigation practices under limited water availability. *Frontiers in Sustainable Food Systems*, 4, 17. https://doi.org/10.3389/fsufs.2020.00017.

Cardoso, I. M., Cardoso, R. M., & da Silva, J. C. E. (2021). Advanced oxidation processes coupled with nanomaterials for water treatment. *Nanomaterials*, 11(8), 2045. https://doi.org/10.3390%2Fnano11082045.

Cescon, A., & Jiang, J. Q. (2020). Filtration process and alternative filter media material in water treatment. *Water*, 12(12), 1–20. https://doi.org/10.3390/w12123377.

Chaplin, B. P. (2018). Advantages, disadvantages, and future challenges of the use of electrochemical technologies for water and wastewater treatment. In *Electrochemical Water and Wastewater Treatment* (pp. 451–494). Butterworth-Heinemann. https://doi.org/10.1016/B978-0-12-813160-2.00017-1.

Dores, R., Hussain, A., Katebah, M., & Adham, S. (2012). Using advanced water treatment technologies to treat produced water from the petroleum industry. In *SPE International Production and Operations Conference & Exhibition*. OnePetro. https://doi.org/10.2118/157108-MS.

Doyle, M. E M., Archer, J, Kaspar, C. W., & Weiss, R. (2006). Human illness caused by E. coli O157:H7 from Food and Non-food Sources. *Food Research Institute, UW-Madison*. Retrieved from https://fri.wisc.edu/docs/pdf/FRIBrief_EcoliO157H7humanillness.pdf. Accessed on September 20, 2023.

Environmental Protection Authority. (1993). Western Australian water quality guidelines for fresh and marine waters. EPA Bulletin, 711.

EPA. (2023). Partnerships with the EPA. *EPA*. Retrieved from https://www.epa.nsw.gov.au/working-together/partnerships-with-the-epa. Accessed on September 22, 2023.

Ezeuko, A. S., Ojemaye, M. O., Okoh, O. O., & Okoh, A. I. (2022). The effectiveness of silver nanoparticles as a clean-up material for water polluted with bacteria DNA conveying antibiotics resistance genes: Effect of different molar concentrations and competing ions. *OpenNano*, 7, 100060. https://doi.org/10.1016/j.onano.2022.100060.

FAO. (2017). The future of food and agriculture – Trends and challenges. Rome. Retrieved from https://www.fao.org/3/i6583e/i6583e.pdf. Accessed on September 20, 2023.

FAO & IWMI. (2017). Water pollution from agriculture: A global review. Retrieved from https://www.fao.org/3/i7754e/i7754e.pdf. Accessed September 21, 2023.

Gilbert, A. (2022). Adaptation and water resources management: Examining adaptive governance in Montana (Doctoral dissertation, Montana State University-Bozeman, College of Letters & Science).

Grizzetti, B., Liquete, C., Antunes, P., Carvalho, L., Geamănă, N., Giucă, R., … Woods, H. (2016). Ecosystem services for water policy: Insights across Europe. *Environmental Science & Policy*, 66, 179–190. https://doi.org/10.1016/j.envsci.2016.09.006.

Hashmi, I., Farooq, S., & Qaiser, S. (2009). Chlorination and water quality monitoring within a public drinking water supply in Rawalpindi Cantt (Westridge and Tench) area, Pakistan. *Environmental Monitoring and Assessment*, 158, 393–403. https://doi.org/10.1007/s10661-008-0592-z.

Hussaini, A. B., Cao, J., Yateh, M., & Wu, B. (2023). Effects of industrial wastewater effluents on irrigation water quality around Bompai Industrial Area, Kano State, Nigeria. *Open Access Library Journal*, 10(7), 1–14. https://doi.org/10.4236/oalib.1110395.

International Water Association. What are IWA communities? Retrieved from https://iwa-network.org/iwa-groups/. Accessed on September 22, 2023.

Jadhav, S. A., Garud, H. B., Patil, A. H., Patil, G. D., Patil, C. R., Dongale, T. D., & Patil, P. S. (2019). Recent advancements in silica nanoparticle-based technologies for removal of dyes from water. *Colloid and Interface Science Communications*, 30, 100181. https://doi.org/10.1016/j.colcom.2019.100181.

Jain, K., Patel, A. S., Pardhi, V. P., & Flora, S. J. S. (2021). Nanotechnology in wastewater management: A new paradigm towards wastewater treatment. *Molecules*, 26(6), 1797. https://doi.org/10.3390/molecules26061797.

Jarvie M. (2014). History of food safety in the US – Part 1. University of Michigan. Retrieved from https://www.canr.msu.edu/news/history_of_food_safety_in_the_us_part_1. Accessed on September 20, 2023.

KC, B., Schultz, B., McIndoe, I., Rutter, H., Dark, A., Prasad, K., … Paudel, K. (2020). Impacts of dairy farming systems on water quantity and quality in Brazil, Ethiopia, Nepal, New Zealand and the USA. *Irrigation and Drainage*, 69(4), 944–955. https://doi.org/10.1002/ird.2486.

Lebelo, K., Malebo, N., Mochane, M. J., & Masinde, M. (2021). Chemical contamination pathways and the food safety implications along the various stages of food production: A review. *International Journal of Environmental Research and Public Health*, 18(11), 5795. https://doi.org/10.3390%2Fijerph18115795.

Lin, C., & Lee, C. (2020). Adsorption of ciprofloxacin in water using Fe3O4 nanoparticles formed at low temperature and high reactant concentrations in a rotating packed bed with co-precipitation. *Materials Chemistry and Physics*, 240, 122049. https://doi.org/10.1016/j.matchemphys.2019.122049.

Linderhof, V., De Lange, T., & Reinhard, S. (2021). The dilemmas of water quality and food security interactions in low-and middle-income countries. *Frontiers in Water*, 3, 736760. https://doi.org/10.3389/frwa.2021.736760.

Malakar, A., Snow, D. D., & Ray, C. (2019). Irrigation water quality-A contemporary perspective. *Water*, 11(7): 1482. https://doi.org/10.3390/w11071482.

Manetu, W. M., & Karanja, A. M. (2021). Waterborne disease risk factors and intervention practices: A review. *Open Access Library Journal*, 8(5), 1–11. https://doi.org/10.4236/oalib.1107401.

Martínez-Huitle, C. A., Rodrigo, M. A., Sirés, I., & Scialdone, O. (2023). A critical review on latest innovations and future challenges of electrochemical technology for the abatement of organics in water. *Applied Catalysis B: Environmental*, 328, 122430. https://doi.org/10.1016/j.apcatb.2023.122430.

Mensah, M. B., Lewis, D. J., Boadi, N. O., & M. Awudza, J. A. (2021). Heavy metal pollution and the role of inorganic nanomaterials in environmental remediation. *Royal Society Open Science*, 8(10), 201485. https://doi.org/10.1098/rsos.201485.

Mondejar, M. E., Avtar, R., Diaz, H. L. B., Dubey, R. K., Esteban, J., Gómez-Morales, A., … Garcia-Segura, S. (2021). Digitalization to achieve sustainable development goals: Steps towards a Smart Green Planet. *Science of the Total Environment*, 794, 148539. https://doi.org/10.1016/j.scitotenv.2021.148539.

Nagar, A., & Pradeep, T. (2020). Clean water through nanotechnology: Needs, gaps, and fulfillment. *ACS Nano*, 14(6), 6420–6435. https://doi.org/10.1021/acsnano.9b01730.

National Research Council, & Safe Drinking Water Committee. (1977). *Drinking Water and Health: Volume 1*. The National Academies Press.

Nicholas, N. (2019). Benefits and disadvantages of the advanced oxidation process. *Genesis Water Technologies*. Retrieved from https://www.wateronline.com/doc/benefits-and-disadvantages-of-the-advanced-oxidation-process-0001. Accessed on September 22, 2023.

Nizamuddin, S., Siddiqui, M., Mubarak, N., Baloch, H. A., Abdullah, E., Mazari, S. A., Griffin, G., Srinivasan, M., & Tanksale, A. (2019). Iron oxide nanomaterials for the removal of heavy metals and dyes from wastewater. *Nanoscale Materials in Water Purification*, 447–472. https://doi.org/10.1016/B978-0-12-813926-4.00023-9.

Nõlvak, H., Truu, M., Tiirik, K., Devarajan, A. K., Peeb, A., & Truu, J. (2023). The effect of synthetic silver nanoparticles on the antibiotic resistome and the removal efficiency of antibiotic resistance genes in a hybrid filter system treating municipal wastewater. *Water Research*, 237, 119986. https://doi.org/10.1016/j.watres.2023.119986.

Okorogbona, A. O., Denner, F. D., Managa, L. R., Khosa, T. B., Maduwa, K., Adebola, P. O., ... Macevele, S. (2018). Water quality impacts on agricultural productivity and environment. *Sustainable Agriculture Reviews*, 27, 1–35. https://doi.org/10.1007/978-3-319-75190-0_1.

Our World in Data Team. (2023). Ensure access to water and sanitation for all. *Our World in Data*. Retrieved from 'https://ourworldindata.org/sdgs/clean-water-sanitation. Accessed on September 22, 2023.

Ramírez-Castillo, F. Y., Loera-Muro, A., Jacques, M., Garneau, P., Avelar-González, F. J., Harel, J., & Guerrero-Barrera, A. L. (2015). Waterborne pathogens: Detection methods and challenges. *Pathogens*, 4(2), 307–334. https://doi.org/10.3390%2Fpathogens4020307.

Rather, I. A., Koh, W. Y., Paek, W. K., & Lim, J. (2017). The sources of chemical contaminants in food and their health implications. *Frontiers in Pharmacology*, 8, 830. https://doi.org/10.3389%2Ffphar.2017.00830.

Ringler, C., Agbonlahor, M., Baye, K., Barron, J., Hafeez, M., Lundqvist, J., ... Uhlenbrook, S. (2023). Water for food systems and nutrition. *Science and Innovations for Food Systems Transformation*, 497. https://doi.org/10.1007/978-3-031-15703-5_26.

Salem Attia, T. M., Hu, X. L., & Yin, D. Q. (2013). Synthesized magnetic nanoparticles coated zeolite for the adsorption of pharmaceutical compounds from aqueous solution using batch and column studies. *Chemosphere*, 93(9), 2076–2085. https://doi.org/10.1016/j.chemosphere.2013.07.046.

Sarkar, S., Gil, J. D. B., Keeley, J., & Jansen, K. (2021). The use of pesticides in developing countries and their impact on health and the right to food. European Union. https://doi.org/10.2861/28995.

Saroa, A., Singh, A., Jindal, N., Kumar, R., Singh, K., Guleria, P., & Kumar, V. (2023). Nanotechnology-assisted treatment of pharmaceuticals contaminated water. *Bioengineered*, 14(1), 2260919. https://doi.org/10.1080/21655979.2023.2260919.

Sharma, S., & Bhattacharya, A. J. A. W. S. (2017). Drinking water contamination and treatment techniques. *Applied Water Science*, 7(3), 1043–1067. https://doi.org/10.1007/s13201-016-0455-7.

Singh, D., Goswami, R. K., Agrawal, K., Chaturvedi, V., & Verma, P. (2022). Bio-inspired remediation of wastewater: A contemporary approach for environmental clean-up. *Current Research in Green and Sustainable Chemistry*, 5, 100261. https://doi.org/10.1016/j.crgsc.2022.100261.

Singh, J., Saharan, V., Kumar, S., Gulati, P., & Kapoor, R. K. (2018). Laccase grafted membranes for advanced water filtration systems: A green approach to water purification technology. *Critical Reviews in Biotechnology*, 38(6), 883–901. https://doi.org/10.1080/07388551.2017.1417234.

Tarhan, T., Tural, B., Boga, K., & Tural, S. (2019). Adsorptive performance of magnetic nano-biosorbent for binary dyes and investigation of comparative biosorption. *SN Applied Sciences*, 1, 1–11. https://doi.org/10.1007/s42452-018-0011-1.

Tatarchuk, T., Soltys, L., & Macyk, W. (2023). Magnetic adsorbents for removal of pharmaceuticals: A review of adsorption properties. *Journal of Molecular Liquids*, 384, 122174. https://doi.org/10.1016/j.molliq.2023.122174\.

Thompson, W. R. (2016). Climate, water, and political-economic crises in Ancient Mesopotamia and Egypt 1. In *The World System and the Earth System* (pp. 163–179). Routledge. https://doi.org/10.4324/9781315416854.

Tiemann, M. (2014). Safe drinking water act (SDWA): A summary of the act and its major requirements (pp. 7–5700). Washington, DC: Congressional Research Service.

Tudi, M., Daniel Ruan, H., Wang, L., Lyu, J., Sadler, R., Connell, D., ... Phung, D. T. (2021). Agriculture development, pesticide application and its impact on the environment. *International Journal of Environmental Research and Public Health*, 18(3), 1112. https://doi.org/10.1007%2Fs40201-020-00506-8.

Umar, S., Munir, M. T., Azeem, T., Ali, S., Umar, W., Rehman, A., & Shah, M. A. (2014). Effects of water quality on productivity and performance of livestock: A mini review. *Veterinaria*, 2(2), 11–15.

Vilakazi, N., Nyirenda, K., & Vellemu, E. (2019). Unlocking water issues towards food security in Africa. *Food Security in Africa*, 1–17. https://doi.org/10.5772/intechopen.86788.

Wang, Z., Wang, S., Liu, W., Su, Q., Tong, H., Xu, X., ... Liu, J. (2020). Hydrochemical characteristics and irrigation suitability evaluation of groundwater with different degrees of seawater intrusion. *Water*, 12(12), 3460. https://doi.org/10.3390/w12123460.

Wichelns, D. (2015). Achieving water and food security in 2050: Outlook, policies, and investments. *Agriculture*, 5(2), 188–220. https://doi.org/10.3390/agriculture5020188.

World Health Organization. (2019). Safer Water, Better Health. WHO. Retrieved from https://iris.who.int/bitstream/handle/10665/329905/9789241516891-eng.pdf. Accessed on September 22, 2023.

World Health Organization. (2021). Towards stronger food safety systems and Global Cooperation. World Health Organization, 2022–2030.

Wu, S., Shi, W., Li, K., Cai, J., & Chen, L. (2022). Recent advances on sustainable bio-based materials for water treatment: Fabrication, modification and application. *Journal of Environmental Chemical Engineering*, 10(6), 108921. https://doi.org/10.1016/j.jece.2022.108921.

Yaqoob, A. A., Parveen, T., Umar, K., & Mohamad Ibrahim, M. N. (2020). Role of nanomaterials in the treatment of wastewater: A review. *Water*, 12(2), 495. https://doi.org/10.3390/w12020495.

Yin, L., Tao, F., Chen, Y., & Wang, Y. (2022). Reducing agriculture irrigation water consumption through reshaping cropping systems across China. *Agricultural and Forest Meteorology*, 312, 108707. https://doi.org/10.1016/j.agrformet.2021.108707.

10 Nanotechnology in Food Processing

Ahmed Raza Sidhu, Sarfaraz Ahmed Mahesar,
Zahid Hussain Laghari, Hadia Shoaib, and
Syed Tufail Hussain Sherazi

10.1 INTRODUCTION

The food industry is a worldwide network involved in the distribution of food products on a global scale. The food industry encompasses a range of industrial activities that involve the production, processing, conversion, distribution, preservation, and processing of various food products. Similarly, the proper storage and secure distribution of processed food products are crucial, as many food items are lost during these stages (Onyeaka et al., 2022). Chemical preservatives and outdated processing techniques reduce the food industry's profits, pose serious health risks, and impair the food's nutritional content. The proliferation of health-related assertions and counterarguments on particular food items can be attributed to the advent of contemporary food processing techniques. Determining the reliability of various food sources can often pose a challenge. Nutrient loss can occur during the processing and subsequent storage of food. The degradation of essential nutrients can occur due to various processing methods and their respective intensities (Kumar et al., 2021). In addition, plastic containers have revolutionized our capacity to prolong food longevity while imposing minimal weight (a crucial factor when considering the energy required for transportation). However, the production and disposal of plastic containers incur substantial environmental costs (García-Pinilla et al., 2019). Technological advancements are imperative in the food market to maintain a dominant position in the food processing industry. These advancements enable the production of fresh, authentic, convenient, and flavorful food products while extending their shelf life and freshness. Furthermore, they enhance the overall quality of food (Mehmood et al., 2021).

Food safety is significantly emphasized due to the swift evolution of food recipes and dietary practices. Foodborne pathogens, toxins, and other contaminants pose substantial risks to human health. The conventional techniques employed for identifying pathogens and their toxins are characterized by a high degree of manual labor and a lengthy time requirement. The progress made in nanotechnology (NT) has accelerated efforts to tackle food safety concerns related to microbial contaminants. Additionally, it has led to enhancements in the detection of toxins and improved shelf-life and packaging techniques (Singh et al., 2023). NT exhibits numerous favorable effects within the food industry, encompassing lowered costs, decreased risks of pollution, disease prevention for loss mitigation, and enhanced agricultural and food chain management practices.

10.2 NANOTECHNOLOGY IN FOOD

The global food industry faces increasing pressure to satisfy consumer demands for safe, healthy, and fresh food. Additionally, it is confronted with the challenge of complying with updated and stringent food safety regulations. The ease of accessing digital media has led to increased consumer awareness, which in turn has driven the demand for food products that are fresh, nutritious, safe, least processed, ready-to-eat, and have clearly defined labels. To ensure the safety and integrity of food products along the entire food supply chain, various stakeholders such as food manufacturers,

DOI: 10.1201/9781003514039-10

traders, buyers, and regulatory authorities are actively seeking an innovative, cost-effective, efficient, and reliable method to monitor the quality of processed food (Janjarasskul & Suppakul, 2018; Sarkar et al., 2017). Numerous food scientists have acknowledged the significant potential of NT to revolutionize various sectors within the food processing industry. However, it is essential to note that the current applications of NT in this domain remain somewhat limited (Singh et al., 2017). NT has emerged as a transformative field in the food industry, offering novel opportunities and advantages. It enables the exploration of innovative flavors and sensations, the creation of unique textures in food components, the encapsulation of food additives, and the enhancement of nutritional values in various food products (Mlalila et al., 2016).

NT is a scientific discipline that works at the nanometer scale, focusing on manipulating and studying atoms, molecules, and macromolecules. Its primary objective is to develop and utilize materials with unique properties, typically ranging in size from approximately 1 to 100 nm (Naseer et al., 2018). On a large scale, it has been seen that these materials have unique properties that are different from those of their counterparts. This can be attributed to their high surface-to-volume ratio and other unique physiochemical characteristics such as color, strength, solubility, toxicity, diffusivity, and magnetic, optical, and thermodynamic properties (Rai et al., 2009; Gupta et al., 2016). According to Qureshi et al. (2012), the emergence of NT has sparked a new era of industrial transformation, attracting the attention and investment of both industrialized and developing nations. Hence, NT presents diverse prospects for advancing and utilizing novel structures, materials, or systems exhibiting unique qualities across several domains, such as agriculture, food, and medicine.

The increasing consumer apprehensions regarding food quality and its impact on health motivate researchers to explore methods to improve food quality without significantly compromising the product's nutritional content. The demand for nanoparticle-based materials in the food business has experienced an increase due to their inclusion of critical nutrients and their demonstrated non-toxicity (Roselli et al., 2003). In addition, it has been observed that they exhibit stability even under conditions of elevated temperature and pressure (Sawai, 2003). NT provides comprehensive solutions for the whole food industry, encompassing manufacturing, processing, and packaging. Nanomaterials significantly impact the quality and safety of food and the health advantages associated with its consumption. Numerous entities, including organizations, researchers, and enterprises, actively develop innovative approaches, procedures, and products that directly apply NT in food science (Dasgupta et al., 2015).

10.3 FOOD PROCESSING

Food processing can be characterized as applying various procedures and techniques to preserve and convert food into a state suitable for consumption. The strategies employed in this context are specifically formulated to protect the food's flavor and quality while safeguarding it against microbial infection, a primary cause of food spoilage (Neethirajan & Jayas, 2011). Food safety is an increasingly significant issue in the realm of public health on a global scale. The principal objective of food safety is to guarantee that food does not pose any potential harm to individuals throughout its preparation and consumption (Pal, 2017). Safeguarding food from potential physical, chemical, and biological contaminants throughout manufacturing, handling, and distribution is imperative. The food business has significantly transformed due to recent advancements in NT. These advancements have profoundly impacted several aspects of the industry, including food processing, safety, and security.

Additionally, it has made significant progress in improving the nutritional content of food, prolonging its shelf-life, and reducing packaging waste (Wesley et al., 2014). The primary objective of food processing is to preserve the integrity of the food and extend its longevity. Processed foods facilitate long-distance transportation by minimizing the possibility of spoilage, benefiting the producer. Including micronutrients in processed food has become a significant aspect of addressing the objective of promoting healthier food options, providing substantial advantages to consumers.

Food processing encompasses a diverse range of unit processes employed to transform the raw materials into the final product. Consequently, the utilization of NT-based solutions can significantly enhance numerous methods. Based on process technology, foodstuff NT can be utilized in several applications, such as packaging, antimicrobials, and food components. These applications can be categorized as "direct" or "indirect" uses. Direct applications encompass the incorporation of nanostructured materials into the food matrix and the detection of their existence. Other applications include aromas, preservatives, coloring agents, antioxidants, and bioactive constituents such as omega-3 fatty acids, polyphenols, vitamins, and diverse food components. According to McClements and Xiao (2012), the indirect applications encompass utilizing nanostructured particles through innovative packaging techniques for lipid hydration or implementing meticulously nanostructure catalysts. The field of NT has presented novel approaches to food processing, which aim to enhance the physicochemical properties of food and promote the stability and availability of nutrients (Moraru et al., 2009; Omerović et al., 2021). The utilization of NT in food systems has yielded a diverse range of innovative products that exhibit enhanced features related to food quality, including texture, taste, sensory aspects, and stability (Anandharamakrishnan & Parthasarathi, 2019). Scientists and business organizations, particularly those in agriculture and food processing, have recognized NT as a viable application across various domains (Figure 10.1) within the food sector (Pathakoti et al., 2017; Sahoo et al., 2021).

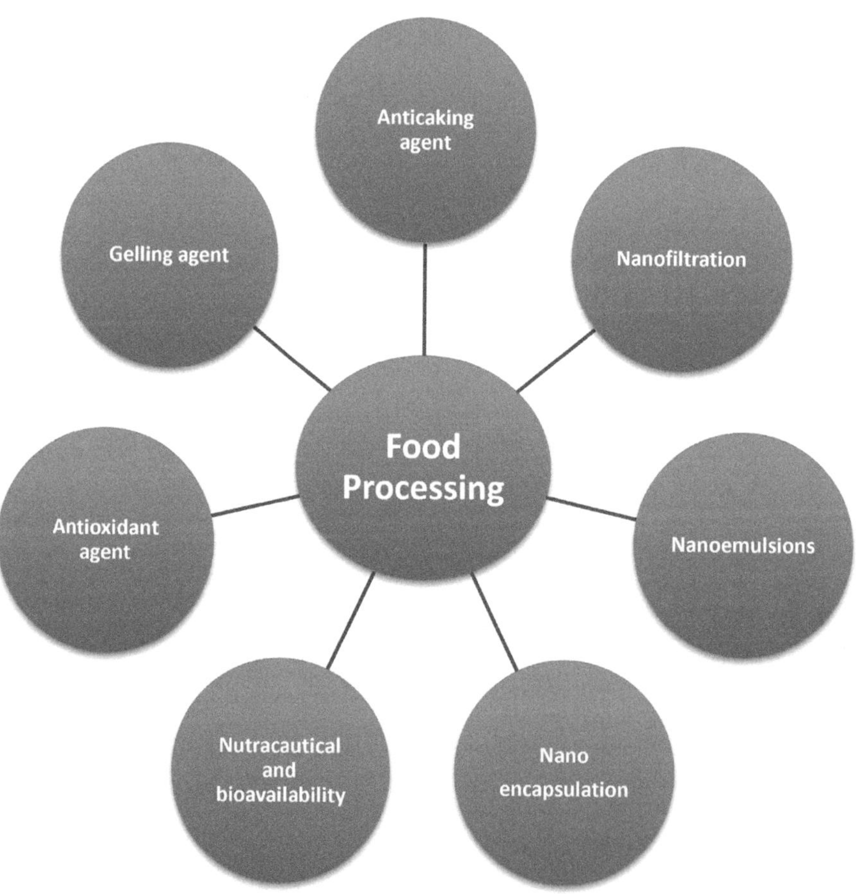

FIGURE 10.1 Applications of nanotechnology in the field of food science.

10.4 ANTICAKING AGENTS

The caking phenomenon in food powders commonly arises during their processing, handling, and storage stages. This occurrence can detrimentally impact the quality and functionality of products due to the creation of clumps and clusters. Additionally, caking adversely influences the ability of food to rehydrate and disperse, leading to degradation in its sensory characteristics and a reduction in shelf life (Cozzolino et al., 2016; Fitzpatrick et al., 2010). The initiation of caking in crystalline ingredients due to environmental moisture involves forming liquid bridges that connect the crystals. This is caused by either partial deliquescence or capillary condensation (Chen et al., 2018). Anticaking agents function through multiple mechanisms, which include competing with the host powder for moisture, is forming moisture-resistant layers on particle surfaces, providing smooth surfaces to minimize friction between particles, and impeding crystal growth (Belton et al., 2019). Anticaking additives are frequently employed in powder processes to delay or prevent caking, yet there is limited understanding of their influence on powdered substances' structural and chemical durability. No specialized analytical approach is available for assessing the efficacy of anticaking compounds. This challenge stems from numerous independent variables impacting powders' caking propensity, characteristics of the base powder, temperature, moisture, etc. Furthermore, additives like calcium silicate, calcium carbonate, aluminum silicate, calcium phosphate (E341), magnesium compounds (E340), and various others are utilized in both granular and powdered food products to inhibit caking (Pui et al., 2020; Akins, 2002).

10.4.1 CALCIUM SILICATE

Calcium silicate is a frequently used anticaking agent in sodium chloride and other substances. It demonstrates the capacity to absorb hydrophobic (oil) and hydrophilic (water) substances. Despite their classification as food additives, anticaking agents offer a range of additional functionalities (Yapıcı et al., 2021). Numerous anticaking compounds can absorb excess moisture or envelop particles in a water-repellent layer. These additives, available in powdered and granulated forms, are frequently used across diverse food items, a notable example being calcium silicate. In powdered form, it is employed in various things such as eggs, soups, *S. cerevisiae* (yeast), spices, flavored salts, vending machine ingredients (like cocoa, dairy, and cream), sauces, confectionery, and shredded cheese, among others (EFSA, 2011).

10.4.2 CALCIUM CARBONATE

Calcium carbonate is a versatile compound widely utilized in the food industry due to its multifunctional properties. It is derived from natural sources such as limestone, marble, and chalk and is recognized as a safe and effective food ingredient. It finds application as an anticaking agent in powdered or granulated food items, effectively preventing particles' undesirable aggregation or adhesion. This preserves the smooth, free-flowing texture of products such as powdered spices, baking mixes, and instant beverages. It's commonly added to dairy alternatives, cereals, and nutritional supplements to enhance their calcium content (EFSA, 2016).

10.4.3 ALUMINUM SILICATE

Precipitation of soluble aluminum salts with a compatible metal is a standard method for producing aluminum silicates. These minerals are finely ground powders utilized within the food industry as free-flowing additives. Compared to other anticaking agents, aluminum silicates are frequently used due to their cost-effectiveness and capacity to enhance the flow of powdered substances (Pandey et al., 2012). A powdery substance characterized by its smooth, white, and free-flowing nature is known as sodium aluminum silicate, containing a SiO_2 concentration that ranges from 66 to 88 g/100 g. Another frequently utilized variation is potassium aluminum silicate, which is even more widespread than its sodium counterpart. Furthermore, potassium aluminum silicate is used to lower sodium content in products whenever necessary (Afzaal et al., 2022).

10.4.4 PHOSPHATE OF CALCIUM E341 AND MAGNESIUM E340

Tri-calcium phosphate (E341) is a versatile food additive within the food industry and serves multiple purposes. Its prominent roles include acting as a calcium source, a buffering agent, and a texturizer. This additive commonly applies to dairy alternatives, beverages, and baked goods. Notably, its significance extends to processed meats, canned vegetables, and sauces. Similarly, magnesium compounds, particularly magnesium hydroxide, find utility as pH regulators in food products. They play a pivotal role in adjusting and controlling the acidity or alkalinity of various food and beverage items. This function holds particular importance in sectors like confectionery, baking, and dairy products (Silva & Lidon, 2016).

10.5 ANTICAKING AGENTS IN FOOD NANOTECHNOLOGY

10.5.1 POLYDIMETHYLSILOXANE

Extensive industrial utilization of the polymer polydimethylsiloxane (PDMS) has spurred broad research into composite materials enhanced with nanoscale inorganic fillers. Incorporating minute quantities of nano-oxides from transition metals or rare earth metals into polymers is acknowledged for enhancing specific attributes such as thermal resilience, conductivity, visual appeal, hydrophobicity, and interfacial behavior. A prime example of such a composite is PDMS/silica, a nanocomposite polymer blend currently manufactured and employed on a large scale (Clarson, 2010; Kulyk et al., 2015). Notably, PDMS/silica plays a crucial role as an anticaking agent in confectionery and flour products and an antifoaming agent in edible oils within food-related processes. The customarily used organ silicon polymer PDMS constitutes approximately 7% of the composition in these applications. PDMS has gained approval for usage across a wide range of consumables, adhering to the maximum allowable concentrations outlined in the general standard for food additives (GSFA), which span from 10 to 100 mg kg^{-1} in various food items, including vegetable oils and fats that are commonly employed in cooking and frying (Valerini et al., 2018).

On the one hand, numerous studies, particularly those centered on polymer applications in the industrial context, point to the suitability of polydimethylsiloxane-based polymers for stabilizing foam in food manufacturing. This suitability stems from their heat stability and chemical inertness. The occurrence of silanones as precursors during thermal interactions involving organo-silicon compounds, low-molecular-weight which encompass silenes, alkoxyvinylsilanes, alkoxysilanes, hydridosilylketenes, and polysilylated compounds. Dimethylsilanone has been identified as a product arising from linear and cyclic PDMS. Moreover, various studies, especially those focusing on polymer applications within the industrial context, indicate the appropriateness of polydimethylsiloxane-based polymers for stabilizing foam in food production. This appropriateness is attributed to their heat resistance and chemical inertness (Rücker & Kümmerer, 2015).

10.5.2 SILICON DIOXIDE

Silicon dioxide (SiO$_2$) is utilized as a food ingredient primarily for its role as an anti-clumping agent, effectively preventing powdery food components from sticking together or causing blockages. The size of these agglomerates directly impacts the effectiveness of E551 as an anticaking agent within food products (EFSA, 2018; Lee et al., 2021). SiO$_2$ NPs function as an anticaking agent, enhancing the overall quality of food products. They are categorized as food additives in European and American countries under the code E551. Combining SiO$_2$ NPs with Zn renders them non-toxic to humans. These particles exhibit remarkable antifungal and antimicrobial properties against gram-positive and gram-negative bacteria (He & Hwang, 2016). The nano-encapsulation of food items and active ingredients plays a crucial role in maintaining sensory attributes like taste and aroma in a wide array of food products, thereby stimulating appetite. SiO_2 NPs have been credited with preserving the fragrance and natural extracts in food. Their incorporation has been demonstrated to enhance the consistency, flow characteristics, plumpness, and aromatic qualities of aqueous food ingredients (Zhao et al., 2018b). In another previous study Chen et al. (2013) converted the paper into a water lily-inspired hydrophobic surface using R812S silica NPs and polydimethylsiloxane silicon oil, rendering it exceptionally water-repellent. Additionally, SiO_2 is widely employed as an anticaking agent in various powdered foods. SiO_2 and carbon with a diameter of hundreds of nanometers possess valuable properties that make them suitable for application as food additives and packaging materials.

10.5.3 SYNTHETIC AMORPHOUS SILICA

Synthesized amorphous silica, also known as synthetic amorphous SiO_2, has been an important food additive for many years. The substantial production volumes and widespread utilization of synthetic amorphous silica across various sectors raise concerns about potential environmental, industrial, and consumer exposure (Ruan et al., 2007). This versatile substance finds application as an anticaking agent, thickener, adsorbent material, stabilizer, free-flow additive, dispersion stabilizer, and carrier in diverse industries, including pharmaceuticals, cosmetics, food, and animal feed. The estimations for the average consumption of SiO_2 as the food ingredient (E551) spanned from 0.28 to 4.53 mg kg^{-1} bw day^{-1} (Dekant et al., 2017).

10.5.4 TITANIUM DIOXIDE

The US Food and Drug Administration formally approved TiO_2 NPs as a food additive and component in food interactions in 1996 (Belton et al., 2019). TiO_2 NPs find widespread application in food packaging due to their roles as a bleaching agent, effective photocatalyst, antimicrobial agent, and anticaking agent. These TiO_2 NPs are employed in food packaging for color enhancement and as a protective layer, effectively extending the shelf life by preventing clumping. By interacting with the unsaturated polyphospholipid segment of bacterial cell membranes, TiO_2 NPs can trigger oxidation, leading to a bactericidal effect (Tankhiwale & Bajpai, 2012). This mechanism involves the emission of oxygen ions such as superoxide ions, hydrogen oxide, and hydroxyl radicals upon exposure to ultraviolet or natural light. These ions swiftly target bacterial cell walls, impeding microbial growth and development (Kumar et al., 2014). Bright white TiO_2 NPs effectively disperse visible light in applications such as pasteurized milk and readily available chicken meat, showcasing antimicrobial capabilities similar to those of ZnO against *Staphylococcus aureus*. Mainly, TiO_2 is employed as a food coloring ingredient without inducing adverse side effects. Significant research has explored the impact of TiO_2 on human tissues, revealing an increase in intracellular radical oxygen molecules without causing considerable damage to DNA or the endoplasmic reticulum (Kamat, 2012).

10.6 GELLING AGENTS

Food additives like jellies and candies rely on gelling agents (GA) to thicken and stabilize the ingredients. Heteropolysaccharides and hydrocolloids constitute a significant proportion of polysaccharides that possess gelling properties. These GAs find widespread application in various products, including sweets, jams, jellies, marmalade, salad dressings, yogurts, and more (Dutta et al., 2012; Nazir et al., 2017). Different proteins (Figure 10.2), both animal-derived and plant-derived, like zein from maize and whey proteins, contribute to the formation of gels (Dekant et al., 2017). In a broader sense, colloidal proteins and polysaccharides sourced from microorganisms or plants establish a continuous three-dimensional structure, acting as medium solidifiers or stabilizers. Through the incorporation of GAs, the diffusion properties of the gel medium are altered, resulting in its firming. The viscosity of the medium, governed by the concentration and physicochemical attributes of the GAs, dictates the diffusion rate. Some gelling substances can transition between liquid and gel

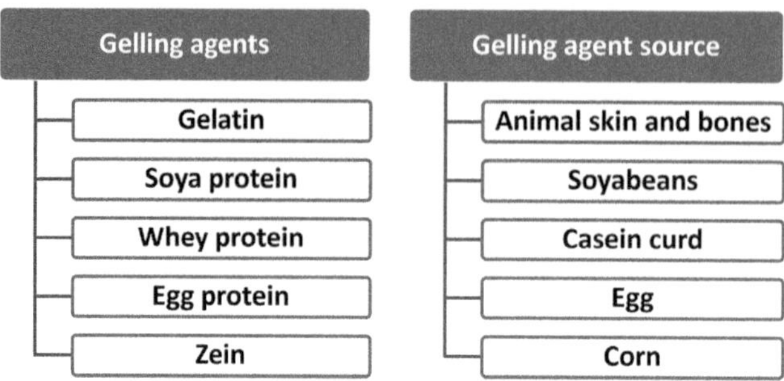

FIGURE 10.2 Use of proteins as GA.

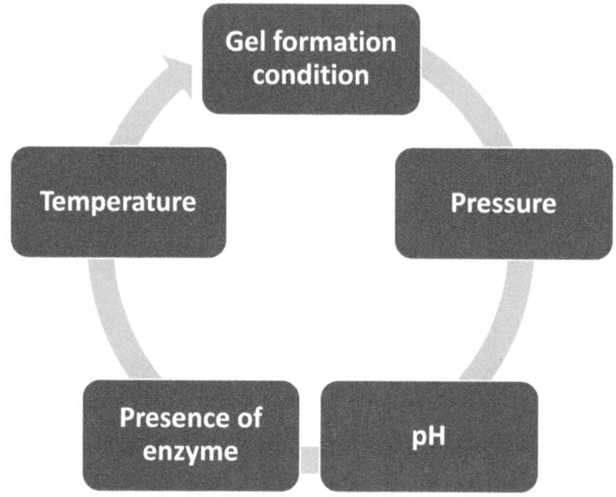

FIGURE 10.3 Typical parameters influence the gel formation.

states based on temperature changes, adding versatility (Eom et al., 2018). The GA is characterized by its ability to maintain a controlled temperature and pH range. Additionally, removing different gelling substances from GA may necessitate the utilization of diverse groups of microorganisms, hence demanding the usage of various types of GA.

Over recent years, the availability of traditional GA sources has diminished, spurring the innovation of new GAs. Gelation research advances, allowing tolerance for a wide range of pressures and temperatures, have facilitated the culture of novel microorganisms and extremophiles that were previously difficult to grow. Despite their importance in microbiology, essential GAs are not produced in one area. Commonly used GAs encompass natural gums, agar-agar, pectin, starches, and proteins, broadly categorized into proteins and polysaccharides (Nielsen et al., 2020).

10.6.1 Gel Formation Conditions

The formation of a gel occurs when a fundamental polymer dispersion or particle suspension interacts with controlled external factors like temperature or solution composition. The sol-gel conversion process often involves aggregating particles or macromolecules, forming a network that extends over the whole volume of the container (Yadav et al., 2020). Gelation processes can be broadly classified into chemically induced and physically induced gelation. In the context of protein gels, the elucidation of the native protein structure necessitates the presence of a motivating factor. This is subsequently followed by an aggregation mechanism, forming a three-dimensional configuration including clustered molecules interconnected through covalent or noncovalent connections (Thakur & Thakur, 2014). As indicated in Figure 10.3, a range of physicochemical properties primarily influence the factors that promote gel formation.

10.7 NANOAPPLICATIONFOR GELLING AGENTS

Nanostructured materials can be used as GA in food processing to enhance texture. The food industry aims to provide high-quality, tender, and healthful products. Food items that undergo texture modification often exhibit increased demand. NT contributes significantly to the creation of tenderized foods (Kiss, 2020). Traditional methods like high-pressure processing, pulsed electric fields, enzyme impregnation, sonication, and freeze-thawing are utilized to soften foods. At the same time, innovative techniques like micro fluidics, 3D printing, electro spinning, and electro spraying are employed to preserve color and flavor (Nowacka et al., 2019). Also, the essential components of texture-modified food include lipids, proteins, and carbohydrates. Globular proteins undergo unfolding and denaturation during heating, increasing fluid viscosity. Subsequent heating leads to the assembly of these proteins, forming aggregates, fibrils, and gel network chains. Polysaccharides serve as condensing and stabilizing agents in fluid systems, acting as GA. At the same time, gums and starches function as thickening agents, contributing to the viscosity and stability of enzymes and colorants (Jiang et al., 2019).

10.7.1 MICROGEL

Microgels consist of polymer chains internally linked within small dimensions, ranging from hundreds of nanometers to a few micrometers. These microgels are dispersed within colloidal solutions and exhibit a structure akin to solid particles, mainly when their surface is well-defined. Characterized by substantial water retention capacity, extensive surface area, and an internal network, these gels hold significant promise for applications such as drug delivery systems. Microgels derived from biopolymers garner considerable attention in food, drug delivery, and tissue engineering domains. This is attributed to their advantageous characteristics like biodegradability, lack of toxicity, relatively affordable production, and abundance in natural sources (Barroso et al., 2021).

10.7.2 NANOGEL

Nanogels represent groundbreaking nanometer-scale systems with immense potential across bioNT, nanomedicine, pharmaceuticals, and nutraceuticals. While sharing a similar internal structure with microgels, nanogels stand out due to their size variation (up to 100 nm) and heightened responsiveness, resulting in numerous advantages (Barroso et al., 2021). The diminutive scale enhances the solubility of hydrophobic drugs, amplifies drug accumulation within tumors, ensures robust stability against enzymatic and chemical degradation for therapeutic agents, and reduces cytotoxic side effects. Nanogels are pivotal in drug encapsulation, boasting a substantial surface area and a well-maintained interior network structure (Li et al., 2017).

10.8 APPLICATIONS OF MICROGELS AND NANOGELS IN THE FOOD INDUSTRY

10.8.1 MICROGELS

Microgels find diverse and significant applications within food systems, depending on the intended objectives. They offer value in areas such as texture management, encapsulation, acting as delivery systems, and safeguarding agents. Microgels' key attribute is their ability to enhance viscosity by swelling within suitable solvents (Zhang et al., 2016; Stokes, 2012). Given the imperative to reduce excessive fats in food, microgels have gained traction in the food industry due to their ability to enhance food products' texture and sensory experience. Additionally, microgels are emerging as encapsulation agents, with recent experimentation focused on their capacity to safeguard and gradually release bioactive compounds like minerals, antioxidants, phytochemicals, vitamins, antimicrobials, probiotics, nutraceuticals, enzymes, macronutrients, and flavors in a controlled manner (McClements, 2017a).

The prominence of microgels as encapsulation agents can be attributed to addressing sensory issues associated with numerous bioactives. Many of these compounds possess an unstable chemical nature that could lead to physical or chemical transformations in the gastrointestinal tract, potentially compromising their bioavailability and efficacy. Microgels can be tailored to serve as efficient delivery systems or preventive measures against food spoilage (McClements, 2017b).

In pharmaceutical and biomedical contexts, microgels become a focus when stability, degradation, or controlled dissolution are vital aspects tailored to the specific requirements of applications like drug delivery systems, tissue engineering, contact lenses, and wound dressings (Yadav et al., 2020). Among the myriad applications of microgels, a particularly captivating area of research revolves around their potential to deliver chemotherapy for cancer treatment. While chemotherapy exhibits efficacy in combating tumors, it grapples with several constraints, chiefly its lack of specificity, which is the primary cause of its toxicity (Shewan & Stokes, 2013).

10.8.2 Nanogels

Like microgels, nanogels have established a foothold in the food industry, particularly in encapsulating, delivering, and safeguarding bioactive compounds within food products. An example of a widely adopted nanogel is protein nanogel, harnessed from milk-derived proteins such as casein and whey. The utilization of nanogels is gaining momentum due to the imperative to craft products that can effectively resist degradation or spoilage and offer the advantageous attribute of controlled release for precisely targeted delivery of bioactive. Nanogels prove particularly effective for delivering poorly water-soluble substances owing to their singular structure encompassing a hydrophobic core, micellar design, and hydrophilic outer layer. Research conducted by Hu et al. (2021) focused on the utilization of acylated ovalbumin nanogels (AOVA) as a novel delivery vehicle for encapsulating curcumin. The AOVA nanogels were created using acylation modification and heat-induced self-assembly. The results indicated that under gastrointestinal conditions, curcumin encapsulated within AOVA nanogels exhibited significantly higher encapsulation efficiency of 93.64% and a slower sustained release pattern than native ovalbumin nanogels. This study is crucial due to the hydrophobic nature of curcumin, which leads to poor absorption. In the fields of pharmaceutical and biomedical applications, nanogels hold immense potential across a spectrum of roles, including disease diagnosis and chemotherapy, controlled release of bioactive agents, contrast agents, vaccines, cell culture systems, antimicrobials, biocatalysis, as well as the creation of bioactive frameworks within regenerative medicine (Lee et al., 2013).

Furthermore, nanogels exhibit versatility, extending their utility as nanodevices, sensors, super absorbents, nanoreactors, and biomimetic mechanical components. Their interpenetrating network structure significantly enhances drug encapsulation, rendering them effective for various routes of administration, including nasal, oral, intraocular, and pulmonary pathways. Additionally, nanogels can simultaneously encapsulate two drugs, a vital attribute for co-administrating multiple anticancer agents (Khan et al., 2021).

10.9 NANOTECHNOLOGY IN MODERN APPLICATIONS

10.9.1 Nanofiltration

Nanofiltration is a technique through which small-molecule solutes can be isolated from a liquid environment. Nanofiltration membranes are an internationally recognized membrane technology with promising applications in the food sector due to reliability, cost-effectiveness, and standardized performance. Due to the demand for lower pressure-driven membranes than reverse and ultra-filtration membranes, it has also drawn the interest of several industries (Yadav et al., 2022). The effectiveness of a nanofiltration membrane is primarily determined by three factors: effective pore radius, effective ratio of membrane thickness to porosity, and effective charge density (Dasgupta et al., 2015). In the food processing industry, nanofiltration membranes are widely used in various products, including vegetable oils, dairy products, drinks, and other foodstuffs. These membranes are used for operations such as separation, microbial reduction, deacidification, demineralization, and concentration and are specially used in the purification step or to obtain specific solutes.

On a commercial scale, nanofiltration finds practical use in several domains, such as the desalination of seawater, the concentration of juices, demineralization processes, and the extraction of color from water. The technique is also valuable for wastewater treatment, water purification, and cheese production (Warczok et al., 2004). Another commercial application of nanofiltration is lactic acid separation, a critical food sector product that requires costly purifying equipment (Duke et al., 2008). Within the dairy industry, nanofiltration serves dual purposes: enhancing product quality and facilitating the separation of mineral salts from lactose. This separation occurs after the removal of proteins via ultrafiltration. In addition, nanofillers have been used to make milk suitable for lactose-intolerant people by eliminating lactose and replacing it with alternative sugars (Shah et al., 2018). Nanoscale filters have also been used to eradicate bacterial species from milk or water without boiling it. Nanomaterials can create nanosieves that act as milk and beer filters. Moreover, NT produces better foods with less salt, sugar, and fat to prevent foodborne illnesses (Nile et al., 2020).

10.9.2 Nanoadditives

Nano additives facilitate the enhanced dispersion of water-insoluble additives, such as pigments, flavors, additives, and supplements, without using a surfactant. These nano additives elevate flavor and nutritional value and boost the body's absorption of these constituents due to their vast surface area. Nanostructured bioactive compounds, which hold nutraceutical value, are effectively delivered through a nanoparticle delivery system termed nanofood additives. This system effectively disrupts the delivery of these nanosized particles for improved bioavailability Nile et al. (2020). Vitamins, antioxidants, colors, flavors, and preservatives stand as notable instances of nano additives that are currently available. Within the food industry, nano additives such as calcium, selenium, iron, silica, human-sized silver, and magnesium are actively employed to enhance various food products. Nano additives based on TiO_2 NPs find practical application as ingredients in specific food products, including confectionery, certain types of cheeses, and sauces. These additives confer distinctive optical characteristics such as heightened delicacy and brightness, thus enhancing the visual appeal of the products (Krishna et al., 2022). Nano additives based on AgNPs have found utility in coloring the surfaces of various food products, including items like sugar, confectionery, and bakery goods. These additives enhance the visual appeal of these products through effective coloring techniques (Kothari & Wani, 2021). SiO_2 NPs are incorporated into specific powdered food products as antimicrobial agents. These NPs are used in salts, icing sugar, flavors, dried milk, and dry mixes to enhance their antimicrobial properties and improve food safety (Go et al., 2017). Magnesium, zinc, and iron oxide NPs hold promise for application within the food processing sector as potent antimicrobial agents, enabling a reduction in the necessary pasteurization temperature for milk. Lipid-based NPs are being developed as colloidal delivery systems. These systems encapsulate, shield, and release hydrophobic bioactive substances, such as flavors, antimicrobials, supplements, and nutraceuticals (Blanco-Llamero et al., 2022; Mirhosseini & Dehestani, 2020).

10.9.3 Nanocarriers

In the food industry, nanocarriers are currently employed as delivery systems to transport food additives within food products while maintaining their fundamental structure. This approach safeguards food additives from thermal degradation and conceals their flavors (Srivastava et al., 2022). NT offers diverse avenues to elevate food quality and enhance flavor profiles. Rutin, a prevalent dietary flavonoid known for its significant pharmacological properties, faces limitations in food industry applications due to its poor solubility. However, through the encapsulation of ferritin nanocages, serving as nanocarriers, the thermal stability, solubility, and resistance to UV radiation of ferritin-trapped rutin have been notably improved when contrasted with free rutin (Nakagawa, 2014).

Compared to larger particles, which typically release encapsulated substances more slowly and over more extended periods, NPs offer an effective way to enhance the bioavailability of nutraceutical compounds because of their sub-cellular size. Common metallic oxides, like SiO_2 and TiO_2, have traditionally found application as colorants or flow agents in food products. Among these, SiO_2 nanomaterials stand out as extensively employed carriers of fragrances or flavors in various food items (Dekkers et al., 2011).

10.9.4 NANOENCAPSULATION

Nanoencapsulation is a technology that involves encapsulating substances at a minuscule scale, resulting in end products with enhanced functionality, notably controlled release of the enclosed core material. The application of nanoencapsulation holds promise in addressing the food industry's challenges related to the efficient delivery of beneficial functional ingredients and the precise release of flavor compounds (Maqsoudlou et al., 2022).

In the field of food products, nanoencapsulation provides a range of advantages. These include improving the stability and solubility of bioactive compounds, safeguarding against degradation during manufacturing, transportation, and storage, minimizing undesirable tastes, achieving heightened activity levels of the encapsulated ingredients, and controlling the release of these compounds (Taouzinet et al., 2023). One common application involves masking unappealing odors or flavors in food formulations or converting liquid phases into solid states (Pateiro et al., 2021).

Numerous technologies are available for creating nanoencapsulation systems, such as emulsification, inclusion complexation, coacervation, emulsification solvent evaporation, nanoprecipitation, extraction, electrospraying (spray drying), and electrospinning. These are the most common processes to produce nanoencapsulation systems (Siddiqui et al., 2022; Jayaprakash et al., 2023; Khairnar et al., 2022; Buljeta et al., 2022; Bensid et al., 2022).

10.9.5 NANOEMULSION

An emulsion is a mixture of two liquid phases that cannot blend, with one phase being dispersed as droplets within the other. Nanoemulsions are characterized by having oil droplets at the nanoscale, typically measuring between 10 and 100 nm (Azmi et al., 2019). Generally, a nanoemulsion comprises three essential constituents: an oil phase, an aqueous phase, and an emulsifier. The emulsifier holds a pivotal role within nanoemulsions. Even in minute quantities, its presence is crucial for creating nanoemulsions by effectively diminishing the interfacial tension between the water and oil phases.

Furthermore, the emulsifier dramatically contributes to stabilizing nanoemulsions (Espitia et al., 2019). The oil phase utilized to create food-grade nanoemulsions consists of diverse nonpolar molecules, including free fatty acids, waxes, mineral oils, monoacylglycerols, di and tri glycerol, or various lipophilic nutraceuticals. Triglyceride-rich oils derived from sunflower, soybean, corn, safflower, olive, flaxseed, fish, and algae are commonly employed in nanoemulsions due to their cost-effectiveness and high nutritional value (Azmi et al., 2019). On the other hand, the aqueous phase in food-grade nanoemulsion preparations can encompass various polar molecules such as minerals, acids, proteins, carbohydrates, or alcoholic co-solvents. Stabilizers heavily influence the stability of nano emulsions over extended periods. As a result, nanoemulsions' formation depends on the choice of appropriate stabilizers. These stabilizers can serve multiple functions, including modifying texture, acting as emulsifiers, retarding ripening processes, enhancing weight, and aiding emulsification.

Nanoemulsion is a pivotal nano-vehicle employed within the food industry to enhance food products' physical characteristics and effectiveness. It plays a crucial role in ensuring food safety through proper packaging and preservation and maintaining the qualities of food materials by incorporating antibacterial, antifungal, and antioxidant agents. Importantly, nanoemulsion delivers a wide array of food components efficiently (Shahidi & Zhong, 2010).

Numerous bioactive food constituents are susceptible to degradation during food processing and oxidative spoilage during storage due to enzymatic, microbiological, chemical, and physical alterations. These changes can adversely affect sensory attributes, nutritional value, and physicochemical properties (Ahmed, 2022). Some bioactive exhibit low solubility and rapid metabolism, diminishing their bioavailability, while others are volatile and sensitive to processing conditions. These challenges can be addressed by employing nanoemulsions to encapsulate bioactive compounds within the food matrix.

Encapsulating bioactive compounds within an oil phase or emulsifier through nanoencapsulation ensures stability, heightened bioavailability, and a controlled release rate. The utilization of nanoencapsulation with natural compounds not only enhances food processing efficiency but also augments food safety (Guía-García et al., 2022). The primary enhancements are directed toward altering food product texture, encapsulating edible substances or additives, creating original flavors, and improving the bioaccessibility and bioavailability of food constituents (Pateiro et al., 2021). A range of applications of nanoemulsions in the food industry involving bio-actives and their outcomes are detailed in Table 10.1.

Polyphenols encompass diverse molecules, such as phenolic acids, flavonoids, and others. The utilization of NT for encapsulating these compounds has emerged as a potential solution to address numerous challenges while enhancing the functional attributes of food products. As an example, a phenolic extract derived from guabiroba fruit encapsulated at the nanoscale within poly (D, L-lactic-co-glycolic) acid through a modified emulsion-evaporation technique (Pereira et al., 2018). This approach effectively safeguarded the phenolic content and its bioactivity, allowing them to be retained until consumption and enabling an extended-release over time. A study by Guan et al. (2019) employed nanoencapsulation to encapsulate caffeic acid phenethyl ester (a significant hydrophobic bioactive constituent of propolis extract) within an aqueous propylene glycol solution. This was achieved using sucrose fatty acid ester and a temperature-cycle method. This innovative approach enhanced physiological characteristics and increased effectiveness against cancer cells.

In recent times, vitamins have recently gained popularity due to their dual functions as essential nutrients and agents for disease control. Elbarbary et al. employed chitosan/vitamin C complexes encompassing a range of NPs in a study focused on meat lipid peroxidation. This complex demonstrated notable antioxidant potency, resulting in a remarkable 75% reduction in lipid peroxidation (Elbarbary et al., 2015).

In a study by Zhang and coworkers (2022), vitamin D_3 was encapsulated using Tween 80 and Span 80. This approach enhanced storage stability and resulted in sustained release during simulated gastrointestinal digestion. Baek et al. (2021) nano-encapsulated vitamin C within cellulose/chitosan nanocapsules, this novel approach was reported to improve stability, antimicrobial activity, and the release of vitamin C. In previous study, Zhao and coworkers (2018a) generated a nanoemulsion based on lutein, utilizing whey protein. Their study explored its morphological characteristics, storage conditions, and stability.

Similarly, Tan et al. (2014) developed carotenoid-loaded nano-liposomes using egg yolk phosphatidylcholine. This research delved into understanding how carotenoids exhibited antioxidant activity following encapsulation. The study revealed increased lipid peroxidation inhibition capacity after encapsulation. Min et al. (2016) formulated chitosan-based NPs containing iron to enhance iron's bioavailability and stability during storage. In another study, Sharifi and coworkers (2013) developed nano-encapsulation involving Fe/Zn-loaded alginate for food fortification. This encapsulation method was applied to fortify ice cream, and the study assessed the sensory and rheological aspects of the fortified ice cream. Gülseren et al. (2012) produced zinc NPs based on whey protein and evaluated their efficiency and stability. This study focused on incorporating minerals into food products through innovative methods, allowing for the creation of acidic dairy beverages containing nutritionally valuable mineral content.

TABLE 10.1

An Overview of Nanoemulsion of Bioactive Compounds for Food Applications

Bioactive Compounds	Purpose	Nanoemulsion Formation	Applications	References
Citral	Flavoring compound	Nanoemulsions were prepared with lecithin-stabilized palm kernel lipid in pH three buffer 1:1 ratio of citral and antioxidants such as β-carotene, tanshinone, and black tea extract	Improve stability of citral by decreasing the formation of off-flavor compounds such as α, p-dimethyl styrene, and p-methyl acetophenone	Yang et al. (2011)
β-Carotene	Coloring compound	β-carotene nanoemulsions have been prepared by coating with starch caseinate and chitosan-epigallocatechin-3-gallate conjugates	Improve physiochemical properties and stability of β-Carotene by degradation	Wei and Gao (2016)
β-Carotene	Coloring compound	The o/w carotenoid nanoemulsions encapsulated in porcine gelatin and whey protein isolate had an average particle size of 70–160 nm	Good physiochemical properties and improved stability of yogurt for up to 60 days	Medeiros et al. (2019)
Resveratrol	Nutraceuticals	Encapsulation of resveratrol was formulated by spontaneous emulsification using a 10% oil phase (grape seed oil and orange oil), 10% surfactant (Tween 80), and 80% aqueous phase had 100 nm droplet size and could carry 120 ± 10 µg mL^{-1} of resveratrol	Improved chemical stability against UV-light degradation	Davidov-Pardo and McClements (2015)
Vitamin D3		Emulsifiers polysorbate 20, soybean lecithin and their mixtures, and dispersed oil phase of soybean oil or combinations of the oil with cocoa butter were used to prepare nanoemulsions of mean diameters <200 nm with high-pressure homogenizer	Fortified whole-fat milk showed stability	Golfomitsou et al. (2018)
Astaxanthin		Nanoemulsions of astaxanthin have been formulated by coating chitosan and carrageenan	Protected from UV light and photodegradation	Alarcón-Alarcón et al. (2018)
Clove and oregano essential oil	Preservatives	Nanoemulsions of clove bud and oregano essential oil with 180–250 nm droplet size have been prepared by incorporating edible methylcellulose films	Extending the self-life of the sliced bread for 15 days prevented the growth of yeasts and molds.	Otoni et al. (2014)
Oregano essential oil		Oregano oil nanoemulsions of 148 nm size have been prepared by mixing with tween 20 through ultrasonication	Inhibited growth of foodborne bacteria *Listeria monocytogenes*, *Salmonella* Typhimurium, and *E. coli* O157:H7 on fresh lettuce	Bhargava et al. (2015)
Ginger essential oil		Nanoemulsions of ginger essential oil have been prepared by coating with sodium caseinate	Significantly decrease refrigerated chicken fillets' total aerobic psychrophilic bacteria during 12 days.	Noori et al. (2018)
capsaicin (oleoresin capsicum)		High-pressure homogenization and ultrasonication prepared capsaicin nanoemulsions with Tween 80 as the aqueous phase	Increase the bioavailability of lipophilic compounds	Akbas et al. (2018)

10.10 NANOTECHNOLOGY IN FOOD PRESERVATION AND TECHNIQUES

Recent developments in NT have shown interest in developing nano-based products for efficient nutrient delivery. NT-based methods can be employed for food processing involving the development of nanostructured food ingredients, which improve consistency, taste, texture, color, flow properties, and stability during processing or shelf life extension (Singh, 2018). Using various food processing techniques, one food quality can be kept at a required level and maximize its nutritional value. Food preservation methods include growing, harvesting, processing, packaging, and distributing food. In the 20th century, a variety of techniques for food preservation were developed (Hamad et al., 2018). There exists a diverse array of cutting-edge technologies encompassing physical methods such as high hydrostatic pressure and high-pressure homogenization, electromagnetic techniques like pulsed electric fields, ohmic heating, microwaves, radio waves, and UV light, as well as other approaches such as membrane technology and compact phase CO_2. There exists a diverse array of cutting-edge technologies encompassing physical methods such as high-pressure homogenization and high hydrostatic pressure and electromagnetic techniques like pulsed electric fields, ohmic heating, UV light, microwaves, and radio waves as well as other approaches such as membrane technology and compact phase CO_2. Due to the increasing demand from customers for convenient food options that offer optimal nutritional value, pleasant aroma, and natural taste, there is currently a significant emphasis placed on the degree to which minimally processed products resemble fresh meals (Sharifi et al., 2013). Some of the typical applications of NT in food preservation are shown in Table 10.2.

TABLE 10.2

Use of Nanotechnology in Food Industry

Nanotechnology	Examples	Uses/Applications	References
Nanocomposites	Aegis	Oxygen scavenger	Yotova et al. (2013)
		Reduces CO_2 leaks from drinks (carbonated)	
	Durethan (polyamide)	Paperboard containers (fruit juices)	Davis et al. (2013)
	Top Screen DS13 Guard in Fresh	Absorption of ethylene gas to ripen fruits and vegetables	Burdo (2005)
	Imperm (nylon)	Oxygen scavenger	Thirumurugan et al. (2013)
	PEG-coated with garlic oil nanocomposites	Helps manage pests at stores that contaminate packaged foods	Mason et al. (2006)
	NanoCeram PAC	Quick absorption of components (e.g., unpleasant odors and taste)	Burdo (2005)
	Nanolaminates	Ccoating of bakery products, cheese, meats, vegetables and fruits	Flanagan and Singh (2006)
Nanosensors	Array biosensors and nano-test strips	Changes in color in food material	Biswal et al. (2012)
	Polyaniline and Carbon black	Detection of foodborne pathogens, microorganisms, and carcinogens in the food material	
	Cerium oxide immunosensors and chitosan-based nanocomposites	Toxin detection (ochratoxin A) in the food	Mousavi and Rezaei (2011)
	Gold, platinum, and palladium	Detecting color change & toxins (aflatoxin B_1) in the food material	Pradhan et al. (2015)
	Nano-biosensors	Detection of viruses and bacteria	Coles and Frewer (2013)
	Nano-barcodes	Quality detection of food produced	

(Continued)

TABLE 10.2 (*Continued*)

Use of Nanotechnology in Food Industry

Nanotechnology	Examples	Uses/Applications	References
Nanoparticles	TiO_2	Acts as a food colorant	Jones et al. (2008)
		Photocatalytic disinfecting agent	
	Polymeric	Act as an efficient bactericide	Bouwmeester et al. (2009)
	SiO_2	Drying and anticaking agent	Raghupathi et al. (2011)
	Magnetic	Detection of microbes & toxins	Zhao et al. (2008)
	Silver	Detection of food components	Bouwmeester et al. (2009)
		Act as an antimicrobial agent	
		Improve shelf life	
	Zinc oxide	Detection of toxins & chemicals	
		Decreases the flow of oxygen in packaged food.	
Nanofilms	Ag-based cellulose film	Used as antibacterial agents	Gu et al. (2019)
	TiO_2-based polyacrylonitrile film	Ethylene scavengers	Zhu et al. (2019)
	TiO_2 based chitosan film	Protect fresh fruits due to high antimicrobial activity	Zhang et al. (2017)
	ZnO based chitosan film	Antibacterial agent and antioxidant	Yadav et al. (2021)

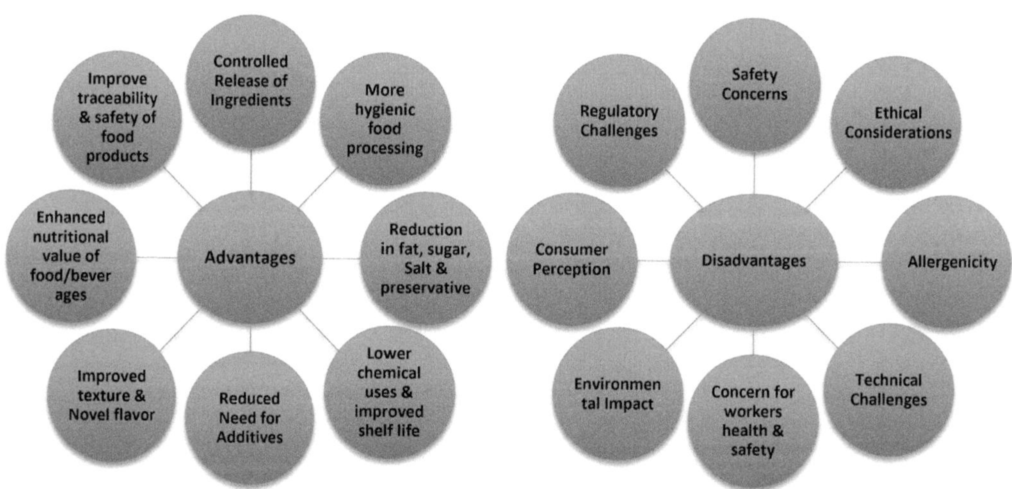

FIGURE 10.4   Advantages and disadvantages of nanotechnology applications in food and related products.

10.11   ADVANTAGES AND DISADVANTAGES OF NANOPARTICLES IN FOOD PROCESSING

10.11.1   ADVANTAGES

Due to their unique properties, NPs have shown great promise in various fields, including food processing. However, they also have potential disadvantages and challenges, particularly regarding food safety, regulation, and environmental impact (Prakash et al., 2019). Figure 10.4 shows the advantages and disadvantages of using NT in food processing. NT offers vast export potential for producers through increasing the local processing of essential commodities such as sugar, coffee, tea, rice, spices, and bananas, thus growing volumes of exports and, subsequently, profit margins.

Implementing intelligent labels in the food processing industry might additionally assist businesses in guaranteeing the genuineness, traceability, and safety of their food items.

Additionally, there are prospects for developing innovative food and beverage products that exhibit enhanced nutritional content, visual appeal, taste, and texture while catering to individual consumer preferences and specific dietary and health needs. For instance, using NT enables the enhancement of food products by incorporating fruits and vegetables, increasing the nutritional value and density of such foods (Vasilache et al., 2011). Nanoencapsulation technologies allow the incorporation of additional nutrients into food and beverage items while preserving their original flavor and quality characteristics. Some nutrients and additives can be delivered to a specific target place in the body, providing customers with additional health benefits. Nanoencapsulation technologies offer numerous advantages to businesses, primarily by safeguarding food ingredients during processing and extending the shelf life of products. These benefits mitigate food waste and enhance profits (Alfadul & Elneshwy, 2010; Prerna & Ratan, 2021).

10.11.2 DISADVANTAGES

NT employs the manipulation of nanomaterials for various applications across various domains, including but not limited to food, agriculture, and human health. The utilization of NT in food processing has raised several considerations on human health, safety, and interconnected toxicity (Parveen et al., 2016). There have been reports of NPs migrating through processed foods and harming humans. Increasing scientific evidence has linked exposure to NPs (titanium dioxide, iron oxide, carbon black, and silicates) to oxidative damage and inflammatory reactions in the gastrointestinal tract. Moreover, prolonged and continuous exposure has been associated with acute toxic reactions such as kidney and liver lesions and various types of cancer (Chellaram et al., 2014). NPs' toxicological and safety properties suggest that nanomaterials incorporation into food may pose new risks to consumers. The primary mode of NP entrance into the gastrointestinal tract is typically via ingesting food and beverages. The biggest health concerns about NPs are their increased toxicological effects at lower concentrations due to their larger surface area, greater access to the body, improved bioavailability, compromised immune system response, and possible more prolonged pathological effects (Divya & Jisha, 2018).

BIBLIOGRAPHY

fzaal, M., Saeed, F., Ahmed, A., Islam, F., Khalid, M. A., Hussain, M., & Afzal, A. (2022). Anticaking agents in food nanotechnology. In *Application of Nanotechnology in Food Science, Processing and Packaging*, Cham: Springer International Publishing, 141–151. doi: 10.1007/978-3-030-98820-3_9.

Ahmed, T. M. K. (2022). Side effects of preservatives on human life. *Journal of Pharmaceutical Sciences and Research*, 2(5), 1–14.

Akbas, E., Soyler, B., & Oztop, M. H. (2018). Formation of capsaicin loaded nanoemulsions with high pressure homogenization and ultrasonication. *LWT – Food Science and Technology*, 96, 266–273. doi: 10.1016/j.lwt.2018.05.043.

Akins, M. L. (2002). Effects of starch-based anticaking agents on the functional properties of shredded mozzarella cheese (*Doctoral dissertation, Virginia Tech*). https://hdl.handle.net/10919/35783.

Alarcón-Alarcón, C., Inostroza-Riquelme, M., Torres-Gallegos, C., Araya, C., Miranda, M., Sánchez-Caamaño, J. C., & Oyarzun-Ampuero, F. A. (2018). Protection of astaxanthin from photodegradation by its inclusion in hierarchically assembled nano and microstructures with potential as food. *Food Hydrocolloids*, 83, 36–44. doi: 10.1016/j.foodhyd.2018.04.033.

Alfadul, S. M., & Elneshwy, A. A. (2010). Use of nanotechnology in food processing, packaging and safety-review. *African Journal of Food, Agriculture, Nutrition and Development*, 10(6), 2719–2739. doi: 10.4314/ajfand.v10i6.58068.

Anandharamakrishnan, C., & Parthasarathi, S. (2019). *Food Nanotechnology: Principles and Applications* (1st Ed.). CRC Press, 454.

Azmi, N. A. N., Elgharbawy, A. A., Motlagh, S. R., Samsudin, N., & Salleh, H. M. (2019). Nanoemulsions: Factory for food, pharmaceutical and cosmetics. *Processes*, 7(9), 617. doi: 10.3390/pr7090617.

Baek, J., Ramasamy, M., Willis, N. C., Kim, D. S., Anderson, W. A., & Tam, K. C. (2021). Encapsulation and controlled release of vitamin C in modified cellulose nanocrystal/chitosan nanocapsules. *Current Research in Food Science, 4*, 215–223. doi: 10.1016/j.crfs.2021.03.010.

Barroso, L., Viegas, C., Vieira, J., Ferreira-Pego, C., Costa, J., & Fonte, P. (2021). Lipid-based carriers for food ingredients delivery. *Journal of Food Engineering, 295*, 110451. doi: 10.1016/j.jfoodeng.2020.110451.

Belton, D. J., Hickman, G. J., & Perry, C. C. (2019). Traditional materials from new sources-conflicts in analytical methods for calcium carbonate. *Food Additives & Contaminants: Part A, 36*(3), 366–373. doi: 10.1080/19440049.2019.1571288.

Bensid, A., El Abed, N., Houicher, A., Regenstein, J. M., & Özogul, F. (2022). Antioxidant and antimicrobial preservatives: Properties, mechanism of action and applications in food-a review. *Critical Reviews in Food Science and Nutrition, 62*(11), 2985–3001. doi: 10.1080/10408398.2020.1862046.

Bhargava, K., Conti, D. S., da Rocha, S. R. P., & Zhang, Y. (2015). Application of an oregano oil nanoemulsion to the control of foodborne bacteria on fresh lettuce. *Food Microbiology, 47*, 69–73. doi: 10.1016/j.fm.2014.11.007.

Biswal, S. K., Nayak, A. K., Parida, U. K., & Nayak, P. L. (2012). Applications of nanotechnology in agriculture and food sciences. *International Journal of Science Innovations and Discoveries, 2*(1), 21–36.

Blanco-Llamero, C., Fonseca, J., Durazzo, A., Lucarini, M., Santini, A., Señoráns, F. J., & Souto, E. B. (2022). Nutraceuticals and food-grade lipid nanoparticles: From natural sources to a circular bioeconomy approach. *Foods, 11*(15), 2318. doi: 10.3390/foods11152318.

Bouwmeester, H., Dekkers, S., Noordam, M. Y., Hagens, W. I., Bulder, A. S., De Heer, C., & Sips, A. J. (2009). Review of health safety aspects of nanotechnologies in food production. *Regulatory Toxicology and Pharmacology, 53*(1), 52–62. doi: 10.1016/j.yrtph.2008.10.008.

Buljeta, I., Pichler, A., Šimunović, J., & Kopjar, M. (2022). Polysaccharides as carriers of polyphenols: Comparison of freeze-drying and spray-drying as encapsulation techniques. *Molecules, 27*(16), 5069. doi: 10.3390/molecules27165069.

Burdo, O. G. (2005). Nanoscale effects in food-production technologies. *Journal of Engineering Physics and Thermophysics, 78*(1), 90–96. doi: 10.1007/s10891-005-0033-6.

Chellaram, C., Murugaboopathi, G., John, A. A., Sivakumar, R., Ganesan, S., Krithika, S., & Priya, G. (2014). Significance of nanotechnology in food industry. *APCBEE Procedia, 8*, 109–113. doi: 10.1016/j.apcbee.2014.03.010.

Chen, M., Wu, S., Xu, S., Yu, B., Shilbayeh, M., Liu, Y., & Gong, J. (2018). Caking of crystals: Characterization, mechanisms and prevention. *Powder Technology, 337*, 51–67. doi: 10.1016/j.powtec.2017.04.052.

Chen, W., Wang, X., Tao, Q., Wang, J., Zheng, Z., & Wang, X. (2013). Lotus-like paper/paperboard packaging prepared with nano-modified overprint varnish. *Applied Surface Science, 266*, 319–325. doi: 10.1016/j.apsusc.2012.12.018.

Clarson, S. J. (2010). Advanced materials containing the siloxane bond. In *Advances in Silicones and Silicone-Modified Materials. American Chemical Society, Chapter 1*, Clarson, S., Owen, M., Smith, S., & Van Dyke, M. (Eds.), 3–10. doi: 10.1021/bk-2010-1051.ch001.

Coles, D., & Frewer, L. J. (2013). Nanotechnology applied to European food production – A review of ethical and regulatory issues. *Trends in Food Science & Technology, 34*(1), 32–43. doi: 10.1016/j.tifs.2013.08.006.

Cozzolino, C. A., Castelli, G., Trabattoni, S., & Farris, S. (2016). Influence of colloidal silica nanoparticles on pullulan-coated BOPP film. *Food Packaging and Shelf Life, 8*, 50–55. doi: 10.1016/j.fpsl.2016.03.003.

Dasgupta, N., Ranjan, S., Mundekkad, D., Ramalingam, C., Shanker, R., & Kumar, A. (2015). Nanotechnology in agro-food: From field to plate. *Food Research International, 69*, 381–400. doi: 10.1016/j.foodres.2015.01.005.

Davidov-Pardo, G., & McClements, D. J. (2015). Nutraceutical delivery systems: Resveratrol encapsulation in grape seed oil nanoemulsions formed by spontaneous emulsification. *Food Chemistry, 167*, 205–212. doi: 10.1016/j.foodchem.2014.06.082.

Davis, D., Guo, X., Musavi, L., Lin, C. S., Chen, S. H., & Wu, V. C. (2013). Gold nanoparticle-modified carbon electrode biosensor for the detection of Listeria monocytogenes. *Industrial Biotechnology, 9*(1), 31–36. doi: 10.1089/ind.2012.0033.

Dekant, W., Fujii, K., Shibata, E., Morita, O., & Shimotoyodome, A. (2017). Safety assessment of green tea based beverages and dried green tea extracts as nutritional supplements. *Toxicology Letters, 277*, 104–108. doi: 10.1016/j.toxlet.2017.06.008.

Dekkers, S., Krystek, P., Peters, R. J., Lankveld, D. P., Bokkers, B. G., van Hoeven-Arentzen, P. H., & Oomen, A. G. (2011). Presence and risks of nanosilica in food products. *Nanotoxicology, 5*(3), 393–405. doi: 10.3109/17435390.2010.519836.

Divya, K., & Jisha, M. S. (2018). Chitosan nanoparticles preparation and applications. *Environmental Chemistry Letters, 16*, 101–112. doi: 10.1007/s10311-017-0670-y.

Duke, M. C., Lim, A., da Luz, S. C., & Nielsen, L. (2008). Lactic acid enrichment with inorganic nanofiltration and molecular sieving membranes by pervaporation. *Food and Bioproducts Processing*, *86*(4), 290–295. doi: 10.1016/j.fbp.2008.01.005.

Dutta, J., Tripathi, S., & Dutta, P. K. (2012). Progress in antimicrobial activities of chitin, chitosan and its oligosaccharides: A systematic study needs for food applications. *Food Science and Technology International*, *18*(1), 3–34. doi: 10.1177/1082013211399195.

EFSA. Panel on Food Additives and Nutrient Sources (ANS). (2016). Scientific Opinion on the reevaluation of benzoic acid (E 210), sodium benzoate (E 211), potassium benzoate (E 212) and calcium benzoate (E 213) as food additives. *EFSA Journal*, *14*(3), 4433. doi: 10.2903/j.efsa.2016.4433.

EFSA. Panel on Food Additives and Nutrient Sources added to Food. (2011). Scientific opinion on reevaluation of calcium carbonate (E 170) as a food additive. *EFSA Journal*, *9*(7), 2318. doi: 10.2903/j.efsa.2011.2318.

EFSA. Panel on Food Additives and Nutrient Sources added to Food (ANS), Younes, M., Aggett, P., Aguilar, F., Crebelli, R., Dusemund, B., & Lambré, C. (2018). Reevaluation of silicon dioxide (E 551) as a food additive. *EFSA Journal*, *16*(1), e05088. doi: 10.2903/j.efsa.2018.5088.

Elbarbary, A. M., El-Sawy, N. M., & Hegazy, E. S. A. (2015). Antioxidative properties of irradiated chitosan/vitamin C complex and their use as food additive for lipid storage. *Journal of Applied Polymer Science*, *132*(24), 42105. doi: 10.1002/app.42105.

Eom, S. H., Chun, Y. G., Park, C. E., Kim, B. K., Lee, S. H., & Park, D. J. (2018). Application of freeze-thaw enzyme impregnation to produce softened root vegetable foods for elderly consumers. *Journal of Texture Studies*, *49*(4), 404–414. doi: 10.1111/jtxs.12341.

Espitia, P. J., Fuenmayor, C. A., & Otoni, C. G. (2019). Nanoemulsions: Synthesis, characterization, and application in bio-based active food packaging. *Comprehensive Reviews in Food Science and Food Safety*, *18*(1), 264–285. doi: 10.1111/1541-4337.12405.

Fitzpatrick, J. J., Descamps, N., O'Meara, K., Jones, C., Walsh, D., & Spitere, M. (2010). Comparing the caking behaviours of skim milk powder, amorphous maltodextrin and crystalline common salt. *Powder Technology*, *204*(1), 131–137. doi: 10.1016/j.powtec.2010.07.029.

Flanagan, J., & Singh, H. (2006). Microemulsions: A potential delivery system for bioactives in food. *Critical Reviews in Food Science and Nutrition*, *46*(3), 221–237. doi: 10.1080/10408690590956710.

García-Pinilla, S., Villalobos-Espinosa, J. C., Cornejo-Mazón, M., & Gutiérrez-López, G. F. (2019). Nanotechnology in food processing. *In Advances in Processing Technologies for Bio-based Nanosystems in Food*, CRC Press, 259–276.

Go, M. R., Bae, S. H., Kim, H. J., Yu, J., & Choi, S. J. (2017). Interactions between food additive silica nanoparticles and food matrices. *Frontiers in Microbiology*, *8*, 1013. doi: 10.3389/fmicb.2017.01013.

Golfomitsou, I., Mitsou, E., Xenakis, A., & Papadimitriou, V. (2018). Development of food grade O/W nanoemulsions as carriers of vitamin D for the fortification of emulsion based food matrices: A structural and activity study. *Journal of Molecular Liquids*, *268*, 734–742. doi: 10.1016/j.molliq.2018.07.109.

Gu, R., Yun, H., Chen, L., Wang, Q., & Huang, X. (2019). Regenerated cellulose films with amino-terminated hyperbranched polyamic anchored nanosilver for active food packaging. *ACS Applied Bio Materials*, *3*(1), 602–610. doi: 10.1021/acsabm.9b00992.

Guan, Y., Chen, H., & Zhong, Q. (2019). Nanoencapsulation of caffeic acid phenethyl ester in sucrose fatty acid esters to improve activities against cancer cells. *Journal of Food Engineering*, *246*, 125–133. doi: 10.1016/j.jfoodeng.2018.11.008.

Guía-García, J. L., Charles-Rodríguez, A. V., Reyes-Valdés, M. H., Ramírez-Godina, F., Robledo-Olivo, A., García-Osuna, H. T., & Flores-López, M. L. (2022). Micro and nanoencapsulation of bioactive compounds for agri-food applications: A review. *Industrial Crops and Products*, *186*, 115198. doi: 10.1016/j.indcrop.2022.115198.

Gülseren, İ., Fang, Y., & Corredig, M. (2012). Zinc incorporation capacity of whey protein nanoparticles prepared with desolvation with ethanol. *Food Chemistry*, *135*(2), 770–774. doi: 10.1016/j.foodchem.2012.04.146.

Gupta, A., Eral, H. B., Hatton, T. A., & Doyle, P. S. (2016). Nanoemulsions: Formation, properties and applications. *Soft Matter*, *12*(11), 2826–2841. doi: 10.1039/c5sm02958a.

Hamad, A. F., Han, J. H., Kim, B. C., & Rather, I. A. (2018). The Intertwine of nanotechnology with the food industry. *Saudi Journal of Biological Sciences*, *25*(1), 27–30. doi: 10.1016/j.sjbs.2017.09.004.

He, X., & Hwang, H. M. (2016). Nanotechnology in food science: Functionality, applicability, and safety assessment. *Journal of Food and Drug Analysis*, *24*(4), 671–681. doi: 10.1016/j.jfda.2016.06.001.

Hu, G., Batool, Z., Cai, Z., Liu, Y., Ma, M., Sheng, L., & Jin, Y. (2021). Production of self-assembling acylated ovalbumin nanogels as stable delivery vehicles for curcumin. *Food Chemistry*, *355*, 129635. doi: 10.1016/j.foodchem.2021.129635.

Janjarasskul, T., & Suppakul, P. (2018). Active and intelligent packaging: The indication of quality and safety. *Critical Reviews in Food Science and Nutrition*, *58*(5), 808–831. doi: 10.1080/10408398.2016.1225278.

Jayaprakash, P., Maudhuit, A., Gaiani, C., & Desobry, S. (2023). Encapsulation of bioactive compounds using competitive emerging techniques: Electrospraying, nano spray drying, and electrostatic spray drying. *Journal of Food Engineering*, *339*, 111260. doi: 10.1016/j.jfoodeng.2022.111260.

Jiang, Y., Liu, L., Wang, B., Yang, X., Chen, Z., Zhong, Y., & Sui, X. (2019). Polysaccharide-based edible emulsion gel stabilized by regenerated cellulose. *Food Hydrocolloids*, *91*, 232–237. doi: 10.1016/j.foodhyd.2019.01.028.

Jones, N., Ray, B., Ranjit, K. T., & Manna, A. C. (2008). Antibacterial activity of ZnO nanoparticle suspensions on a broad spectrum of microorganisms. *FEMS Microbiology Letters*, *279*(1), 71–76. doi: 10.1111/j.1574-6968.2007.01012.x.

Kamat, P. V. (2012). TiO2 nanostructures: Recent physical chemistry advances. *The Journal of Physical Chemistry C*, *116*(22), 11849–11851. doi: 10.1021/jp305026h.

Khairnar, S. V., Pagare, P., Thakre, A., Nambiar, A. R., Junnuthula, V., Abraham, M. C., & Dyawanapelly, S. (2022). Review on the scale-up methods for the preparation of solid lipid nanoparticles. *Pharmaceutics*, *14*(9), 1886. doi: 10.3390/pharmaceutics14091886.

Khan, J., Rudrapal, M., Bhat, E. A., Ali, A., Alaidarous, M., Alshehri, B., & Egbuna, C. (2021). Perspective insights to bio-nanomaterials for the treatment of neurological disorders. *Frontiers in Bioengineering and Biotechnology*, *9*, 724158. doi: 10.3389/fbioe.2021.724158.

Kiss, É. (2020). Nanotechnology in food systems: A review. *Acta Alimentaria*, *49*(4), 460–474. doi: 10.1556/066.2020.49.4.12.

Kothari, R., & Wani, K. A. (2021). Environmentally friendly slow release nano-chemicals in agriculture: A synoptic review. In *Research Anthology on Synthesis, Characterization, and Applications of Nanomaterials*, Mehdi Khosrow-Pour, D. B. A. (Ed.), IGI Global, 409–425. doi: 10.4018/978-1-7998-8591-7.ch019.

Krishna, A. R., Gurumoorthy, S., Elayappan, P., Sakthivadivel, P., Kumaran, S., & Pushparaj, P. (2022). A review on the application of nanotechnology in food industries. *Current Research in Nutrition & Food Science*, *10*(3), 871–883. doi: 10.12944/CRNFSJ.10.3.5.

Kulyk, K., Borysenko, M., Kulik, T., Mikhalovska, L., Alexander, J. D., & Palianytsia, B. (2015). Chemisorption and thermally induced transformations of polydimethylsiloxane on the surface of nanoscale silica and ceria/silica. *Polymer Degradation and Stability*, *120*, 203–211. doi: 10.1016/j.polymdegradstab.2015.07.004.

Kumar, A., Pratush, A., & Bera, S. (2021). Significance of nanoscience in food microbiology: Current trend and future prospects. *Nanotechnology for Advances in Medical Microbiology*, 249–267. doi: 10.1007/978-981-15-9916-3_10.

Kumar, P. S. M., Francis, A. P., & Devasena, T. (2014). Biosynthesized and chemically synthesized titania nanoparticles: Comparative analysis of antibacterial activity. *Journal of Environmental Nanotechnology*, *3*(3), 73–81. doi: 10.13074/jent.2014.09.143098.

Lee, J. H., You, S. M., Luo, K., Ko, J. S., Jo, A. H., & Kim, Y. R. (2021). Synthetic ligand-coated starch magnetic microbeads for selective extraction of food additive silicon dioxide from commercial processed food. *Nanomaterials*, *11*(2), 532. doi: 10.3390/nano11020532.

Lee, W. H., Loo, C. Y., Bebawy, M., Luk, F., Mason, R. S., & Rohanizadeh, R. (2013). Curcumin and its derivatives: Their application in neuropharmacology and neuroscience in the 21st century. *Current Neuropharmacology*, *11*(4), 338–378. doi: 10.2174/1570159X11311040002.

Li, D., van Nostrum, C. F., Mastrobattista, E., Vermonden, T., & Hennink, W. E. (2017). Nanogels for intracellular delivery of biotherapeutics. *Journal of Controlled Release*, *259*, 16–28. doi: 10.1016/j.jconrel.2016.12.020.

Maqsoudlou, A., Assadpour, E., Mohebodini, H., & Jafari, S. M. (2022). The influence of nanodelivery systems on the antioxidant activity of natural bioactive compounds. *Critical Reviews in Food Science and Nutrition*, *62*(12), 3208–3231. doi: 10.1080/10408398.2020.1863907.

Mason, T. G., Wilking, J. N., Meleson, K., Chang, C. B., & Graves, S. M. (2006). Nanoemulsions: Formation, structure, and physical properties. *Journal of Physics: Condensed Matter*, *18*(41), R635. doi: 10.1088/0953-8984/18/41/R01.

McClements, D. J. (2017a). Designing biopolymer microgels to encapsulate, protect and deliver bioactive components: Physicochemical aspects. *Advances in Colloid and Interface Science*, *240*, 31–59. doi: 10.1016/j.cis.2016.12.005.

McClements, D. J. (2017b). Recent progress in hydrogel delivery systems for improving nutraceutical bioavailability. *Food Hydrocolloids*, *68*, 238–245. doi: 10.1016/j.foodhyd.2016.05.037.

McClements, D. J., & Xiao, H. (2012). Potential biological fate of ingested nanoemulsions: Influence of particle characteristics. *Food & Function*, *3*(3), 202–220. doi: 10.1039/C1FO10193E.

Medeiros, A. K. D. O. C., de Carvalho Gomes, C., de Araújo Amaral, M. L. Q., de Medeiros, L. D. G., Medeiros, I., Porto, D. L., & Passos, T. S. (2019). Nanoencapsulation improved water solubility and color stability of carotenoids extracted from Cantaloupe melon (Cucumis melo L.). *Food Chemistry*, *270*, 562–572. doi: 10.1016/j.foodchem.2018.07.099.

Mehmood, T., Ahmed, A., & Ahmed, Z. (2021). Food-grade nanoemulsions for the effective delivery of β-carotene. *Langmuir*, *37*(10), 3086–3092. doi: 10.1021/acs.langmuir.0c03399.

Min, K. A., Cho, J. H., Song, Y. K., & Kim, C. K. (2016). Iron casein succinylate-chitosan coacervate for the liquid oral delivery of iron with bioavailability and stability enhancement. *Archives of Pharmacal Research*, *39*, 94–102. doi: 10.1007/s12272-015-0684-6.

Mirhosseini, M., & Dehestani, R. (2020). Increasing the shelf life of milk by metal oxide nanoparticles and mild heat. *Journal of Nutrition and Food Security*, *5*(3), 227–235. doi: 10.18502/jnfs.v5i3.3795.

Mlalila, N., Kadam, D. M., Swai, H., & Hilonga, A. (2016). Transformation of food packaging from passive to innovative via nanotechnology: Concepts and critiques. *Journal of Food Science and Technology*, *53*, 3395–3407. doi: 10.1007/s13197-016-2325-6.

Moraru, C., Huang, Q., Takhistov, P., Dogan, H., & Kokini, J. (2009). Food nanotechnology: Current developments and future prospects. In *Global Issues in Food Science and Technology*, Barbosa-Cánovas, G., Mortimer, A., Lineback, D., Spiess, W., Buckle, K., & Colonna, P. (Eds.), Pullman, WA: Washington State University, 369–399. doi: 10.1016/B978-0-12-374124-0.00021-1.

Mousavi, S. R., & Rezaei, M. (2011). Nanotechnology in agriculture and food production. *Journal of Applied Environmental and Biological Sciences*, *1*(10), 414–419.

Nakagawa, K. (2014). Chapter 10: Nano- and micro-encapsulation of flavor in food systems. In *Nano- and Microencapsulation for Foods*, H.- S. Kwak, (Ed.), Oxford: John Wiley & Sons, 249–272. doi: 10.1002/9781118292327.ch10.

Naseer, B., Srivastava, G., Qadri, O. S., Faridi, S. A., Islam, R. U., & Younis, K. (2018). Importance and health hazards of nanoparticles used in the food industry. *Nanotechnology Reviews*, *7*(6), 623–641. doi: 10.1515/ntrev-2018-0076.

Nazir, A., Asghar, A., & Maan, A. A. (2017). Food gels: Gelling process and new applications. In *Advances in Food Rheology and Its Applications*, Woodhead Publishing, 335–353. doi: 10.1016/B978-0-08-100431-9.00013-9.

Neethirajan, S., & Jayas, D. S. (2011). Nanotechnology for the food and bioprocessing industries. *Food and Bioprocess Technology*, *4*, 39–47.

Nielsen, A. V., Beauchamp, M. J., Nordin, G. P., & Woolley, A. T. (2020). 3D printed microfluidics. *Annual Review of Analytical Chemistry*, *13*, 45–65. doi: 10.1146/annurev-anchem-091619-102649.

Nile, S. H., Baskar, V., Selvaraj, D., Nile, A., Xiao, J., & Kai, G. (2020). Nanotechnologies in food science: Applications, recent trends, and future perspectives. *Nano-micro Letters*, *12*, 1–34. doi: 10.1007/s40820-020-0383-9.

Noori, S., Zeynali, F., & Almasi, H. (2018). Antimicrobial and antioxidant efficiency of nanoemulsion-based edible coating containing ginger (Zingiber officinale) essential oil and its effect on safety and quality attributes of chicken breast fillets. *Food Control*, *84*, 312–320. doi: 10.1016/j.foodcont.2017.08.015.

Nowacka, M., Wiktor, A., Dadan, M., Rybak, K., Anuszewska, A., Materek, L., & Witrowa-Rajchert, D. (2019). The application of combined pre-treatment with utilization of sonication and reduced pressure to accelerate the osmotic dehydration process and modify the selected properties of cranberries. *Foods*, *8*(8), 283. doi: 10.3390/foods8080283.

Omerović, N., Djisalov, M., Živojević, K., Mladenović, M., Vunduk, J., Milenković, I., Knezevic, N. Z., Gadjanski, I., & Vidić, J. (2021). Antimicrobial nanoparticles and biodegradable polymer composites for active food packaging applications. *Comprehensive Reviews in Food Science and Food Safety*, *20*(3), 2428–2454. doi: 10.1111/1541-4337.12727.

Onyeaka, H., Passaretti, P., Miri, T., & Al-Sharify, Z. T. (2022). The safety of nanomaterials in food production and packaging. *Current Research in Food Science*, *5*, 763–774. doi: 10.1016/j.crfs.2022.04.005.

Otoni, C. G., Pontes, S. F., Medeiros, E. A., & Soares, N. D. F. (2014). Edible films from methylcellulose and nanoemulsions of clove bud (Syzygium aromaticum) and oregano (Origanum vulgare) essential oils as shelf life extenders for sliced bread. *Journal of Agricultural and Food Chemistry*, *62*(22), 5214–5219. doi: 10.1021/jf501055f.

Pal, M. (2017). Nanotechnology: A new approach in food packaging. *Journal of Food: Microbiology, Safety & Hygiene*, *2*(2), 121.

Pandey, S., Baker, S. N., Pandey, S., & Baker, G. A. (2012). Fluorescent probe studies of polarity and solvation within room temperature ionic liquids: A review. *Journal of Fluorescence, 22*, 1313–1343. doi: 10.5772/34455.

Parveen, K., Banse, V., & Ledwani, L. (2016, April). Green synthesis of nanoparticles: Their advantages and disadvantages. In *AIP conference proceedings* (Vol. 1724, No. 1), AIP Publishing. doi: 10.1063/1.4945168.

Pateiro, M., Gómez, B., Munekata, P. E., Barba, F. J., Putnik, P., Kovačević, D. B., & Lorenzo, J. M. (2021). Nanoencapsulation of promising bioactive compounds to improve their absorption, stability, functionality and the appearance of the final food products. *Molecules, 26*(6), 1547. doi: 10.3390/molecules26061547.

Pathakoti, K., Manubolu, M., & Hwang, H. M. (2017). Nanostructures: Current uses and future applications in food science. *Journal of Food and Drug Analysis, 25*(2), 245–253. doi: 10.1016/j.jfda.2017.02.004.

Pereira, M. C., Oliveira, D. A., Hill, L. E., Zambiazi, R. C., Borges, C. D., Vizzotto, M., & Gomes, C. L. (2018). Effect of nanoencapsulation using PLGA on antioxidant and antimicrobial activities of guabiroba fruit phenolic extract. *Food Chemistry, 240*, 396–404. doi: 10.1016/j.foodchem.2017.07.144.

Pradhan, N., Singh, S., Ojha, N., Shrivastava, A., Barla, A., Rai, V., & Bose, S. (2015). Facets of nanotechnology as seen in food processing, packaging, and preservation industry. *BioMed Research International, 2015*, 365672. doi: 10.1155/2015/365672.

Prakash, J., Vignesh, K., Anusuya, T., Kalaivani, T., Ramachandran, C., Sudha, R. R., & DevanandVenkatasubbu, G. (2019). Application of nanoparticles in food preservation and food processing. *Journal of Food Hygiene and Safety, 34*(4), 317–324. doi: 10.13103/JFHS.2019.34.4.317.

Prerna, D. A., & Ratan, G. (2021). Nanoparticles: An overview. *Drugs and Cell Therapies in Haematology, 10*(1), 1487–1497.

Pui, L. P., Karim, R., Yusof, Y. A., Wong, C. W., & Ghazali, H. M. (2020). Anticaking agent effects on the properties of spray-dried 'cempedak' fruit powder. *Pertanika Journal of Tropical Agricultural Science, 43*(4), 621–635.

Qureshi, M. A., Karthikeyan, S., Karthikeyan, P., Khan, P. A., Uprit, S., & Mishra, U. K. (2012). Application of nanotechnology in food and dairy processing: An overview. *Pakistan Journal of Food Science, 22*(1), 23–31.

Raghupathi, K. R., Koodali, R. T., & Manna, A. C. (2011). Size-dependent bacterial growth inhibition and mechanism of antibacterial activity of zinc oxide nanoparticles. *Langmuir, 27*(7), 4020–4028. doi: 10.1021/la104825u.

Rai, M., Yadav, A., & Gade, A. (2009). Silver nanoparticles as a new generation of antimicrobials. *Biotechnology Advances, 27*, 76–83. doi: 10.1016/j.biotechadv.2008.09.002.

Roselli, M., Finamore, A., Garaguso, I., Britti, M. S., & Mengheri, E. (2003). Zinc oxide protects cultured enterocytes from the damage induced by Escherichia coli. *The Journal of Nutrition, 133*(12), 4077–4082. doi: 10.1093/jn/133.12.4077.

Ruan, R., Choi, Y. J., & Chung, M. S. (2007). Caking in food powders. *Food Science and Biotechnology, 16*(3), 329–336.

Rücker, C., & Kümmerer, K. (2015). Environmental chemistry of organosiloxanes. *Chemical Reviews, 115*(1), 466–524. doi: 10.1021/cr500319v.

Sahoo, M., Vishwakarma, S., Panigrahi, C., & Kumar, J. (2021). Nanotechnology: Current applications and future scope in food. *Food Frontiers, 2*(1), 3–22. doi: 10.1002/fft2.58.

Sarkar, P., Choudhary, R., Panigrahi, S., Syed, I., Sivapratha, S., & Dhuma, C. V. (2017). Nano-inspired systems in food technology and packaging. *Environmental Chemistry Letters, 15*, 607–622. doi: 10.1007/s10311-017-0649-8.

Sawai, J. (2003). Quantitative evaluation of antibacterial activities of metallic oxide powders (ZnO, MgO and CaO) by conductimetric assay. *Journal of Microbiological Methods, 54*(2), 177–182. doi: 10.1016/S0167-7012(03)00037-X.

Shah, M. A., Mir, S. A., & Bashir, M. (2018). Nanoencapsulation of food ingredients. In *Food Science and Nutrition: Breakthroughs in Research and Practice EDS*, D. B. A. Mehdi Khosrow-Pour (Ed.), IGI Global, 218–234. doi: 10.4018/978-1-5225-5207-9.ch011.

Shahidi, F., & Zhong, Y. (2010). Lipid oxidation and improving the oxidative stability. *Chemical Society Reviews, 39*(11), 4067–4079. doi: 10.1039/b922183m.

Sharifi, A., Golestan, L., & Sharifzadeh Baei, M. (2013). Studying the enrichment of ice cream with alginate nanoparticles including Fe and Zn salts. *Journal of Nanoparticles, 2013*, 754385. doi: 10.1155/2013/754385.

Shewan, H. M., & Stokes, J. R. (2013). Review of techniques to manufacture micro-hydrogel particles for the food industry and their applications. *Journal of Food Engineering, 119*(4), 781–792. doi: 10.1016/j.jfoodeng.2013.06.046.

Siddiqui, S. A., Bahmid, N. A., Taha, A., Abdel-Moneim, A. M. E., Shehata, A. M., Tan, C., & Jafari, S. M. (2022). Bioactive-loaded nanodelivery systems for the feed and drugs of livestock; purposes, techniques and applications. *Advances in Colloid and Interface Science, 308*, 102772. doi: 10.1016/j. cis.2022.102772.

Silva, M. M., & Lidon, F. C. (2016). An overview on applications and side effects of antioxidant food additives. *Emirates Journal of Food and Agriculture, 28*(12), 823–832. doi: 10.9755/ejfa.2016-07-806.

Singh, P. (2018). Nanotechnology in food preservation. *Food Science, 9*(2), 435–441. doi: 10.15740/HAS/ FSRJ/9.2/435-441.

Singh, R., Dutt, S., Sharma, P., Sundramoorthy, A. K., Dubey, A., Singh, A., & Arya, S. (2023). Future of nanotechnology in food industry: Challenges in processing, packaging, and food safety. *Global Challenges, 7*(4), 2200209. doi: 10.1002/gch2.202200209.

Singh, T., Shukla, S., Kumar, P., Wahla, V., Bajpai, V. K., & Rather, I. A. (2017). Application of nanotechnology in food science: Perception and overview. *Frontiers in Microbiology, 8*, 1501. doi: 10.3389/ fmicb.2017.01501.

Srivastava, S., Bhargava, A., Srivastava, S., & Bhargava, A. (2022). Nanobiotechnology: Present Status and Future Prospects. *Green Nanoparticles: The Future of Nanobiotechnology*, 345–352. doi: 10.1007/978-981-16-7106-7_16.

Stokes, J. R. (2012). Food biopolymer gels, microgel and nanogel structures, formation and rheology. In *Food Materials Science and Engineering*, B. Bhandari, & Y. H. Roos (Eds.),West Sussex, UK: Wiley-Blackwell, 151–176. doi: 10.1002/9781118373903.ch6.

Tan, C., Xue, J., Abbas, S., Feng, B., Zhang, X., & Xia, S. (2014). Liposome as a delivery system for carotenoids: Comparative antioxidant activity of carotenoids as measured by ferric reducing antioxidant power, DPPH assay and lipid peroxidation. *Journal of Agricultural and Food Chemistry, 62*(28), 6726– 6735. doi: 10.1021/jf405622f.

Tankhiwale, R., & Bajpai, S. K. (2012). Preparation, characterization and antibacterial applications of ZnO-nanoparticles coated polyethylene films for food packaging. *Colloids and Surfaces B: Biointerfaces, 90*, 16–20. doi: 10.1016/j.colsurfb.2011.09.031.

Taouzinet, L., Djaoudene, O., Fatmi, S., Bouiche, C., Amrane-Abider, M., Bougherra, H., & Madani, K. (2023). Trends of nanoencapsulation strategy for natural compounds in the food industry. *Processes, 11*(5), 1459. doi: 10.3390/pr11051459.

Thakur, V. K., & Thakur, M. K. (2014). Recent advances in graft copolymerization and applications of chitosan: A review. *ACS Sustainable Chemistry & Engineering, 2*(12), 2637–2652. doi: 10.1021/sc500634p.

Thirumurugan, A., Ramachandran, S., & Shiamala Gowri, A. (2013). Combined effect of bacteriocin with gold nanoparticles against food spoiling bacteria-an approach for food packaging material preparation. *International Food Research Journal, 20*(4), 1909–1912.

Valerini, D., Tammaro, L., Di Benedetto, F., Vigliotta, G., Capodieci, L., Terzi, R., & Rizzo, A. (2018). Aluminum-doped zinc oxide coatings on polylactic acid films for antimicrobial food packaging. *Thin Solid Films, 645*, 187–192. doi: 10.1016/j.tsf.2017.10.038.

Vasilache, V., Popa, C., Filote, C., Cretu, M. A., & Benta, M. (2011, February). Nanoparticles applications for improving the food safety and food processing. In 7th *International Conference on Materials Science and Engineering-BRAMAT Braşov, ebruary*, 24–26.

Warczok, J., Ferrando, M., Lopez, F., & Güell, C. (2004). Concentration of apple and pear juices by nanofiltration at low pressures. *Journal of Food Engineering, 63*(1), 63–70. doi: 10.1016/S0260-8774(03)00283-8.

Wei, Z., & Gao, Y. (2016). Physicochemical properties of β-carotene bilayer emulsions coated by milk proteins and chitosan-EGCG conjugates. *Food Hydrocolloids, 52*, 590–599. doi: 10.1016/j.foodhyd.2015.08.002.

Wesley, S. J., Raja, P., Raj, A. A., & Tiroutchelvamae, D. (2014). Review on-nanotechnology applications in food packaging and safety. *International Journal of Engineering Research, 3*(11), 645–651.

Yadav, D., Karki, S., & Ingole, P. G. (2022). Nanofiltration (NF) membrane processing in the food industry. *Food Engineering Reviews, 14*(4), 579–595. doi: 10.1007/s12393-022-09320-4.

Yadav, S., Mehrotra, G. K., Bhartiya, P., Singh, A., & Dutta, P. K. (2020). Preparation, physicochemical and biological evaluation of quercetin based chitosan-gelatin film for food packaging. *Carbohydrate Polymers, 227*, 115348. doi: 10.1016/j.carbpol.2019.115348.

Yadav, S., Mehrotra, G. K., & Dutta, P. K. (2021). Chitosan based ZnO nanoparticles loaded gallic-acid films for active food packaging. *Food Chemistry, 334*, 127605. doi: 10.1016/j.foodchem.2020.127605.

Yang, X., Tian, H., Ho, C. T., & Huang, Q. (2011). Inhibition of citral degradation by oil-in-water nanoemulsions combined with antioxidants. *Journal of Agricultural and Food Chemistry, 59*(11), 6113–6119. doi: 10.1021/jf2012375.

Yapıcı, E., Karakuzu-İkizler, B., & Yücel, S. (2021). Anticaking additives for food powders. In *Food Powders Properties and Characterization*. Food Engineering Series, E. Ermiş (Ed.), Cham: Springer. doi: 10.1007/978-3-030-48908-3_6.

Yotova, L., Yaneva, S., & Marinkova, D. (2013). Biomimetic nanosensors for determination of toxic compounds in food and agricultural products. *Journal of Chemical Technology & Metallurgy, 48*(3), 215–227.

Zhang, H., Zhai, Y., Wang, J., & Zhai, G. (2016). New progress and prospects: The application of nanogel in drug delivery. *Materials Science and Engineering: C, 60*, 560–568. doi: 10.1016/j.msec.2015.11.041.

Zhang, X., Song, R., Liu, X., Xu, Y., & Wei, R. (2022). Fabrication of vitamin D3 nanoemulsions stabilized by Tween 80 and Span 80 as a composite surface-active surfactant: Characterization and stability. *Colloids and Surfaces A: Physicochemical and Engineering Aspects, 645*, 128873. doi: 10.1016/j.colsurfa.2022.128873.

Zhang, X., Xiao, G., Wang, Y., Zhao, Y., Su, H., & Tan, T. (2017). Preparation of chitosan-TiO2 composite film with efficient antimicrobial activities under visible light for food packaging applications. *Carbohydrate Polymers, 169*, 101–107. doi: 10.1016/j.carbpol.2017.03.073.

Zhao, C., Shen, X., & Guo, M. (2018a). Stability of lutein encapsulated whey protein nano-emulsion during storage. *PLoS One, 13*(2), e0192511. doi: 10.1371/journal.pone.0192511.

Zhao, J., Niu, Y., Ren, B., Chen, H., Zhang, S., Jin, J., & Zhang, Y. (2018b). Synthesis of Schiff base functionalized superparamagnetic Fe3O4 composites for effective removal of Pb (II) and Cd (II) from aqueous solution. *Chemical Engineering Journal, 347*, 574–584. doi: 10.1016/j.cej.2018.04.151.

Zhao, R., Torley, P., & Halley, P. J. (2008). Emerging biodegradable materials: Starch-and protein-based bio-nanocomposites. *Journal of Materials Science, 43*, 3058–3071.

Zhu, Z., Zhang, Y., Zhang, Y., Shang, Y., Zhang, X., & Wen, Y. (2019). Preparation of PAN TiO2 nanofibers for fruit packaging materials with efficient photocatalytic degradation of ethylene. *Materials, 12*(6), 896. doi: 10.3390/ma12060896.

11 Nanotechnology in Food Packaging

*Jamil A. Buledi, Arfana Mallah, Ali Hyder,
and Hassan Karimi Maleh*

11.1 INTRODUCTION

The domain of food packaging represents a continuously evolving industry, permanently under pressure to originate and enhance its contributions. Food packaging materials serve as crucial barriers, safeguarding the integrity and healthfulness of consumable products for end-users. Consequently, the field of processed food products places significant emphasis on the quality of packaging. The beginning of the Industrial Revolution amplified the mission for novel packaging materials endowed with numerous qualities, including aesthetics, robustness, healthy properties, visual application, and environmental sustainability, thereby highlighting the significance of material innovation (Enescu, Cerqueira, Fucinos, & Pastrana, 2019; Vasile & Baican, 2021). Furthermore, a primary objective in the packaging field is enhancing food quality safety and extending its shelf life, aligning with the demands of modern, fast-paced lifestyles. Alterations in food characteristics, such as fragrance and flavor, stemming from aroma absorption and the transfer of flavors from packaging materials, constitute pivotal mechanisms implicated in the deterioration of packaged food products (Porta, 2019; Yaris & Sezgin, 2015). Plastic materials are widely used in the food industry for packaging due to their cost-effectiveness, lightweight nature, and ability to provide a barrier against gas and moisture infiltration (Porta, 2019; Realini & Marcos, 2014). However, many plastic packaging materials exhibit limited resistance to high temperatures, contain potentially harmful plasticizers, and are non-biodegradable, leading to environmental pollution concerns (Ahari & Soufiani, 2021). Consequently, there is an urgent demand within the food industry to identify alternative packaging materials and to develop active and intelligent packaging solutions that can mitigate the drawbacks associated with traditional plastic packaging.

Furthermore, nanotechnology represents a valuable source of innovations that can address contemporary challenges related to food safety and sustainability (Alfei, Marengo, & Zuccari, 2020; Khatoon et al., 2022; Kuswandi & Moradi, 2019; Shah et al., 2023). Recently, significant advancements have been made in packaging technologies to enhance the durability, quality, safety, and freshness of food products (Dobrucka & Cierpiszewski, 2014; Kuswandi et al., 2011). Notably, the utilization of biodegradable polymers, bio-nanocomposites, nanostructured materials, and engineered nanoparticles has become increasingly prevalent in the development of active and intelligent materials for food packaging (Buledi et al., 2022; Reichert et al., 2020). Additionally, innovative methodologies and equipment have been devised to enable rapid in situ analysis and the production of environmentally friendly and degradable food packaging (Bakhsh, Buledi, Ghumro, et al., 2021; Firouz, Mohi-Alden, & Omid, 2021; Geueke, Groh, & Muncke, 2018). In addition, nanomaterials, including metal nanoparticles, nanoemulsions, and nanoclays, exhibit substantial potential for packaging applications. The incorporation of these nanomaterials into packaging enhances the stability and susceptibility of compounds, ultimately promoting the preservation and safeguarding of packaged food items (Mihindukulasuriya & Lim, 2014; Ncube, Ude, Ogunmuyiwa, Zulkifli, & Beas, 2020). However, it is worth noting that the application of nanotechnology in food packaging is still in its nascent stages. Presently, there is a growing trend in the packaging industry toward introducing various functional nanoparticles to extend the shelf life of products and ensure the safety of packaged food items (Pathakoti, Manubolu, & Hwang, 2017).

Over the past decade, there has been substantial progress in packaging materials and packaging technology. This chapter provides insights into various packaging technologies, elucidating their inherent limitations. Additionally, it examines the potential of nanomaterials, incorporating metal nanoparticles, nanoemulsions, and nanoclays within the area of food packaging while addressing societal concerns. This information sheds light on prevailing trends and the potential applications of nanomaterials in food packaging.

11.2 ANTIMICROBIAL PACKING AGENTS

In the last few decades, the packaging industry has experienced consistent growth and technological advancements, particularly in the types of materials used for packaging and the packaging systems utilized to extend food preservation and decrease food waste (Vermeiren, Devlieghere, van Beest, de Kruijf, & Debevere, 1999). Food packaging is crucial in the food production industry and is at the heart of manufacturing businesses. It is vital to guarantee food quality for proper growth and to maintain the human body for physical work (Robertson, 2005; Saadat, Pandey, Tharmavaram, Braganza, & Rawtani, 2020). Moreover, preventing waste, minimizing preservatives, and safeguarding against spoilage are vital functions of packaging materials employed in the food industry to enhance the self-life of food products for long-term uses. Therefore, food packaging systems are designed to safeguard food from a range of potential threats such as environmental contamination, odors, physical impacts, dust, temperature variations, mechanical forces, breakage, moisture, gases, physical damage, light exposure, microorganisms, and humidity (Bakhsh, Buledi, Khand, et al., 2021; Braga, Rangel, Suarez, & Machado, 2018; Buledi et al., 2021). However, these protections are essential during various stages, such as transportation, processing, storage, and marketing of food products. Also, food packaging plays a dynamic part in maintaining the essential qualities of food, including its color, temperature, taste, texture, and overall quality of food products. In addition, a well-designed food packaging system can effectively mitigate the primary causes of food deterioration, including oxidation, microbial spoilage, and metabolic processes. This, in turn, leads to improved food quality and extended shelf life. Oxidation, for example, can diminish food products' nutritional value, energy content, flavor, and color, ultimately compromising their overall quality. To avoid the deterioration nature of food products, a vast number of food-protecting agents such as gourds, shells, grass, wood, paper, glass, metal, plastic, and biopolymer are widely employed to maintain the food quality and protect from pathogen substances (Buledi et al., 2022; Risch, 2009). However, these food-packing substances have drawbacks such as high production volume, short usage time, and non-biodegradable nature. Therefore, it is urgently necessary to develop a cost-effective and efficient material to protect food products from pathogens and enhance their shelf-life for long-term use.

Nanotechnology is a powerful interdisciplinary tool that promotes the manufacture of innovative products. It was predicted that nanotechnology would substantially impact at least $3 trillion in the global economy in 2020 and generate demand for six million jobs in various industries worldwide (Buledi et al., 2022; Duncan, 2011). Due to rapid development and advancement in the field of nanotechnology, several nanoscale materials such as metal and metal oxide nanoparticles, nanofiber materials, carbon nanotubes, core-shell nanomaterials, clay nanomaterials, graphene oxide-based composites materials, and organic-inorganic nanohybrids were employed to enhance the reinforcement of gas and moisture barriers in food packaging which improve the processability of polymers and impart specialized capabilities such as antimicrobial activity and gas sensing features (Garcia, Shin, & Kim, 2018; Hyder, Buledi, et al., 2023; Pandey, Munguambe, Tharmavaram, Rawtani, & Agrawal, 2017). Because of their nanosized dimensions, these materials have an exceptionally high surface-to-volume ratio and surface activity (Buledi, Hyder, et al., 2023; Hyder, Memon, Buledi, Memon, Memon, Rajpar, et al., 2023; Hyder, Memon, Buledi, Memon, Memon, Shaikh, et al., 2023;

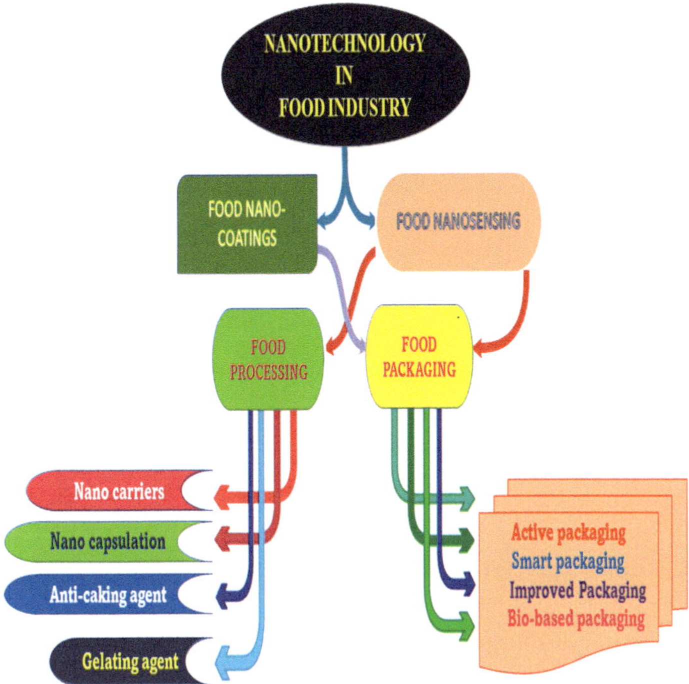

FIGURE 11.1 Schematic representation applications of nanotechnology in the food industry.

Yang & Theboc, 2023). When incorporated into compatible polymers, nanomaterials can significantly improve the material properties of the resulting nanocomposites. These enhancements can include increased mechanical strength, enhanced thermal stability, elevated electrical conductivity, and other desirable characteristics of the food products (Buledi, Solangi, et al., 2023; Uskoković, 2007). Consequently, nanomaterials hold promise for enhancing the mechanical and barrier properties of food packaging and for creating advanced structures suitable for active and intelligent applications in the food industry. These classical functions can be summarized in Figure 11.1.

Numerous studies examining the physical, microbiological, and chemical effects of nano-food packaging have been reported in recent years due to their increasing significance in food packing. In this section, we have provided the nanomaterials' antimicrobial application in food packing.

Moreover, many researchers have been working on food packing by exploiting nanomaterials, such as Li and coworkers (2016) have fabricated the core-shell nanocomposite material using N-halamine, i.e., 3-ally 5,5-dimethylhydantoin (ADMH) precursor to produce TiO_2-ADMH nanocomposite through miniemulsion polymerization via employing 3-allyl-5,5-dimethylhydantoin (ADMH) and methyl methacrylate (MMA) to create TiO_2@poly(ADMH-co-MMA nanocomposite material as shown in Figure 11.2. The fabricated core-shell TiO_2@poly(ADMH-co-MMA nanocomposite material showed antimicrobial effectiveness against both Gram-positive *S. aureus* and Gram-negative bacteria *E. coli*. The nanocomposite exhibited a 100% neutralization efficacy against *S. aureus* and *E. coli* within 10–30 minutes. Moreover, the consistent transfer of effective oxidizing agent chlorine from the surface of N-halamine into microbial cells explained the mechanism behind the solid antimicrobial action of the nanocomposite material. Furthermore, the in vitro cytocompatibility results indicated that the nanocomposite was compatible with mammalian cells. Therefore, these results suggested that the prepared N-halamine modified core-shell titanium dioxide core-shell nanoparticles are anticipated to have excellent applications in various antibacterial fields such as healthcare materials, paints, food packaging, and coating industries.

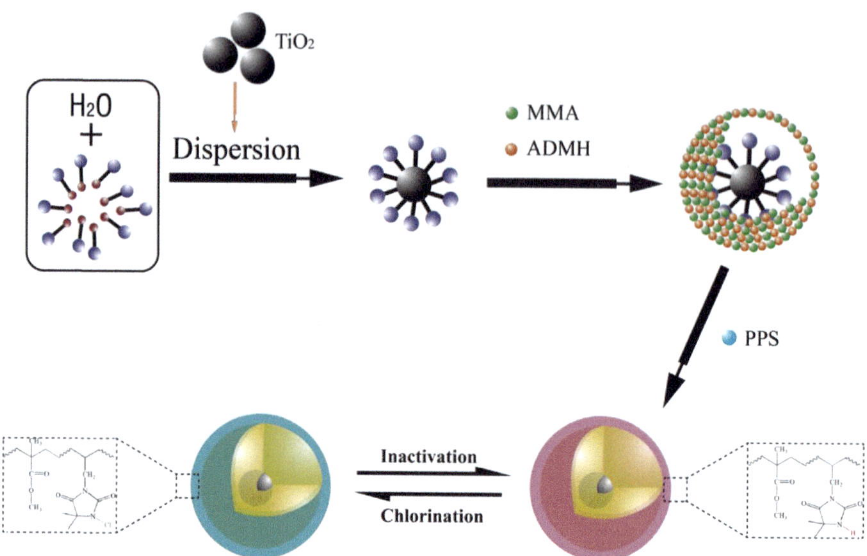

*MMA: Methyl methacrylate; **ADMH**: 3-allyl-5,5-dimethylhydantoin; **PPS**: Potassium persulfate*

FIGURE 11.2   The schematic illustration of TiO₂@poly (ADMH-co-MMA)-Cl N.P.s preparation procedure. Copyright © 2023 with permission from Elsevier (Li et al., 2016).

Similarly, Zhao and coworkers prepared silica nanoparticles functionalized with new *N*-halamine precursor material through the surface immobilization of 3-(4′-epoxyethyl-benzyl)-5,5-dimethylhydantoin (EBDMH) of amino-self-modified silica nanoparticles (EBDMH-SiO₂). Then, the EBDMH-SiO₂ nanoparticles were incorporated into the PLA medium using the melt blending functionalization technique. Moreover, the effectiveness of the EBDMH-SiO₂/PLA nanocomposite as an antimicrobial material was assessed against *S. aureus* and *E. coli* bacteria, respectively. Antimicrobial evaluations against the microbial strains were conducted through contact killing. The minimum inhibition concentration (MIC) was determined for EBDMH-SiO₂ nanoparticles alone, and the MIC values against both *S. aureus* and *E. coli* were found to be 60 mg mL^{-1} against these bacteria. It was observed that the oxidative chlorine content in this nanocomposite was responsible for the enhanced antimicrobial activity of the EBDMH-SiO₂/PLA nanocomposite against both *S. aureus* and *E. coli*. However, the EBDMH-SiO₂/PLA nanocomposite displayed remarkable bactericidal effectiveness. With a contact time of 10 minutes, it neutralized approximately 90.2% of *S. aureus* and 89.4% of *E. coli*. When the contact time was extended to 180 minutes (3 hours), a kill efficiency of 99.97% was achieved against *S. aureus* and 99.91% against *E. coli* bacterial species.

Due to their tremendous antimicrobial properties, low bio-toxicity, and outstanding biocompatible nature, EBDMH-SiO₂/PLA nanocomposite material was utilized in drinking water bottle packaging to enhance their self-life period and protect them from bacterial attack. Also, the prepared EBDMH-SiO₂/PLA nanocomposite material found versatile applications in food packaging, packaging hygiene products, filters, and medical and textile devices (Zhao et al., 2020). In addition, Liu et al. have engineered a novel synthetic protocol to synthesize a food packaging material. This method involved the functionalization of polymer poly(vinyl alcohol) (PVA) onto dimethylol-5,5-dimethylhydantoin (DMDMH). Moreover, the entire synthesis process was environmentally friendly, and the PVA nanofibrous substrate was covalently attached to the surface of DMDMH nanoparticles. However, the resulting DMDMH/PVA nanocomposite was chlorinated using a sodium hypochlorite solution. The prepared DMDMH/PVA nanocomposite showed excellent features, including sustainable antimicrobial properties and robust mechanical strength. Therefore, these remarkable properties of the DMDMH/PVA nanocomposite make it a strong aspirant for applications like antimicrobial food packaging substances in the food industry. The presence of a chlorinated functional group on the surface of DMDMH/PVA nanocomposite is responsible for enhancing antimicrobial performance against both *S. aureus* and *E. coli*, respectively. Also, the DMDMH/PVA nanocomposite material significantly reduced the colony-forming unit (CFU) log count. Initially, 108 CFU mL^{-1} concentrations were used for *S. aureus*, and 4×10^8 CFU mL^{-1} was used for *E. coli*. However, the chlorinated DMDMH/PVA nanocomposite material achieved a 6 log CFU mL^{-1} reduction in both *S. aureus* and *E. coli* within 1 minute. This biocidal activity is equivalent to a 99.99% kill efficiency, which led to the utilization of the chlorinated functionalized DMDMH/PVA nanocomposite for packaging nectarine fruits. When this chlorinated functionalized DMDMH/PVA nanocomposite material was employed in food packaging, it showed excellent antimicrobial features. It extended the shelf life of fruit and prevented spoilage for up to 7 days. However, nectarine fruit packaged with conventional materials had a shorter shelf life, typically lasting less than 3 days. Therefore, the findings of this study have contributed to enhancing the shelf life and protecting the fruit significantly from bacterial attack (Liu et al., 2020).

In another work, Yousef and his coworkers fabricated the cellulose acetate functionalized copper (CA-Cu) nanocomposite through the reduction method. The antimicrobial activity of the CA-assisted copper nanocomposite was evaluated against both *S. aureus* and *P. aeruginosa*, respectively. Experiments were carried out to determine the inhibition zones by loading the 2% weight of CuNPs onto the surface of cellulose acetate (CA). However, this method resulted in the highest bactericidal activity against *S. aureus* and *P. aeruginosa*. Any load exceeding 2% led to the irregular distribution of CuNPs on the CA surface, resulting in reduced antimicrobial activity. The inhibition zone diameter was 21 mm against *S. aureus* and 18 mm against *P. aeruginosa*. So, the higher susceptibility of *S. aureus* toward the Cu-CA nanocomposite was recognized due to differences in the composition of the microbial membranes between Gram-negative and Gram-positive bacteria. Thus, Cu-CA nanocomposite material enhanced the shelf life of packaged food products and effectively inhibited microbial growth within the packaging materials (Abou-Yousef, Saber, Abdel-Aziz, & Kamel, 2018).

In addition, Tóth and coworkers have synthesized the copper-humate cellulose nanocomposite material through a simple and cost-effective method. The synthesized copper-humate cellulose nanocomposite material was used as an effective antimicrobial agent for the *E. faecalis* and *L. monocytogenes* bacteria. The synthesized copper-humate cellulose nanocomposite material efficiency was determined to be approximately 100%. Additionally, the Cu-humate cellulose nanocomposite film was suitable for preserving and prolonging the shelf life of blackberry fruit and vegetables (Tóth, Balázs, & Halász, 2020). Aline et al. and their colleagues prepared using low-density polyethylene (LDPE) functionalized silver nanoparticle films through a cost-effective method. The developed functionalized is employed as an efficient antimicrobial agent against *S. aureus* and *E. coli*. Ag-LDPE nanocomposite material's enhanced antimicrobial performance is due to the high surface-to-volume ratio. In addition, the Ag-LDPE nanocomposite material interacted with the sulfhydryl component of the bacterial membrane, which is responsible for inhibiting the respiration process and ultimately causing the death of bacterial cells from the inhibition zones. Also, the prepared Ag-LDPE nanocomposite material showed excellent antimicrobial performance toward the Gram +ve bacteria *S. aureus* than Gram –ve bacteria *E. coli*. Moreover, when the Ag-LDPE nanocomposite material is incorporated with food packaging ingredients, it primarily enhances the shelf life of food products, including chicken fillets, by improving their freshness and sustainability period (Becaro et al., 2015).

In another study, Volova and coworkers (2018) engineered cellulose functionalized silver nanoparticles using a facile and eco-friendly method. The cellulose molecules contain various functional groups responsible for stabilizing the silver nanoparticles and modifying their surface through these groups. Furthermore, the engineered Ag-Cellulose nanocomposite was used as an efficient antimicrobial agent against *S. aureus* and *E. coli*. Also, the antimicrobial efficiency of the Ag-Cellulose nanocomposite was qualitatively checked out through the film inhibition zone method.

The composite material was brought into contact with bacteria species, and it was determined that the composite exhibited a kill efficiency of 99.9% against *E. coli*, and a 10% kill efficiency was observed for *S. aureus* bacterial cell death. However, the extraordinary antimicrobial efficiency of Ag-Cellulose nanocomposite material is due to the abundant silver nanoparticles that are uniformly distributed on the surface of the functionalizing agent cellulose and also cellulose molecule possessing the different functional groups which efficiently stabilized the silver nanoparticles and controlled the size of silver nanoparticles. Additionally, the uniform distribution allows silver nanoparticles to readily attach to bacterial cells, ultimately leading to their sterilization. Therefore, this study shows that nanocomposites like the Ag-Cellulose nanocomposite have applications in food packaging, particularly for meat and melon cuts, where they can help extend the shelf life and maintain the quality of these products by providing antimicrobial protection (Volova et al., 2018).

Mayorga and coworkers have synthesized functionalized copper oxide (CuO) nanoparticles by using poly (3-hydroxybutyrate-co-hydroxyvalerate) (PHBV) as functionalizing and stabilizing through a straightforward method. The functionalized CuO-PHBV nanocomposite material showed tremendous antimicrobial activity against *S. enterica* and *L. monocytogenes*. However, the CuO-PHBV nanocomposite material successfully reduced the bacterial concentration by about 3 log CFU mL^{-1} for both types of bacteria. The effectiveness in reducing bacterial counts shows the potential of CuO-PHBV nanocomposite material, which can be attributed to enhanced dispersion and distribution of CuO nanoparticles within the PHBV polymer matrix and effectively utilized in food packaging, including fruits, vegetables, beverage containers, and films (Castro Mayorga, Fabra Rovira, Cabedo Mas, Sánchez Moragas, & Lagarón Cabello, 2018).

Moreover, using a cost-effective and eco-friendly synthetic protocol, Nouri and coworkers fabricated Montmorillonite-copper oxide (MMT-CuO) nanocomposite material. The developed MMT-CuO nanocomposite material introduced a chitosan polymer matrix to escalate its optical, mechanical, thermal stability, and tensile properties. The refined CuO-MMT/Chitosan nanocomposite material was an effective antimicrobial agent against *S. aureus*, *B. cereus*, *E. coli*, *and P. aeruginosa*, as shown in Figure 11.3. Thus, this study revealed that the CuO-MMT/Chitosan nanocomposite material is well-suited for integration into food packaging materials. Hence, applying this nanocomposite can be harnessed for active packaging purposes for pecans and meat products (Nouri, Yaraki, Ghorbanpour, Agarwal, & Gupta, 2018).

Similarly, Rokbani and his coworkers have developed antibacterial low-density polyethylene (LDPE) films fabricated by coating zinc oxide nanoparticles (ZnO NPs) onto their surface using a facile and cost-effective method. The produced ZnO-LDPE nanocomposite films were applied as an efficient and selective antimicrobial agent against *S. aureus* and *E. coli*. Moreover, antimicrobial agent ZnO-LDPE nanocomposite was positioned close to the bacteria to ensure effective bactericidal activity against both bacteria. Also, the developed nanocomposite material effectively suppressed approximately 99.9% of bacterial cell deaths. However, this study displayed that the ZnO-LDPE nanocomposite films possess outstanding characteristics when applied as an active coating material in food packaging (Rokbani, Daigle, & Ajji, 2019).

Kumar and coworkers developed chitosan and gelatin-functionalized zinc oxide nanoparticles (ZnO NPs) films using a cost-effective and straightforward method. The developed nanocomposite film shows excellent antimicrobial activity toward *S. aureus* and *E. coli*. The disc diffusion assays were employed to determine the antimicrobial activity against *S. aureus* and *E. coli*. Also, the nanocomposite film possessed outstanding thermal stability and mechanical strength due to the incorporation of nanomaterial on the surface of the polymer substrate. In addition, the developed nanocomposite film in this study has the potential to serve as a biodegradable alternative for post-harvest packaging of fresh fruits and vegetables (Kumar et al., 2020).

In addition, Ibrahim and coworkers also prepared polystyrene (PS) functionalized zinc oxide nanoparticles using an eco-friendly and facile synthetic route. The prepared ZnO-PS nanocomposite films exhibited tremendous antimicrobial activity against *S. aureus* and *P. aeruginosa*. However, the assessments were examined via contact killing tests and measurement of inhibition zones to

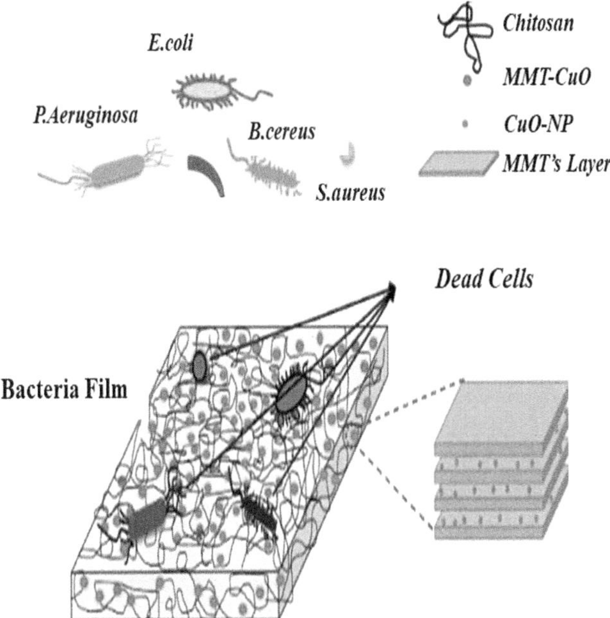

FIGURE 11.3   The antimicrobial mechanism against the targeted bacterial species. Copyright © 2023 with permission from Elsevier (Nouri et al., 2018).

evaluate the antimicrobial properties of ZnO-PS nanocomposite films against the *S. aureus* and *P. aeruginosa* bacteria. Therefore, the ZnO-PS nanocomposite films are recommended as a promising candidate for use as active coating substrate in the food packaging industry due to their small size, well-defined shape control, thermal stability, antimicrobial effectiveness, and mechanical strength (Ibrahim, El-Naggar, Youssef, & Abdel-Aziz, 2020).

Besides, Pina and his coworkers have developed novel nanocomposite films using zinc oxide nanoparticles employed on poly (ε-caprolactone) (PCL) surfaces through a simple method. The produced ZnO-PCL nanocomposite films were utilized as an effective antimicrobial agent for the *S. aureus* bacteria. Therefore, the developed ZnO-PCL nanocomposite films successfully inhibited the growth of bacterial cells on the surface of the packaging substrate and served as a preventive measure against foodborne diseases with packaging substrate (Pina et al., 2020).

Venkatesan and coworkers synthesized biodegradable poly (butylene adipate-co-terephthalate) (PBAT) functionalized titanium oxide (TiO₂) nanoparticles through a facile method. The synthesized TiO₂-PBAT nanocomposite films possess excellent properties such as tensile strength, thermal stability, and outstanding mechanical strength. The TiO₂-PBAT nanocomposite films exhibited profound antimicrobial activity against foodborne pathogenic bacteria such as *S. aureus* and *E. coli*. The remarkable antimicrobial properties of the TiO₂-PBAT nanocomposite films make them a suitable candidate for food packaging applications (Venkatesan & Rajeswari, 2017).

Cheng and coworkers engineered a double-layered hydroxide (LDHs) by intercalating with the para-hydro benzoic acid (PHBA) cost-effectively. The prepared ZnAl-PHBA-LDH composite material possessed outstanding mechanical strength, thermal stability, and tensile strength. Moreover, the engineered ZnAl-PHBA-LDH composite material was evaluated as an antimicrobial agent against S. aureus bacteria. The measurement of the inhibition zone diameter yielded a value of 15.12 mm, indicating vigorous antimicrobial activity toward *S. aureus* bacteria. However, the intercalating of PHBA with the surface of Zn-Al LDH significantly reduces contact between food and PHBA, thereby establishing a founding for the utilization of Zn-Al LDH nanocomposite-based film as an antimicrobial agent in food packaging (Cheng et al., 2019). In another study, a group researcher developed novel food packaging material utilizing Starch/halloysite/nisin nanocomposite material. The developed nanocomposite material exhibited good bactericidal activity toward *L. monocytogenes* bacteria in cheese, *C. perfringens,* and *S. aureus.* The measured inhibition zone diameters were approximately 8–10 mm when tested *L. monocytogenes,* 15–17.5 mm against *C. perfringens* and *S. aureus.* This study demonstrated its effectiveness in reducing *L. monocytogenes,* making it a valuable choice for application in cheese packaging (Meira, Zehetmeyer, Scheibel, Werner, & Brandelli, 2016).

Additionally, Lin and his colleagues have developed novel antibacterial packaging material using chitosan nanofibers loaded within the chrysanthemum essential oil (CHEO) through electrospinning. Also, the antibacterial activity of the CHEO/CS nanofibers revealed a minimum inhibitory concentration (MIC) value of 2.5 mg mL^{-1} against *L. monocytogenes.* However, the antibacterial activity of the CHEO/CS nanofibers against *L. monocytogenes* was assessed on beef resulting in inhibition rates of 99.91%, 99.97%, and 99.95% at temperatures of 4°C, 12°C, and 25°C, respectively, following 7 days of storage. Therefore, the obtained results of this study indicate that CHEO/CS/ nanofiber can enhance the shelf lifetime of beef for long-term uses and implying a promising application in food packaging for other food materials (Lin, Mao, Sun, Rajivgandhi, & Cui, 2019).

Thus, using nanocomposites incorporated with biomaterials has a broader scope in food packaging. These materials are continuously evolving and opening up new possibilities for addressing challenges related to food products' health, safety, and sustainability.

11.3   NANOMATERIALS TO IMPROVE PHYSICAL TEXTURE AND STABILITY OF FOOD

In recent years, nanotechnology has the potential for both the mechanical strength and antimicrobial properties of food packaging materials, effectively preventing microorganisms' infiltration into the food (Bumbudsanpharoke, Choi, & Ko, 2015). This improvement in the food barrier properties makes nanomaterials an auspicious choice for food packaging applications. However, utilizing nanomaterials in packaging design can contribute to preserving food freshness for long-term use. Moreover, nanomaterials find applications in diverse areas, including environmental protection, the pharmaceutical industry, the development of new materials, agriculture, food processing, and packaging (Carbone, Donia, Sabbatella, & Antiochia, 2016). The food industry commonly utilizes metal and metal oxide nanoparticles, graphene, carbon nanotubes, metal-organic framework, organic nanoparticles, nanoscaled cellulose, and inorganic hybrid material for packaging materials, storage enhancement to improve migration properties, employing nanopore filters for purification, and utilizing nano encapsulated additives and nanosized food and nutrients as dietary supplements (Sharma, Jafari, & Sharma, 2020).

In this connection, Yu and coworkers have fabricated proanthocyanin-loaded nanoparticles by using chitosan cross-linker via chondroitin sulfate (CS) were employed as nanofillers within the chitosan film matrix to create an innovative edible bioactive packaging material. Moreover, the electrostatic interactions and hydrogen bonding between the nanoparticles and the chitosan film matrix resulted in denser film, significantly enhancing both its mechanical strength and barrier properties. Additionally, the nanoparticles notably enhanced the film's antioxidant properties and mitigated the adverse effects of ultraviolet rays and thermal exposure on the biological activities of proanthocyanins (PC). Also, the chitosan-loaded nanoparticle films proved to be highly effective in preserving the texture of fish fillets and preventing the elevation of TVB-N and biogenic amine levels during storage. Its potent antioxidant, antibacterial, and oxygen isolation properties achieved this. Notably, using the CH-NPs film did not negatively impact the color or pH of the fish fillets (Yu et al., 2022).

Similarly, Yang and his coworkers have developed casein and octenyl succinic anhydride-modified starch (OSAS) nanocomposite material to improve food products' physical texture. However, the interaction between casein and oxidized starch (OSAS) was evident through alterations in protein structure and hydrogen-bonding interactions. Nanoemulsions stabilized by combinations of casein and oxidized starch (OSAS), or casein alone, exhibited superior oxidative stability compared to emulsions solely stabilized by OSAS. Thus, using casein-OSAS combinations can enhance various physical properties of protein-based emulsions and elevate the oxidative stability of modified starch-based emulsions. This suggests that using protein-modified starch combinations holds more promise in the emulsion-based food industry when compared to employing each of the two emulsifiers independently in the food packaging products (Yang, Qin, Kan, Liu, & Zhong, 2019).

In addition, Voon and coworkers (2012) have fabricated novel bio-composite films based on bovine gelatin, incorporating halloysite nanosclay and nano-SiO_2 by using the casting method. The developed bio-composite films exhibited better mechanical strength and water stability than the pristine gelatin films. However, the bovine gelatin bio-composite films potentially enhance the food packaging substances' barrier and mechanical properties. It also improves the stability and texture of food products by preventing microbial attacks (Voon, Bhat, Easa, Liong, & Karim, 2012). Joye and coworkers (Joye, Nelis, & McClements, 2015) have developed gliadin nanoparticles, which can be protected from aggregation in specific conditions by encapsulating them with anionic polysaccharides, such as pectin. Coating gliadin nanoparticles with pectin resulted in a notable enhancement in their stability when exposed to variations in ionic strength and temperature. Therefore, the gliadin nanoparticles can be used in the food industry to improve and enhance the stability of food packaging.

In addition, Qi and his colleagues have employed soybean oil emulsion stabilized by nanocrystalline cellulose (NCCs) as a replacement for animal fat in sausage production. They found that the NCC-based emulsion effectively preserved moisture content and maintained the compact structure of the sausages. This resulted in a product with favorable stability and texture characteristics. This outcome can likely be attributed to the strong emulsifying capacity, water-holding capacity (WHC), and gel-forming properties of the NCCs after interacting with meat proteins, other NCC particles, or a combination of both (Qi et al., 2020).

11.4 INTELLIGENT PACKAGING

In the enhancement of food packaging through the integration of smart or intelligent types, nanomaterials are employed to determine the chemical, biochemical, and microbial variations occurring within the food and its surroundings, a concept often referred to as smart or intelligent Packaging. For instance, specialized nanosensors have been designed to detect particular pathogens and gases by identifying food spoilage. In this scenario, nanomaterials can serve as responsive components within the packaging to relay and convey the state of the enclosed product, referred to as nanosensors (Kerry, O'grady, & Hogan, 2006; Toydemir et al., 2020). Moreover, the nanosensors can react to alterations in internal or external parameters, facilitating the communication and detection of changes within the food product. The objective is to ensure its quality and safety from the microbial attack. Also, the nanomaterials are responsible for enhancing food product packaging via incorporating intelligent properties, including freshness nanosensors, oxygen nanosensors, and nano-devices designed for tracking and authenticating food products. However, the nanosensors integrated into food packaging are also being advanced to detect the conditions of different food products or their containers, whether internal or external, as they assign the supply chain. For instance, packaging can be designed to be responsive to changes in humidity or temperature over a specific duration, subsequently conveying pertinent information about the food's condition through alterations in color. The nanosensors incorporated into plastic packaging can detect the gases emanating from spoiled food. In response, the packaging can undergo a color change and serve as a visual alter for customers. Thus, nanosensors can respond to alterations in the micro-environment within the food package, including humidity, temperature, and oxygen exposure levels (Bouwmeester et al., 2009).

Consequently, they can signal food deterioration or changes in microbial levels. In addition, Huang et al. synthesized iridium oxide (IrO x) nanoparticles using the sol-gel method. The fabricated iridium oxide nanoparticles were deposited on the surface of the glassy carbon electrode to construe an efficient electrochemical device for food quality monitoring. A wireless pH sensing system was evaluated for its ability to continuously monitor spoilage processes in fish meats in real time for 18 hours. However, the practicability of remotely monitoring pH values in fish meats through wireless means can serve as an indicator of spoilage in food products (Huang et al., 2011). Guinovart and his coworkers introduced an electrode created through inkjet printing in another study. However, the layers of printed electrodes consisting of positively charged polymer and negatively charged iridium oxide nanoparticles onto an ITO/PET substrate, achieving a noble pH sensitivity of 59 mV^{-1}. Also, the sensor exhibits excellent durability when subjected to mechanical stress, in addition to demonstrating superior repeatability and reproducibility (Guinovart, Valdés-Ramírez, Windmiller, Andrade, & Wang, 2014).

Moreover, Zabala and coworkers have developed a Time-Temperature Indicator (TTI) using a rotary printer, wherein water-based ink was applied to a pure cellulose substrate to create a flat and flexible TTI electrode. However, the manufacturing process involves steam sterilization, followed by vigorous mechanical stirring of all TTI materials for approximately 3 hours. Subsequently, the pH of the solution is regulated, and the bacterial source is thoroughly mixed with a small quantity of water before being introduced into the mixture. The outcomes indicate that rotary printing is versatile for creating TTIs with diverse shapes, sizes, and colors. Therefore, these intelligent labels can be effortlessly incorporated into food packaging and serve as valuable tools for effective cold chain management (Zabala, Castán, & Martínez, 2015).

Liu et al. (2017). They used an intelligent starch/polyvinyl alcohol film infused with anthocyanin (ANT) and limonene (LIM), which could detect pH alterations and prevent undesirable microbial proliferation in food products. There was a noticeable and distinct color change when the film was submerged in solutions with pH values from 1.0 (resulting in a red-orange color) to 14.0 (yielding an intense blue-green hue). Ultimately, the film exhibited effective color indication and demonstrated antimicrobial properties when applied to pasteurized milk. In addition, Mills and Hazafy (2009) have developed nanocrystalline SnO_2 nanoparticles as photosensitizers in colorimetric O_2 indicator, where the film's color changed in response to varying levels of O_2 exposure. Beniwal and coworkers have fabricated a Ni-SnO_2 nano thin film sensor to detect ethylene gas in apple fruit at ambient conditions. The developed sensor detected a low concentration of ethylene gas (5–50 mg L^{-1}), which shows the sensor's high sensitivity (Beniwal & Sunny, 2019).

Galstyan and his coworkers synthesized novel Nb-doping of TiO_2 nanotubes through a facile and cost-effective method. The developed nanocomposite material is exploited as a sensor for the selective and sensitive detection of dimethylamine from food products. The sensor exhibited excellent sensing properties at ambient temperature and concentration of dimethylamine (10–50 ppm). Moreover, the results demonstrate the significant potential of Nb-TiONTs for their utilization in monitoring and ensuring the quality of seafood products (Galstyan et al., 2020).

Finally, The more this technology advances, the closer we are to avoiding petroleum-based packaging and reducing food waste by using nanotechnology directly on consumer products. For example, Tanpichai and collaborators (2023) have just demonstrated that it is possible to extend shelf life and inhibit bacterial growth, prevent moisture loss, and delay the growth of *S. Typhimurium* from fresh cucumber using chitin (from shrimp shell waste) nanofiber coatings.

11.5 DELIVERY OF FUNCTIONAL AND BIOACTIVE INGREDIENTS (INCLUDING VITAMINS, PROBIOTICS, BIOACTIVE PEPTIDES, AND ANTIOXIDANTS)

A growing trend is to incorporate health-promoting components like phenolic compounds, vitamins, and minerals into food and beverages to create "functional foods." These are specially formulated to enhance human health, well-being, and performance. Nevertheless, there are often challenges associated with integrating these nutraceuticals into foods. This is due to their limited solubility, potential for introducing unfavorable flavor attributes, susceptibility to chemical instability, or issues related to low bioavailability. This issue can be overcome by encapsulating the bioactive compounds within nanoparticle-based delivery systems. Encapsulating bioactive agents often leads to improved bioavailability because the reduced particle size allows for quicker digestion, improved mucus layer penetration, and direct cell uptake. Therefore, the nanomaterials can be designed in a way that enables them to withstand the challenges posed by specific parts of the gastrointestinal tract and subsequently release their contents at targeted locations, thereby maximizing their potential health-enhancing effects.

Dewapriya and Kim prepared bioactive proteins, and protein hydrolysates are widely recognized terms in contemporary nutritional supplements. However, it's important to note that not all high-protein sources are suitable for supplement development, as their biological value represents the proportion or percentage of protein the body can effectively absorb and must be considered (Dewapriya & Kim, 2014). In another study, Fathi and coworkers reviewed protein-based nanoencapsulation strategies, which examined physicochemical properties. They found that protein modification techniques were valuable for enhancing the functionality and performance of bioactive agents within diverse protein-based nanocarriers. Consequently, proteins emerge as a primary choice for creating nanoparticles, nanogels, and nanofibers from various origins (Fathi, Donsi, & McClements, 2018). Guan and coworkers employed a nanoencapsulation method to encapsulate caffeic acid phenethyl ester, a significant hydrophobic bioactive compound in propolis extract. They utilized aqueous propylene glycol and sucrose fatty acid ester in conjunction with a temperature-cycle process to enhance this compound's physiological characteristics and anti-cancer activity (Guan, Chen, & Zhong, 2019).

Bezerra and coworkers have evaluated the use of yellow passion fruit albedo flour, which contains significant pectin content, as an encapsulating agent in developing nanodispersions containing carotenoid extract from Spirulina sp. LEB 18. These nanodispersions were characterized, assessing their stability, antioxidant capacity, physicochemical properties, and the retention of carotenoids over a 60-day storage period at 4°C. The study concluded that flour could serve as a natural alternative to commercial polymers, offering the potential for incorporation into various food formulations (Bezerra et al., 2019).

Similarly, Chapeau and coworkers explored the co-assembly process to bind lactoglobulin and lactoferrin, utilizing them as biocarriers for vitamin B_9. They concluded that natural food components possess significant potential as carriers for functional ingredients (Chapeau et al., 2016). In addition, Hategekimana and coworkers also employed nanoemulsion technology to create nanocapsules, using octenyl succinic anhydride starches as emulsifiers and wall materials. These nanocapsules were loaded with vitamin E and found potential applications in the pharmaceutical and beverage sectors (Hategekimana, Chamba, Shoemaker, Majeed, & Zhong, 2015). Albadran and coworkers discovered that chitosan coating had a substantial positive impact on enhancing the protective properties of gelatin-agar particles for probiotics. This enhanced protective effect was attributed to the robust interaction between chitosan and gelatin, leading to an improved survival rate of encapsulated probiotics when exposed to gastric acid. Furthermore, chitosan also enhanced the adhesion of probiotic agents, further contributing to their effectiveness (Albadran, Monteagudo-Mera, Khutoryanskiy, & Charalampopoulos, 2020). Moreover, Lin and coworkers (2021) observed that probiotics encapsulated within a yeast membrane exhibited the ability to transport probiotics to lymphoid follicles through oral administration. This process facilitated a mucosal immune response, helped maintain the balance of intestinal microorganisms, and reduced the occurrence of damage to the intestinal barrier.

Besides, Xu and coworkers (2015) developed tumor-homing peptides as systematic nanoparticles to enhance receptor-mediated cell penetrability. These nanoparticulate peptide structures proved notably more efficient as mediators of receptor-dependent uptake than their free counterparts. This outcome emphasizes an additional advantage of nanostructured materials built using repetitive building blocks, particularly regarding the multivalent presentation of cell ligands. This multivalency could enhance cell penetration, especially in drug delivery applications.

11.6 CONCLUSION

In conclusion, incorporating nanotechnology into food packaging represents a model-shifting development in the food industry, evolving in a new era of innovation and enhancement. This incorporation has yielded substantial progress in diverse aspects of food packaging, including incorporating antimicrobial agents, utilizing nanomaterials to expand food items' physical characteristics and stability, creating intelligent packaging systems, and efficiently delivering functional and bioactive constituents. Antimicrobial agents integrated into packaging materials have demonstrated their capacity to extend food products' shelf life by inhibiting deleterious microorganisms' proliferation. This ensures food safety, diminishes food wastage, and elevates overall product quality.

Conversely, nanomaterials have emerged as invaluable tools in restoring foodstuffs' physical characteristics and stability. They can be customized to bestow barrier properties, mechanical robustness, and controlled release features, thereby facilitating the preservation of the packaged food items' sensory features and nutritional value.

Intelligent packaging systems equipped with sensors and indicators have stimulated a revolution in monitoring food freshness and safety. These systems supply real-time information concerning the state of the packaged product, thereby empowering consumers to make informed choices and mitigating the probability of consuming spoiled or contaminated food. Additionally, transporting functional and bioactive constituents, such as vitamins, probiotics, bioactive peptides, and antioxidants, through nanotechnology-incorporated packaging affords exceptional precision and efficacy. This ensures that consumers receive these constituents' intended health benefits and sensory experiences. Nonetheless, it is imperative to acknowledge that while the application of nanotechnology in food packaging is rife with promise, it also elicits concerns regarding nanomaterials' potential health and environmental effects. Imposing rigorous regulations and undertaking comprehensive safety assessments are indispensable to reducing associated risks.

In summation, the deployment of nanotechnology in food packaging represents a transformative influence on the food industry, discussing an excess of advantages, ranging from prolonged shelf life and elevated food quality to increased food safety and heightened consumer engagement. As research and development in this domain continue to advance, it is imperative to strike a delicate balance between innovation and safety to connect nanotechnology's full potential in the food packaging sector.

BIBLIOGRAPHY

Abou-Yousef, H., Saber, E., Abdel-Aziz, M. S., & Kamel, S. (2018). Efficient alternative of antimicrobial nanocomposites based on cellulose acetate/Cu-NPs. *Soft Materials, 16*(3), 141–150.

Ahari, H., & Soufiani, S. P. (2021). Smart and active food packaging: Insights in novel food packaging. *Frontiers in Microbiology, 12,* 657233.

Albadran, H. A., Monteagudo-Mera, A., Khutoryanskiy, V. V., & Charalampopoulos, D. (2020). Development of chitosan-coated agar-gelatin particles for probiotic delivery and targeted release in the gastrointestinal tract. *Applied Microbiology and Biotechnology, 104,* 5749–5757.

Alfei, S., Marengo, B., & Zuccari, G. (2020). Nanotechnology application in food packaging: A plethora of opportunities versus pending risks assessment and public concerns. *Food Research International, 137,* 109664.

Bakhsh, H., Buledi, J. A., Ghumro, T., Khand, N. H., Ameen, S., Solangi, A. R., … Shah, Z. U. (2021). Sodium dodecyl sulfate stabilized NiO nanoseeds: A potential procedure for ultra-sensitive determination of bentazone in vegetables. *Journal of Materials Science: Materials in Electronics, 32*(12), 15917–15929.

Bakhsh, H., Buledi, J. A., Khand, N. H., Junejo, B., Solangi, A. R., Mallah, A., & Sherazi, S. T. H. (2021). NiO nanostructures based functional none-enzymatic electrochemical sensor for ultrasensitive determination of endosulfan in vegetables. *Journal of Food Measurement and Characterization, 15,* 2695–2704.

Becaro, A. A., Puti, F. C., Correa, D. S., Paris, E. C., Marconcini, J. M., & Ferreira, M. D. (2015). Polyethylene films containing silver nanoparticles for applications in food packaging: Characterization of physico-chemical and antimicrobial properties. *Journal of Nanoscience and Nanotechnology, 15*(3), 2148–2156.

Beniwal, A., & Sunny. (2019). Apple fruit quality monitoring at room temperature using sol-gel spin coated Ni-SnO$_2$ thin film sensor. *Journal of Food Measurement and Characterization, 13*(1), 857–863.

Bezerra, P. Q. M., de Matos, M. F. R., Ramos, I. G., Magalhães-Guedes, K. T., Druzian, J. I., Costa, J. A. V., & Nunes, I. L. (2019). Innovative functional nanodispersion: Combination of carotenoid from Spirulina and yellow passion fruit albedo. *Food Chemistry, 285*, 397–405.

Bouwmeester, H., Dekkers, S., Noordam, M. Y., Hagens, W. I., Bulder, A. S., De Heer, C., … Sips, A. J. (2009). Review of health safety aspects of nanotechnologies in food production. *Regulatory Toxicology and Pharmacology, 53*(1), 52–62.

Braga, L. R., Rangel, E. T., Suarez, P. A. Z., & Machado, F. (2018). Simple synthesis of active films based on PVC incorporated with silver nanoparticles: Evaluation of the thermal, structural and antimicrobial properties. *Food Packaging and Shelf Life, 15*, 122–129.

Buledi, J. A., Buledi, P. A., Batool, M., Solangi, A. R., Mallah, A., Ameen, S., … Maleh, H. K. (2021). Exploring electrocatalytic proficiencies of CuO nanostructure for simultaneous determination of bentazone and mexacarbate pesticides. *Applied Nanoscience, 11*(6), 1889–1902.

Buledi, J. A., Hyder, A., Khand, N. H., Memon, S. A., Batool, M., & Solangi, A. R. (2023). Potential mitigation of dyes through Mxene composites. In *Handbook of Functionalized Nanostructured MXenes: Synthetic Strategies and Applications from Energy to Environment Sustainability* (pp. 283–300): Springer.

Buledi, J. A., Solangi, A. R., Hyder, A., Khand, N. H., Memon, S. A., Mallah, A., … Behzadpour, M. (2022). Selective oxidation of amaranth dye in soft drinks through tin oxide decorated reduced graphene oxide nanocomposite based electrochemical sensor. *Food and Chemical Toxicology, 165*, 113177.

Buledi, J. A., Solangi, A. R., Mallah, A., Shah, Z. U., Sherazi, S. T., Shah, M. R., … Ali, S. (2023). Electrochemical monitoring of isoproturon herbicide using NiO/V$_2$O$_5$/rGO/GCE. *Journal of Food Measurement and Characterization, 17*(2), 1628–1639.

Buledi, J. J. A., Solangi, A. R., Hyder, A., Batool, M., Mahar, N., Mallah, A., … Ghalkhani, M. (2022). Fabrication of sensor based on polyvinyl alcohol functionalized tungsten oxide/reduced graphene oxide nanocomposite for electrochemical monitoring of 4-aminophenol. *Environmental Research, 212*, 113372.

Bumbudsanpharoke, N., Choi, J., & Ko, S. (2015). Applications of nanomaterials in food packaging. *Journal of Nanoscience and Nanotechnology, 15*(9), 6357–6372.

Carbone, M., Donia, D. T., Sabbatella, G., & Antiochia, R. (2016). Silver nanoparticles in polymeric matrices for fresh food packaging. *Journal of King Saud University-Science, 28*(4), 273–279.

Castro Mayorga, J. L., Fabra Rovira, M. J., Cabedo Mas, L., Sánchez Moragas, G., & Lagarón Cabello, J. M. (2018). Antimicrobial nanocomposites and electrospun coatings based on poly (3-hydroxybutyrate-co-3-hydroxyvalerate) and copper oxide nanoparticles for active packaging and coating applications. *Journal of Applied Polymer Science, 135*(2), 45673.

Chapeau, A.-L., Tavares, G. M., Hamon, P., Croguennec, T., Poncelet, D., & Bouhallab, S. (2016). Spontaneous co-assembly of lactoferrin and β-lactoglobulin as a promising biocarrier for vitamin B9. *Food Hydrocolloids, 57*, 280–290.

Cheng, H.-M., Gao, X.-W., Zhang, K., Wang, X.-R., Zhou, W., Li, S.-J., … Yan, D.-P. (2019). A novel antimicrobial composite: ZnAl-hydrotalcite with p-hydroxybenzoic acid intercalation and its possible application as a food packaging material. *New Journal of Chemistry, 43*(48), 19408–19414.

Dewapriya, P., & Kim, S.-K. (2014). Marine microorganisms: An emerging avenue in modern nutraceuticals and functional foods. *Food Research International, 56*, 115–125.

Dobrucka, R., & Cierpiszewski, R. (2014). Active and intelligent packaging food research and development-A Review. *Polish Journal of Food and Nutrition Sciences, 64*(1), 7–15.

Duncan, T. V. (2011). Applications of nanotechnology in food packaging and food safety: Barrier materials, antimicrobials and sensors. *Journal of Colloid and Interface Science, 363*(1), 1–24.

Enescu, D., Cerqueira, M. A., Fucinos, P., & Pastrana, L. M. (2019). Recent advances and challenges on applications of nanotechnology in food packaging. A literature review. *Food and Chemical Toxicology, 134*, 110814.

Fathi, M., Donsi, F., & McClements, D. J. (2018). Protein-based delivery systems for the nanoencapsulation of food ingredients. *Comprehensive Reviews in Food Science and Food Safety, 17*(4), 920–936.

Firouz, M. S., Mohi-Alden, K., & Omid, M. (2021). A critical review on intelligent and active packaging in the food industry: Research and development. *Food Research International, 141*, 110113.

Galstyan, V., Ponzoni, A., Kholmanov, I., Natile, M. M., Comini, E., & Sberveglieri, G. (2020). Highly sensitive and selective detection of dimethylamine through Nb-doping of TiO2 nanotubes for potential use in seafood quality control. *Sensors and Actuators B: Chemical, 303*, 127217.

Garcia, C. V., Shin, G. H., & Kim, J. T. (2018). Metal oxide-based nanocomposites in food packaging: Applications, migration, and regulations. *Trends in Food Science & Technology, 82*, 21–31.

Geueke, B., Groh, K., & Muncke, J. (2018). Food packaging in the circular economy: Overview of chemical safety aspects for commonly used materials. *Journal of Cleaner Production, 193*, 491–505.

Guan, Y., Chen, H., & Zhong, Q. (2019). Nanoencapsulation of caffeic acid phenethyl ester in sucrose fatty acid esters to improve activities against cancer cells. *Journal of Food Engineering, 246*, 125–133.

Guinovart, T., Valdés-Ramírez, G., Windmiller, J. R., Andrade, F. J., & Wang, J. (2014). Bandage-based wearable potentiometric sensor for monitoring wound pH. *Electroanalysis, 26*(6), 1345–1353.

Hategekimana, J., Chamba, M. V., Shoemaker, C. F., Majeed, H., & Zhong, F. (2015). Vitamin E nanoemulsions by emulsion phase inversion: Effect of environmental stress and long-term storage on stability and degradation in different carrier oil types. *Colloids and Surfaces A: Physicochemical and Engineering Aspects, 483*, 70–80.

Huang, W.-D., Deb, S., Seo, Y.-S., Rao, S., Chiao, M., & Chiao, J. (2011). A passive radio-frequency pH-sensing tag for wireless food-quality monitoring. *IEEE Sensors Journal, 12*(3), 487–495.

Hyder, A., Buledi, J. A., Memon, R., Qureshi, A., Niazi, J. H., Solangi, A. R., … Thebo, K. H. (2023). Modified electrochemical sensor via supramolecular structural functionalized graphene oxide for ultra-sensitive detection of gallic acid. *Diamond and Related Materials, 139*, 110357.

Hyder, A., Memon, S. S., Buledi, J. A., Memon, S., Memon, Z.-U., Rajpar, D. B., & Sirajuddin. (2023). A highly selective sensor based on p-tetranitrocalix [4] arene-capped copper nanoparticles for colorimetric and bare-eye detection of cyclophosphamide. *Analytical Sciences, 39*, 1981–1992.

Hyder, A., Memon, S. S., Buledi, J. A., Memon, S., Memon, Z.-U., Shaikh, S. G., & Rajpar, D. B. (2023). Strategic fabrication of para diethylamine calix [4] arene-functionalized CuO nanostructures: A metal-organic framework-based sensor for quantification of antipyrine. *Chemical Papers, 77*(12), 7737–7748.

Ibrahim, S., El-Naggar, M. E., Youssef, A. M., & Abdel-Aziz, M. S. (2020). Functionalization of polystyrene nanocomposite with excellent antimicrobial efficiency for food packaging application. *Journal of Cluster Science, 31*, 1371–1382.

Joye, I. J., Nelis, V. A., & McClements, D. J. (2015). Gliadin-based nanoparticles: Stabilization by post-production polysaccharide coating. *Food Hydrocolloids, 43*, 236–242.

Kerry, J., O'grady, M., & Hogan, S. (2006). Past, current and potential utilisation of active and intelligent packaging systems for meat and muscle-based products: A review. *Meat Science, 74*(1), 113–130.

Khatoon, A., Syed, J. A., Buledi, J. A., Shakeel, S., Mallah, A., Solangi, A. R., … Shah, M. R. (2022). Bio-green fabrication of bell pepper mediated silver nanoparticles: An efficient material for electrochemical sensing of arbutin in cosmetics. *Journal of the Iranian Chemical Society, 19*(8), 3659–3672.

Kumar, S., Mudai, A., Roy, B., Basumatary, I. B., Mukherjee, A., & Dutta, J. (2020). Biodegradable hybrid nanocomposite of chitosan/gelatin and green synthesized zinc oxide nanoparticles for food packaging. *Foods, 9*(9), 1143.

Kuswandi, B., & Moradi, M. (2019). Improvement of food packaging based on functional nanomaterial. In *Nanotechnology: Applications in Energy, Drug and Food* Siddiquee, S., Hong Melvin, G. J., & Rahman, M. M. (Eds.) (pp. 309–344). Springer.

Kuswandi, B., Wicaksono, Y., Jayus, Abdullah, A., Heng, L. Y., & Ahmad, M. (2011). Smart packaging: Sensors for monitoring of food quality and safety. *Sensing and Instrumentation for Food Quality and Safety, 5*, 137–146.

Li, L., Ma, W., Cheng, X., Ren, X., Xie, Z., & Liang, J. (2016). Synthesis and characterization of biocompatible antimicrobial N-halamine-functionalized titanium dioxide core-shell nanoparticles. *Colloids and Surfaces B: Biointerfaces, 148*, 511–517.

Lin, L., Mao, X., Sun, Y., Rajivgandhi, G., & Cui, H. (2019). Antibacterial properties of nanofibers containing chrysanthemum essential oil and their application as beef packaging. *International Journal of Food Microbiology, 292*, 21–30.

Lin, S., Mukherjee, S., Li, J., Hou, W., Pan, C., & Liu, J. (2021). Mucosal immunity-mediated modulation of the gut microbiome by oral delivery of probiotics into Peyer's patches. *Science Advances, 7*(20), eabf0677.

Liu, B., Xu, H., Zhao, H., Liu, W., Zhao, L., & Li, Y. (2017). Preparation and characterization of intelligent starch/PVA films for simultaneous colorimetric indication and antimicrobial activity for food packaging applications. *Carbohydrate Polymers, 157*, 842–849.

Liu, M., Wang, F., Liang, M., Si, Y., Yu, J., & Ding, B. (2020). In situ green synthesis of rechargeable antibacterial N-halamine grafted poly(vinyl alcohol) nanofibrous membranes for food packaging applications. *Composites Communications, 17*, 147–153. https://doi.org/10.1016/j.coco.2019.11.018.

Meira, S. M. M., Zehetmeyer, G., Scheibel, J. M., Werner, J. O., & Brandelli, A. (2016). Starch-halloysite nanocomposites containing nisin: Characterization and inhibition of Listeria monocytogenes in soft cheese. *LWT-Food Science and Technology, 68*, 226–234.

Mihindukulasuriya, S., & Lim, L.-T. (2014). Nanotechnology development in food packaging: A review. *Trends in Food Science & Technology, 40*(2), 149–167.

Mills, A., & Hazafy, D. (2009). Nanocrystalline SnO2-based, UVB-activated, colourimetric oxygen indicator. *Sensors and Actuators B: Chemical, 136*(2), 344–349.

Ncube, L. K., Ude, A. U., Ogunmuyiwa, E. N., Zulkifli, R., & Beas, I. N. (2020). Environmental impact of food packaging materials: A review of contemporary development from conventional plastics to polylactic acid based materials. *Materials, 13*(21), 4994.

Nouri, A., Yaraki, M. T., Ghorbanpour, M., Agarwal, S., & Gupta, V. K. (2018). Enhanced Antibacterial effect of chitosan film using montmorillonite/CuO nanocomposite. *International Journal of Biological Macromolecules, 109*, 1219–1231.

Pandey, G., Munguambe, D. M., Tharmavaram, M., Rawtani, D., & Agrawal, Y. (2017). Halloysite nanotubes–An efficient 'nano-support' for the immobilization of α-amylase. *Applied Clay Science, 136*, 184–191.

Pathakoti, K., Manubolu, M., & Hwang, H.-M. (2017). Nanostructures: Current uses and future applications in food science. *Journal of Food and Drug Analysis, 25*(2), 245–253.

Pina, H. d . V., Farias, A. J. A. d., Barbosa, F. C., William de Lima Souza, J., de Sousa Barros, A. B., Batista Cardoso, M. J., … Wellen, R. M. R. (2020). Microbiological and cytotoxic perspectives of active PCL/ZnO film for food packaging. *Materials Research Express, 7*(2), 025312.

Porta, R. (2019). *The Plastics Sunset and the Bio-Plastics Sunrise* (Vol. 9, p. 526). MDPI.

Qi, W., Wu, J., Shu, Y., Wang, H., Rao, W., Xu, H.-N., & Zhang, Z. (2020). Microstructure and physiochemical properties of meat sausages based on nanocellulose-stabilized emulsions. *International Journal of Biological Macromolecules, 152*, 567–575.

Realini, C. E., & Marcos, B. (2014). Active and intelligent packaging systems for a modern society. *Meat Science, 98*(3), 404–419.

Reichert, C. L., Bugnicourt, E., Coltelli, M.-B., Cinelli, P., Lazzeri, A., Canesi, I., … Agostinis, L. (2020). Bio-based packaging: Materials, modifications, industrial applications and sustainability. *Polymers, 12*(7), 1558.

Risch, S. J. (2009). Food packaging history and innovations. *Journal of Agricultural and Food Chemistry, 57*(18), 8089–8092.

Robertson, G. L. (2005). *Food Packaging: Principles and Practice*. CRC Press.

Rokbani, H., Daigle, F., & Ajji, A. (2019). Long-and short-term antibacterial properties of low-density polyethylene-based films coated with zinc oxide nanoparticles for potential use in food packaging. *Journal of Plastic Film & Sheeting, 35*(2), 117–134.

Saadat, S., Pandey, G., Tharmavaram, M., Braganza, V., & Rawtani, D. (2020). Nano-interfacial decoration of halloysite nanotubes for the development of antimicrobial nanocomposites. *Advances in Colloid and Interface Science, 275*, 102063.

Shah, S.-Z., Taqvi, I. H., Ameen, S., Mallah, A., Buledi, J. A., Khand, N. H., & Solangi, A. R. (2023). Plant based fabrication of CuO/NiO nanocomposite: A green approach for low-level quantification of vanillin in food samples. *Pure and Applied Chemistry, 95*(5), 501–511.

Sharma, R., Jafari, S. M., & Sharma, S. (2020). Antimicrobial bio-nanocomposites and their potential applications in food packaging. *Food Control, 112*, 107086.

Tanpichai, S., Pumpuang, L., Srimarut, Y., Woraprayote, W., & Malila, Y. (2023). Development of chitin nanofiber coatings for prolonging shelf life and inhibiting bacterial growth on fresh cucumbers. *Scientific Reports, 13*, 13195.

Tóth, A., Balázs, B., & Halász, K. (2020). Antimicrobial activity of copper-humate-cellulose sheets. *Packaging Technology and Science, 33*(3), 123–137.

Toydemir, G., Cekic, S. D., Ozkan, G., Uzunboy, S., Avan, A. N., Capanoglu, E., & Apak, R. (2020). Nanosensors for foods. *Nano-Food Engineering, 1*, 327–375.

Uskoković, V. (2007). Nanotechnologies: What we do not know. *Technology in Society, 29*(1), 43–61.

Vasile, C., & Baican, M. (2021). Progresses in food packaging, food quality, and safety-controlled-release antioxidant and/or antimicrobial packaging. *Molecules, 26*(5), 1263.

Venkatesan, R., & Rajeswari, N. (2017). TiO$_2$ nanoparticles/poly (butylene adipate-co-terephthalate) bionanocomposite films for packaging applications. *Polymers for Advanced Technologies, 28*(12), 1699–1706.

Vermeiren, L., Devlieghere, F., van Beest, M., de Kruijf, N., & Debevere, J. (1999). Developments in the active packaging of foods. *Trends in Food Science & Technology, 10*(3), 77–86.

Volova, T. G., Shumilova, A. A., Shidlovskiy, I. P., Nikolaeva, E. D., Sukovatiy, A. G., Vasiliev, A. D., & Shishatskaya, E. I. (2018). Antibacterial properties of films of cellulose composites with silver nanoparticles and antibiotics. *Polymer Testing, 65*, 54–68.

Voon, H. C., Bhat, R., Easa, A. M., Liong, M., & Karim, A. (2012). Effect of addition of halloysite nanoclay and SiO$_2$ nanoparticles on barrier and mechanical properties of bovine gelatin films. *Food and Bioprocess Technology, 5*, 1766–1774.

Xu, Z., Unzueta, U., Roldán, M., Mangues, R., Sánchez-Chardi, A., Ferrer-Miralles, N., … Vázquez, E. (2015). Formulating tumor-homing peptides as regular nanoparticles enhances receptor-mediated cell penetrability. *Materials Letters, 154*, 140–143.

Yang, J., & Theboc, K. H. (2023). MXene-based nanocomposites: An emerging aspirant for removal of antibiotics, dyes, and heavy metal ions. *Materials Chemistry Frontiers, 7*, Article 100202.

Yang, L., Qin, X., Kan, J., Liu, X., & Zhong, J. (2019). Improving the physical and oxidative stability of emulsions using mixed emulsifiers: Casein-octenyl succinic anhydride modified starch combinations. *Nanomaterials, 9*(7), 1018.

Yaris, A., & Sezgin, A. (2015). Food packaging: Glass and plastic. *Researches on Science and Art in* 21st Century. Chapter 81, Gece Kitapligi, Türkiye (pp.735–740).

Yu, Z., Jiang, Q., Yu, D., Dong, J., Xu, Y., & Xia, W. (2022). Physical, antioxidant, and preservation properties of chitosan film doped with proanthocyanidins-loaded nanoparticles. *Food Hydrocolloids, 130*, 107686.

Zabala, S., Castán, J., & Martínez, C. (2015). Development of a time-temperature indicator (TTI) label by rotary printing technologies. *Food Control, 50*, 57–64. https://doi.org/10.1016/j.foodcont.2014.08.007.

Zhao, Y., Wei, B., Wu, M., Zhang, H., Yao, J., Chen, X., & Shao, Z. (2020). Preparation and characterization of antibacterial poly (lactic acid) nanocomposites with N-halamine modified silica. *International Journal of Biological Macromolecules, 155*, 1468–1477.

12 Nanotechnology Principles for the Detection of Foodborne Bacterial Pathogens and Toxins

Mauricio Redondo-Solano

12.1 INTRODUCTION

World Health Organization (WHO) defines unsafe food and water as food or water containing pathogenic microorganisms or chemical substances (World Health Organization, 2013, 2015). While foodborne illnesses are also caused by heavy metals, harmful chemicals, parasites, fungi, and viruses, a majority of food poisoning cases are caused by pathogenic bacteria, such as *Escherichia coli* O157:H7 (Rani et al., 2021), *Salmonella enterica* (Al-Rifai et al., 2020), *Listeria monocytogenes* (Kallipolitis et al., 2020) *Staphylococcus aureus* (Oniciuc et al., 2017) and *Campylobacter jejuni* (Thomas et al., 2020). Sometimes, the pathogenic microorganisms cause food poisoning, and sometimes, the toxins they release may pose health hazards to humans. Natural toxicants are those produced naturally by living organisms and do not cause any harm to the organisms themselves. Some of the natural toxicants are pyrrolizidine alkaloids (PAs), poisonous mushrooms, solanines and chaconine (glycoalkaloids), mycotoxins, goitrogens, furocoumarins, cyanogenic glycosides (phytotoxins), aquatic biotoxins (algal toxins), and so on (Girigoswami et al., 2021).

Factors such as climate change, the globalization of the food market, and consumer lifestyle habits (including a shift toward healthy, minimally processed foods) impact the likelihood that larger quantities of persons may come directly in contact with contaminated food and water. Further exacerbating these trends is an increase in the human population with a demographic shift toward an aging population composed of a growing number of immunocompromised people more susceptible to foodborne diseases.

Pathogenic microbial infections have posed an enormous threat to global public health, like in the case of the pandemic coronavirus disease 2019 (COVID-19) (Hajipour et al., 2021), and food safety remains a significant issue in the field of public health (especially under the current situation that biosafety is given great concern). The epidemiology of foodborne disease is changing parallel due to some pathogenic microorganisms' emergence, reemergence, and world spread. The WHO (Chen and Park, 2016) estimated that globally 2.2 million people die each year from food- and waterborne diarrheal diseases alone; approximately 30% of death occurs among children (5 years) due to foodborne illness (World Health Organization, 2020). In the United States, an estimated 48 million people are sickened, 128,000 are hospitalized, and 3,000 die of foodborne diseases yearly (Centers for Disease Control and Prevention, 2011). The severity of diseases is also increasing because of the emergence of pathogens and the spread of antibiotic resistance, primarily by unwise use of antibiotics in human and animal medicine. In addition to direct impacts on human health, illness associated with contaminated food has wide-ranging effects on economic development, healthcare costs, and international trade. These burdens disproportionately affect developing countries (Yang and Duncan, 2021). The annual medical and industrial costs of the illnesses are over \$15.5 billion, including a loss of \$10 million to \$30 million per product recall for the food industry (Centers for Disease Control and Prevention, 2015). Economic losses due to the recall of products in cases of sanitary alarms are frequent, and it is estimated that a combination of lost

DOI: 10.1201/9781003514039-12

wages, insurance claims, and medical bills costs between \$7.7 and \$23 billion annually (Centers for Disease Control and Prevention, 2017). The majority of the illnesses were attributed to pathogens, and according to the U.S. Food and Drug Administration (FDA) (2015), bacterial and viral agents are responsible for the second largest number of food product recalls in the United States, following by undeclared allergens.

Because food and water are among the most important living necessities, tampering with food by adding infectious agents could lead to disastrous outcomes. Food operators should consider that foodborne pathogenic microorganisms or their metabolites may contaminate a food matrix at every food chain step (from farm to fork) and may proliferate to become responsible for foodborne infections. Thus, timely detection and identification of pathogens in food samples become crucial for public health and food industries' financial stability. Early and accurate detection of foodborne illness is pivotal for preventing, controlling, and mitigating the impact of potential outbreaks.

Accurate and rapid detection of pathogenic microorganisms at the first opportunity is significant in public health and food safety, environmental monitoring, clinical diagnosis, anti-bioterrorism, etc. (Huang et al., 2023; Ji et al., 2020; Riu and Giussani, 2020). While extremely sensitive and effective, traditional culture enrichment techniques for pathogen detection are laborious and can extend over several days (Mandal et al., 2011). From a food manufacturer's standpoint, although routine testing for potential contamination is needed to comply with regulations and avoid costly product recalls, the high cost, lack of robustness, and long waiting periods associated with existing detection techniques can be discouraging. Detection challenges are due to the complexity of the food matrix, the potential lack of homogeneity, and the potential for a low number of pathogens within the matrix. As the location of pathogens varies significantly between products and types of contamination, homogenization is required to obtain a complete product analysis. Food products may be subjected to processing and environmental factors so that the resulting microbiota may include viable, dead, stressed, or injured microbial cells (Fusco et al., 2015). Enumerating and detecting the viable, dead, stressed, or injured microbial cells in a food matrix is of utmost importance, especially if those organisms are pathogenic (Quero et al., 2014).

Current gold standards of pathogen detection in food and agricultural products are mainly based upon the FDA Bacteriological Analytical Manual (Hammack et al., 2015), the U.S. Department of Agriculture, Food Safety and Inspection Service Microbiology Laboratory Guidebook (U.S. Department of Agriculture, Food Safety and Inspection Service, 2015), or International Organization for Standardization standards (Chen and Park, 2016). These microbiological tests consist of pre-enrichment and selective enrichment steps followed by confirmation through selective plating and biochemical tests (Lopez-Campos et al., 2012). Most established microbial culture methods may help detect a specific bacterium. Still, this procedure can be quite lengthy (turnover time of several days to more than 1 week for most foodborne pathogens), laborious, and resource-demanding with limited throughput. Later, the isolated pathogens must be further identified to serotype with a panel of antisera. Their antimicrobial resistance profiles may also be obtained by assessing growth on plates with antibiotic disks. In addition, standard microbiological enumeration using selective media may vastly underestimate microbial numbers (Speck, 1970; Hameed et al., 2018). Regarding the matrix effect, mechanical methods are usually performed to homogenize the food sample, ranging from simple blending to more advanced techniques such as ultrasound (Rohde et al., 2015). The various matrix compounds, such as proteins, vitamins, minerals, saccharides, and lipids, must also be eliminated to prevent interference (Lisalova et al., 2016). Separating and concentrating pathogens has been performed using physical-, chemical-, and biological-based (e.g., antibody) methods. Potential cell damage during homogenization, in combination with the typically low numbers of pathogens often present in contaminated samples, necessitates long microbial enrichment periods, hindering the development of more rapid techniques.

In the case of molecular methods, they are not dependent on the growth of the targeted microorganism and can circumvent the underestimation of injured cells (Salazar et al., 2015; Lee et al., 2019); however, the methods need to have some means to differentiate injured from truly dead cells

(Norton and Batt, 1999). Typically, this involves the use of a target, such as mRNA, whose presence is correlated with viability or DNA intercalating dyes, such as EMA (ethidium monoazide) and PMA (propidium monoazide) (Fusco and Quero, 2014). Significant advancements have been made in reducing detection time by developing enzyme-based nucleic acid amplification methods, such as multiplex polymerase chain reaction (mPCR), which provide accurate results in about 24 hours (Law et al., 2015). Using mPCR, shorter processing times are achieved by replacement culture enrichment procedures with those that amplify specific nucleic acid sequences. The use of genetic material also leads to highly specific identification, particularly important for microorganisms that cannot be readily cultured. Quantitative or real-time PCR provides real-time data (fluorescent reading) with less processing. However, intensive sample preparation and expensive thermal cyclers still limit this technique to some laboratory settings (Zhang et al., 2011). However, these PCR-based screening methods' relative ease, speed, and accuracy have made them one of the USDA's and the United States Food and Drug Administration's preferred laboratory procedures for microbial analysis in foods (USDA, 2017).

Other principles include immunological detection such as enzyme-linked immunosorbent assay (ELISA) and lateral flow immunoassay (Wang et al., 2014; Wu et al., 2019), or chromatography methods for detection of toxins and chemical contaminants, chromatographic techniques (high-performance liquid chromatography (HPLC), and mass spectrometry (MS) (Vazquez-Roig and Pico, 2012) are frequently used for the detection of agricultural pollutants. Although most of these methods have high sensitivity and selectivity, they require complex steps, are time-consuming, and need expensive equipment and highly trained operators, which hinder their application, especially in low-resource regions (Law et al., 2015; Zhao et al., 2014), which make them incompatible for point-of-care. Some "rapid" techniques include miniaturized biochemical kits that use biochemical characteristics for pathogen identification (Chen et al., 2013). The requirement of complex sample preparation and sophisticated instruments also hinders these techniques (Cifuentes, 2012) from implementation in point-of-use (POU) applications.

Alternatively, achieving the simultaneous detection of multiple pathogenic bacteria by an intelligent sensing platform is desirable, which is significant for controlling the pathogenic bacteria and safeguarding human health accurately (Lv et al., 2020). However, because of the coexistence of environmental complexity of multiple pathogenic bacteria and the similarity of pathogens and non-pathogens, as well as the interference of proteins, acids, and salts in the sample matrix to the detection sensitivity, the precise detection of multiple pathogens in the synchronization way is still a considerable challenge. In general, two critical aspects of detection are the capacity to correctly identify and quantify pathogenic bacteria, which should comprehensively consider the analytical performances of response sensitivity, selectivity, stability, time, cost, etc. (Cesewski and Johnson, 2020).

From the information discussed above, the need to establish more rapid and sensitive detection methods for early determination of bacterial and toxical contamination is clear. New methods necessitate the development of reliable, sensitive, multiplexed, high-throughput, portable, and automated platforms to detect foodborne illness-causing agents and their toxins in foods (Fusco and Quero, 2014). The current trend is to make use of micro- and nanofabrication technologies to achieve the miniaturization of these platforms into lab-on-a-chip (LOC) devices, also referred to as biomedical nano- and microelectromechanical systems (BioNEMS or BioMEMS), for integrated and automated sample preparation and (quantitative) detection.

12.2 NANOTECHNOLOGY APPLICATION FOR THE DETECTION OF MICROBIAL CONTAMINANTS IN FOOD

Nanostructures have been used as innovative sensing platforms because of advancements in nanotechnology and biotechnology due to their high specific surface area, size-dependent electrical characteristics, and potential for machine downsizing (Sertova, 2015). Various nanomaterials have frequently been utilized to establish efficient chemo/biosensing platforms for different target analytes, resulting from their unique magnetic, optical, electronic, and other relevant properties (Sai-Anand et al., 2019). In particular, the sizeable surface-volume ratio of nanomaterials makes them favorable for attaching a variety of recognition moieties through covalent or noncovalent binding modes, allowing for their properties to be altered and controlled (Muniandy et al., 2019). When a nano sensing platform was developed using nanomaterials like carbon nanomaterials (nanotubes and graphene), nanofibers, nanotubes, and nanostructured metal oxide nanoparticles for pathogen and mycotoxin identification, such nanostructures reduce the time it takes to identify pathogens.

An unprecedented opportunity for pathogenic bacteria detection has emerged by employing suitable nanomaterials as a scaffold (Ye et al., 2020). Most bacteria or natural toxins exist in a contaminated matrix at low concentrations, so it is difficult to detect them effectively (Kim and Oh, 2021). Using nanotechnology-based methods, the pathogenic bacteria and toxins present in complex food products can be detected with high sensitivity and specificity compared to conventional methods. In particular, due to the excellent controllability in size and surface structure, the analytical performances (e.g., cost, reaction kinetics, accuracy, repeatability, etc.) of assay-based nanomaterials can be significantly improved, making it easier to achieve the rapid on-site detection of pathogenic bacteria which requires only small sample volumes, satisfying the requirements of actual public safety (Alamer et al., 2018). Compared with the traditional pathogen detection methods, the nanomaterials-based assays have the advantages of low-cost, rapid, and high accuracy, especially in high- throughput screening, label-free, and real-time response, which can adequately satisfy the needs of food industries, hospitals, and regulatory agencies (Zhang et al., 2020). Technologies that utilize nanoparticles with electrochemical or optical detection modalities offer rapid, sensitive, and cost-effective methods that permit miniaturization and automation for point-of-care testing. Advancements in micro- and nanofabrication techniques may lead to miniaturized devices suitable for field applications. From an economic perspective, reduced reagent consumption and lower overall cost are expected due to more efficient chemistry. Most nanomaterial-based techniques are not used alone, and the high level of compatibility with existing techniques may result in significant improvements to current methods.

12.3   NANO/BIO SENSOR FOR FOOD APPLICATIONS

Biosensors encompass a wide range of analytical techniques that translate specific biorecognition events into measurable signals (Singh et al., 2013). A biosensor is a device with a biological element that recognizes an analyte. It then transduces the event to produce a signal proportional to the analyte concentration and conveys it to the detector. The recognition element should be highly specific and can combine with only the analyte of interest and not with any other interfering species. These chemical or mechanical sensing devices containing nanoscale materials can be called nanosensors (Mohammad et al., 2022). No official definition must be adhered to when deciding whether a technology qualifies as a nanosensor. Still, the standard definitions of the nanoscale (length range of approximately 1–100 nm) and nanomaterials (materials with at least one dimension on the nanoscale) may be used as a general guide to the usage of the term, where nanomaterials include both one-dimensional (1D), 2D and 3D nano-objects (nanoplates, nanotubes, nanoparticles, and so on) as well as materials with nanostructured surfaces (Sundar et al., 2010).

Nanosensors convert physical quantities into signals, which can be easily detected and analyzed. In response to different external stimuli, nanomaterials can change shape, altering physical and chemical properties, one of nanomaterials' most prominent sensing mechanisms. Thus, it has great potential to recognize pathogens more quickly and accurately than other currently available plant pathogen diagnosis methods (Joseph and Morrison, 2006). Nanosensors have emerged as novel nanotechnology in food safety and can be used to detect pathogens or other types of contaminants rapidly and as surface biocidal techniques (Eleftheriadou et al., 2017). This approach is already used to detect foodborne pathogens, food spoilage, and toxins as nanosensors have immobilized bioreceptor probes that are specific for intended molecules in the sample and have the advantages of being small, portable, responsive to real-time screening, accurate, computational, reliable, stable, repeatable, and resilient to detect food contamination issues. Nanosensors attract considerable attention from academic research scientists for all these characteristics, and interest in food applications is growing. A survey of peer-reviewed literature published over the past 15 years reveals a steady growth in articles related to all types of nanosensors.

Nanoparticle-based nanosensors have great potential in providing quality assurance and tracing the chemical, physical, and biological modifications during food processing and preservation. (Berekaa, 2015). Contaminants can be detected using nanosensors for in-lab methodological applications or direct in-product applications, like smart packaging systems. Smart packaging systems consist primarily of nanosensors to detect food contaminants, and nanoparticles are mostly used for their development (Nile et al., 2020); they respond to environmental stimuli and alert consumers to product contamination or the presence of pathogens (Fuertes et al., 2016; Neethirajan and Jayas, 2011). This is interesting as packaging is the most active food nanoscience research and development area. This is likely connected to the fact that some studies have shown that the public is more willing to embrace nanotechnology in "out of food" applications than those directly adding nanoparticles to foods (Department of Industry, Innovation, Science and Research, Australia, 2009). Also, nanosensors may open the possibility for real-time monitoring of contaminants at every stage of the production chain; this is especially critical in the early identification of rapid contamination events such as bioterrorism attacks aimed at the food supply (Pedrero et al., 2009). From a preventative standpoint, POU pathogen detection devices would also be beneficial to overcome quality assurance challenges faced by restaurants and grocery stores. Rapid monitoring of food goods would allow for detecting pathogens before their distribution.

12.4 GENERAL SCHEMES FOR DETECTION OF MICROBIAL CONTAMINANTS

According to Carlson et al. (2017), to effectively enable pathogen monitoring at any stage in the food supply chain, an ideal POU sensor would have the following attributes: (i) minimal or no sample preparation, (ii) real-time detection of low cell counts, (iii) identification of multiple pathogens, (iv) multiple uses (extended use, resistance to fouling), (v) inexpensive, (vi) ease of operation, (vii) ability to function in real-world conditions, and (viii) extended shelf life. Ultimately, a POU device would also be portable with integrated detection and analysis compact enough to be easily carried by a person or deployed to a sensor network. Overcoming the technological complexities of developing such a sensor will require a multifaceted approach, potentially utilizing various platforms for performance. The following sections discuss a variety of methods that have the potential to satisfy some of the desirable attributes of an ideal sensor.

Nanosensors represent an exciting option for developing new state-of-the-art POU devices. At least in the basic research phase, nanosensor principles are diverse in the analytes they have been designed to detect. The types of transducer platforms they leverage for detection (optical, electrochemical, surface-enhanced Raman spectroscopy (SERS), etc.) (Figure 12.1). Small molecules are typically the most frequently detected analytes and (Ameta et al., 2020) are the most common transducer platform (SERS is a sensing technique based on the enhancement of inelastic molecular light scattering in the presence of nanostructured metallic surfaces).

The analyte that can be detected can be an antigen, toxin, DNA, RNA, enzyme substrate, protein, amino acid, microRNA, cell-specific antigen, and various disease biomarkers. It means that nano/biosensors can be used to detect foodborne pathogenic bacteria or their products (toxins). The recognition element can be protein, enzyme, antibody, aptamer, microorganism, cell, ssDNA/RNA, etc. Nanomaterials in food nanosensors are most frequently used as transducers (e.g., fluorescent nanoparticles or conductive nanotubes), but in some nanosensor designs, they may be employed as recognition elements (molecularly imprinted polymers, MIPs), capture agents (magnetic nanoparticles), signal amplifiers (nanopatterned SERS substrates) and more (Girigoswami et al., 2021). Different types of sensing techniques used in biosensors are fluorescence, DNA microarray, surface

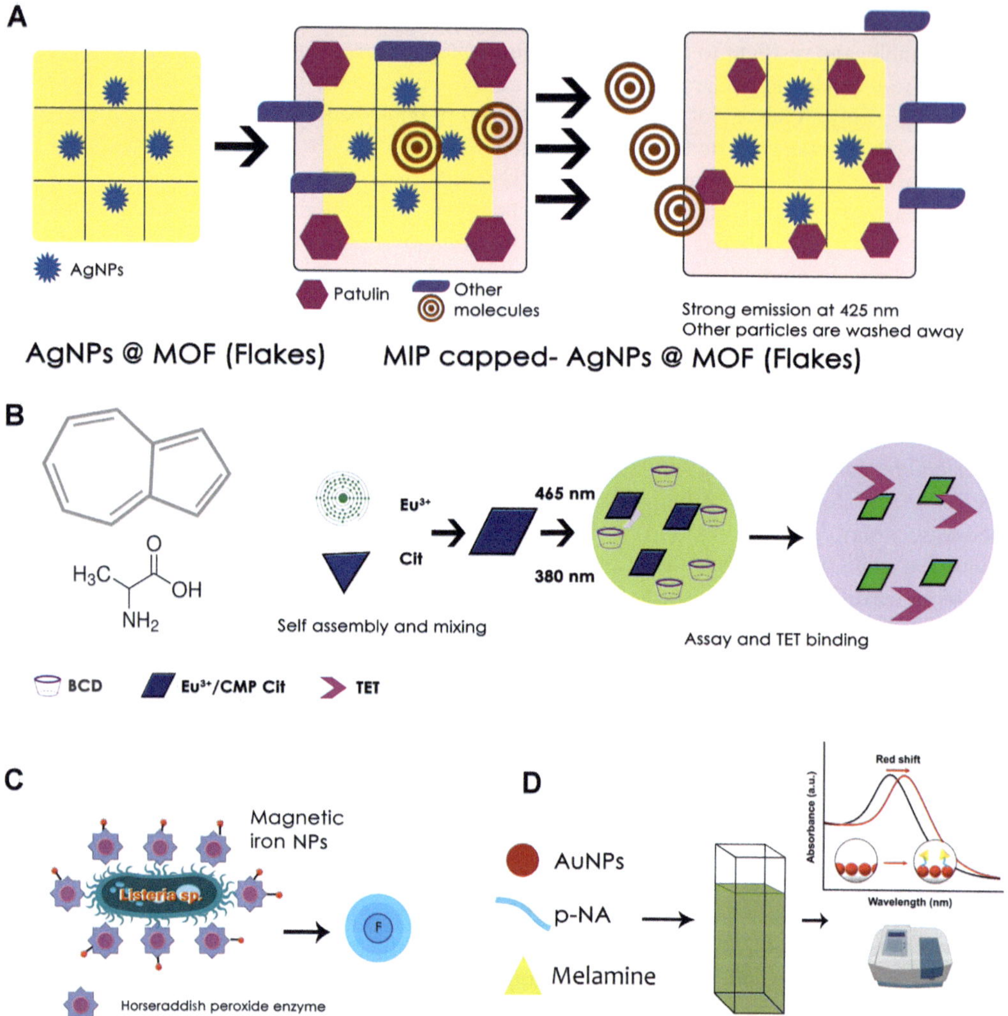

FIGURE 12.1 Details for some principles for nano-based detection of various food contaminants. (a) Mimetic AgNPs@ ZnMOF capped with molecularly imprinted polymer for the selective detection of patulin; (b) detection of tetracycline by a double-signal response fluorescent nanohybrid; (c) detection of *Listeria monocytogenes* through lateral-flow enzyme immunoconcentration for rapid detection; (d) LSPR sensor for sensitive detection of Melamine in infant formulas.

plasmon resonance (SPR), impedance spectroscopy, SPM (Scanning probe microscopy, AFM, STM), Quartz crystal microbalance (QCM), SERS, electrochemical-potentiometric, amperometric, and conductimetric.

Food nanosensors also vary substantially, both in complexity and design: they may take the form of simple nanoparticle probes that improve the sensitivity or speed of basic laboratory-scale food analysis techniques, or they may be more sophisticated, self-contained devices that monitor safety parameters in marketed food products constantly over periods of weeks and send that information wirelessly to users' smartphones. Micro- and nanotechnologies have further enabled miniaturized, chip-based formats for performing nucleic acid analysis, dramatically enhancing portability and improving multiplexing capability in some cases (Cady et al., 2017). Still, the distribution of food nanosensor designs has not changed appreciably in the past 5 years, except for a slight growth in the proportion of chip-based nanosensor platforms, including microfluidic systems, paper test strips, and electrochemical devices. This change coincides with a recent uptick in the number of reported nanosensors intended to integrate with smartphones for data recording, reflecting a growing interest in bringing the laboratory to the sample rather than the other way around.

The following sections will explain the main principles of detecting pathogens and toxins in food nanosensors. Table 12.1 summarizes the main detection principles, characteristics, applications, and general performance.

12.4.1 Antigen-Based Detection

The immunosensing phenomenon has numerous applications, including developing a new class of biosensors. Therefore, a sensor that detects analytes or antigens of interest with the concept of immunoassay utilizing specific antibodies to generate stable immunocomplex is termed an immunosensor. The amount of stable immunocomplex is measured by combining this reaction with the surface of specialized transducers. After the detection of immunocomplex, the transducer converts it to the electrical signal to be processed. The ideal immunosensors should be able to identify the target antigens quickly and generate stable immunocomplex without supplementary reagents. The high reproducibility and ability to detect the target analytes in actual samples are also required for the ideal immunochemical biosensors (Mahato et al., 2018).

For pathogen detection, antigen-binding agents can be tethered to a surface to remove the target from the bulk sample or attached to a nano- or microparticle, enabling detection or enhanced removal from the sample. The antigen-binding event can be coupled with a unique transduction mechanism for direct detection (such as changes in optic or electric properties), or the binding event can be used for immobilization (Cady et al., 2017) and performance of a secondary detection assay (such as a secondary labeling step to produce fluorescence, chemiluminescence, etc.). For instance, whole cells can be immobilized on a chip and directly detected through micro-optic methods (St John et al., 1998) or via microscale separation methods, followed by the more conventional ELISA described by et al. (2013a).

Some nanoparticles (NP) that have SPR properties can be modified with targeting species (e.g., antibodies) as their recognition element to allow them to target and attach to receptors/proteins/epitopes expressed on the surface of the target bacteria. This receptor interaction induces NP aggregation of the nearby targeted pathogen and changes the SPR/colorimetric response (Wilson, 2008). SPR is a unique property of metallic NPs.

Recognition elements such as antibodies have high affinity and avidity to their corresponding antigens expressed on bacteria; therefore, one approach to improve the sensitivity of immunosensors is to enhance the localized surface plasmon resonance (LSPR) signal or promote the detection of SPR signals that are representative of binding target bacteria. NP-based colorimetric sensors show different sensitivities and performances depending on the size, shape, particle and binding site number, and aggregation state of gold NPs (You et al., 2020). As many types of antibodies targeting many pathogens are available, and the surface functionalization of gold NPs is feasible, the design of colorimetric immune sensors is adaptable to isolate many types of pathogens. Different types of bacteria can be detected depending on the kind of antibodies immobilized on the NP or platform–biosensor surface.

12.4.2 Aptasensors

Although monoclonal antibodies are still the most common type of antigen-binding agent employed and have sufficient sensitivity and selectivity for bacterial recognition, their purification and production are expensive and time-consuming. In addition, issues with solubility and stability remain to be addressed. Therefore, alternative biorecognition elements have been used for the development of nanosensors. In this sense, aptamers are growing in popularity. Aptamers are short sequences

TABLE 12.1

Summary of main principles used to develop nano/biosensors to detect foodborne pathogenic bacteria and toxins (adapted from Mohammad et al., 2022 and Sheng et al., 2022)

Nanomaterial	Targets	Detection methods	Recognition elements	Signal readout	Time	LOD Ref range	Type of matrix
MNPs	*E. coli,* *E. coli* O157:H7, *L. monocytogenes,* *S.* Typhimurium, *S. enterica,* *S. agalactiae,* *S. uberis,* *S. aureus,* *V. parahaemolyticus*	Sandwich immunoassay and enzymatic catalysis Sandwich immunoassay and fluorescence analysis Magnetoresistive cytometer Magnetic capture and bacterial growth IMS and nucleic acid amplification	Antibody Mannose-binding lectin	Colorimetry Electrochemistry Pressure Fluorimetry Magnetoresistivity Standard culture plating Turbidity	45 min–6 h (most of the methods have a two-hour detection time)	2–10⁴ CFU/ mL	Culture broth, raw milk, and egg yolk.
Metal nanomaterials	*B. cereus,* *C. jejuni,* *E. coli,* *E. coli* O157:H7, *E. faecium* *L. monocytogenes,* Salmonella, *S.* Typhimurium, *S. aureus,* *Shigella,* *V. parahaemolyticus* Mycotoxins (ochratoxin A, zearalenone, aflatoxin B1)	AuNPs aggregation AuNPs SERS AuNPs label Reduction of gold ions PtNPs catalysis MOFs catalysis MOF/PtNPs catalysis	Antibody Aptamer Polyethylenimine Oligonucleotides Target gene	Colorimetry Raman spectroscopy Pressure Thermal sensor	20–40 min	10–93 CFU/ mL For mycotoxins 0.08–10 ng/mL	Culture broth and fresh-cut vegetables Mycotoxins in corn, corn oil, milk, peanuts, red wine.
QDs	*A. baumannii,* DNA, *E. coli,* *E. coli* O157:H7, *S.* Enteritidis, *S.* Typhimurium, *S. aureus,* *S. pneumoniae*	QDs immunolabelling IMS FRET QDs LFSA Cation exchange reaction	Antibody Aptamer Oligonucleotides	Fluorimetry	12 min–2 h	8–10² CFU/ mL	Culture broth

(Continued)

TABLE 12.1 (*Continued*)

Summary of main principles used to develop nano/biosensors to detect foodborne pathogenic bacteria and toxins (adapted from Mohammad et al., 2022 and Sheng et al., 2022)

Nanomaterial	Targets	Detection methods	Recognition elements	Signal readout	Time	LOD Ref range	Type of matrix
Carbon nanomaterials	DNA, E. coli, E. coli O157:H, P. putida, S. enterica, S. Typhimurium S. Enteritidis, S. epidermidis, S. aureus, Y. enterocolitica	CDs immunolabelling IMS SWCNT immunolabelling SWCNT FET SWCNT immune recognition and nucleic acid amplification MWCNT hybridization Graphene immune recognition GO FRET Graphene SERS Graphene LFSA rGO LFSA Graphene FET Graphene/AuNPs love wave Au/Graphene impedimetric detection	Vancomycin Amikacin Antibody Oligonucleotides Aptamer Electrostatic adsorption	Fluorimetry Electrochemistry Turbidity Raman spectroscopy Colorimetry	10 min–1.5 h	10^0–10^7	Culture broth, kimchi.
UCNPs	B. cereus B. pseudomallei, C. sakazaki, E. coli, E. coli O157:H7, L. monocytogenes S. aureus, S. Typhimurium, Shigella flexneri, V. parahaemolyticus Y. pestis	UCNPs SERS UCNPs label UCNPs LFSA UCNPs LRET Nucleic acid amplification	Electrostatic Adsorption Antibody Aptamer Target gene	Raman spectroscopy Fluorimetry Electrochemistry	5 min–1 h	8–10^3 CFU/ mL	Culture broth, tap water, pasteurized milk, pork

MNPs: Magnetic nanoparticles; QDs: Quantum dots; UCNPs. lanthanide-doped upconversion nanoparticles; IMS: Immunomagnetic separation; AuNPs: gold nanoparticles; SERS: Surface enhanced Raman scatering; PtNPs: platinum nanoparticles; MOF: Metal-organic frameworks; FRET: Fluorescence resonance energy transfer; LFSA: Lateral flow strip assay; CDs: carbon dots; SWCNT: single-walled carbon nanotubes; FET: field-effect transistor; MWCNT: multi-walled carbon nanotube; GO: Graphene oxide; LRET: Luminiscence resonance energy transfer.

of peptide or single-strand DNA or RNA (20–60 nucleotides) that can fold into different structures and target bacteria, proteins, and enzymes with high affinity and specificity (Zhou and Rossi, 2017). The binding of aptamers with their specific molecules generates distinct conformations such as G-quadruplex and hairpin. These conformational changes alter the local environment and are crucial to avoid non-specific interaction, which offers a great opportunity and feasibility for designing aptamer-based biosensors.

The aptamers could be considered nucleotide alternatives to antibodies, and they are inexpensive, readily modifiable with reporters, and show neither immunogenicity nor toxicity, even at high doses. Aptamers are generated by a process known as the "systematic evolution of ligands by exponential enrichment" or SELEX (Potyrailo et al., 1998); this means that aptamers are designed by exponential enrichment, enabling the identification of a matching sequence from a vast database of random sequences to target many different types of receptors/epitopes expressed explicitly on the bacterial species (Sefah et al., 2010). This approach selects DNA, RNA, or polypeptide libraries for their binding ability to target antigens. Multiple washing and reselection procedures then enrich them to identify unique, high-affinity sequences. Once a high-affinity sequence is identified, it can be integrated with a wide range of antigen-based detection approaches (Wu et al., 2015). Since aptamers are nucleic acids, they can more easily be modified using simple chemistry for signal improvements and to attach electrochemical probes, optical probes, or quenchers.

The aptamer can specifically bind to the chosen target molecule by folding into a sequence-defined unique structure in target bacteria. The captured bacteria are then enriched and isolated using an external magnetic field or another separation technique. This versatile platform can isolate and detect many types of bacteria by selecting an aptamer for the epitope of interest. DNA and RNA aptamers are the most commonly used for biosensing applications, mainly due to their resistance to protease degradation and ability to withstand thermal denaturation and refolding cycles. One of the other advantages of using nucleic acid-based aptamers is the relative ease of producing large quantities of these aptamers (using modern DNA and RNA synthesis methods). Aptamers are widely used as recognition elements in SERS- and fluorescent-based sensors developed for bacterial detection (Tang et al., 2018).

12.4.3 Magnetic Nanoparticles

Magnetic nanoparticles (MNPs) consist of either magnetic metals, such as iron, cobalt, and nickel, or their oxides and mixed oxides, such as Fe_3O_4, Fg-Fe_2O_3 (Dadfar et al., 2019), $CoFe_2O_4$ (Ojha and Kant, 2019), and $Mn_{0.6}Zn_{0.4}Fe_2O_4$ (Sunil et al., 2021). MNPs are small particles that, at sizes <20 nm, harbor an unusual magnetic property called "superparamagnetism." Specifically, superparamagnetic nanoparticles exhibit single magnetic domains with a rapid response to applied fields but no longer show magnetic properties upon the removal of the external magnetic field (Samrot et al., 2021). Magnetic nanoparticles have recently gained increasing attention for their great practical potential for biochemical analysis, attributable to the merits of superparamagnetic, excellent monodispersity, great separation speed, and so on (Karakoti et al., 2015). Considering such unique features, magnetic nanoparticles have usually been employed as an excellent matrix to identify various pathogens efficiently (Wang et al., 2021).

Most commercially available immunomagnetic beads are in the 1–100 μm size range. Still, more recently, research groups have demonstrated that nanoscale magnetic particles can also be used for fast separation, particularly for bacterial detection, through immunomagnetic separation (IMS). In brief, IMS technology can be regarded as a nanomachine to selectively enrich the bacteria by employing the magnetic responsiveness and immunological reaction of immunomagnetic beads (IMBs). In this approach, microscale magnetic beads are surface-modified and coated with antibodies, aptamers, and oligonucleotides to enable selective binding of target bacteria or their nucleic acids. Modified MNPs are first mixed with the target sample to form MNP-target complexes. Then, a magnetic field is applied to orient, capture, and concentrate the magnetic targets, which can be analyzed directly or after isolating them from the mixture (Qiu et al., 2009; Srisa-Art et al., 2018). Target pathogens that are removed from the sample can be detected by using conventional culturing on selective media, or by a variety of modern methods, for example, fluorescent labeling (Zhu et al., 2011; Wang et al., 2013b), PCR detection (Wang et al., 2013b), electrochemical detection (Gehring and Tu, 2005), flow cytometry (Seo et al., 1998), or even labeling with fluorescent nanomaterials/quantum dots (Su and Li, 2004; Xue et al., 2018), all post capture.

Based on magnetic nanoparticles, IMS has the characteristics of stronger target specificity, convenient operation, and high separation efficiency, thus effectively reducing the cross-contamination process. This is possible given that nanoparticles can achieve higher degrees of binding (i.e., more particles bound to each target cell). Since microparticles are the same size as bacterial cells, typically, only one or two microparticles can bind to a target cell simultaneously. When using high concentrations of nanoparticles, hundreds can attach to a single cell. This can enhance the total magnetic activity of the target cell post attachment. This can also enable the targeting of multiple antigens to a single cell. One of the primary reasons for using IMS is that immunomagnetic micro- or nanobeads can be directly added to the food matrix (often with some dilution) and then be captured via magnetic separation (Ogden et al., 2001). This obviates the need for complicated filtration and concentration methods, which are more typically employed for enrichment and growth-based approaches. This has enabled the detection of bacterial pathogens in complex samples, such as ground beef, apple juice, and milk (Gehring and Tu, 2005; Seo et al., 1998).

Enlightened by the impressive merits of IMS above and that MNPs have intense color and can separate the target material from the complex matrix, some groups have concentrated on the simultaneous isolation of multiple pathogenic bacteria in a single microbial fouling system using functional IMBs, which can be assembled by coupling different bacteria recognition antibodies into magnetic nanoparticles (Cai et al., 2018). The combination of MNPs and microfluidic systems can be utilized for the efficient continuous flow separation of targets. Various researchers have used magnetic beads to develop flow assays for detecting pathogenic bacteria (Qiao et al., 2017).

MNPs can be used not only for whole bacterial detection but also for nucleic acid assays. As an example, target bacteria were first separated by mannose-binding lectin-modified magnetic beads and then analyzed directly on a microfluidic chip on which cell lysis, nucleic acid amplification, and optical detection were also performed (Cai et al., 2018). In more recent research, Garrido-Maestu et al. (2020) applied immunomagnetic separation with magnetic nanoparticles in combination with q-PCR to detect *L. monocytogenes* in spiked food samples. The optimized methodology resulted in a very low limit of detection (9.7 CFU/25 g) with values for sensitivity, specificity, accuracy, and positive and negative predictive values of 100%, resulting in a kappa index of concordance of 1.00. These results confirmed the usefulness of MNPs to improve current detection methods.

Alternatively, combining IMS and immunofluorescent assays might improve the detection sensitivity. The bacteria can be labeled with MNPs in solution and then transferred onto microfluidic chips for further analysis. To make the system more portable and automated, MNPs can be pre-fixed into the microsystem. To overcome the deficiencies of conventional organic fluorophores, IMBs are permanently bonded with luminescent nanomaterials, especially for QDs, to form sandwich nanocomposite structures for reporting the bacteria fouling levels by measuring its fluorescence emission intensity changes (Cheng et al., 2016). However, in the actual case, the size of IMBs is usually much larger than that of QDs, blocking the efficient luminescent space QDs within the formed sandwich nanocomposites, which affects the efficient output of luminescent QDs fluorescence signals and thus reduces the detection sensitivity.

Although the IMS technologies-based fluorescence assays have made some positive progress in multiplex pathogens' simultaneous isolation and identification, the broad emission spectra of the reported luminescent materials (e.g., around 50–100 nm emission spectra for fluorescent dyes & proteins and about 30–50 nm emission spectra for UNCPs & semiconductor nanoparticles) essentially limit the number of resolvable colors on a single sample because of the adverse spectral cross-talk (Fan et al., 2019), thereby making two to three pathogens detection at most in a same time. To address the limitations of aforementioned IMS technologies-based optical assays, some groups have demonstrated the promising potential by employing magnetic nanoparticles as the efficiently encodable elements to achieve multiplex pathogenic bacteria detection as much as possible through their superparamagnetic features (Lee et al., 2010).

From such in-depth analysis above, it could be easy to find that magnetic nanoparticles, in combination with fluorescence materials and immunological techniques, exhibit promising application prospects, but some drawbacks are also unavoidable. Firstly, the sensitivity of one assay based on IMS technology usually suffers the restriction from antibody titer, and the analytical selectivity for pathogenic bacteria largely depends on the antibody activity. Antibodies are generally less active (Chen and Yang, 2015). In addition, the high preparation cost of immune nanoparticles also seriously limits their widespread use, failing to meet the demands of the daily on-site checking of pathogenic bacteria in food or agricultural-related industries. Considering such two points, it is of great significance to screen the higher activity of pathogenic bacteria recognition antibodies in a cheaper way for improving the multiplexing detection capabilities of pathogenic bacteria, which is conducive to the daily use of ordinary people, finally safeguarding human health.

According to Shang et al. (2022), regarding the integration of MNPs in microfluidic systems, specific problems must be addressed. (i) The magnetic property of MNPs should be improved to meet the application requirements of a fast external magnetic field. (ii) The stability of MNPs could be improved so that they may, for example, be stored at room temperature. (iii) Preparation of multifunctional MNPs with improved surface functionalization and ligand bonding efficiency would be beneficial.

12.4.4 QUANTUM DOTS

Quantum Dots (QDs) are fluorescent semiconductor nanocrystals made of semiconducting materials, such as cadmium selenide and lead selenide, that can transport electrons which in bulk form are nonfluorescent but in their nanoscale form are susceptible to quantum effects and become highly fluorescent materials (Farzin and Abdoos, 2021), are resistant to photobleaching, and have a significant Stokes shift (difference between the maximum optic excitation and emission wavelengths) (Cady et al., 2007). Notably, the significant Stokes shift of QDs can effectively circumvent the adverse effect of excitation and emission spectral overlap on sensing accuracy (Chandan et al., 2018). In addition, QDs exhibit higher fluorescence intensity and photostability than traditional organic fluorescent dyes (nearly 20 times and 2 orders of magnitude higher, respectively, than those of rhodamine 6G) (Bajorowicz et al., 2018). Notably, their broad excitation and narrow emission spectra allow several QDs to be excited from one source without emission signal overlap. Additionally, their large stokes shift can reduce interference from background auto-fluorescence, which is also valuable for multiplexing (Stanisavljevic et al., 2015). QDs can be easily functionalized with various compounds via electrostatic or covalent conjugation (Karakoti et al., 2015). Given such advantageous features, many groups have employed QDs in microfluidic systems to detect pathogenic bacteria. Several studies have used IMS to separate target bacteria from samples, subsequently using QDs to label them, effectively increasing the detection sensitivity.

The quantum dots are fabricated following colloidal synthesis, plasma synthesis, viral assembly, electrochemical assembly, etc. These are used in photovoltaic devices (solar cells, hybrid solar cells), light-emitting diodes, photocatalysts, photodetector devices, etc. Durán et al. (2015) developed a β-cyclo dextrin modified Cd Se/ZnS QDs sensor to determine vanillin in milk and sugar. Carbon quantum dots (CDs) are microscopic artificial semiconductor particles with sizes normally less than 10 nm. These nanoparticles are extensively used in research due to their high luminescence properties, high solubility, and biocompatibility (Sun et al., 2006). They are quasi-sphere nanoparticles (smaller than 10 nm) formed from crystalline sp^2 hybridization graphite cores and amorphous aggregations used in bioanalytics and biolabeling. These nanostructures based on carbon are two distinct allotropes. Both are functionalized with oxygen-related complex surface group molecules such as carboxylates or hydroxylate derivatives that improve optical features and particle solubility (Ding et al., 2014).

On the other hand, graphene quantum dots (GDs) consist of single or very few graphene lattices (<10). Due to more extensive conjugated domains and periodic structure, the GDs are generally more crystalline than CDs. The variability in the fabrication of these materials gives rise to diverse surface functionalization and more complex hybridization in biosensor applications. Additionally, graphene quantum dots are used as sensors for detecting phenols in olive oil, as an optical sensor for glucose, a photoluminescent sensor for pesticide detection, visual detection of copper ions, and further used as multifunctional sensors (Ghosh et al., 2022).

The multicolor properties and photostability of QDs compared to conventional fluorescent dyes for detecting pathogens make it a better choice for biosensing applications. They have several advantages over organic dyes due to their widespread excitation spectra. With about 10–100 times greater molar extinction coefficient, distinctive adjustable fluorescence peak, and longer fluorescence lifespan, quantum dots resist photobleaching. These properties of quantum dots allow for the activation of many colors of quantum dots from a single source without the emission signals overlapping, making quantum dots brighter than standard fluorophores (Zhao and Zeng, 2015). In particular, their distinctive features, such as high emission quantum yield, excellent photostability, and size-tunable narrow and symmetrical emission wavelengths, make them to be used for fluorescence sensing and bioimaging both in vitro and in vivo (Pandey and Bodas, 2020)

These benefits led to the adoption of QDFRET-based nanosensors in agricultural and related industries. These sensors are most commonly used in enzyme and nucleic acid activity detection (Stanisavljevic et al., 2015). The ability of QD-based nanosensors to exhibit several enzyme activities has been demonstrated by Knudsen et al. (2013). The surface functionalization of QDs with different recognition elements targeting viruses, bacteria, and protozoa has been frequently reported (Gilmartin and O'Kennedy, 2012). The fundamental reason is that QDs usually exhibit a broad absorption band, which could also be favored to establish a versatile analytical platform for achieving the simultaneous detection of multi-channel and multi-target objects by utilizing the same excitation source to simultaneously excite multicolor QDs (Ruedas-Rama et al., 2012). Because of this, it could be an excellent choice to employ multicolor QDs as multicolor fluorescent indicators for simultaneous reporting of the fouling levels of multiple pathogenic bacteria with acceptable results.

Fluorescent quantum dots, fluorescent silica nanoparticles, and carbon nanotubes have been used in various applications, including DNA detection and the development of immunoassays for detecting bacteria and toxins (Edgar et al., 2006). For bacterial detection, a method for possible quantitation of bacteria based on the emission wavelength shifts of quantum dots (Qdots) on binding with bacteria has been reported by Dwarakanath et al. (2004). In nucleic acid-based detection schemes, quantum dots are used instead of traditional dye-based fluorophores. For instance, quantum-dot molecular beacons were synthesized to detect unique DNA sequences for *Salmonella* Typhimurium, *L. monocytogenes*, and *Bacillus thuringiensis* (Cady et al., 2007). In this approach, short- DNA hairpins were attached to a quantum dot at one end and a fluorescent quencher (or gold particle) at the opposite. In this configuration, the fluorophore does not emit light (does not fluoresce) because of a phenomenon known as fluorescent resonant energy transfer (FRET). Optimal energy transfer (and subsequent quenching) occurs when the fluorophore and quencher are in very close proximity (typically around 5 nm distance). When the DNA hairpin is exposed to target DNA, it can unravel and hybridize with the target DNA, physically separating the fluorophore and quencher, resulting in a sharp increase in fluorescence. This results in a high signal-to-noise assay because there is no fluorescent (Figure 12.2).

QDs can be variously modified without difficulty, which is conducive to connecting with various specific biocognitive molecules to achieve the purpose of particular detection. However, there are still some problems and challenges when using these inorganic semiconductor QDs as fluorescent indicators. First, the traditional inorganic semiconductor QDs containing heavy metal elements (*e.g.,* Cd & Pb) have certain toxicity and potentially harm the environment, human body, and microorganisms (Manshian et al., 2017). Some alternative semiconductor QDs based on harmless metallic elements (e.g., Cu) have been reported with low toxicity but still do not seem ideal (Li et al., 2013). Therefore, developing safer semiconductor QDs with high fluorescence quantum yield for constructing multiple pathogenic bacteria-responsive assays may be more significant. Secondly, due to the wide absorption band of traditional inorganic semiconductor QDs, multicolor QDs may cause fluorescence resonance energy transfer between each other and weaken the fluorescence emission efficiency, thereby impairing the response sensitivity of coexistence pathogenic bacteria. The practical alternative is to develop novel QDs with a narrow enough fluorescence emission peak, allowing remarkable performance for high-throughput analysis. Finally, to improve the accuracy of pathogenic bacteria assays based on QDs fluorescence emission intensity, the development of dual-emission ratiometric fluorescence assay based on QDs can effectively circumvent the interference from various target-independent factors (*e.g.,* instrumental parameters & variations in bacterial microenvironment around QDs) through built-in self-calibration, achieving an improved detection sensitivity and a wide dynamic linear range for pathogenic bacteria assays (Hao et al., 2016). In a word, by exploring new synthesis methods of QDs or performing the ingenious engineering designs of functionalized QDs, QDs still have more broad research space for practical applications in the simultaneous detection of multiple pathogenic bacteria (Figure 12.2).

As discussed above, considerable progress has been made in developing QD-based biosensors for detecting foodborne bacteria. Still, certain reasons prevent their conversion into commercial products: (i) Their sensitivity has a high potential for improvement, as the LODs have not reached a satisfactory level. The detectable bacteria levels in food samples are usually very low; signal

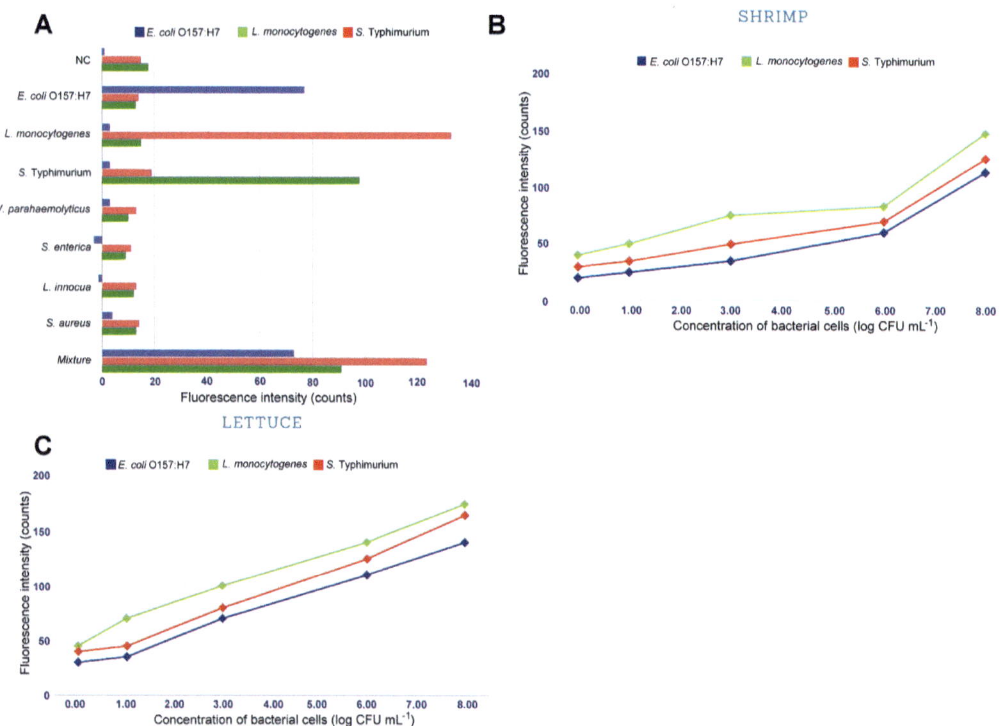

FIGURE 12.2 (a) Specific response for a portable fluorescent biosensing system for *E. coli* O157:H7, *L. monocytogenes*, and *S. typhimurium*. (b & c) Calibration curves obtained after field detection of three bacteria in shrimp and lettuce, respectively. A schematic diagram for the simultaneous tracing of *E. coli* O157:H7, *L. monocytogenes*, and *S. Typhimurium* in food samples based on QDs.

(Continued)

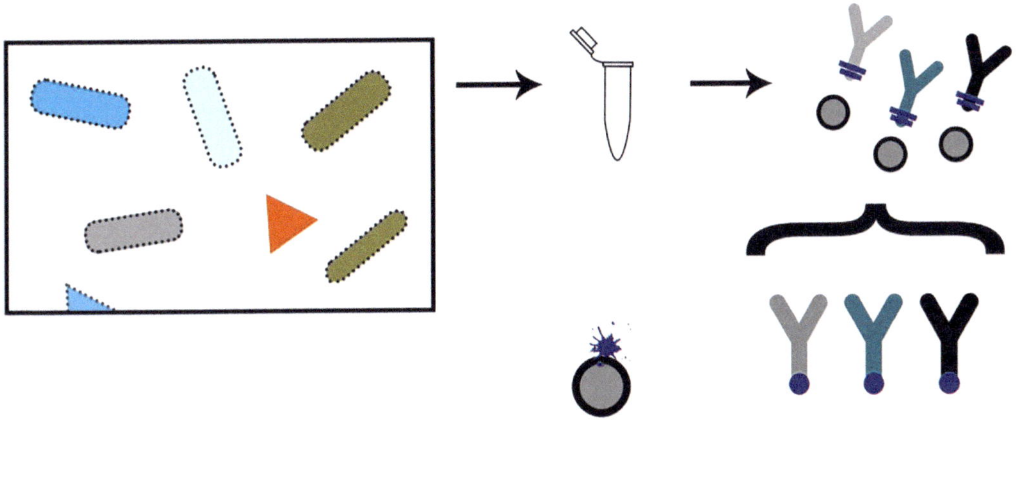

SEPARATION

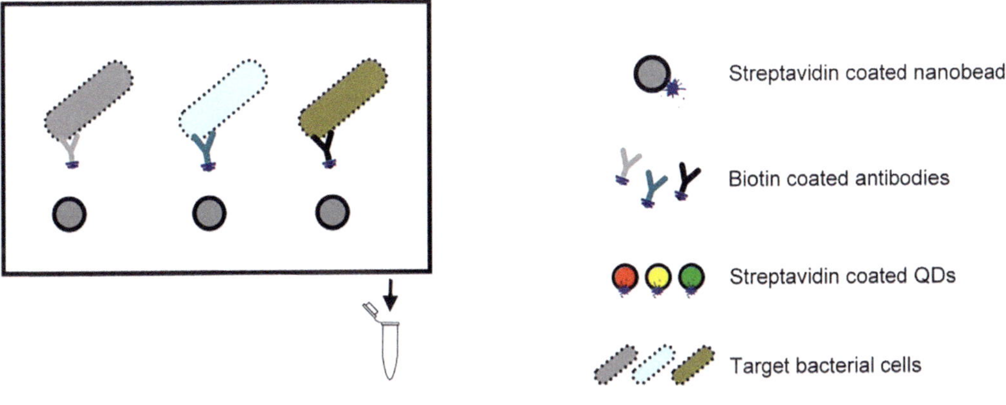

DETECTION

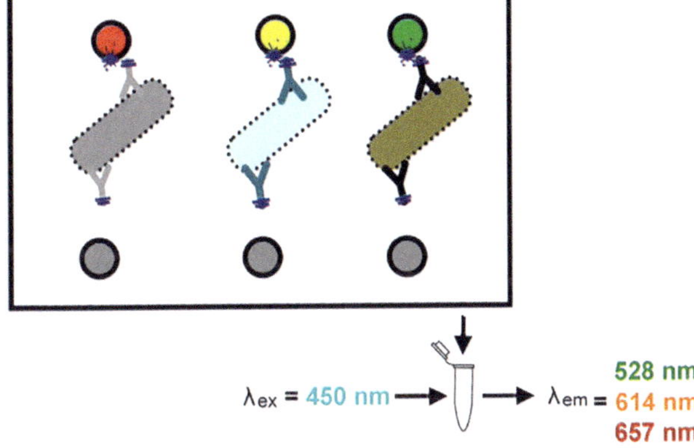

FIGURE 12.2 (Continued)   (a) Specific response for a portable fluorescent biosensing system for *E. coli* O157:H7, *L. monocytogenes*, and *S. typhimurium*. (b & c) Calibration curves obtained after field detection of three bacteria in shrimp and lettuce, respectively. A schematic diagram for the simultaneous tracing of *E. coli* O157:H7, *L. monocytogenes*, and *S. Typhimurium* in food samples based on QDs.

amplification strategies in this aspect need further study. (ii) High stability and lower costs are more attractive. The ideal QDs must be reusable, where binding with and separation from bacteria does not affect their performance. This requires further research. (iii) New synthesis methods of QDs are required. QD toxicity remains an obstacle to its practical application. Therefore, green synthesis or biosynthesis methods are needed to produce QDs to reduce their toxicity (Shang et al., 2022).

12.4.5 Carbon-Based Materials

Carbon can form different structures of the same element because of its valency. The allotrope of carbon is available in crystalline form or amorphous forms. The other allotropes of carbon are diamond, lonsdaleite, graphene, graphite, carbine, fullerenes, carbon nanotubes, carbon dots, etc. The zero-dimensional carbon allotropes are fullerene, carbon dot, graphene dot, and nanodiamond. The one-dimensional carbon allotropes are single-walled carbon nanotubes, multi-walled carbon nanotubes, and carbon nanohorns. The two-dimensional carbon allotropes are graphene and multi-layered graphite sheets. The three-dimensional carbon allotropes are diamond and graphite (Liu and Speranza, 2019).

On the other hand, carbon nanotubes are tubular structures made of carbon and are arranged in graphene sheet form, which is rolled to get a cylinder shape. Carbon-based nanotubes are classified as single-walled nanotubes (SWNTs) and multi-walled nanotubes (MWNTs). SWNTs comprise a single graphene cylinder, a single sheet of graphene rolled into a seamless molecular cylinder, with a diameter and length ranging between 0.75–3 and 1–50 nm, respectively. MWNTs are made up of more than two concentric cylindrical shells of graphene sheets around a central hollow core (Kierkowicz et al., 2018) with a diameter usually in the range of 2–30 nm and sometimes even more than 100 nm (Alim et al., 2018). As a result of the high surface area, which allows immobilization of a substantial number of recognition elements (nucleic acids, aptamers, antibodies, oligonucleotides, and enzymes), CNT-based microchips can achieve high detection sensitivity (Rathinavel et al., 2021). In the last decade, carbon nanotubes (CNTs) have emerged as one of the most extensively used nanomaterials in drug delivery and biomolecular techniques. They have unique structures, high electrical and mechanical properties, chemical stability, lightweight, high thermal conductivity, unique electrocatalytic action, minimal surface fouling, and a high surface-to-volume ratio (Pandit et al., 2016). MWCNTs are commonly used to modify electrodes and simultaneously detect various pathogenic bacteria, showing significantly superior performance. Due to the simple preparation of the screen printing process, convenient manual operation, and low cost, it is often used to prepare disposable electrodes (Niu et al., 2012). SWCNTs have been employed in a DNA sensor application for the detection of *Salmonella* using N-ethyl-N'-(3-dimethylaminopropyl) carbodiimide hydrochloride (3-dimethylaminopropyl) covalently bonded to the nanotubes (Weber et al., 2011).

Unlike QDs-based assays that inherently rely upon toxic metal ions (e.g., Cd^{2+}), carbon-based nanomaterials such as carbon nanotubes, carbon dots (CDs), and graphene were primarily composed of the hypotoxic nonmetallic carbon, which has attracted tremendous attention in the construction of the efficiently biocompatible assays for various chemical or biological compounds. Furthermore, carbon-based nanomaterials have many other advantages over conventional QDs and organic fluorescent dyes, including robust chemical inertness, easy functionalization, and high resistance to photobleaching; these features have attracted more attention in the development of versatile biosensing platforms (Chen et al., 2020). In particular, most carbon-based nanomaterials usually exhibit a π-rich surface, which can be endowed with a strong affinity toward various target analytes through the noncovalent π-π stacking and hydrophobic interaction. Apart from these, some carbon-based nanomaterials (*e.g.*, graphene & carbon nanotubes) also exhibit promising electrochemical properties with good electrical conductivity and high efficiency of electrocatalysis and, therefore, have potential applications in the development of multifarious electrochemical sensors (Adhikari et al., 2020). Generally, these carbon-based nanomaterials can be used as excellent ion-electronic converters to detect the conversion of biological information into electrical signals. The typical representative is a carbon nanotube that can act as a superb accelerator to speed electron transfer after modifying the surface of electrodes.

Carbon dots, a type of fluorescent carbon nanomaterial, are spherical or quasi-spherical nanoparticles with sizes less than 10 nm and are comprised of sp^2/sp^3 carbon skeletons and abundant oxygen/nitrogen-based groups or polymeric aggregations. CDs have superior fluorescence and can be utilized for bioanalysis as they have the most fascinating photoluminescence feature, probably enabling them to present multicolor for marking different pathogenic bacteria on the same sample. Several studies have reported using CDs as fluorescent probes for bioimaging and identification of metal ions (Mehta et al., 2021). However, only a few studies have focused on the selective detection of bacteria using CDs, and even fewer have focused on its implementation in microfluidic systems. However, developing surface-modified CDs with high quantum yield is still challenging; all CD-based assays are performed in solution and not on microchips. Future efforts in this field must be encouraged.

Other nanoscale materials for food pathogen detection include carbon-based materials, graphene, and fullerene nanotubes. Fullerenes, or C_{60} or carbon nanospheres, are spheres of single-layer π-bonded carbon with unique electric properties. However, among available carbon allotropes, graphene is considered "the mother of all graphitic forms of carbon," consisting of an atomically thin, sheet-like nanomaterial consisting of π-bonded carbon or atoms in a single layer with 2D honeycomb lattice. Graphene has many beneficial properties, including high mobility concerning charge carriers, impermeable to gasses, high thermal conductivity, exceptionally stiff, high thermal conductivity, and highly flexible thin films. Compared with carbon nanotubes, graphene oxide (GO) is usually considered a prominent quencher to construct different fluorescence assays because of its broad absorption band from far-red to near-infrared regions (Anju and Renuka, 2019). Graphene-based nanomaterials can be used as a quencher to generate fluorescent transducer-based biosensors, as reduced graphene oxide (rGO), graphene (G), and GO possess a very high fluorescent quenching efficiency (Batir et al., 2018). In a fine energy transfer sensing system, GO can effectively accept the energy from its adsorbed light-emitting materials such as organic fluorescent dye, QDs, CDs, etc., leading to efficient fluorescence quenching of these luminophores.

Moreover, as an amphipathic nanomaterial, GO possesses a strong adsorption capacity to bind with hydrophobic molecules through noncovalent π-π stacking interaction and assembly with hydrophilic substances via hydrogen bonds. The electron mobility in graphene is not highly affected by temperature changes, and the carboxyl, hydroxyl, and epoxy groups on the surface also provide vast binding sites for other molecules (Bokhari et al., 2020). As with CNTs, graphene can also be used directly on the electrode surface. Moreover, graphene can improve the optical performance of some sensors. It has been reported that SPR and SERS signals are increased due to the charge transfer between graphene and the probe molecules (Rao et al., 2009).

Graphene is the most commonly used nanomaterial for different biosensor designs. It is generally used for converting targeted and receptor molecules for detectable measurement using EDC/NHS chemistry (Pumera, 2011) with varying transduction modes. Graphene-based electric sensors have been used to detect *E. coli* via impedance measurements (Wang et al., 2013b). Graphene was embedded in a filter material in this measurement approach and modified by adding gold nanoparticles. This formed a high surface area electrode surface that could then be used to detect *E. coli* cells via affinity-based binding and subsequent electric impedance measurement. As with other impedance-based measures, the added capacitance and resistance of the bacterial cells change the frequency-dependent impedance of the electrode (in this case, the graphene-modified paper). Fullerenes have also been used for electrochemical antigen-based detection of *E. coli* (Li et al., 2013). Other carbon-based electric sensors have been developed beyond using nanoscale carbon-based materials for micro and nanobacterial sensors. These include screen-printed carbon electrodes, such as those used by Alfonso et al. (2013) to detect *Salmonella*. Screen printing is essentially a low-resolution form of pattern transfer that enables printing liquid-based materials through a previously patterned porous "screen" material. For microscale electrodes, screen printing is a fast and easy way of generating electrodes in a high throughput manner. This aids in reducing the final cost of the sensor and can enable the precise patterning of materials (such as carbon paste) that would be difficult to achieve with other methods.

Through the above examples, various carbon-based nanomaterials exhibit broad development prospects in the simultaneous detection of multiple pathogenic bacteria, attributable to their outstanding optical (*e.g.*, high brightness, color tunability, and tunable excitation) and electrochemical (*e.g.*, prominent electrical conductivity, chemical stability, and large surface area) features. However, it must be admitted that some challenges still need to be overcome in simultaneously detecting multiple pathogenic bacteria for practical applications. One of the significant challenges is to clearly understand the exact luminescence mechanism of carbon-based nanomaterials, which might be in favor of controlling the width of emission spectra as narrow as possible. A clear understanding of the properties of carbon nanomaterials, allowing them to maintain high stability without changing the structure and function of the biomolecules or the material itself, is also vital. In this circumstance, it is conducive to developing a versatile optical sensing platform for simultaneously detecting more pathogenic bacteria on a sample. Besides this, the current reports only focus on constructing various sensing strategies based on carbon-based nanomaterials. Their actual attainment from scientific research to practical applications, such as food safety, environment monitoring, and clinical diagnosis, would need more effort. In the future, the development of multiplex pathogenic bacteria stimuli-responsive sensing devices based on carbon nanomaterials with high sensitivity, selectivity, stability, and reproducibility should be the primary focus of efforts, which can be integrated into microfluidics and portable sensor chips for endowing the capability of point-of-care testing, thus establishing the early warning mechanism of pathogenic bacteria fouling and ultimately safeguarding human health.

12.4.6 COLORIMETRIC ASSAYS AND PHOTON-BASED METHODS

Optical detection correlates an observed optical signal to the interaction between a specified bioprobe and the target pathogen. A unique advantage of photon-based measurements over electron-based sensors (e.g., electrochemical) is their immunity to electromagnetic interference, which may be necessary in POU field applications. Among the many detection and quantification mechanisms, the most common pathogen detection method is based on fluorescence. Fluorescence sensing platforms are based on interactions between fluorescently tagged target pathogens and their associated biorecognition molecules (Martins et al., 2013). Environmental conditions leading to interferences include variability in luminescent response with pH, susceptibility to photobleaching, short excited state lifetimes, luminescent emission shift, and short-term stability in aqueous mediums. These factors would make these sensors best suited for production facilities where testing could be performed in a controlled environment.

Colorimetric assays are among the most prominent and affordable nanotechnology-based toxin detection venues that utilize nanoparticle probes. For instance, using thiolated lactose derivatives and gold nanoparticles, researchers were able to detect cholera toxin through molecular mimicry, as the nanoparticles' coating resembled the extracellular matrix terminal portion of GM1 ganglioside which is found in the apical membrane of intestinal epithelial cells (Schofield et al., 2007). Detection and quantification were achieved visually and spectrophotometrically through color changes in the nanoparticle suspension (red to deep purple) and shifts in the nanoparticles' plasmonic band, as increases in the toxin concentration facilitated concomitant redshifts in the UV–visible absorption spectra (Schofield et al., 2007). Since portability is crucial for detecting toxins, antibody-carrying gold nanoparticles have been immobilized in immunochromatography strips, allowing the visual detection of aflatoxins in grain samples (Shim et al., 2007).

An emerging class of label-free biosensors appears to have the greatest potential for deployment in a wide range of POU applications. Label-free methods detect target molecules in their natural state, reducing processing complexity and eliminating potential interference effects from fluorophores (Massad-Ivanir et al., 2011). Quantitative and kinetic measurements can be performed with high sensitivity depending on the detection mechanism, as the detection signal does not scale down with sample volume. SPR (Lisalova et al., 2016) affinity biosensors are considered one of the most versatile and powerful label-free techniques, relying on the excitation of surface plasmons in a resonant manner with incident light (Martins et al., 2013). One of the main challenges to real-world implementation of SPR biosensors is signal interference from the sample matrix; nonspecific adsorption of nontarget molecules leads to false positive signals and biofouling.

On the other hand, Raman Spectroscopy (RS) is an inelastic scattering that can be used to get information about a molecule's unique molecular bond vibrations. When a sample is irradiated with incident light, light scatters either elastically or inelastically. Most light is elastically scattered with the same frequency as the incident light, called Rayleigh scattering. Although RS provides a vibrational fingerprint of the molecule, it is feeble because only a small part of the incident photons is inelastically scattered. Thus, detecting low-abundant molecules in complex media is not reasonably possible without any improvement. With plasmonic nanostructures, enhancement of the Raman signal ranging from 107 to 1,014 can be achieved, and this phenomenon is called surface-enhanced Raman scattering (SERS) (Zhao et al., 2006).

With the nanotechnology idea, SERS has become a powerful vibrational technique that can give information from the fingerprint of the molecules. It provides high sensitivity, low water interference, and short detection time and does not require complex and lengthy sample preparation steps. The obtained signal can be tailored with changing substrate properties, including size, shape, composition, and surface chemistry (Sharma et al., 2012). SERS is a good alternative for on-site detection systems, which can be integrated with developed SERS probes and portable Raman spectrometers. Many different detection platforms or SERS probes were fabricated for on-site detection, such as microfluidic devices, paper-based, and swab-based systems (Krafft et al., 2021). With such systems, sample volume can be decreased, sample handling can be automatized, and the whole system can be turned into a portable device. Many excellent review papers have been published in the literature based on the application of SERS in detecting agricultural pollutants. As an on-site detection system, SERS is a widely used powerful technique that can provide rapid, specific, sensitive, and on-site detection (Petersen et al., 2021). As a different microfluidic-based system, Dina et al. (2020) reported the detection of *E. coli, Pseudomonas aeruginosa, S. aureus,* and *Enterococcus faecalis* from a microfluidic flowcell platform coupled with a portable Raman spectrometer.

Jeanmaire et al. proposed electromagnetic field enhancement (EM) theory, resulting from the surface plasmon polaritons (SPPs). Surface plasmons are the collective oscillation of free electrons in the conduction band of the noble metal. When surface plasmons oscillate with the incoming laser at a frequency, it creates an electromagnetic field gradient formed in a very thin zone (~10 nm). When a molecule approaches the SERS substrate (plasmonic metal structure), the localized electromagnetic field enhances Raman scattering from the molecule. EM enhancement depends on the resonance between plasmons, excitation, and scattered field.

Moreover, when excitation light and Raman scattering resonate with the surface plasmon frequency, the SERS signal is maximized by a factor (enhancement factor, EF) of about 105–106 (Schatz and Van Duyne, 2006). Albrecht and Creighton (1977) proposed a chemical enhancement (CE) theory based on the charge transfer mechanism. When a molecule is adsorbed on a metal surface, chemisorption causes a new electronic state formation, mostly called the charge transfer state. This newly formed state could serve as a resonant intermediate state in Raman scattering and increase the Raman probability of the adsorbate, enhancing the Raman signals.

12.4.7 Inorganic Nanomaterials

Inorganic nanomaterials comprise nanoparticles comprising inorganic material or metallic nanostructures such as QDs (Yu et al., 2018). Gold, silver, iron oxide, titanium dioxide, silicon dioxide, and zinc oxides are the inorganic materials used to make inorganic nanoparticles. These particles can be crystalline or amorphous solids with spherical or nonspherical shapes at ambient temperature. Metallic nanoparticles composed of gold or silver have many optical and electronic properties derived from their size and composition (Nath et al., 2009). When coupled with affinity ligands, these nanomaterials find essential applications as chemical sensors. Metallic nanoparticles in colloidal solution appear cultured. For example, colloidal GNPs generally appear ruby red, purple, blue, and orange depending on their shape, size, and synthesis conditions.

Similarly, titanium (Ti) and platinum (Pt) nanoparticles give blue and dark brown colors. The different colors of metal nanoparticles arise from SPR and their confinement (Cady et al., 2017). Silver (Ag), Gold (Au), Silicon (Si), and other metal oxide nanoparticles are typical examples of inorganic nanomaterials that are used in the standard class of nanosensors. They employ signal transducers and recognition ligands (such as antibodies or DNA oligos) to identify and measure target molecules (Darr et al., 2017). These nanoparticles can be used as distinctive sensing platforms to examine interactions in vitro or in original plant systems between nanoparticles and bioanalytes that are specifically relevant to plant diseases (Kwak et al., 2017). Changes in the surface-enhanced optical characteristics resulting from the LSPR of nanoparticles are frequently used to record distinctive sensor responses. To build sensitive and target-specific nano-biosensors, a wide range of self-assembled nanoparticles containing oligonucleotides (Lau and Botella, 2017) or proteins (Li et al., 2018) have been investigated using SERS.

The properties of gold nanoparticles include chemically inert, biological compatibility, high stability, high surface area to volume ratio, high dispersity, non-cytotoxicity, plasmonic nanoparticles, scattering solid properties, and high electrical and heat conductivity (Yusoff et al., 2015). GNPs have distinguished themselves in biodiagnostics due to their size-dependent optical properties, variety of surface coatings, and biocompatibility (Pandey et al., 2008). Owing to the prominent optical and electrochemical properties, gold nanomaterials, such as gold nanoparticles (AuNPs), nanorods (AuNRs), nanoshells (AuNSs), nanoclusters (AuNCs), nanostars (AuNTs), and gold nanowires (AuNWs), have been a hot candidate to develop various chem/biosensors (Pissuwan et al., 2020). The sensor based on AuNPs depends on the color change of the solution or aggregation/disaggregation of particles, and they can directly or indirectly sense specific targets when combined with tools such as fluorescence labels, nucleic acid hybridization, and enzyme-mimicking activity. These NPs detect hazardous agents, toxins, pesticide drugs, and heavy metals. Gold nanoparticles (AuNPs) are extensively used in several ways, such as drug delivery, radiotherapy, photothermal therapy, imaging agents, nanoenzymes, cancer therapy, biosensors, tissue engineering, and enzyme immobilization. Bhardwaj.

Spherical GNPs exhibit SPR-related optical absorption at 520 nm, which depends on particle size and morphology. The rich surface chemistry of GNPs allows surface modification with various biofunctional groups, such as nucleic acids, sugars, and proteins, via the strong affinity of a gold surface with thiol ligands, creating multifunctionality to tailor the needs of biomedical applications including imaging, diagnostics, and therapy. The conjugation of nanoparticles with biomolecules, such as proteins DNA, can be done either by direct covalent linkage or noncovalent interactions. Biomolecules are often covalently linked to ligands on the nanoparticle surface via traditional coupling strategies such as carbodiimide-mediated amidation and esterification. The high surface-to-volume ratio of GNPS makes surface electrons sensitive to minor changes in the medium's dielectric (refractive index) constant. Therefore, changes to the surface chemistry and environment of these particles (surface modification, aggregation, medium refractive index) lead to colorimetric changes in the dispersions that become the basis for detecting any analyte of interest. This has further facilitated bio-detection applications via numerous detection methods (Boyer et al., 2002).

Gold nanoparticles are frequently used in nanotechnology-based applications, especially for detecting food contaminants. For example, gold nanoparticles conjugated with specific oligonucleotides can sense complementary DNA strands, detectable by color changes (Mirkin et al., 1996). Gold nanoparticle (AuNPs) synthesis in organic or aqueous solvents requires a stabilizing mediator (surfactant) for stability. It can be achieved by chemical binding or adsorbing the appropriate mediator on the gold NPs. The biological compatibility, exceptional conducting capability, and high surface-to-volume ratio are the few characteristics of AuNPs, which make them the nanoparticle of interest (Guo and Wang, 2007). They also have SPR characteristics, and the redox activity is an interesting characteristic of gold NPs, enhancing the sensitivity of electrochemical biosensors in analyzing foodborne pathogens.

Gold nanoparticles possess extraordinary optical properties at their nanoscale, wine red in color of gold nanoparticles (GNPs), which become blue when aggregating. This property has been exploited to design many biosensors for dairy toxins detection. Worthy of special mention is the fact that the optical features of these Au nanomaterials are closely related to their own structural composition and LSPR effects, allowing them to be used as an excellent signal transduction unit for rational designing of versatile optical analytical methods (Rakshit et al., 2017). In general, the LSPR effects of Au nanomaterials could be effectively activated by a light-induced free electron oscillation on their surface, manifesting a strong LSPR absorption peak, which is easily changed by the alterations of nanomorphology, composition, distance between nanoparticles as well as surrounding permittivity of nanoparticles (Yu et al., 2020). Driven by this fact, by mediating the distance of nanoparticles, their LSPR absorption bands will be effectively changed, along with a noticeable color change, thus making them prominent candidate nanoplatforms for colorimetric detection of various targeting analytes. Because of such enjoyable features, many groups have also devoted themselves to employing the intuitive color changes of Au nanomaterials for performing the colorimetric assay of various pathogenic bacteria. The basic design strategy mainly relies on the aggregation or anti-aggregation phenomena, observing obvious color change. Although the aggregation-based colorimetric platforms exhibit relatively simple and sensitive, one nerve-wracking issue of the self-aggregation of nanoparticles often causes false positive/negative results. In contrast, the anti-aggregation colorimetric strategy is more reliable and precise.

Gold and silver nanoparticles can enhance the signal intensity for RS, also known as surface-enhanced Raman spectroscopy (SERS). The SPR effect of gold nanomaterials is one of the most useful sensing principles for designing the simultaneous colorimetric assay for multiple pathogenic bacteria into a single matrix with fast kinetics. In particular, by employing more specific markers as the specific signal molecules to indicate various bacteria, the simultaneous detection of more pathogenic bacteria can be effectively realized as well, guaranteeing no spectral cross-talk between the different detection channels to affect the accuracy of the assay. Unfortunately, the current Raman spectral equipment is relatively expensive and bulky, which is only suitable for laboratory operation, failing to meet the requirements of on-site detection of bacteria. To overcome this defect, it is necessary to develop a portable Raman spectrometer, and some efforts are underway. Recently, the lateral flow detection strip (LFA) has attracted more and more attention in on-site assay because of its advantages of simple, fast, and long-term stability (Wang et al., 2020a). However, the relatively low sensitivity and selectivity are also the inevitable issues.

In contrast, the SERS assay based on the LSPR effect of gold nanomaterials generally shows a higher sensitivity for realizing the simultaneous assay of multiple pathogenic bacteria through the enhanced characteristic Raman scattering signals. Unlike fluorescence assay, SERS can usually avoid the headaches of photobleaching and photodegradation, which is more suitable for long-term monitoring. More importantly, the SERS characteristic peak frequency width is very narrow (10–100 times thinner than the width of the fluorescence peak excited by traditional organic dyes and even QDs), allowing it to use a single laser for realizing the multi-path detection of various pathogenic bacteria at the same time (Wang et al., 2020b).

In addition to the visual colorimetric detection technology, applying the local plasmon resonance characteristics of gold nanomaterials can prepare spectroscopic biosensors for detecting pathogenic bacteria. The basic principle is that the gold nanomaterial is combined with the target analyte through the coupled identification element, which effectively triggers the change of the refractive index of the gold nanomaterial, thus affecting the LSPR effects of gold nanomaterials, ultimately outputting the response signal for the detection of pathogenic bacteria on an ultraviolet absorption spectroscopy (Kaushik et al., 2019). Colorimetric characteristics of gold nanoparticles are beneficial for toxin detection in foods. A recent review by Geleta (2022) illustrates how the principle of gold nanoparticle detection can be coupled with aptamers to increase the sensitivity of bacterial (*C. botulinum, C. difficile, S. aureus*) and fungal (mycotoxins) toxins detection in foods. Additional and specific details on nanosensors for toxin detection can be seen in Section 12.4.12.

Silver nanoparticles like GNPs exhibit peculiar properties of localized SPR. As described for GNPs, the colloidal solutions of AgNPs have high extinction coefficients and color change properties in the visible region of the spectrum depending on the intermolecular distance. The silver nanoparticles are antiviral, antibacterial, anti-inflammatory, anti-angiogenic, antitumor, and anti-oxidative. The antibacterial application of silver NPs includes footwear, apparel, paints, plastics, and appliances. Furthermore, AgNP-based sensors are extensively used, such as (i) Detection of specific toxic dyes, (ii) Prohibited food colorants, (iii) Detection of chloramphenicol in food, (iv) Monitoring meat spoilage in intelligent packaging, (v) Detection of Vitamin B1 in food and water samples, (vi) Detection of glyphosate in food samples (Ghosh et al., 2022).

Titanium oxide is a transition metal oxide with anatase, rutile, and brookite phase structures. It is an inexpensive, non-toxic, inert metal with potential action against various microorganisms (Kshirod et al., 2019). The anatase and rutile have a tetragonal structure, and brookite has an orthorhombic structure. The nanoform of TiO_2 has the properties of biocompatibility, nontoxicity, cost-effectiveness, high UV absorbance, high photochemical activity, large surface area, and remarkable chemical and electronic properties. The TiO_2-based sensors are used in sensing volatile chemicals and detecting soluble organics and biological substances. Besides as a biosensor, titanium dioxide is widely used as a sensor in several ways, such as (i) Detection of pork freshness, (ii) Evaluate the shelf life of mango, egg, and fish, (iii) Detection of histamine in Salmon fillet, (iv) As gas sensor in determining fish freshness.

Similarly, zinc oxide NPs generally have diameters less than 100 nm, whereas the NPs have high catalytic activity and surface area. Sensing food properties during storage, zinc oxide, and its nanoforms are used in detecting seafood odors, vapor sensing properties for monitoring quality during storage, biogenic amine detection, and trace detection of sunset yellow in soft drinks (Ghosh et al., 2022).

A special mention is made for metal-organic frameworks (MOFs), a type of crystalline nanomaterials composed of metal ions and organic ligands that have garnered much attention because of their large specific areas, stable structures, and adjustable pore sizes (Cheng et al., 2021). MOFs containing different metal and organic ligand compositions offer considerable functionality, including catalytic, electrochemical, and optical activity. About 70,000 MOFs have been reported (Ryu et al., 2021). Although the unique properties of metal nanomaterials have pushed the development of microchip platforms, further studies would be required to develop better applications. Primarily, it is necessary to design synthetic routes for metal nanomaterials with high homogeneity and stability in large quantities. In addition, the potential biological toxicity of the chip materials for cells, tissues, and the environment requires careful study. Moreover, the complete integration of metal nanomaterials with microfluidic systems for point-of-care testing (POCT) in complex food matrices.

12.4.8 LAB-ON-CHIP-BASED DETECTION

A microfluidic chip, or "lab-on-a-chip" (LOC) device, can integrate different units, which offers several advantages, including miniaturization, high throughput, low sample consumption, and cost-effectiveness (Xing et al., 2021a). In recent years, microfluidic chips have been used in food production and have shown some promise in detecting foodborne bacteria (Xing et al., 2021b). However, as most bacteria exist in a contaminated matrix at low concentrations, it is difficult to detect them (Kim and Oh, 2021) effectively. Nanomaterials can improve the analytical performance of microfluidic chips regarding selectivity and sensitivity, thus expanding their applications.

They are also referred to as biomedical nano- and microelectromechanical systems (BioNEMS or BioMEMS) for integrated and automated sample preparation and (quantitative) detection.

LOC systems have been developed for PCR-based detection of various food pathogens (Tachibana et al., 2015). In these detection formats, total nucleic acids are extracted from a sample, followed by microliterscale PCR reactions, typically in a chip-based form. The small size of the reaction chamber enables exceptionally rapid heat transfer, which can reduce the PCR cycling time and even improve the efficiency of PCR amplification. Most microchip-based PCR systems employ real-time PCR assays to quantify successful PCR amplification. These technologies have been further enhanced by adding multiplexing capability via multiplex real-time PCR and microarray-based methods (Tortajada-Genaro et al., 2015). Mu et al. (2015) have recently developed a surface acoustic wave (SAW) microfluidic device to concentrate bacterial cells in suspensions. In a different approach, but still exploiting microfluidics, Clime et al. (2015) optimized a tool that could remove the debris particles from beef samples and extract the pathogens without the need for membranes.

Integrating biosensors for pathogen detection with microfluidic components for sample pretreatment opens the possibility of progress and innovation in the LOC and micro total analysis systems (μTAS). The rationale is to avoid the need for off-chip procedures and specialized personnel to increase the throughput and reduce the assay cost. However, LOC systems enable the determination of other analytes of food significance, such as pesticide contaminations, vitamins, antioxidants, and allergens. Two main approaches have been developed to detect foodborne pathogens on-chip: (i) methods based on nucleic acid amplification and (ii) methods based on detecting whole cells without the need for cell lysis. Multiple-analyte detection is one of the most essential improvements possible by exploiting the possibility offered by miniaturization and LOC platforms. In this respect, Brandão et al. (2015) have proposed a new platform for the multiplexed detection of foodborne pathogens. All the systems previously described require the external contribution of an operator to perform some steps of the process. Still, the final goal of a LOC platform would be to obtain a fully automatic assay for "sample-to-results" analysis. In this respect, centrifugal microfluidic can accelerate the development of new analytical devices because fluidic handling and sample processing can be achieved without external pumping but with a rotational motor to create the necessary forces to drive the flow.

Despite significant advances in nanoscale LOC devices, there are substantial challenges to directly detecting food pathogens. Interfacing the large, complicated, and different food samples with the microscale fluid interrogation volumes requires unique solutions. Other approaches can be used for bacterial concentration in microfluidic devices by exploiting various physical mechanisms, such as gravitational force, electric force, dielectrophoretic force, magnetic force, and acoustic-primary radiation force.

12.4.9 UCNP-Based Assays

In recent years, the lanthanide-doped upconversion nanoparticles (UCNPs) have become a considerable research interest in the development of chemical/biosensors, attributable to their unparalleled optical merits such as relatively sharp emission wavelength, large anti-Stokes shift luminescence, insignificant photobleaching, easy functionalized modifications, and long luminescence lifetimes (Zheng et al., 2019). Usually, the UCNPs are synthesized by embedding lanthanide rare earth ions into an inorganic matrix of around 10–100 nm. The luminescence features of UCNPs are primarily dependent on the stepped energy level of the dopped rare earth ions (*e.g.*, Pr^{3+}, Nd^{2+}, Dy^{3+}, Er^{3+}, Ho^{3+}, and Tm^{3+}) (Wen et al., 2018). Compared with traditional organic fluorescent dyes and QDs, UNCPs present excellent optical attributes, including significant anti-Stokes shifts, resistance to photobleaching, low background auto-fluorescence, high chemical stability, and low toxicity (Kumar et al., 2021). The unique anti-Stokes shift luminescence signals of UCNPs can distinguish the auto-fluorescence of complex real matrices such as food, environment, and biological tissues, thereby reducing the background interference and improving the sensitivity during sensing. Moreover, UCNPs show narrower emission spectra, which could be more helpful for accurately detecting multiplex pathogenic bacteria simultaneously [174,175 Review pathogen detection 2021]. Upconversion nanoparticles (UCNPs) are gaining considerable interest, mainly in improving and expanding upon their applications (Mendez-Gonzalez et al., 2017). Given that UCNPs NaYF4:Yb3fl, Er3 fl can convert near near-infrared (NIR) light into visible light to promote LSPR frequencies of AuNPs, LFSAs that incorporate UCNP-based probes could also be used to detect bacteria. However, the low photoluminescence efficiency, unfavorable hydration modification, and poor antibody modification of UCNPs pose severe challenges to their practical application in LFSA platforms. Recently, some studies have reported solutions to these limitations.

Furthermore, a few approaches based on luminescence resonance energy transfer (LRET) have been established for the rapid and sensitive detection of pathogens, where the UCNPs serving as donors are hybridized with inorganic/organic LRET acceptors such as QDs, AuNPs, MOFs, carbon-based structures, MoS_2, and MnO_2. The LRET offers advantages over FRET, such as strong energy transfer even at relatively large distances, increased sensitivity, and reduced background due to the longer donor lifetime. Three main strategies are used: energy transfer between donor and acceptor followed by excitation of the acceptor, quenching of donor luminescence induced by resonance energy transfer, and recovery of donor luminescence induced by restoring the original properties of UCNPs. For this purpose, AuNPs, gold nanorods, and WS2 (Shang et al., 2022) were bonded with UCNPs through aptamer-cDNA pairing or electrostatic interaction, quenching upconversion fluorescence. In the presence of target bacteria in the system, its specific aptamer preferentially combined with the bacteria and dissociated donor UCNPs from the acceptor, leading to fluorescence recovery.

Further exploiting the design of UNCP- and MNP-conjugated aptasensors for simultaneous magnetic separation and bacterial detection would be helpful for practical applications. In addition, the promising potentials of UCNP-based anti-Stokes shift luminescence properties in the simultaneous optical detection of multiplex pathogenic bacteria are substantial. However, whether the Stokes shift or anti-Stokes shift luminescence feature inevitably suffered from the limitation of signal cross-talk and spectral overlap in multiplexing detection. To address the trouble of spectral overlap, an alternative method might be developing a dual-excitation sensing strategy by integrating anti-Stoke shift luminescent UCNPs with other Stokes shift luminescent nanoparticles (*e.g.,* QDs & CDs) into a single matrix.

From the discussed examples, it can be concluded that although the development of UCNPs-based optical analysis has been successfully applied to achieve the simultaneous detection of multiplex pathogenic bacteria in the same sample, the fabrication of the multifunctional identification platform based on UCNPs is relatively complicated, which requires the strict reaction condition, long preparation time as well as high reaction temperature. Additionally, a literature survey demonstrates that the anti-Stocks luminescence emission peak and color of UCNPs can be effectively adjusted by doping with various rare earth elements or transition metal ions, thus achieving multiple UCNPs with different colors. This point seems to offer an effective way to design an efficient assay based on UCNPs for expanding the simultaneous detection number of pathogenic bacteria. However, the width of the luminescence spectra of rare earth elements incorporated into UCNPs can only be controlled to ~30 nm. The smaller spectral widths are almost impossible to achieve. As a result, the simultaneous detection number of pathogenic bacteria based on UCNPs assay is typically four to five at most. Finally, despite tremendous progress being made in the development of UCNP-based analytical probes, some concerns must still be addressed concerning synthesizing smaller UCNPs (~10 nm), increasing the solubility of the particles in water, minimizing nonspecific adsorption, and improving upconversion emission efficiency. Unless addressed, these issues will hinder any progress in the practical applications of UCNPs.

12.4.10 Electrochemical Sensors

Electrochemical detection is another popular method by which nanomaterial-based sensors with applications in the food industry function. Compared to optical (colorimetric or fluorimetric) methods, electrochemical approaches may be more beneficial for food matrices because the problem of light scattering and absorption from the various food components can be avoided. Many electrochemical sensors bind selective antibodies to a conductive nanomaterial (e.g., carbon nanotube) and then monitor changes to the material's conductivity when the target analyte binds to the antibodies.

The focus on electrochemical techniques has recently increased due to speed, sensitivity, specificity, and cost advantages. The nature of the detection mechanism leads to detection times on the order of minutes, compared with hours for other platforms, and minimal instrumentation makes automation and field deployment feasible. Data collection can also be controlled through compact devices, such as smartphones, allowing remote sensing and real-time data distribution. The idea behind this methodology is that binding events between target molecules and the biosensing platform are converted into measurable electrical signals that can be used for pathogen identification and quantification. Various electrochemical analysis techniques are utilized in the formation and characterization of the bioprobes and subsequent pathogen detection; Cyclic voltammetry (CV) is often used for controlled growth onto probe substrates (Miodek et al., 2015) or to examine the effects of different experimental conditions. In contrast, differential pulse voltammetry (DPV) (Chen et al., 2015) and impedance spectroscopy (Tian et al., 2016) have been used for direct pathogen detection.

Electrochemical detection is often enhanced through single or multi-walled carbon nanotubes (SWCNTs or MWCNTs). The nm-sized CNTs are ideal for sensing due to their low charge carrier density. When binding events occur, charge accumulation or depletion ensues throughout the structure's bulk, leading to easily measurable electrical property changes. These analytical devices transduce energy from biochemical reactions such as antigen-antibody or enzyme-substrate reactions to electrical signals. The electrical signals include current, voltage, impedance, frequency, etc. An electrode is a crucial component in electrochemical biosensors that supports biomolecules' immobilization, like an antibody, enzymes, and nucleic acids. The journey of electrochemical biosensors started with the development of the first version of the blood glucose sensor. Presently, many electrochemical biosensors are introduced and commercialized for diverse applications. A common electrochemical biosensor consists of a bioreceptor that can bind to a specific target to generate the signal as an electron that will be measured electronically using a transducer and data analysis devices (Sharma et al., 2017b). The biorecognition element can be antibodies, enzymes, nucleic acids, aptamers, proteins, etc. The addition of nanomaterials as a part of biosensors has given a new direction toward better sensitivity due to their high surface-to-volume ratio. The capacity to load biomolecules or immobilize biorecognition elements has also improved for the same reason.

Another critical factor in the electrochemical biosensors is the transducer. The transducer combines electrodes and electrolytes (Vikrant et al., 2019). The sensitivity depends on the specific materials used to design the transducers, and the signal translation rate is related to the properties of the surface. In this regard, gold electrodes, carbon-based electrodes, conducting polymers, metal oxides, and nanocomposites are well known (Abdulbari and Basheer, 2017). The electrochemical biosensors' operating principle depends on the specific transducers that convert biochemical signals to detectable electronic responses. It can be potentiometric, amperometric, or impedimetric. Potentiometric transducers generally work on the ion-selective principle, and they are widely used to detect several bioanalytes like glucose, triglycerides, pesticides, etc. Potentiometric transducers generally have poor sensitivities. Typically, noble metals are used to design the working electrodes of the amperometric transducers (Jia et al., 2010). After applying a particular potential, the current produced at the working electrode is measured concerning a reference electrode.

The current produced is directly proportional to the degree of oxidation or reduction of the analytes to be identified. These types of biosensors are comparatively simple and cost-effective. The impedimetric biosensors are label free and highly sensitive biosensors. In impedimetric biosensors, the receptor-analyte interaction is related to the impedance change. The working electrode's surface plays a crucial role in this type of electrochemical biosensor (Gupta et al., 2021). The difference in impedance is monitored to analyze the frequency change concerning the time.

It should be noted that the impedance is related to two significant factors – resistance and capacitance. This type of electrochemical biosensor's major drawbacks are slow response time and complicated models to calculate the response. The electrochemical biosensors can be employed to detect toxins like food additives present in foods. Several synthetic dyes, like metanil yellow, are illegally added to food items to generate bright colors and attract food consumers. Food regulatory bodies ban these artificial dyes even though they are available and regularly added to sauces, soft drinks, spices, juices, and so on.

A brief comment is necessary for electrochemical impedance spectroscopy (EIS), extensively applied for biosensor fabrication to quantify foodborne bacteria. Impedance can be used to measure the release of ionic metabolites induced by bacterial metabolism and cell growth, distinguishing viable and dead cells (Wawerla et al., 1999). Other electrochemical methods can be used for the sensitive detection of foodborne bacteria. Research groups exploit two leading solutions to improve sensor performance: using nanomaterials for electrodes and discovering new functionalization strategies to obtain a better electric response. In this respect, DNA nanostructures have attracted great interest for their unique capacity to self-assemble in a predictable form and the possibility of easy coupling with surface and other functionalities.

12.4.11 Additional Principles

Several other nanotechnology-based approaches are being studied to develop new biosensors to detect foodborne bacteria and toxins. Many rely on still experimental principles or a combination of the abovementioned methods.

For example, dendrimers (DEN) are complex globular-shaped-branched structures of 2–20 nm in size. The structural properties like monodispersity, manageable size, easily amendable surface functionalities, hydrophilicity, and high mechanical and chemical strength make them the preferred synthetic nanoparticle for developing biosensors (Bahadir and Sezgintürk, 2016). The polyamido-amine (PAMAM) dendrimer has gained significant attention as it provides large surface areas with many functional groups to allow the easy binding of biological entities. It also contains mono-disperse and hyper-branched polymers with active functional groups at the end of the dendrimer structure. These functional groups aid in immobilizing the bio-recognizing molecules by acting as a bio-conjugating moiety and play diverse roles in biosensor technology. Electrochemical techniques like amperometric, electrochemiluminescence, impedimetric, and potentiometric are generally used for estimating specific molecules using high selectivity and sensitivity dendrimers.

With advances in nanotechnology, a new method called "bio-barcoding" is developed to recognize enzyme-free ultrasensitive proteins and DNAs. Compared to traditional ELISA-based assays that depend on target and sample density, protein barcode assays would be more intricate, sensitive, and extensive. Compared with other conventional techniques like ELISA, qPCR, and so on. The nanoparticle-based biobarcode test is sensitive to pathogen identification (Bao et al., 2006), helping detect early plant disease. The bio barcode method uses two probes: (i) Magnetic microbeads (MMB), which carry an antibody or DNA as a biological probe, are designed for target recognition. (ii) Au-NPs containing polyclonal antibodies or an oligonucleotide (Bio barcode). It is a responsive method for detecting enzyme- and PCR-free proteins and DNA. Bio barcode technology is evolving as a result of advances in nanotechnology. Without enzymes or PCR, identifying protein complexes and nucleic acids is a very efficient approach. Improvements have been made in several areas to increase the utility of bio barcodes in various fields. To identify fungi, DNA barcoding has been recommended (Xu, 2016). The oligo-AuNP sensors hybridize with oligonucleotide-functionalized magnetic microparticle (MMP) probes when molecular targets employ the target gene as a linker.

Other sensing techniques are based on recognition by bacteriophages. Bacteriophages can express proteins bound to bacterial receptors, recognize live and dead cells, and proliferate in live cells (Flores et al., 2011). They are cheap, quickly produced, and infect only bacteria. Moreover, bacteriophages demonstrate enhanced stability compared to their antibody counterparts, which are more sensitive to environmental stress (i.e., pH, temperature, etc.). Genetic engineering can also produce distinct bacteriophages or receptor-binding proteins (RBPs) targeting the desired bacterial species. Tail fibers (TF) of bacteriophages can also be genetically engineered and used for targeting bacterial species (Bai et al., 2019). Phage-immobilized NPs enhanced bacterial detection and discrimination in dark-field microscopy, the most sensitive and effective light-scattering imaging tool for bacterial detection. A plasmon-scattering nanoprobe was developed through immobilization.

On the other hand, array-based sensing, without needing any recognition element or molecular labels, was developed to detect and discriminate bacteria and analytes (Geng et al., 2019). As bacterial species or analytes have different affinities to varying plasmonic NPs with different sizes, compositions, and surface chemistry, they show different levels of light scattering, fluorescent emission, or UV absorbance signal after exposure to plasmonic NPs. Based on those characteristics, gold–silver alloy nanoclusters were developed to detect and discriminate sulfur-producing bacteria.

Conducting polymers with distinctive characteristics has made these an effective alternative for some materials currently used in biosensor fabrication. Polymers are good insulators, and some polymers are found to have good conducting properties due to their combination of metallic and semiconductor characteristics. Various conducting polymers are used in different applications (Faridbod et al., 2008). Out of these, polyaniline, polythiophene, and polypyrrole are used as nanomaterials and show biocompatibility and can reduce the leading disturbances affecting the working environment and these also help in preventing the electrodes from fouling (Oh et al., 2013). Only polyaniline and polypyrrole are extensively used for the detection of foodborne pathogens. Conducting polymers are an excellent immobilizing platform with biomolecules at electrodes to deliver better signal transduction, high sensitivity, selectivity, durability, biocompatibility, and flexibility.

Nanoscale electric wires (nanowires) and three-dimensional field-effect transistors (FETs) are relatively recent technologies employed for biosensing applications. A FET is a semiconductor device sensitive to the amount of charge (or electric field) at its surface. In electric devices, FETs are configured with two terminals (a source and a drain) connected within an electric circuit. A third portion of the device, the "gate," lies between the source and drain and can be perturbed by a metal electrode or by adjusting the charge/electric field environment in its proximity. For biosensing applications, this is useful because many biological molecules and even whole cells have an inherent electric charge. Although FETs have been used in semiconductor devices for many years, they have recently gained popularity for biosensing applications.

The optoelectronic nose is an array-based bacterial-detecting method consisting of cross-reactive sensors that interact with bacteria or their analytes and generate a unique pattern response, in essence, a distinct optical "fingerprint" for different bacteria (Svechkarev et al., 2020). It can act as an olfactory system by enabling bacterial detection based on a generated array response and by collating to a predefined library of answers. In the literature, several other authors have described effective applications of biosensors, nano-biosensors, biomimetic sensors, etc., in various food industry segments (Coles and Frewer, 2013). Biomimetic sensors are developed using protein and biomimetic membranes to detect mycotoxins and other toxic compounds. Reflective interferometry sensors that detect *E. coli* contaminations in packaged foods have also been reported in the literature (Wanekaya et al., 2006). This particular sensor works on the principle of scattering of light by the mitochondria, and the scattering of light is detected by analyzing digital images with reflective interferometry. The *E. coli* protein is placed on the silicon chip, which binds a similar protein in the presence of contamination.

Finally, a newly developed approach has been identified as a nanopore sequencing platform. Nanopore sequencing must solve two serious challenges: (i) identifying the nucleotides as the strand travels through the nanopore and (ii) regulating the pace of the DNA strand as it passes through the nanopore. The protein nanopore and enzyme are designed to handle a single strand of DNA, and the DNA is subjected to a direct electrical inspection as it travels through the nanopore. The protein nanopore is placed over the top of a microwell in a polymer bilayer membrane. A sensor chip in each microwell detects the ionic current when a single molecule passes through the nanopore.

However, the DNA strand moves too quickly through the nanopore for proper detection. To determine the sequence, four different-sized tags generated from 5′-phosphate-modified nucleotides were discovered using nanopore-based sequencing (Nano-SBS), which differentiates four DNA bases. The extended frames of RNA and DNA can be sequenced using nano-scanning blotting. IBM and Roche unveiled the "DNA transistor," an advanced sequencing technique that could record the nucleotide array by dragging the template through a nanopore sensor (Shivashakarappa et al., 2022).

12.4.12 Advancements in Foodborne Toxin Detection

Mycotoxins are toxic compounds produced by particular fungal species as secondary metabolites with a range of molecular weights up to 700 Da, produced by the fungi, commonly found in food crops, cereals, and so on, during harvesting and accumulate toxins during the storage of the products. These mycotoxins render the food unpalatable, contaminate about 25% of the global food production, and render it unsuitable for human consumption. Approximately 300 mycotoxins have been recorded, but researchers have focused on toxins that are proven to be carcinogenic or toxic, like aflatoxins, ochratoxins, fumonisins, zearalenone, patulin, tremorgenic toxins, trichothecenes, and ergot alkaloids (Zain, 2011). It is mandatory and crucial to detect mycotoxins to ensure food safety across the national and international markets. Although all the principles described so far apply to bacterial pathogens and toxin detections, particular progress has been made with some technologies aimed at sensing the presence of chemical contaminants in food.

Current toxin detection methods have been intervened with nanotechnology-based methods to expedite the process at minimal costs, portability, and user-friendly least detection limits to eliminate long-term bioterrorism issues. Nanoparticle-based colorimetric assays and molecular mimicry have been explored for toxin detection and produced positive results in cholera toxin detection (Tark et al., 2010). For example, nanozymes have accelerated the detection of contaminants in food spoilage and stand as potential alternatives to overcome the shortcomings of biological enzymes. Overall, they have excellent advantages such as higher selectivity, precise sensitivity, focused target recognition, reduced detection time, and enhanced signal readout compared to conventional methods such as chromatographic approaches and immunological assays (Mohamad et al., 2020). Modern research focused on the attributes of these nanozyme-based biosensors considers them the best-suited strategy for detecting mycotoxins, heavy metal ions, food contaminants, and intentional adulterants in the packed food industry.

Nanomaterials like gold nanoparticles, dendrimers, quantum dots, and GO possess excellent optical and electronic properties, due to which they have been integrated into fluorescent biosensing devices for mycotoxin sensing. More interestingly, aptamers that are synthetic oligonucleotides, i.e., single-stranded DNA or RNA (approximately 80 bases), have appreciable specificity and binding affinity to the target. The significant milestone in nano-based sensors was achieved through fluorescence resonance energy transfer in detecting Aflatoxin M_1, Ochratoxin A, and Fumonisin B_1 in food commodities. This technique was constructed using the immobilization of aptamers or antibodies arranged on the fluorescent nanoparticles-GO, quantum dots-AuNPs, or nanogold-strips to form thin films which enable low detection limits up to 0.02 ng mL^{-1} (Pushparaj et al., 2022). Surface Plasmon Resonance Sensors (SPRS) involve the integration of response from fluorescence or modifications of the sensor chip with metallic nanoparticles (Figure 12.1).

Currently, sensors such as electrochemical and biosensors based on metallic nanoparticles, supermagnetic nanoparticles, and newly developed nanomaterials (carbon nanotubes) are used to detect different toxins in food products. Metallic nanoparticles such as silver, gold, copper, platinum, and zinc are commonly used to form sensors. Some studies have observed that aflatoxins, a group of toxin and carcinogenic compounds, are widely found in various foodstuffs. For example, Sharma and coworkers (2017a) noted the potential application of functionalized gold nanoparticles in detecting aflatoxin B_1 from food samples. Similarly, Radoi et al. (2008) used superparamagnetic nanoparticles and gold nanoprobes to detect aflatoxin M_1 in milk samples. Moreover, in another study (Mohammad et al., 2022), gold-based nanoprobe immunochromatographic assays were also used to detect brevetoxins in fishery product samples. Similarly, gold-based immunochromatographic assays have been used to detect botulinum neurotoxin type B from processed foods. Some studies detected palytoxin from contaminated seafood using carbon nanotube-based electro-chemiluminescent sensors. In the literature, microfluidic sensors have also been used to detect toxic substances in aqueous samples. Usually, these sensors are very sensitive and effective even when detecting contaminants in the microliter range.

Mak et al. (2010) reported a highly sensitive magnetic nanoparticle immunoassay for recognizing various mycotoxins. The immobilization of the reactants and their uniform distribution across the whole volume of the reaction media were accomplished by employing the magnetic nanoparticle as the solid phase, which offered a significantly higher surface area. Using a magnetic field separated the reactants quickly and easily, simplifying the wash procedures required for conventional microplate-based ELISA. Operating these advantages, the MNP-based immunoassay approach was developed to detect aflatoxin B_1, zearalenone, and HT-2 mycotoxins. Colorimetric biosensors are interesting optical biosensors because they provide rapid and certain visual confirmation of harmful microbes in the sample through a color change, even without the need for any additional equipment or chemical substances. The solution-based colorimetric sensor functions similarly to the lateral flow test. When a colloidal gold nanoparticle-mounted receptor reacts with the target pathogens, nanoparticle aggregation changes color from red to purple.

12.5 CONCLUSIONS

A wide range of approaches to detect food pathogens using micro- and nanotechnologies have been presented. These include using nanomaterials as enhancers for analytical assays, microchip-based sensors for small volumes, rapid diagnostic assays, and unique LOC systems that incorporate multiple sample processing and analytical steps onto a single platform. One of the great advantages of using micro- and nanotechnologies for food-pathogen detection is the promise of more rapid and portable detection platforms. Incorporation of multiple laboratory steps into a single, automated platform also reduces the likelihood of human error and sets the stage for remote (operator-free) detection of pathogens in many environments. We also anticipate individuals' eventual use of such devices for pathogen detection at home. In the future, we hope to see more cross-talk between the food science and medical fields within the context of pathogen detection. Although the sample matrices can be quite different, the core technology for many of these biosensing platforms is the same, and thus, technology sharing between these fields is paramount.

Nanotechnology is often described as 'science-based' or 'knowledge-intensive.' This means that fundamental advances are driven by basic research usually generated at universities. The transfer of technology from academic research to marketed products or implemented processes is generally visualized as a roughly linear pathway. The linearity of technology development models suggests that the progression of a nanosensor from the initial university concept to the finished product at a partner corporation should be conceptually straightforward. Yet, as we have seen, few examples of commercialized or implemented nanosensors exist, particularly for foods. This is especially true for technologies and studies performed in developing countries.

The rapid and accurate detection of foodborne bacteria is the need of the hour to ensure food safety and reduce the number of foodborne illnesses. Microfluidic technology has been used to study bacteria for nearly 10 years. It possesses many attractive advantages, such as low sample consumption, high integration, small size, portability, and high throughput. However, improving signal intensity or exploiting other sensing mechanisms that depend on nanomaterials are avenues to be explored moving forward. Microfluidic chips are better than traditional detection methods as they can detect foodborne bacteria much faster (in hours; enrichment not included) with LODs ranging from several CFU mL^{-1} to 10^3 CFU mL^{-1}. Thus, it is a potential platform for alternative analysis compared to traditional techniques.

Despite the progress that has been made in developing nanomaterial-based microfluidic chips for bacterial detection, some challenges and limitations must be addressed. (i) It is desirable to achieve the simultaneous detection of multiple bacteria, as they may coexist in contaminated food. One possible approach is using different segmented regions or multiple probes. Therefore, it is crucial to combine nanomaterials with other properties (e.g., fluorescent, magnetic, or plasma properties) in a single nanostructure entity or to explore new luminescent nanomaterials with very narrow emission spectra, which may help in avoiding spectral cross-talk between different detection channels. (ii) Integration and simplicity of sample pretreatment are needed. Current studies largely depend on pure cultures rather than raw samples for bacterial identification, mainly because of the low microbial concentration and high interference from the food matrix. Although different combinations of nanomaterials could be used for sample enrichment, pretreatment still accounts for most of the detection time and, thus, is quite far from being implemented in practical applications. Further development should focus on improving the sample pretreatment ability or limiting this procedure using ultrasensitive methods. The strategies for cell analysis in the microfluidic system also provide a reference for separating bacteria (Zhang et al., 2021). (iii) Continuous improvements are needed to achieve more sensitivity, high selectivity, and robust stability of detection abilities.

The application of sensors in food processing industries has also changed the current trend as these can identify the various contaminants formed within the food chain with high sensitivity. The advancements in diagnostics have increased the demand for portable devices for robust and precise detection in food industries. Nanosensors have the potential to meet both the need for miniaturization and low-cost analytical devices. In the past few years, applications of e-nose technologies have come through advances in sensor design, material improvements, software innovations, and progress in micro-circuitry design and systems integration. The extensive research progress in nanotechnology for nanomaterial exploration and the development of new mechanisms in the future will enable researchers to develop highly sensitive, specific, and unobtrusive nanosensors for analyzing foodborne microbes at an affordable cost.

The potential risks arising from nanoscience and nanotechnologies in food and feed are observed by international regulatory agencies, who consider the current risk assessment paradigm appropriate for nanomaterials. However, there is limited data on oral exposure to nanomaterials and any consequent toxicity; there are limited methods to characterize, detect, and measure nanomaterials in food/feed. Toxicological and toxicokinetic profiles of nanomaterials cannot be determined by extrapolation from data on their equivalent non-nano forms. A case-by-case approach is needed.

BIBLIOGRAPHY

Abdulbari H.A.; Basheer E.A. 2017. Electrochemical biosensors: Electrode development, materials, design, and fabrication. *Chem. Bio. Eng.* 4:92–105.

Adhikari J.; Rizwan M.; Keasberry N.A.; Ahmed M.U. 2020. Current progresses and trends in carbon nanomaterials-based electrochemical and electrochemiluminescence biosensors. *J. Chin. Chem. Soc.* 67:937–960.

Alamer S.; Eissa S.; Chinnappan R.; Herron P.; Zourob M. 2018. Rapid colorimetric lactoferrin-based sandwich immunoassay on cotton swabs for the detection of foodborne pathogenic bacteria. *Talanta.* 185:275–280.

Albrecht M.G.; Creighton J.A. 1977. Anomalously intense Raman spectra of pyridine at a silver electrode. *J. Am. Chem. Soc.* 99(15):5215–5217.

Alfonso A.S.; Perez-Lopez B.; Faria R.C.; Mattoso L.H.; Hernandez-Herrero M.; Roig-Sagues A.X.; Maltez-da Costa M.; Merkoci A. 2013. Electrochemical detection of *Salmonella* using gold nanoparticles. *Biosens. Bioelectron.* 40:121–126.

Alim S.; Vejayan J.; Yusoff M.M.; Kafi A.K.M. 2018. Recent uses of carbon nanotubes & gold nanoparticles in electrochemistry with application in biosensing: A review. *Biosens. Bioelectron.* 121:125–136.

Al-Rifai R.H.; Chaabna K.; Denagamage T.; Alali W.Q. 2020. Prevalence of nontyphoidal *Salmonella enterica* in food products in the Middle East and North Africa: A systematic review and meta-analysis. *Food Control.* 109:106908.

Ameta S.K., et al. 2020. Use of nanomaterials in food science. In: *Biogenic Nano-particles and Their Use in Agro-ecosystems*. Singapore: Springer.

Anju M.; Renuka N.K. 2019. Grapheneedye hybrid optical sensors. *Nano-Struct. Nano-Obj.* 17:194–217.

Bahadir E.B.; Sezgintürk M.K. 2016. Poly (amidoamine) (PAMAM): An emerging material for electrochemical bio(sensing) applications. *Talanta.* 148:427–438.

Bai Y.L.; Shahed-Al-Mahmud M.; Selvaprakash K.; Lin N.T.; Chen Y.C. 2019 Tail fiber protein-immobilized magnetic nanoparticle-based affinity approaches for detection of *Acinetobacter baumannii. Anal. Chem.* 91(15):10335–10342.

Bajorowicz B.; Kobylanski M.P.; Gołabiewska A.; Nadolna J.; Zaleska-Medynska A.; Malankowska A. 2018. Quantum dot-decorated semiconductor micro and nanoparticles: A review of their synthesis, characterization and application in photocatalysis. *Adv. Colloid. Interface Sci.* 256:352–372.

Bao Y.P.; Wei T.F.; Lefebvre P.A.; An H.; He L.; Kunkel G.T.; Müller U.R. 2006. Detection of protein analytes via nanoparticle-based bio bar code technology. *Anal. Chem.* 78(6):2055–2059.

Batır G.G.; Arık M.; Caldıran Z.; Turut A.; Aydogan, S. 2018. Synthesis and characterization of reduced graphene oxide/rhodamine 101 (rGO-Rh101) nanocomposites and their heterojunction performance in rGORh101/p-Si device configuration. *J. Electron. Mater.* 47:329–336.

Berekaa M.M. 2015. Nanotechnology in food industry; advances in food processing, packaging and food safety. *Int. J. Curr. Microbiol. Appl. Sci.* 4(5):345–357.

Bokhari S.W.; Siddique A.H.; Sherrell P.C.; Yue X; Karumbaiah K.M.; Wei S.; Ellis A.V.; Gao W. 2020. Advances in graphene-based supercapacitor electrodes. *Energy Rep.* 6:2768–2784.

Boyer D.; Tamarat P.; Maali A.; Lounis B.; Orrit M. 2002. Photothermal imaging of nanometer-sized metal particles among scatterers. *Science.* 297:1160–1163.

Brandão D.; Liébana S.; Campoy S.; Cortés M.P.; Alegret S.; Pividori M.I. 2015. Simultaneous electrochemical magnetogenosensing of foodborne bacteria based on triple-tagging multiplex amplification. *Biosens. Bioelectron.* 74:652–659.

Cady N.C.; Fusco V.; Maruccio G.; Primiceri E.; Batt C.A. 2017. Micro- and nanotechnology-based approaches to detect pathogenic agents in food. In: Alexandru Mihai Grumezescu (ed). *Nanobiosensors.* Academic Press, pp. 475–510.

Cady N.C.; Strickland A.D.; Batt C.A. 2007. Optimized linkage and quenching strategies for quantum dot molecular beacons. *Mol. Cell. Probes.* 2:116–124.

Cai G.; Wang S.; Zheng L.; Lin J. 2018. A fluidic device for immunomagnetic separation of foodborne bacteria using self-assembled magnetic nanoparticle Chains. *Micromachines.* 9:624.

Carlson K.; Misra M.; Mohanty S. 2017. Developments in micro- and nanotechnology for foodborne pathogen detection. *Foodborne. Pathog. Dis.* 15(1) 16–25.

Centers for Disease Control and Prevention. 2011. Estimates of foodborne illness in the United States. Available at: https://www.cdc.gov/foodborneburden/2011-foodborne-estimates.html#annual. Accessed 1 October 2023.

Centers for Disease Control and Prevention. CDC. List of Selected Multistate Foodborne Outbreak Investigations. 2017. Available at: www.cdc.gov/foodsafety/outbreaks/multistateoutbreaks/outbreaks-list.html. Accessed 30 September 2023.

Centers for Disease Control and Prevention Foundation. 2015. CDC fights foodborne diseases, protects America's businesses and consumers. Available at: https://www.cdcfoundation.org/businesspulse/food-safety. Accessed 1 October 2023.

Cesewski E.; Johnson B.N. 2020. Electrochemical biosensors for pathogen detection. *Biosens. Bioelectron.* 159:112214.

Chandan H.R.; Schiffman J.D.; Balakrishna R.G. 2018. Quantum dots as fluorescent probes: Synthesis, surface chemistry, energy transfer mechanisms, and applications. *Sens. Actuators B Chem.* 258:1191–1214.

Chen A.; Yang S. 2015. Replacing antibodies with aptamers in lateral flow immunoassay. *Biosens. Bioelectron.* 71:230–242.

Chen J.; Jiang Z.; Ackerman J.; Yazdani M.; Hou S.; Nugen S.; Rotello V. 2015. Electrochemical nanoparticle-enzyme sensors for screening bacterial contamination in drinking water. *Analyst* 140:4991–4996.

Chen J.; Park B. 2016. Recent advancements in nanobioassays and nanobiosensors for foodborne pathogenic bacteria detection. *J. Food Prot.* 79(6):1055–1069.

Chen P.; Li Y.; Cui T.; Ruan R. 2013. Nanoparticles based sensors of rapid detection of foodborne pathogens. *Int. J. Agric. Biol. Eng.* 6:28–35.

Chen, S., Qui, L., & Cheng, H-M. (2020). Carbon-Based Fibers for Advanced Electrochemical Energy Storage Devices. *Chem. Rev.* 120: 2811–2878.

Cheng D.; Yu M.; Fu F.; Han W.; Li G.; Xie J.; Song Y.; Swihart M.T.; Song E. 2016. Dual recognition strategy for specific and sensitive detection of bacteria using aptamer-coated magnetic beads and antibiotic-capped gold nanoclusters. *Anal. Chem.* 88:820–825.

Cheng W.; Tang X.; Zhang Y.; Wu D.; Yang W. 2021. Applications of metal-organic framework (MOF)-based sensors for food safety: Enhancing mechanisms and recent advances. *Trends Food Sci. Technol.* 112:268–282.

Cifuentes A. 2012. Food analysis: Present, future, and foodomics. *ISRN Anal. Chem.* 2012:16.

Clime L.; Hoa X.Y.D.; Corneau N.; Morton K.J.; Luebbert C.; Mounier M.; Brassard D.; Geissler M.; Bidawid S.; Farber J.; Veres T. 2015. Microfluidic filtration and extraction of pathogens from food samples by hydrodynamic focusing and inertial lateral migration. *Biomed. Microdev.* 17(1):1–14.

Coles D.; Frewer L.J. 2013. Nanotechnology applied to European food production – A review of ethical and regulatory issues. *Trends Food Sci. Technol.* 34(1):32–43.

Dadfar S.M.; Roemhild K.; Drude N.I.; von Stillfried S.; Knüchel R.; Kiessling F.; Lammers T. 2019. Iron oxide nanoparticles: Diagnostic, therapeutic and theranostic Applications. *Adv. Drug Deliv. Rev.* 138:302–325.

Darr J.A.; Zhang J.; Makwana N.M.; Weng X. 2017. Continuous hydrothermal synthesis of inorganic nanoparticles: Applications and future directions. *Chem. Rev.* 117(17):11125–11238.

Department of Industry, Innovation, Science and Research, Australia. 2009. Market attitude research services, Australian community attitudes about nanotechnology – 2005–2009.

Dina N.E.; Marconi D.; Colniță A.; Gherman A.M.R. 2020. Microfluidic portable device for pathogens 'rapid SERS detection. *Proceedings of the 1st International Electronic Conference on Biosensors* 7089.

Ding C.; Zhu A.; Tian Y. 2014. Functional surface engineering of C-dots for fluorescent biosensing and in vivo bioimaging. *Acc. Chem. Res.* 47:20–30.

Durán G.M.; Contento A.M.; Ríos A. 2015. β-Cyclodextrin coated CdSe/ZnS quantum dots for vanillin sensoring in food samples. *Talanta.* 131:286–291.

Dwarakanath S.; Bruno J.G.; Shastry, A.; Phillips T.; John A.; Kumar A.; Stephenson L.D. 2004. Quantum dot-antibody and aptamer conjugates shift fluorescence upon binding bacteria. *Biochem. Biophys. Res. Commun.* 325(3):739–743.

Edgar R.; McKinstry M.; Hwang J.; Oppenheim A.B.; Fekete R.A.; Giulian G.; Merril C.; Nagashima K.; Adhya S. 2006. High-sensitivity bacterial detection using biotin-tagged phage and quantum-dot nanocomplexes. *Proc. Natl. Acad. Sci. USA.* 103(13):4841–4845.

Eleftheriadou M.; Pyrgiotakis G.; Demokritou P. 2017. Nanotechnology to the rescue: Using nano-enabled approaches in microbiological food safety and quality. *Curr. Opin.* 44:87–93.

Fan Y.; Wang S.; Zhang F. 2019. Optical multiplexed bioassays for improved biomedical diagnostics. *Angew. Chem. Int. Ed.* 58:13208–13219.

Faridbod F.; Norouzi P.; Dinarvand R.; Ganjali M.R. 2008. Developments in the field of conducting and non-conducting polymer based potentiometric membrane sensors for ions over the past decade. *Sensors.* 8:2331–2412.

Farzin M.A.; Abdoos H. 2021. A critical review on quantum dots: From synthesis toward applications in electrochemical biosensors for determination of disease-related biomolecules. *Talanta.* 224:121828.

Flores C.O.; Meyer J.R.; Valverde S.; Farr L.; Weitz J.S. 2011. Statistical structure of host-phage interactions. *Proc. Natl. Acad. Sci. U S A.* 108(28):E288–E297.

Fuertes G., et al. 2016. Intelligent packaging systems: Sensors and nanosensors to monitor food quality and safety. *J. Sens.* https://doi.org/10.1155/2016/4046061.

Fusco V.; den Besten H.M.; Logrieco A.F.; Rodriguez F.P.; Skandamis P.N.; Stessl B.; Teixeira P. 2015. Food safety aspects on ethnic foods: Toxicological and microbial risks. *Curr. Opin. Food Sci.* 6:24–32.

Fusco V.; Quero G.M. 2014. Culture-dependent and -independent nucleic acid based methods used in the microbial safety assessment of milk and dairy products. *Comp. Rev. Food Sci. Food Saf.* 13:493–537.

Garrido-Maestu A.; Azinheiro S.; Carvalho J.; Espiña B.; Prado M. 2020. Evaluation and implementation of commercial antibodies for improved nanoparticle-based immunomagnetic separation and real-time PCR for faster detection of *Listeria monocytogenes*. *J. Food Sci. Technol.* 57(11):4143–4151.

Gehring A.G.; Tu S.I. 2005. Enzyme-linked immunomagnetic electrochemical detection of live *Escherichia coli* 0157: H7 in apple juice. *J. Food Protect.* 68:146–149.

Geleta G.S. 2022. A colorimetric aptasensor based on gold nanoparticles for detection of microbial toxins: An alternative approach to conventional methods. *Anal. Bioanal. Chem.* 414(24):7103–7122

Geng Y.; Peveler W.J.; Rotello V.M. 2019. Array-based "chemical nose" sensing in diagnostics and drug discovery. *Angew. Chem. Int. Ed. Engl.* 58(16):5190–5520.

Ghosh T.; Bhagya Raj G.V.S.; Kumar Dash K. 2022. A comprehensive review on nanotechnology based sensors for monitoring quality and shelf life of food products. *Measurement: Food.* 7:100049.

Gilmartin N.; O'Kennedy R. 2012. Nanobiotechnologies for the detection and reduction of pathogens. *Enzym. Microb. Technol.* 50:87–95.

Girigoswami A.; Mitra Ghosh M.; Pallavi P.; Ramesh S.; Girigoswami K. 2021. Nanotechnology in detection of food toxins –Focus on the dairy products. *Biointerface Res. Appl. Chem.* 11(6):14155–14172.

Guo S; Wang E. 2007. Synthesis and electrochemical applications of gold nanoparticles. *Anal. Chim. Acta.* 598:181–192.

Gupta R.; Raza N.; Bhardwaj S.K.; Vikrant K.; Kim K.H.; Bhardwaj N. 2021. Advances in nanomaterial-based electrochemical biosensors for the detection of microbial toxins, pathogenic bacteria in food matrices. *J. Hazard Mater.* 401:121379.

Hajipour M.J.; Saei A.A.; Walker E.D.; Conley B.; Omidi Y.; Lee K.B.; Mahmoudi M. 2021. Nanotechnology for targeted detection and removal of bacteria: Opportunities and challenges. *Adv. Sci.* 8:2100556.

Hameed S.; Xie L.; Ying Y. 2018. Conventional and emerging detection techniques for pathogenic bacteria in food science: A review. *Trends Food Sci. Technol.* 81:61–73.

Hammack T.; Davidson M.; Feng P.; Gharst G.; Ge B.; Jinneman K.; Regan P.M.; Kase J.; Orlandi P.; Burkhardt W. (ed.). 2015. *Bacteriological Analytical Manual*. Rockville, MD: U.S. Food and Drug Administration.

Hao T.; Wei X.; Nie Y.; Xu Y.; Lu K.; Yan Y.; Zhou Z. 2016. Surface modification and ratiometric fluorescence dual function enhancement for visual and fluorescent detection of glucose based on dual-emission quantum dots hybrid. *Sens. Actuators B Chem.* 230:70–76.

Huang, T.; Zhang, R.; Li, J. 2023. CRISPR-Cas-based techniques for pathogen detection: Retrospect, recent advances, and future perspectives. J. Adv. Res. 50:69–82.

Ji T.; Liu Z.; Wang G.; Guo X.; Akbar Khan S.; Lai C.; Chen H.; Huang S.; Xia S.; Chen B.; Jia H.; Chen Y.; Zhou Q. 2020. Detection of COVID-19: A review of the current literature and future perspectives. *Biosens. Bioelectron.* 166:112455–112455.

Jia W.-Z.; Wang K.; Xia X.-H. 2010. Elimination of electrochemical interferences in glucose biosensors. *TrAC Trends Anal. Chem.* 29:306–318.

Joseph T.; Morrison M. 2006. Nanotechnology in Agriculture and Food. London: Institute of Nanotechnology.

Kallipolitis B.; Gahan C.G.M.; Piveteau P. 2020. Factors contributing to *Listeria monocytogenes* transmission and impact on food safety. *Curr. Opin. Food Sci.* 36:9–17.

Karakoti A.S.; Shukla R.; Shanker R.; Singh S. 2015. Surface functionalization of quantum dots for biological applications. *Adv. Colloid. Interface Sci.* 215:28–45.

Kaushik S; Tiwari U.K; Pal S.S; Sinha R.K. 2019. Rapid detection of *Escherichia coli* using fiber optic surface plasmon resonance immunosensor based on biofunctionalized Molybdenum disulfide (MoS2) nanosheets. *Biosens. Bioelectron.* 126:501–509.

Kierkowicz M Pach E.; Santidrián A.; Sandoval S.; Gonçalves G.; Tobías-Rossell E.; Kalbáč M.; Ballesteros B; Tobias G.. 2018. Comparative study of shortening and cutting strategies of single-walled and multi-walled carbon nanotubes assessed by scanning electron microscopy. *Carbon.* 139:922–932.

Kim J.-H.; Oh S.-W. 2021. Pretreatment methods for nucleic acid-based rapid detection of pathogens in food: A review. *Food Control.* 121:107575.

Knudsen B.R.; Jepsen M.L.; Ho Y.P. 2013. Quantum dot-based nanosensors for diagnosis via enzyme activity measurement. *Expert Rev. Mol. Diagn.* 13(4):367.

Krafft B.; Tycova A.; Urban R.D.; Dusny C.; Belder D. 2021. Microfluidic device for concentration and SERS-based detection of bacteria in drinking water. *Electrophoresis.* 42(1–2):86–94.

Kshirod K.; Dash N.; Afzal A.; Dipannita D.; Mohanta D. 2019. Thorough evaluation of sweet potato starch and lemon-waste pectin based-edible films with nano-titania inclusions for food packaging applications. *Int. J. Biol. Macromol.* 139:449–458.

Kumar B.; Malhotra K.; Fuku R.; Van Houten J.; Qu G.Y.; Piunno P.A.E.; Krull U.J. 2021. Recent trends in the developments of analytical probes based on lanthanide-doped upconversion nanoparticles. *TrAC-Trends Anal. Chem.* 139:116256.

Kwak S.Y.; Wong M.H.; Lew T.T.S.; Bisker G.; Lee M.A.; Kaplan A.; Strano M.S. 2017. Nanosensor technology applied to living plant systems. *Annu. Rev. Anal. Chem.* 10:113–140.

Lau H.Y.; Botella J.R. 2017. Advanced DNA-based point-of-care diagnostic methods for plant disease detection. *Front. Plant. Sci.* 8:2016.

Law J.; Mutalib N.; Chan K.; Lee L. 2015 Rapid methods for the detection of foodborne bacterial pathogens: Principles, applications, advantages and limitations. *Front. Microbiol.* 5:1–19.

Lee H.; Kim J.; Kim H.; Kim J.; Kwon S. 2010. Colour-barcoded magnetic microparticles for multiplexed bioassays. *Nat. Mater.* 9:745–749.

Lee S.H.; Park S.-m.; Kim B.N.; Kwon O.S.; Rho W.-Y.; Jun B.-H. 2019. Emerging ultrafast nucleic acid amplification technologies for next-generationmolecular diagnostics. *Biosens. Bioelectron.* 141:111448.

Li S.; Chen Y.; Huang L.; Pan D. 2013. Simple continuous-flow synthesis of Cu-In-Zn-S/ZnS and Ag-In-Zn-S/ZnS core/shell quantum dots. *Nanotechnology.* 24:395705

Li Z.; Askim J.R.; Suslick K.S., 2018. The optoelectronic nose: Colorimetric and fluorometric sensor arrays. *Chem. Rev.* 119(1):231–292.

Lisalova H.; Visova I.; Ermini M.; Springer T., Song X.; Mrazek J.; Lamacova J.; Lynn N.; Sedivak P.; Homola J. 2016. Low-fouling surface plasmon resonance biosensor for multi-step detection of foodborne bacterial pathogens in complex food samples. *Biosens. Bioelectron.* 80:84–90.

Liu W.; Speranza G. 2019. Functionalization of carbon nanomaterials for biomedical applications. *J. Carbon Res.* 5(4):72.

Lopez-Campos G., et al. 2012. Detection, identification, and analysis of foodborne pathogens. In: López-Campos, G., et al. (Eds.), *Microarray Detection and Characterization of Bacterial Foodborne Pathogens.* Boston, MA: Springer US, pp. 13–32.

Lv W.; Ye H.; Yuan Z.; Liu X.; Chen X.; Yang W. 2020. Recent advances in electrochemiluminescence-based simultaneous detection of multiple targets. *TrAC Trends Anal. Chem.* 123:115767.

Mahato, K.; Kumar, S.; Srivastava, A.; Maurya, P.K.; Singh, R.; Chandra, P. 2018. Electrochemical immunosensors: Fundamentals and applications in clinical diagnostics. In: *Handbook of Immunoassay Technologies.* Vashist, S. K., Luong, J. H. T. (eds.). Academic Press, pp. 359–414.

Mak A.C.; Osterfeld S.J.; Yu H.; Wang S.X.; Davis R.W.; Jejelowo O.A.; Pourmand, N. 2010. Sensitive giant magnetoresistive-based immunoassay for multiplex mycotoxin detection. *Biosens. Bioelectron.* 25(7):1635–1639.

Mandal P.; Biswas A.; Choi K.; Pal U. 2011. Methods for rapid detection of foodborne pathogens: An overview. *Am. J. Food Technol.* 6:87–102.

Manshian B.B.; Martens T.F.; Kantner K.; Braeckmans K.; De Smedt S.C.; Demeester J.; Jenkins G.J.S.; Parak W.J.; Pelaz B.; Doak S.H.; Himmelreich U.; Soenen S.J. 2017. The role of intracellular trafficking of CdSe/ZnS QDs on their consequent toxicity profile. *J. Nanobiotechnol.* 15:45.

Martins T.; Ribeiro A.; Camargo H.; Filho P.; Cavalcante H.; Dias D. 2013. New insights on optical biosensors: Techniques, construction and application. In Rinken, T (ed.): *State of the Art in Biosensors.* Rijeka, Croatia: InTech.

Massad-Ivanir N.; Shtenberg G.; Tzur A.; Krepker M.; Segal E. 2011. Engineering nanostructured porous SiO2 surfaces for bacteria detection via "direct cell capture".' *Anal. Chem.* 83:3282–3289.

Mehta V.N.; Desai M.L.; Basu H.; Kumar R.; Kailasa S.K. 2021. Recent developments on fluorescent hybrid nanomaterials for metal ions sensing and bioimaging applications: A review. *J. Mol. Liq.* 333:115950.

Mendez-Gonzalez D.; Lopez-Cabarcos E.; Rubio-Retama J.; Laurenti M. 2017. Sensors and bioassays powered by upconverting materials. *Adv. Colloid. Interface Sci.* 249:66–87.

Miodek A.; Mejri N.; Gomgnimbou M.; Sola C.; Youssoufi H. 2015. EDNA sensor of *Mycobacterium tuberculosis* based on electrochemical assemble of nanomaterials (MWCNTs/PPy/PAMAM). *Anal Chem.* 87:9257–9264.

Mirkin C.A.; Letsinger R.L.; Mucic R.C.; Storhoff J.J. 1996. A DNA-based method for rationally assembling nanoparticles into macroscopic materials. *Nature.* 382:607–609.

Mohamad A.; Rizwan M.; Keasberry N.A.; Nguyen A.S.; Lam T.D.; Ahmed M.U. 2020. Gold-microrods/ Pd-nanoparticles/polyaniline-nanocomposite-interface as a peroxidase-mimic for sensitive detection of tropomyosin. *Biosens. Bioelectron.* 155:112108.

Mohammad Z.; Ahmad F.; Ibrahim S.A.; Zaidi S. 2022. Application of nanotechnology in different aspects of the food industry. *Discover. Food.* 2:12.

Mu C.; Zhang Z.; Min Lin Dai Z.; Cao X. 2015. Development of a highly effective multi-stage surface acoustic wave SU-8 microfluidic concentrator. *Sens. Actuat. B.* 215:77–85.

Muniandy S.; The S.J; Thong K.L.; Thiha A.; Dinshaw I.J.; Lai C.W.; Ibrahim F.; Leo B.F. 2019. Carbon nanomaterial-based electrochemical biosensors for foodborne bacterial detection. *Crit. Rev. Anal. Chem.* 49:510–533.

Nath S.; Kaittanis C.; Ramachandran V.; Dalal N.S.; Perez J.M. 2009. Synthesis, magnetic characterization, and sensing applications of novel dextrancoated iron oxide nanorods. *Chem. Mater.* 21(8):1761–1767.

Neethirajan S.; Jayas D.S. 2011. Nanotechnology for the food and bioprocessing industries. *Food Bioproc. Technol.* 4(1):39–47.

Nile S.H; Baskar V.; Selvaraj D.; Nile A.; Xiao J.; Kai G. 2020. Nanotechnologies in food science: Applications, recent trends, and future perspectives. *Nano-Micro Lett.* 12(1):1–34.

Niu X.; Chen C.; Zhao H.; Tang J.; Li Y.; Lan M. 2012. Porous screen-printed carbon electrode. *Electrochem. Commun.* 22:170–173.

Norton D.M.; Batt C.A. 1999. Detection of viable *Listeria monocytogenes* with a 5′ nuclease PCR assay. *Appl. Environ. Microbiol.* 65:2122–2127.

Ogden I.D.; Hepburn N.F.; MacRae M. 2001. The optimization of isolation media used in immunomagnetic separation methods for the detection of *Escherichia coli* O157 in foods. *J. Appl. Microbiol.* 91:373–379.

Oh W.H.; Kwon O.S.; Jang J. 2013. Conducting polymer nanomaterials for biomedical applications: Cellular interfacing and biosensing. *Polym. Rev.* 53:407–442.

Ojha V.H.; Kant K.M. 2019. Temperature dependent magnetic properties of superparamagnetic $CoFe_2O_4$ nanoparticles. *Phys. B.* 567:87–94.

Oniciuc E.A.; Nicolau A.I.; Hernandez M.; Rodríguez-Lázaro D. 2017. Presence of methicillin-resistant *Staphylococcus aureus* in the food chain. *Trends Food Sci. Technol.* 61:49–59.

Pandey P.; Singh S.P.; Arya S.K.; Sharma A.; Datta M.; Malhotra B.D. 2008. Gold nanoparticle-polyaniline composite films for glucose sensing. *J. Nanosci. Nanotechnol.* 8:3158–3163.

Pandey S.; Bodas D. 2020. High-quality quantum dots for multiplexed bioimaging: A critical review. *Adv. Colloid Interface Sci.* 278:102137.

Pandit S.; Dasgupta D.; Dewan N.; Ahmed P. 2016. Nanotechnology based biosensors and its application. *Pharam. Innov. J.* 5:18–25.

Pedrero M.; Campuzano S.; Pingarron J. 2009. Electroanalytical sensors and devices for multiplexed detection of foodborne pathogen microorganisms. *Sensors.* 9:5503–5520.

Petersen M.; Yu Z.; Lu X. 2021. Application of Raman spectroscopic methods in food safety: A review. *Biosensors* 11(6):1–22.

Pissuwan D.; Gazzana C.; Mongkolsuk S.; Cortie M.B. 2020. Single and multiple detections of foodborne pathogens by gold nanoparticle assays. *WIREs Nanomed. Nanobiotechnol.* 12:e1584.

Potyrailo R.A.; Conrad R.C.; Ellington A.D.; Hieftje G.M. 1998. Adapting selected nucleic acid ligands (aptamers) to biosensors. *Anal. Chem.* 70:3419–3425.

Pumera M. 2011. Graphene in biosensing. *Mater. Today.* 14:308–315.

Pushparaj K.; Liu W.C.; Meyyazhagan A.; Orlacchio A.; Pappusamy M.; Vadivalagan C.; Robert A.A.; Arumugam V.A.; Kamyab H.; Klemeš J.J.; Khademi T. 2022. Nano- from nature to nurture: A comprehensive review on facets, trends, perspectives and sustainability of nanotechnology in the food sector. *Energy.* 240:122732.

Qiao Z.H; Lei C.Y; Fu Y.C; Li Y.B. 2017. Rapid and sensitive detection of E. *coli* O157:H7 based on antimicrobial peptide functionalized magnetic nanoparticles and urease-catalyzed signal amplification. *Anal. Methods.* 9:5204–5210.

Qiu J.; Zhou Y.; Chen H.; Lin J.-M. 2009. Immunomagnetic separation and rapid detection of bacteria using bioluminescence and microfluidics. *Talanta.* 79:787–795.

Quero G.M.; Santovito E.; Visconti A.; Fusco V. 2014. Quantitative detection of *Listeria monocytogenes* in raw milk and soft cheeses: Culture-independent versus liquid- and solid-based culture-dependent real time PCR approaches. *LWT Food Sci. Technol.* 58:11–20.

Radoi A.; Targa M.; Prieto-Simon B.; Marty J.L. 2008. Enzyme-linked immunosorbent assay (ELISA) based on superparamagnetic nanoparticles for aflatoxin M1 detection. *Talanta.* 77(1):138–143.

Rakshit S.; Moulik S.P.; Bhattacharya S.C. 2017. Understanding the effect of size and shape of gold nanomaterials on nanometal surface energy transfer. *J. Colloid Interface Sci.* 491:349–357.

Rani A.; Ravindran V.B.; Surapaneni A.; Mantri N.; Ball A.S. 2021. Review: Trends in point-of-care diagnosis for *Escherichia coli* O157:H7 in food and water. *Int. J. Food Microbiol.* 349:109233.

Rao C.N.R.; Sood A.K.; Subrahmanyam K.S.; Govindaraj A. 2009. Graphene: The new two-dimensional nanomaterial. *Angew. Chem. Int. Ed.* 48:7752–7777.

Rathinavel S.; Priyadharshini K.; Panda D. 2021. A review on carbon nanotube: An overview of synthesis, properties, functionalization, characterization, and the application. *Mater. Sci. Eng. B-Adv.* 268:115095.

Riu J.; Giussani B. 2020. Electrochemical biosensors for the detection of pathogenic bacteria in food. *TrAC Trends Anal. Chem.* 126:115863.

Rohde A.; Hammerl J.; Appel B.; Dieckmann R.; Dahouk S. 2015. Sampling and homogenization strategies significantly influence the detection of foodborne pathogen in meat. *Biomed. Res. Int.* 2015(1):145437.

Ruedas-Rama M.J.; Walters J.D.; Orte A.; Hall E.A.H. 2012. Fluorescent nanoparticles for intracellular sensing: A review. *Anal. Chim. Acta* 751:1–23.

Ryu U.; Jee S.; Rao P.C.; Shin J.; Ko C.; Yoon M.; Park K.S.; Choi K.M. 2021. Recent advances in process engineering and upcoming applications of metaleorganic frameworks. *Coord. Chem. Rev.* 426:213544.

Sai-Anand G.; Sivanesan A.; Benzigar M.R.; Singh G.; Gopalan A.I.; Baskar A.V.; Ilbeygi H.; Ramadass K.; Kambala V.; Vinu A. 2019. Recent progress on the sensing of pathogenic bacteria using advanced nanostructures. *BCSJ.* 92:216–244.

Salazar J.K.; Wang Y.; Yu S.; Wang H.; Zhang W. 2015. Polymerase chain reaction-based serotyping of pathogenic bacteria in food. *J. Microbiol. Methods.* 110:18–26.

Samrot A.V.; Sahithya C.S.; Selvarani J.A.; Purayil S.K.; Ponnaiah P. 2021. A review on synthesis, characterization and potential biological applications of superparamagnetic iron oxide nanoparticles. *Curr. Res. Green Sustain. Chem.* 4:100042.

Schatz G.C.; Van Duyne R.P. 2006. Electromagnetic mechanism of surface-enhanced spectroscopy. *Handb. Vib. Spectrosc.* 1:759–774.

Schofield C.L.; Field R.A.; Russell D.A. 2007. Glyconanoparticles for the colorimetric detection of cholera toxin. *Anal. Chem.* 79(4):1356–1361.

Sefah K.; Shangguan D.; Xiong X.; O'Donoghue M.B.; Tan W. 2010. Development of DNA aptamers using Cell-SELEX. *Nat. Protoc.* 5(6):1169–1185.

Seo K.H.; Brackett R.E.; Frank J.F. 1998. Rapid detection of *Escherichia coli* O157:H7 using immuno-magnetic flow cytometry in ground beef, apple juice and milk. *Int. J. Food Microbiol.* 44:115–123.

Sertova N.M. 2015. Application of nanotechnology in detection of mycotoxins and in agricultural sector. *J. Cent. Eur. Agric.* 16(2):117–130.

Shang Y.; Xiang X.; Ye Q.; Wu Q.; Zhang J.; Lin J.M. 2022. Advances in nanomaterial-based microfluidic platforms for on-site detection of foodborne bacteria. *Trends Anal. Chem.* 147:116509.

Sharma A.; Goud K.Y.; Hayat A.; Bhand S.; Marty J.L. 2017. Recent advances in electrochemical-based sensing platforms for aflatoxins detection. *Chemosensors.* 5:1.

Sharma B.; Frontiera R.R.; Henry A.I.; Ringe E.; Van Duyne R.P. 2012. SERS: Materials, applications, and the future. *Mater. Today.* 15(1–2):16–25.

Sharma C; Dhiman R.; Rokana N.; Panwar H. 2017. Nanotechnology: An untapped resource for food packaging. *Front. Microbiol.* 8:1735.

Shim W.B.; Yang Z.Y.; Kim J.S.; Kim J.Y.; Kang S.; Gun-Jo W.; Chung Y.C.; Eremin S.A.; Chung D.H. 2007. Development of immunochromatography strip-test using nanocolloidal gold-antibody probe for the rapid detection of aflatoxin B1 in grain and feed samples. *J. Microbiol. Biotechnol.* 17(10):1629–1637.

Shivashakarappa K.; Reddy V.; Krishna V.; Ali Farnian T.; Vuppula A.; Gunnaiah R. 2022. Nanotechnology for the detection of plant pathogens. *Plant Nano Biol.* 2 100018.

Singh A.; Poshitban S.; Evoy S. 2013. Recent advances in bacteriophage-based biosensors for food-borne pathogen detection. *Sensors.* 13:1763–1786.

Speck M. 1970. Selective culture of spoilage and indicator organisms. *J. Milk Food Technol.* 33:163–167.

Srisa-Art M.; Boehle K.E.; Geiss B.J.; Henry C.S. 2018. Highly sensitive detection of *Salmonella* Typhimurium using a colorimetric paper-based analytical device coupled with immunomagnetic separation. *Anal. Chem.* 90:1035–1043.

Stanisavljevic M.; Krizkova S.; Vaculovicova M.; Kizek R.; Adam V. 2015. Quantum dots-fluorescence resonance energy transfer-based nanosensors and their application. *Biosens. Bioelectron.* 74:562–574.

St John P.M.; Davis R.; Cady N.; Czajka J.; Batt C.A.; Craighead H.G. 1998. Diffraction-based cell detection using a microcontact printed antibody grating. *Anal. Chem.* 70:1108–1111.

Su X.L.; Li Y. 2004. Quantum dot biolabeling coupled with immunomagnetic separation for detection of *Escherichia coli* O157:H7. *Anal. Chem.* 76:4806–4810.

Sun Y.P.Zhou B.; Lin Y.; Wang W.; Fernando K.S.; Pathak P.; Meziani M.J.; Harruff B.A.; Wang X.; Wang H.; Luo P.G. 2006. Quantum-sized carbon dots for bright and colorful photoluminescence. *J. Am. Chem. Soc.* 28:7756–7757.

Sundar S.; Kundu J.; Kundu SC. 2010. Biopolymeric nanoparticles. *Sci. Technol. Adv. Mater.* 11:014104

Sunil K.C.; Utsav S.; Nairy R.K.; Chethan G.; Shenoy S.P.; Mustak M.S.; Yerol N. 2021. Synthesis and characterization of Zn0.4Co0.6Fe$_2$O$_4$ superparamagnetic nanoparticles as a promising agent against proliferation of colorectal cancer cells. *Ceram. Int.* 47:19026–19035.

Svechkarev D.; Sadykov M.R.; Houser L.J.; Bayles K.W.; Mohs A.M. 2020. Fluorescent sensor arrays can predict and quantify the composition of multicomponent bacterial samples. *Front. Chem.* 15(7):916.

Tachibana H.; Saito M.; Shibuya S.; Tsuji K.; Miyagawa N.; Yamanaka K.; Tamiya E. 2015. On-chip quantitative detection of pathogen genes by autonomous microfluidic PCR platform. *Biosens. Bioelectron.* 74:725–730.

Tang H.; Zhu C.; Meng G.; Wu N. 2018. Review-surface-enhanced Raman scattering sensors for food safety and environmental monitoring. *J. Electrochem. Soc.* 165(8):B3098–B3118.

Tark S.H.; Das A.; Sligar S.; Dravid V.P. 2010. Nanomechanical detection of cholera toxin using microcantilevers functionalized with ganglioside nanodiscs. *Nanotechnology.* 21(43):435502.

Thomas K.M.; de Glanville W.A.; Barker G.C.; Benschop J.; Buza J.J.; Cleaveland S.; Davis M.A.; French N.P.; Mmbaga B.T.; Prinsen G.; Swai E.S.; Zadoks R.N.; Crump J.A. 2020. Prevalence of *Campylobacter* and *Salmonella* in African food animals and meat: A systematic review and meta-analysis. *Int. J. Food Microbiol.* 315:108382.

Tian F.; Lyu J.; Shi J.; Tan F.; Yang M. 2016. A polymeric microfluidic device integrated with nanoporous alumina membranes for simultaneous detection of multiple foodborne pathogens. *Sens. Actuators B Chem.* 225:312–318.

Tortajada-Genaro L.A.; Rodrigo A.; Hevia E.; Mena S.; Ninoles R.; Maquieira A. 2015. Microarray on digital versatile disc for identification and genotyping of *Salmonella* and *Campylobacter* in meat products. *Anal. Bioanal. Chem.* 407:7285-7294.

USDA. Microbiology Laboratory Guidebook. United States Department of Agriculture. 2017. Available at: www.fsis.usda.gov/wps/portal/fsis/topics/science/laboratories-and-procedures/guidebooks-and-methods/microbiology-laboratory-guidebook/microbiology-laboratory-guidebook. Accessed 2 October 2023.

U.S. Department of Agriculture, Food Safety and Inspection Service. 2015. Microbiology laboratory guidebook. Available at: https://www.fsis.usda.gov/wps/portal/fsis/topics/science/laboratoriesandprocedures/guidebooks-and-methods/microbiology-laboratoryguidebook/microbiology-laboratory-guidebook. Accessed 1 October 2023.

U.S. Food and Drug Administration. 2015. Recalls, market withdrawals, & safety alerts. Available at: https://www.fda.gov/Safety/Recalls/. Accessed 4 October 2023.

Vazquez-Roig P.; Pico Y. 2012. 2-gas chromatography and mass spectroscopy techniques for the detection of chemical contaminants and residues in foods A2-Schrenk, D. In: *Chemical Contaminants and Residues in Food*. Woodhead Publishing, pp. 17–61.

Vikrant K.; Bhardwaj N.; Bhardwaj S.K.; Kim K.-H.; Deep A. 2019. Nanomaterials as efficient platforms for sensing DNA. *Biomaterials*. 214.

Wanekaya A.K.Chen W.; Myung N.V.; Mulchandani A. 2006. Nanowire-based electrochemical biosensors. *Electroan. Int. J. Devot. Fund Pract. Asp. Electroanal.* 18(6):533–550.

Wang J.; Liang D.; Jin Q.; Feng J.; Tang X. 2020a. Bioorthogonal SERS nanotags as a precision theranostic platform for in VivoSERS imaging and cancer photothermal therapy. *Bioconjugate Chem.* 31:182–193.

Wang L., et al. 2014. A bare-eye-based lateral flow immunoassay based on the use of gold nanoparticles for simultaneous detection of three pesticides. *Microsc. Acta.* 181(13):1565–1572.

Wang L.; Shen X.; Wang T.; Chen P.; Qi N.; Yin B.C.; Ye B.C. 2020b. A lateral flow strip combined with Cas9 nickase-triggered amplification reaction for dual food- borne pathogen detection. *Biosens. Bioelectron.* 165:112364.

Wang S.; Shen W.; Zheng S.; Li Z.; Wang C.; Zhang L.; Liu Y. 2021. Dual-signal lateral flow assay using vancomycin-modified nanotags for rapid and sensitive detection of *Staphylococcus aureus*. *RSC Adv.* 11:13297–13303.

Wang Z.; Wang J.; Yue T.; Yuan Y.; Cai R.; Niu C. 2013a. Immunomagnetic separation combined with polymerase chain reaction for the detection of *Alicyclobacillus acidoterrestris* in apple juice. *PLoS One.* 8:e82376.

Wang Z.; Yue T.; Yuan Y.; Cai R.; Niu C.; Guo C. 2013b. Development and evaluation of an immunomagnetic separation-ELISA for the detection of *Alicyclobacillus* spp. in apple juice. *Int. J. Food Microbiol.* 166:28–33.

Wawerla M.; Stolle A.; Schalch B.; Eisgruber H. 1999. Impedance microbiology· Applications in food hygiene. *J. Food Prot.* 62:1488–1496.

Weber J.E.; Pillai S.; Rama M.K.; Kumar A.; Singh S.R. 2011. Electrochemical impedance-based DNA sensor using a modified single walled carbon nanotube electrode. *Mater. Sci. Eng. C.* 31:821–825.

Wen S.; Zhou J.; Zheng K.; Bednarkiewicz A.; Liu X.; Jin D. 2018. Advances in highly doped upconversion nanoparticles. *Nat. Commun.* 9:2415.

Wilson R. 2008. The use of gold nanoparticles in diagnostics and detection. *Chem. Soc. Rev.* 37:2028.

World Health Organization. 2020. Estimating the burden of foodborne diseases. Available at: https://www.who.int/activities/estimating-the-burden-of-foodborne-diseases. Accessed 2 October 2023.

World Health Organization, Food safety food safety, World Health Organization. 2013. Available at: https://www.who.int/health-topics/food-safety. Accessed 2 October 2023.

World Health Organization, Water, sanitation and hygiene, World Health Organization. 2015. Available at: https://www.who. int/ health- topics/water- sanitation- and- hygiene-wash. Accessed 2 October 2023.

Wu L.; Li G.; Xu X.; Zhu L.; Huang R.; Chen X. 2019. Application of nano-ELISA in food analysis: Recent advances and challenges. *TrAC-Trends Anal. Chem.* 113:140–156.

Wu W.; Zhao S.; Mao Y.; Fang Z.; Lu X., Zeng L. 2015. A sensitive lateral flow biosensor for *Escherichia coli* O157:H7 detection based on aptamer mediated strand displacement amplification. *Anal. Chim. Acta.* 861:62–68.

Xing G.; Zhang W.; Li N.; Pu Q.; Lin J.M. 2021a. Recent progress on microfluidic biosensors for rapid detection of pathogenic bacteria. *Chin. Chem. Lett.* 33:1743-1751

Xing Y.; Zhao L.; Cheng Z.; Lv C.; Yu F.; Yu F. 2021b. Microfluidics-based sensing of Biospecies. *ACS Appl. Bio Mater.* 4:2160–2191.

Xu J. 2016. Fungal DNA barcoding. *Genome.* 59(11):913–932.

Xue L.; Zheng L.; Zhang H.; Jin X.; Lin J. 2018. An ultrasensitive fluorescent biosensor using high gradient magnetic separation and quantum dots for fast detection of foodborne pathogenic bacteria. *Sens. Actuators B Chem.* 265:318–325.

Yang T.; Duncan T.V. 2021. Challenges and potential solutions for nanosensors intended for use with foods. *Nat. Nanotechnol.* 16:251–265.

Ye Y.; Zheng L.; Wu T.; Ding X.; Chen F.; Yuan Y.; Fan G.C.; Shen Y. 2020. Size-dependent modulation of polydopamine nanospheres on smart nanoprobes for detection of pathogenic bacteria at single-cell level and imaging-guided photothermal bactericidal activity. *ACS Appl. Mater. Interfaces.* 12:35626–35637.

You Y.; Lim S.; Gunasekaran S. 2020. Streptavidin-coated Au nanoparticles coupled with biotinylated antibody-based bifunctional linkers as plasmon-enhanced immunobiosensors. *ACS Appl. Nano Mater.* 3:1900–1909.

Yu H.; Park J.; Y. Kwon; C.W. Hong; S. C. Park; K. M.; Chang P. S. 2018. An overview of nanotechnology in food science: Preparative methods, practical applications, and safety. *J. Chem.* 5427978:1–10.

Yu L.; Song Z.; Peng J.; Yang M.; Zhi H.; He H. 2020. Progress of gold nanomaterials for colorimetric sensing based on different strategies. *TrAC Trends Anal. Chem.* 127:115880.

Yusoff N.; Pandikumar A.; Ramaraj R.; Lim H.N.; Huang N.M. 2015. Gold nanoparticle based optical and electrochemical sensing of dopamine. *Microchim. Acta.* 182:2091–2114.

Zain M.E. 2011. Impact of mycotoxins on humans and animals. *J. Saudi Chem. Soc.* 15:129–144.

Zhang G.; Brown E.; Escalona N. 2011. Comparison of real-time PCR, reverse transcriptase real-time PCR, loop-mediated isothermal amplification, and the FDA conventional microbiological method for the detection of Salmonella spp. in produce. *Appl. Environ. Microbiol.* 77:6495–6501.

Zhang Q.; Feng S.; Lin L.; Mao S.; Lin J.M. 2021. Emerging open microfluidics for cell manipulation. *Chem. Soc. Rev.* 50:5333–5348.

Zhang R.; Belwal T.; Li L.; Lin X.; Xu Y.; Luo Z. 2020. Nanomaterial-based biosensors for sensing key foodborne pathogens: Advances from recent decades. *Compr. Rev. Food Sci. Food Saf.* 19:1465–1487.

Zhao L.; Jensen L.; Schatz G.C. 2006. Pyridine-Ag20 cluster: A model system for studying surface-enhanced Raman scattering. *J. Am. Chem. Soc.* 128(9):2911–2919.

Zhao M.X.; Zeng E.Z. 2015. Application of functional quantum dot nanoparticles as fluorescence probes in cell labeling and tumor diagnostic imaging. *Nanoscale Res.* Lett. 10(1):1–9.

Zhao X. Lin C.W.; Wang J.; Oh D.H. 2014. Advances in rapid detection methods for foodborne pathogens. *J. Microbiol. Biotechnol.* 24(3):297–312.

Zheng L.; Qi P.; D. Zhang. 2019. Identification of bacteria by a fluorescence sensor array based on three kinds of receptors functionalized carbon dots. *Sens. Actuators B Chem.* 286:206–213.

Zhou J; Rossi J. 2017. Aptamers as targeted therapeutics: Current potential and challenges. *Nat. Rev. Drug Discov.* 16(3):181–202.

Zhu P.; Shelton D.R.; Li S.; Adams D.L.; Karns J.S.; Amstutz P.; Tang C.M. 2011. Detection of E. coli O157:H7 by immunomagnetic separation coupled with fluorescence immunoassay. *Biosens. Bioelectron.* 30:337-341.

13 Nanotechnology in the Detection of Food Contaminants

*Madeeha Batool, Hafiz Muhammad Junaid,
and Amber Rehana Solangi*

13.1 INTRODUCTION

Food is part and parcel of our lives and essential to our survival. Contamination of food with substances having adverse effects on the health of living beings is one of this era's burning questions. Therefore, food quality and safety are the subject of much concern for researchers today. Food quality is based on two key parameters: intrinsic quality, like nutritional characteristics, functional importance, taste, and tang, while external authenticity deals with illegal and contaminants in food. Subsequently, food safety can be split into two categories, i.e., exogenous safety and endogenous safety. The exogenous safety concerns include the manifestation of deposits of pesticides, drugs, and heavy metals. Endogenous safety problems comprise the occurrence of biotoxins and the injurious species produced as a result of thermal treatment and fermentation in food.

Food can be contaminated during production, packaging, transportation, or storage. New industrial processes, agricultural practices, environmental pollution, and climate change also increase the toxicant residues in the food.

13.2 FOOD CONTAMINANTS

13.2.1 NATURAL CONTAMINANTS

These are the substances that are naturally present in crops but produce negative impacts on the health of living organisms. The most common natural contaminants are aflatoxins and ochratoxins.

13.2.1.1 Aflatoxins

Aflatoxins are fungi-based toxic entities naturally found in crops such as cotton seeds, peanuts, corn, etc. Mainly, two fungal species, i.e., *Aspergillus parasiticus* and *Aspergillus flavus*, are responsible for producing aflatoxins as these usually exist in humid and warm areas of the world. These environmental conditions are ideal for the growth of fungus. These fungal species can grow on the crop during its development as well as during its storage. Aflatoxins possess mutagenic, teratogenic, immune-suppressing, and carcinogenic properties. International agency for cancer research has declared aflatoxins as carcinogens for humans.

13.2.1.2 Ochratoxins

Like aflatoxins, ochratoxin belongs to the mycotoxins family, chiefly secondary metabolites of *Aspergillus ochraceus* and *Penicillium verrucosum*. Ochratoxins are known for their hepatotoxic, neurotoxic, nephrotoxic, immunotoxic, teratogenic, and carcinogenic behaviors. Ochratoxins are common in cereal grain crops.

DOI: 10.1201/9781003514039-13

13.2.2 Chemical Contaminants

Chemical contamination refers to toxic chemical species in food materials in a concentration that is lethal to living beings. These contaminants are one of the major causes of foodborne diseases. Chemical contaminants may come from multiple sources, including air, water, soil, environment, personal care, medicinal products, and packaging materials, from the farm to the dinner plate.

13.2.2.1 Pesticides

Pesticides are the chemical compounds used to eradicate the pests. These may include herbicides (Grillo et al. 2014; Ahmadian et al. 2020; Kurniadie et al. 2022), fungicides (Jing et al. 2022), or insecticides (Ma et al. 2022) depending upon the species to be removed, i.e., herbs, fungi, or insects. Pesticides are composed of toxic persistent chemicals like lead arsenate and phenylmercuric acetate etc. However, these pesticides increase the crop yield but also act as contaminants.

13.2.3 Persistent Organic Pollutants

There are large numbers of organic compounds that contaminate the foodstuff due to different activities. Various organic contaminants can be produced during food processing by heating, drying, baking, smoking, frying, and microwave treatment. Such organics include acrylamide, polycyclic aromatic compounds, heterocyclic aromatic amines, N-nitroso species, etc. Moreover, many portable and food storage containers are made from organic polymeric species like polystyrene, melamine, microplastics, etc. These organics can also contaminate the food when there is temperature storage followed by microwave treatment before usage. Furthermore, artificial sweeteners and food colors are considered emerging food contaminants, complex organic molecules. Subsequently, veterinary drug contaminants like antibiotics are also a source of many foods-based allergic reactions.

13.2.4 Heavy Metals

Heavy metals contamination of edible stuff occurs in many ways, including agricultural activities, i.e., excessive use of pesticides and fertilizers, irrigation of crops with heavy metals polluted water, uptake of heavy metals by aquatic animals used as food, and many environmental processes for examples transportation of heavy metals through wind, smoke, water, etc. Moreover, storage of food in metal containers for an extensive period is also a cause of heavy metal contamination. Therefore, remediation of heavy metals is very crucial. The heavy metal contamination of crops can be controlled using industrial and municipal wastewater after proper treatment. The number of nanomaterials has been reported in this regard, summarized in Table 13.1.

13.2.5 Biocontaminants

Biological contaminants usually comprise bacterial species. The bacterial contamination of edible materials may occur due to contaminated raw materials or bacterial exposure during the processing or storage of food. Pathogenic bacteria, such as *Staphylococcus aureus, Bacillus cereus, Salmonella, Escherichia coli, Campylobacter,* and *Listeria monocytogenes,* are responsible for mainstream foodborne ailment outbursts.

13.3 DETECTION OF FOOD CONTAMINANTS BY NANOMATERIALS

Conventionally, hi-tech instruments have been employed for food analysis using GC-FID (Hunter Jr et al. 2010), HPLC-UV (Yoshioka and Ichihashi 2008), GC-MS (Fu et al. 2016), HPLC-MS

TABLE 13.1

Nanomaterials for Heavy Metal Remediation from Food Materials

Role in Heavy Metal Remediation	Nanomaterials	References
Sensors	Clay-based nanomaterials	El Mhammedi et al. (2009)
	MOF based nanomaterials	Singhet al. (2020)
	Organic-inorganic based nano-hybrids	Batool and Junaid (2022)
	Metal NPs	Zhu et al. (2014)
	CNTs	Afkhamiet al. (2012)
	Graphene-based nanomaterials	Ting et al. (2015)
	Nanoporous silica	Javanbakht et al. (2009)
	Carbon Dots	Batool et al. (2022a)
	Nano-polymeric materials	Huang et al. (2014)
Adsorbents	Chitosan-based nano-adsorbents	Haripriyan et al. (2022)
	Magnetic nano-adsorbents	Haripriyan et al. (2022)
	Clay nano-adsorbents	Zhang et al. (2021)
	Metal oxide nano-adsorbents	Dhiman and Kondal (2021)
	CNTs	Bankole et al. (2019)
	Nano-zeolites	Zhang et al. (2020b)
	MXenes	Batool and Junaid (2023)

(Zhou et al. 2019), NIR (Fu et al. 2019), Raman spectroscopy (Luo et al. 2016), fluorescence spectrometry (Pereira et al. 2018), UV/Vis. spectroscopy (Martins et al. 2017), supercritical fluid chromatography (SFC), solid phase extraction (SPE), supercritical fluid extraction (SFE), headspace (HS), and flow injection analysis (FIA) followed by complex and laborious sample preparation procedures. The primary limitation of these methods is that they are not commercially designed specifically for food analysis. Higher separation and detection efficiencies characterize chromatographic methods, but these are very costly, and suffer from retention time shifts and inaccurate peak extraction issues. Spectroscopic techniques, i.e., NIR, Raman spectroscopy, fluorescence spectroscopy, and UV/Vis. spectrometers are simple and easy to operate but over-lapping peaks cause poor separation efficiency. Due to these drawbacks, chromatographic and spectroscopic procedures are not methods of choice for food analysis. Furthermore, biological methods, e.g., immunoassay, enzyme inhibition procedures, and biosensing, are also broadly applied for quickly detecting food contaminants with a high sensitivity and selectivity. However, these strategies may show false-positive results because of matrix effects and thermal instability (Zhang et al. 2019). Therefore, selective, sensitive, quick, and on-site detection of food contaminants is crucial.

To address these limitations, nano-structured materials can be used due to their marvelous properties, i.e., surface effects, quantum sizes, quantum tunnels, and dielectric limits (Lv et al. 2018). The application of nanomaterials as an active component of sensing systems offers rapid, economical, simple, easy, and green detection of food contaminants with higher selectivities and sensitivities. Such nanomaterial-based sensors include various quantum and carbon dots (Wang et al. 2019b; Yang et al. 2020b), nano-porphyrin (Paolesse et al. 2017), gold nanoparticles (Au-NPs), silver nanoparticles (Ag-NPs) (Haes et al. 2004), up-conversion nanoparticles (UCNPs) (Wang et al. 2019a), nanozymes (Huang et al. 2019), metal oxide nanomaterials (Sebők and Dékány 2015) and organic fluorescent molecules-based nanomaterials (Svechkarev and Mohs 2019).

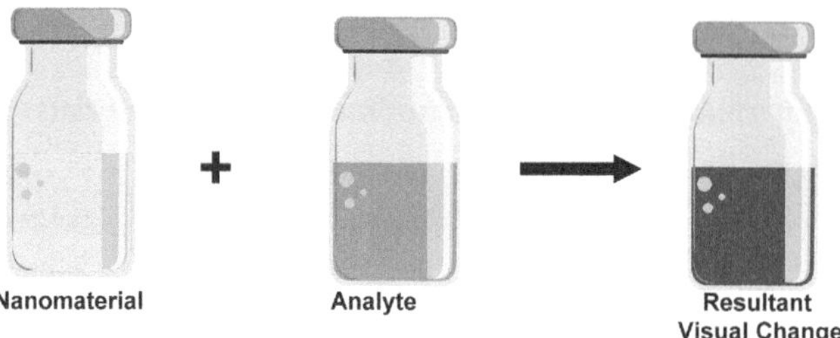

FIGURE 13.1 Colorimetric sensing model.

13.4 TYPES OF NANOMATERIALS-BASED SENSORS FOR FOOD CONTAMINANTS

13.4.1 COLORIMETRIC SENSORS

Colorimetric detection systems are quick, sensitive, selective, low cost, and robust for detecting a particular analyte with the aid of the naked eye without using any complex instrument (Junaid et al. 2021; Junaid et al. 2022; Junaid and Batool 2023) as shown in Figure 13.1. Detection using a colorimetric sensor is attributed to a color change produced by the sensor and analyte interaction. Colorimetric methods have gained the interest of food technologists for an on-site food safety inspection as observable color changes can readily produce the selective chemical reaction without using any sophisticated instrument. Colorimetric sensors based on nanomaterials have improved sensitivity and selectivity for detecting food contaminants. AuNPs, AgNPs, and Au-nanorods have been reported for their sensing abilities toward various food contaminants.

13.4.2 FLUORESCENT SENSORS

Fluorescent sensors are of great importance because of their higher sensitivities, selectivities, rapid response, low detection limits with low cost, and sample requirements in very small amounts (Batool et al., 2022a, b). With the emergence of nanomaterials, a progressive development in fluorescent sensors has also been acknowledged. Regarding food safety, fluorescent sensors based on nanomaterials have revolutionized food quality control due to their robust, reproducible, and rapid response to various food contaminants. Quantum dots (Zrazhevskiy et al. 2010) and metal nanoclusters (Li et al. 2014a) are appreciable nanomaterials in this respect. The fluorescent sensing model is given in Figure 13.2.

13.4.3 ELECTROCHEMICAL SENSORS

The advancement in electrochemical sensing has great importance because of its capability to rectify many analytical issues and challenges, including agriculture, medicine, environmental monitoring, and food safety. Owing to the large surface area and small size, nanomaterials are the ideal option for electrode materials in sensitive and selective electrochemical sensing of particular analytes. Electrochemical sensors based on nanomaterials provide an excellent platform for detecting and quantifying various contaminants in water and food. The general electrochemical sensing scheme is summarized in Figure 13.3.

13.4.4 BIOSENSORS

Biosensors are a combination of physicochemical sensors and biological species. This approach is used to detect analytes neither detected by a chemical sensor nor a biosensor. The higher sensitivities, selectivities, and lowest detection limits are some striking features of biosensors. Advancements

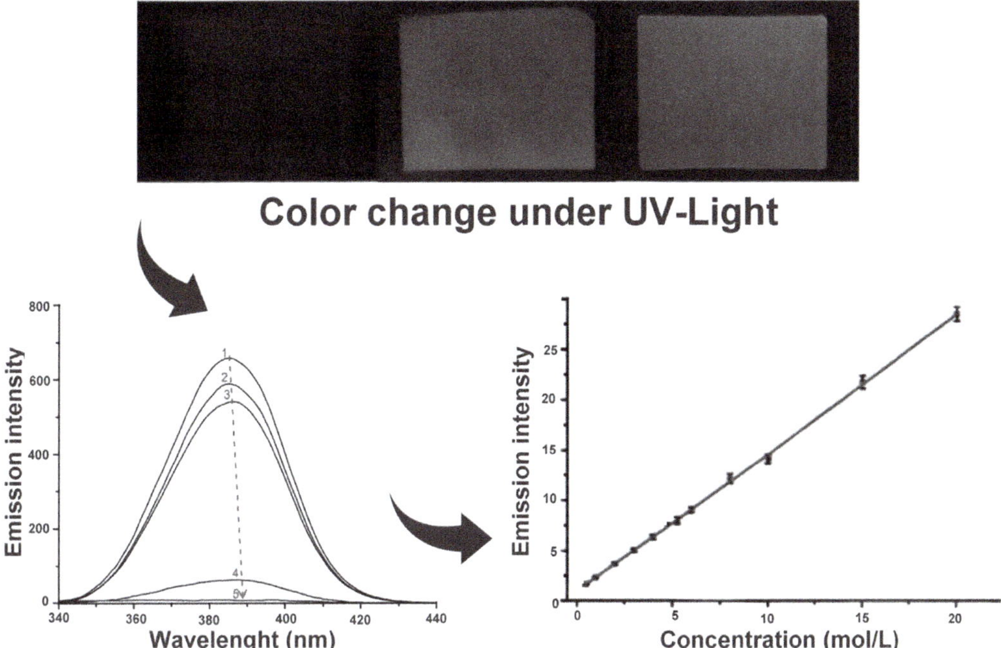

FIGURE 13.2 Fluorescent sensing model.

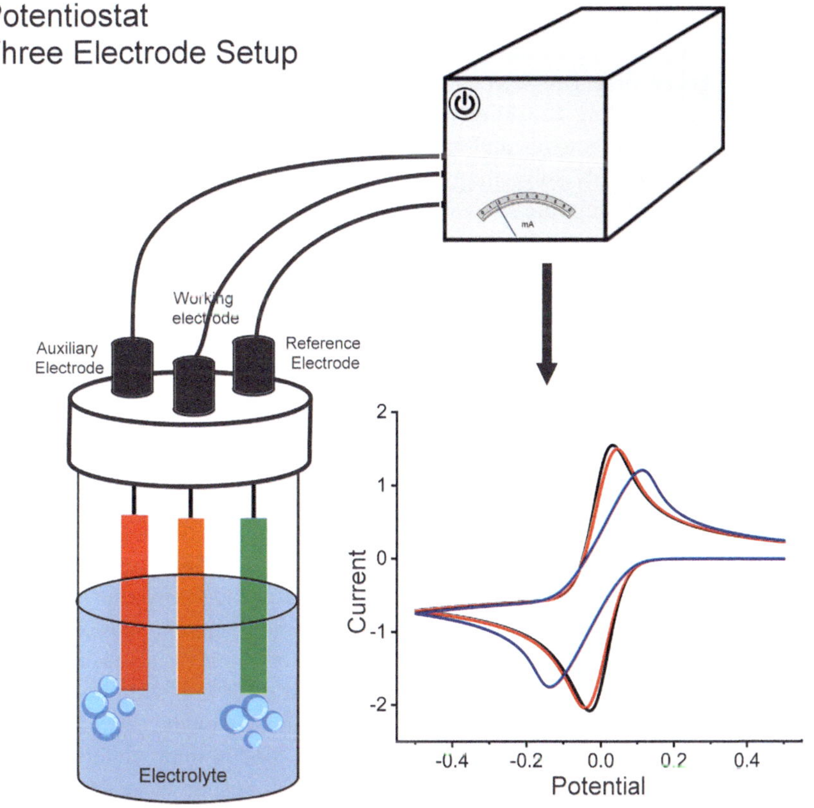

FIGURE 13.3 Electrochemical sensing model.

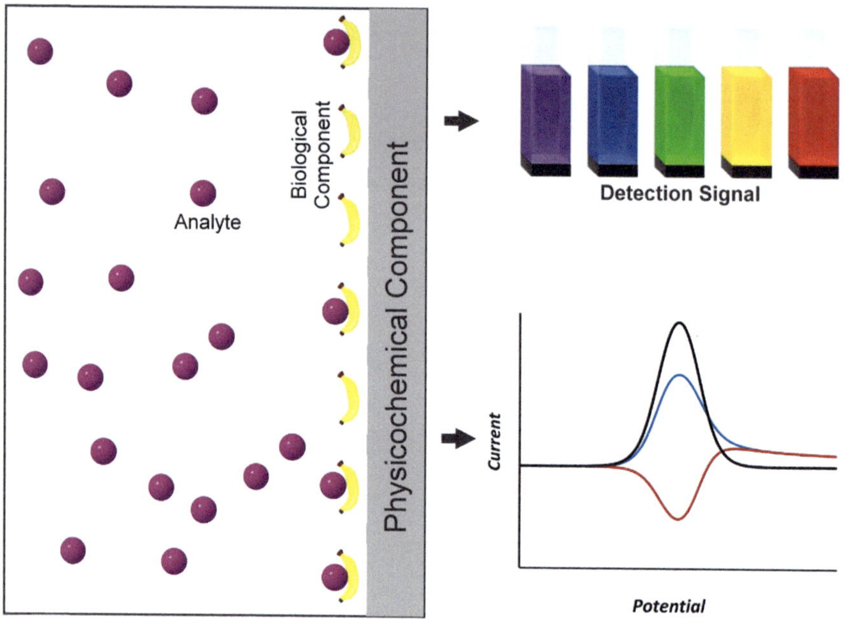

FIGURE 13.4 Biosensing model.

in nanomaterials have expedited the progress of biosensing to detect contaminants associated with edible stuff (Warriner et al. 2014; Bülbül et al. 2015; Sharma et al. 2015). Some critical properties of nanomaterials-based biosensors include selective recognition of targeted analyte, elevation in the output signal, enhancement in selectivity and sensitivity, and decrease in the analysis time. Many nanomaterials, including zero-dimensional (0D) nanoparticles (NPs, including nanodots), 1D nano-rods (containing nanowires and nanotubes), 2D nanosheets, and even 3D metal-organic frameworks (MOFs) effectively solve the problems of developing advanced nanobiosensing systems. Schematic biosensing steps are shown in Figure 13.4.

13.5 SENSING APPLICATIONS OF NANOMATERIALS FOR FOOD CONTAMINANTS

13.5.1 Detection of Organic Contaminants

13.5.1.1 Detection of Acrylamide

Acrylamide is manufactured commercially by acrylonitrile through a catalytic hydrolytic reaction and is used to produce various materials such as plastics, pigments, adhesives, and cosmetics. It also acts as a coagulant in sewage or wastewater management plants. On a laboratory scale, acrylamide is frequently used in organic synthesis with selective modification of –SH groups, gel electrophoresis, and binding of biological products. These chemicals harm the environment and human health and are expected to cause muscle weakness, skin and mucosa irritation, nausea, drowsiness, brain damage, and carcinoma (Oracz et al. 2011). Unfortunately, acrylamide is also widely found in edible stuff like potato chips and French fries due to reduced sugar's reaction with the amino acid asparagine at elevated cooking temperatures (Krajewska et al. 2008). Given the above, there is a dire need to evaluate, mitigate, and control the amount of acrylamide food to improve food safety.

Quick, selective, and sensitive detection of acrylamide using electrochemical techniques (Stobiecka et al. 2007; Krajewska et al. 2008; Silva et al. 2011; Batra et al. 2013, Sun et al. 2013) and quartz crystal microbalance (QCM) (Kleefisch et al. 2004) have been reported in the literature. However, various nanomaterials have also been found for their applications in the fabrication of biosensors. Batra and coworkers (2011) developed an amperometric biosensor for acrylamide detection using hemoglobin-functionalized MWCNT-copper nanoparticles-PANI with a detection limit as low as $0.2\,nmol\,L^{-1}$ having an excellent sensitivity of $72.5\,mA\,nmol^{-1}L^{-1}cm^{-2}$, in a more comprehensive linear range of 5–$75\,m\,mol\,L^{-1}$ (rapid response in <2 seconds and a longer stability *ca.* 100 days stability). Another nanomaterial-based biosensor consists of a monolayer of hemoglobin on SWCNTs to generate an electrode with an efficient LOD of $1\,nmol\,L^{-1}$ (Krajewska et al. 2008). Similarly, nanocomposites containing graphene oxide and AuNPs have been used for their biosensing properties toward acrylamide when conjugating with pheochromocytoma cells (Sun et al. 2013). However, the LOD of this detection system is high, i.e., $40\,mmol\,L^{-1}$. An ion-selective electrode-based biosensor has been presented by Silva and coworkers (2011) for the sensing of acrylamide. The best results were obtained using a polyethersulfone bi-membrane system with a reaction time of 6 minutes, stability of up to 27 days, and LODs in the $\mu g\,kg^{-1}$ range.

13.5.1.2   Detection of Polycyclic Aromatic Hydrocarbons

Polycyclic aromatic hydrocarbons (PAHs) are a family of organic compounds containing fused aromatic rings. PAHs originate from fossil fuels such as gasoline, coal, and tar deposits and are released as byproducts during the combustion of these fossil fuels. Therefore, the most hydrophobic compounds are also released into the air. PAHs have low solubility in water but high solubility in lipids; for example, the solubility of anthracene in water at $25°C$ is $<0.086\,mg\,L^{-1}$ ($4.8\times10^{-7}mol\,L^{-1}$). Thus, PAHs can interact with lipids such as cell membranes and bioaccumulate in the lipids of bacteria (Engraff et al. 2011). PAHs can act as teratogens, carcinogens, mutagens, and endocrine disrupters. It is essential to detect, quantify, and remove PAHs in food and drinking water due to their potential to cause environmental pollution. Various methods to detect PAHs include immunoassays (Meimaridou et al. 2010) and bioassays using microbes (Engraff et al. 2011). However, these techniques are time-consuming and have low stabilities with high EC 50 values ($>1\,mg\,L^{-1}$). Many nanomaterials, i.e., SWCNTs (Carrillo-Carrión et al. 2009), AuNPs (Mailu et al. 2010), graphene oxide (Shen et al. 2012), etc. based sensors have been developed for the detection of PAHs. Cerrilo-Carion and workers (2009) have introduced a fluorescent nanosensor based on SWCNT-doped cadmium selenium quantum dots (QD) for sensing pyrene, benzopyrene, benzo anthracene, and perylene. This optical nanosensor exhibits a high sensitivity to pyrene, having a broad linear range with LOD as low as $<1\,nmol\,L^{-1}$. Duong and coworkers (2006) have presented a similar fluorescence probe based on sole-gel functionalized cadmium selenium/zinc sulfide QDs for sensing anthracene, phenanthrene, and pyrene. Trace levels of PAHs can elevate the fluorescence intensity of the CdSe/ZnS QD-entrapped membranes with an LOD of *ca.* $5\,nmol\,L^{-1}$. The electrochemical sensors, e.g., polypyrrole-silver-AuNPs (Mailu et al. 2010), humic acid-Fe_3O_4 (Lin et al. 2010), graphene-GCE (Shen et al. 2013), antibody-gold electrode (Ahmad et al. 2011), and thiolated DNA/gold electrode (Del Carlo et al. 2012) have got popularity for the detection of PAHs these days. These sensors have been applied for the detection of anthracene (Mailu et al. 2010), benzo pyrene (Lin et al. 2010; Ahmad et al. 2011; Del Carlo et al. 2012), and 1-hydroxypyrene (Shen et al. 2013). These exhibit good selectivities with efficient sensitivities, low LODs, wider linear range, and good reproducibility in results.

13.5.1.3   Detection of Malamine

Melamine (1, 3, 5-triazine-2, 4, 6-triamine) is synthesized from urea on an industrial scale, producing cyanic acid as a byproduct (Hau et al. 2009). Animal models have shown that melamine can cause acute toxicity and stone formation in the urinary tract (Hau et al. 2009). Edible material containing melamine contaminants has been found to cause the death of hundreds of animals. The World Health Organization (WHO) recommends that daily melamine intake be less than $0.2\,mg\,kg^{-1}$ of body weight. Optical sensors-based gold nanoparticles (Au-NPs) have been reported for the sensing of melamine in raw milk (Li et al. 2010). In this detection method, a selective naked eye color change from red to blue has been observed upon the interaction of Au-NPs with melamine in milk.

Similarly, fluorescent sensors have also been known for detecting melamine in foodstuffs. For example, Water-dispersable thioglycolic acid-capped cadmium telluride (CdTe) QD have been introduced as fluorescent melamine sensors (Zhang et al. 2012). Likewise, single-wall carbon nanotubes (SW-CNTs) modified using glassy carbon have also been applied to detect melamine in electrochemical methods (Liu et al. 2011).

13.5.1.4 Detection of Antibiotics

Ampicillin belongs to the penicillin class of beta-lactam antibiotics and is widely used in medicine and agriculture to treat bacterial infections and improve animal growth. It successfully treats infectious diseases caused by various bacterial species, i.e., *E. coli, Neisseria gonorrhoeae, Haemophilus influenzae, Shigella* sp., and *Salmonella* sp. However, overuse of ampicillin in livestock and many household products can lead to food contamination, causing health problems harmful to humans, such as allergies, breathing disorders, and seizures. Song and coworkers (2012) have developed a dual-functioning probe, i.e., colorimetric and fluorimetric, based on Au-NPs for the sensing of ampicillin.

Similarly, oxytetracycline belongs to the tetracycline class of antibiotics used to cure uncomplicated bacterial infections in humans. Its residues in wastewater released from pharmaceuticals are a source of its entrance into the food chain. Its contamination in food causes liver toxicity and gastro-intestinal problems. Li et al. (2020) have introduced La-MOF strips for the on-spot detection of oxytetracycline with an LOD of $1.95\,nmol\,L^{-1}$.

Likewise, quinolone antibiotics have extensively been employed to treat infectious ailments caused by gram-positive bacteria. These antibiotics may enter the food chain through pharmaceutical wastewater, and their contamination in food may lead to bacterial resistance in our bodies, which is very harmful to us. Eu-MOF-based nano-fluorescent sensors have been reported for selective detection of three quinolones, namely ciprofloxacin, norfloxacin, and enrofloxacin, with LODs of 0.75, 0.52, and $0.12\,\mu mol\,L^{-1}$, respectively (Wang et al. 2022a).

13.5.1.5 Detection of Carbofuran

Carbofuran is an extensively used pesticide for eradicating insects and nematodes in various crops because of its broad-range biological activity and relatively less persistent nature than organochlorine pesticides (Pogačnik and Franko 2003). Environmental pollution due to the intensive use of pesticides for agricultural and non-agricultural purposes directly causes many problems in terms of human health. Carbofuran toxicity includes nausea, vomiting, abdominal pain, sweating, diarrhea, hypersalivation, weakness, blurred vision, shortness of breath, increased blood pressure, and urinary incontinence. Its high doses can cause death from respiratory failure associated with carbofuran exposure (Kuswandi et al. 2017).

An electrochemical-immunosensor based on Au-NPs, multiwalled carbon nano tubes-chitosan (MWCNTs-CTS) nanocomposite film, and a protein (SPA) has been reported for the selective detection of carbofuran (Sun et al. 2012). In these films, Au-NPs and MWCNTs-CTS have been incorporated to elevate the electrical activity and stability of this immunosensing probe. This 3D MWCNTs-CTS nanocomposite film provides various amino and carboxyl groups to cross-link SPA, generating a large surface area. These self-assembled SPA layers have been integrated into the electrode surface to enhance the binding ability of the antibody (Sun et al. 2012).

13.5.1.6 Detection of DDT

1,1,1-trichloro-2,2-bis (4-chlorophenyl) ethane, commonly known as DDT, has broadly used as an organochlorine pesticide throughout the world since 1939 after the recognition of its pesticidal properties by Muller (Anfossi et al. 2004). Due to its effectiveness, about 1 million tons of DDT are used annually around the globe (Botchkareva et al. 2002). DDT has been banned in various countries due to its dangerous effects and environmental risks to wildlife and human beings. An immune-assay technique based on Au-NPs has been established to detect DDT in the nano-gram range (Lisa et al. 2009). In this technique, Au-NPs of a particular size are synthesized and conjugated to anti-DDT antibodies, which act as the detecting medium.

13.5.1.7 Detection of 2,4-Dinitrophenol

2,4-Dinitrophenol (DNP) was previously used to control body weight to enhance fat metabolism. DNP was used widely used in weight loss pills in the 1930s. As late as 1938, it was proved to be a fatal cause of influenza in the United States, which kicked down its use. Excessive use of DNP can cause death due to increased body temperature, possibly due to losing energy in heat down the proton gradient rather than producing ATP (Sonawane et al. 2014).

Ko and coworkers (2010) have presented a colorimetric nano-biosensor for detecting DNP based on the hybridization of the chromogenic effect of latex microspheres with Au-NPs. A DNP analog, 2, 4-dinitrophenol–bovine serum albumin (DNP–BSA), has been attached with Au-NPs to allow the hybridization of DNP with an anti-DNP antibody on latex microspheres, which results in the development of pinkish-red color. Moreover, a similar probe has also detected and quantified DNP–glycine. Upon displacement of Au-NPs from host latex microspheres in the presence of the DNP, an optical change from pinkish-red to white has been observed.

13.5.1.8 Detection of Dioxins

Dioxins are chemically polychlorodibenzo-p-dioxins (PCDDs) and polychlorodibenzofurans (PCDFs). These are highly stable carcinogenic environmental pollutants that cannot be degraded naturally. Dioxin enters the food chain when it contaminates various crops and can mount up in the adipose tissues of living beings (Chobtang et al. 2011). The toxicity of dioxin toward life is due to the chemical reaction of its aryl hydrocarbon with xenobiotic-reactive substances. It causes mutations, which may lead to carcinoma and immune defects (Kasai et al. 2006; Chobtang et al. 2011).

An oligopeptide-cysteine-QCM-based electro-static sensor has been reported for quick, sensitive, and selective sensing of dioxins (Mascini et al. 2004, 2005) with a ca. 1 µg kg^{-1} detection limit.

13.5.1.9 Detection of Bisphenol A

(2, 2-bis(4-hydroxyphenyl)propane) also known as Bisphenol A, is an organic molecule used as raw material for the manufacturing of transparent plastic items (Choi and Lee 2017; Bahmani et al. 2020). Its large-scale applications lead to its dispersion in the environment as a pollutant. Plastic packing of edible materials is a source of food contamination from Bisphenol A. It is responsible for various human medical issues, such as endocrine malfunctioning, reproductive diseases, and tumors (Li et al. 2016; Mirzajani et al. 2017); thus, its sensing is critical.

Like other persistent organic pollutants, nanomaterials have also been reported to detect Bisphenol A. For example, ZnO and biochar nano-hybrid (Hu et al. 2022) and AuNPs-poly amide composites (Mercante et al. 2021) have been reported as selective electrochemical sensors of Bisphenol A with LODs in the order of 10^{-7} mol L^{-1}.

13.5.1.10 Detection of Furan

Furan is an organic compound composed of a five-membered aromatic ring (Hashemi-Moghaddam and Ahmadifard 2016) with four corners occupied by carbon atoms and an oxygen atom at the fifth corner. Because of its broad range of applications in organic and polymeric synthesis, non-degradable nature, and carcinogenic properties, it is considered a pollutant (Loui and Chiang 2018) and a food contaminant. Nanomaterials have also found their sensing applications for the detection of furan. For example, CNTs (Torres and Correa 2021) and SWCNTs doped with Ag, Rh, and Pd (Yuksel et al. 2022) have been described for the detection of furan based on first principle and DFT calculations.

13.5.1.11 Detection of Allergens

The food contaminants that can harmfully affect the immunity of a living being and cause various diseases are called food allergens (Burks et al. 2001). Nanomaterials have also been known for their allergen detection properties. AuNPs (Momeni et al. 2022) and AuNPs/AgNPs (Anfossi et al. 2019) based lateral flow-immuno assay systems have been reported for the detection of food allergens like gluten, ovalbumin, and hazelnut.

13.5.2 Detection of Heavy Metals

Heavy metals are one of the most lethal contaminants of food, which pose a severe danger to human health, even in minute amounts (Gilani et al. 2015). Commonly, heavy metals contaminate food in several ways, i.e., particulate matter in the air, irrigation of crops from industrial and sewage wastewater, agricultural activities, gaseous discharge from industries, automobile discharge, etc. Heavy metal contamination in the food due to industrial waste effluents is one of today's biggest problems. For example, wastewater released from tanneries contains chromium in hexavalent and trivalent forms. The crops cultivated in the areas present in the neighborhood of tanneries can accumulate chromium ions. The heavy metal ions are tolerable by living beings up to a certain threshold level. However, these are highly toxic beyond that tolerance limit (Ntihuga 2006). These are supposed to cause diseases in the various organ systems, such as the nervous system, liver, skin, bone, and teeth (Aragay and Merkoci 2012; Soldatkin et al. 2012).

Many classical, instrumental, and modern techniques have existed for detecting and determining heavy metals; however, recently, many nanomaterials (Batool and Junaid 2022, 2023; Junaid et al. 2023) have been reported for the rapid, sensitive, and selective sensing of heavy metals. For instance, a potentiometric sensor for sensing Pb^{2+} based on MWCNT modified with 2-aminothiophenol has been reported (Guo et al. 2011a, b). The selectivity of this sensor for Pb^{2+} is attributed to covalent bonding between the $-NH_2$ moieties of aminothiophenol and the carbonyl functionalities of MWCNTs to form a coordination complex with Pb^{2+} ions. Similarly, a colorimetric probe based on Au-NPs has been presented for the sensing of mercuric (Hg^{2+}) ions, which exhibits a color change from ruby red to royal purple with a detection limit in nano-molar range that is much lower than the WHO allowable limit for Hg^{2+} in drinking water (Ding et al. 2012; Li et al. 2014b; Zhou et al. 2014; Chen et al. 2015). Electrochemical nanosensors based on AuNP-based electrochemical sensors have also been reported (Gong et al. 2010; Wan et al. 2015) for the selective sensing of Pb^{2+} and Cu^{2+} metal ions with LODs in ppb ranges.

13.5.3 Detection of Biotoxins

13.5.3.1 Detection of Aflatoxins

Lateral flow strip sensors based on metallic particles have been used to detect Aflatoxin AFB_1 in food samples (Liao and Li 2010). This immuno-sensor can immobilize the monoclonal antibody on the surface of NPs with a silver core and a gold shell (AgAu), which acts as the detection reagent. Subsequently, membrane-based immuno-dipsticks have also been known for the sensing of AFB_1, which consists of a test line comprising AFB_1 conjugated to bovine serum albumin (BSA) and a control line. In the strip, one to two colored lines are formed on the membrane using red AgAu nanoparticles coated with anti-AFB_1 as the sensing reagent for naked-eye detection of AFB_1. For an extensive discussion on nanotechnological approaches applied to ameliorate and detect aflatoxins, the reader is suggested to the work by Zhang and coworkers (2020a) and Yan and coworkers (2020).

13.5.3.2 Detection of Ochratoxins

A selective and sensitive immuno-logical technique has been devised for sensing ochratoxin A (OTA) (Wu et al. 2011). Aspergillus species commonly produce this mycotoxin. The method is comprised of a fluorescent immunoassay for the simultaneous detection of AFB_1 and OTA in food-stuffs utilizing antigen-modified magnetic nanoparticles (MNPs) as immuno-sensing probes and antibody-functionalized yttrium doped nanoparticles have been applied as polychromic signal probes. The fluorescent intensity of the sensor decreases with the increase in OTA concentration. The advantage of using MNPs is that these provide rapid detection, reducing the time for the detection process.

13.5.3.3 Detection of Bacterial Toxins

Food poisoning characterized by diarrhea, fever, and abdominal cramps is usually due to *Salmonella* species. At the same time, the one caused by *E. coli* shows the symptoms of bloody diarrhea, stomach spasms, nausea, and vomiting. A nano-biosensor has been synthesized by incorporating carbon nanotubes (CNTs) into monoclonal antibodies to select *Salmonella* sp. (Jain et al. 2012). The large surface area of CNTs is covalently coupled with monoclonal antibodies of *Salmonella* under a diimide-activated imidation coupling process followed by immobilization onto a glassy carbon electrode (GCE). The resistance, impedance, and charge transfer alterations before and after forming an antigen-antibody complex are considered detection signals. Likewise, an integrated multi-stage sensor has been developed for quick and selective detection of *E. coli* (Maurer et al. 2012). The assembly was accomplished by growing CNTs on a graphite substrate, adding AuNPs directly to the nanotube surface, and adding *E. coli*-specific thiolated RNA to the nanoparticles. Composite nanomaterials have the unique advantage of protecting the electrical properties of CNTs from AuNPs. The method's sensitivity was enhanced by increasing the surface area using carbon nanotubes as additional connection points. Subsequently, RNA-coated AuNPs increase the chances of *E. coli* detection by 189% compared to bare AuNPs.

13.5.4 Miscellaneous

Most regulated undesirable substances in feed and food can be found in 2002/32/EC and 2023/915/EC, respectively. Both documents establish maximum limits on several of these substances. Nanobiotechnological-based detection devices for a myriad of emerging contaminants and other substances that are to be avoided in food have been recently constructed (Table 13.2). For example, nanotechnology has made strides in the detection of allergenic biogenic amines, whose presence in foods (especially meat) that are related to food spoilage. Nanozymes and enzyme/bio/chemosensors have recently been used to detect such compounds *in situ* (Ahangari et al., 2021; Givanoudi et al., 2023) (Figure 13.5).

13.6 CONCLUSION

Food contamination is one of the severe threats of this era. Food contamination can occur at various stages, from raw material to finished packed products. The need of the hour is to detect and eradicate these contaminants from edible stuff to improve public health. For these purposes, nanomaterials have emerged as rising stars. However, a space exists in this field; only a few quick, selective, and on-site food contaminant sensors have been reported for highly hazardous species like arsenic, chromium, cyanides, fluorides, nitrophenols, organic dyes, etc. Research must be further conducted to fill this gap so that food safety can be ensured for all.

TABLE 13.2

Nanobiotechnological Applications of Additional Undesired Substances, Contaminants, and Residues

Detected Analyte	Food Matrix	Nanoparticle Used	Principle	Detection Limit	References
Terpenoid Aldehyde					
Gossipol	Cotton seeds	BSA-stabilized copper nanoclusters	Fluorescence	25 nmol L^{-1}	Xu et al. (2018)
Gossipol	Cotton seeds	Photoluminescence of a Yb^{3+} MOF, Yb-NH$_2$-TPDC	Near-infrared emitting lanthanide	25 µg mL^{-1}	Luo et al. (2020)
Allergens/Spoilage					
Biogenic amines	Wine	AuNPs prepared by laser ablation in liquid	Colorimetry	0.2 µmol L^{-1}	Lapenna et al. (2020)
Biogenic amines	Wine	Gold nanoarray islands	Competitive reaction in an evanescent field. Fluorescence of BA-conjugated quantum dots	63 a mol L^{-1}	Nguyen et al. (2021)
Biogenic amines	Meat	Spherical Au/Ag plasmonic nanoparticles	Colorimetry	0.1 µmol L^{-1}	Abbasi-Moayed et al. (2023)
Biogenic amines	Pork and meat samples (spoilage)	Polydiacetylene-based hydrogel beads	Colorimetric sensors		Jang et al. (2023)
Biogenic amines and sulfides	Raw and preserved meat, fish, crustaceans	*Bacillus subtilis* spore-based biosensor	Colorimetry. Inhibit electron transfer between Cu ion sites within CotA-laccase on the spore.	0.17 mg L^{-1}	Qin et al. (2023)
Parvalbumin	Fish	Gold aptasensor	Colorimetric and fluorescent	2.38 µg mL^{-1}	Wang et al. (2022c)
Detection of Virus and Bacteria					
White Sot Syndrome Virus (WSSV)	Shrimp	Gold aptasensor	Colorimetric	10^4 copies of WSSV	Etedali et al. (2022)
Salmonella enterica, *Listeria monocytogenes*, and *Escherichia coli*	Not reported	SiN micropore with polydimethylsiloxane	Microfluidic chip potential change	Not reported	Yang et al. (2022a)
Antimicrobials					
Sulfonamides	Meat	Iron oxide β-cyclodextrin-gold	Magnetic solid phase extraction/mass spectrometry	0.8–1.6 µg kg^{-1}	Yang et al. (2020a)
Ampicillin	Milk	Gold nanoparticles and CdTe QDs	Fluorimetry	18 nmol L^{-1}	Yu et al. (2022)
Nisin	Milk and egg white	tris(2,2-bipyridyl) dichlororuthenium (II) encapsulated in liposomes	Electrochemical	9.3 ng mL^{-1}	Luo et al. (2022)

(Continued)

TABLE 13.2 (*Continued*)

Nanobiotechnological Applications of Additional Undesired Substances, Contaminants, and Residues

Detected Analyte	Food Matrix	Nanoparticle Used	Principle	Detection Limit	References
Toxic Anions					
Cyanide	Water, cassava flour and bitter almonds	Anthraquinone	Colorimetric	29.48 μmol L^{-1}	Zhu et al. (2020)
Sulfite ion (SO_3^{2-})	Beer and fruit and vegetable juices and apple cider vinegar	Au metal-organic frameworks	Rayleigh scattering and basic fuchsin adsorption	0.0800 μmol L^{-1}	Lv et al. (2022)
Mycotoxins					
Aflatoxin B_1 and B_2	Vegetable oil	Iron oxide graphene nanocomposite	Magnetic solid phase extraction/fluorimetry	0.01 μg kg^{-1}	Yu et al. (2019)
Ochratoxin A	Water, corn and groundnut	Gold aptamer	Colorimetric	242, 545.45, and 95.69 ng mL^{-1}	Shahdeo et al. (2022)
Metallic Ions					
Silver	Mineral drinks	Erythrosine and Ag(I) aggregates	Rayleigh scattering	0.12 ng mL^{-1}	Wang et al. (2022b)
Lead	Water and milk	MXeneTi$_3$C$_2$ nanosheet	Fluorimetry	0.05 nmol L^{-1}	Zhi et al. (2022)
Lead	Noodles/serving i.e., noodles, seasoning spices, water	Nanodiamond@Bi$_2$MoO$_6$ composite	Preconcentrating before FAAS	1.75 μg L^{-1}	Arain et al. (2023)
Lead and manganese	Water, tea, cinnamon	NiCo$_2$O$_4$@ZnCo$_2$O$_4$	Solid phase micro-extraction	1.7 and 4.0 μg L^{-1}	Tokalıoğlu et al. (2023)
PAHs					
Polycyclic Aromatic hydrocarbons	Smoked pork, wild fish, grilled fish, smoked bacon, coffee, and water.	Iron oxide Benzidine and 1,3,5-triformylphloroglucinol	Magnetic solid phase extraction/diode array	0.83–11.7 ng L^{-1}	Li et al. (2018)
Xanthine/Methylxanthines					
Guanine	Beer	tris(2,2-bipyridyl) dichlororuthenium (II)	Electrochemical, mesoporous silica film	0.058 μmol L^{-1}	Yang et al. (2022b)
Caffeine, theobromine, and paraxanthine	Tea, coffee and chocolate/human serum	Silver and gold citrate and borohydride colloids	surface-enhanced Raman scattering	Not reported	Alharbi et al. (2015)
Colorants					
Tartrazine	Beverages	N,Cl doped quantum dots	Fluorescence	48 nmol L^{-1}	Yang et al., 2020b

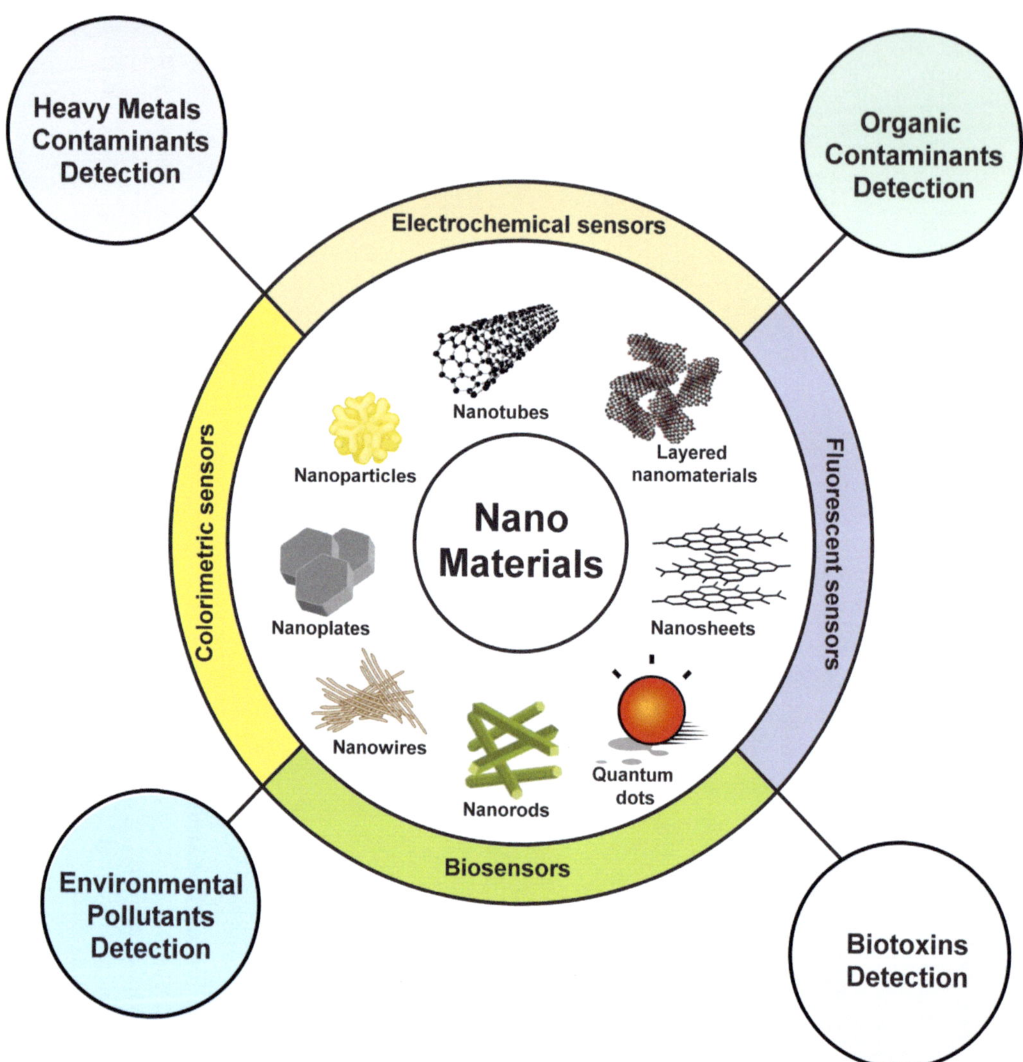

FIGURE 13.5 Types of the sensors.

BIBLIOGRAPHY

Abbasi-Moayed, S., A. Orouji and M. Hormozi-Nezhad (2023). Multiplex detection of biogenic amines for meat freshness monitoring using nanoplasmonic colorimetric sensor array. *Biosensors* **13**(8): 803.

Afkhami, A., H. Bagheri, H. Khoshsafar, M. Saber-Tehrani, M. Tabatabaee and A. Shirzadmehr (2012). Simultaneous trace-levels determination of Hg (II) and Pb (II) ions in various samples using a modified carbon paste electrode based on multiwalled carbon nanotubes and a new synthesized Schiff base. *Analytica Chimica Acta* **746**: 98–106.

Ahangari, H., S. Kurbanoglu, A. Ehsani and B. Uslu (2021). Latest trends for biogenic amines detection in foods: Enzymatic biosensors and nanozymes applications. *Trends in Food Science & Technology* **112**: 75–87.

Ahmad, A., A. Paschero and E. Moore (2011). Amperometric immunosensors for screening of polycyclic aromatic hydrocarbons in water. *Journal of Physics: Conference Series*, IOP Publishing.

Ahmadian, E., A. Eftekhari, T. Kavetskyy, A. Y. Khosroushahi, V. A. Turksoy and R. Khalilov (2020). Effects of quercetin loaded nanostructured lipid carriers on the paraquat-induced toxicity in human lymphocytes. *Pesticide Biochemistry and Physiology* **167**: 104586.

Alharbi, O., Y. Xu and R. Goodacre (2015). Simultaneous multiplexed quantification of caffeine and its major metabolites theobromine and paraxanthine using surface-enhanced Raman scattering *Analytical and Bioanalytical Chemistry* **407**(27): 8253–8261.

Anfossi, L., F. Di Nardo, A. Russo, S. Cavalera, C. Giovannoli, G. Spano, S. Baumgartner, K. Lauter and C. Baggiani (2019). Silver and gold nanoparticles as multi-chromatic lateral flow assay probes for the detection of food allergens. *Analytical and Bioanalytical Chemistry* **411**: 1905–1913.

Anfossi, L., G. Giraudi, C. Tozzi, C. Giovannoli, C. Baggiani and A. Vanni (2004). Development of a non-competitive immunoassay for monitoring DDT, its metabolites and analogues in water samples. *Analytica Chimica Acta* **506**(1): 87–95.

Aragay, G. and A. Merkoci (2012). Nanomaterials application in electrochemical detection of heavy metals. *Electrochimica Acta* **84**: 49–61.

Arain, M. B., H. E. H. Ahmed and M. Soylak (2023). Dispersive solid phase microextraction (DSP-µE) by using nanodiamond@Bi_2MoO_6 composite for the separation-preconcentration of Pb(II) in food and water samples. *Microchemical Journal*: 109495.

Bahmani, R., D. Kim, M. Modareszadeh, A. J. Thompson, J. H. Park, H. H. Yoo and S. Hwang (2020). The mechanism of root growth inhibition by the endocrine disruptor bisphenol A (BPA). *Environmental Pollution* **257**: 113516.

Bankole, M. T., A. S. Abdulkareem, I. A. Mohammed, S. S. Ochigbo, J. O. Tijani, O. K. Abubakre and W. D. Roos (2019). Selected heavy metals removal from electroplating wastewater by purified and polyhydroxylbutyrate functionalized carbon nanotubes adsorbents. *Scientific Reports* **9**(1): 4475.

Batool, M. and H. M. Junaid (2022). Organic-inorganic nanohybrid-based sensors for metal ions sensing. In *Hybrid Nanomaterials: Biomedical, Environmental and Energy Applications*, Springer: 201–225.

Batool, M. and H. M. Junaid (2023). Environmental remediation of heavy metals through MXene composites. In *Handbook of Functionalized Nanostructured MXenes: Synthetic Strategies and Applications from Energy to Environment Sustainability*, Springer: 229–248.

Batool, M., H. M. Junaid, S. Tabassum, F. Kanwal, K. Abid, Z. Fatima and A. T. Shah (2022a). Metal ion detection by carbon dots-a review. *Critical Reviews in Analytical Chemistry* **52**(4): 756–767.

Batool, M., Z. Afzal, H. M. Junaid, A. R. Solangi and A. Hassan (2022b). Sulfonamides as optical chemosensors. *Critical Reviews in Analytical Chemistry*: 1–28.

Batra, B., S. Lata, M. Sharma and C. Pundir (2013). An acrylamide biosensor based on immobilization of hemoglobin onto multiwalled carbon nanotube/copper nanoparticles/polyaniline hybrid film. *Analytical Biochemistry* **433**(2): 210–217.

Botchkareva, A. E., F. Fini, S. Eremin, J. V. Mercader, A. Montoya and S. Girotti (2002). Development of a heterogeneous chemiluminescent flow immunoassay for DDT and related compounds. *Analytica Chimica Acta* **453**(1): 43–52.

Bülbül, G., A. Hayat and S. Andreescu (2015). Portable nanoparticle-based sensors for food safety assessment. *Sensors* **15**(12): 30736–30758.

Burks, W., R. Helm, S. Stanley and G. A. Bannon (2001). Food allergens. *Current Opinion in Allergy and Clinical Immunology* **1**(3): 243–248.

Carrillo-Carrión, C., B. M. Simonet and M. Valcárcel (2009). Carbon nanotube-quantum dot nanocomposites as new fluorescence nanoparticles for the determination of trace levels of PAHs in water. *Analytica Chimica Acta* **652**(1–2): 278–284.

Chen, H., W. Hu and C. M. Li (2015). Colorimetric detection of mercury (II) based on 2, 2′-bipyridyl induced quasi-linear aggregation of gold nanoparticles. *Sensors and Actuators B: Chemical* **215**: 421–427.

Chobtang, J., I. J. De Boer, R. L. Hoogenboom, W. Haasnoot, A. Kijlstra and B. G. Meerburg (2011). The need and potential of biosensors to detect dioxins and dioxin-like polychlorinated biphenyls along the milk, eggs and meat food chain. *Sensors* **11**(12): 11692–11716.

Choi, Y. J. and L. S. Lee (2017). Aerobic soil biodegradation of bisphenol (BPA) alternatives bisphenol S and bisphenol AF compared to BPA. *Environmental Science & Technology* **51**(23): 13698–13704.

Chow, C-F. (2020). Biogenic amines- and sulfides-responsive gold nanoparticles for real-time visual detection of raw meat, fish, crustaceans, and preserved meat. *Food Chemistry* **311**: 125908.

Del Carlo, M., M. Di Marcello, M. Giuliani, M. Sergi, A. Pepe and D. Compagnone (2012). Detection of benzo (a) pyrene photodegradation products using DNA electrochemical sensors. *Biosensors and Bioelectronics* **31**(1): 270–276.

Dhiman, V. and N. Kondal (2021). ZnO nanoadsorbents: A potent material for removal of heavy metal ions from wastewater. *Colloid and Interface Science Communications* **41**: 100380.

Ding, N., H. Zhao, W. Peng, Y. He, Y. Zhou, L. Yuan and Y. Zhang (2012). A simple colorimetric sensor based on anti-aggregation of gold nanoparticles for Hg2+ detection. *Colloids and Surfaces A: Physicochemical and Engineering Aspects* **395**: 161–167.

Dong, C., H. Qian, N. Fang and J. Ren (2006). Study of fluorescence quenching and dialysis process of CdTe quantum dots, using ensemble techniques and fluorescence correlation spectroscopy. *The Journal of Physical Chemistry B* **110**(23): 11069–11075.

El Mhammedi, M., M. Achak, M. Hbid, M. Bakasse, T. Hbid and A. Chtaini (2009). Electrochemical determination of cadmium (II) at platinum electrode modified with kaolin by square wave voltammetry. *Journal of Hazardous Materials* **170**(2–3): 590–594.

Engraff, M., C. Solere, K. E. Smith, P. Mayer and I. Dahllöf (2011). Aquatic toxicity of PAHs and PAH mixtures at saturation to benthic amphipods: Linking toxic effects to chemical activity. *Aquatic Toxicology* **102**(3–4): 142–149.

Etedali, P., M. Behbahani, H. Mohabatkar and G. Dini (2022). Field-usable aptamer-gold nanoparticles-based colorimetric sensor for rapid detection of white spot syndrome virus in shrimp. *Aquaculture* **548**(2): 737628.

Fu, H., O. Hu, L. Xu, Y. Fan, Q. Shi, X. Guo, W. Lan, T. Yang, S. Xie and Y. She (2019). Simultaneous recognition of species, quality grades, and multivariate calibration of antioxidant activities for 12 famous green teas using mid-and near-infrared spectroscopy coupled with chemometrics. *Journal of Analytical Methods in Chemistry* **2019**: 4372395.

Fu, H.-Y., H.-D. Li, B. Wang, J.-L. Cai, J.-W. Guo, H.-P. Cui, X.-B. Zhang and Y.-J. Yu (2016). Quantification of acid metabolites in complex plant samples by using second-order calibration coupled with GC-mass spectrometry detection to resolve the influence of seriously overlapped chromatographic peaks. *Analytical Methods* **8**(4): 747–755.

Gilani, S. R., M. Batool, S. R. Ali Zaidi, Z. Mahmood, A. A. Bhatti and A. I. Durrani (2015). Central nervous system (CNS) toxicity caused by metal poisoning: Brain as a target organ. *Pakistan Journal of Pharmaceutical Sciences* **28**(4).

Givanoudi, S., M. Heyndrickx, T. Depuydt, M. Khorshid, J. Robbens and P. Wagner (2023). A review on bio- and chemosensors for the detection of biogenic amines in food safety applications: The status in 2022. *Sensors* **23**: 613.

Gong, J., T. Zhou, D. Song and L. Zhang (2010). Monodispersed Au nanoparticles decorated graphene as an enhanced sensing platform for ultrasensitive stripping voltammetric detection of mercury (II). *Sensors and Actuators B: Chemical* **150**(2): 491–497.

Grillo, R., A. E. Pereira, C. S. Nishisaka, R. De Lima, K. Oehlke, R. Greiner and L. F. Fraceto (2014). Chitosan/tripolyphosphate nanoparticles loaded with paraquat herbicide: An environmentally safer alternative for weed control. *Journal of Hazardous Materials* **278**: 163–171.

Guo, J., Y. Chai, R. Yuan, Z. Song and Z. Zou (2011a). Lead (II) carbon paste electrode based on derivatized multiwalled carbon nanotubes: Application to lead content determination in environmental samples. *Sensors and Actuators B: Chemical* **155**(2): 639–645.

Guo, Z., J. Ren, J. Wang and E. Wang (2011b). Single-walled carbon nanotubes based quenching of free FAM-aptamer for selective determination of ochratoxin A. *Talanta* **85**(5): 2517–2521.

Haes, A. J., S. Zou, G. C. Schatz and R. P. Van Duyne (2004). Nanoscale optical biosensor: Short range distance dependence of the localized surface plasmon resonance of noble metal nanoparticles. *The Journal of Physical Chemistry B* **108**(22): 6961–6968.

Haripriyan, U., K. Gopinath and J. Arun (2022). Chitosan based nano adsorbents and its types for heavy metal removal: A mini review. *Materials Letters* **312**: 131670.

Hashemi-Moghaddam, H. and M. Ahmadifard (2016). Novel molecularly-imprinted solid-phase microextraction fiber coupled with gas chromatography for analysis of furan. *Talanta* **150**: 148–154.

Hau, A. K.-c., T. H. Kwan and P. K.-t. Li (2009). Melamine toxicity and the kidney. *Journal of the American Society of Nephrology* **20**(2): 245–250.

Hu, J., D. Mao, P. Duan, K. Li, Y. Lin, X. Wang and Y. Piao (2022). Green synthesis of ZnO/BC nanohybrid for fast and sensitive detection of Bisphenol A in water. *Chemosensors* **10**(5): 163.

Huang, L., D. W. Sun, H. Pu and Q. Wei (2019). Development of nanozymes for food quality and safety detection: Principles and recent applications. *Comprehensive Reviews in Food Science and Food Safety* **18**(5): 1496–1513.

Huang, M.-R., Y.-B. Ding and X.-G. Li (2014). Combinatorial screening of potentiometric Pb (II) sensors from polysulfoaminoanthraquinone solid ionophore. *ACS Combinatorial Science* **16**(3): 128–138.

Hunter Jr, R. E., A. M. Riederer and P. B. Ryan (2010). Method for the determination of organophosphorus and pyrethroid pesticides in food via gas chromatography with electron-capture detection. *Journal of Agricultural and Food Chemistry* **58**(3): 1396–1402.

Jain, S., S. R. Singh, D. W. Horn, V. A. Davis, M. Ram and S. Pillai (2012). Development of an antibody functionalized carbon nanotube biosensor for foodborne bacterial pathogens. *Journal of Biosensors and Bioelectronics* **11**: 002.

Jang, S., S. U. Son, J. Kim, H. Kim, J. Lim, S. B. Seo, B. Kang, J. Jung, S. Seo and E.-K. Lim (2022). Polydiacetylene-based hydrogel beads as colorimetric sensors for the detection of biogenic amines in spoiled meat. *Food Chemistry* **403**: 134317.

Javanbakht, M., F. Divsar, A. Badiei, M. R. Ganjali, P. Norouzi, G. M. Ziarani, M. Chaloosi and A. A. Jahangir (2009). Potentiometric detection of mercury (II) ions using a carbon paste electrode modified with substituted thiourea-functionalized highly ordered nanoporous silica. *Analytical Sciences* **25**(6): 789–794.

Jing, X., X. Huang, Y. Zhang, M. Wang, H. Xue, X. Wang and L. Jia (2022). Cyclodextrin-based dispersive liquid-liquid microextraction for the determination of fungicides in water, juice, and vinegar samples via HPLC. *Food Chemistry* **367**: 130664.

Junaid, H. M., A. R. Solangi and M. Batool (2021). Carbon dots as naked eye sensors. *Analyst* **146**(8): 2463–2474.

Junaid, H. M. and M. Batool (2023). Optical chemosensing of arsenite and ferrum ions using 4-(Benzylideneamino)-1,5-dimethyl-2-phenyl-1,2-dihydropyrazol-3-one. Chemical Papers.

Junaid, H. M., M. Batool, F. W. Harun, M. S. Akhter and N. Shabbir (2022). Naked eye chemosensing of anions by Schiff bases. *Critical Reviews in Analytical Chemistry* **52**(3): 463–480.

Junaid, H. M., S. Munir and M. Batool (2023). Unraveling the chemistry of ionic liquid mediated carbon dots as sensing probe – A review. *Trends in Environmental Analytical Chemistry*: e00214.

Kasai, A., N. Hiramatsu, K. Hayakawa, J. Yao, S. Maeda and M. Kitamura (2006). High levels of dioxin-like potential in cigarette smoke evidenced by in vitro and in vivo biosensing. *Cancer Research* **66**(14): 7143–7150.

Kleefisch, G., C. Kreutz, J. Bargon, G. Silva and C. A. Schalley (2004). Quartz microbalance sensor for the detection of acrylamide. *Sensors* **4**(9): 136–146.

Ko, S., S. Gunasekaran and J. Yu (2010). Self-indicating nanobiosensor for detection of 2, 4-dinitrophenol. *Food Control* **21**(2): 155–161.

Krajewska, A., J. Radecki and H. Radecka (2008). A voltammetric biosensor based on glassy carbon electrodes modified with single-walled carbon nanotubes/hemoglobin for detection of acrylamide in water extracts from potato crisps. *Sensors* **8**(9): 5832–5844.

Kurniadie, D., U. Umiyati, R. Widianto and H. Kato-Noguchi (2022). Effect of chitosan molecules on paraquat herbicidal efficacy under simulated rainfall conditions. *Agronomy* **12**(7): 1666.

Kuswandi, B., D. Futra and L. Heng (2017). Nanosensors for the detection of food contaminants. In *Nanotechnology Applications in Food*, Elsevier: 307–333.

Lapenna, A., M. Dell'Aglio, G. Palazzo and A. Mallardi (2020). "Naked" gold nanoparticles as colorimetric reporters for biogenic amine detection. *Colloids and Surfaces A: Physicochemical and Engineering Aspects* **600**: 124903.

Li, C., C. Zeng, Z. Chen, Y. Jiang, H. Yao, Y. Yang and W.-T. Wong (2020). Luminescent lanthanide metal-organic framework test strip for immediate detection of tetracycline antibiotics in water. *Journal of Hazardous Materials* **384**: 121498.

Li, J., J.-J. Zhu and K. Xu (2014a). Fluorescent metal nanoclusters: From synthesis to applications. *TrAC Trends in Analytical Chemistry* **58**: 90–98.

Li, L., B. Li, D. Cheng and L. Mao (2010). Visual detection of melamine in raw milk using gold nanoparticles as colorimetric probe. *Food Chemistry* **122**(3): 895–900.

Li, N., D. Wu, N. Hu, G. Fan, X. Li, J. Sun, X. Chen, Y. Suo, G. Li and Y. Wu (2018). Effective enrichment and detection of trace polycyclic aromatic hydrocarbons in food samples based on magnetic covalent organic framework hybrid microspheres. *Journal of Agricultural and Food Chemistry* **66**(13): 3572–3580.

Li, S., G. Zhang, P. Wang, H. Zheng and Y. Zheng (2016). Microwave-enhanced Mn-Fenton process for the removal of BPA in water. *Chemical Engineering Journal* **294**: 371–379.

Li, Y.-L., Y.-M. Leng, Y.-J. Zhang, T.-H. Li, Z.-Y. Shen and A.-G. Wu (2014b). A new simple and reliable Hg2+ detection system based on anti-aggregation of unmodified gold nanoparticles in the presence of O-phenylenediamine. *Sensors and Actuators B: Chemical* **200**: 140–146.

Liao, J.-Y. and H. Li (2010). Lateral flow immunodipstick for visual detection of aflatoxin B 1 in food using immuno-nanoparticles composed of a silver core and a gold shell. *Microchimica Acta* **171**: 289–295.

Lin, H., Z. Chen, Q. Lu, Q. Cai and C. A. Grimes (2010). A wireless and sensitive sensing detection of polycyclic aromatic hydrocarbons using humic acid-coated magnetic Fe$_3$O$_4$ nanoparticles as signal-amplifying tags. *Sensors and Actuators B: Chemical* **146**(1): 154–159.

Lisa, M., R. Chouhan, A. Vinayaka, H. Manonmani and M. Thakur (2009). Gold nanoparticles based dipstick immunoassay for the rapid detection of dichlorodiphenyltrichloroethane: An organochlorine pesticide. *Biosensors and Bioelectronics* **25**(1): 224–227.

Liu, F., X. Yang and S. Sun (2011). Determination of melamine based on electrochemiluminescence of Ru (bpy) 3 2+ at bare and single-wall carbon nanotube modified glassy carbon electrodes. *Analyst* **136**(2): 374–378.

Loui, A. and S. Chiang (2018). An experimental study of furan adsorption and decomposition on vicinal palladium surfaces using scanning tunneling microscopy. *Surface Science* **670**: 13–22.

Luo, H., Y. Huang, K. Lai, B. A. Rasco and Y. Fan (2016). Surface-enhanced Raman spectroscopy coupled with gold nanoparticles for rapid detection of phosmet and thiabendazole residues in apples. *Food Control* **68**: 229–235.

Luo, T-Y., P. Das, D. L. White, C. Liu, A. Star and N. L. Rosi (2020). Luminescence "turn-on" detection of gossypol using Ln³⁺-based metal-organic frameworks and Ln³⁺ salts. *Journal of the Amercian Chemical Society* **142**(6): 2897–2904.

Luo, X., T. Zhang, H. Tang and J. Liu (2022). Novel electrochemical and electrochemiluminescence dual-modality sensing platform for sensitive determination of antimicrobial peptides based on probe encapsulated liposome and nanochannel array electrode. *Frontiers in Nutrition* **9**: 962736.

Lv, M., Y. Liu, J. Geng, X. Kou, Z. Xin and D. Yang (2018). Engineering nanomaterials-based biosensors for food safety detection. *Biosensors and Bioelectronics* **106**: 122–128.

Lv, X., Y. Liu, S. Zhou, M. Wu, Z. Jiang and G. Wen (2022). A stable and sensitive Au metal organic frameworks resonance Rayleigh scattering nanoprobe for detection of SO_3^{2-} in food based on fuchsin addition reaction. *Frontiers in Nutrition* **9**: 1019429.

Ma, W., J. Li, X. Li and H. Liu (2022). Enrichment of diamide insecticides from environmental water samples using metal-organic frameworks as adsorbents for determination by liquid chromatography tandem mass spectrometry. *Journal of Hazardous Materials* **422**: 126839.

Mailu, S. N., T. T. Waryo, P. M. Ndangili, F. R. Ngece, A. A. Baleg, P. G. Baker and E. I. Iwuoha (2010). Determination of anthracene on Ag-Au alloy nanoparticles/overoxidized-polypyrrole composite modified glassy carbon electrodes. *Sensors* **10**(10): 9449–9465.

Martins, A. R., M. Talhavini, M. L. Vieira, J. J. Zacca and J. W. B. Braga (2017). Discrimination of whisky brands and counterfeit identification by UV-Vis spectroscopy and multivariate data analysis. *Food Chemistry* **229**: 142–151.

Mascini, M., A. Macagnano, D. Monti, M. Del Carlo, R. Paolesse, B. Chen, P. Warner, A. D'Amico, C. Di Natale and D. Compagnone (2004). Piezoelectric sensors for dioxins: A biomimetic approach. *Biosensors and Bioelectronics* **20**(6): 1203–1210.

Mascini, M., A. Macagnano, G. Scortichini, M. Del Carlo, G. Diletti, A. d'Amico, C. Di Natale and D. Compagnone (2005). Biomimetic sensors for dioxins detection in food samples. *Sensors and Actuators B: Chemical* **111**: 376–384.

Maurer, E. I., K. K. Comfort, S. M. Hussain, J. J. Schlager and S. M. Mukhopadhyay (2012). Novel platform development using an assembly of carbon nanotube, nanogold and immobilized RNA capture element towards rapid, selective sensing of bacteria. *Sensors* **12**(6): 8135–8144.

Meimaridou, A., W. Haasnoot, L. Noteboom, D. Mintzas, J. Pulkrabova, J. Hajslová and M. W. Nielen (2010). Color encoded microbeads-based flow cytometric immunoassay for polycyclic aromatic hydrocarbons in food. *Analytica Chimica Acta* **672**(1–2): 9–14.

Mercante, L. A., L. E. Iwaki, V. P. Scagion, O. N. Oliveira Jr, L. H. Mattoso and D. S. Correa (2021). Electrochemical detection of bisphenol a by tyrosinase immobilized on electrospun nanofibers decorated with gold nanoparticles. *Electrochem* **2**(1): 41–49.

Mirzajani, H., C. Cheng, J. Wu, J. Chen, S. Eda, E. N. Aghdam and H. B. Ghavifekr (2017). A highly sensitive and specific capacitive aptasensor for rapid and label-free trace analysis of Bisphenol A (BPA) in canned foods. *Biosensors and Bioelectronics* **89**: 1059–1067.

Momeni, A., M. Rostami-Nejad, R. Salarian, M. Rabiee, E. Aghamohammadi, M. R. Zali, N. Rabiee, F. R. Tay and P. Makvandi (2022). Gold-based nanoplatform for a rapid lateral flow immunochromatographic test assay for gluten detection. *BMC Biomedical Engineering* **4**(1): 5.

Nguyen, H. T. T., S. Lee, J. Lee, J-H. Ha and S. H. Kang (2021). Ultrasensitive biogenic amine sensor using an enhanced multiple nanoarray chip based on competitive reactions in an evanescent field. *Sensors and Actuators B: Chemical* **345**: 130354.

Ntihuga, J. (2006). Biosensor to detect heavy metals in waste water. In Proceedings from the International Conference on Advances in Engineering and Technology, Elsevier.

Oracz, J., E. Nebesny and D. Żyżelewicz (2011). New trends in quantification of acrylamide in food products. *Talanta* **86**: 23–34.

Paolesse, R., S. Nardis, D. Monti, M. Stefanelli and C. Di Natale (2017). Porphyrinoids for chemical sensor applications. *Chemical Reviews* **117**(4): 2517–2583.

Pereira, C. G., J. Andrade, T. Ranquine, I. N. de Moura, R. A. da Rocha, M. A. M. Furtado, M. J. V. Bell and V. Anjos (2018). Characterization and detection of adulterated whey protein supplements using stationary and time-resolved fluorescence spectroscopy. *LWT* **97**: 180–186.

Pogačnik, L. and M. Franko (2003). Detection of organophosphate and carbamate pesticides in vegetable samples by a photothermal biosensor. *Biosensors and Bioelectronics* **18**(1): 1–9.

Qin, Y., W. Ke, A. Faheem, Y. Ye and Y. Hu (2023). A rapid and naked-eye on-site monitoring of biogenic amines in foods spoilage. *Food Chemistry* **404**: 134581.

Sebők, D. and I. Dékány (2015). ZnO$_2$ nanohybrid thin film sensor for the detection of ethanol vapour at room temperature using reflectometric interference spectroscopy. *Sensors and Actuators B: Chemical* **206**: 435–442.

Shahdeo, D., A. Ali Khan, A. M. Alanazi, V. K. Bajpar, S. Shukla and S. Gandhi (2022). Molecular diagnostic of ochratoxin A with specific aptamers in corn and groundnut via fabrication of a microfluidic device. *Frontiers in Nutrition* **9**: 851787.

Sharma, R., K. Ragavan, M. Thakur and K. Raghavarao (2015). Recent advances in nanoparticle based aptasensors for food contaminants. *Biosensors and Bioelectronics* **74**: 612–627.

Shen, H., Y. Huang, R. Wang, D. Zhu, W. Li, G. Shen, B. Wang, Y. Zhang, Y. Chen and Y. Lu (2013). Global atmospheric emissions of polycyclic aromatic hydrocarbons from 1960 to 2008 and future predictions. *Environmental Science & Technology* **47**(12): 6415–6424.

Shen, X., Y. Cui, Y. Pang and H. Qian (2012). Graphene oxide nanoribbon and polyhedral oligomeric silsesquioxane assembled composite frameworks for pre-concentrating and electrochemical sensing of 1-hydroxypyrene. *Electrochimica Acta* **59**: 91–99.

Silva, N., M. J. Matos, A. Karmali and M. M. Rocha (2011). An electrochemical biosensor for acrylamide determination: Merits and limitations. *Portugaliae Electrochimica Acta* **29**(5): 361–373.

Singh, S., A. Numan, Y. Zhan, V. Singh, T. Van Hung and N. D. Nam (2020). A novel highly efficient and ultrasensitive electrochemical detection of toxic mercury (II) ions in canned tuna fish and tap water based on a copper metal-organic framework. *Journal of Hazardous Materials* **399**: 123042.

Soldatkin, O., I. Kucherenko, V. Pyeshkova, A. Kukla, N. Jaffrezic-Renault, A. El'Skaya, S. Dzyadevych and A. Soldatkin (2012). Novel conductometric biosensor based on three-enzyme system for selective determination of heavy metal ions. *Bioelectrochemistry* **83**: 25–30.

Sonawane, S. K., S. S. Arya, J. G. LeBlanc and N. Jha (2014). Use of nanomaterials in the detection of food contaminants. *European Journal of Food Research & Review* **4**(4): 301.

Song, K.-M., E. Jeong, W. Jeon, M. Cho and C. Ban (2012). Aptasensor for ampicillin using gold nanoparticle based dual fluorescence-colorimetric methods. *Analytical and Bioanalytical Chemistry* **402**: 2153–2161.

Stobiecka, A., H. Radecka and J. Radecki (2007). Novel voltammetric biosensor for determining acrylamide in food samples. *Biosensors and Bioelectronics* **22**(**9–10**): 2165–2170.

Sun, X., J. Ji, D. Jiang, X. Li, Y. Zhang, Z. Li and Y. Wu (2013). Development of a novel electrochemical sensor using pheochromocytoma cells and its assessment of acrylamide cytotoxicity. *Biosensors and Bioelectronics* **44**: 122–126.

Sun, X., S. Du and X. Wang (2012). Amperometric immunosensor for carbofuran detection based on gold nanoparticles and PB-MWCNTs-CTS composite film. *European Food Research and Technology* **235**: 469–477.

Svechkarev, D. and A. M. Mohs (2019). Organic fluorescent dye-based nanomaterials: Advances in the rational design for imaging and sensing applications. *Current Medicinal Chemistry* **26**(21): 4042–4064.

Ting, S. L., S. J. Ee, A. Ananthanarayanan, K. C. Leong and P. Chen (2015). Graphene quantum dots functionalized gold nanoparticles for sensitive electrochemical detection of heavy metal ions. *Electrochimica Acta* **172**: 7–11.

Tokalıoğlu, Ş., S. T. H. Moghaddam, Y. Yılmaz and Ş. Patat (2023). NiCo2O4@ZnCo2O4 nanomaterial for selective and fast dispersive solid phase micro-extraction of manganese and lead in water, tea and cinnamon samples followed by FAAS determination. *Microchemical Journal* 195: 109515.

Torres, A. M. and J. Correa (2021). First-principles calculation of volatile organic compound adsorption on carbon nanotubes: Furan as case of study. *Carbon Letters* 31: 1061–1070.

Wan, H., Q. Sun, H. Li, F. Sun, N. Hu and P. Wang (2015). Screen-printed gold electrode with gold nanoparticles modification for simultaneous electrochemical determination of lead and copper. *Sensors and Actuators B: Chemical* **209**: 336–342.

Wang, C.-Y., C.-C. Wang, X.-W. Zhang, X.-Y. Ren, B. Yu, P. Wang, Z.-X. Zhao and H. Fu (2022a). A new Eu-MOF for ratiometrically fluorescent detection toward quinolone antibiotics and selective detection toward tetracycline antibiotics. *Chinese Chemical Letters* **33**(3): 1353–1357.

Wang, J., S. Liu and W. Shen (2022b). Absorption and resonance Rayleigh scattering spectra of Ag(I) and erythrosin system and their analytical application in food safety. *Frontiers in Nutrition* **9**: 900215.

Wang, P., H. Li, M. M. Hassan, Z. Guo, Z.-Z. Zhang and Q. Chen (2019a). Fabricating an acetylcholinesterase modulated UCNPs-Cu^{2+} fluorescence biosensor for ultrasensitive detection of organophosphorus pesticides-diazinon in food. *Journal of Agricultural and Food Chemistry* **67**(14): 4071–4079.

Wang, Q., Q. Yin, Y. Fan, L. Zhang, Y. Xu, O. Hu, X. Guo, Q. Shi, H. Fu and Y. She (2019b). Double quantum dots-nanoporphyrin fluorescence-visualized paper-based sensors for detecting organophosphorus pesticides. *Talanta* **199**: 46–53.

Wang, Y., H. Li, J. Zhou, Q. Qi and L. Fu (2022c). A colorimetric and fluorescent gold nanoparticle-based dual-mode aptasensor for parvalbumin detection. *Microchemical Journal* **159**: 105413

Warriner, K., S. M. Reddy, A. Namvar and S. Neethirajan (2014). Developments in nanoparticles for use in biosensors to assess food safety and quality. *Trends in Food Science & Technology* **40**(2): 183–199.

Wu, S., N. Duan, C. Zhu, X. Ma, M. Wang and Z. Wang (2011). Magnetic nanobead-based immunoassay for the simultaneous detection of aflatoxin B1 and ochratoxin A using upconversion nanoparticles as multicolor labels. *Biosensors and Bioelectronics* **30**(1): 35–42.

Xu, S., K. Zhou, D. Fan and L. Ma (2018). Highly sensitive and selective fluorescent detection of gossypol based on BSA-stabilized copper nanoclusters. *Molecules* **24**(1): 95.

Yan, C., Q. Wang, Q. Yang and W. Wu (2020). Recent advances in aflatoxins detection based on nanomaterials. *Nanomaterials* **10**(9): 1626.

Yang, L., T. Zhang, H. Zhou, F. Yan and Y. Liu (2020a). Silica nanochannels boosting Ru(bpy)$_3^{2+}$-mediated electrochemical sensor for the detection of guanine in beer and pharmaceutical samples. *Frontiers in Nutrition* **9**: 987442.

Yang, T., Z. Luo, R. A. Wu, L. Li, Y. Xu, T. Ding and X. Lin (2022a). Rapid and label-free identification of single foodborne pathogens using microfluidic pore sensors. *Frontiers in Nutrition* **9**: 959317.

Yang, X., J. Xu, N. Luo, F. Tang, M. Zhang and B. Zhao (2020b). N, Cl co-doped fluorescent carbon dots as nanoprobe for detection of tartrazine in beverages. *Food Chemistry* **310**: 125832.

Yang, Y., G. Li, D. Wu, A. Wen, Y. Wu and X. Zhou (2022b). β-Cyclodextrin-/AuNPs-functionalized covalent organic framework-based magnetic sorbent for solid phase extraction and determination of sulfonamides. *Microchimica Acta* **187**: 278.

Yoshioka, N. and K. Ichihashi (2008). Determination of 40 synthetic food colors in drinks and candies by high-performance liquid chromatography using a short column with photodiode array detection. *Talanta* **74**(5): 1408–1413.

Yu, L., F. Ma, L. Zang and P. Li (2019). Determination of aflatoxin B1 and B2 in vegetable oils using Fe^{3}O^4/rGO magnetic solid phase extraction coupled with high-performance liquid chromatography fluorescence with post-column photochemical derivatization. *Toxins* **11**(11): 621.

Yu, W., A. Hao, Y. Mei, Y. Yang and C. Dai (2022). A turn-on fluorescent aptasensor for ampicillin detection based on gold nanoparticles and CdTe QDs. *Microchemical Journal* **179**: 107454.

Yuksel, N., A. Kose and M. F. Fellah (2022). Pd, Ag and Rh doped (8, 0) single-walled carbon nanotubes (SWCNTs): A DFT study on furan adsorption and detection. *Surface Science* **715**: 121939.

Zhang, M., H. Ping, X. Cao, H. Li, F. Guan, C. Sun and J. Liu (2012). Rapid determination of melamine in milk using water-soluble CdTe quantum dots as fluorescence probes. *Food Additives & Contaminants: Part A* **29**(3): 333–344.

Zhang, T., W. Wang, Y. Zhao, H. Bai, T. Wen, S. Kang, G. Song, S. Song and S. Komarneni (2021). Removal of heavy metals and dyes by clay-based adsorbents: From natural clays to 1D and 2D nanocomposites. *Chemical Engineering Journal* **420**: 127574.

Zhang, X., D. Wu, X. Zhou, Y. Yu, J. Liu, N. Hu, H. Wang, G. Li and Y. Wu (2019). Recent progress in the construction of nanozyme-based biosensors and their applications to food safety assay. *TrAC Trends in Analytical Chemistry* **121**: 115668.

Zhang, X., G. Li, D. Wu, J. Liu and Y. Wu (2020a). Recent advances on emerging nanomaterials for controlling the mycotoxin contamination: From detection to elimination. *Food Frontiers* **1**(4): 360–381.

Zhang, X., T. Cheng, C. Chen, L. Wang, Q. Deng, G. Chen and C. Ye (2020b). Synthesis of a novel magnetic nano-zeolite and its application as an efficient heavy metal adsorbent. *Materials Research Express* **7**(8): 085007.

Zhi, S., Q. Wei, C. Zhang, C. Yi, C. Li and Z. Jiang (2022). MXene catalytic amplification-fluorescence/absorption dimode aptamer sensor for the detection of trace Pb2+ in milk. *Frontiers in Nutrition* **9**: 1008620.

Zhou, P., O. Hu, H. Fu, L. Ouyang, X. Gong, P. Meng, Z. Wang, M. Dai, X. Guo and Y. Wang (2019). UPLC-Q-TOF/MS-based untargeted metabolomics coupled with chemometrics approach for Tieguanyin tea with seasonal and year variations. *Food Chemistry* **283**: 73–82.

Zhou, Y., H. Dong, L. Liu, M. Li, K. Xiao and M. Xu (2014). Selective and sensitive colorimetric sensor of mercury (II) based on gold nanoparticles and 4-mercaptophenylboronic acid. *Sensors and Actuators B: Chemical* **196**: 106–111.

Zhu, L., L. Xu, B. Huang, N. Jia, L. Tan and S. Yao (2014). Simultaneous determination of Cd (II) and Pb (II) using square wave anodic stripping voltammetry at a gold nanoparticle-graphene-cysteine composite modified bismuth film electrode. *Electrochimica Acta* **115**: 471–477.

Zhu, T., Z. Li, C. Fu, L. Chen, X. Chen, C. Gao, S. Zhang and C. Liu (2020). Development of an anthraquinone-based cyanide colorimetric sensor with activated C-H group: Large absorption red shift and application in food and water samples. *Tetrahedron* **76**(38): 131479.

Zrazhevskiy, P., M. Sena and X. Gao (2010). Designing multifunctional quantum dots for bioimaging, detection, and drug delivery. *Chemical Society Reviews* **39**(11): 4326–4354.

14 Nanotechnology Approaches in Food Adulterant and Spoilage Detection

Mehrunnisa Tanzil and Huma Shaikh

14.1 FOOD ADULTERATION

Any substance consumed by an organism for nutritional support is called food, which is necessary for living beings to gain strength and growth. Food preservation or adulteration of food has been knuckled down since the start of human evolution, as it is the prime source of energy for living beings. Food fraud or adulteration is a method of threatening public health that is economically fueled. The vile and straightforward means of obtaining enormous sums of money seriously endanger people's lives. Individuals mix impurities into food items to satisfy their greed for profit, aiming to generate more financial gain by selling lower-quality goods at elevated costs. Food adulteration has indeed turned into a beneficial enterprise in recent times. Various factors, such as vendors, shopkeepers, and other traders, jeopardize ordinary people's health for their illicit gains. Virtually all food products are susceptible to adulteration, including grains, dairy items, oils, seafood, honey, alcoholic beverages, and more. Even vegetables and fruits in the market are not immune, often tainted with harmful pesticides and chemicals. There are numerous methods employed to compromise the worth of food.

A food item is assumed to be adulterated if its precious components are removed, substandard foods are concealed within a genuine food product, or if an entirely different food item is substituted for the original. Food quality can also deteriorate when unidentified substances are added or when its container contains toxic materials. Sometimes, unsafe pesticides or other chemical substances are inadvertently introduced. It's important to note that food contamination can occur naturally, such as through the action of microorganisms on vegetables and fruits or through the spoilage of dairy products, unpreserved beverages, and other food items.

Traditional food safety methods alone are insufficient to address this issue effectively. Therefore, there's a need for innovative and advanced approaches that can be quickly adopted by the general public or individuals with limited training in this field to monitor food quality. These methods should be user-friendly, and affordable tools must be developed for evaluating food quality and achieving the desired objectives.

The primary food safety law administered by the Food and Drug Administration (FDA), and according to the Federal Food, Drug, and Cosmetic Act (FFDCA), food can be considered contaminated if:

a. An additional matter that is detrimental to health.
b. Addition of inferior or cheaper items.
c. Any precious constituent haul out from the primary food course.
d. Below quality standard food.
e. Summation of contamination to enhance bulk or weight.
f. To improve its appearance.

DOI: 10.1201/9781003514039-14

Foods are frequently susceptible to fraudulent activities. Food fraud encompasses intentional acts involving the addition, substitution, misrepresentation of food ingredients, tampering or packaging, and misleading or bogus product information, all driven by the desire for financial gain. One particular form of fraud involves sellers fraudulently adding counterfeit substances to food, removing or replacing authentic ingredients without the purchaser's knowledge, all in pursuit of economic benefits.

While traditionally, people prepared and consumed homemade meals with a focus on health and safety, modern times have witnessed shifts in lifestyles and increases in income levels. These factors have progressively led individuals to opt for ready-to-eat foods at restaurants regularly. The food in these outlets may prioritize taste and flavor over nutritional value. These foods may satisfy the palate but not provide a wholesome, balanced meal. However, as suggested by the contributors, the majority of individuals typically are unaware that these items are tainted by 25%–30%. This has a significant negative impact on the health of these customers.

Food quality and safety, next to the various prospects that impact, have emerged as a significant part of concern through the food supply chain. Researchers, government agencies, and regulatory bodies have increasingly focused on such critical issues. Food adulteration can involve the calculated addition of non-food substances to food products for various reasons, including increasing quantity, preserving the product, or enhancing its appearance. These actions can occur deliberately or accidentally and pose significant risks to consumer health (Figure 14.1).

   i. **Dilution with Water**: Adding water to milk, fruit juices, or other beverages to increase volume.
  ii. **Substituting Inferior Ingredients**: Replacing high-quality ingredients with cheaper or lower-quality alternatives and substituting a different type of meat for what is advertised.
 iii. **Addition of Harmful Substances**: Introducing harmful or toxic substances, such as chemicals, pesticides, or contaminants, into food products.
 iv. **Foreign Objects**: Including non-food items like sand, gravel, stones, insects, or animal hairs in grain products or other food items.

These adulterations can have profound health implications for consumers. Consuming contaminated food can lead to foodborne illnesses, allergic reactions, or long-term health issues. In extreme cases, it can even be life-threatening. To protect themselves, consumers should be aware of standard

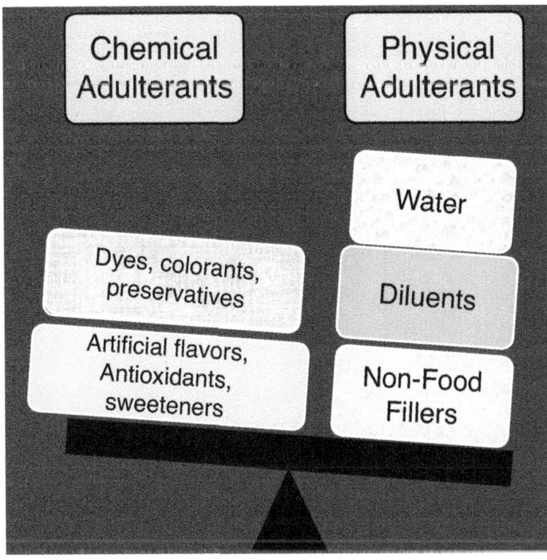

FIGURE 14.1   Adulterants based on physical and chemical classification.

variants of adulterants in the food products they purchase, understand their potential health risks, and take precautions to guarantee the food safety and authenticity they consume. This includes being vigilant about product labels, sourcing food from reputable sources, and reporting suspicious or unscrupulous practices to relevant authorities. Additionally, government agencies and regulatory bodies play a significant role in monitoring and enforcing food safety standards to safeguard consumer interests.

Certainly, adulterants can indeed be chemical substances that are not supposed to be present in a particular food or beverage. These substances may be intentionally added for various deceptive or malicious reasons, often to cut manufacturing costs, increase the apparent quantity of the product, or achieve some other fraudulent objective. This practice can compromise the food or beverage's quality and safety and could negatively affect customers. It underscores the importance of strict regulations and monitoring to prevent such deceptive practices in the food industry. Since there was inadequate or nonexistent government oversight, food adulteration was more prevalent in nations from very early times. The act of adulteration might occasionally even involve extremely harmful substances and toxins. For example, it was stated that, in the People's Republic of China (Chinese milk scandal case with melamine), Chinese dairy companies added melamine to their milk and infant formula products to boost their protein content readings during quality testing artificially. Unfortunately, this led to severe health consequences, especially in young children and infants who consumed the unhygienic products. Tragically, several children died, and thousands suffered from kidney stones and other health issues due to melamine exposure. Consequently, food adulteration continues to rank among the major public health concerns in various nations. The improper use of hazardous chemicals in food products can severely affect the consumer's health and the nutritional value of the food.

14.2 NATURE OF ADULTERATION

Indeed, food adulteration has been one of the major global concerns in recent decades. Many researchers have come upon the cross to remove adulterants from food products due to its fatal impact on consumer health. Screening and ensuring the authenticity of food products is crucial for daily life to safeguard public health and well-being. Examining contaminants or adulterants solely visually is challenging, as they are present in traces at mg L^{-1} or mg kg^{-1} levels. The following categories of adulteration can be differentiated based on the type of contamination, the producer's intention, and the processing method:

14.2.1 Intended Adulterants

Color adulteration is indeed one of the most common types of food adulteration. This deceptive practice involves adding substances to food products to manipulate their color for various reasons, often to make the product appear fresher, more appealing, or of higher quality than it is. This type of adulteration can be complicated for consumers to detect without proper screening and testing. Examples of intentional adulteration include accumulating irrelevant stuff to ground spices, water to liquid milk, or substituting or removing milk solids from the natural product. It includes adding sand, stones, marble chips, mad, other filth, chalk powder, mineral oil, water, and harmful colors to earn some unfair profit. Such adulterants harm the human body. Efforts to combat intentional food adulteration involve stringent regulations, quality control measures, and testing procedures. Public awareness and consumer vigilance are crucial in holding food producers and suppliers accountable for the safety and authenticity of the products they sell. Food safety authorities in many countries work tirelessly to detect and prevent food adulteration, but it remains a persistent challenge in some regions.

According to the literature reported, various chemical substances and practices are used in agriculture that can impact the quality and safety of fruits and vegetables. For example, using calcium carbide, copper sulfate, oxytocin, wax coatings, and excessive pesticides and herbicides can negatively affect the health and safety of these products if not adequately regulated and communicated to consumers. Consumers must be aware of these potential issues, support regulatory efforts that ensure food safety, and make informed choices when purchasing these products. Ensuring transparency in the supply chain and adhering to food safety standards are vital steps in mitigating the risks associated with food adulteration and fraudulent practices.

14.2.2 METALLIC CONTAMINATION

Combining substandard and superior-quality materials like metals, including arsenic tin present in pesticides, mercury from effluents, lead from water, and similar contaminants are also reasons for metallic contamination of food products. During processing. These adulterants accidentally end up in food or through environmental pollution, and they can be present in food in minimal amounts.

14.2.3 ACCIDENTAL ADULTERATION

As the name signifies, accidental food adulteration happens unintentionally (inadvertently) and without our knowledge. These include inadvertent adulterants, including larvae in food, mouse droppings, pesticide residues, and others. Lead, mercury, or arsenic is conceivable as accidental metal contamination. Addressing unintentional adulteration and contamination in food products is crucial for ensuring food safety and quality. Food producers and processors must adhere to good agricultural practices, proper storage, processing techniques, and regular inspections to mitigate these risks. Regulatory agencies also set standards and conduct inspections to enforce food safety regulations and protect consumers from unintentional contamination and adulteration. The most common accidental adulterants are pesticides, DDT, and residues on the plant product. For example, between September 1st and October 4th of this 2023 alone, the FDA included $n = 15$ product recalls (Table 14.1, https://www.fda.gov/safety/recalls-market-withdrawals-safety-alerts) in food and beverage items.

TABLE 14.1
Most Relevant Recalls Made by the US FDA

Product Description	Recall Reason Description
Bacterial Disease	
Dog food	Salmonella sp. contamination
Ice cream products (×2)	Potential foodborne illness
Diced organic butternut squash	Potential *Escherichia coli* O45 contamination
Whole cantaloupe	Potential foodborne illness
Physical Hazards	
Cheese slices	A packaging defect causes a potential for film to remain adhered to the cheese slice after the wrapper has been removed.
Undesired Substances	
Nuez de la India (×2)	Possible health risks include cardiac glycosides.
Undeclared Allergens	
Mooncakes	Undeclared eggs
Homemade hamburger buns	Undeclared allergen/milk
Chocolate cake	Undeclared peanuts
Sweet corn pancakes (×2)	Undeclared wheat and soy allergens; undeclared yellow #5
Ginger snap milk chocolate bar	Undeclared peanuts
Brady street cheese sprinkle	Undeclared sesame seeds
Chocolate flavored protein powder	Undeclared sesame

14.2.4 Natural Adulteration

This type may occur when different chemicals are involved, like organic compounds or radicals that are hazardous to health but are naturally present in foods and not consciously or accidentally added. It may be the addition of poisonous mushrooms, varieties of pulses, green and other vegetables, fish, and shellfish, which are a few examples of numerous natural adulterations, which in some cases could be called anti-nutrients. There are many edible marine fish species, some of which have deadly variants. Adding, replacing, or removing substances is another description of food adulteration.

A food ingredient or important authentic constituent may be partially or entirely replaced with a low-cost alternative to misrepresent the ingredient's origin and the process as a whole. On the other side, addition uses a small number of unauthenticated ingredients to cover up inferior ones. The removal form of adulteration involves removing vital and authentic ingredients without the consumers' knowledge. The purpose of this is to gain a financial advantage. Foods produced using organic, genetically modified, or radioactive ingredients may be adulterated.

14.2.5 Adulteration in Organic Foods

If a product is labeled as organic, it must have been grown and produced in line with local regulations. In organic farming, synthetic pesticides are not released into the environment. The most widely used insecticides allowed for restricted use by most organic standards are pyrethrum and rotenone. Therefore, any food advertised as organic but not complying with the specifications is adulterated.

14.2.6 Adulteration during Food Irradiation

Food is irradiated with ionizing radiation, such as the radioisotope Cobalt-60, to eliminate any bacteria, viruses, insects, or other pathogens present, stop sprouting, delay ripening, and encourage rehydration. The amount of radiation used when irradiating food is crucial. Radiation ionization causes DNA cleavage, alterations to the internal metabolism of cells, and free radical production. Chemical bonds are broken during the process, which may lead to the appearance of resistant microbes, insufficient analytical techniques to identify irradiation in food and public opposition. If this irradiation is used at a level where it might be an overdose, there will be substantial health concerns. As a result, all of these actions taken without the consumer's knowledge are considered adulteration.

14.2.7 Adulteration during the Production of Genetically Modified Foods

Genetically modified (GM) foods are foods produced from organisms whose DNA has undergone alterations through genetic engineering techniques. In the context of GM foods, adulteration typically refers to situations where GM foods are improperly labeled or misrepresented. Here are some key considerations related to adulteration during the production of GM foods:

Labeling and Transparency: Many countries have regulations to ensure that GM foods are appropriately labeled, allowing consumers to make informed choices. Adulteration can occur when GM ingredients are not adequately disclosed on food labels or when non-GM foods are falsely labeled as GM. Labeling and transparency are essential to avoid such adulteration.

Cross-Contamination: Adulteration can also occur through unintentional cross-contamination during GM and non-GM foods' production, processing, or distribution. This is especially relevant for foods like grains, where pollen or seeds from GM crops can mix with non-GM crops, leading to unintended GM content in non-GM products.

Misrepresentation: Deliberate misrepresentation of a food product as GM when it is not, or vice versa, is another form of adulteration. This can occur for various reasons, including economic gain or the desire to capitalize on consumer preferences for or against GM foods.

Regulatory Oversight: To prevent adulteration in the production of genetically modified foods, regulatory agencies often establish guidelines and standards for the labeling and safety assessment of GM products. These regulations ensure that consumers can access accurate information about GM ingredients in their food.

Traceability and Testing: Traceability systems and testing methods are essential tools for verifying the authenticity of GM foods and preventing adulteration. Food producers and regulatory authorities use these methods to track the presence of GM ingredients and ensure compliance with labeling requirements.

14.3 IMPACT OF ADULTERANTS

Nowadays, it is widespread with news articles describing tainted food products widely promoted and consumed by consumers, posing various health dangers. According to news sources, milk and milk products are being contaminated by adding urea, detergent, and other hazardous compounds. To make crops grow overnight, they are also being injected. Steroids were injected into chickens to quickly transform them into hens, which causes several health problems in people. A few health risks are anemia, paralysis, rise in tumors, diseases in vital organs, anomalies of the eyes and skin, and body and stomach aches. Thus, Food adulteration should be taken seriously because it significantly impacts the general public's health.

14.4 FOOD PRODUCTS PRONE TO ADULTERATION

Taste and appearance significantly influence food product commercial value. Artificial ripening, sweetening, colorants, and preservatives enhance food palatability, appearance, and shelf life, providing financial benefits. Although food adulteration affects people's health, it is clear that it should be appropriately taken. According to reports from many authors around the world, there are a variety of beverages and foods that are vulnerable to adulteration. Therefore, it could be complex for the general public to find food products like wheat, pulses, oil, fruit, vegetables, milk, sweets, spices, tea, coffee, honey, bakery goods, fruit juice, chocolate, and similar items that are free from one or more types of adulterants. For example;

i. Fruit vendors select the fruits well before they are fully mature and use chemicals to artificially ripen them just before retailing to avoid the financial loss caused by the breakage of climacteric fruits during harvest, processing, and transportation. Artificial ripening becomes necessary if fruit vendors want to sell their produce ahead of schedule to increase profits. Artificially ripening fruits and vegetables involves using ethanol, ethylene, methanol, propylene, methyl jasmonate, ethylene glycol, ethephon, and calcium carbide.

ii. Farmers use growth hormones to promote growth, including oxytocin, alpha naphthyl acetic acid, and gibberellic acid on fruits and vegetables. Being a veterinary medicine and a hormone found in mammals, oxytocin is inappropriate for use on vegetable crops. However, it is frequently used to increase the size and color of bitter, bottle gourds, pumpkins, and cucumbers.

iii. The utilization of artificial sweeteners, such as food sweetness, is a crucial factor in demand and marketability. As a result, dealers constantly desire to raise the sweetness of particular food items artificially. Sweet foods include fruits and vegetables, drinks, sweeteners, and confectionery. On one side of the fruit, artificial sweeteners are injected using injector pumps to change the fruit's inherent sweetness. Vendors inject

The sweeteners into the fruit at various locations to guarantee uniform distribution. It was discovered that a saccharine combination was being intentionally introduced into melons and watermelons to increase their sweetness. One typical method of adulterating fruit juice is adding exogenous sugar or sugar solution. If the concentrated fruit juice is being exported, external water and sugar are added to it to give it natural qualities similar to those of natural juice. To enhance the composition and increase the Brix value, high-fructose corn syrup, partially hydrolyzed cane syrup, and medium invert sugar from beets are added. Add sucrose, glucose, fructose, maltose, beet sugar, maize sugar, and cane sugar to honey.

iv. The fructose-to-glucose ratio is altered when honey and confectionery items are tainted with fructose or glucose. An indirect kind of adulteration known as feeding honey bees syrup and synthetic sugar after the hovers have been naturally available is known to occur and is exceedingly challenging to identify. Providing low-quality honey to honeybees is also reported. Different confectionery and bakery products are sweetened with Acesulfame-K and Aspartame. Sugar, coloring agents, aromatizing agents, synthetic red dyes, and sweeter foreign wines are often added to wine for quality enhancement. Glycerol boosts sweetness, decreases acidity, and slows fermentation, but diethylene glycol gives wines a relish-like flavor. Reports have also been of cane or beet ethanol added to whiskey and root or cane sugar added to tequila.

v. Synthetic Coloring Agents' use tends consumers to choose their preferred food items based on the products' color, texture, and look. Foods with appealing colors are more profitable and marketable. As a result, numerous natural and synthetic colors are added to various food items in accordance with this trend. The majority of foods with vibrant colors are vulnerable to such misconduct. Fruits, vegetables, and foods produced from eggs, spices, sweets, and confectionery are among the best options.

vi. Additionally, coloring substances are added to processed foods, meat, and fish. In food products made with meat, paprika oleoresin is frequently employed as a natural coloring ingredient. Natural hues in fruits and vegetables, like chlorophyll, annatto, and caramel, have been noted numerous times. The sellers prefer synthetic colors since they outperform natural ones in many ways. These are more stable, bright, and highly effective because they were chemically created. They are also widely available and inexpensive, which are important aspects. Cut fruits and vegetables are colored with rhodamine B, auramine, Metanil yellow, Congo red, Orange II, malachite green, and other allowed and prohibited colors. Azo dyes are used to colorize eggs artificially. Because egg yolk color can be used to determine nutritional content and freshness, the insertion of unlawful synthetic colors motivates traders. Sudan dyes are a kind of synthetic azo dye used in printing and industries. Sudan dye I is chemically similar to another azo dye called Para red, used in printing. Customers want eggs that are yellow or orange. To improve the color of the egg yolk, the dealers frequently give the hens food laced with dyes.

vii. Red pepper chili powder is colored with Sudan I, Sudan IV, Metanil Yellow, Sudan III, Oil Orange SS, Rhodamine B, Auramine O, coal tar red, Para Red, etc. Paprika powder is combined with Sudan I, Sudan IV, Acid Black I, Annatto, and other colors—the addition of Sudan I, Metanil Yellow, Lead Chromate, and other chemicals to turmeric powder. Sumac is colored with Basic Red 46 and Amaranth Red. Curry powder is combined with auramine O and chysoidin. Saffron flower is used with several dyes, including Acid Orange II, Metanil Yellow, Ponceau 4R, Gardenia Yellow, and dye made from Buddleja officinalis Maxim flowers. Cayenne pepper is combined with crystal violet. Colored paper and wood are frequently utilized in addition to synthetic dyes. Dark honey, sometimes known as "forest honey," has a higher commercial value than light honey and is richer in minerals and nutrients. Because of this, light honey is frequently portrayed as dark honey after being polluted with sulfite ammonia caramel.

Consumers are very concerned about the quality and origin of their food since specific foods grown in particular places and by certain species have higher economic worth because of their superior quality. Environmental damage caused by geographic origins is a significant contributing component to this. Food from high-quality batches, species, and cultures is frequently swapped with low-quality items and, on purpose, mislabeled by traders to increase profits. Such conduct infringes upon the rights of consumers, lessens the advantages enjoyed by regional farmers, and fosters unfair commercial rivalry. Only labeling and administrative documents can guarantee traceability, which makes it vulnerable to frequent fraud. Scientific studies provide the most evidence for these problems with fish, shellfish, meat, processed meat, and staple foods.

Additionally, there have been allegations of tea, fruit juice, vinegar, honey, and alcoholic beverages having incorrect labels and origins. The botanical origin of expensive honey varieties, including linden, pine, thyme, orange flower, chestnut, heather, manuka, acacia, and litchi, is frequently mislabeled. The protected designation of origin (PDO) plant food "Fava Santorinis" is commonly substituted with subpar yellow split peas. 'Eglouvis' lentils cultivated in the Ionian Islands are often subject to origin fraud. Most often, olive oils and cocoa butter are misrepresented in terms of their geographical and botanical origins. Olive oil quality and composition vary depending on certain regional traits, environmental factors, and manufacturing processes, retaining various commercial values. For example, due to breeding and selection techniques, Olea europaea L. cultivars of some olives are known to have higher quality. Labeling designates the origin of the olive oils from those areas or varietals. They are frequently misrepresented because of their high market value. Due to the high concentration of monounsaturated fatty acids, vitamins, and antioxidants, extra virgin olive oil has been given a PDO. High commercial value leads to frequent labeling errors.

14.5 FOOD SPOILAGE

One of the significant challenges humans are facing today is to secure or maintain the safety of food. An excellent food proportion gets spoiled during the supply chain before or after reaching the consumer's hand. Any undesirable change in the quality of food that renders it unwanted and unfit for consumers, either by animals or humans, is considered as spoilage. Agriculture and Food Organization of the United Nations has reported that one-third of the food produced for the consumption of humans is either wasted or spoiled. Food spoilage has, therefore, constituted a global concern that needs the attention of experienced stakeholders.

Food spoilage is considered quite complex process with underlying causes like attacking microbiological, chemical or physical changes, which has incurred substantial economic losses to producers, retailers, and consumers. Most countries' legislation has tended to reinforce the concept of food spoilage. The two areas cannot be distinguished according to a microbiological and ecological perspective. Despite considerable efforts, even in advanced countries, microbiological security assures a remote criterion. Suffering, economic losses, death, and civil claims on behalf of victims due to foodborne diseases are matched by the financial losses caused by food spoilage.

To ensure the safety and food quality, it is essential to prevent food from spoilage by adopting specific strategies, like storing perishable foods like dairy products, meat, seafood, and so on in the refrigerator at a temperature below 40°F or keeping dry storage, or vacuum sealing and maintaining cleanliness or consumer having self-awareness regarding food spoilage. A consumer should have excellent knowledge regarding food safety to reduce food waste. Educating consumers on proper food handling and storage practices can help prevent food spoilage and ensure the safety and freshness of food.

For consumer health requirements, food safety issues have prompted the public to construct fast, cost-effective, specific, and sensitive analytical methods and techniques for food analysis. Nanomaterials, with a size from 1 to 100 nm, can be quickly developed into different forms for different detection needs, such as zero dimension (0D) of nanoparticles/nano-clusters, I dimension (1D) of nanowire/nanorod, II dimension (2D) of nanosheet, and III dimension (3D) of nano-net/ nano-flower/nano-bulk. Due to their unique light, electrical, and mechanical properties, their large surface area, good biocompatibility, catalytic activity, and rich bonding sites, nanomaterials have been widely used in various fields, especially for the production of sensing surface elements that can not only increase the sensitivity but also provide lower limits of detection (LOD). Nanomaterial-based sensors and biosensors are being explored to reduce food safety problems.

14.6   GENERAL PRINCIPLES UNDERLYING FOOD SPOILAGE

The core concepts governing food rotting are connected to food degradation after removal from its natural source. Rancidity, fermentation, and putrefaction are all parts of the deterioration process. Different types of food spoilage result in modifications to the meals' appearance, flavor, texture, production of slime, and discharge of fluids. The pH, air, temperature, nutrition, and the occurrence of diverse compounds are the leading causes of food spoilage. The general principles for food deterioration are as follows:

i. Microbial Activity
   Activity or growths of microorganisms, or higher forms, sometimes multiple organisms are involved, such as yeast, molds, bacteria, and so on.
ii. Enzymatic reactions in food spoilage
   Enzymes naturally present in food can cause spoilage by catalyzing chemical reactions that produce undesirable color, flavor, and texture changes. Enzymatic browning, for example, can cause fruits and vegetables to turn brown when cut or bruised.
iii. The intrinsic characteristics influencing microbial deterioration in foods include substrates, endogenous enzymes, light sensitivity, and oxygen. The inherent food qualities affect the type and rate of microbial deterioration and its projected shelf life or perishability.
iv. Oxygen exposure of food (oxidation), Exposed to oxygen (oxidation) can lead to the spoilage of certain foods, especially those containing fats and oils. Oxygen promotes the development of off-flavors and can cause rancidity.
v. Temperature is a critical factor in food spoilage. Most perishable foods spoil more rapidly at higher temperatures. Refrigeration and freezing slow down microbial growth and enzymatic reactions, helping to extend the shelf life of food.
vi. Moisture Content is another essential factor. High moisture levels can promote microbial growth and lead to the growth of molds and yeasts. Foods with low moisture content, such as dried goods, are less susceptible to spoilage.
vii. **pH Levels**: A food's acidity (pH) can influence its susceptibility to spoilage. Some microorganisms thrive in acidic conditions, while others prefer alkaline or neutral pH. The pH of a food product can be adjusted to control spoilage.
viii. **Food Packaging**: Proper packaging can help prevent spoilage by protecting food from exposure to oxygen, moisture, and contaminants. Vacuum-sealed packaging and airtight containers are standard methods to extend the shelf life of products.
ix. Cross-contamination occurs when hazardous microbes spread from one food product to another. Proper food handling, storage, and sanitation practices are essential to prevent cross-contamination and foodborne illnesses.
x. **Time**: All foods will eventually spoil over time, even if stored under ideal conditions. The spoilage rate depends on the food's composition, storage conditions, and other factors.

14.7   NANO-BIOTECHNOLOGY IN THE FIELD OF DETECTION OF ADULTERATION AND SPOILAGE OF FOOD DETECTION

Different kinds of nanomaterials are used in the food sector for various purposes. The methods for identifying adulterants in foods depend on chemical, physical, biochemical, and other techniques. Food adulteration is an ongoing emerging new issue, which has replaced the early organoleptic and further empirical testing. For most adulteration situations, many analytical processes are typically used so that the analyst can select the best one (a variety of samples to be tested, level of sensitivity needed, etc). Nanotechnology in the food industry helps to enhance the safety of food products through its application to pathogens and toxin detection and by providing information on nutritional status.

14.7.1 DETECTION OF CHEMICAL CONTAMINANTS

14.7.1.1 Nanotechnology Incorporated with Calorimetric Analysis

14.7.1.1.1 Detection of Heavy Metals Using Calorimetric Analysis

To analyze the analysts including heavy metals, calorimetric analysis is used based on a dye's color change. Colorimetric analysis typically works in conjunction with nanotechnology to identify and measure chemical contaminations in the food industry. Metal nanoparticles were found to be one of the most successful nanotechnologies commonly used to determine heavy metals. The presence of heavy metals in foods can be detected through a color shift of metal nanoparticles responding to the aggregation level of metal nanoparticles (Table 14.2).

14.7.1.1.2 Calorimetric Analysis Concerning Antibiotic Detection

Kanamycin is an aminoglycoside antibiotic commonly used in agriculture, aquaculture, and livestock care to treat bacterial infections. Tang and coworkers developed tungsten disulfide (WS_2) nanosheets coupled with aptamers and 3,3′,5,5′ -tetramethylbenzidine (TMB). In the presence of kanamycin, the aptamer can no longer enhance the peroxidase-mimicking activity of WS_2 nanosheets because of the specific binding of kanamycin to the aptamer, leading to the retention of colorless TMB. The limit detection for kanamycin was as low as 0.06 µM.

Similarly, Ma et al. developed a colorimetric detection of tobramycin in milk and chicken eggs using AuNPs coupled with aptamers. The presence of tobramycin induced the aggregation of AuNPs, changing the color of the solution from red to blue, and this approach could detect as low as 23.3 nM from other antibiotics, such as sulfamethoxazole, sulfadimethoxine, streptomycin, kanamycin, and sulphachlorpyridazine.

TABLE 14.2
Examples of Various Metals Detection Via Calorimetric Analysis

Metals	Types of Nanoparticles	Detection
Pb^{2+}	AuNPs	NPs, coated with valine, work as a ligand for the attachment of Pb^{2+}. Valine-AuNPs only change from red to blue in the presence of Pb^{2+}, while the presence of all other metal ions retained valine-AuNPs as red, and the limit of detection was one ppm for detection of Pb^{2+}.
Triple detection of Hg^{2+}, Cu^{2+}, and Ag^+	AuNPs coupled with o-phenylenediamine (OPDA)	This technique worked based on the oxidizability of Hg^{2+}, Cu^{2+}, and Ag^+ ions toward OPDA and the ability of OPDA to trigger the aggregation of AuNPs. Among the tested ions (Fe^{3+}, Al^{3+}, Cr^{3+}, Co^{2+}, Ni^{2+}, Zn^{2+}, Mn^{2+}, Mg^{2+}, Ca^{2+}, Pb^{2+}, Cd^{2+}, Fe^{2+}, K^+, Na^+, SO_4^{2-}, CO_3^{2-}, PO_4^{3-}, NO_3^-, NO^{2-}, and Cl^-, Hg^{2+}, Cu^{2+}, and Ag^+), only Hg^{2+}, Cu^{2+}, and Ag^+ could retain the red color of AuNPs, while the presence of other ions led to the formation of the blue aqueous solution. Suitable masking reagents were required to distinguish Hg^{2+}, Cu^{2+}, or Ag^+ from each other, and the detection limit was 19.5 nmol L^{-1} by the naked eye.

14.7.1.2 SERS Analysis

14.7.1.2.1 SERS Analysis for Heavy Metal Detection

A facile SERS substrate for heavy metals detection was fabricated by inkjet printing an AgNPs ink layer (about 400 nm thick) on a silicon wafer. The AgNPs-printed SERS substrates could increase the Raman scattering for the detection of cadmium sulfide (CdS), zinc oxide (ZnO), and mercury sulfide (HgS) by three, four, and five times, respectively, in comparison to the bare silicon substrates (Table 14.3).

14.7.1.2.2 SERS Analysis for Detection of Food Allergens

When the body's immune system is supposed to react against some substances because of an immunological mechanism involving IgE antibodies in the human body, it is referred to as food allergens. Food allergens are generally harmless. Food allergies can have significant effects on a vulnerable person. Therefore, identifying food allergies is crucial so that consumers can be warned. The unique interaction between ligands coupled to the SERS substrate and the allergens allows SERS technology to detect food allergens. Aptamers and antibodies are the two main classes of ligands for detecting food allergies.

14.7.1.2.3 SERS Analysis for Detection of Antibiotics

SERS-based detection of antibiotics in foods was recently reported (Table 14.4).

14.7.1.3 Electrochemical Analysis with Assistance of Nanomaterials

Nanomaterials can significantly improve electrochemical performance when used in electrochemical analysis. Nanomaterials' enormous surface area-to-volume ratios make the electro-catalytic

TABLE 14.3
Reported Synthesized SERS Nanomaterials for Heavy Metal Detection

Metals	Types of Nanoparticles	Reported by	Detection Limit
Arsenic (As)	Cuprous oxide (Cu$_2$O) nanoparticles and AgNPs	Barimah et al.	5.61 µg L^{-1}
Mercury (Hg)	Au@AgPt nanoparticles	Song et al.	0.52 µmol L^{-1}
Lead (Pb)	The SERS-based sensor was synthesized by depositing a nanolayer of Ag and Au and a monolayer of graphene on a porous gallium nitride substrate, and the thiolated probe (Cy3-DNAzyme) for hybridization of single-stranded DNA	He et al.	4.31 pmol L^{-1}

TABLE 14.4
Different Synthesized Methods Reported in the Literature for Antibiotic Detection

Method	Reported by	Antibiotic Compounds	Detection Limit
Aptamer conformation cooperated enzyme-assisted SERS method	Fang et al.	Chloramphenicol (CAP)	15 fmol L^{-1}
AgNP-decorated TiO$_2$	Jing et al.	2-mercapto-5-methyl-1, 3,4-thiadiazole (MMT)	0.11 µmol L^{-1}
SPME-SERS substrate with co-deposition of reduced graphene oxide and silver on silver-copper alloy fibers	Cui et al.	sulfadiazine and sulfamethoxazole analysis	10 ng mL^{-1}
Bimetallic Au@Ag core-shell nanorods	Tian et al.	levofloxacin	0.37 ng L^{-1} (10^{-9} mol L^{-1})

TABLE 14.5

Few Examples of Electrode Modifiers Reported in the Literature

Electrode Modifiers	Reported by	Compounds	Detection Limit
Carbon nanotubes-coated electrodes	Chen et al.	Carbofuran	$0.1\ \mu g\ L^{-1}$
Quantum dots@porous carbon platform	Lin et al.	Oxytetracycline	$3.23 \times 10^{-9} mol\ L^{-1}$
Nano cobalt (II, III) oxide-modified carbon paste	Fekry et al.	Caffeine	$0.016\ \mu mol\ L^{-1}$

processes possible, significantly improving the sensitivity of more extensive materials. Therefore, various nanomaterials were developed to amplify electrochemical signals for ultrasensitive detection. Two main strategies for enhancing electrochemical signals using nanomaterials are electrode modifiers and signal tags.

14.7.1.3.1 Nanomaterials as Electrode Modifiers

By expanding the electrode's effective surface area and speeding up the rate of electron transfer on the electrode surface, nanomaterials with high conductivity have the potential to improve electrochemical signals significantly. Due to their conductive qualities and biocompatibility, various nanomaterials, including carbon nanotubes, quantum dots, graphene, metal, and oxide nanoparticles, can be included in an electrode to enhance analytical performance. The electroanalytical sensors can be modified with single nanomaterials, a binary composite, or triple and multiple nanocomposites (Table 14.5).

14.7.1.3.2 Nanomaterials as Signal Tags

The combination of signal tags can enhance the electrochemical signals. Signal tags are labeled components affixed to the electrode's surface to generate electrochemical signals for analyte analysis. Excellent electrochemical activity in nanomaterials makes them potential catalysts for metabolic activities that provide catalytic signals. Wei and coworkers synthesized thymine-functionalized silver nanoparticles (Ag-T) as the signal tags for mercury detection. Typically, mercury caused Ag-T nanoparticles to aggregate, which enhanced electrochemical signals. The decrease in electrochemical signals was caused by the retention of Ag-T dispersion in the absence of mercury. This mercury electrochemical sensor's detection threshold was $5\,pmol\ L^{-1}$.

14.7.2 DETECTION OF BIOLOGICAL CONTAMINANTS

A wide range of analytical techniques were developed to detect biological contaminations. Still, along with significant advances in nanotechnology, the limitations of current techniques for biological contaminant detection in foods can be improved. It can be described as (Figure 14.2);

14.7.3 DETECTION OF MICRO-NANO PLASTICS

Extensive use of plastics has become a public threat, causing micro/nanoplastics (MP/NPs)- related environmental contamination. MNPs can enter the animal or human body by ingesting contaminated foods or packaged beverages. There is an urgent need for a novel method that is highly accurate, straightforward, and quick to quantify MP/ NPs. The promising approaches for MP/NPs in food are mainly vibrational spectroscopy (FITR, Raman) and electrochemical analysis toward nanostructure design and development. Using Raman spectroscopy, several studies were adapted for MNP detection in food samples using the combination of nanostructure-SERS.

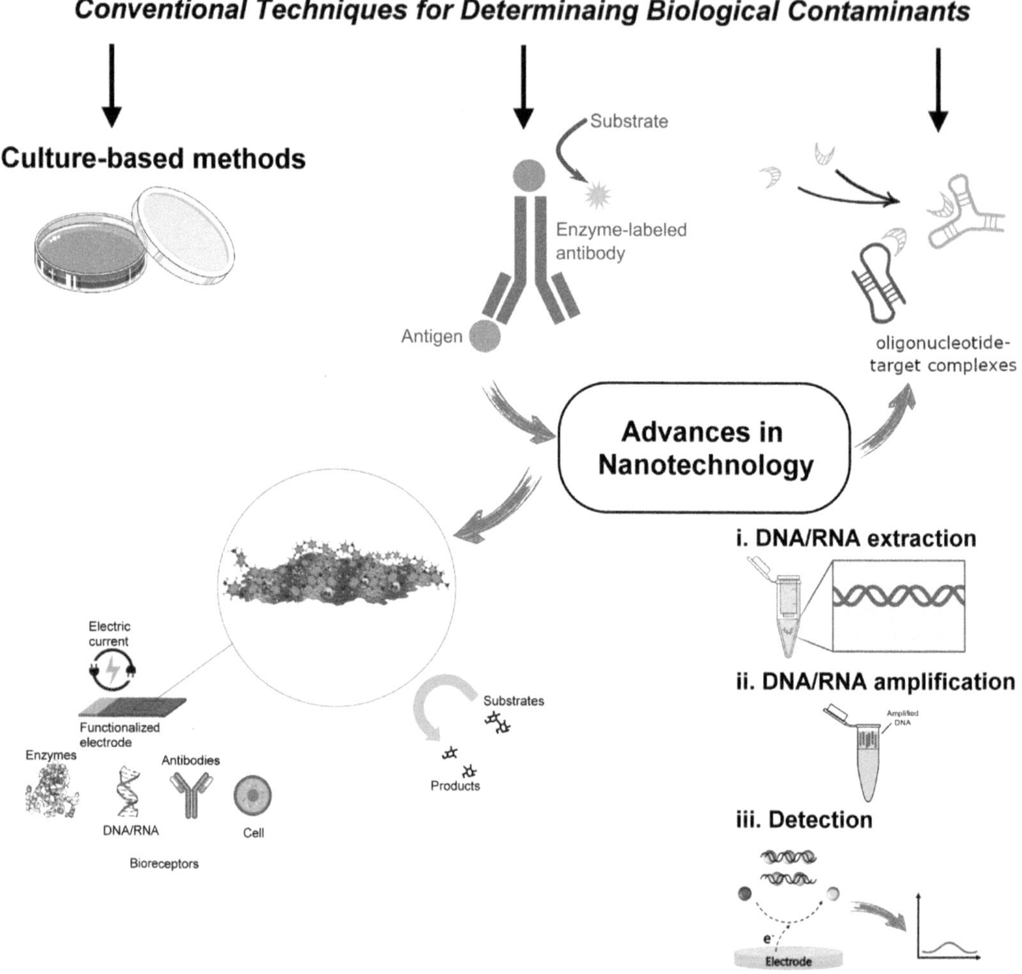

FIGURE 14.2   Conventional techniques for determining biological contaminants.

The electrochemical sensors for MNPs were widely applied to detect MP/NPs and extended to the infield testing of different types of samples without prior purification or isolation. Typical electrochemical behavior sensing of MNPs is based on label-free electrochemical impedance spectroscopy (EIS), amperometry, and voltammetry.

14.8   ROLE OF NANOPARTICLES IN FOOD ADULTERATION/SPOILAGE DETECTION

"Nanotechnology is the future, not just a part of it." Nanotechnology offers creative solutions to problems faced by many industries, including the food industry. The high surface area of nanomaterials allows for substantial molecular interactions. Nanomaterials include metal and metal oxide NPs, carbon nanotubes (CNTs), graphene and its derivatives, carbon nanofibers (CNFs), magnetic NPs, and electrospun nanofibers play a vital role in the design of sensors and biosensors to detect adulteration and spoilage in food (Table 14.6).

Researchers have improved nanosensors' technology by creating more sensitive and selective sensors, allowing for the quick examination of complicated mixtures in food industries.

TABLE 14.6

Different NP Applications for Food Detection

Nanoparticles	Functionality
Metal NPs, such as gold (Au) and silver (Ag) NPs, and metal oxide NPs, such as titanium dioxide (TiO_2), zinc oxide (ZnO), and silica (SiO_2) NPs	Sensors, absorbent materials, drug delivery systems, and antimicrobial materials
Carbon NTs	Sensors
Graphene and its derivatives, including mechanical and chemical exfoliation of graphite and CVD graphene oxide (GO), reduced graphene oxide (rGO), and graphene quantum dots	Conjugated nanoplatelets

14.8.1 Need for Artificial Intelligence

14.8.1.1 In Detection of Food Adulteration

Artificial intelligence (AI) plays a crucial role in addressing food adulteration by offering advanced tools and techniques for detection, prevention, and mitigation. Detection AI presents a chance for the food industry. Technology plays a big part in supporting our food system because it helps form precessions and different aspects of the consumption and production of food. It serves as a quality control matrix in food industries. As a result of AI, concepts such as food production, quality, distribution, and others are changing, and the introduction of intelligent mobile apps has dramatically facilitated this transformation. Artificial intelligence can be used to create and analyze food databases efficiently. It could lead to a cheaper, healthier food business for customers and workers. The advanced methods used in this field are electronic noses, electronic tongues, computer vision, spectral imaging, spectroscopy, and so on, which have been widely used to detect food quality. These techniques may collect a substantial amount of digital information about the make-up and characteristics of food, but it is essential to analyze this information and extract pertinent details.

Nevertheless, implementing these techniques in practical settings is a complex undertaking. The quality of the data will be assessed, and the data will be analyzed using AI-based algorithms. There are various methods to deal with a large amount of data, such as partial least squares, artificial neural network (ANN), support vector machine (SVM), random forest, k-nearest neighbor (KNN), and so on. For feature extraction, principal component analysis (PCA), wavelet transform (WT), independent component correlation algorithm (ICA), scale-invariant feature transform, histogram of oriented gradient, and so on. These techniques are beneficial for handling this kind of data. Analyzing the current situation of food quality degradation has inspired researchers to investigate this field of study.

14.8.1.2 In the Detection of Food Spoilage

An AI approach can detect and analyze food spoilage. Researchers are moving toward new methodologies to maintain food quality and increase shelf life. Machine learning can speed up the monitoring and analysis of food spoilage while delivering precise and reliable results.

A prototype has been developed to monitor food quality and control storage setups at home. The system uses convolutional neural network (CNN) modeling to identify the different types of fruits and vegetables. The suggested system then uses sensors and actuators to check food spoiling levels by monitoring the temperature, humidity, and gas emission levels of fruits and vegetables. Based on the freshness and condition of the food, a message alerting the client to the food's level of deterioration is delivered to their registered mobile phones.

AI and biosensors can be used for food shelf-life detection. For example, a pH sensor can detect when food gets spoiled. Machine learning can be used to detect food spoilage. The cause of food spoilage can be due to factors such as temperature, humidity, and precipitation.

Overall, AI and machine learning can be used to detect food spoilage and maintain food quality. Using sensors and actuators can help monitor fruits and vegetables' temperature, humidity, and gas emission levels to check the food spoilage level. AI and biosensors can also help detect food spoilage and augment the shelf life of food.

14.9   CONCLUSION

Nanotechnology is an emerging technology that utilizes the characteristics of atomic and molecular structures and their interaction principles at the nanoscale. The data has demonstrated that nano/biotechnology can significantly improve the analytical performance of food analytical methods. We believe that nano/biotechnology has great potential in developing food safety and spoilage detection technology. It is noteworthy that sensor technology is also playing its role in the food industry. Various types of nanosensors have already been developed that are robust and cheap. These include electrochemical sensors, aptamer-based electrochemical sensors, immunosensors, optical sensors, etc. The validated analytical methods have been developed for the analysis of food. These methods should be implemented as food testing methods to save labor and cost. However, the standardization of nanosensor-based food testing methods is essential and should be given priority.

BIBLIOGRAPHY

Adukia, Rajkumar S. Food processing industry in India. Bhupendra Kumar and Sohan Lal. History of food processing industry and recent development in India. *Int. J. Res. Finance Manage* 5 no. 2 (2022): 216-221.

Agarwal, Yagini, Varsha Gupta, Mansvi Yadav, Deepesh K Neelam, and Ravi Kant. A review of food adulteration and its impact on human health. *J Coast Life Med* 11 (2023): 1649–1659.

Akramzadeh, Naeimeh, Hedayat Hosseini, Zahra Pilevar, Nader Karimian Khosroshahi, Kianoush Khosravi-Darani, Rozita Komeyli, Francisco J Barba, Alessandro Pugliese, Mahesha Manjunatha Poojary, and Amin Mousavi Khaneghah. Physicochemical properties of novel non-meat sausages containing natural colorants and preservatives. *J Food Process Preserv* 42, no. 9 (2018): e13660.

Alauddin, Shafqat. Global research analysis international. *Food Adult Soc* 1, no. 7 (2012): 3–5.

Amit, Sadat Kamal, Md Mezbah Uddin, Rizwanur Rahman, S M Rezwanul Islam Islam, and Mohidus Samad Khan. A review on mechanisms and commercial aspects of food preservation and processing. *Agric Food Secur* 6, no. 1 (2017): 1–22.

Anita, Gautam, and Singh Neetu. Hazards of new technology in promoting food adulteration. *IOSR J Environ Sci Toxicol Food Technol* 5, no. 1 (2013): 08–10.

Awasthi, Shruti, Kirti Jain, Anwesha Das, Raza Alam, Ganesh Surti, and N. Kishan. Analysis of food quality and food adulterants from different departmental & local grocery stores by qualitative analysis for food safety. *IOSR JESTFT* 8 (2014): 22–26.

Ayalew, Henok, Amare Birhanu, and Biruhtesfa Asrade. Review on food safety system: Ethiopian perspective. *Afr J Food Sci* 7, no. 12 (2013): 431–440.

Ayza, Asrat, and Zelalem Yilma Yilma. Patterns of milk and milk products adulteration in Boditti town and its surrounding, South Ethiopia. *J Agric Sci* 4, no. 10 (2014): 512–516.

Ayza, Asrat, and Ermias Belete. Food adulteration: Its challenges and impacts. *Food Sci Qual Manag* 41 (2015): 50–56.

Ballin, Nicole Z. Authentication of meat and meat products. *Meat Sci* 86, no. 3 (2010): 577–587.

Bansal, Sangita, Apoorva Singh, Manisha Mangal, Anupam K. Mangal, and Sanjiv Kumar. Food adulteration: Sources, health risks, and detection methods. *Critic Rev Food Sci Nutr 57,* no. 6 (2017): 1174–1189.

Banti, Misgana. Food adulteration and some methods of detection, review. Int J Nutr Food Sci 9, no. 3 (2020): 86–94.

Barimah, Alberta Osei, Zhiming Guo, Akwasi A Agyekum, Chuang Guo, Ping Chen, Hesham R El-Seedi, Xiaobo Zou, and Quansheng Chen. Sensitive label-free Cu2O/Ag fused chemometrics SERS sensor for rapid detection of total arsenic in tea. Food Control 130 (2021): 108341

Bertelli, Davide, Massimo Lolli, Giulia Papotti, Laura Bortolotti, Giorgia Serra, and Maria Plessi. Detection of honey adulteration by sugar syrups using one-dimensional and two-dimensional high-resolution nuclear magnetic resonance. J Agric Food Chem 58, no. 15 (2010): 8495–8501.

Bjørklund, Geir, Maryam Dadar, Joachim Mutter, and Jan Aaseth. The toxicology of mercury: Current research and emerging trends. *J Environ Res* 159 (2017): 545–554.

Bogdanov, Stefan, Anton Imdorf, Jean-Daniel Charrière, Peter Fluri, and Verena Kilchenmann. *Bulgarian J Vet Med* 6, no, 2 (2003): 59–70. The Contaminants of the Bee Colony. Swiss Bee Research Centre. Bern, Switzerland. 12 (2003).

Chachan, Satyam, Anand Kishore, Khushbu Kumari, and Arun Sharma. Trends of food adulteration in developing countries and its remedies. *J Food Chem* (2021): 165–187.

Chattopadhyay, Sanchari, Utpal Raychaudhuri, and Runu Chakraborty. Artificial sweeteners – A review. *J Food Sci Technol* 51 (2014): 611–621.

Chen, Haode, Xiaolei Zuo, Shao Su, Zhuzhao Tang, Aibo Wu, Shiping Song, Dabing Zhang, and Chunhai Fan. An electrochemical sensor for pesticide assays based on carbon nanotube-enhanced acetycholinesterase activity. *Analyst* 133, no. 9 (2008): 1182–1186.

Choudhary, Ankita, Neeraj Gupta, Fozia Hameed, and Skarma Choton. An overview of food adulteration: Concept, sources, impact, challenges and detection. *Int J Chem Stud* 8, no. 1 (2020): 2564–2573.

Cui, Jingcheng, Shichao Chen, Xicheng Ma, Hua Shao, and Jinhua Zhan. Galvanic displacement-induced codeposition of reduced-graphene-oxide/silver on alloy fibers for non-destructive Spme@ Sers analysis of antibiotics. *Microchimica Acta* 186 (2019): 1–8.

Dais, Photis, and Emmanuel Hatzakis. Quality assessment and authentication of virgin olive oil by NMR spectroscopy: A critical review. *Analytica Chimica Acta* 765 (2013): 1–27.

Di Renzo, Laura, Carmen Colica, Alberto Carraro, Beniamino Cenci Goga, Luigi Tonino Marsella, Roberto Botta, Maria Laura Colombo, et al. Food safety and nutritional quality for the prevention of non communicable diseases: The nutrient, hazard analysis and critical control point process (Naccp). *J Transl Med* 13 (2015): 1–13.

Drivelos, Spiros A, Georgios P Danezis, Serkos A Haroutounian, and Constantinos A Georgiou. Rare Earth elements minimal harvest year variation facilitates robust geographical origin discrimination: The case of Pdo "Fava Santorinis". *Food Chem 213* (2016): 238–245.

Durazzo, Alessandra, Márcio Carocho, Sandrina Heleno, Lillian Barros, Eliana B Souto, Antonello Santini, and Massimo Lucarini. Food dyes and health: Literature quantitative research analysis. *Measurement: Food* 7 (2022): 100050.

El-Loly, Mohamed Mansour, Ali Ibrahim Ali Mansour, and Ramadan Omar Ahmed. Evaluation of raw milk for common commercial additives and heat treatments. *Int J Food Safety* 15, no. 10 (2013).

Ali Eshkeiti, Binu Baby Narakathu, Avuthu Sai Guruva Reddy , Akhil Moorthi, Massood Zandi Atashbar, Erika Rebrosova , Marián Rebroš, Margaret Kehoe Joyc, and Margaret Joyce. Detection of heavy metal compounds using a novel inkjet printed surface enhanced Raman spectroscopy (Sers) substrate. *Sens Actuat B: Chem* 171 (2012): 705–711.

Essuman, Edward Ken, Ernest Teye, Rosemond Godbless Dadzie, and Livingstone K Sam-Amoah. Consumers' knowledge of food adulteration and commonly used methods of detection. *J Food Qual* 2022 (2022): 1–10.

Fang, Qianqian, Yingying Li, Xinxing Miao, Yiqiu Zhang, Jun Yan, Tainrong Yu, and Jian Liu Sensitive detection of antibiotics using aptamer conformation cooperated enzyme-assisted sers technology. *Analyst* 144, no. 11 (2019): 3649–3658.

Asim Faraz, Muhammad Amir Lateef, Muhammad Iqbal Mustafa, Pervez Akhtar , Muhammad Yaqoob Shahid ur-rehman. Detection of adulteration, chemical composition and hygienic status of milk supplied to various canteens of educational institutes and public places in Faisalabad. *J Anim Plant Sci* 23, no. Suppl 1 (2013): 119–124.

Amany Mohammed Fekry, Mohammed Shehata, Shereen M. Azab, and Alain Walcarius. Voltammetric detection of caffeine in pharmacological and beverages samples based on simple nano-Co (Ii, Iii) oxide modified carbon paste electrode in aqueous and micellar media. *Sens Actuat B: Chem* 302 (2020): 127172.

Galvin-King, Pamela, Simon A Haughey, and Christopher T Elliott. Herb and spice fraud; the drivers, challenges and detection. *Food Contr* 88 (2018): 85–97.

Ganopoulos, Ioannis, Christos Bazakos, Panagiotis Madesis, Panagiotis Kalaitzis, Athanasios Tsaftaris. Barcode DNA high-resolution melting (Bar-Hrm) analysis as a novel close-tubed and accurate tool for olive oil forensic use. *J Sci Food Agric* 93, no. 9 (2013): 2281–2286.

Ghidini, Sergio, Maria Olga Varrà, and Emanuela Zanardi. Approaching authenticity issues in fish and seafood products by qualitative spectroscopy and chemometrics. *Molecules* 24, no. 9 (2019): 1812.

Gizaw, Zemichael. Public health risks related to food safety issues in the food market: A systematic literature review. *Environ Health Prevent Med* 24 (2019): 1–21.

Goonatilake, Ruchitha. Effects of diluted ethylene glycol as a fruit-ripening agent. *Glob J Biotechnol Biochem* 3, no. 1 (2008): 8–13.

Gu, Xinzhe, Li Zhang, Liting Li, Nan Ma, Kang Tu, Lijun Song, and Leiqing Pan. Multisource fingerprinting for region identification of walnuts in Xinjiang combined with chemometrics. *J Food Proc Eng* 41, no. 4 (2018): e12687.

He, Limin, Yijuan Su, Xiangguang Shen, Zhenling Zeng, and Yahong Liu. Determination of Sudan dye residues in eggs by liquid chromatography and gas chromatography-mass spectrometry. *Analytica Chimica Acta* 594, no. 1 (2007): 139–146.

He, Qihang, Yingkuan Han, Yuzhen Huang, Jianwei Gao, Yakun Gao, Lin Han, and Yu Zhang. Reusable dual-enhancement Sers sensor based on graphene and hybrid nanostructures for ultrasensitive lead (II) detection. *Sens Actuat B: Chem* 341 (2021): 130031.

He, Yong, Xiulin Bai, Qinlin Xiao, Fei Liu, Lei Zhou, Chu Zhang. Detection of adulteration in food based on nondestructive analysis techniques: A review. *Crit Rev Food Sci Nutr* 61, no. 14 (2021): 2351–2371.

Hegazi, Nesrine M, Ghada E Abd Elghani, Mohamed A Farag. The super-food manuka honey, a comprehensive review of its analysis and authenticity approaches. *J Food Sci Technol* 59, no. 7 (2022): 2527–2534.

Islam, Md Nazibul, Mollik Yousuf Imtiaz, Sabrina Shawreen Alam, Farrhin Nowshad, Swarit Ahmed Shadman, Mohidus Samad Khan. Artificial ripening on banana (Musa Spp.) samples: Analyzing ripening agents and change in nutritional parameters. *Cogent Food Agric* 4, no. 1 (2018): 1477232.

Islam, Md Shahedul, Nusrat Jahan Jhily, Md Mahedi Hasan Shampad, Jilon Hossain, Sujan Chandra Sarkar, Md Maminur Rahman, and Md Ashiqul Islam. Dreadful practices of adulteration in foods: Their frequent consequences on public health. *Asian J Health Sci* 8, no. 1 (2022): 32.

Jamwal, Rahul, Shivani Kumari, Sushma Sharma, Simon Kelly, Andrew Cannavan, and Dileep Kumar Singh. Recent trends in the use of Ftir spectroscopy integrated with chemometrics for the detection of edible oil adulteration. Vibratio Spectrosc 113 (2021): 103222.

Jing, Mengyu, Hui Zhang, Ming Li, Zhu Mao, and Xiumin Shi. Silver nanoparticle-decorated Tio$_2$ nanotube array for solid-phase microextraction and Sers detection of antibiotic residue in milk. *Spectrochim Acta Part A: Mol Biomol Spectrosc* 255 (2021): 119652.

Kamiloglu, Senem. Authenticity and traceability in beverages. *Food Chemistry* 277: 12–24. https://doi.org/10.1016/j.foodchem.2018.10.091

Kathirvelan, Jayaraman, Vijayaraghavan, Rajagopalan. An infrared based sensor system for the detection of ethylene for the discrimination of fruit ripening. *J Infrared Phys* Technol 85 (2017): 403–409.

Meenakshi Khapre, Abhay Mudey, Sonali Chaudhary Vasant Govind Wagh, Ajay Dawale. Buying practices and prevalence of adulteration in selected food items in a rural area of Wardha district: A cross-sectional study. *Online J Health Allied* Sci 10, no. 3 (2011): 4.

Koester, Ulrich. Food loss and waste as an economic and policy problem. In *World Agricultural Resources and Food Security*, 275–288: Emerald Publishing Limited, 2017.

Youji Kohda, Iffat Haque. Knowledge and perception about food adulteration problem among school children in Bangladesh: Iffat Haque. *Eur J Publ Health* 27, no. suppl_3 (2017): ckx189. 69.

Dirk W. Lachenmeier. Advances in the detection of the adulteration of alcoholic beverages including unrecorded alcohol. In *Advances in Food Authenticity Testing*, 565–584: Elsevier, 2016.

Lachenmeier, Dirk W, Eva-Maria Sohnius, Rainer Attig, and Mercedes G López. Quantification of selected volatile constituents and anions in Mexican agave spirits (Tequila, Mezcal, Sotol, Bacanora). *J Agric Food Chem* 54, no. 11 (2006): 3911–3915.

Nychas, George John È. , Panagou, Efstathios Z., Lianou, Alexandra. Microbiological spoilage of foods and beverages. In *The Stability and Shelf Life of Food*, 3–42: Elsevier, 2016.

Lin, Jinyu, Junchao Qian, Yaping Wang, Yue Yang, Yuxuan Zhang, Jianping Chen, Xingwang Chen, and Zhigang Chen. Quantum dots@ porous carbon platform for the electrochemical sensing of oxytetracycline. *Microchem J* 167 (2021): 106341.

Tharaka S. D. Maduwanthi, Rajapaksha Arachchilage Upul Janapriya Marapana, and Rauj Marapana. Induced ripening agents and their effect on fruit quality of banana. *Int J Food Sci* 2019 (2019).

Majumdar, Arnab, Neha Pradhan, Jibin Sadasivan, Ananya Acharya, Nupur Ojha, Swathy Babu, and Sutapa Bose. Food degradation and foodborne diseases: A microbial approach. In *Microbial Contamination and Food Degradation*, 109–148: Elsevier, 2018.

Manasha, S, M Janani. Food adulteration and its problems (intentional, accidental and natural food adulteration). *Int J Res Finance* Market 6, no. 4 (2016): 131–140.

Momtaz, Mysha, Saniya Yesmin Bubli, and Mohidus Samad Khan. Mechanisms and Health Aspects of Food Adulteration: A Comprehensive Review. Foods 12 no. 1 (2023): 199. Study of Adulterants in Foods, Delhi, India, 2011.

Martic, Sanela, Meaghan Tabobondung, Stephanie Gao, and Tyra Lewis. Emerging electrochemical tools for microplastics remediation and sensing. *Front Sens* 3 (2022): 958633.

Martínez Montero, Cristina Martínez Montero, María del Carmen Rodríguez-Dodero, Dominico Guillén-Sánchez, Carmelo García-Barroso. Analysis of low molecular weight carbohydrates in food and beverages: A review. *Chromatographia* 59 (2004): 15–30.

Mohammadi, Zahra, and Seid Mahdi Jafari. Nanoparticles/nanofibers for checking adulteration/spoilage of food products. In *Handbook of Food Nanotechnology*, 459–492: Elsevier, 2020.

Mursalat, Mehnaz, Asif Hasan Rony, AHMS Rahman, MN Islam, and Mohidus Samad Khan. A critical analysis of artificial fruit ripening: Scientific, legislative and socio-economic aspects. *Che Thoughts* 3, no. 1 (2013): 1–7.

Nadimi, Mohammad, Eric Hawley, Jing Liu, Kurt Hildebrand, Elaine Sopiwnyk, Jitendra Paliwal. Enhancing traceability of wheat quality through the supply chain. *Comprehens Rev Food Sci Food Saf* 22, no. 4 (2023): 2495–2522.

Naila, Aishath, Steve H Flint, Ahmad Ziad Sulaiman, Azilah Ajit, and Zuben Weeds. Classical and novel approaches to the analysis of honey and detection of adulterants. *Food Control* 90 (2018): 152–165.

Suresh Neethirajan, Xuan Weng, Krishnamoorthy Vasanth Ragavan. Nano-biosensor platforms for detecting food allergens-new trends. *Sens Bio-sens Res* 18 (2018): 13–30.

Niharika, M, D Madhavi, and Sireesha Guttapalam. Evaluation of Kap and detection of adulterants in spices by physical and chemical methods. *Ind J Appl* Pure Bio Vol 37, no. 3 (2022): 855–863.

Nilghaz, Azadeh, Seyed Mahdi Mousavi, Miaosi Li, Junfei Tian, Rong Cao, Xungai Wang. Based microfluidics for food safety and quality analysis. *Trends Food Sci Technol* 118 (2021): 273–284.

Panghal, Anil, DN Yadav, Bhupender S Khatkar, Himanshu Sharma, Vikas Kumar, Navnidhi Chhikara. Post-harvest malpractices in fresh fruits and vegetables: Food safety and health issues in India. *Nutr Food Sci* 48, no. 4 (2018): 561–578.

Pereira, Leonor, Sonia Gomes, Sara Barrias, Jose Ramiro Fernandes, and Paula Martins-Lopes. Applying high-resolution melting (Hrm) technology to olive oil and wine authenticity. *Food Res Int* 103 (2018): 170–181.

Priyadarshini, Eepsita, and Nilotpala Pradhan. Metal-induced aggregation of valine capped gold nanoparticles: An efficient and rapid approach for colorimetric detection of Pb2+ ions. *Sci Rep* 7, no. 1 (2017): 9278.

Qian, Lanting, Sharmila Durairaj, Scott Prins, and Aicheng Chen. Nanomaterial-based electrochemical sensors and biosensors for the detection of pharmaceutical compounds. *Biosens Bioelectron* 175 (2021): 112836.

Raieta, Katia, Livio Muccillo, and Vittorio Colantuoni. A novel reliable method of DNA extraction from olive oil suitable for molecular traceability. *Food Chem* 172 (2015): 596–602.

Revel, Messika, Amélie Châtel, and Catherine Mouneyrac. Micro (nano) plastics: A threat to human health? *Curr Opin Environ Sci Health* 1 (2018): 17–23.

Rinke, Peter. Tradition meets high tech for authenticity testing of fruit juices. In *Advances in Food Authenticity Testing*, 625–665: Elsevier, 2016.

Rout, Prangya Ranjan, Anee Mohanty, Ana Sharma, Mehak Miglani, Dezhao Liu, and Sunita Varjani. Micro-and nanoplastics removal mechanisms in wastewater treatment plants: A review. *J Hazard Mater Adv* 6 (2022): 100070.

Saadat, Sacida, Hardi Pandya, Aayush Dcy, and Deepak Rawtani. Food forensics: Techniques for authenticity determination of food products. *Forens Sci Int* 333 (2022): 111243.

Sagandykova, Gulyaim N, Mereke B Alimzhanova, Yenglik T Nurzhanova, and Bulat Kenessov. Determination of semi-volatile additives in wines using Spme and Gc-Ms. *Food Chem* 220 (2017): 162–167.

Everstine, Karen D., Henry B. Chin, Fernando A. Lopes, and Jeffrey C. Moore. Database of Food Fraud Records: Summary of Data from 1980 to 2022. *J Food Protect* 87 no. 3 (2024): 100227.

Sezgin, Aybuke Ceyhun, and Sibel Ayyıldız,. Food additives: Colorants. In *Science within Food: Up-to-Date Advances on Research and Educational Ideas,* 87–94, 2017.

Sharma, Ameeta, Neha Batra, Anjali Garg, and Ankita Saxena. Food adulteration: A review. *Int J Res Appl Sci Eng Technol (IJRASET)* 5, no. 11 (2017): 686–689.

Sharma, Kajal, DC Singh, Suresh Chaubey, and Ramesh Chandra Tiwari. World J *Pharm Res* 11, no. 5 (2022): 342–350. *Adulteration in Food and Drug and Drug Evaluation for Identification, Determination of Quality-Purity and Detection of Adulteration,* 2022.

Shukla, Seema, Ravi Shankar, and Surya Prakash Singh. Food safety regulatory model in India. *Food Control* 37 (2014): 401–413.

Siddiqui, Amna Jabbar, Syed Ghulam Musharraf, and Mohammed Iqbal Choudhary. Application of analytical methods in authentication and adulteration of honey. *Food Chem* 217 (2017): 687–698.

Snyder, Abigail B, and Randy W Worobo. Fungal spoilage in food processing. *J Food Protec* 81, no. 6 (2018): 1035–1040.

Song, Chunyuan, Jinxiang Li, Youzhi Sun, Xinyu Jiang, Jingjing Zhang, Chen Dong, and Lianhui Wang. Colorimetric/Sers dual-mode detection of mercury ion via Sers-active peroxidase-like Au@ Agpt Nps. *Sens Actuat B: Chem* 310 (2020): 127849.

Song, Weiran, Zhiyuan Song, Jordan Vincent, Hui Wang, and Zhe Wang. Quantification of extra virgin olive oil adulteration using smartphone videos. *Talanta* 216 (2020): 120920.

Sonwani, Ekta, Urvashi Bansal, Roobaea Alroobaea, Abdullah M Baqasah, and Mustapha Hedabou. An artificial intelligence approach toward food spoilage detection and analysis. *Front Public Health* 9 (2022): 816226.

Tajik, Somayeh, Zahra Dourandish, Fariba Garkani Nejad, Hadi Beitollahi, Peyman Mohammadzadeh Jahani, and Antonio Di Bartolomeo. Transition metal dichalcogenides: Synthesis and use in the development of electrochemical sensors and biosensors. *Biosens Bioelectron* 216 (2022): 114674.

Tang, Yue, Yang Hu, Pei Zhou, Chunxiao Wang, Han Tao, and Yuangen Wu. Colorimetric detection of kanamycin residue in foods based on the aptamer-enhanced peroxidase-mimicking activity of layered Ws2 nanosheets. *J Agric Food Chem* 69, no. 9 (2021): 2884–2893.

Techane, Tolcha. Effect of adulterants on quality and safety of cow milk: A review. *Int J Diabetes MetabDisord* 8, no. 1 (2023): 277–287.

Tian, Yue, Guanhua Li, Hua Zhang, Linlin Xu, Anxin Jiao, Feng Chen, and Ming Chen. Construction of optimized Au@ Ag core-shell nanorods for ultralow Sers detection of antibiotic levofloxacin molecules. *Opt Exp* 26, no. 18 (2018): 23347–23358.

Tsanakas, Georgios F, Photini V Mylona, Katerina Koura, Anthoula Gleridou, and Alexios N Polidoros. Genetic diversity analysis of the Greek lentil (Lens Culinaris) Landrace' Eglouvis' using morphological and molecular markers. *Plant Genet Res* 16, no. 5 (2018): 469–477.

Tsopelas, Fotios, Dimitris Konstantopoulos, and Anna Tsantili Kakoulidou. Voltammetric fingerprinting of oils and its combination with chemometrics for the detection of extra virgin olive oil adulteration. *Anal Chim Acta* 1015 (2018): 8–19.

Vemireddy, Lakshminarayana R, Valluri V. Satyavathi, Ebrahimali Abubacker Siddiq, Javaregowda Nagaraju. Review of methods for the detection and quantification of adulteration of rice: Basmati as a case study. *J Food Sci Technol* 52 (2015): 3187–3202.

von Engelhardt, Nikolaus, Rie Henriksen, and Ton G.G. Groothuis. Steroids in chicken egg yolk: Metabolism and uptake during early embryonic development. *General Groothuis Compar Endocrinol* 163, no. 1–2 (2009): 175–183.

Wei, Tianxiang, Tingting Dong, Zhaoyin Wang, Jianchun Bao, Wenwen Tu, and Zhihui Dai. Aggregation of individual sensing units for signal accumulation: Conversion of liquid-phase colorimetric assay into enhanced surface-tethered electrochemical analysis. *J Am Chem Soc* 137, no. 28 (2015): 8880–8883.

Whitehouse, Christina R, Joseph Boullata, and Linda A McCauley. The potential toxicity of artificial sweeteners. *Aaohn J* 56, no. 6 (2008): 251–261.

Woldemariam, Henock Woldemichael, and Biresaw Demelash Abera. The extent of adulteration of selected foods at Bahir Dar, Ethiopia. *Int J Interdisciplin Res* 1, no. 6 (2014): 1–6.

Wu, Liming, Bing Du, Yvan Vander Heyden, Lanzhen Chen, Liuwei Zhao, Miao Wang, and Xiaofeng Xue. Recent advancements in detecting sugar-based adulterants in honey – A challenge. *TrAC Trends Anal Chem* 86 (2017): 25–38.

Wu, Xijun, Baoran Xu, Renqi Ma, Shibo Gao, Yudong Niu, Xin Zhang, Zherui Du, Hailong Liu, and Yungang Zhang. Botanical origin identification and adulteration quantification of honey based on Raman spectroscopy combined with convolutional neural network. *Vibration Spectrosc* 123 (2022): 103439.

Yang, Zhe-Han, Shirong Ren, Ying Zhuo, Ruo Yuan, and Ya-Qin Chai. Cu/Mn double-doped Ceo2 nanocomposites as signal tags and signal amplifiers for sensitive electrochemical detection of procalcitonin. *Anal Chem* 89, no. 24 (2017): 13349–13356.

Zábrodská, Blanka, and Lenka Vorlová. Adulteration of honey and available methods for detection – A review. *Acta Veterinaria Brno* 83, no. 10 (2015): 85–102.

Zhang, Jianyi, Jiyun Nie, Lixue Kuang, Youming Shen, Haidong Zheng, Hui Zhang, Saqib Farooq, Syed Asim. Geographical origin of Chinese apples based on multiple element analysis. *J Sci Food Agric* 99, no. 14 (2019): 6182–6190.

15 Nanosensors Applied within the Food Industry
The Role of Volatile Organic Compounds Sensing

Gabriela Flores-Rangel, Lorena Díaz de León-Martínez,
Boris Mizaikoff, and José Alejandro Roque-Jiménez

15.1 INTRODUCTION

Food safety is the absence -at acceptable and safe levels- of hazards in food that can generate adverse effects on consumers' health (Akbari et al. 2022). This refers to the care and attention in consuming a product so that it does not cause damage to health by chemical, physical, and biological agents (FAO 2022). If food is not safe, there can be no food safety, and in a world where the food supply chain has become more complex, any adverse food safety incident can adversely affect public health, trade, and the economy (Akbari et al. 2022). Ensuring food safety is a complex process that begins at the farm and ends with the consumer (Verbeke 2005). Ensuring that food is safe, nutritious, and sufficient at any supply chain level is vital. However, some chemical-biological agents contaminate food, which cannot be easily perceived or detected (Perkowski et al. 2012). These include bacteria (Thakali and MacRae 2021), fungi (Pandey et al. 2023), viruses (Petrović and D'Agostino 2016), parasites (Jhu and Sinha 2022), pesticides (Fletcher and Netzel 2020), biological toxins (Kumar Verma et al. 2022), and additives (Thakali and MacRae 2021). More than 200 diseases, from diarrhea to cancer, originate in contaminated foods that transmit viruses, bacteria, parasites, or dangerous chemical compounds (FAO 2022). Due to the growing concerns about food safety, hygiene, and adulteration in the food chain, the demand for rapid and accurate quality inspection of food products has steadily increased (Siddiqui et al. 2022).

In recent decades, food production and consumer concerns related to food quality and safety have substantially increased international regulations of this sector (Loutfi et al. 2015). This approach has led to food inspections on the final product and food quality being assured at every level of supply and production to meet regulatory and consumer requirements. In this regard, analytical chemistry plays a vital role in the food industry, both in the control of its quality and in its safety (Valdés et al. 2009). Evaluating food quality and safety involves applying different instruments/techniques and procedures such as cell culture, molecular techniques like PCR, infrared spectroscopy, chromatography, and mass spectrometry. However, these instruments are associated with high costs of analysis, arduous sample pretreatment, require highly trained personnel for their development and application, and some cannot be performed on-site and in real-time (Valdés et al. 2009). Therefore, the research community has focused on developing analytical techniques to detect and extract these compounds (Jagtiani 2022). Public health concerns about food quality and security have driven the need to develop rapid, sensitive, specific, portable instrumentation to detect food hazards and minimal pre-sample treatment. Most of these methods rely on developing and applying nanotechnology to screen, detect, and extract these contaminants from foods. Nanotechnology'sanosensors play an essential role in achieving these goals (Nile et al. 2020).

DOI: 10.1201/9781003514039-15

Nanosensors are nanoscale devices that measure physical, chemical, biological, or environmental quantities and convert these into signals that can be detected and analyzed (He and Hwang 2016). These have been recognized as promising tools for several applications within the food industry; they offer advantages such as selectivity, sensitivity, and speed compared to other analytical techniques for detecting contaminants and microorganisms and evaluating freshness in food (Singh et al. 2023). Therefore, nanosensors are increasingly employed for on-site and fast analysis to develop detection techniques for sensing contamination, adulteration, and freshness of food materials to assess the traceability and detection of food contamination (Valdés et al. 2009).

To this extent, volatile organic compounds (VOCs) have been analyzed in foods as biomarkers. Food products generate VOCs during production, processing, storage, and transportation due to their chemical reactions or the influence of external microorganisms, which are inextricably linked to food quality and safety (Lin et al. 2022). Currently, the gold standard for evaluating VOCs is gas chromatography coupled with mass spectrometry (GC-MS); the results are generally accurate and reliable; however, as mentioned above, they are costly and specialized techniques, which makes their application in this sector complicated (Bai and Shi 2007). In this regard, equipment based on nanosensors has been developed, called electronic noses (E-nose); these types of equipment are designed to imitate the sense of smell of mammals; they are equipped with arrays of nanosensors used to detect VOCs and are trained with chemometric methods for pattern recognition, to distinguish odors in complex samples at a low cost accurately, they are non-destructive, susceptible and specific, and with a fast response time (Karakaya et al. 2020). The E-nose has been applied in various scenarios within the food sector, including for quality control of raw and manufactured products: process, freshness and maturity monitoring, shelf-life investigations, authenticity assessments of premium products, denomination of origin, and microbial pathogen detection (Khatib and Haick 2022).

Due to the background information presented, the main objective of this chapter is to give a state-of-the-art overview of nanodevices and the processes in which this technology plays a vital role in the food industry, focusing on applying E-noses for VOC detection in different approaches.

15.2 NANOSENSORS IN THE FOOD INDUSTRY

Nanosensors are novel devices that harness the unique properties of nanomaterials to detect and analyze various substances at the nanoscale (Shawon et al. 2020). Subsequently, the acquired nanoscale information is translated into data for comprehensive analysis. Within this framework, nanosensors have gained significant attention in the food industry over the past decade, owing to their potential for on-site analysis (see Figure 15.1).

Nanosensors represent devices that transmute physical stimuli, such as thermal, magnetic, and luminescent cues, into electric signals (Ramezani et al. 2020). The main characteristics of nanosensors encompass sensitivity, selectivity, resolution, stability, and calibration.

According to Siddiqui et al. (2022), nanosensors can be categorized into two overarching functional groups. The initial group includes nano-dimensional attributes, delineated by tridimensional characteristics within nanoscale proportions, such as quantum dots (QD) resembling fluorescent nanosized semiconductor crystals and thin films of various metallic oxides employed for gas sensor applications. The alternate group incorporates measurements void of nanoscale dimensions (Siddiqui et al. 2022).

Several authors state that nanosensors can be classified into active and passive (John et al. 2022). First, active sensors, such as thermistors, encompass those that require an energy source. In contrast, passive sensors operate without power, such as piezoelectric sensors (Aylott 2003). Nanosensors can also be categorized according to the detection of signals from physical, chemical, and biological effects (Ramezani et al. 2020). Physical nanosensors comprise mechanical, optical, and electromagnetic sensors that quantify physical properties such as force, density, absorbance, pressure, and viscosity. Chemical nanosensors determine the presence and

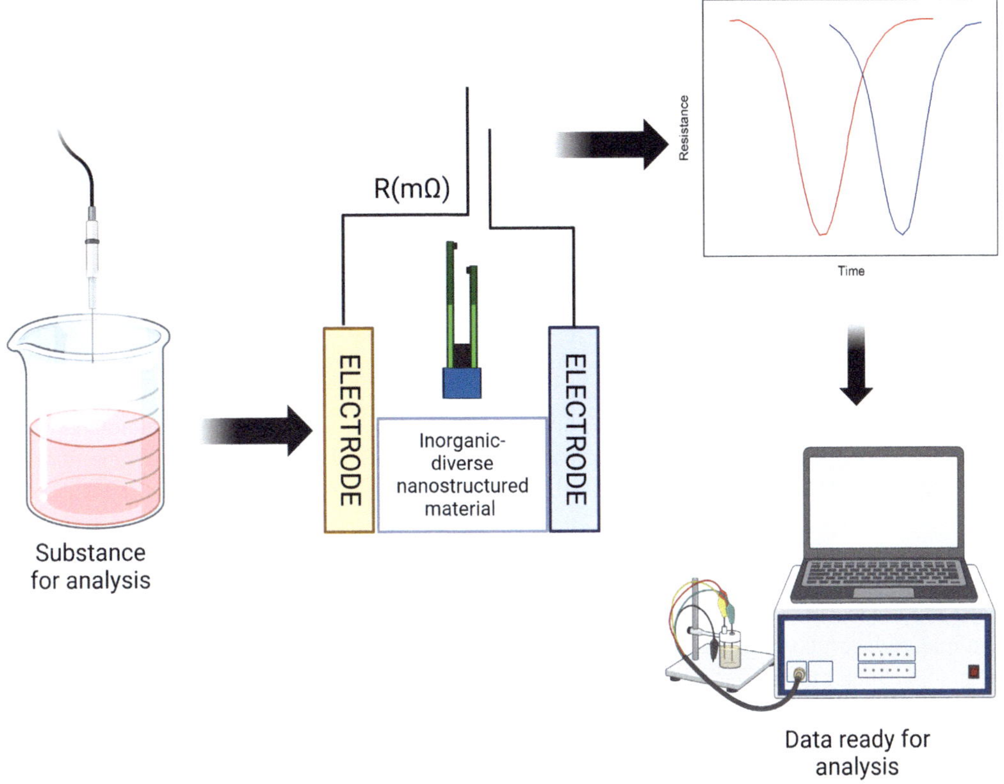

FIGURE 15.1 Schematic illustration of device-based sensing of diverse nanostructures.

concentration of various chemicals, including pH, while biological sensors are designed to interact with microorganisms (Cullum and Vo-Dinh 2000).

15.2.1 Optical Nanosensors

Optical sensors take advantage of the interaction of light with materials to detect changes in optical properties in the presence of analytes with dimensions smaller than 1,000 nm (Ramezani et al. 2020). This detection is based on modifying light signals such as absorbance, emission, or refraction due to the interaction with the analyte. Subsequently, in the optical nanosensor, an optical receiver is incorporated, typically an inert or functionalized matrix, which allows the detection and quantification of the specific properties of the assigned analyte (Aylott 2003). These sensors are susceptible and enable real-time analysis, making them ideal for food industry applications.

Implementing optical nanosensors in the food industry is particularly valuable in this context (Servarayan et al. 2023). These devices can measure critical parameters such as food products' pH, acidity, and other spoilage indicators. For example, the first optical nanosensor developed aimed to measure pH using fluorescence trapped in polyacrylamide nanoparticles (Aylott 2003). This strategy opened the door for further applications in the field, enabling more accurate and detailed monitoring of foods' physical and chemical properties.

15.2.2 Fiber Optic Nanosensors

Fiber optic nanosensors are structured in layers of chemical or biological recognition components, which possess specific chemical or biological activity (Vo-Dinh and Kasili 2005). These layers are bonded to an optical transducer (Lakshmipriya and Gopinath 2019). The interaction between the target analyte (Vo-Dinh and Kasili 2005) and its corresponding receptor triggers a physicochemical modification that can be translated into a discernible electrical signal. Subsequently, this signal is captured by the optical probe and channeled to a database for its meticulous quantification. This procedure is designed to ensure high levels of precision and accuracy.

Siddiqui et al. (2022) outlined two different detection methods for fiber optic nanosensors. The first method involves a direct measurement, where the intrinsic optical properties of the target analyte are evaluated. The second method involves an indirect approach, which governs optically detectable biosensors. These sensors are classified into fiber optic chemical nanosensors and fiber optic biosensors. The first fiber optic nanosensors developed for food science were adapted to detect Escherichia coli in the food matrix. For example, the Analyte 2000 device, a core-type fiber optic sensor created by Research International in the United States, was designed for discerning pathogens by a sandwich assay using fluorophore-labeled antibodies directed against various pathogenic bacteria. After food packaging, the fiber optic sensors use a sensitive coating of syndiotactic polystyrene thermoplastic material with high heat resistance and chemical stability, boasting outstanding dielectric properties (Vo-Dinh and Kasili 2005).

15.2.3　Nanoparticle-Based Nanosensors

In the food industry context, nanosensors are conceived from precisely engineered nanomaterials. Nanomaterials structured in dimensions ranging below 100 nm are the foundation of this applied technology. Such materials, such as gold nanoparticles, quantum dots, and carbon nanotubes, comprise one or more essential components. In this application, the interaction between these nanoparticles and analytes of interest in food is strategically explored to enable high-accuracy detection and quantification in the food industry (Sharma et al. 2023).

15.2.4　Nanoparticles

Nanoparticles (NPs) are characterized by their dimensions on the nanometer scale (up to 100 nm) and exhibit surface properties that allow interactions with metal ions, small molecules, polymers, and surfactants. These NPs are organic, inorganic, and carbon-based (You et al. 2019). Metallic NPs are distinguished by their localized surface plasmon resonance (LSPR), generating unique electrical and optical properties. Ceramic NPs, on the other hand, as non-metallic inorganic particles, exhibit the ability to absorb oxygen due to their photocatalytic behavior. Nanoscale materials, such as nano-cellulose, nanoparticles, and nanoclays, show promise in food packaging, although migration studies are imperative to safeguard consumer safety (Sharma et al. 2023).

As far as nanosensors in food packaging are concerned, they make use of nanoparticles such as zinc oxide (ZnO NPs), silver (Ag NPs), titanium dioxide (TiO_2 NPs), nanoclays, carbon nanotubes (CNTs), and cellulose nanofibers as functional agents. These components, with specific chemical and physical attributes, are used for various purposes in packaging (Shawon et al. 2020).

15.2.5　Electrochemical Sensors

Electrochemical sensors (ECs) are innovative devices that detect alterations in electrical properties resulting from interactions between nanomaterials and target analytes. These sensors operate through ionic conduction mechanisms, which gives them remarkable sensitivity and rapid analysis capabilities (Wojnowski et al. 2017). As a result, ECs are widely used to detect various contaminants in food samples. ECs comprise several categories according to their different electrical behaviors, including potential sensors, conductivity sensors, electric current sensors, polarographic sensors, and electrolytic sensors. These sensors are versatile tools that evaluate gaseous, liquid, or solid components dissolved in liquids. ECs can accurately measure parameters such as liquid pH,

TABLE 15.1

Nanosensors Applied in Food Analysis

Type of Nanosensor	Matrix	Objective Compound	Applicability	References
Gold nanoparticles	Milk	*E. coli*	Rapid bacterial detection	Guo et al. (2021)
Quantum dots	Fruit	Pesticides	Pesticide levels	Desai et al. (2019)
Carbon nanotubes	Seafood	Histamine	Deterioration detection	Billing et al. (2022)
Graphene-based	Fruit	Pesticides	Residue analysis	Maciel et al. (2020)
Carbon nanofibers	Seafood	Heavy metals	Contaminant detection	Rasheed et al. (2019)
Conductive polymers	Cereals	Mycotoxins	Evaluation of mycotoxin levels	Machungo et al. (2022)
Plasmonic	Beverages	pH	Acidity monitoring	Steinegger et al. (2020)
Fluorescence	Beverages and food	Lactose	Detection of lactose intolerance products	Mei et al. (2021)
Resonance	Meat	Ammonia	Deterioration indicator	Liu et al. (2023), Siribunbandal et al. (2023)

conductivity, and oxidation-reduction potential, providing valuable information on the composition and quality of food products (Khatib and Haick 2022).

The integration of nanotechnology has significantly boosted the field of ECs, improving these devices' accuracy, selectivity, distinctiveness, and sensitivity. These transformative advances have immense potential for the food industry and other applications that rely heavily on accurate and reliable sensor technology (Table 15.1).

The origin of the applicability of EC sensors has not been free of ambiguities. However, according to the historical review by Simões and Xavier (2017), ECs began to be used in the 1950s to monitor and measure oxygen (O_2) in industrial environments. These devices were conceived by Leland C. Clark, who designed a cell with two electrodes separated by an oxygen-permeable membrane and immersed in an electrolyte solution (Simões and Xavier 2017). Oxygen diffused through the membrane was reduced at the indicator electrode, resulting in a current proportional to the O_2 concentration. Subsequently, Clark's sensor found various applications in the medical and environmental fields. However, significant milestones in the applicability of Electrochemical Sensors emerged in the years 1981 and 1986 with the introduction of the Scanning Tunneling Microscope (STM) and the Atomic Force Microscope (AFM) by the teams of Binnig and Rohrer (Binning and Quate 1986). These advances in scientific instrumentation enabled the visualization of material structures at the minuscule scale of 1.0×10^{-9} m (1 nm), laying the foundation for the nascent technology known as nanoscience and nanotechnology. Currently, ECs measure gases, generating an electrical signal proportional to the gas concentration. This application offers a valuable tool for monitoring and quantifying gases in different contexts (Simões and Xavier 2017).

The operation of ECs is based on reactions that allow measuring the concentration of a specific gas, generating, as a result, an electrical signal proportional to the detected concentration (Figure 15.2). The process starts when the gas impinges on the sensor through the small capillary-shaped opening and then passes through the hydrophobic barrier layer until it reaches the ends of the electrode surface. This carefully designed procedure ensures enough gas interacts with the sensor electrode, triggering a reaction that translates into an electrical signal. At the same time, electrolyte leakage from the sensor is prevented. The gas passes through a barrier and encounters the sensor electrode, where an oxidation or reduction process occurs depending on the gas type present. The electrode materials are designed to catalyze these reactions according to the measured gas. Once the electrodes are connected by a resistor, an electric current is generated between the positive and negative electrodes. The magnitude of this current is a direct indicator of the concentration of the gas being analyzed. ECs are often called 'gas stacks' or 'micro fuels' because of the generation of this electrical current during the measurement process (Simões and Xavier 2017).

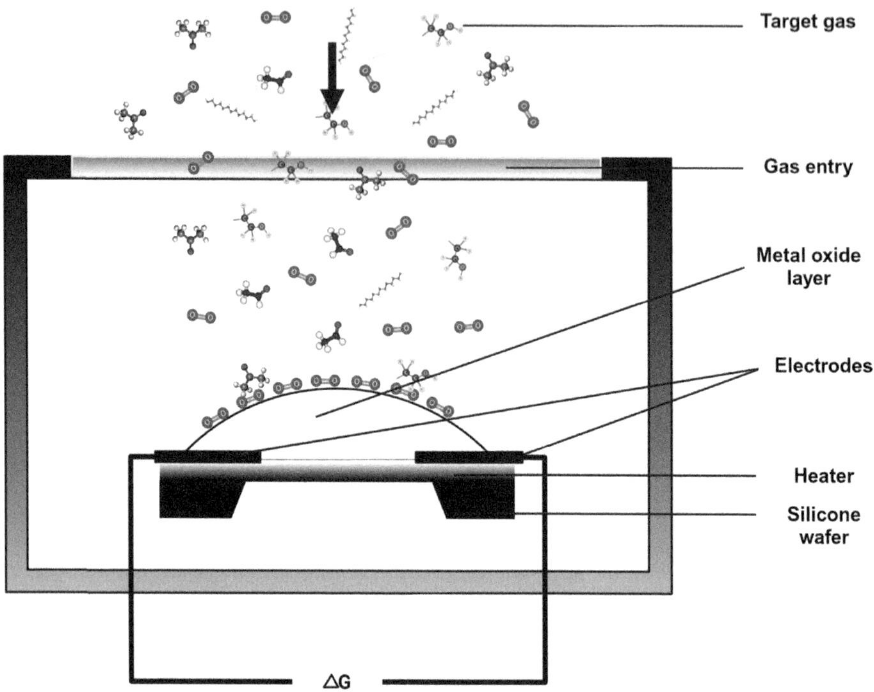

FIGURE 15.2   Electrochemical sensor structure diagram.

15.3 APPLICATIONS OF NANOSENSORS IN FOOD SAFETY

15.3.1 DETECTION OF PATHOGENIC MICROORGANISMS

It is essential to note the impact of various microorganisms such as bacteria, viruses, parasites, and fungi on food, leading to foodborne diseases (Yang and Wang 2008). Food spoilage is the process by which microorganisms alter the biochemical properties of food, reducing its suitability for consumption. The International Association for Food Protection (IAFP), World Resources Institute (WRI), World Food Programme (WFP), Food and Agriculture Organization (FAO), and International Food Information Council (IFIC) monitor food safety, considering factors such as sustainability, climate change, and population growth. Weakened immune systems contribute significantly to the severity of foodborne illnesses (Yang and Wang 2008).

Determining microorganisms in food through nanosensors is a crucial field of study in the search for advanced solutions for food safety control. Based on highly sensitive and selective nanomaterials, these nanosensors offer a promising way to detect and quantify pathogenic microorganisms in food products more efficiently and accurately than conventional methods.

In the context of detecting foodborne bacterial pathogens such as *Salmonella* (Perçin et al. 2017), *Escherichia coli* (Guo et al. 2021), and *Listeria* sp. (Servarayan et al. 2023), nanosensors provide the ability to identify the presence of these bacteria at extremely low concentrations. Early and reliable detection is achieved by exploiting specific interactions between the nanomaterials' surfaces and the pathogens' biomarkers. This reduces the time required for analysis and helps minimize the risk of foodborne illness by rapidly identifying bacterial contamination in the food supply chain (Saravanan et al. 2021).

In the case of fungi and molds, nanosensors have also been proven to be effective in detecting and quantifying their metabolites and breakdown products in food (Perçin et al. 2017). These spoilage agents can adversely affect food quality and safety, underscoring the importance of their early detection. Nanosensors enable continuous, real-time monitoring of food quality, which helps prevent the growth of fungi and molds and ultimately reduces food waste. A highlight of nanosensors is their ability to operate under field and on-site conditions, which facilitates practical application in the food industry. In addition, the nanomaterials' versatility in these sensors allows adaptation to various food matrices and processing environments. The detection of microorganisms in food safety determination using nanosensors is based on multiple detection methods and principles. Some of the standard techniques used are as follows (Saravanan et al. 2021):

- **Optical Detection**: Many nanosensors use optical detection, where changes in fluorescence, absorbance, or light scattering indicate the presence of microorganisms. Nanomaterials such as QD and metal nanoparticles can emit specific optical signals when bound to microbial biomarkers.
- **Electrochemical Sensing**: Nanosensors can exploit changes in electrical properties, such as resistance or electrical current, when interacting with microorganisms. Carbon nanotubes and conductive polymers are examples of nanomaterials used in electrochemical sensing.
- **Detection of Mechanical Changes**: Some nanosensors use changes in the mechanical properties of nanomaterials when they bind to microorganisms. These changes can be detected by techniques such as microcantilever or mass resonance.
- **Magnetic Detection**: Functionalized magnetic nanoparticles can bind to microorganisms and then be detected by magnetic methods, such as nuclear magnetic resonance (**NMR**) or magnetoresistance.
- **Impedance Change Detection**: Changes in electrical impedance caused by the interaction between nanosensors and microorganisms can be measured and used for detection.
- **Detection of Conductivity Changes**: Changes in the electrical conductivity of nanosensors due to the binding of microorganisms can be used to detect their presence.
- **Amplification Techniques**: Some nanosensors can use amplification techniques, such as PCR or loop-mediated isothermal amplification (**LAMP**), to increase the detection signal and improve sensitivity.

15.3.2 Monitoring and Elimination of Chemical Contaminants in Foods

The determination or extraction of contaminants in the food industry, such as additives, pesticides, and mycotoxins, through nanosensors has revolutionized how the safety and quality of food products are monitored and ensured. Based on highly sensitive and selective nanomaterials, these nanosensors enable more accurate, rapid, and efficient detection of various contaminants (Martinović et al. 2016), essential to protect consumers' health and complying with regulatory standards. For food additives, nanosensors can detect and quantify even small concentrations of these compounds used to improve the taste, texture, or appearance of food. Nanosensor technology enables real-time monitoring of the presence of additives, helping to ensure that levels comply with established regulations and prevent consumer overexposure to these substances (Kögler et al. 2016).

In the case of pesticides, nanosensors offer a more efficient and sensitive alternative to traditional detection methods. These sensors can identify the presence of pesticides in food early and at minimal concentrations, helping to avoid exposure to harmful chemical residues and maintain the integrity of agricultural products (Shawon et al. 2020).

Detecting mycotoxins, toxic metabolites produced by fungi, is another area where nanosensors prove highly effective (Pandey et al. 2023). These sensors can identify mycotoxins in foods, such as grains and processed products, helping to prevent the ingestion of contaminated food and ensure food safety. In addition to the techniques listed for detecting microorganisms that can be used for mycotoxins, pesticides, and additives, we can name the specific detection of biomarkers and the monitoring of changes in chemical properties such as redox potential and response to contaminants (Cheli et al. 2023).

15.3.3 Assessment of Food Quality and Freshness

Assessing food quality and freshness is essential to ensure food safety and consumer satisfaction (Sharma et al. 2023). Nanosensors have emerged as promising tools to address this challenge, enabling accurate, real-time monitoring of the physical, chemical, and biological changes that occur in food as it ages or deteriorates. These nanosensors take advantage of the unique properties of nanomaterials and their high sensitivity to detect even slight variations in food characteristics (Siddiqui et al. 2022).

One of the main focuses of nanosensors in evaluating food quality and freshness is the detection of ripening and spoilage gases. During the ripening process of fruits and vegetables, gases such as ethylene are released, influencing their flavor and texture (Shafiq et al. 2020). Nanosensors based on porous nanomaterials, such as metal oxides and polymers (Bai and Shi 2007), can selectively adsorb these gases and generate detection signals, allowing estimation of the degree of ripening and prediction of the remaining shelf life of the products. Nanosensors can also evaluate physical properties such as food texture and moisture (James et al. 2005). Sensors based on conductive nanomaterials, such as carbon nanotubes or polymer nanofibers, can detect changes in electrical resistance caused by changes in texture. In addition, moisture nanosensors can measure the amount of water in food, which is crucial for preventing microorganism growth and spoilage (Sharma et al. 2023).

15.4 ROLE OF ELECTROCHEMICAL SENSORS

15.4.1 Principles of Electrochemical Sensing in the Food Industry

Sensory evaluation of food materials relies on various techniques, ranging from the traditional involvement of human experts to the analysis of chemicals by gas chromatography-mass spectrometry (GC-MS) and the overall assessment of volatile scent intensity using advanced sensor technologies (Deisingh et al. 2004; Loutfi et al. 2015). However, the inherent complexity of most food flavors leads to high costs in their characterization by conventional methods, such as GC or GC-olfactometry. In parallel, sensory analysis carried out by a panel of experts also represents a significant investment, as it requires highly trained personnel who can only work for relatively short intervals. In addition, this approach poses challenges, such as the subjectivity of human responses to odors and the inherent variability between individuals (James et al. 2005; Karakaya et al. 2020).

The ECs have been widely used, but they also experience limited accuracy and problems with long-time stability. An ideal gas sensor should exhibit reliability, robustness, sensitivity, selectivity, and reversibility. Achieving high selectivity and reversibility is a complex challenge (Roy and Yadav 2022). Persaud and Dodd (1982) addressed this problem by creating a set of reversible but partially selective ID layers, each with distinct chemical properties. Selectivity was achieved by applying pattern recognition techniques to the responses generated by this sensor array. The evolution of chemical sensors has benefited from constant advances in computing power, integrated electronics, innovative designs, and processing tools (Persaud and Dodd 1982). This progress is reflected in the growing presence of miniaturized, affordable, portable chemical sensors with static and continuous measurement capabilities, highlighting their usefulness in food, biomedical, industrial, and clinical applications.

15.4.2 Electrochemical Sensor for the Identification of Volatile Compounds

In this context, the need arose to develop an instrument based on the applicability of ECs, taking advantage of their remarkable benefits regarding high sensitivity and correlation with data obtained by human sensory panels in various specific food control applications. For this purpose, the E-nose was created for the efficient and economical analysis of VOCs, dispensing with the need for highly specialized technical personnel. E-noses are devices composed of several gas sensors -mostly

ECs- each semi-selective and reversible, whose outputs are analyzed by pattern recognition software (Loutfi et al. 2015). These methodologies allow us to classify and quantify chemical compounds responsible for sensory perceptions. In this context, Persaud and Dodd (1982) presented the pioneering design of an electronic nose (E-nose) designed to detect and differentiate between complex aromas trained with various biological or chemical odor-sensitive substances (Gonzalez Viejo and Fuentes 2022).

15.5 APPLICATION OF E-NOSE IN THE FOOD INDUSTRY

15.5.1 INTRODUCTION OF E-NOSE TECHNOLOGY

Previous reviews have summarized numerous publications and reports concerning the different methods of VOC determination in the food industry. These methods play a significant role in production, providing information about food quality and its relationship to consumer preferences (Khatib and Haick 2022). Given that, as previously described, various chemical groups such as alcohols, aldehydes, acids, esters, terpenes, pyrazines, and furans contribute to aroma formation during food processing, such as thermal treatment, fermentation, and storage (Karakaya et al. 2020). Thus, analyzing scent composition in food can be challenging due to the mixture of various molecules and the interpretation of the results (Figure 15.3).

Based on scientific reports, the four primary steps in assessing food VOCs are isolation and concentration, separation, identification, and sensory characterization (Lin et al. 2022; Khatib and Haick 2022). The most common are as follows: (i) Solid-phase micro-extraction (SPME) and stir bar sportive extraction (SBSE) are the most used techniques to separate a fraction of volatiles from non-volatiles. (ii) Solvent-assisted flavor evaporation (SAFE) is used to study the active components of food aroma by gas chromatography with an olfactometry detector (GC-O). (iii) Volatiles are mainly separated using GC systems such as GC or comprehensive two-dimensional GC×GC, with mass spectrometry (MS, MS/MS, ToF–MS) used for chemical compound identification. (iv) E-nose technology capable of detecting and discriminating between complex mixtures of scents-associated volatile metabolites without identifying them in foods.

Since its description in 1982, the E-nose has been developed to measure olfaction by replicating and mimicking natural abilities for detecting and decoding VOCs as scent and gas from fermentation (Karakaya et al. 2020). The olfactory systems in nature are highly sensitive and can

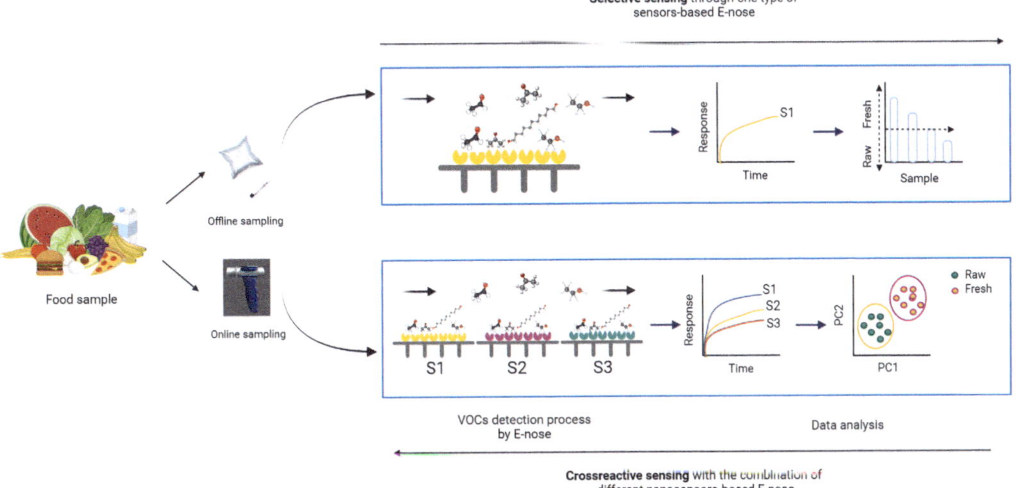

FIGURE 15.3 Functionality and application of electronic nose in the food industry.

distinguish between scents better than other sensory systems. For instance, the nose in mammals can discriminate more than 1 trillion different olfactory stimuli starting at a 0.01 nmol L^{-1} odor concentration (James et al. 2005; Karakaya et al. 2020). Specifically, in humans, 400 intact olfactory receptors have been identified. Thus, it is impossible to relate this ability to the "lock and key" model, in which each receptor responds to only one odor. Therefore, natural olfactory systems absorb and interact with VOCs and scents through a limited number of odor receptors to interpret as computational processes inside the brain, using different ligand binding affinities for the odors. Comparably, the E-nose is designed based on a cross-reactive chemical sensor array, incorporating other sensing materials adapted to algorithms that generate distinct sensor responses to classify the VOCs (Karakaya et al. 2020). These procedures are usually followed by an affinity training and learning phase to identify the correct pattern in the target of pre-classified odor samples under different conditions (Figure 15.4).

Achieving the correct pattern in the target of odors and scents from different food samples requires the same principle of the biological olfactory system in mammals. The E-nose comprises hardware and software components (Banerjee et al. 2019; Deisingh et al. 2004). The software part can be considered the "brain," and the hardware part can be seen as the "olfactory receptors." The software part mainly contains a data processing unit that identifies and classifies each scent detected using digital signatures of the sensed chemicals. The hardware part is a sensor array. Since the main objective of the E-nose is detecting and classifying multiple scents, the sensing array should encompass different types of individual sensors, where each sensor is responsible for detecting another VOC (Karakaya et al. 2020). Selecting suitable sensors for a given specific identity is critical in E-nose. Choosing appropriate hardware components and efficient software analysis is essential in designing and implementing a successful E-nose for a particular problem in the food industry (Mavani et al. 2022).

The components of an E-nose start by detecting ambient gas absorbed by the sensor array. After, the section of the input signal occurs according to the variation in voltage, frequency, and resistance parameters (Deisingh et al. 2004). For example, food contamination is activated by the activity of microorganisms, enzymes, and fat oxidation. This decline in quality may lead to the formation of harmful bacteria that release chemical gases, which are sensed in changes in resistance by an E-nose. However, the E-nose requires artificial neuronal networks (ANNs) to connect sensor signals to the process data. Rather than an algorithm, ANNs have constituted a "machine learning framework" (Kohavi and Provost 1998) to process data and present highly accurate results during VOC recognition for the food industry and other purposes. Therefore, when ANNs apply to the E-nose training using datasets, different chemical mixtures, and diverse aromas, the ANNs increase their extension and construction of pattern recognition (Wilson and Baietto 2011).

Pattern recognition is a general definition for recognizing specific or individual patterns and regular patterns in the data (Persaud and Dodd 1982). Whereas classification is an example of pattern recognition in which a trained model separates the data into classes. Later, these classes are modeled using the collected data during training data of E-nose to identify the correct class label of the object of interest. Several learning-based approaches are used by the ANNs to pattern recognition algorithms in E-nose, increasing the extension of datasets of patterns (Karakaya et al. 2020). The most frequently used data analysis methods include linear discriminate analysis

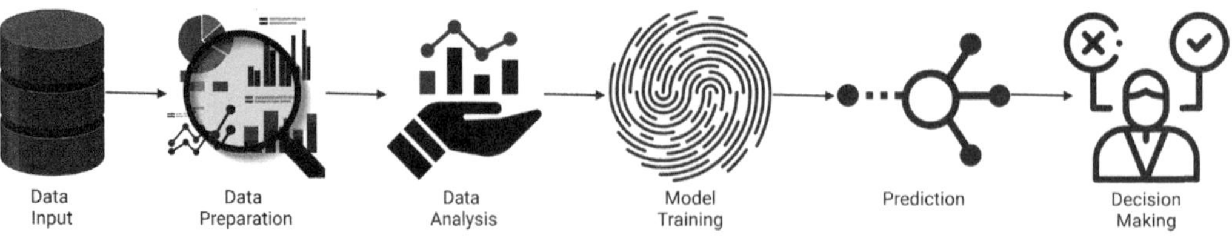

FIGURE 15.4 General block diagram of machine learning algorithm.

(LDA), discriminant function analysis (DFA), stepwise discriminant analysis (SDA), partial least squares discrimination analysis (PLS-DA), generalized least squares regression (GLSR), multiple linear regression (MLR), principal component analysis (PCA), support vector machines (SVMs), and ANNs that may be fed of different model analysis to expend their system. Karakaya et al. (2020) summarized the most frequently utilized data analysis algorithms for the responses by the different E-nose sensor types and the data analysis methods with their characteristics (Table 15.2).

15.5.2 Use of E-nose in the Identification of Alcoholic Beverages

The E-nose has been used to evaluate and classify wine according to its aroma and quality. These noses can identify specific volatile compounds in wine that contribute to its aroma and flavor (Wang et al. 2022). This is especially useful for ensuring product consistency and detecting possible defects in the wine. Electronic noses are devices used to detect adulteration in alcoholic and non-alcoholic beverages by analyzing changes in aroma profiles. These devices employ various sensors to differentiate between different spirits and drinks. Pattern recognition techniques such as PCA, DFA, LDA, ANN, and SVM are applied to classify pure and adulterated samples (Roy and Yadav 2022).

In the wine industry, the characteristics of wine depend on its chemical composition, influenced by grape variety, winemaking, and aging processes. The E-nose and electronic tongues (e-tongues)

TABLE 15.2

Multivariate Statistical Algorithms Reported in E-nose Data Treatment

Algorithm	Advantages	Disadvantages	Sensor Type
PCA	Reduce the data dimensionality. Provide a set of uncorrelated components. Measure probability estimations of high-dimensional data	High computational time with a large amount of data	Metal oxide Conducting polymer Quartz crystal Microbalance Acoustic wave Catalytic bead Optical
Discriminant analysis	Easy to use. Fast in classification applications. Linear decision boundary.	High computational time for training Complex matrix operations Gaussian assumption	Metal oxide Conducting polymer Quartz crystal microbalance Acoustic wave
Regression	Easy to use. Usage in several cases. Great with a small amount of data.	Outliers can cause problems. Overfitting occurs	Electrochemical
ANNs	Can work with incomplete knowledge. Fault tolerance is high. Parallel processing. Once trained, predictions are fast. Practical with a large amount of data. For both regression and classification.	Hardware-dependent computational time. It is hard to find the optimal network structure. Overfitting may occur.	Metal oxide Conducting polymer Quartz crystal microbalance Acoustic wave Electrochemical
SVMs	Effective in high-dimensional spaces. Relatively memory efficient. Fast in both binary and multi-class classification. Works excellent with non-linear data. Effective even in cases when the number of dimensions is greater than the number of data samples.	High computational time with large amounts of data. Noisy data can cause overlapping classes.	Electrochemical

have been used at various stages of wine production for quality control, from the evaluation of grape quality to the analysis of the organoleptic properties of bottled wine (Deisingh et al. 2004; Rodríguez-Méndez et al. 2016). In evaluating grape quality and crush, e-tongues based on ISFETS sensors and voltammetric biosensors have been used to distinguish between grape varieties and vintages. These biosensors have shown correlations between ripe grapes' sugar and phenol content (Banerjee et al. 2019).

During wine fermentation, the conversion of sugars to alcohol is monitored. The E-nose and e-tongues have been used to a lesser extent to monitor wine fermentation, as it is a turbulent process that affects the performance of the sensors. Electronic devices have demonstrated the usefulness of digital technologies such as near-infrared spectroscopy (NIR) and a low-cost electronic nose (E-nose) to detect and assess wine defects (Rodríguez-Méndez et al. 2016). These methods showed high accuracy in detecting defects in red and white wines, which could benefit the wine industry regarding decision-making and product quality in a changing environment (Loutfi et al. 2015). In summary, E-nose and sensor-based technologies have an essential role in the alcoholic and non-alcoholic beverage industry, particularly in detecting adulteration and assessing wine quality at different stages of production (Wilson and Baietto 2011).

15.5.3 Use of E-nose in the Mycotoxin and Fungi Contamination in Foods

Mycotoxins, toxic secondary metabolites of certain fungi, significantly threaten food safety and human health. Although effective, mycotoxin detection methods often require extensive sample preparation and lengthy analytical procedures. In recent years, electronic noses (E-noses) have emerged as promising tools for mycotoxin detection due to their rapid and non-invasive nature (Camardo Leggieri et al. 2021). Accurate and rapid detection of mycotoxins is imperative for ensuring food safety and quality. E-noses offer several advantages over traditional analytical methods, such as reduced analysis time and minimal sample preparation (Mota et al. 2021). E-noses can rapidly screen large quantities of grains, nuts, and other agricultural products for mycotoxin contamination. This early warning system enables producers to take corrective measures and prevent contaminated products from reaching consumers.

Several sensor devices for E-noses to detect fungi and mycotoxin contamination in foods have been developed using different types of detection: optical, thermal, electrochemical, and gravimetric (James et al. 2005). The presence of fungi contamination in foods can produce VOCs; it is essential to note that the numbers and the amounts of individual VOCs vary depending on several factors such as fungi species and strains and growth conditions, such as substrate, nutrients, pH, humidity, and temperature (Cheli et al. 2023). The main VOCs found in cultures of fungi grown on cereals and dry fruits belong to categories such as alcohols, aldehydes and ketones, benzene derivatives, hydrocarbons, and terpenes. Some authors state that Aflatoxin B_1 maize contamination is correlated with the level of VOCs such as E-2-octenal-M, benzene, acetaldehyde, (E-hept-2-enal-M, 2-heptanone-D, and 2-pentyl furan (Li et al. 2021). Other authors primarily correlate the presence of VOCs in wheat, maize, barley, and rice (Gu et al. 2019) with the contamination by mycotoxigenic fungi such as *Aspergillus candidus* (Perkowski et al. 2012), *Aspergillus fumigatus*, *Fusarium graminearum* (Li et al. 2021); and *Aspergillus oryzae* (Pandey et al. 2023).

Despite their promise, E-nose applications in mycotoxin detection face challenges such as sensor drift, interference from background VOCs, and the need for standardized calibration procedures. Future research efforts should focus on sensor stability, data fusion techniques, and developing portable, field-ready E-nose devices. As technology continues to evolve, E-noses hold great potential for revolutionizing the field of mycotoxin monitoring, improving food safety, and protecting public health (Pandey et al. 2023). Addressing current challenges and exploring innovative solutions will pave the way for broader adoption of E-nose technology in this critical application (Cheli et al. 2023).

15.5.4 Application of E-noses in Food Dehydration

Food dehydration is a well-established method for preserving food by reducing its moisture content, preventing microbial growth, and maintaining flavor and nutritional quality (Santonico et al. 2010). Monitoring dehydration is crucial to achieving consistent product quality, optimizing energy consumption, and ensuring food safety (Lopez de Lerma et al. 2012). E-noses respond to VOCs emitted during dehydration, allowing for real-time moisture loss monitoring and detecting off-flavors or spoilage (Santonico et al. 2010). Corrective actions can be taken promptly if the sensor pattern indicates over-drying or developing off-flavors. Adjustments in drying parameters, such as temperature, humidity, or airflow, can be made to optimize the process and ensure the final product meets quality standards.

Santonico et al. (2010) applied this methodology to assess Montepulciano grapes dehydration in postharvest; they reported that VOCs changed significantly with the temperature during the dehydration process. Some reported VOCs altered during this process were ethanol, acetaldehyde, terpenols, and ethylacetate. They mentioned that using the E-nose was successfully achieved to monitor the progressive change of aroma during this process to monitor metabolic changes. Other authors have used the E-nose to establish wine grapes' optimum off-vine dehydration time (Lopez de Lerma et al. 2012). Pei et al. (2016) evaluated two different drying approaches in button mushrooms; the changes in volatile composition were investigated. Results showed that the content of C8 compounds decreased during freeze drying/microwave vacuum drying methodology; they also found that alkanes and heterocyclic compounds were detected in the latter drying periods. The E-nose analysis could discriminate button mushroom samples and different drying periods, having better results through this approach rather than with GC-MS (Pei et al. 2016).

Applying E-noses in food dehydration processes represents a transformative advancement in food preservation and quality control. It offers real-time monitoring that enables precise control over this process, ensuring consistent product quality and optimizing energy consumption.

15.5.5 E-noses in Food Quality Assessment

Food quality assessment ensures consumer satisfaction, safety, and adherence to industry standards (Grassi et al. 2019). Food quality assessment is a multifaceted process involving the evaluation of sensory attributes, shelf life, safety, and adherence to industry specifications. The ability to accurately and efficiently evaluate these aspects of food quality is critical for producers, distributors, and consumers alike (Xiong et al. 2023). Traditional methods for assessing food quality often involve time-consuming and labor-intensive sensory evaluations or analytical laboratory techniques, which may only sometimes provide rapid and cost-effective results. In this context, E-noses have emerged as revolutionary tools that bridge the gap between human perception and analytical chemistry (Stephen et al. 2016). E-noses provide insights into food freshness, detecting subtle changes in aroma that may signal spoilage or shelf-life limitations. This capability is particularly valuable for perishable products, where maintaining quality and safety is paramount.

Some of the applications of the E-noses in freshness assessment are summarized below:

i. **Perishable Foods**: E-noses are highly effective in evaluating the freshness of perishable products like fruits, vegetables, seafood, and dairy items. They can detect subtle alterations in aroma caused by microbial contamination, temperature abuse, or package integrity breaches (Grassi et al. 2019).

ii. **Meat and Poultry**: E-noses are instrumental in assessing the freshness of meat and poultry products. They can identify the onset of spoilage, the presence of pathogens, and the effects of storage conditions on meat quality (Loutfi et al. 2015).

   iii. **Bakery Products**: In the bakery industry, E-noses help monitor the freshness of bread, pastries, and baked goods by detecting changes in VOC profiles related to staling, rancidity, or mold growth (Pulluri and Kumar 2022).

   iv. **Beverages**: E-noses play a role in assessing the freshness of drinks, including juices and wines. They can detect off-flavors or spoilage indicators, improving quality control (Banerjee et al. 2019).

Several examples of E-nose applications in freshness assessment in foods are in the literature; nevertheless, we chose two representative examples for this chapter. Grassi et al. (2019) applied a portable E-nose system to assess meat and fish freshness -Mastersense-; this system was used to test beef and poultry slices and plaice and salmon fillets during their shelf life at 4 °C from the day of packaging and beyond the expiration date. All the obtained models to predict spoilage presented sensitivity and prediction percentages higher than 80%; they also reported that their system would be able to distinguish between different meat species and cuts and differentiate between fish. Another example is the one by Xiong et al. (2023), in which they applied the E-nose to assess chicken freshness. Their prediction models reached more than 94% accuracy; it is essential to note that these authors highlight the need for excellent knowledge and, therefore, application of chemometric analysis to accurately select the features that better explain the variability between samples, to offer a reliable predictive model, which in every case of application of this technology is of paramount importance (Xiong et al. 2023).

15.5.6 Application of E-nose in Food Authenticity Assessment

Food fraud, including mislabeling and adulteration, can seriously affect public health and trust in the food supply chain. The authenticity of food products, ensuring they meet their claimed origin, composition, and quality, is a vital concern for consumers, producers, and regulatory authorities. Detecting and preventing food fraud requires sophisticated analytical techniques to identify subtle product composition and quality variations. Authentic food items possess unique aroma profiles influenced by geographical origin, ingredients, and production methods. E-noses can detect these patterns and have been studied for this specific purpose.

   E-noses find application in a wide range of food authenticity assessments:

   i. **Geographical Origin Verification**: E-noses can discern the geographical origin of products by detecting unique aroma profiles associated with specific regions or terroirs, providing a tool for verifying origin claims.

   ii. **Ingredient Authentication**: They assist in confirming the presence of specific ingredients or identifying unauthorized substitutions, enabling the detection of adulteration.

   iii. **Quality and Age Verification**: E-noses can assess the quality and age of products, identifying freshness deviations, counterfeit products, or improperly stored items.

   iv. **Process Verification**: They validate adherence to specific production methods or certification standards, ensuring the authenticity of specialty products.

One example that describes geographical origin classification through E-noses is the one reported by Yu et al. (2022), where they analyze ginger varieties and geographical origins from seven major production areas in China; a distinct separation between the two types of ginger was achieved. Also, some potential flavor components were selected as essential factors for origin certification (Yu et al. 2022). Miao et al. (2015) applied the E-nose to assess the geographical origin of rums of distinct places; they could distinguish between rums of China, Cuba, Jamaica, the Philipines, and Guatemala (Liu et al. 2023). As for ingredient authentication, Bougrini et al. (2014) reported the detection of adulteration in argan oil through E-nose; they trained their system with different proportions of argan oil adulterated with sunflower oil; their predictive models achieved excellent success classification rates (91.67% and 83.34%) in the recognition of comestible and cosmetic argan oil (Ordukaya and Karlik 2017).

Despite their potential, challenges such as the need for robust reference databases, sensor drift management, and data variability due to various factors must be addressed. As the technology matures, ongoing research and development efforts are expected to enhance further E-noses' reliability and applicability in food authenticity assessment. It is important to note that E-noses offer an efficient means of screening large quantities of products, aiding in the early detection of authenticity issues. The E-noses represent a transformative tool in the fight against food fraud, promoting transparency and trust in the food supply chain (Bougrini et al. 2014). Their role in objectively and rapidly assessing food authenticity is instrumental in safeguarding consumers, upholding industry standards, and maintaining the integrity of food products (Lee-Rangel et al. 2022). As E-nose technology continues to evolve and researchers work to overcome existing challenges, these devices are poised to play an increasingly pivotal role in ensuring the authenticity and safety of food products worldwide. Their adoption promises to strengthen consumer confidence, support industry sustainability, and ultimately contribute to a safer and more reliable food supply chain.

15.5.7 Current E-nose Applications in the Food Industry

Food products generate VOCs during production, processing, storage, and transportation due to their chemical reactions or the influence of external microorganisms, which are inextricably linked to food quality and safety (Table 15.3; Magan and Evans 2000).

It is essential to highlight that VOCs comprise aliphatic and aromatic substances with low molecular weights and boiling points. Solvents, dry cleaning compounds, degreasers, combustion byproducts, paints, water chlorination, chemical intermediates, microwaving processes, and various industrial products are all VOCs. Some VOCs are even permitted as indirect food additives derived from commercial packaging components (Singh et al. 2023). Specifically in the food industry, VOCs arise from various sources, including the ingredients used in the recipe, cooking methods, storage conditions, and microbial activity. However, not all volatile compounds in food are desirable. Some can indicate spoilage or deterioration, such as producing off-flavors and off-odors in meat, dairy, or vegetable products. Certain volatile compounds can also affect food safety, such as making toxic compounds like acrylamide in high heat cooking or storing starchy foods (Pandey et al. 2023). Therefore, it is essential to monitor the volatiles in food to ensure product quality, safety, and consistency (Table 15.4).

TABLE 15.3
Volatiles Reported in Foods

Volatiles in Food	Sources	Examples	References
Scents and flavors.	Ingredient sources Cooking procedures Storage Microbial activity	Aldehydes Ketones Furans Pyrazines	Kumar Verma et al. (2022)
Off-flavors and off-odors.	Spoilage	Rancid oils Ammonia Putrid odors	Siribunbandal et al. (2023)
Toxic compounds.	Storage Heat temperature Time	Acrylamide Biogenic amines Polycyclic aromatic hydrocarbons	Khatib and Haick (2022)

TABLE 15.4

Examples of VOCs That Have Been Reported in Foods

Family Group	Compounds Example	Aroma Description	Examples of Food Sample	References
Alcohols	Hexanol	Bitter, floral	Watermelon	Wang et al. (2022)
Terpenes	Myrcene	Fruity	Lemongrass, thyme, basil, clove	Desai et al. (2019)
	α- and β-pinene	Woody, green, pine-like	Black pepper	Klein et al. (2016)
	Limonene	Citrus	Pistachio, gin, nutmeg	Nazir et al. (2022)
	Linalool	Floral	Saffron, oregano, basil	Raguso (2016)
Aldehydes	Hexanal	Freshly cut grass	Almonds, chestnut peanut, lamb meat	Chen et al. (2019)
	Octanal	Fruity	Orange juice, citrus pulp	Rey-Serra et al. (2022)
Esters	Ethyl octanoate	Fruity, sweet, apricot	Wine, beer, dried fish	Rodríguez-Méndez et al. (2016)
	Ethyl hexanoate	Fruity, floral	Alcoholic beverages	Banerjee et al. (2019)
Furans	Furan	Roast, coffee-line	Peanut, cookies	Karakaya et al. (2020)
	Furfural	Almond, sweet	Bread	Khatib and Haick (2022)
	2,5-dimethyl-4-hydroxy-3(2H)-furanone (furaneol)	Fruity	Kiwifruit, strawberry	Lin et al. (2022)

15.6 CONCLUSIONS AND FUTURE PERSPECTIVES

Nanosensors represent a technological frontier with extraordinary potential to revolutionize the food industry. Maintaining rigorous control over food quality, safety, and authenticity is paramount, as these are critical concerns for consumers, producers, and regulatory agencies. Nanosensors are poised to address pressing challenges such as quality control, food safety assurance, and traceability thanks to their nanoscale dimensions, fantastic sensitivity, and selectivity. As highlighted in this review, the application of nanosensors in the food sector ranges from freshness, flavor, and texture assurance to protection against pathogens, allergens, and contaminants. They contribute significantly to authenticating food products, extending shelf life, and improving supply chain monitoring. In addition, nanosensors play an essential role in reducing food waste, which has direct economic and environmental implications.

One of the most promising fields for integrating nanosensors in the food industry is their use in electronic systems based on E-nose. Due to their increased sensitivity, reduced size, and increased selectivity, nanosensors offer a transformative advantage in this context. By mimicking the olfactory system and detecting VOCs, E-nose with nanosensors becomes a formidable tool for assessing food quality, freshness, and authenticity. These systems provide objective and highly accurate real-time measurements, overcoming traditional sensory and chemical analysis limitations. Despite their potential, it is essential to note that adopting nanosensors in the food industry still faces challenges, such as scalability, stability, and standardization.

In summary, E-noses are gaining popularity due to their automated, non-destructive nature and ability to characterize food flavors efficiently. They are easy to build and affordable, making them an attractive option for the food industry. They provide valuable information in a short period, which is essential in an industry where speed and accuracy are critical.

ACKNOWLEDGMENTS

This study has been in part supported by the Ministerium für Wissenschaft, Forschung und Kunst (MWK) in Baden-Württemberg, Germany, within the program "Sonderförderlinie COVID-19", the EU Horizon 2020 project PHOTONFOOD (No. 101016444) which is part of the Photonics Public Private Partnership, the Research Council of Norway vias project SFI Digital Food Quality (DIGIFOODS) No. 309259 (grant numbers 101016444 and 309259) and the EU Horizon 2022 project M3NIR (No. 101093008).

BIBLIOGRAPHY

Akbari M, Foroudi P, Shahmoradi M, Padash H, Parizi ZS, Khosravani A, Ataei P, Cuomo MT. 2022. The evolution of food security: Where are we now, where should we go next? *Sustainability*. 14(6):3634. https://doi.org/10.3390/su14063634.

Aylott JW. 2003. Optical nanosensors-an enabling technology for intracellular measurements. *Analyst*. 128(4):309–312. https://doi.org/10.1039/b302174m.

Bai H, Shi G. 2007. Gas sensors based on conducting polymers. *Sensors*. 7(3):267–307. https://doi.org/10.3390/s7030267.

Banerjee R, Tudu B, Bandyopadhyay R, Bhattacharyya N. 2019. Application of electronic nose and tongue for beverage quality evaluation. In: *Engineering Tools in the Beverage Industry*. Elsevier; p. 229–254. https://doi.org/10.1016/B978-0-12-815258-4.00008-1.

Billing BK, Verma M, Chaudhary M. 2022. Functionalized carbon nanotube based cyanide detection and degradation. *ChemistrySelect*. 7(2). https://doi.org/10.1002/slct.202104014.

Binning G, Quate CF, Gerber Ch. 1986. Atomic Force Microscope. *Physical Review Letters*. 56: 930.

Bougrini M, Tahri K, Haddi Z, Saidi T, El Bari N, Bouchikhi B. 2014. Detection of adulteration in argan oil by using an electronic nose and a voltammetric electronic tongue. *Journal of Sensor*. 2014:1–10. https://doi.org/10.1155/2014/245831.

Camardo Leggieri M, Mazzoni M, Fodil S, Moschini M, Bertuzzi T, Prandini A, Battilani P. 2021. An electronic nose supported by an artificial neural network for the rapid detection of aflatoxin B1 and fumonisins in maize. *Food Control*. 123:107722. https://doi.org/10.1016/j.foodcont.2020.107722.

Cheli F, Ottoboni M, Fumagalli F, Mazzoleni S, Ferrari L, Pinotti L. 2023. E-nose technology for mycotoxin detection in feed: Ready for a real context in field application or still an emerging technology? *Toxins (Basel)*. 15(2):146. https://doi.org/10.3390/toxins15020146.

Chen W, Wang Z, Gu S, Wang J. 2019. Detection of hexanal in humid circumstances using hydrophobic molecularly imprinted polymers composite. *Sensors and Actuators B: Chemical*. 291:141–147. https://doi.org/10.1016/j.snb.2019.04.065.

Cullum BM, Vo-Dinh T. 2000. The development of optical nanosensors for biological measurements. *Trends in Biotechnology*. 18(9):388 393. https://doi.org/10.1016/S0167-7799(00)01477-3.

Deisingh AK, Stone DC, Thompson M. 2004. Applications of electronic noses and tongues in food analysis. *International Journal of Food Science & Technology*. 39(6):587–604. https://doi.org/10.1111/j.1365-2621.2004.00821.x.

Desai ML, Jha S, Basu H, Singhal RK, Park T-J, Kailasa SK. 2019. Acid oxidation of muskmelon fruit for the fabrication of carbon dots with specific emission colors for recognition of Hg^{2+} ions and cell imaging. *ACS Omega*. 4(21):19332–19340. https://doi.org/10.1021/acsomega.9b02730.

FAO. 2022. *Food Outlook – Biannual Report on Global Food Markets*. FAO. https://doi.org/10.4060/cc2864en.

Fletcher MT, Netzel G. 2020. Food safety and natural toxins. *Toxins (Basel)*. 12(4):236. https://doi.org/10.3390/toxins12040236.

Gonzalez Viejo C, Fuentes S. 2022. Digital assessment and classification of wine faults using a low-cost electronic nose, near-infrared spectroscopy and machine learning modelling. *Sensors*. 22(6):2303. https://doi.org/10.3390/s22062303.

Grassi S, Benedetti S, Opizzio M, Nardo E, Buratti S. 2019. Meat and fish freshness assessment by a portable and simplified electronic nose system (mastersense). *Sensors*. 19(14):3225. https://doi.org/10.3390/s19143225.

Gu S, Wang J, Wang Y. 2019. Early discrimination and growth tracking of Aspergillus spp. contamination in rice kernels using electronic nose. *Food Chemistry*. 292:325–335. https://doi.org/10.1016/j.foodchem.2019.04.054.

Guo Y, Luo Y, Tang M, Zhang M, Yuan M, Chen S, Tu Q, Wang J. 2021. Gold nanosensor for the selective identification of Escherichia coli in foodstuff and its antibacterial ability. *Sensors and Actuators B: Chemical.* 344:130191. https://doi.org/10.1016/j.snb.2021.130191.

He X, Hwang H-M. 2016. Nanotechnology in food science: Functionality, applicability, and safety assessment. *Journal of Food and Drug Analysis.* 24(4):671–681. https://doi.org/10.1016/j.jfda.2016.06.001.

Jagtiani E. 2022. Advancements in nanotechnology for food science and industry. *Food Frontiers.* 3(1):56–82. https://doi.org/10.1002/fft2.104.

James D, Scott SM, Ali Z, O'Hare WT. 2005. Chemical sensors for electronic nose systems. *Microchimica Acta.* 149(1–2):1–17. https://doi.org/10.1007/s00604-004-0291-6.

Jhu M-Y, Sinha NR. 2022. Parasitic plants: An overview of mechanisms by which plants perceive and respond to parasites. *Annual Review of Plant Biology.* 73(1):433–455. https://doi.org/10.1146/annurev-arplant-102820-100635.

John SA, Chattree A, Ramteke PW, Shanthy P, Nguyen TA, Rajendran S. 2022. Nanosensors for plant health monitoring. In: *Nanosensors for Smart Agriculture.* Elsevier; p. 449–461. https://doi.org/10.1016/B978-0-12-824554-5.00012-4.

Karakaya D, Ulucan O, Turkan M. 2020. Electronic nose and its applications: A survey. *International Journal of Automation and Computing.* 17(2):179–209. https://doi.org/10.1007/s11633-019-1212-9.

Khatib M, Haick H. 2022. Sensors for volatile organic compounds. *ACS Nano.* 16(5):7080–7115. https://doi.org/10.1021/acsnano.1c10827.

Klein F, Farren NJ, Bozzetti C, Daellenbach KR, Kilic D, Kumar NK, Pieber SM, Slowik JG, Tuthill RN, Hamilton JF, et al. 2016. Indoor terpene emissions from cooking with herbs and pepper and their secondary organic aerosol production potential. *Scientific Reports.* 6(1):36623. https://doi.org/10.1038/srep36623.

Kögler M, Zhang B, Cui L, Shi Y, Yliperttula M, Laaksonen T, Viitala T, Zhang K. 2016. Real-time Raman based approach for identification of biofouling. *Sensors & Actuators, B: Chemical.* 230:411–421. https://doi.org/10.1016/j.snb.2016.02.079.

Kohavi R, Provost F. 1998. Machine learning-special issue on applications of machine learning and the knowledge discovery process. *Machine Learning.* 30(2/3):271–274. https://doi.org/10.1023/A:1017181826899.

Kumar Verma D, Thyab Gddoa Al-Sahlany S, Kareem Niamah A, Thakur M, Shah N, Singh S, Baranwal D, Patel AR, Lara Utama G, Noe Aguilar C. 2022. Recent trends in microbial flavour compounds: A review on chemistry, synthesis mechanism and their application in food. *Saudi Journal of Biological Sciences.* 29(3):1565–1576. https://doi.org/10.1016/j.sjbs.2021.11.010.

Lakshmipriya T, Gopinath SCB. 2019. An introduction to biosensors and biomolecules. In: *Nanobiosensors for Biomolecular Targeting.* Elsevier; p. 1–21. https://doi.org/10.1016/B978-0-12-813900-4.00001-4.

Lee-Rangel HA, Mendoza-Martinez GD, Diaz de León-Martínez L, Relling AE, Vazquez-Valladolid A, Palacios-Martínez M, Hernández-García PA, Chay-Canul AJ, Flores-Ramirez R, Roque-Jiménez JA. 2022. Application of an electronic nose and HS-SPME/GC-MS to determine volatile organic compounds in fresh Mexican cheese. *Foods.* 11(13):1887. https://doi.org/10.3390/foods11131887.

Li H, Kang X, Wang S, Mo H, Xu D, Zhou W, Hu L. 2021. Early detection and monitoring for Aspergillus flavus contamination in maize kernels. *Food Control.* 121:107636. https://doi.org/10.1016/j.foodcont.2020.107636.

Lin H, Jiang H, Adade SY-SS, Kang W, Xue Z, Zareef M, Chen Q. 2022. Overview of advanced technologies for volatile organic compounds measurement in food quality and safety. *Critical Reviews in Food Science and Nutrition*:1–23. https://doi.org/10.1080/10408398.2022.2056573.

Liu Z, Huang Y, Kong S, Miao J, Lai K. 2023. Selection and quantification of volatile indicators for quality deterioration of reheated pork based on simultaneously extracting volatiles and reheating precooked pork. *Food Chemistry.* 419:135962. https://doi.org/10.1016/j.foodchem.2023.135962.

Lopez de Lerma N, Bellincontro A, Mencarelli F, Moreno J, Peinado RA. 2012. Use of electronic nose, validated by GC-MS, to establish the optimum off-vine dehydration time of wine grapes. *Food Chemistry.* 130(2):447–452. https://doi.org/10.1016/j.foodchem.2011.07.058.

Loutfi A, Coradeschi S, Mani GK, Shankar P, Rayappan JBB. 2015. Electronic noses for food quality: A review. *Journal of Food Engineering.* 144:103–111. https://doi.org/10.1016/j.jfoodeng.2014.07.019.

Machungo C, Berna AZ, McNevin D, Wang R, Trowell S. 2022. Comparison of the performance of metal oxide and conducting polymer electronic noses for detection of aflatoxin using artificially contaminated maize. *Sensors & Actuators, B: Chemical.* 360:131681. https://doi.org/10.1016/j.snb.2022.131681.

Maciel EVS, Mejía-Carmona K, Jordan-Sinisterra M, da Silva LF, Vargas Medina DA, Lanças FM. 2020. The current role of graphene-based nanomaterials in the sample preparation arena. *Frontiers in Chemistry*. 8. https://doi.org/10.3389/fchem.2020.00664.

Magan N, Evans P. 2000. Volatiles as an indicator of fungal activity and differentiation between species, and the potential use of electronic nose technology for early detection of grain spoilage. *Journal of Stored Products Research*. 36(4):319–340. https://doi.org/10.1016/S0022-474X(99)00057-0.

Martinović T, Andjelković U, Gajdošik MŠ, Rešetar D, Josić D. 2016. Foodborne pathogens and their toxins. *Journal of Proteomics*. 147:226–235. https://doi.org/10.1016/j.jprot.2016.04.029.

Mavani NR, Ali JM, Othman S, Hussain MA, Hashim H, Rahman NA. 2022. Application of artificial intelligence in food industry – A guideline. *Food Engineering Reviews*. 14(1):134–175. https://doi.org/10.1007/s12393-021-09290-z.

Mei J, Bao J, Cheng X, Ren D, Xu G, Wei F, Sun Y, Hu Q, Cen Y. 2021. Novel dual-emissive fluorescent silicon nanoparticles for detection of enzyme activity in supplements associated with lactose intolerance. *Sensors & Actuators, B: Chemical*. 329:129164. https://doi.org/10.1016/j.snb.2020.129164.

Miao L, He S, Mo J, Lv S. 2015. Application of electronic nose analysis in rum classification and wine base discrimination. *China Brewing*, 8:106-110. https://manu61.magtech.com.cn/zgnz/EN/10.11882/j.issn.0254-5071.2015.08.022.

Mota I, Teixeira-Santos R, Cavaleiro Rufo J. 2021. Detection and identification of fungal species by electronic nose technology: A systematic review. *Fungal Biology Reviews*. 37:59–70. https://doi.org/10.1016/j.fbr.2021.03.005.

Nazir NU, Abbas SR, Nasir H, Hussain I. 2022. Electrochemical sensing of limonene using thiol capped gold nanoparticles and its detection in the real breath sample of a cirrhotic patient. *Journal of Electroanalytical Chemistry*. 905:115977. https://doi.org/10.1016/j.jelechem.2021.115977.

Nile SH, Baskar V, Selvaraj D, Nile A, Xiao J, Kai G. 2020. Nanotechnologies in food science: Applications, recent trends, and future perspectives. *Nanomicro Letters*. 12(1):45. https://doi.org/10.1007/s40820-020-0383-9.

Ordukaya E, Karlik B. 2017. Quality control of olive oils using machine learning and electronic nose. *Journal of Food Quality*. 2017:1–7. https://doi.org/10.1155/2017/9272404.

Pandey AK, Samota MK, Kumar A, Silva AS, Dubey NK. 2023. Fungal mycotoxins in food commodities: Present status and future concerns. *Frontiers in Sustainable Food Systems*. 7. https://doi.org/10.3389/fsufs.2023.1162595.

Pei F, Yang W, Ma N, Fang Y, Zhao L, An X, Xin Z, Hu Q. 2016. Effect of the two drying approaches on the volatile profiles of button mushroom (Agaricus bisporus) by headspace GC-MS and electronic nose. *LWT – Food Science and Technology*. 72:343–350. https://doi.org/10.1016/j.lwt.2016.05.004.

Perçin I, Idil N, Bakhshpour M, Yılmaz E, Mattiasson B, Denizli A. 2017. Microcontact imprinted plasmonic nanosensors: Powerful tools in the detection of Salmonella paratyphi. *Sensors*. 17(6):1375. https://doi.org/10.3390/s17061375.

Perkowski J, Stuper K, Buśko M, Góral T, Kaczmarek A, Jeleń H. 2012. Differences in metabolomic profiles of the naturally contaminated grain of barley, oats and rye. *Journal of Cereal Science*. 56(3):544–551. https://doi.org/10.1016/j.jcs.2012.07.012.

Persaud K, Dodd G. 1982. Analysis of discrimination mechanisms in the mammalian olfactory system using a model nose. *Nature*. 299(5881):352–355. https://doi.org/10.1038/299352a0.

Petrović T, D'Agostino M. 2016. Viral contamination of food. In: *Antimicrobial Food Packaging*. Elsevier; p. 65–79. https://doi.org/10.1016/B978-0-12-800723-5.00005-X.

Pulluri KK, Kumar VN. 2022. Qualitative and quantitative detection of food adulteration using a smart E-nose. *Sensors*. 22(20):7789. https://doi.org/10.3390/s22207789.

Raguso RA. 2016. More lessons from linalool: Insights gained from a ubiquitous floral volatile. *Current Opinion in Plant Biology*. 32:31–36. https://doi.org/10.1016/j.pbi.2016.05.007.

Ramezani M, Esmaelpourfarkhani M, Taghdisi SM, Abnous K, Alibolandi M. 2020. Application of nanosensors for food safety. In: *Nanosensors for Smart Cities*. Elsevier; p. 369–386. https://doi.org/10.1016/B978-0-12-819870-4.00021-9.

Rasheed T, Nabeel F, Adeel M, Rizwan K, Bilal M, Iqbal HMN. 2019. Carbon nanotubes-based cues: A pathway to future sensing and detection of hazardous pollutants. *Journal of Molecular Liquids*. 292:111425. https://doi.org/10.1016/j.molliq.2019.111425.

Rey-Serra P, Mnejja M, Monfort A. 2022. Inheritance of esters and other volatile compounds responsible for the fruity aroma in strawberry. *Frontiers in Plant Science*. 13. https://doi.org/10.3389/fpls.2022.959155.

Rodríguez-Méndez ML, De Saja JA, González-Antón R, García-Hernández C, Medina-Plaza C, García-Cabezón C, Martín-Pedrosa F. 2016. Electronic noses and tongues in wine industry. *Frontiers in Bioengineering and Biotechnology*. 4. https://doi.org/10.3389/fbioe.2016.00081.

Roy M, Yadav BK. 2022. Electronic nose for detection of food adulteration: A review. *Journal of Food Science and Technology*. 59(3):846–858. https://doi.org/10.1007/s13197-021-05057-w.

Santonico M, Bellincontro A, De Santis D, Di Natale C, Mencarelli F. 2010. Electronic nose to study post-harvest dehydration of wine grapes. *Food Chemistry*. 121(3):789–796. https://doi.org/10.1016/j.foodchem.2009.12.086.

Saravanan A, Kumar PS, Hemavathy RV, Jeevanantham S, Kamalesh R, Sneha S, Yaashikaa PR. 2021. Methods of detection of foodborne pathogens: A review. *Environmental Chemistry Letters*. 19(1):189–207. https://doi.org/10.1007/s10311-020-01072-z.

Servarayan KL, Krishnamoorthy G, Sundaram E, Karuppusamy M, Murugan M, Piraman S, Vasantha VS. 2023. Optical immunosensor for the detection of *Listeria monocytogenes* in food matrixes. *ACS Omega*. 8(18):15979–15989. https://doi.org/10.1021/acsomega.2c07848.

Shafiq M, Anjum S, Hano C, Anjum I, Abbasi BH. 2020. An overview of the applications of nanomaterials and nanodevices in the food industry. *Foods*. 9(2):148. https://doi.org/10.3390/foods9020148.

Sharma A, Ranjit R, Pratibha, Kumar N, Kumar M, Giri BS. 2023. Nanoparticles based nanosensors: Principles and their applications in active packaging for food quality and safety detection. *Biochemical Engineering Journal*. 193:108861. https://doi.org/10.1016/j.bej.2023.108861.

Shawon ZBZ, Hoque ME, Chowdhury SR. 2020. Nanosensors and nanobiosensors: Agricultural and food tech-nology aspects. In: *Nanofabrication for Smart Nanosensor Applications*. Elsevier; p. 135–161. https://doi.org/10.1016/B978-0-12-820702-4.00006-4.

Siddiqui S, Chattree A, Higgins P, Kavipriya K, Nguyen TA, Rajendran S. 2022. Food products safety. In: *Nanosensors for Smart Agriculture*. Elsevier; p. 757–768. https://doi.org/10.1016/B978-0-12-824554-5.00031-8.

Simões FR, Xavier MG. 2017. Electrochemical sensors. In: *Nanoscience and Its Applications*. Elsevier; p. 155–178. https://doi.org/10.1016/B978-0-323-49780-0.00006-5.

Singh R, Dutt S, Sharma P, Sundramoorthy AK, Dubey A, Singh A, Arya S. 2023. Future of nanotechnology in food industry: Challenges in processing, packaging, and food safety. *Global Challenges*. 7(4). https://doi.org/10.1002/gch2.202200209.

Siribunbandal P, Osotchan T, Kim Y-H, Jaisutti R. 2023. Highly sensitive colorimetric ammonia sensors based on polydiacetylene/zinc oxide nanopellet-embedded PDMS films for meat spoilage detection. *ACS Applied Polymer Materials*. https://doi.org/10.1021/acsapm.3c00993.

Steinegger A, Wolfbeis OS, Borisov SM. 2020. Optical sensing and imaging of pH values: Spectroscopies, materials, and applications. *Chemical Reviews*. 120(22):12357–12489. https://doi.org/10.1021/acs.chemrev.0c00451.

Stephen Inbaraj B, Chen BH. 2016. Nanomaterial-based sensors for detection of foodborne bacterial pathogens and toxins as well as pork adulteration in meat products. *Journal of Food and Drug Analysis*. 24(1):15–28. https://doi.org/10.1016/j.jfda.2015.05.001.

Thakali A, MacRae JD. 2021. A review of chemical and microbial contamination in food: What are the threats to a circular food system? *Environmental Research*. 194:110635. https://doi.org/10.1016/j.envres.2020.110635.

Valdés MG, Valdés González AC, García Calzón JA, Díaz-García ME. 2009. Analytical nanotechnology for food analysis. *Microchimica Acta*. 166(1–2):1–19. https://doi.org/10.1007/s00604-009-0165-z.

Verbeke W. 2005. Consumer acceptance of functional foods: Socio-demographic, cognitive and attitudinal deter-minants. *Food Quality and Preference*. 16(1):45–57. https://doi.org/10.1016/j.foodqual.2004.01.001.

Vo-Dinh T, Kasili P. 2005. Fiber-optic nanosensors for single-cell monitoring. *Analytical and Bioanalytical Chemistry*. 382(4):918–925. https://doi.org/10.1007/s00216-005-3256-7.

Wang A, Zhu Y, Qiu J, Cao R, Zhu H. 2022. Application of intelligent sensory technology in the authentication of alcoholic beverages. *Food Science and Technology*. 42. https://doi.org/10.1590/fst.32622.

Wilson AD, Baietto M. 2011. Advances in electronic-nose technologies developed for biomedical applications. *Sensors*. 11(1):1105–1176. https://doi.org/10.3390/s110101105.

Wojnowski W, Majchrzak T, Dymerski T, Gębicki J, Namieśnik J. 2017. Portable electronic nose based on electrochemical sensors for food quality assessment. *Sensors*. 17(12):2715. https://doi.org/10.3390/s17122715.

Xiong Y, Li Y, Wang C, Shi H, Wang S, Yong C, Gong Y, Zhang W, Zou X. 2023. Non-destructive detection of chicken freshness based on electronic nose technology and transfer learning. *Agriculture*. 13(2):496. https://doi.org/10.3390/agriculture13020496.

Yang H, Wang Y. 2008. Application of atomic force microscopy on rapid determination of microorganisms for food safety. *Journal of Food Science*. 73(8):N44–N50. https://doi.org/10.1111/j.1750-3841.2008.00918.x.

You Q, Zhang X, Wu F-G, Chen Y. 2019. Colorimetric and test stripe-based assay of bacteria by using vancomycin-modified gold nanoparticles. *Sensors and Actuators B: Chemical*. 281:408–414. https://doi.org/10.1016/j.snb.2018.10.103.

Yu D, Zhang X, Guo S, Yan H, Wang J, Zhou J, Yang J, Duan J-A. 2022. Headspace GC/MS and fast GC e-nose combined with chemometric analysis to identify the varieties and geographical origins of ginger (Zingiber officinale Roscoe). *Food Chemistry*. 396:133672. https://doi.org/10.1016/j.foodchem.2022.133672.

16 Safety Issues and Regulatory Challenges of Nanoproducts in Food

Fabio Granados-Chinchilla,
Eduardo Alberto López Maldonado,
and Amber R. Solangi

16.1 REGULATORY FRAMEWORK FOR FOOD NANOMATERIALS

Nanotechnology is a field of applied sciences and technologies involving manipulating matter at the atomic and molecular scale, usually <100 nm. These materials possess different properties than their non-nano counterparts as their high surface area increases their reactivity (Gatoo et al., 2014).

Nanotechnology products could substantially impact the food and feed sector in the future, potentially offering benefits for industry and the consumer. However, possible risks need to be considered (Falk et al., 2002). Both public and private funded research worldwide are developing applications in fields such as treating food's mechanical and sensory properties (e.g., to achieve changed taste or texture) and modified nutritional value. For instance, nanotechnology may also be used in packaging to ensure better food protection or detect its freshness (Ashfaq et al., 2022).

The food industry has already incorporated nanomaterials into foods, dietary supplements, and food contact materials (FCMs), including cutting boards, plastic containers, and sandwich bags. Companies also use nanotechnologies as food additives, flavor/taste modifiers, and for preservation through nano antimicrobials (Lugani et al., 2021). However, for potential health risks, nanomaterials' specific properties and characteristics must be considered. For example, food or food packaging nanoparticles can access the human body via ingestion, inhalation, or skin penetration (Ashfaq et al., 2022). When ingested, their small size allows them to circulate through the body and access sensitive organs [e.g., bone marrow, lymph nodes, the spleen, the brain, the liver, and the heart (Ashfaq et al., 2022)]. Furthermore, some nanomaterials can travel deeper into the nucleus of cells and damage the DNA (Onyeaka et al., 2022).

As the use of nanobiotechnology has increased and new technologies have been developed, further questions have emerged from consumers and scientists. At the same time, countries/regions have begun to advance, refine, or articulate regulatory approaches for nanofoods and advance in regulatory science and other research efforts to support the responsible expansion of nanotechnology in these areas. Interestingly, the potential food safety problems related to new uses of biotechnology were discussed early on (National Research Council, 1988).

16.1.1 LEGISLATION FRAMEWORK IN EUROPE: EFSA

Since 2006, EFSA has been ensuing nanotechnology advances within its remit, which includes reviewing the current state of knowledge and the latest developments in nanotechnology regarding food and feed. Consequently, it has been within the EFSA Scientific Committee's responsibilities to advise how to assess applications from food operators to use (manufactured) nanomaterials in food additives, enzymes, flavorings, FCMs, novel foods, food supplements, feed additives, and pesticides.

DOI: 10.1201/9781003514039-16

The work considers the risks of nanomaterials and nanoparticles that might be present in the food chain for human and animal health. In this regard, EFSA recently released a draft opinion on nanotechnologies in food and feed (EFSA, 2021a, Figure 16.1). They found that current approaches to risk assessment can be applied to nanomaterials but should be carried out on a case-by-case basis.

As a result of this work, the two guidance documents were released on August 3, 2021, by the agency will help to clarify further how the European scientific community has approached the evaluation of nanomaterials within the food and feed chain:

i. **Guidance on Risk Assessment of Nanomaterials in the Food and Feed Chain**: animal and human health – presents a blueprint to put in place and gradually assist the valuation of nanomaterials. It uses the most relevant scientific studies on the properties and potential hazards of nanomaterials to detail aspects relating to exposure assessment, hazard identification, and characterization to consider, introduces a tiered framework for toxicological testing, and induces attention to contemplations concerning *in vitro/in vivo* toxicologic studies specific for nanomaterials (EFSA, 2021a).
ii. Guidance on technical requirements for regulated food and feed product applications to establish the presence of small particles, including nanoparticles, relates criteria for assessing to conventional materials that contain a fraction of small particles but do not necessarily meet the definition of engineered nanomaterials and define the requirements for submissions in the regulated food and feed product expanses. It primarily focuses on which materials require assessment and how to perform such evaluation. Still, it also sketches how and when a claimant can incorporate 'existing scientific studies' into any application (EFSA, 2021b).

Both documents also set down data and information requirements for claimants when submitting materials for assessment as part of EU market authorization procedures, e.g., for use as food additives or FCMs.

In addition to the above interpretations, according to the Guidance on Risk Assessment, measuring the exposure to nanomaterials from food is "essentially the same as for non-nanomaterials and will require consideration of the likely exposure scenarios." A subsection specific to FCMs is included, which asserts that nanomaterials incorporated in an FCM may "structurally differ from the pristine nanomaterial." Therefore, in addition to the characterization of the nanomaterial used to manufacture an FCM, it becomes necessary to characterize the nanomaterial as present in the FCM and possibly when released from the FCM.

Moreover, the Guidance on Technical Requirements applies to "chemical materials either as substances or mixtures to be assessed by EFSA" unless the claimant can confirm they do not contain small particles in suspension. This includes non-intentionally added substances in the food that may transfer from FCMs. However, if the levels of a substance migrating into food are <60 mg L^{-1} and can be solubilized before ingestion of the packed food, a conventional FCM risk assessment may be sufficient because the migrant complies with EU Regulation 10/2011 (i.e., on plastic materials and articles intended to come into contact with food) and is not a particle.

On the other hand, titanium dioxide nanoparticles in the food sector are widespread as a white food coloring. Nevertheless, the European Food Safety Authority (EFSA, 2021c) has stated that TiO$_2$ can no longer be considered safe as a food additive. A Commission Regulation (2022/63), released in January, amended EC 1333/2008 regarding TiO$_2$ withdrawn authorization (Blaznik et al., 2022; Boutillier et al., 2022).

Regulation 2015/2283 on novel foods, released in November, defines novel foods as "newly developed, innovative food, food produced using new technologies and production processes, as well as food which is or has been traditionally eaten outside of the European Union" (Pisanello and Caruso, 2018). More specifically, there were ten categories considered novel foods produced from new sources of ingredients and processing technologies. Moreover, novel food may also cover food consisting of certain micelles or liposomes and manufactured nanomaterials. The European

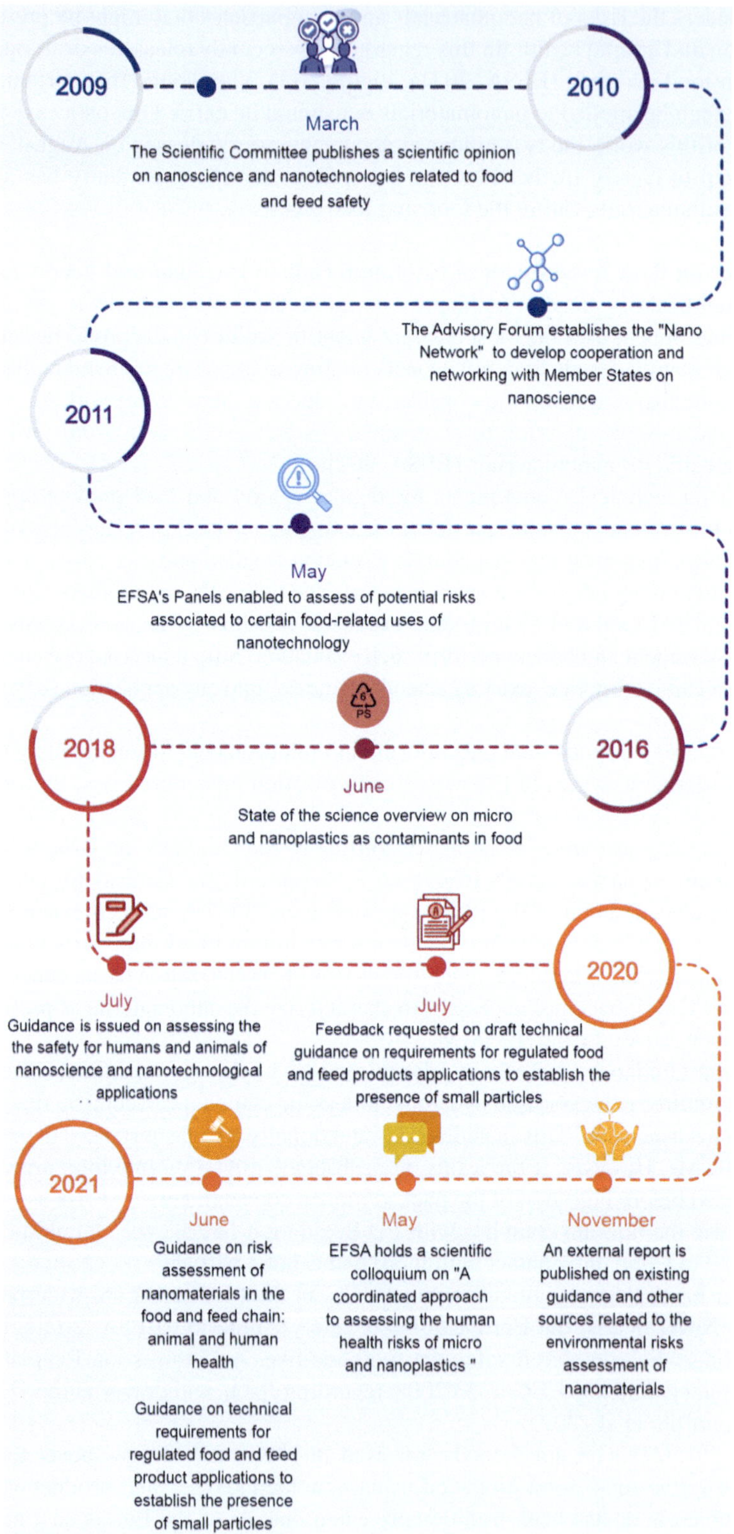

FIGURE 16.1　EFSA contributions to the development of the regulatory framework. Timeline prepared with Visme (Easy WebContent, Inc. Rockville, MD, USA.)

Commission has clarified that the provisions of the Novel Food legislation extend to feed food additives and nutrient sources (EFSA, 2021a).

The "Active and Intelligent Materials and Articles Regulation" (EC) No 450/2009 forecasts instituting an inventory of substances sanctioned for manufacturing active and intelligent materials. Such types of materials (e.g., production machines, conveyor belts, packaging materials, and products like tableware, cutlery, and cutting machines) serve to extend food shelf life by retaining or improving the condition of packaged food by releasing or absorbing substances to or from the food or the environment surrounding it.

An additional regulatory instrument worth mentioning to be updated includes the use of nano forms of pesticides is regulated under Regulation (EU) 2018/1881 on the Registration, Evaluation, Authorization, and Restriction of Chemicals (REACH), linked to the Commission Recommendation 2011/696/EU on the definition of a nanomaterial. Finally, a comprehensive overview of the regulatory provisions for nanotechnologies in food in the EU can be found in the recent work by Schoonjans and coworkers (2023).

16.1.2 LEGISLATION FRAMEWORK IN THE USA: FDA

The US Food and Drug Administration (FDA) polices a wide range of products, including foods, some of which may employ nanotechnology or encompass nanomaterials. As early as 2006, a task force was created to determine the regulatory approaches that foster the persistent development of innovative, safe, and effective regulated products that use nanotechnology or nanomaterials (Institute of Medicine, 2009). It is also commended with identifying and recommending any knowledge or policy gaps within the nanotechnology research community.

The Guidance for the Food Industry laid by the FDA, as of July 2014, includes the following:

- Use of Nanomaterials in Food Animals # 220
- Considering whether an FDA-regulated product concerns the usage of nanotechnology
- Reviewing the consequences of substantial manufacturing process amendments, including evolving technologies, on the safety and regulatory stages of food ingredients (including color additives) and FCMs

On August 12, 2020, the Nanotechnology Industries Association (NIA) announced the publication of the report "Nanotechnology-Over a Decade of Progress and Innovation" by the US Food and Drug Administration (FDA).

The report provides an update on the FDA's advancements in the field of nanotechnology since its last report in 2007, hence providing the current state of nanomaterials science and nanotechnology (https://nanotechia.org/sites/default/files/ntf_report_2020_july.pdf), it also highlights the agency's role in advancing the public health through the regulation of products within its jurisdiction that involve the application of nanotechnology and references industry guidance for use of nanomaterials in human and animal food and in drug products strengthening the regulatory science research in nanotechnology. Finally, it summarizes collaborations with federal agencies, academic institutions, and international partners.

The FDA guidelines also rationalize that nanochemicals constitute a new food technology and recommend that corporations consult with the agency beforehand taking products to market and warn that nanotechnology products may require supplementary safety appraisal on a case-by-case basis. Within the current guidelines is also affirmed that nano foods would not be allowed through a fast-track process known as "generally recognized as safe" (GRAS); far too little is known about the human health and environmental implications of this technology to allow it to be transferred into the food chain without rigorous assessment (Institute of Medicine, 2009; Najahi-Missaoui et al., 2021).

Hence, until now, the FDA has maintained a product-focused and science-based regulatory policy to appropriately regulate products using this emerging technology. The industry remains liable for warranting that its products meet all applicable legal obligations, including safety standards, notwithstanding the emerging nature of the technology involved in manufacturing a product.

To provide sound guidelines to its stakeholders and ensure the proper allocation of resources, the FDA works in conjunction with other organizations that advance nanotechnology regulatory science, for example, the Centre for Food Safety and Applied Nutrition (CFSAN), whose goal of using systematic approaches to characterize nanomaterials, inspecting the likelihood of nanomaterial leaching from FCMs and to determine if there is a safety apprehension related to these packaging materials, and investigating diverse approaches to study the latent toxicity of nanomaterials used in foods. To put contact materials migrating from food packaging into perspective, Canada alone's *per capita* sales of ultra-processed foods were estimated in 2016 at 275 per year, the fourth highest among 80 countries (Polsky et al., 2020).

As a significant increase in the use of nanoscale materials in drugs, devices, biologics, cosmetics, and food has been manifested, other institutions that have supported the FDA review and approve nanotechnology-based products, including the National Centre for Toxicological Research (NCTR)/Office of Regulatory Affairs (ORA) Nanotechnology Core Facility, NCTR and Arkansas Human and Animal Food Laboratory (ARLHAF). Together, they have translated the methodologies to detect nanoscale materials in toxicological studies to biological samples of FDA-regulated products. For example, NCTR has developed three lipid quantitation standards already published by ASTM (i.e., E297-21, E3323-21, and E3324-22).

It is worth mentioning that the FDA is not alone in emitting regulations and guidelines, more recent than efforts from other agencies; the last report emitted by the OECD regarding nanotechnological applications in foods was in 2013, "Regulatory Frameworks for Nanotechnology in Foods and Medical Products." The OECD has been constantly developing assay guidelines for regulatory testing of nanomaterials to ensure mutual acceptance of data (Rasmussen et al., 2019).

16.1.3 SHORTCOMINGS IN THE REGULATORY WORK

An improved understanding of nanomaterials enables control agencies to evaluate possible adverse health effects from regulated nanotechnology products. However, much research is still needed to comprehend how nanoparticles move around an organism and what tests should be conducted to determine their toxicity (Savage et al., 2020; Fröhlich, 2017).

There is still information lacking regarding the mechanisms of absorption, distribution, and excretion of these nanomaterials, which must be, in turn, well characterized. Furthermore, there is still an unclear definition of nanotechnology or nanomaterials, as in food, you have natural nanomaterials (e.g., droplets of homogenized milk, mayonnaise, carbohydrate-based food caramels, such as bread, cornflakes, sugar, and biscuits; Rogers, 2016). Notwithstanding, most of the literature solves this using "manufactured or engineered nanomaterials" to avoid confusion with natural nano-scaled materials.

Aside from some aspects of EU regulation that directly link nanomaterials with existing legislation, most nanobiotechnological applications are regulated in the agricultural, feed, and food sectors by mainly building on guidance for industry (Amenta et al., 2015). Although the recommendation for the sector primarily relies on the producers and manufacturers to abide by these standards, agencies should take a more active approach. Under these guidelines, companies will consult the corresponding agencies, but the products are not reviewed for safety. Considering the novel risks of nanotechnology, agencies must issue mandatory regulations. In fact, in the USA, the Center for Food Safety petitioned the FDA for compulsory nanotechnology regulations in food and later sued the agency for its failure to respond.

Regulatory agencies still use a case-by-case approach for assessing nanotechnology products, employing the combination product framework to determine the product type and resulting supervisory requirements (Paradise, 2019). Simultaneous with the debate about whether the existing body of law is sufficient, pronounced interrogatives have emerged regarding the inherent risks of nanotechnology and nanoparticle products. Recurring themes for concern include nanoparticle toxicity and human health impacts of exposure (Sahu and Hayes, 2017), especially the effects of exposure or administration routes, inadvertent repercussions of nanoparticles' capability to traverse the blood-brain barrier, and chronic effects of nanoparticles.

On the other hand, traditional definitional distinctions and accompanying legal requirements for review, approval, and post-market surveillance and assessment may not be ideal for evaluating such nano-based products. The existing legal framework may function for the prevailing products. Nonetheless, the expanding intricacy of nanotechnology and its junction with other fields will likely extend their reach (Paradise, 2019).

Regulation must consider if the labeling of foods for consumers is sufficient to advise them that products include nanotechnology or nanomaterials. Also, consumer education and engagement may be warranted (Chuah et al., 2018; Siegrist and Keller, 2011), and determining the public's perception and opinion is crucial (Rothen-Rutishauser et al., 2021). As recently as December of 2022, AVICENN, a French specialist association that monitors nanomaterials (https://veillenanos.fr/avicenn/), surveyed food products and found that $n=20$ out of the 23 items tested contained unlabelled nanomaterials. The products ranged from infant formula, ham, and vitamin supplements to dog food. The Food Information to Consumers Regulation establishes the rules for labeling ingredients, with conditions concerning manufactured nano foods. Under this directive, all manufactured nanomaterials must be undoubtedly specified in the list of ingredients. The names of such constituents must be followed by the word "nano" in brackets (Figure 16.2).

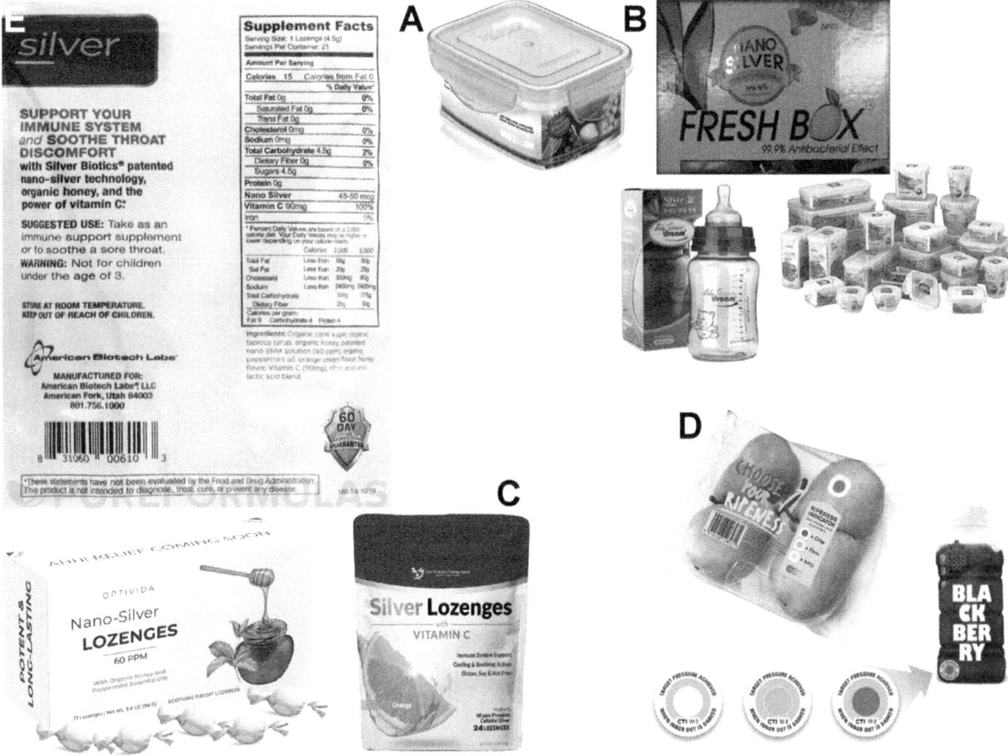

FIGURE 16.2   (a) Examples of food containers sold commercially using nano silver technology. (b) Feeding bottles and mug cups developed with antibacterial technology (Bumbudsanpharoke and Ko, 2015). (c) Organic honey and vitamin C mineral supplement are sold in the commercial market with declared use of silver nanoparticles. (d) Smart packaging in pears using ripeSense→technology (Alam et al., 2021). (e) Guaranteed label declaring the use of nanoparticles of silver in food.

Finally, despite some efforts, no internationally recognized standard protocols for toxicity testing nanomaterials in food or feed are currently available. A consistent regulatory framework for evaluating nanotechnology is necessary for food and animal feed.

16.2 NANOTOXICITY

16.2.1 COMMON USES OF NANOPARTICLES AND TOXICITY OUTCOMES

Nanotechnology poses the food industry with many new approaches for improving food quality, shelf life, safety, and health. Yet, there is apprehension from consumers, regulatory agencies, and the food industry regarding potential adverse effects (toxicity) accompanying the application of nanotechnology in foods.

Several characteristics intervene in the toxicity outcome of nanoparticles, including nature, chemical composition, particle size, interfacial properties, surface area, shape, aspect ratio, crystallinity, dissolution (Egbuna et al., 2021), ionic potential (Sardoiwala et al., 2018), the capacity of accumulation or aggregation states, biocorona (a layer(s) of biomolecules that adsorbs to the surface of a nanoparticle in a biological environment) formation/evolution (Liu et al., 2022), the food matrix and food ingredients (Moradi et al., 2022; Youn and Choi, 2022; Chang et al., 2023), their coating (even when considered biocompatible, Oliveira et al., 2020) and even their use (i.e., how they interact with food).

The most common nanoparticle uses include (i) food ingredients for color, texture, and flavor (ii) food production and packaging (i.e., improved mechanical barriers, detection of microbial contamination, and potentially enhanced bioavailability of nutrients) (iii) Nutrients and Dietary Supplements (e.g., ingredients and additives, such as vitamins, antimicrobials, antioxidants in nutrients, and health supplements for enhanced absorption and bioavailability) (iv) safe food storage and transport (v) construction of food nanosensors (vi) they are improving food quality safety (Table 16.1).

16.2.2 SOME RECENT EXAMPLES OF TOXICITY IN ANIMALS

There is still insufficient information regarding nano foods, their fate, and possible toxic effects in the long term after human exposure. However, there is extensive research focused on the materials themselves. Most of these toxicological tests are *in vitro* (mono-, co-, and three-dimensional culture, and microfluidic chips), and some are *in vivo* (mice models, Caenorhabditis elegans models, zebrafish models, human volunteers, Kumar et al., 2017; Hu et al., 2022a). However, despite advances in nanotoxicology models, neither *in vitro* nor *in silico* (computer-simulated toxicity models) testing tools can dependably predict adverse outcomes or replace *in vivo* testing (Hu et al., 2022a). Innovative testing tactics are required to foretell the potential toxicity of nanoparticles before, during, and after their manufacturing process to aid the introduction of safer materials into our lives.

Most research regarding toxicity has been done in plants and bacteria. However, nanoparticles' ionic properties (cations, anions) positively influence the toxicity pattern toward bacteria (de Oliveira Mallia et al., 2022). Interestingly, Dudefoi and coworkers (2017) demonstrated that TiO_2 nanoparticles do not impact the bacterial human intestinal community. Also, crustaceans, usually reserved to evaluate ecotoxicological effects, have been used to assess nanoparticle toxicity (Sardoiwala et al., 2018). For example, two different crustaceans (i.e., *Tigriopus fulvus* and *Corophium insidiosum*) have served as tools to evaluate the toxicity of Zn-based nanoparticles (Vimercati et al., 2020). The authors confirmed that the toxic effect could be mainly attributed to the Zn ions, ensuring that the dissolution processes play a crucial role in the toxicity of the ZnO nanoparticles. On the other hand, in humans, the research has focused on the toxicity of specific cellular lines (Sardoiwala et al., 2018). However, there is definite evidence regarding toxicity and allergenicity in humans due to occupational exposure (Di Gioacchino et al., 2022; Mohammadi and Galera, 2023).

TABLE 16.1

Selected Applications of Nanoparticles in Food Industries

Nanoparticle [Additive Code]	Nanoparticle Sample[a]	Common Uses in the Food Industry	Relevant References
Calcium carbonate, $CaCO_3$, [E170]		Pickering emulsifiers, plastic food contact materials, plant nutrition, and pest control.	EFSA (2022a, b), Guo et al. (2021)
Titanium oxide, TiO_2, [E171]		Whitening and brightening agent, active food packaging.	EFSA (2021c), Zhang and Rhim (2022)
Iron oxide or hydroxide, Fe_2O_3 or $Fe(OH_2)$ [E172]		Antimicrobial coatings, food packaging, enzyme immobilization, protein purification, and food analysis as a pigment	Góral et al. (2023)
Silver, Ag [E174]		Antimicrobial agent in food packaging technologies.	EFSA (2021d), Ahmad et al. (2021)
Gold, Au [175]		Antibacterial activity, improving barrier properties, using as biosensors, scavenging.	Ahari et al. (2022)
Microcellulose [E460i]		Texturizer, an anti-caking agent, a fat substitute, an emulsifier, an extender, and a bulking agent.	EFSA (2018b)
Manganese oxide, MnO, [E530]		Seed priming, growth promoter (Mn source)	Kasote et al. (2021), Matuszewski et al. (2020)
Silica, SiO_2, [E551]		Keeping fruit fresh, carriers for fragrances and flavors, anticoagulant, antifoam, thickener, auxiliary filter, and clarifier.	EFSA (2018a)
Copper/copper oxide or carbonate, CuO or $CuCO_3$		Improve biopolymers' mechanical features and barrier properties—antimicrobial and antioxidant activities in food packaging.	Mesgari et al. (2022)
Zinc/zinc oxide, Zn/ ZnO		Zn supplement, nutrient fortifier, and agricultural fertilizer, active packaging.	Youn and Choi (2022)
Carbon nanostructures		Food analysis, safety, and packaging materials. Food preservative. Antibacterial. Improving seed and sprout germination/growth	Baz et al. (2020), Raul et al. (2022), Halawani et al. (2023)
Lipid nanoparticles		Release of bioactive substances, beverage emulsions	McClements and Xiao (2017)

[a] Images acquired using scanning (SEM) or transmission electron microscope (TEM) at 100 μm.

Focused on the toxicity of the nanoparticle itself, for example, recent attention has been devoted to nanometals such as selenium (Bano et al., 2022), zinc oxide (Chong et al., 2021; Youn and Choi, 2022; Rahman et al., 2022), iron oxide (Wu et al., 2022), and carbon-based nanoparticles (Zhu, 2022). There is also evidence of the toxicity of nanoparticles targeting specific organs in animal models, for example, the reproductive system (Ajdary et al., 2021), intestine and liver (Voss et al., 2021), bone marrow and liver (Alghriany et al., 2022), spleen (Ilmiawan Sutomo et al., 2022), and respiratory system (Deng et al., 2023).

However, rodents are the most reliable model (Hu et al., 2022b). Several murine models have been used to assess the toxicity of diverse nanoparticles, including silver (Iyola et al., 2017; Cho et al., 2018), gold (Bahamonde et al., 2018), magnesium oxide (Mazaheri et al., 2019), iron oxide (Talash et al., 2019; Wu et al., 2022), aluminum oxide (Alghriany et al., 2022), and, more recently, nickel oxide (Lyon-Darden et al., 2023). In an exciting study, Wu and coworkers (2022) used magnetic resonance imaging to assess the toxicity of Fe_3O_4 particles of 2.3, 4.2, and 9.3 nm in multiple organs. They observed toxicity of small nanoparticles (2.3 and 4.2 nm), especially in the hearts of mice, where they observed the production of reactive oxygen species, especially the ·OH radical. Additionally, no significant toxicity was observed for SiO_2 and gold nanoparticles. They attributed the toxicity of the iron nanoparticles to both the iron element and size. Size-dependent toxicity was already described for silver in mice by Cho and coworkers (2018).

In terms of fate and toxicity outcomes, dissolution rates of nanoparticles can be increased by one of two mechanisms. Firstly, there is aqueous complexation, where aqueous species complex free ions are released from the material's surface. The second mechanism is ligand-enhanced dissolution, extracting surface metal atoms from nanoparticle surfaces. Dissolution properties within food matrixes have been reported previously, for example, for ZnO nanoparticles (Youn and Choi, 2022). Additionally, even during biocorona formation under physiological conditions, the transformation and dissolution of cobalt nanoparticles have been demonstrated to be influenced by phosphate molecules (Mei et al., 2019). Biocorona formation initially imparts nanoparticles with properties that aid them in escaping immunologic detection; there is evidence that indicates that they can trigger an inflammatory response in human HeLa cells and keratinocytes (García-Hevia et al., 2021; Li et al., 2022).

16.2.3 PRO VS CONS

Benefits and toxicity should always be evaluated together. For example, Badgar and Prokisch (2021) using a paramecium model, showed that selenium nanoparticles are less toxic than non-nano selenium selenite. The authors proved that biosynthesized selenium nanoparticles function as antidotes against silver and bulk selenium-induced toxicity, increasing the tolerance for these toxicants two-fold after treatment with the nanoparticles. Similarly, zinc oxide nanoparticles have been demonstrated to improve animal health, quality, and quantity in milk and meat production (Rahman et al., 2022). Similarly, Alghriany and coworkers (2022) demonstrated that curcumin nanoparticles increased the therapeutic efficiency of the natural pigment against the harmful effects of oxidative stress associated with Al_2O_3 nanoparticles. In this regard, most available evidence points to oxidative stress as the primary device of toxicity of nanoparticles. Food additives have been demonstrated to induce gastrotoxicity, hepatotoxicity, gut microbiota alterations, and animal behavior alterations (Medina-Reyes et al., 2020; Campos et al., 2021).

16.2.4 INTESTINAL EPITHELIAL CELL TOXICITY INDUCED BY NANOPARTICLES

As food nanoparticles' main avenue of exposure is ingestion, it is logical to assess the impact and burden these substances impart on the gastrointestinal system and its cellular conglomerate. Additionally, nanoparticle-loading drugs or other biologics research is centered on developing nanoparticles for the oral administration route rather than the intravenous one (Vitolo et al., 2022). The digestive tract's singular pH, the intestinal mucus layer, the intestinal absorption capacity, or the gastrointestinal tract microbiota will allow for some complex potential interactions with the nanoparticles (Vitolo et al., 2022). For example, Shabbir and coworkers (2023) recently demonstrated that silica nanoparticles significantly regulate and modify the gut microbiota configuration in immunodeficient mice, affecting the abundance of different bacterial communities. Analogously, Ren and coworkers (2023) have demonstrated that oral exposure to silver nanoparticles generates intestinal cytotoxicity, specifically damaging the epithelial structure, reducing the mucosal layer's thickness, and altering the microbiota. Notably, the reduced thickness of the mucosal layer increased the phagocytosis of silver nanoparticles by dendritic cells, which generates reactive oxygen species and induces uncontrolled apoptosis.

Further experiments have demonstrated that the chronically inflamed intestine is still more susceptible to silver nanoparticles' necrotic/apoptotic effect (Kämpfer et al., 2020). Finally, Abudayyak and coworkers (2020) demonstrated using Caco-2 cells (human intestinal epithelial cells) that nickel oxide nanoparticles substantially reduced cell viability. They caused DNA and oxidative damage at concentrations as low as 30–150 µg mL^{-1}. Again, apoptosis might be a primary cell death mechanism for the exposed intestinal cells.

Geppert and coworkers (2021) use rainbow trout cells to create an *in vitro* intestinal barrier model and test the toxicity of several nanoparticles. They demonstrated that the toxicity order as follows: polyvinylpyrrolidone coated Ag < uncoated Ag < CuO < ZnO < TiO$_2$ nanoparticles [from the most toxic (lowest EC$_{50}$) to the least harmful (highest EC$_{50}$)]. They also demonstrated basolateral translocation of Ag, Cu, and Zn-based nanoparticles or ions liberated from them through the epithelial cell layer. Similarly, Agans and coworkers (2019) used a human gut simulator to assess that TiO$_2$ nanoparticles elicit a modest effect over the microbiota compared to their Ag counterparts.

On the flip side, 5-fluorouracil and polyethyleneglycol-loaded silica and ceria nanoparticles have been used to treat colon cancer and inflammatory bowel syndrome, respectively (Perumal et al., 2021; Vitulo et al., 2022). Similarly, as the small intestine is unique, it has also been targeted as an excellent candidate for localized drug delivery. Compared to the other parts of the gastrointestinal tract, it has (i) more agreeable conditions than the stomach, (ii) offers comparatively longer stay times, and (iii) possesses a greater surface area. The internalization of nanoparticles can be achieved by crossing intestinal barriers such as the mucus layer, the epithelial layer, and tight junctions. The physical and chemical properties of nanoparticles also affect their internalization and fate, various factors that contribute to nanocarriers' cellular uptake, including their surface chemistry, surface morphology, and functionalization of nanoparticles (Madni et al., 2020).

16.2.5 PARTICULAR CASE OF MICRO(NANO)PLASTICS

Finally, a recent area of research that has come to light and now draws considerable interest is the effect on micro(nano)plastics. Nowadays, it is recognized that these microplastics travel along the food chain and that this pollution is directly and indirectly generated through human food systems following the order of food production, food processing, and food consumption (Wen et al., 2022). There is already evidence of residues of micro(nano) plastics in edible fish species and fishery products (Alberghini et al., 2023) and of bioaccumulation in different organs (Guerrera et al., 2021). These nanoparticles are of interest from the risk management and hazard analysis point of view since they may represent three types of hazards (physical, chemical, and biological). However, despite their relevance, there is still limited information on the presence of nanoplastics in food (Molina and Benedé, 2022).

Plastic particles have been reported to induce physical stress and damage, apoptosis, necrosis, inflammation, oxidative stress, and immune responses, which could contribute to advancing diseases such as cancer, metabolic disorders, and neurodevelopmental conditions. Also, they may affect food allergies, where they could act by altering the digestibility of food allergens, increasing intestinal permeability, promoting an intestinal inflammatory milieu, or dysbiosis, which could encourage food allergen sensitization (Molina and Benedé, 2022). In fact, within the gastrointestinal system, nanoplastics have been explicitly reported to cause particle translocation (Hu et al., 2022b), neurobehavioral toxicity (Huang et al., 2022), cytotoxicity, damage to the gut barrier, intestinal inflammation, and microbial alteration (Wen et al., 2022).

16.2.6 RISK AND HAZARD ASSESSMENT FOR NANOPARTICLES

Risk management of nanoparticles is needed as nanoparticles have diversified effects on humans and the environment and can be used in industrial applications. Technical reports such as TR12855 and TR13121 for providing a framework related to risk assessment for nanomaterials have been established by the International Organization for Standardization (ISO). However, no accurate risk assessment of nanomaterials can be performed until the complete identification of hazards, exposure, and severity associated are considered. Still today, there is uncertainty around the transport, partitioning, degradation, transformation, and mutual interaction (environmental fate) of a combination of nanoparticles and their interaction with living organisms. Hence, insufficient data is available to associate the relationship between exposure assessment and hazard characterization. Therefore, limited studies are available on exposure assessments only.

In contrast, hazard characterization or toxicity studies have been extensively performed. Risk characterization is the final step toward completing a risk assessment in which scenario and sensitivity analysis of several nanoparticles must be performed before being included in food applications. Developing a more unified risk assessment methodology with further enhanced food-based nanoparticle modeling techniques and data sets would help advance knowledge in this area (de Oliveira Mallia et al., 2022). However, several steps and strategies for hazard assessment and quantitation have been taken (Fernández-Cruz et al., 2018).

16.3 FOODS, NANOMATERIALS, AND ALLERGENICITY

Nanotechnology has rapidly transformed food production, manufacturing, and processing in recent years, intending to enhance food quality. However, their use in this continuous effort to improve foods could have unexpected health consequences as they can act as haptens or allergens and trigger antigen response. This may contribute to the epidemic of immune-related maladies in children, such as food allergies, a significant public health concern (Chen et al., 2023).

Though food allergies generally impact a considerable percentage of adults, children seem to be the most affected, with their prevalence continuously increasing in recent decades (Chen et al., 2023). So far, allergy development has been linked with environmental factors, but the higher prevalence in children suggests that environmental factors are essential during the earliest stages of life. Dietary practices and the environment impact gut health in young children, which may ultimately affect food tolerance.

On the other hand, antigen presentation is a crucial immunological process for developing an adaptive immune response. Dysfunction of the innate [cellular, e.g., macrophages, dendritic cells, neutrophils and non-cellular, e.g., the complement system, inflammatory mediators] or adaptive (acquired, composed by lymphocytes: B and T cells) immune systems could be associated with activation, suppression, or modulation (Leite-de-Moraes et al., 2012). However, overstimulation of the immune system could result in unfavorable outcomes, collectively called immune hypersensitivity reactions. These reactions can be triggered by harmful pathogens or their products, as well as innocuous antigens (Leite-de-Moraes et al., 2012).

Allergic diseases include asthma, food allergy, allergic rhinitis, allergic conjunctivitis, atopic dermatitis, atopic eczema, and life-threatening anaphylaxis (Leite-de-Moraes et al., 2012). An allergic form of atopic disease develops following the extension of T helper 2 (Th2) cells and immunoglobulin (Ig) class switching to IgE by plasma cells following first exposure to an allergen (e.g., dust mite, molds, animal dander, insect stings) (Leite-de-Moraes et al., 2012; Ellenbogen et al., 2018). Food allergies and other food hypersensitivities bear on millions of people. The primary food allergens are milk, eggs, fish, crustacean shellfish, tree nuts, peanuts, wheat, and soybeans, but more than $n=160$ food allergens are recognized (in fact, US FDA recently released the guidance titled "Evaluating the Public Health Importance of Food Allergens Other Than the Major Food Allergens Listed in the Federal Food, Drug, and Cosmetic Act"). In this regard, there is considerable evidence that can link nanomaterials with type I allergic hypersensitivity disorders (atopy), including asthma (including effects on established and *de novo* airway disease) (Meldrum et al., 2017; Alsaleh and Brown, 2020).

Though nanomaterials are generally considered poorly immunogenic, even in the company of stark adjuvants, specific antibodies have been produced against liposomes, synthetic polymers, and fullerenes (Dobrovolskaia, 2022). Once again, due to their unique physicochemical properties (compared to their non-nano counterparts), nanomaterials (specifically, nanoparticles) have shown wide biodistribution and tissue accumulation (see examples below), which also contribute to their toxicological outcomes (Kumar et al., 2023). For example, extended access to tissue locations could mean the direct interface with Th2-type effector cells.

16.3.1 Allergenicity of Metal Nanoparticles in Murine Models

As reflected by most examples herein, murine models of food allergy are currently considered the best predictor of food allergenicity for assessing the potential allergenicity of novel food (Castan et al., 2020). In a recent study, Tsuchida and coworkers (2023) found that nickel nanoparticle administration in BALB/c mice showed intestinal epithelial tissue damage, elevated serum interleukin (IL)-17 and IL-1β levels, and higher nickel accumulation in the liver and kidney. After intraperitoneal lipopolysaccharide in conjunction with intradermal dermal nickel nanoparticles administration to mice, an allergic reaction to nickel was induced. The nanoparticles induced tissue accumulation, significant lymphocytic infiltration, and increased serum IL-6 and IL-17 levels. Nickel nanoparticles caused a heightened sensitization to the allergy reactions (as demonstrated by an increase in Th17 cells). Hence, oral exposure to the nickel nanoparticles results in biotoxicity and accumulation in tissues (even when compared to microparticles and the naked ion) and suggests a higher probability of allergy development.

In agreement with the above results, Hirai and coworkers (2016) already demonstrated that nickel nanoparticles in the presence of lipopolysaccharides trigger the onset of metal allergy in mice. The authors also showed that mice pretreated with Ag nanoparticles and lipopolysaccharides but not with the silver ions (thought to cause allergies) developed CD4+ T cells and IL-17A-mediated allergic inflammation in response to Ag. Remarkably, Au and Si nanoparticles, which are minimally ionizable, did not cause sensitization. Quantitative analysis of the Ag distribution suggested that small Ag nanoparticles (i.e., ≤10 nm) transferred to the draining lymph node and freed ions more readily than a large counterpart. These results suggest that metal nanoparticles served as ion carriers to enable metal sensitization.

In contrast, Johnson and coworkers (2022) demonstrated that the interaction of the birch pollen allergen with silica nanoparticles will not worsen allergic sensitization, a state of type 2 inflammation, but instead seems to decrease it by skewing toward a T helper 1(Th1)-dominated immune response (a type 1 inflammatory profile). Lastly, current knowledge regarding other metal nanomaterials and their potential to induce/exacerbate dermal and respiratory allergy are summarized in the work by Roach and coworkers (2019).

16.3.2 Milk Allergenicity and Its Relationship with Nanoparticles

One of the main issues with nanoparticles in foods is that their exposure could be amplified as not only food represents a direct way into the organism, but in general terms, the amount of food consumed by a mammal is considerable (World, Europe, and North America average of food consumption for 2018 was 2.13, 2.36, 1.85 kg food per person per day, Collins and Sanders, 2017). Hence, chronic dietary exposure to nanomaterials through foodstuffs should be a consideration. Issa and coworkers (2022) have found evidence that nanoparticles (i.e., SiO_2, TiO_2, and Ag), used as food additives, can cross the placenta during pregnancy and reach the fetus, putting infants at higher risk of developing life-threatening food allergies after birth. Excretion in milk is also suggested, hence continuing to expose the neonate during a critical window of susceptibility. Nanoparticle exposure may disrupt the host-intestinal microbiota and interfere with the intestinal barrier and gut-associated immune system development in the fetus and neonate. Dorier and coworkers (2019) had already determined that exposure to TiO_2 (as a food additive E171) to intestinal cells (i.e., Caco-2 and HT29-MTX cells) causes an inflammatory profile, together with increased mucus secretion demonstrating dysregulation on several features that contribute to the protective function of these cells within the intestine.

In an exciting study, Phue and coworkers (2022) demonstrated that food additive nanomaterials (based on Si and Ti) increased the antigenicity and allergenicity of milk proteins (β-lactoglobulin and casein) and skimmed milk. Mast cell degranulation (a proxy for allergenicity) was higher when exposed to particle-interacted skim milk, where nanomaterials of Ti showed the highest effect, and this tendency was retained even after subjecting to simulated gut digestion. Particles induced alterations in the structure of milk proteins, exposing epitopes that increase the allergenicity of milk proteins.

16.3.3 Lipid Nanoparticle Allergenicity

Food allergy can occur through several immunological pathways, including lipids (López-Fandiño, 2020). They are essential ingredients of some nano liposome foods (López-Fandiño, 2020). In the case of skin exposure to nanomaterials, immunomodulatory effects may induce allergic reactions (Yoshioka et al., 2017). On the other hand, glycolipids have the potential allergenicity due to their carbohydrate epitopes (Hils et al., 2020). Also, lipids can interact with proteins. Such contacts shift the structural properties of the protein allergen, which may expose linear epitopes inside proteins and affect protein immunogenicity (Meng et al., 2020).

On the other hand, Xu and coworkers (2023) have advanced the means to impart extended relief from allergies by inducing a dynamic immune tolerance state by using a lipid nanoparticle platform to deliver mRNA-encoded peanut allergen epitopes to liver sinusoidal endothelial cells. Those proteins may initiate an allergic response in new organs in the body. Still, in the liver, they cause the targeted cells to activate a tolerant immune reaction that deactivates the allergic response.

Additionally, the authors performed several experiments using egg protein-sensitized mice and liver-targeting nanoparticles. After being exposed to the allergen through inhalation, the mice generated regulatory T cells that are programmed to overturn the allergic response to the egg protein, as well as specific fragments of the egg protein with the liver-targeting nanoparticle boosted immune tolerance, and it did so better than targetted transport of the complete protein. Finally, nanoparticle-treated mice dramatically reduced life-threatening responses to the allergen when modeling anaphylaxis.

16.3.4 Nanotechnological Approaches for Treatment of Food Allergens

Finally, despite the documented drawbacks regarding immune response triggered by nanomaterials, there is still a myriad of accessible nanostructures that, by controlling their physicochemical properties, would permit procuring a sounder and more effectual compound for clinical applications, either in allergy diagnosis or treatment (e.g., specific immunotherapy) (Pohlit et al., 2017; Johnson et al., 2020).

For example, nanoparticles, carbon nanotubes, liposomes, polymers, dendrimers, and nanogels, among others, can be manufactured by managing the sum of their properties, including their functionalization with ligands, which can afford an anticipated interaction with dendritic cell receptors to stimulate or modulate the response, as well as specific immunoglobulin E (IgE), or effector cell receptors (Boraschi et al., 2018; Mayorga et al., 2021; Rai et al., 2023).

In yet another example, applying dendrimeric antigens, nano allergens, and nanoparticles in allergy diagnosis is highly hopeful since it can improve sensitivity by augmenting specific IgE binding, mimicking carrier proteins, or developing signal detection. Additionally, glycodendrimers, liposomes, polymers, and nanoparticles have shown therapeutic promise in acting as scaffolds of allergenic structures, adjuvants, or protectors of the allergen from degradation (Boraschi et al., 2018; Mayorga et al., 2021; Rai et al., 2023).

As a specific example, Shahgordi and coworkers (2020) used a combination of curcumin and ovalbumin encapsulated into polylactic co-glycolic acid (PLGA) nanoparticles to enhance their sublingual immunotherapy efficiency in a murine model of allergic rhinitis. In this case, treatment with all synthesized PLGA formulations significantly decreases total IgE. The study of nasal lavage fluid also showed significantly reduced total and eosinophil cell count levels in the group treated with the nano-formulation. Hence, the author demonstrated that curcumin and allergen nanoparticles can be potential immune modulatory agents.

Again, in a murine model, polyanhydride nanoparticles loaded with cashew (the second most reported tree nut allergen) were tested as an effective oral immunization treatment. Indeed, the nanoparticles system led to a pro-Th1 and Treg immune response [i.e., a higher Th1/Th2 ratio in comparison with animals immunized with free cashew nut proteins, a decrease in splenic Th2 cytokines (IL-4, IL-5, and IL-13), an enhancement of pro-Th1 (IL-12 and IFN-γ) and regulatory (IL-10) cytokines, and an increased expansion of CD4+ T regulatory cells (e.g., CD4+Foxp3+ and CD4+LAP+) in the mesenteric lymph nodes].

Furthermore, their immunomodulatory properties could be introduced as a new approach to managing cashew nut allergy (Araujo Pereira et al., 2018).

16.3.5 Viral Nanoparticles Are Helpful in Food Allergy Research

In Pazos-Castro and coworkers' (2022) effort, a peach allergen Pru p 3, a lipid transport protein, was exposed by genetic fusion on the external surface of viral nanoparticles derived from Turnip mosaic virus, a potyvirus. The recombinant nanoparticles were produced in plants, purified, and characterized from different standpoints.

Their potential usefulness was highlighted by their ability to induce the proliferation of human immune cells, especially when coupled to the allergen natural lipid ligand, which could be transported together with the nanoparticles through intestinal epithelial cells without affecting the monolayer integrity. When delivered to animal models of food allergy, they substantially dropped down some of the primary markers of the allergic response without the need for external adjuvants. In addition, no macroscopic, nephritic, or hepatic alterations were detected in the recipient mice. Thus, these nanoparticles seem to be excellent candidates for further studies about treating pathologies with an immunologic basis.

16.4 CONCLUSION AND FUTURE PERSPECTIVE

In conclusion, the complex environment surrounding safety concerns and regulatory problems related to nanoparticles in the food sector highlights an evolving complication. The advancement of nanotechnology offers vast potential in enhancing food quality, safety, and nutritional value. However, serious concerns exist regarding nanoparticles' potential toxicity and incorporation into food products. Regulatory bodies such as the European Food Safety Authority (EFSA) in Europe and the Food and Drug Administration (FDA) in the USA are pivotal in ensuring the practical use of nanoproducts in the food sector. Despite significant advancements in establishing legislative guidelines for nanoproducts in food, certain shortcomings persist within the regulatory process. The extensive development of nanotechnology often exceeds the capacity of regulatory entities to evaluate and manage these modern products effectively.

Consequently, an ongoing interaction involving regulatory bodies, researchers, and industry stakeholders is imperative to address emerging safety concerns and dynamically refine regulatory plans. Nanoparticles find diverse applications in food products, including enhancing nutrient delivery and improving packaging materials. However, investigations have clarified potential examples of toxicity, mainly when nanoparticles interact with biological systems. Recent toxicity observed in animals has raised severe concern about nanoparticle exposure, highlighting the significance of rigorous safety evaluations. The advantages and disadvantages of employing nanoparticles in food items demand careful consideration. While nanoparticles present innovative solutions, their possible risks warrant cautious examination. For example, intestinal epithelial cell toxicity triggered by nanoparticles emphasizes the necessity for an in-depth understanding of their mechanisms of operation and probable long-term consequences for human health.

Looking ahead, the widespread exploitation of nanoparticles offers instrumental tools for advancing food allergy research. Their distinctive characteristics encourage researchers to examine the allergic mechanisms and improve targeted therapeutic interventions carefully. By employing the capabilities of nanotechnology, the field of food allergy research continues to acquire substantial benefits.

BIBLIOGRAPHY

Abudayyak, M.; Güzel, E.; Özhan, G. 2020 Cytotoxic, genotoxic, and apoptotic effects of nickel oxide nanoparticles in intestinal epithelial cells. *Turk. J. Pharm. Sci.* 17(4):446–451.

Agans, R. T.; Gordon, A.; Hussain, S.; Paliy, O. 2019 Titanium dioxide nanoparticles elicit lower direct inhibitory effect on human gut microbiota than silver nanoparticles. *Toxicol. Sci.* 172(2):411–416.

Ahari, H.; Fakhrabadipour, M.; Paidari, S.; Goksen, G.; Xu, B. 2022 Role of AuNPs in active food packaging improvement: A review. *Molecules.* 27(22):8027.

Ahmad, S. S.; Yousuf, O.; Islam, R. U.; Younis, K. 2021 Silver nanoparticles as an active packaging ingredient and its toxicity. *Packaging Technol. Sci.* 34(11–12):653–663.

Ajdary, M.; Keyhanfar, F.; Moosavi, M. A.; Shabani, R.; Mehdizadeh, M.; Varma, R. S. 2021 Potential toxicity of nanoparticles on the reproductive system animal models: A review. *J. Reprod. Immunol.* 148:103384

Alam, A. U.; Rathi, P.; Beshai, H.; Sarabha, G. K.; Deen, M. J. 2021 Fruit quality monitoring with smart packaging. *Sensors.* 21:1509.

Alberghini, L.; Truant, A.; Santonicola, S.; Colavita, G.; Giaccone, V. 2023 Microplastics in fish and fishery products and risks for human health: A review. *Int. J. Environ. Res. Public Health.* 20(1):789.

Alghriany A. A. I.; Omar, H. E. M.; Mahmood, A. M.; Atia, M. M. 2022 Assessment of the toxicity of aluminum oxide and its nanoparticles in the bone marrow and liver of male mice: Ameliorative efficacy of curcumin nanoparticles. *ACS Omega.* 7(16):13841–13852.

Alsaleh, N. B.; Brown, J. M. 2020 Engineered nanomaterials and type I allergic hypersensitivity reactions. *Front. Immunol.* 11:222.

Amenta, V.; ASchberger, K.; Arena, M.; Bouwmeester, H.; Botelho, F.; Brandhoff, P.; Gottardo, S.; Marvin, H. J. P.; Mech, A.; Quiros, L.; Rauscher, H.; Schoonjans, R.; Vettori, M. V.; Wiegel, S.; Peters, R. J. 2015 Regulatory aspects of nanotechnology in the agri/feed/food sector in EU and non-EU countries. *Regul. Toxicol. Pharmacol.* 73(1):463–476.

Araujo Pereira, M.; Rebouças, J de S.; Ferraz-Carvalho, R de S.; Luis de Redín, I.; Guerra, P. V.; Gamazo C.; Brodskyn, C. I.; Irache J. M.; Santos-Magalhães, N. S. 2018 Poly(anhydride) nanoparticles containing cashew nut proteins can induce a strong Th1 and Treg immune response after oral administration. *Eur. J. Pharm. Biopharm.* 127:51–60.

Ashfaq, A.; Khursheed, N.; Fatima, S.; Anjum, Z.; Younis, K. 2022 Application of nanotechnology in food packaging: Pros and Cons. *J. Agric. Food Res.* 7:100270.

Badgar, K.; Prokisch, J. 2021 Testing toxicity and antidote effect of selenium nanoparticles with *Paramecium caudatum. Open J. Anim. Sci.* 11(4):532–542.

Bahamonde, J.; Brenseke, B.; Chan, M. Y.; Kent, R. D.; Vikesland, R. J.; Prater, M. R. 2018 Gold nanoparticle toxicity in mice and rats: Species differences. *Toxicologic Pathiol.* 46(4) 431–443.

Bano, I.; Skalickova, S.; Arbab, S.; Urbankova, L.; Horky, P. 2022 Toxicological effects of nanoselenium in animals. *J. Anim. Sci. Biotechnol.* 13:72.

Baz, H.; Creech, M.; Chen, J.; Gong, H.; Bradford, K.; Huo, H. 2020 Water-soluble carbon nanoparticles improve seed germination and post-germination growth of lettuce under salinity stress. *Agronomy.* 10:1192.

Blaznik, U.; Krušič, M. H.; Kušar, A.; Žmitek, K.; Pravst, I. 2022 Use of food additive titanium dioxide (E171) before the introduction of regulatory restrictions due to concern for genotoxicity. *Foods.* 10(8):1910.

Boraschi, D.; Swartzwelter, B. J.; Italiani, P. 2018 Interaction of engineered nanomaterials with the immune system: Health-related safety and possible benefits. *Curr. Opinion Toxicol.* 10:74–83.

Boutillier, S.; Fourmentin, S.; Laperche, B. 2022 History of titanium dioxide regulation as a food additive: A review. *Environ. Chem. Lett.* 20:1017–1033.

Bumbudsanpharoke, N.; Ko, S. 2015 Nano-food packaging: An overview of market, migration research, and safety regulations. *J. Food Sci.* 80(5):R910–R923.

Campos, D.; Gómez-García, R.; Oliveira, D.; Madureira, A. R. 2021 Intake of nanoparticles and impact on gut microbiota: In vitro and animal models available for testing. *Gut Microbiome.* 3:e1.

Castan, L.; Bøgh, K. L.; Maryniak, N. Z.; Epstein, M. M.; Kazemi, S.; O'Mahony, L.; Bodinier, M.; Smit, J. J.; van Bilsen, H. M.; Blanchard, C.; Głogowski, R.; Kozáková, H.; Schwarzer, M.; Noti, M.; de Wit, N.; Bouchaud, G.; Bastiaan-Net, S. 2020 Overview of in vivo and ex vivo endpoints in murine food allergy models: Suitable for evaluation of the sensitizing capacity of novel proteins? *Allergy.* 75:289–301.

Chang, T.; Khort, A.; Saeed, A.; Blomberg, E.; Nielsen, M. B.; Hansen, S. F.; Odnevall, I. 2023 Effects of interactions between natural organic matter and aquatic organism degradation products on the transformation and dissolution of cobalt and nickel-based nanoparticles in synthetic freshwater. *J. Hazard. Mater.* 445:130586.

Chen, Z.; Liu, W.; Wang, J.; Yan, D.; Feng, H.; Wu, Y.; Wu, Y.; Chen, H. 2023 Overview of allergenic risk of novel foods. *Food Innov. Adv.* 2(2):115–123.

Cho, Y-M.; Mizuta, Y.; Akagi, J-I.; Toyoda, T.; Sone, M.; Ogawa, K. 2018 Size-dependent acute toxicity of silver nanoparticles in mice. *J. Toxicol. Pathol.* 31(1):73–80.

Chong, C. L.; Fang, C. M.; Pung, S. Y.; Ong, C. E.; Pung, Y. F.; Kong, C.; Pan, Y. 2021 Current updates on the in vivo assessment of zinc oxide nanoparticles toxicity using animal models. *BioNanoScience.* 11:590–620.

Chuah, A. S. F.; Leong, A. D.; Cummings, C. L.; Ho, S. S. 2018 Label it or ban it? Public perceptions of nano-food labels and propositions for banning nano-food applications. *J. Nanoparticle Res.* 20(2): 36.

Collins, K. L.; Sanders, G. 2017 Food consumption by children and adults. *Am. Fam. Phys.* 96(7):469A.

Deng, R.; Zhu, Y.; Wu, X.; Wang, M. 2023 Toxicity and mechanisms of engineered nanoparticles in animals with established allergic asthma. *Int. J. Nanomed.* 18:3489–3508.

de Oliveira Mallia, J.; Galea, R.; Nag, R.; Cummins, E.; Gatt, R.; Valdramidis, V. 2022 Nanoparticle food applications and their toxicity: Current trends and needs in risk assessment strategies. *J. Food Protect.* 85(2):355–372.

Di Gioacchino, M.; Di Giampaolo, L.; Mangifesta, R.; Gangemi, S.; Petrarca, C. 2022 Exposure to nanoparticles and occupational allergy. *Curr. Opin. Allergy Clin. Immunol.* 22:55–63.

Dobrovolskaia, M. A. 2022. Lessons learned from immunological characterization of nanomaterials at the Nanotechnology Characterization Laboratory. *Front. Immunol.* 13 984252.

Dorier, M.; Béal, D.; Tisseyre, C.; Marie-Desvergne, C.; Dubsson, M.; Barreau, F.; Houdeau, E.; Herlin-Biome, N.; Rabilloud, T.; Carriere, M. 2019 The food additive E171 and titanium dioxide nanoparticles indirectly alter the homeostasis of human intestinal epithelial cells *in vitro. Environ. Sci.: Nano.* 6:1549–1561.

Dudefoi, W.; Moniz, K.; Allen-Vercoe, E.; Ropers, M-H.; Walker, V. K. 2017 Impact of food grade and nano-TiO2 particles on a human intestinal community. *Food Chem. Toxicol.* 106(A): 242–249.

EFSA. 2018a Re-evaluation of silicon dioxide (E 551) as a food additive. *EFSA J.* 16(1):e05088.

EFSA. 2018b Re-evaluation of celluloses E 460(i), E 460(ii), E 461, E 462, E 463, E 464, E 465, E 466, E 468 and E 469 as food additives. *EFSA J.* 16(1):e05047.

EFSA. 2021a Guidance on risk assessment of nanomaterials to be applied in the food and feed chain: Human and animal health. *EFSA J.* 9(8):6768.

EFSA. 2021b Guidance on technical requirements for regulated food and feed product applications to establish the presence of small particles including nanoparticles. *EFSA J.* 19(8):6769.

EFSA. 2021c Safety assessment of titanium dioxide (E171) as a food additive. *EFSA J.* 19(5):6585.

EFSA. 2021d Scientific opinion on the safety assessment of the substance silver nanoparticles for use in food contact materials. *EFSA J.* 19(8), 6790.

EFSA. 2022a Safety assessment of the substance nano precipitated calcium carbonate for use in plastic food contact materials. *EFSA J.* 20(2):e07135.

EFSA. 2022b Safety assessment of the substance fatty acid coated nano precipitated calcium carbonate for use in plastic food contact materials. Contact materials. *EFSA J.* 20(2):e07136.

Egbuna, C.; Parmar, V. K.; Jeevanandam, J.; Ezzat, S. M.; Patrick-Iwuanyanwu, K. C.; Adetunji, C. O.; Khan, J.; Onyeike, E. N.; Uche, C. Z.; Akram, M.; Ibrahim, M. S.; El Mahdy, N. M. 2021 Toxicity of nanoparticles in biomedical application: Nanotoxicology. *J. Toxicol.* 2021:9954443.

Ellenbogen, Y.; Jiménez-Saiz, R.; Spill, P.; Chu, D. K.; Waserman, S.; Jordana, M. 2018 The initiation of Th2 immunity towards food allergens. *Int. J. Mol. Sci.* 19(5):1447.

Falk, M. C.; Chassy, B. M.; Harlander, S. K.; Hoban, T. J.; McGloughlin M. N.; Akhlaghi, A. R. 2002 Food biotechnology: Benefits and concerns. *J. Nutr.* 132(6):1384– 1390.

Fernández-Cruz, M. L.; Hernández-Moreno, D.; Ctalán, J.; Cross, R. K.; Stockman-Juvala, H.; Cabellos, J.; Lopes, V. R.; Matzke, M.; Ferraz, N.; Izquierdo, J. J.; Navas, J. M.; Park, M.; Sverdsen, C.; Janer, G. 2018 Quality evaluation of human and environmental toxicity studies performed with nanomaterials – The GUIDEnano approach. *Environ Sci: Nano.* 5:381–397.

Fröhlich, E. 2017 Role of omics techniques in the toxicity testing of nanoparticles. *J. Nanobiotechol.* 15(84).

García-Hevia, L.; Saramiforoshani, M.; Monge, J.; Iturrioz-Rodríguez, N.; Padín-González, E.; González, F.; González-Legarreta, L.; González, J.; Fanarraga, M. L. 2021 The unpredictable carbon nanotube biocorona and a functionalization method to prevent protein biofouling. *J. Nanobiotechnol.* 19:129.

Gatoo, M. A.; Naseem, S.; Arfat, M. Y.; Dar, A. M.; Qasim, K.; Zubair, S. 2014 Physicochemical properties of nanomaterials: Implication in associated toxic manifestations. *Biomed. Res. Int.* 2014:498420.

Geppert, M.; Sigg, L.; Schimer, K. 2021 Toxicity and translocation of Ag, CuO, ZnO and TiO$_2$ nanoparticles upon exposure to fish intestinal epithelial cells. *Environ. Sci.: Nano.* 8:2249–2260.

Góral, D.; Marczuk, A.; Góral-Kowalczyk, M.; Koval, I.; Andrejko, D. 2023 Application of iron nanoparticle-based materials in the food industry. *Materials.* 16(2):780.

Guerrera, M. C.; Aragona, M.; Porcino, C.; Fazio, F.; Laura, R.; Levanti, M.; Montalbano, G.; Germanà, G.; Abbate, F.; Germanà, A. 2021 Micro and nano plastics distribution in fish as model organisms: Histopathology, blood response and bioaccumulation in different organs. *Appl. Sci.* 11(13):5768.

Guo, X.; Li, X.; Chan, L.; Huang, W.; Chen, T. 2021 Edible CaCO3 nanoparticles stabilized pickering emulsion as calcium-fortified formulation. *J. Nanobiotechnol.* 19:67.

Halawani, R. F.; AbdElgawad, H.; Aloufi, F. A.; Balkhyour, M. A.; Zrig, A.; Hassan, A. H. A. 2023 Synergistic effect of carbon nanoparticles with mild salinity for improving chemical composition and antioxidant activities of radish sprouts. *Front. Plant Sci.* 14:1158031.

Hils, M.; Wölbing, F.; Hilger, C.; Fischer, J.; Hoffard, N.; Biedermann, T. 2020 The history of carbohydrates in Type I Allergy. *Front. Immunol.* 11:586924.

Hirai, T.; Yoshioka, Y.; Izumi, N.; Ichihashi, K-I.; Handa, T.; Nishijima, N.; Uemura, E.; Sagami, K-I.; Takahashi, H.; Yamaguchi, M.; Nagano, K.; Mukai, Y.; Kamada, H.; Tsunoda, S-I.; Ishi, K. J.; Higashisaka, K.; Tsutsumi, Y. 2016 Metal nanoparticles in the presence of lipopolysaccharides trigger the onset of metal allergy in mice. *Nat. Nanotechnol.* 11(9):808–816.

Hu, L.; Zhou, Y.; Wnag, Y.; Zhang, D.; Pan, X. 2022a Transfer of Micro(nano)plastics in animals: A mini-review and future research recommendation. *J. Hazard. Mater. Adv.* 7:100101.

Hu, W.; Wang, C.; Gao, D.; Liang, Q. 2022b Toxicity of transition metal nanoparticles: A review of different experimental models in the gastrointestinal tract. *J. Appl. Toxicol.* 43(1):32–46.

Huang, J-N.; Wen, B.; Xu, L.; Ma, H-C.; Li, X-X.; Gao, J-Z.; Chen, Z-Z. 2022 Micro/nano-plastics cause neurobehavioral toxicity in discus fish (Symphysodon aequifasciatus): Insight from brain-gut-microbiota axis. *J. Hazard. Mater.* 421:126830.

Ilmiawan Sutomo, D. F.; Ahsani, D. N.; Fidianingsih, I. 2022 Toxicity of nanoparticles on the spleen in animal studies: A scoping review. *Jurnal Ilmu Kefarmasian Indonesia.* 20(2):158–168.

Institute of Medicine (US.) Food Forum. 2009 Nanotechnology in food products: Workshop summary. In *Safety and Efficacy of Nanomaterials in Food Products,* vol. 3. Washington (DC): National Academies Press (US). Available from: https://www.ncbi.nlm.nih.gov/books/NBK32731/

Issa, M.; Rivière, G.; Houdeau, E.; Adel-Patient, K. 2022 Perinatal exposure to foodborne inorganic nanoparticles: A role in the susceptibility to food allergy? *Front. Allergy.* 3:1067281.

Iyola, O. A.; Olafimihan, T. F.; Sulaiman, F. A.; Anifowoshe, A. T. 2017 Genotoxicity and histopathological assessment of silver nanoparticles in Swiss albino mice. *UNED Res. J.* 10(1):102–109.

Johnson, L.; Aglas, L.; Punz, B.; Dang, H-H.; Christ, C.; Pointer, L.; Wenger, M.; Hofstaetter, N.; Hofer, S.; Geppert, M.; Andosh, A.; Ferrerira, F.; Hoejs-Hoeck, J.; Duschl, A.; Himly, M. 2022 Mechanistic insights into silica nanoparticle-allergen interactions on antigen presenting cell function in the context of allergic reactions. *Nanoscale.* 15(5):2262–2275.

Johnson, L.; Duscht, A.; Himly, M. 2020 Nanotechnology-based vaccines for allergen-specific immunotherapy: Potentials and challenges of conventional and novel adjuvants under research. *Vaccines.* 8(2):237.

Kämpfer, A. A. M.; Urbán, P.; La Spina, R.; Jiménez, I. O.; Kanase, N.; Stone, V.; Kinsner-Ovaskainen, A. 2020 Ongoing inflammation enhances the toxicity of engineered nanomaterials: Application of an *in vitro* co-culture model of the healthy and inflamed intestine. *Toxicol. in Vitro.* 63:104738.

Kasote, D. M.; Lee, J. H.; Jayaprakasha, G. K.; Patil, B. S. 2021 Manganese oxide nanoparticles as safer seed priming agent to improve chlorophyll and antioxidant profiles in watermelon seedlings. *Nanomaterials.* 11(4):1016.

Kumar, M.; Kulkarni, P.; Liu, S.; Chemuturi, N.; Shah, D. K. 2023 Nanoparticle biodistribution coefficients: A quantitative approach for understanding the tissue distribution of nanoparticles. *Adv. Drug. Deliv. Rev.* 194:114708

Kumar, V.; Sharma, N.; Maitra, S. S. 2017 *In vitro* and *in vivo* toxicity assessment of nanoparticles. *Int. Nano Lett.* 7:243–256.

Leite-de-Moraes, M.; Hammad, H.; Dy, M. 2012 Crosstalk between innate and adaptive cells on allergic process. *J Allergy (Cairo)*. 2012:720568.

Li, X.; Li, D.; Zhang, G.; Zeng, Y.; Monteiro-Riviere, N. A.; Chang, Y-Z.; Li, Y. 2022 Biocorona modulates the inflammatory response induced by gold nanoparticles in human epidermal keratinocytes. *Toxicol. Lett.* 369:34–42.

Liu, Y.; Zhu, S.; Gu, Z.; Chen, C.; Zhao, Y. 2022 Toxicity of manufactured nanomaterials. *Particuology*. 69:31–48.

López-Fandiño, R. 2020 Role of dietary lipids in food allergy. *Crit. Rev. Food Sci. Nutr.* 60:1797–1814.

Lugani, Y.; Sooch, B. S.; Singh, P.; Kumar, S. 2021 Nanobiotechnology applications in food sector and future innovations. *Microb. Biotechnol. Food Health*. 2021:197–225.

Lyon-Darden, T.; Blum, J. L.; Schooley, M. W.; Ellis, M.; Durando, J.; Marrill, D.; Oller, A. R. 2023 An assessment of the oral and inhalation acute toxicity of nickel oxide nanoparticles in rats. *Nanomaterials*. 13(2):261.

Madni, A.; Rehman, S.; Sultan, H.; Khan, M. M.; Ahmad, F.; Raza, M. R.; Rai, N.; Parveen, F. 2020 Mechanistic approaches of internalization, subcellular trafficking, and cytotoxicity of nanoparticles for targeting the small intestine. *AAPS PharmSciTech*. 22:3.

Matuszewski, A.; Łukasiewicz, M.; Łozicki, A.; Niemiec, J.; Zielińska-Górska, M.; Scott, A.; Chwalibog, A.; Sawosz, E. 2020 The effect of manganese oxide nanoparticles on chicken growth and manganese content in excreta. *Animal Feed Sci. Technol.* 114597.

Mayorga, C.; Perez-Inestrosa, E.; Rojo, J.; Ferrer, M.; Montañez, M. I. 2021 Role of nanostructures in allergy: Diagnostics, treatments and safety. *Allergy*. 76(11):3292–3306.

Mazaheri, N.; Naghsh, N.; Karimi, A.; Salavati, H. 2019 In vivo toxicity investigation of magnesium oxide nanoparticles in rat for environmental and biomedical applications. *Iranian J. Biotech*. 17(1):e1543.

McClements, D. J.; Xiao, H. 2017 Is nano safe in foods? Establishing the factors impacting the gastrointestinal fate and toxicity of organic and inorganic food-grade nanoparticles. NPJ *Sci. Food*. 1:6.

Medina-Reyes, E. I.; Rodríguez-Ibarra, C.; Déciga-Alcaraz, A.; Díaz-Urbina, D.; Chirino, Y. I.; Pedraza-Chaverri, J. 2020 Food additives containing nanoparticles induce gastrotoxicity, hepatotoxicity and alterations in animal behavior: The unknown role of oxidative stress. *Food Chem. Toxicol.* 146:111814.

Mei, N.; Hedberg, J.; Wallinder, I. O.; Blomberg, E. 2019 Influence of biocorona formation on the transformation and dissolution of cobalt nanoparticles under physiological conditions. *ACS Omega*. 4(26):21778–21791.

Meldrum, K.; Guo, C.; Marczylo, E. L.; Gant, T. W.; Smith, R.; Leonard, M. O. 2017 Mechanistic insight into the impact of nanomaterials on asthma and allergic airway disease. *Part Fibre Toxicol*. 14(1):45.

Meng, X.; Zeng, Z.; Gao, J.; Tong, P.; Wu, Y.; Li, X.; Chen, H. 2020 Conformational changes in bovine α-lactalbumin and β-lactoglobulin evoked by interaction with C18 unsaturated fatty acids provide insights into increased allergic potential. *Food Funct*. 11:9240–9251.

Mcsgari, M.; Aalami, A. H.; Sathyapalan, T.; Sahebkar, A. 2022 A comprehensive review of the development of carbohydrate macromolecules and copper oxide nanocomposite films in food nanopackaging. *Bioinorg. Chem. Appl.* 2022:7557825, 28 pages.

Mohammadi, P.; Galera, A. 2023 Occupational exposure to nanomaterials: A bibliometric study of publications over the last decade. *Int. J. Hygiene Environ. Health*. 249:114132.

Molina, E.; Benedé, S. 2022 Is there evidence of health risks from exposure to micro- and nanoplastics in foods? *Front. Nutr.* 9:910094.

Moradi, M.; Razavi, R.; Omer, A. K.; Farhangfar, A.; McClements, D. J. 2022 Interactions between nanoparticle-based food additives and other food ingredients: A review of current knowledge. *Trends Food Sci. Technol.* 120:75–87.

Najahi-Missaoui, W.; Arnold, R. D.; Cummings, B. S. 2021 Safe nanoparticles: Are we there yet? *Int. J. Mol. Sci.* 22(1):385.

National Research Council (US.) Commission on Life Sciences. 1988 Biotechnology and the food supply: Proceedings of a symposium. In *Potential Food Safety Problems Related to New Uses of Biotechnology*. Washington (DC): National Academies Press (US). Available from: https://www.ncbi.nlm.nih.gov/books/NBK235027/

Oliveira, E. M. N.; Selli, G. I.; von Smunde, A.; Miguel, C.; Laurent, S.; Vianna, M. R. M. Papaléo, R. M. 2020 Developmental toxicity of iron oxide nanoparticles with different coatings in zebrafish larvae. *J. Nanopart. Res.* 22:87.

Onyeaka, H.; Passaretti, P.; Miri, T.; Al-Shrify, Z. T. 2022 The safety of nanomaterials in food production and packaging. *Curr. Res. Food Sci.* 5:763–774.

Paradise, J. 2019 Regulating nanomedicine at the Food and Drug Administration. *AMA J. Ethics.* 21(4):E347–355.

Pazos-Castro, D.; Margain, C.; Gonzalez-Klein, Z.; Yuste-Calvo, C.; Garrido-Arandia, M.; Zurita, L.; Esteban, V.; Tome-Amat, J.; Diaz-Perales, A.; Ponz, F. 2022 Suitability of potyviral recombinant virus-like particles bearing a complete food allergen for immunotherapy vaccines. *Front. Immunol.* 13:986823.

Perumal, K.; Ahmad, S.; Mohd-Zahud, M. H.; Wan Hanaffi, W. N.; Iksander, Z. A.; Six, J-L.; Ferji, K.; Jaafar, J.; Boer, J. C.; Plebanski, M.; Uskoković, V.; Mohamud, R. 2021 Nanoparticles and gut microbiota in colorectal cancer. *Front. Nanotechnol.* 3: 681760.

Phue, W. H.; Xu, K.; George, S. 2022 Inorganic food additive nanomaterials alter the allergenicity of milk proteins. *Food Chem. Toxicol.* 162:112874.

Pisanello, D.; Caruso, G. 2018 EU regulation on novel foods. In *Novel Foods in the European Union.* Cham: Springer International Press. pp. 1–29.

Pohlit, H.; Bellinghauser, I.; Frey, H.; Saloga, J. 2017 Recent advances in the use of nanoparticles for allergen-specific immunotherapy. *Allergy.* 72(10):1461–1474.

Polsky, J. Y.; Moubarac, J-C.; Garriguet, D. 2020 Consumption of ultra-processed foods in Canada. *Health Rep.* 31(11):3–15.

Rahman, H. S.; Othman, H. H.; Abdullah, R.; Edin, H. Y. A. S.; AL- Haj, N. A. 2022 Beneficial and toxicological aspects of zinc oxide nanoparticles in animals. *Vet. Med. Sci.* 8(4):1769–1779.

Rai, M.; Ingle, A. P.; Alka, Y.; Golińska, P.; Trzcińska-Wencel, J.; Rathod, S.; Bonde, S. 2023 Nanotechnology as a promising approach for detection, diagnosis and treatment of food allergens. *Curr. Nanosci.* 19(1):90–102.

Rasmussen, K.; Rauscher, H.; Kearns, P.; González, M.; Sintes, J. R. 2019 Developing OECD test guidelines for regulatory testing of nanomaterials to ensure mutual acceptance of test data. *Reg. Toxicol. Pharmacol.* 104:74–83.

Raul, P. K.; Thakuria, A.; Das, B.; Devi, R. R.; Tiwari, G.; Yellapa, C.; Kamboj, D. V. 2022 Carbon nanostructures as antibacterials and active food-packaging materials: A review. *ACS Omega.* 7(14):11555–11559.

Ren, Q.; Ma, J.; Li, X.; Meng, Q.; Wu, S.; Xie, Y.; Qi, Y.; Liu, S.; Chen, R. 2023 Intestinal toxicity of metal nanoparticles: Silver nanoparticles disorder the intestinal immune microenvironment. *ACS Appl. Mater. Interfaces.* 15(23):27774–27788.

Roach, K. A.; Stefaniak, A. B.; Roberts, J. R. 2019 Metal nanomaterials: Immune effects and implications of physicochemical properties on sensitization, elicitation, and exacerbation of allergic disease. *J. Immunotoxicol.* 16(1):87–124.

Rogers, M. A. 2016 Naturally occurring nanoparticles in food. *Curr. Opin. Food Sci.* 7:14–19.

Rothen-Rutishauser, B.; Bogdanovich, M.; Harter, R.; Milosevic, A.; Petri-Fink, A. 2021 Use of nanoparticles in food industry: Current legislation, health risk discussions and public perception with a focus on Switzerland. *Toxicol. Environ. Chem.* 103(4):423–437.

Sahu, S. C.; Hayes, A. W. 2017 Toxicity of nanomaterials found in human environment: A literature review. *Toxicol. Res. Application.* 1.

Sardoiwala, M. N.; Kaundal, B.; Choudhury, S. R. 2018 Toxic impact of nanomaterials on microbes, plants and animals. *Environ. Chem. Lett.* 16:147–160.

Savage, D. T.; Hilt, J. Z.; Dziubla, T. D. 2020 *In vitro* methods for assessing nanoparticle toxicity. *Methods Mol. Biol.* 1894:1–29.

Schoonjans, R.; Castenmiller, J.; Chaudhry, Q.; Cubadda, F.; Daskaleros, T.; Franz, R.; Gott, D.; Mast, J:; Mortensen, A.; Oomen, A. G.; Rauscher, H.; Weigel, S.; Astuto, M. C.; Cattaneo, I.; Barthelemy, E.; Rincon, A.; Tarazona, J. 2023 Regulatory safety assessment of nanoparticles for the food chain in Europe. *Trends Food Sci. Technol.* 134:98–111.

Shabbir, S.; Hu, Y.; He, X.; Huang, K.; Xu, W. 2023 Toxicity and impact of silica nanoparticles on the configuration of gut microbiota in immunodeficient mice. *Microorganisms.* 11(5):1183.

Shahgordi, S.; Sankian, M.; Yazdani, Y.; Mashayekhi, K.; Ayati, S. H.; Sadeghi, M.; Saeidi, M.; Hashemi, M. 2020 Immune responses modulation by curcumin and allergen encapsulated into PLGA nanoparticles in mice model of rhinitis allergic through sublingual immunotherapy. *Int. Immunopharmacol.* 84:106525.

Siegrist, M.; Keller, C. 2011 Labeling of nanotechnology consumer products can influence risk and benefit perceptions. *Risk Anal.* 31(11):1762–1769.

Talash, S.; Koohi, M.; Zayerzadeh, E.; Hasan, J.; Shaban, M. 2019 Acute toxicity investigation regarding clinical and pathological aspects following repeated oral administration of iron oxide nanoparticles in rats. *Nanomed. Res. J.* 4(4):228–233.

Tsuchida, D.; Matsuki, Y.; Tsuchida, J.; Iijima, M.; Tanaka, M. 2023 Allergenicity and bioavailability of nickel nanoparticles compared to nickel microparticles in mice. *Materials*. 16(5):1834.

Vimercati, L.; Cavone, D.; Caputi, A.; De Maria, L.; Tria, M.; Prato, E.; Ferri, G. M. 2020 Nanoparticles: An experimental study of zinc nanoparticles toxicity on marine crustaceans. General overview on the health implications in humans. *Front. Public Health*. 8:192.

Vitolo, M.; Gnodi, E.; Meneveri, R.; Barisani, D. 2022 Interactions between nanoparticles and intestine. *Int. J. Mol. Sci.* 23(8):4339.

Voss, L.; Hoché, E.; Stock, V.; Böhmert, L.; Braeuning, A.; Thünemann, A. F.; Sieg, H. 2021 Intestinal and hepatic effects of iron oxide nanoparticles. *Ach. Toxicol*. 95:895–905.

Wen, S.; Zhao, Y.; Wang, M.; Yuan, H.; Xu, H. 2022 Micro(nano)plastics in food system: Potential health impacts on human intestinal system. *Crit. Rev. Food Sci. Nutr.* 64(5):1429–1447.

Wu, L.; Wen, W.; Wang, X.; Huang, D.; Cao, J.; Qi, X.; Shen, S. 2022 Ultrasmall iron oxide nanoparticles cause significant toxicity by specifically inducing acute oxidative stress to multiple organs. *Part. Fibre Toxicol*. 19:24.

Xu, X.; Wang, X.; Liao, Y-P.; Luo, L.; Xia, T.; Nel, A. E. 2023 Use of a liver-targeting immune-tolerogenic mRNA lipid nanoparticle platform to treat peanut-induced anaphylaxis by single- and multiple-epitope nucleotide sequence delivery. *ACS Nano* 17(5):4942–4957.

Yoshioka, Y.; Kuroda, E.; Hirai, T.; Tsutsumi, Y.; Ishii, K. J. 2017 Allergic responses induced by the immuno-modulatory effects of nanomaterials upon skin exposure. *Front. Immunol*. 8:169.

Youn, S-M.; Choi, S-J. 2022 Food additive zinc oxide nanoparticles: Dissolution, interaction, fate, cytotoxicity, and oral toxicity. *Int. J. Mol. Sci.* 23(11):6074.

Zhang, W.; Rhim, J-W. 2022 Titanium dioxide (TiO_2) for the manufacture of multifunctional active food packaging films. *Food Packag. Shelf Life*. 31:100806.

Zhu, C. 2022 The toxicity evaluation of novel carbon-based nanoparticle medicine and how to choose the suitable animal toxicity test models for it. *Adv. Soc. Sci. Educ. Hum. Res*. 666.

17 Environmental Impact of Nanoparticles

Hani Nasser Abdelhamid

17.1 INTRODUCTION

Nanoparticles (NPs) are microscopic particles that measure fewer than 100 nm in length along at least one of their dimensions (Abdelhamid & Badr, 2021). Because of their one-of-a-kind characteristics, they can be utilized effectively in a diverse array of contexts, such as in the fields of medicine, electronics, and energy (Abdelhamid, 2023a–c; Abdelhamid & Mahmoud, 2023; Abdelhamid & Mathew, 2022; Abdellatif et al., 2022; Tran et al., 2022). Despite this, there is a rising concern about the possible effects that nanoparticles could have on the environment. The particle size of nanoparticles governs the material's properties, including physical and chemical characteristics. The melting point of gold can be raised anywhere from 200°C to 1,068°C, depending on the size of the NPs (Link et al., 2000; Tweney, 2006).

One of the primary causes for concern is the simplicity with which nanoparticles can be released into the environment, causing air, water, and soil pollution (Sajid et al., 2015). When nanoparticles are released into the environment, it can be challenging to remove them, and they can remain for extended periods. Additionally, nanoparticles can interact with live creatures in several ways. For instance, they may be absorbed through the lungs via inhalation, the digestive tract via consumption of food or water, or the skin via absorption. As soon as they enter the body, nanoparticles can concentrate in various organs and tissues, which might lead to possible injury. There is a growing amount of research on the potential consequences of nanoparticles on the health of humans and the environment. However, additional research is required to understand the dangers and benefits of these materials thoroughly.

Environmental conditions may cause nanoparticle release. Nanoparticles may undergo photochemical transformation due to light exposure. The decay of nanoparticles under light depends on the wavelength of the incident light, the product's ability to let light through, and the nanomaterial's photosensitivity. Ambient temperature may cause oxidation and reduction of nanoparticles, releasing nanoparticles or their ions into the environment. Therefore, the parameters, e.g., pH, capping, or stabilizer agents, ensure the high stability of nanoparticles should be considered. Nanoparticles can be released into the environment by washing clothes with silver nanoparticles (Figure 17.1). They can also undergo dissolving and precipitation concurrently. For example, dissolving copper or zinc metallic NPs increases their bioavailability and toxicity (Aruoja et al., 2009). Iron NPs may be released from spontaneous coal burning (Silva et al., 2020). Building materials and polyurethane coatings may emit NPs over time (Figure 17.1) (Nowack et al., 2012).

This book chapter summarizes the impact of nanoparticles on the environment. It discusses the reasons for the release of nanoparticles into the environment. It highlights the steps in which nanoparticles cause contamination in the environment. The methods of nanoparticle detection are also discussed.

17.2 ENVIRONMENTAL CHALLENGES

The current environmental predicaments that the globe is dealing with include climate change, air and water pollution, biodiversity contamination, and loss of biodiversity (Al-Shetwi, 2022; Hosseinzadeh-Bandbafha et al., 2022; Olajire, 2020; Shen et al., 2022). The effects of climate change

DOI: 10.1201/9781003514039-17

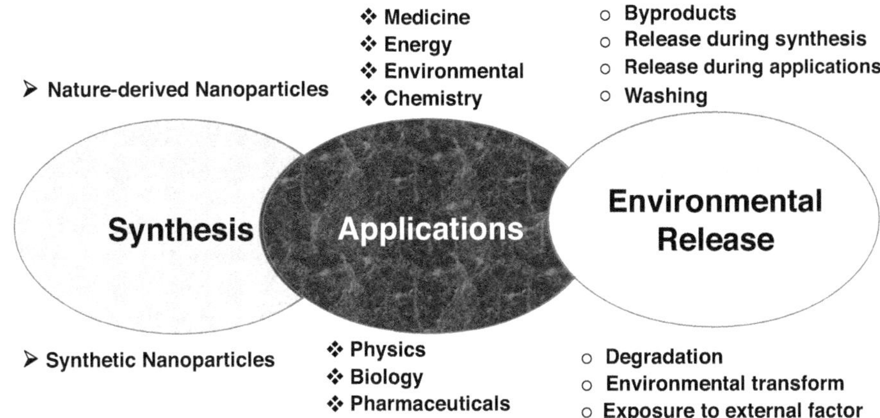

FIGURE 17.1 Synthesis, applications, and environmental release of nanoparticles.

are one of the most significant threats to the environment we face today. The emission of greenhouse gases into the atmosphere, which act as a heat trap and cause the planet to become warmer, is the root cause of this phenomenon. A number of the unfavorable effects of climate change have already begun to manifest in the natural world and human culture. These effects include increasing extreme weather events' frequency and severity, sea levels, and other plant and animal life alterations.

Air pollution is yet another significant problem facing the environment. The emission of pollutants into the atmosphere by sources such as automobiles, factories, and power plants is the root cause of this problem. Air pollution is linked to a wide range of adverse health effects, including but not limited to respiratory infections, cardiovascular disease, and cancer. It is also capable of causing damage to ecosystems and lowering crop harvests.

The term "water pollution" refers to the contamination of bodies of water with pollutants such as human waste, waste from industrial processes, and runoff from agricultural land. Due to the effects of pollution, it is sometimes dangerous to drink the water, swim in it, or fish from it. It can also wreak havoc on ecosystems and wipe out aquatic life.

The loss of plant and animal species, as well as the diversity of those species, is referred to as biodiversity loss. Loss of habitat, excessive fishing, and the effects of climate change are some of the reasons that contribute to this problem. Loss of biodiversity is a significant problem because it endangers the health of ecosystems and the human benefits derived from them.

In addition to these significant problems pertaining to the environment, the globe is also dealing with a variety of other issues, including deforestation, excessive grazing, and the buildup of salt in the soil are all factors that contribute to land degradation, also known as the loss of fertile land. The use of natural resources, including water, minerals, and fossil fuels, can lead to a phenomenon known as resource depletion. Garbage management is disposing of ever-increasing quantities of solid garbage, which may include hazardous and technological waste. There is a connection between each of these environmental problems. For instance, climate change might result in more extreme weather occurrences, which can cause harm to ecosystems and raise the likelihood of natural disasters. Water bodies and soils can also get tainted when polluted air is present. In addition, the degradation of land can be a contributor to climate change as well as the loss of biodiversity.

It is essential to solve these environmental problems to save the earth and ensure everyone has a sustainable future. Several actions can be taken to help address issues relating to the environment, including: (i) Reduce greenhouse gas emissions by switching to renewable energy sources, increasing energy efficiency, and cutting down on deforestation. These are some of the ways this can be accomplished: (ii) Enhance the quality of the air we breathe by lowering the amount of emissions produced by automobiles, factories, and power plants; (iii) Cleaning up polluted water can be accomplished by the treatment of wastewater and the reduction of pollution caused by agriculture and industry; (iv) It is essential to preserve biological diversity, which can be accomplished through limiting overfishing, maintaining habitats, and addressing climate change; (v) Reducing waste can be achieved in many ways, including lowering consumption, recycling, and using compost; and (vi) To effectively address these environmental concerns, it will take a collaborative effort on the part of individuals, corporations, and governments. Protecting the world for the sake of future generations is a shared responsibility that each of us bears.

17.3 NANOPARTICLES

Based on the particle size, the object fills within 1–200 nm (Figure 17.2). Researchers have recorded instances of NPs being used as far back as 4,500 years ago, such as the use of nanofibers in ceramics or the synthesis of "Egyptian blue," the oldest known synthetic pigment, which was created from a mixture of quartz and nanoparticulate glass (Accorsi et al., 2009). Other examples of NP applications include the application of nanofibers in ceramics. In 1959, Richard P. Feynman mentioned manipulating materials on a nanoscale while speaking at a conference of the American Physical Society (Hulla et al., 2015). Over the past 20 years, there has been a gradual increase in the number of studies released and published, as well as their utility and applications.

The following categories can be used to categorize nanoparticles following their place of origin:

1. *In situ* **Synthesized Nanoparticles**: Nanomaterials generated by accident occur as a byproduct of industrial or natural processes such as combustions (for example, the smoke from cigarettes or fires). These nanomaterials can be found in the environment.
2. **Nanomaterials Manufactured Artificially**: These nanomaterials are designed by humans with certain features and characteristics (for example, silver nanoparticles in shampoo).

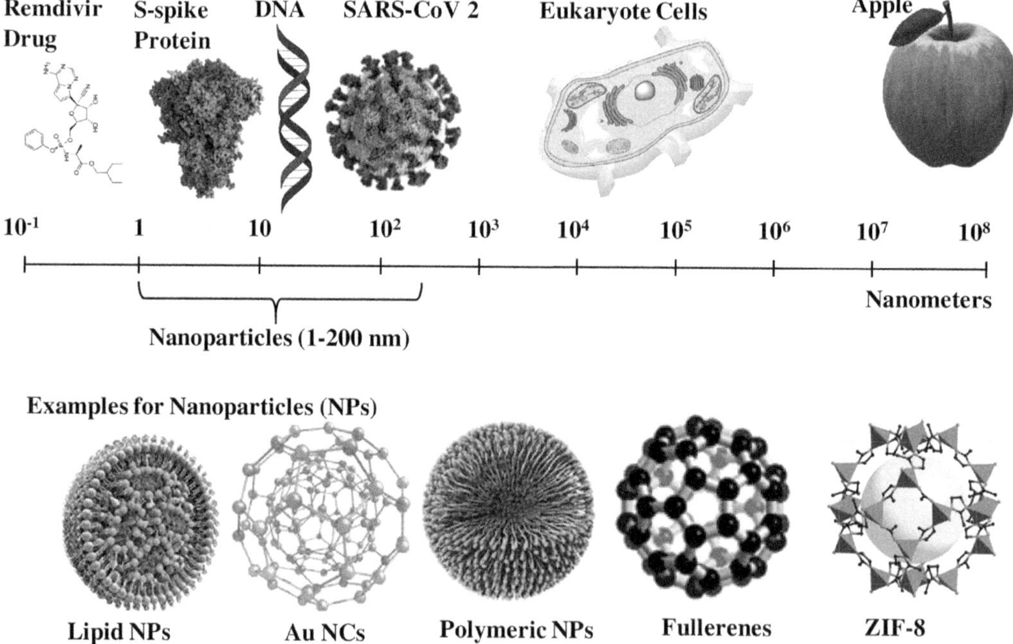

FIGURE 17.2 The size scale of the objects, including examples of each category. (Figure reprinted with permission from Ref. Abdelhamid & Badr, 2021.)

The primary distinction between intentionally generated nanomaterials and unintentionally formed nanomaterials is that the former are designed to be created in specific sizes and compositions, along with other features; the latter, on the other hand, are formed naturally and unplanned.

3. **Nature-Derived Nanoparticles**: Nanomaterials that are the product of natural processes can be discovered in living things and nature, e.g., viruses. In some situations, drawing the line that differentiates raw materials from accidentally formed or synthetic nanoparticles can be challenging.

Green synthesis of nanoparticles was intensively reported in the literature using nature-derived materials. It can be considered green nanotechnology because it uses environmentally friendly processes. Traditional nanosystem manufacture involves hazardous chemicals, high pressure, and temperature. Green synthesis uses natural species and compounds instead of chemicals. Several natural regents for NP synthesis include bacteria, fungi, and vegetable extracts.

The use of microbes to create and assemble NPs was reported. The researchers used extracellular enzymes that make pure, biodegradable NPs with low toxicity. This provides a reliable alternative to conventional synthesis. The cells of some bacteria, fungi, and yeasts may produce biomolecules with metal-attached functional groups, making them ideal for NP production. Microorganisms offer several advantages, including easy handling, high growth rate, low cost, and low environmental toxicity. Rajasekharreddy and coworkers compared biosynthesized and chemically synthesized Ag NP toxicity (Usha Rani & Rajasekharreddy, 2011). They found that protein shells in the Ag core of these nanoparticles reduce their toxicity compared to chemically produced ones.

Vegetable and fruit extracts can be used as chemical reagents for NP synthesis. Seeds, fruits, and leaves include reducing agents that allow vegetable extracts to form NPs. These chemicals create more stable NPs than bacteria. Due to their higher reducing agent concentration, vegetable extracts are chosen for high-scale synthesis (Rajan et al., 2015). Large molecules such as polyphenol inhibited aggregation of the produced NPs at ambient temperature without surfactants or polymer. These molecules are an inexpensive and environmentally friendly resource.

17.4 ENVIRONMENTAL IMPACTS OF NANOPARTICLES

Even though the release of nanoparticles is possible at any stage of the product's life cycle, it causes a significant problem when NPs are used in the final product (Figure 17.3). Washing NP-containing aerosols or fabrics or using washing detergents that contain NPs. This feature makes the task of impact evaluation much more complex. Developing general protocols and rules for dealing with or applying nanoparticles is complicated because several variables, such as insufficient employment and varied climates, cannot be controlled.

During the life cycle of nanoparticles, nanoparticles are released into the environment. We can summarize the steps of nanoparticle release as below:

1. **Nanoparticle Emissions during Product Fabrication**: Nanoparticles (NPs) can be released into the environment by producing products containing NPs. This can happen directly, such as when powder materials are spilled, when open windows allow NPs to escape, or indirectly through improper waste disposal. The release of NPs is exceptionally high during the manufacturing stage when the materials containing NPs are subjected to structural modifications such as cutting, drilling, or high-energy and high-temperature processes to adjust their shape. Workers who handle and produce NPs may also be exposed to them, which could be another route of contamination into the environment via dead bodies after degradation. However, the extent of this exposure is still unknown.

2. **Release during Applications**: The release of nanoparticles can be caused during a product's utilization phase, which might occur either through deliberate actions or as unintended

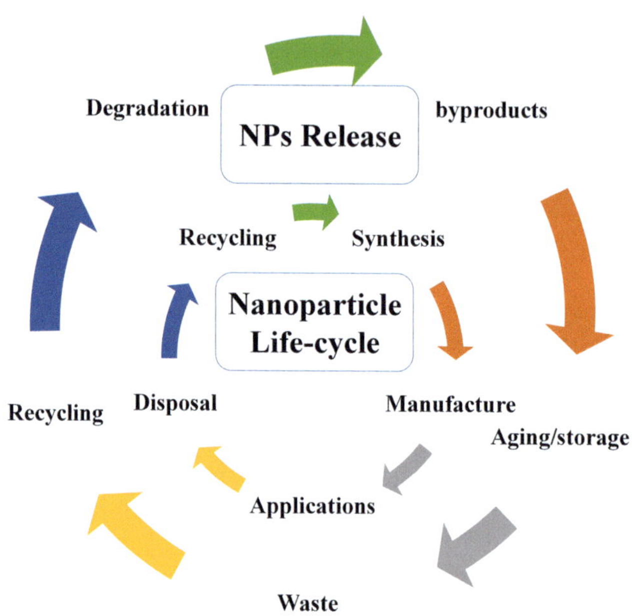

FIGURE 17.3 Life cycle and release of nanoparticles.

occurrences. The precise origin and quantity of NPs released into the environment through purposeful release are well-documented and quantifiable. However, these factors can only be approximated in the case of unintentional release resulting from product degradation and modification. The concentration of nanoparticles varies depending on the specific characteristics and features of the product. According to previous research, most NPs are rapidly discharged when fluid products are utilized.

Nevertheless, in the case of solid items, the nanoparticles they contain are released gradually over time when the product is utilized. For instance, NPs employed in pneumatic tires are released through attrition and friction. The emission of nanoparticles from spray products is instantaneous, but in suspensions, the whole emission occurs within the initial hours, e.g., in cosmetics or sunscreens. Nanoparticles can maintain stability over extended periods when included in textiles and pigments.

3. **Disposal and Recycling**: Disposal and recycling are significant in nanoparticle release into the environment. How waste treatment is conducted impacts both the quantity and how nanoparticles are discharged after applications. In general, nanoparticles are disposed of by dumping into the ground, incineration, or recycling. The landfill is mainly the primary destination for the disposal of a significant portion of nanomaterials. The potential release of nanomaterials into the environment is contingent upon the resilience of the products to degradation and the effectiveness of landfill control technologies—the process of incineration results in the generation of ash particles that become suspended in the atmosphere. Nanoparticles were identified as a significant component of these ash particles. Thus, filters with high removal efficiency are highly required during incineration. The topic of recycling is not only environmentally essential, but it is economically required. The economy and environment are two sides of the same currency.

4. **Release during Storing or Aging**: nanoparticles can be released into the environment during storage. The exposure of nanoparticles to harsh conditions such as acidic environments may cause deterioration or damage of the nanoparticles, causing transformation into toxic nanoparticles or free ions. For instance, metal oxide nanoparticles, e.g., ZnO, Fe_3O_4, CuO, and so on, are dissolved in acidic conditions (Mitrano et al., 2015). Environment

plays a significant role in nanoparticles, causing a variety of transformations due to photochemical transformations, e.g., oxidation/reduction, dissolution, precipitation, adsorption/desorption, combustion, abrasion, and biotransformation. Identifying the gaps in knowledge of the nanoparticle transformation process is essential.

5. **Biological Action on Nanoparticles**: Human myeloperoxidase degrades carbon nanomaterials such as carbon nanotubes, reducing their cytotoxicity (Kagan et al., 2010). Once within the cell, nanoparticles' toxicity level is subjected to the various compartments such as the cell and the type of NPs. For instance, carbon nanotubes typically cause toxicity in mitochondria (Samiei et al., 2020). Still, silver nanoparticles (Ag NPs) usually bind to the membrane, thereby changing the permeability and shape of the membrane (Xiang et al., 2020). Lysosomes are organelles responsible for the excretion of NPs from the cell; nevertheless, the accumulation of NPs in these organelles can also be a source of toxicity. When nanoparticles can get through the nuclear membrane and cause disruptions in the normal functioning of the Golgi complex, this can lead to damage in the nucleus.

Nanoparticles have the potential to interact with ecosystems in a variety of different ways, all of which have the potential to disrupt ecosystems. Nanoparticles, for instance, can bind to enzymes and render them inactive, which might throw crucial biochemical processes off balance. In addition, nanoparticles can cause damage to both DNA and proteins, ultimately resulting in the death of cells. Metallic nanoparticles such as Ag^0 NPs may undergo oxidation into Ag^+ ions (Reed et al., 2016).

The interaction between nanoparticles, the environment, and human beings can be elucidated through many processes. Occupational exposures predominantly arise among those employed in various professional roles, such as engineers, scientists, and technicians, specifically in research-scale synthesis and commercial manufacture of products utilizing nanomaterials. The primary source of this exposure arises from manipulating unprocessed substances during the execution of reactions using the apparatus. The exposure to this form of hazard might also occur from the characterization of the produced material, as well as from its packing and transportation processes. During the subsequent phase, individuals are subjected to said nanomaterial when utilizing and applying it, potentially resulting in adverse and hazardous consequences. In this discourse, we shall examine how nanoparticles (NPs) come into contact with the human body and explore their associated toxicological impacts. The utilization of intricate biological terminology should be minimized to enhance comprehension across a diverse scientific community engaged in nanoparticle applications. The interaction between nanoparticles (NPs) and the human body can manifest through various pathways, including (i) The process of penetrating skin nodes, (ii) The intake via the respiratory system through the process of inhaling, and (iii) The process of nutrient acquisition through the digestive tract occurs through the act of swallowing.

Aquatic environments can be contaminated by nanoparticles (NPs). This is mainly because many consumer goods, like sunscreens and cosmetics, contain NPs, e.g., ZnO or TiO_2. Researchers have found that NPs harm aquatic life, such as fish, daphnia, and single-celled animals. Algae species, dolphins, and zebrafish were exposed to nickel, copper, silver, and aluminum-containing nanoparticles. CuO and TiO_2 nanoparticles are more dangerous to algae than ZnO. Several studies reported that nanoparticles can be harmful to aquatic creatures. The level of toxicity depends on things like the size, type, charge, and species exposed to the NPs.

Nanoparticles have the potential to be bio-persistent, which means that they can survive in their natural environment for extended periods, even decades or centuries. Because nanoparticles are so small, they can easily circumvent the body's biological detoxification processes.

Nanoparticles have the potential to enter the food chain because plants and animals could take up their potential, allowing them to enter the chain. Because of this, people can be exposed to nanoparticles through the food they consume. The presence of nanoparticles such as titanium dioxide (TiO_2) can generate reactive oxygen species (ROS) under sunlight, causing high toxicity toward live organisms (Mulakov et al., 2020; Nel et al., 2006).

To have a complete appreciation of the environmental impact of NP, it is necessary to have an in-depth understanding of the features of these systems, such as their identification, physicochemical qualities, and how they are emitted into the environment, as well as their toxicity to living things. Evaluating and quantifying the amount of released nanomaterial requires conducting exhaustive research on the nanomaterial's entire life cycle, beginning with the processes of producing nanomaterials and ending with the procedures of recycling and disposal, taking into account how they are incorporated into the final products and how they are utilized. The capacity of the NPs to reach and penetrate the various environmental compartments, e.g., soil, water, and air, is a primary factor that plays a crucial role in determining the inherent potential toxicity of the NPs. This effect is proportional to the quantity of NPs dispersed across the biosphere. The specific environmental impacts of nanoparticles can be classified as below:

1. **Impact on Aquatic Ecosystems**: Nanoparticles have the potential to pollute water bodies and cause harm to marine life. This affects aquatic ecosystems. It has been demonstrated that nanoparticles of silver (Ag NPs) and zinc oxide (ZnO), for instance, are hazardous to fish and other aquatic organisms.
2. **Impact on Soil Health**: The accumulation of nanoparticles can harm soil fertility and overall health. In addition, plants can take in nanoparticles, which then make their way into the food chain.
3. **Impact on Human Health**: The inhalation of nanoparticles into the lungs can lead to various respiratory issues and impact human health. Ingestion is another route for nanoparticles to enter the body, which can eventually build up in organs like the liver and kidneys.
4. **Action Transfer**: Nanoparticles can increase the toxicity of other chemicals because they can bind to and transport other species, such as heavy metals and pesticides. This can make the other chemicals more poisonous. Because of this, the toxicity of these compounds can be amplified, making them more dangerous to the creatures they meet.
5. **Constrain Climate Parameters**: Nanoparticles have the potential to change the climate because of their ability to interact with both sunlight and clouds, both of which can affect the local temperature surrounding nanoparticles. For instance, carbon black nanoparticles can soak up sunlight and cause the atmosphere to warm.

It is essential to remember that nanoparticles' effect on the surrounding environment is contingent on several elements, the most important of which are the type of nanoparticle, its size, and its concentration. It is also essential to take into consideration how one encounters nanoparticles. For instance, swallowed nanoparticles are likely to have a different effect on the body than those breathed into the lungs.

Researchers are actively pursuing the development of strategies that will lessen the detrimental effects that nanoparticles have on the surrounding environment. For instance, researchers are creating new nanoparticles that are less harmful and more easily biodegradable.

While NPs may have a negative impact on the environment, they also have potential benefits that should be addressed (Figure 17.4). NPs can improve their surroundings with some applications. NPs can reduce pollution and improve environmental health by promoting energy-efficient production systems or helping remediate water and soil. NPs can minimize soil contamination through remediation and pollutant reduction. Nanotechnology can improve fertilizers, herbicides, insecticides, and growth promoters by reducing their use and increasing their action. This reduces the dose, adverse effects, pollutant discharge, and energy needed for following soil treatments.

The current surge in energy consumption serves as a driving force for scholars to develop environmentally sustainable energy sources. The field of nanotechnology facilitates the advancement of clean and sustainable energy production. Solar photovoltaic energy is often considered a promising energy source due to its notable efficiency, adaptability, and minimal impact on natural resources such as water and soils (Abdelhamid et al., 2019). Photovoltaic cells exhibit a significant drawback

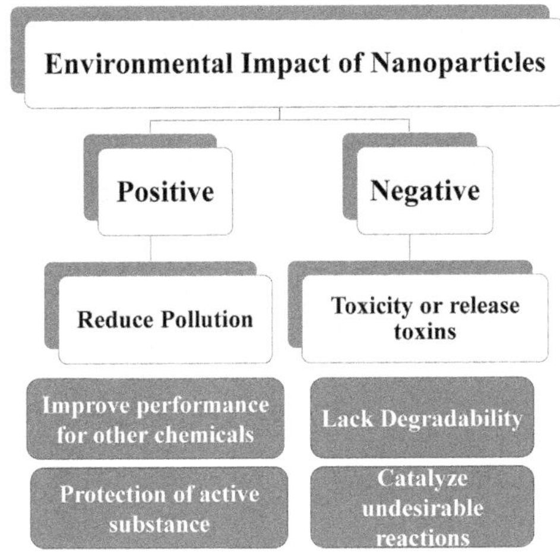

FIGURE 17.4 Environmental impact of nanoparticles.

in terms of their production costs, which are quite high, and their limited efficiency in absorbing solar energy. Additionally, nanoparticles can be fundamental in developing antifreeze coatings, enhancing wind turbines' durability in colder regions. Nanoparticles can effectively lower the depth and temperature required for geothermal energy collecting.

Nanoparticles improved the agricultural environment (Zhou et al., 2020). A star polycation (SPc) synthesized a nanoagent for calcium nutrition (Yan et al., 2022). The assembly of SPc with calcium glycinate is facilitated by hydrogen bonding and Van der Waals forces, resulting in the formation of stable nanoparticles with a nanoscale size of 17.72 nm. The transcriptomic analysis showed that the complex formed by calcium glycinate and SPc can induce the expression of numerous genes associated with transport and disease resistance in tomatoes. This finding suggests that the presence of SPc in calcium glycinate enhances transport processes and the immune response against viral infections. The calcium transport into tomato leaves was shown to be significantly improved by a factor of 3.17 with the assistance of SPc.

Furthermore, the efficacy of calcium glycinate in protecting against tomato mosaic virus exhibited a substantial improvement, reaching 77.40% and 67.31% after the third and fifth applications, respectively, when combined with SPc. Additionally, the utilization of SPc-loaded calcium glycinate has the potential to enhance the photosynthetic rate of leaves and regulate the atypical rapid growth of tomato plants. This study represents a significant achievement as it is the initial successful attempt to utilize a nano-delivery technology to boost calcium transport and promote antiviral immunity. This advancement holds considerable potential for improving nutrient utilization efficiency and exhibits promising prospects for practical implementation in the field (Yan et al., 2022).

Using nanofertilizers from chitosan and oil palm biomass could augment paddy development (Hartoyo et al., 2022). A study reported the growth media possibilities of chitosan nanocomposite films for rice seeds cultivated in tropical peatland environments. The synthesis of chitosan nanocomposites involved the incorporation of activated carbon (AC), nonactivated carbon nanoparticles (n-AC), and lignocellulose nanofibers (LCNFs) into the chitosan matrix. The carbon nanoparticles exhibited reversible aggregation, while the LCNFs did not display an aggregation propensity but exhibited entanglement. The germination test results indicated that the utilization of chitosan nanocomposites yielded the most favorable development patterns for the Dendang paddy variety.

Similarly, the greenhouse test has shown that applying nanocomposites resulted in the most optimal growth patterns for the Indragiri paddy variety. The chitosan/empty fruit bunch ACNP nanocomposites cultivated in a germinator exhibited the highest level of growth normality (100.00%), maximum growth potential (100.00%), and average height (11.27 cm). The greenhouse experiment showed that chitosan/oil palm trunk n-ACNPs exhibited the highest growth natality (16.44%) and growth rate (65.74%). The chitosan nanocomposites exhibited a synergistic biofertilizing impact on both fungi and mycorrhiza. Chitosan nanocomposites have the potential to serve as an effective growth regulator for peatland paddy types, hence facilitating the process of peatland restoration in tropical regions (Hartoyo et al., 2022).

17.5 EVALUATING NANOMATERIAL RELEASE: ENVIRONMENTAL LEGISLATION

Several models that can be used to determine the release of nanoparticles are examples of the approaches that have been taken to evaluate the emission of NPs into the environment using various models and techniques, the methodologies of which vary substantially both in conceptualization and in data collection. Material Flow Analysis (MFA) (Rajkovic et al., 2020) and Environmental Fate Models (EFM) (Dale et al., 2015) are the models that are used the most frequently. MFA can predict the release of nanoparticles in the environment, such as landfills, recycling plants, water treatment plants, etc., and, from there, to the environment; EFM describes the transport and distribution through the various environmental compartments (water, earth, air) (Dale et al., 2015). In other words, the MFA focuses on emissions produced over the product's life cycle. In contrast, the EFM discusses how NPs, once emission has taken place, are transported, changed, and degraded between the various environmental compartments.

Nanoparticles can be detected using several analytical methods (Figure 17.5). Airborne NP measurement systems use different detectors and concepts. Size- and time-resolved devices measure optical characteristics or electrical mobility. Such devices assume all particles are spherical, which is not true at work and biases measurements. The optical particle sizer (OPS) measures particle scattering light. The size of the particles should be at least half the light wavelength for optical detectors. Thus, such detectors cannot measure particles smaller than 200 nm and can make inaccurate size measurements. However, a condensation particle counter (CPC) can bypass this constraint by condensing particles with water or alcohol to a size that optical light can detect. CPC operates from 2.5 nm to 10 μm.

The mobility of an electrical particle can be measured using scanning mobility particle sizers (SPMS). Charged particles are programmed to be exposed to different voltages in a differential

FIGURE 17.5 Common methods used for nanoparticle detection. Abbreviations: APMA, aerosol particle mass analyzer; CPC, condensation particle counter; ELPI, electrical low-pressure impactor; OPS, fast mobility particle sizer; FMPS, optical particle sizer; SEM, scanning electron microscopy; SPMS, scanning mobility particle sizers; TEM, transmission electron microscopy; TEOM, tapered element oscillating microbalance.

mobility analyzer. Electrical mobility separates and counts particles. The charge-to-size ratio affects electrical mobility. Smaller particles with the same charge have more electrical mobility. SMPS uses 3–5 minutes temporal resolutions, while NP size distributions can alter in seconds. Therefore, equipment with lower time resolutions can measure quick size distribution changes. Fast mobility particle sizer (FMPS) measures particles from 5.6 to 560 nm with 1 second resolution. FMPS measurements use electrical mobility like SMPS. Environmental research, toxicological evaluation, and indoor air monitoring use such quick equipment. The condensation particle counter-based SMPS and FMPS devices were compared using ambient and lab-produced particles for particle number concentrations. FMPS measures size distribution faster and gives more significant particle number concentrations. However, other publications say it underestimates particle size. The electrical low-pressure impactor (ELPI) quickly responds with 0.1 second resolution and covers particles from 6 nm to 10 µm. Charged particles enter a low-pressure cascade and are gathered by aerodynamic diameter at distinct impactor stages. Highly sensitive electrical detectors generate a real-time reaction. The electrical signal indicates particle size and concentration. It is most suitable for unstable concentrations or size distributions.

Some real-time sensors detect NP mass as well as particle number concentration. The aerosol particle mass analyzer (APMA) or Tapered element oscillating microbalance (TEOM) measures particles by mass-to-charge ratio, regardless of size or shape. Offline approaches such as electron microscopy gather NPs on filter samplers for morphological or chemical examination. Sample filter media can vary, although analytical methods may determine it. Electron microscopies such as scanning electron microscopy (SEM) and transmission electron microscopy (TEM) can be used for imaging nanoparticles. SEM and TEM are limited in exposure and control studies by tedious sample preparation, high vacuum settings, expensive equipment, and specialized personnel.

Other methods, such as thermal precipitators, can be used for particles under 10 µm in size. It uses a heated wire or sphere between plates. Particles that pass between the heated element and cooler surface settle on the cooler surface and can be examined by any microscope. Dynamic light scattering (DLS) is a non-destructive technique that can be used to measure the size distribution of nanoparticles in suspension. It works by measuring the scattering of light by the nanoparticles. Inductively coupled plasma mass spectrometry (ICP-MS) can be used to measure the elemental composition of nanoparticles. It works by ionizing the nanoparticles and measuring the mass of the ions. Surface plasmon resonance (SPR) is a technique that can be used to measure the size and concentration of nanoparticles in solution. It works by measuring the change in the refractive index of a surface when nanoparticles bind to it.

It is essential to emphasize the fact that several parameters influence these assessments. One of those parameters is how NPs are absorbed in seaweed cells. A study showed that the size of TiO_2 NP for entering the cell wall through their pores was 5–20 nm (Wang & Xia, 2019). The small size of NPs helps the absorption of biomolecules in the cell, such as glycoproteins and polysaccharides, and the membrane modifications caused by the cellular stage. In addition, environmental changes such as pH variations affect absorption (Wang & Xia, 2019).

Nanotechnologies encompass a spectrum of advantages and potential hazards. However, while numerous academics have carefully examined benefits, hazards are not well known. Considering the equilibrium between risks and benefits, it is imperative to assess whether the utilization of NPs is warranted or, conversely, if it should be subject to stringent regulation due to their potential implications for the environment and human health. There exists a pressing imperative to achieve a harmonious equilibrium between the two. Stringent regulations have the potential to impede further advancements and technological progress, whereas lenient laws may result in extensive environmental degradation.

The regulation of NPs is conducted by many organizations, which vary based on the manner and location of their employment or production. In the United States, the Food and Drug Administration (FDA) regulates the utilization of NPs in food, pharmaceuticals, and cosmetics. In contrast, the Environmental Protection Agency (EPA) in the United States assumes responsibility for monitoring and regulating the release of pollutants into the atmosphere, bodies of water, and land, as well as managing the disposal of chemical waste generated by the agricultural sector. This regulation encounters various challenges, including diverse types of non-governmental organizations that hinder the establishment of comprehensive regulations and legislation, the need to monitor all pathways of environmental emissions, including regular employment attrition, and the absence of consolidated data on environmental impact and long-term risks.

Nanodatabase is financially supported by organizations such as the European Research Council and the Danish Consumer Council (Nanodatabase, n.d.). The Nanodatabase categorizes items based on their environmental and human health implications. Utilizing a color-coded system, explicitly employing red, yellow, and green, communicates pertinent information regarding the emission levels and potential toxicity associated with a given substance or entity. In 2019, 3,377 instances were identified, subsequently increasing to almost 3,400 within the initial 4 months of 2020 (Nanodatabase, n.d.).

17.6 CONCLUSIONS

Several steps must be performed to lessen nanoparticles' adverse effects on the surrounding environment. Researchers are putting in a lot of effort to develop new nanoparticles that should be less hazardous and more capable of dissolving in natural surroundings. Any published study should center on methods for eliminating nanoparticles from the surrounding environment. Meanwhile, it is necessary to be aware of the potential threats that nanoparticles offer, not only to the well-being of humans but also to the integrity of the natural world. It is strongly recommended that customers avoid any products on the market that contain nanoparticles whenever it is in any way possible for them to avoid coming into contact with them. The harmful impact of nanoparticles on the environment can be minimized in several ways. One way is encouraging consumers to make well-informed decisions about the items they buy. For instance, consumers can check the labels of products to see if they claim to be "nanoparticle-free." The general public can also lend a hand to enterprises attempting to develop nanotechnology products that are both environmentally friendly and able to withstand the test of time. Because of their size, nanoparticles can be challenging to detect and measure. Thus, determining the exact amount of their impact is a challenging endeavor.

BIBLIOGRAPHY

Abdelhamid, H. N. (2023a). Advanced multifunctional materials (2023). *ChemRevixrxiv*. https://doi.org/10.26434/chemrxiv-2023-w78g3.

Abdelhamid, H. N. (2023b). Biodegradable polymers nanocomposites. *SSRN Electronic Journal*. https://doi.org/10.2139/ssrn.4398109.

Abdelhamid, H. N. (2023c). An introductory review on advanced multifunctional materials. *Heliyon*, e18060. https://doi.org/10.1016/j.heliyon.2023.e18060.

Abdelhamid, H. N., & Badr, G. (2021). Nanobiotechnology as a platform for the diagnosis of COVID-19: a review. *Nanotechnology for Environmental Engineering*. https://doi.org/10.1007/s41204-021-00109-0.

Abdelhamid, H. N., El-Zohry, A. M., Cong, J., Thersleff, T., Karlsson, M., Kloo, L., & Zou, X. (2019). Towards implementing hierarchical porous zeolitic imidazolate frameworks in dye-sensitized solar cells. *Royal Society Open Science*, 6(7), 190723. https://doi.org/10.1098/rsos.190723.

Abdelhamid, H. N., & Mahmoud, G. A. (2023). Antifungal and nanozyme activities of metal-organic framework-derived CuO@C. *Applied Organometallic Chemistry*, 37(3). https://doi.org/10.1002/aoc.7011.

Abdelhamid, H. N., & Mathew, A. P. (2022). Cellulose-based nanomaterials advance biomedicine: a review. *International Journal of Molecular Sciences*, 23(10), 5405. https://doi.org/10.3390/ijms23105405.

Abdellatif, A. B. A., El-Bery, H. M., Abdelhamid, H. N., & El-Gyar, S. A. (2022). ZIF-67 and cobalt-based@ heteroatom-doped carbon nanomaterials for hydrogen production and dyes removal via adsorption and catalytic degradation. *Journal of Environmental Chemical Engineering*, 10(6), 108848. https://doi.org/10.1016/j.jece.2022.108848.

Accorsi, G., Verri, G., Bolognesi, M., Armaroli, N., Clementi, C., Miliani, C., & Romani, A. (2009). The exceptional near-infrared luminescence properties of cuprorivaite (Egyptian blue). *Chemical Communications, 23*, 3392. https://doi.org/10.1039/b902563d

Al-Shetwi, A. Q. (2022). Sustainable development of renewable energy integrated power sector: trends, environmental impacts, and recent challenges. *Science of the Total Environment, 822*, 153645. https://doi.org/10.1016/j.scitotenv.2022.153645

Aruoja, V., Dubourguier, H.-C., Kasemets, K., & Kahru, A. (2009). Toxicity of nanoparticles of CuO, ZnO and TiO2 to microalgae Pseudokirchneriella subcapitata. *Science of the Total Environment, 407*(4), 1461–1468. https://doi.org/10.1016/j.scitotenv.2008.10.053

Dale, A. L., Casman, E. A., Lowry, G. V., Lead, J. R., Viparelli, E., & Baalousha, M. (2015). Modeling nanomaterial environmental fate in aquatic systems. *Environmental Science & Technology, 49*(5), 2587–2593. https://doi.org/10.1021/es505076w

Hartoyo, A. P. P., Octaviani, E. A., Syamani, F. A., Mulsanti, I. W., & Solikhin, A. (2022). Potential of chitosan/carbon nanoparticles and chitosan/lignocellulose nanofiber composite as growth media for peatland paddy seeds. *Environmental Research, 212*, 113235. https://doi.org/10.1016/j.envres.2022.113235

Hosseinzadeh-Bandbafha, H., Nizami, A.-S., Kalogirou, S. A., Gupta, V. K., Park, Y.-K., Fallahi, A., Sulaiman, A., Ranjbari, M., Rahnama, H., Aghbashlo, M., Peng, W., & Tabatabaei, M. (2022). Environmental life cycle assessment of biodiesel production from waste cooking oil: a systematic review. *Renewable and Sustainable Energy Reviews, 161*, 112411. https://doi.org/10.1016/j.rser.2022.112411

Hulla, J., Sahu, S., & Hayes, A. (2015). Nanotechnology. *Human & Experimental Toxicology, 34*(12), 1318–1321. https://doi.org/10.1177/0960327115603588

Kagan, V. E., Konduru, N. V., Feng, W., Allen, B. L., Conroy, J., Volkov, Y., Vlasova, I. I., Belikova, N. A., Yanamala, N., Kapralov, A., Tyurina, Y. Y., Shi, J., Kisin, E. R., Murray, A. R., Franks, J., Stolz, D., Gou, P., Klein-Seetharaman, J., Fadeel, B., … Shvedova, A. A. (2010). Carbon nanotubes degraded by neutrophil myeloperoxidase induce less pulmonary inflammation. *Nature Nanotechnology, 5*(5), 354–359. https://doi.org/10.1038/nnano.2010.44

Link, S., Wang, Z. L., & El-Sayed, M. A. (2000). How does a gold nanorod melt? *The Journal of Physical Chemistry B, 104*(33), 7867–7870. https://doi.org/10.1021/jp0011701

Mitrano, D. M., Motellier, S., Clavaguera, S., & Nowack, B. (2015). Review of nanomaterial aging and transformations through the life cycle of nano-enhanced products. *Environment International, 77*, 132–147. https://doi.org/10.1016/j.envint.2015.01.013

Mulakov, S. P., Gotovtsev, P. M., Gainanova, A. A., Kravchenko, G. V., Kuz'micheva, G. M., & Podbel'skii, V. V. (2020). Generation of the reactive oxygen species on the surface of nanosized titanium(IV) oxides particles under UV-irradiation and their connection with photocatalytic properties. *Journal of Photochemistry and Photobiology A: Chemistry, 393*, 112424. https://doi.org/10.1016/j.jphotochem.2020.112424

Nanodatabase. (n.d.). *Nanodatabase.*

Nel, A., Xia, T., Mädler, L., & Li, N. (2006). Toxic potential of materials at the nanolevel. *Science, 311*(5761), 622–627. https://doi.org/10.1126/science.1114397

Nowack, B., Ranville, J. F., Diamond, S., Gallego-Urrea, J. A., Metcalfe, C., Rose, J., Horne, N., Koelmans, A. A., & Klaine, S. J. (2012). Potential scenarios for nanomaterial release and subsequent alteration in the environment. *Environmental Toxicology and Chemistry, 31*(1), 50–59. https://doi.org/10.1002/etc.726

Olajire, A. A. (2020). The brewing industry and environmental challenges. *Journal of Cleaner Production, 256*, 102817. https://doi.org/10.1016/j.jclepro.2012.03.003

Rajan, R., Chandran, K., Harper, S. L., Yun, S.-I., & Kalaichelvan, P. T. (2015). Plant extract synthesized silver nanoparticles: an ongoing source of novel biocompatible materials. *Industrial Crops and Products, 70*, 356–373. https://doi.org/10.1016/j.indcrop.2015.03.015

Rajkovic, S., Bornhöft, N. A., van der Weijden, R., Nowack, B., & Adam, V. (2020). Dynamic probabilistic material flow analysis of engineered nanomaterials in European waste treatment systems. *Waste Management, 113*, 118–131. https://doi.org/10.1016/j.wasman.2020.05.032

Reed, R. B., Zaikova, T., Barber, A., Simonich, M., Lankone, R., Marco, M., Hristovski, K., Herckes, P., Passantino, L., Fairbrother, D. H., Tanguay, R., Ranville, J. F., Hutchison, J. E., & Westerhoff, P. K. (2016). Potential environmental impacts and antimicrobial efficacy of silver- and nanosilver-containing textiles. *Environmental Science & Technology, 50*(7), 4018–4026. https://doi.org/10.1021/acs.est.5b06043

Sajid, M., Ilyas, M., Basheer, C., Tariq, M., Daud, M., Baig, N., & Shehzad, F. (2015). Impact of nanoparticles on human and environment: review of toxicity factors, exposures, control strategies, and future prospects. *Environmental Science and Pollution Research, 22*(6), 4122–4143. https://doi.org/10.1007/s11356-014-3994-1

Samiei, F., Shirazi, F. H., Naserzadeh, P., Dousti, F., Seydi, E., & Pourahmad, J. (2020). Correction to: toxicity of multi-wall carbon nanotubes inhalation on the brain of rats. *Environmental Science and Pollution Research*, *27*(23), 29699–29699. https://doi.org/10.1007/s11356-020-09667-3

Shen, X., Dai, M., Yang, J., Sun, L., Tan, X., Peng, C., Ali, I., & Naz, I. (2022). A critical review on the phytoremediation of heavy metals from environment: performance and challenges. *Chemosphere*, *291*, 132979. https://doi.org/10.1016/j.chemosphere.2021.132979

Silva, L. F. O., Pinto, D., & Lima, B. D. (2020). Implications of iron nanoparticles in spontaneous coal combustion and the effects on climatic variables. *Chemosphere*, *254*, 126814. https://doi.org/10.1016/j.chemosphere.2020.126814

Tran, H.-V., Ngo, N. M., Medhi, R., Srinoi, P., Liu, T., Rittikulsittichai, S., & Lee, T. R. (2022). Multifunctional iron oxide magnetic nanoparticles for biomedical applications: a review. *Materials*, *15*(2), 503. https://doi.org/10.3390/ma15020503

Tweney, R. D. (2006). Discovering discovery: how Faraday found the first metallic colloid. *Perspectives on Science*, *14*(1), 97–121. https://doi.org/10.1162/posc.2006.14.1.97

Usha Rani, P., & Rajasekharreddy, P. (2011). Green synthesis of silver-protein (core-shell) nanoparticles using Piper betle L. leaf extract and its ecotoxicological studies on Daphnia magna. *Colloids and Surfaces A: Physicochemical and Engineering Aspects*, *389*(1–3), 188–194. https://doi.org/10.1016/j.colsurfa.2011.08.028

Wang, Y., & Xia, Y. (2019). Optical, electrochemical and catalytic methods for in-vitro diagnosis using carbonaceous nanoparticles: a review. *Microchimica Acta*, *186*(1), 50. https://doi.org/10.1007/s00604-018-3110-1

Xiang, Q.-Q., Wang, D., Zhang, J.-L., Ding, C.-Z., Luo, X., Tao, J., Ling, J., Shea, D., & Chen, L.-Q. (2020). Effect of silver nanoparticles on gill membranes of common carp: modification of fatty acid profile, lipid peroxidation and membrane fluidity. *Environmental Pollution*, *256*, 113504. https://doi.org/10.1016/j.envpol.2019.113504

Yan, S., Hu, Q., Wei, Y., Jiang, Q., Yin, M., Dong, M., Shen, J., & Du, X. (2022). Calcium nutrition nanoagent rescues tomatoes from mosaic virus disease by accelerating calcium transport and activating antiviral immunity. *Frontiers in Plant Science*, *13*. https://doi.org/10.3389/fpls.2022.1092774

Zhou, P., Adeel, M., Shakoor, N., Guo, M., Hao, Y., Azeem, I., Li, M., Liu, M., & Rui, Y. (2020). Application of nanoparticles alleviates heavy metals stress and promotes plant growth: an overview. *Nanomaterials*, *11*(1), 26. https://doi.org/10.3390/nano11010026

Index

D

dairy 15, 17, 30, 31, 36, 76, 80, 112, 117, 118, 120, 143, 167, 190, 199, 205, 206, 210, 211, 213, 222, 261, 272, 273, 300, 302, 307, 331, 333
daphnia 365, 372
database 55, 250, 322
deacetylation 10
Debye 3
Decoloration 139
decomposition 2, 170, 192, 295
Decontamination 188, 189, 191, 193, 195, 197, 199, 201
deforestation 361
degradation 8, 11, 78, 105, 119, 144, 145, 149, 152, 153, 154, 159, 177, 178, 192, 193, 195, 198, 202, 205, 209–214, 220, 223, 224, 238, 250, 308, 313, 316, 350, 353, 355, 361, 363, 364, 369, 370
degree 10, 167, 202, 215, 265, 326
dehydration 27, 174, 221, 331, 336, 338
delivery 1, 4, 5, 12, 13, 85, 92, 145, 146, 148, 149, 150, 155, 157–160, 162, 175, 178, 182, 183, 184, 209–212, 215, 218–221, 223, 224, 234–238, 256, 260, 299, 313, 349, 359, 367
dendrimers 155, 266, 268, 352
deoxyglucosone 106, 107
depletion 145, 265, 361
deposition 2, 6, 26, 41, 52, 53, 54, 70, 72, 183, 310
derivatization 80, 91, 114, 298
dermatitis 350
desalination 134, 141
desorption 26, 41, 43, 44, 47, 365
destructive 2, 315, 320, 334, 338, 369
detector 80, 91, 98, 112, 144, 244, 327
detrimental 8, 104, 152, 300, 366
dextran 10, 11, 14, 26, 47
diabetes 19, 20, 31, 45, 65, 66, 69, 126, 318
Diacylglycerol 36
diagnostic 82, 128, 269, 272, 273, 278, 297
diaminopyridine 113
diamond 26, 48, 109, 110, 124, 129, 130, 238, 256, 371
diarrhea 289, 319
dichloromethane 10
dichlorvos 79, 85, 87, 88, 89, 92
dicyandiamide 112
dielectric 4, 62, 260, 281, 322
diesel 174
dietary 17, 19, 21, 41, 46, 47, 51, 59, 68, 74, 75, 101, 104, 108, 121, 124, 128, 129, 140, 202, 211, 217, 232, 340, 346, 350, 351, 357
diffraction 2, 109, 276
diffusion 12, 177, 207, 230
digestibility 172, 185, 349
digestion 18, 32, 164, 171, 172, 182, 184–187, 213, 235, 352
digital 63, 169, 202, 267, 277, 313, 328, 330, 335
dimethylamine 99, 100, 234, 237
dinitrophenol 287, 295
diode 81, 291
dioxide 6, 25, 39, 85, 87, 93, 169, 171, 174, 175, 176, 192, 206, 207, 217, 219, 220, 238, 259, 262, 313, 322, 341, 354, 355, 359, 365
dioxin 287, 293, 295
disaccharides 18, 21, 25, 26

discharge 4, 51, 74, 189, 288, 308, 366
disinfection 191, 194
dots 5, 14, 42, 46, 48, 61, 64, 69, 71, 74, 75, 82, 89, 94, 95, 98, 102, 103, 113, 118, 118, 123–126, 138, 249, 251, 252, 253, 256, 257, 268, 271, 272, 273, 275–278, 281, 282, 285, 290, 291, 293, 295, 297, 298, 299, 311, 313, 316, 320, 322, 323, 335
drinks 20, 25, 26, 39, 44, 141, 210, 215, 237, 262, 266, 291, 298, 305, 329, 332
Dumas 30, 112

E

ecological 68, 71, 73, 74, 102, 153, 155, 159, 171, 195, 196, 307
economic 74, 157, 163–166, 169, 172, 173, 179, 181, 183, 184, 185, 190, 191, 194–197, 200, 241, 244, 301, 305, 307, 316, 317, 334
ecosystems 53, 97, 145, 150, 152, 154, 155, 197, 198, 270, 361, 365, 366
Ecotoxicology 71, 74
edible 8, 13, 14, 15, 54, 55, 56, 68, 69, 72, 73, 175, 182, 184, 185, 206, 213, 214, 220, 221, 232, 273, 280, 284, 285, 287, 289, 304, 316, 349, 356
efsa 101, 104, 112, 120, 121, 124, 125, 126, 205, 206, 219, 340–343, 347, 353, 355
egg 31, 35, 213, 248, 262, 290, 306, 318, 352
eggplant 55, 57
eggshell 174, 182
Egyptian 59, 362, 371
electric 6, 25, 120, 121, 209, 215, 247, 257, 263, 266, 267, 320, 322, 323
electrochemistry 45, 69, 132, 248, 249, 270
electrode 14, 22, 25, 36, 38, 39, 41, 42, 44–48, 61, 62, 71, 74, 84, 85, 88, 92–96, 102, 109, 110, 111, 115, 116, 121, 122, 123, 127-129, 132, 133, 134, 136, 137, 140, 143, 144, 193, 218, 234, 257, 265, 267, 270, 274, 277, 282, 285, 286, 289, 292, 294–297, 299, 311, 315, 323
electrolyte 139, 323
electromagnetic 4, 117, 215, 258, 259, 276, 320
electronics 236, 326, 360
electrospun 10, 11, 12, 14, 15, 16, 42, 237, 296, 312
element 3, 27, 66, 69, 72, 73, 75, 76, 110, 111, 112, 114, 117, 128, 140, 159, 188, 244, 245, 247, 256, 261, 265, 267, 296, 318, 348, 368, 369
emergence 5, 99, 131, 153, 158, 203, 241, 282
emissions 51, 145, 146, 164, 165, 170, 171, 174, 297, 336, 361, 363, 368, 370
emitting 252, 257, 290, 344
emulsifiers 37, 178, 185, 212, 214, 235, 240, 347
emulsion 78, 120, 134, 175, 212, 213, 219, 220, 224, 233, 238, 356
enantiomers 81, 95
encapsulated 87, 147, 149, 150, 154, 155, 160, 210, 212, 213, 214, 224, 232, 235, 290, 296, 353, 358
endosulfan 236
engineered 5, 14, 42, 111, 138, 143, 148, 150, 158, 160, 161, 162, 171, 197, 228, 229, 231, 267, 322, 341, 344, 354, 355, 356, 371
enrofloxacin 286